MEDICAL IMAGING PHYSICS

Fourth Edition

MEDICAL IMAGING PHYSICS

Fourth Edition

William R. Hendee, Ph.D.
Senior Associate Dean and Vice President
Dean of the Graduate School of Biomedical Sciences
Professor and Vice Chair of Radiology
Professor of Radiation Oncology, Biophysics, Bioethics
Medical College of Wisconsin
Professor of Biomedical Engineering
Marquette University

E. Russell Ritenour, Ph.D.
Professor and Chief of Radiology Physics, Medical School
Director of Graduate Studies in Biophysical Sciences and Medical Physics, Graduate School
University of Minnesota

A JOHN WILEY & SONS, INC., PUBLICATION

This book is printed on acid-free paper. ♾

Published simultaneously in Canada.

DISCLAIMER

While the authors, editors, and publisher believe that drug selection and dosage and the specification and usage of equipment and devices, as set forth in this book, are in accord with current recommendations and practice at the time of publication, they accept no legal responsibility for any errors or omissions and make no warranty, express or implied, with respect to material contained herein. In view of ongoing research, equipment modifications, changes in governmental regulations, and the constant flow of information relating to drug therapy, drug reactions, and the use of equipment and devices, the reader is urged to review and evaluate the information provided in the package insert or instructions for each drug, piece of equipment, or device for, among other things, any changes in the instructions or indication of dosage or usage and for added warnings and precautions.

For ordering and customer service information please call 1-800-CALL-WILEY.

Library of Congress Cataloging-in-Publication Data is available.

ISBN 0-471-38226-4

Printed in the United States of America.

10 9 8 7 6 5 4 3 2

Ad hoc, ad loc
and quid pro quo
so little time
so much to know.

Jeremy Hillary Boob, Ph.D.
The Nowhere Man in the Yellow Submarine

CONTENTS IN BRIEF

CONTENTS

PREFACE

Writing and rewriting a text such as *Medical Imaging Physics* over several editions presents two challenges. The first is to keep the information fresh and relevant. This is a particular challenge in medical imaging, because the field is evolving so rapidly. The third edition of this text was published in 1992, just 10 short years ago. Yet in that text no mention was made of topics such as photodiode or direct conversion digital x-ray imagers; digital mammography; digital fluoroscopy; power Doppler ultrasound; functional magnetic resonance imaging; elastography; or helical CT scanning. This is just a partial list of imaging approaches that must be covered today in any text of imaging physics. Being involved in a dynamic and rapidly changing field is one of the more enjoyable aspects of medical imaging. But it places heavy demands on authors trying to provide a text that keeps up with the field.

The second challenge is no less demanding than the first. That challenge is to keep the text current with the changing culture of how people learn, as well as with the educational experience and pedagogical expectations of students. These have changed remarkably over the 30 years since this book first appeared. For maximum effect, information today must be packaged in various ways, including self-contained segments, illustrations, highlights, sidebars, and examples and problems. In addition, it must be presented in a manner that facilitates learning and helps students evaluate their progress. Making the information correct and complete is only half the battle; the other half is using a format that helps the student assimilate and apply it. The latter challenge reflects not only today's learning environment, but also the tremendous amount of information that must be assimilated by any student of medical imaging.

In recognition of these challenges, the authors decided two years ago to restructure *Medical Imaging Physics* into a fourth edition with a fresh approach and an entirely new format. This decision led to a total rewriting of the text. We hope that this new edition will make studying imaging physics more efficient, effective, and pleasurable. It certainly has made writing it more fun.

Medical imaging today is a collaborative effort involving physicians, physicists, engineers, and technologists. Together they are able to provide a level of patient care that would be unachievable by any single group working alone. But to work together, they must all have a solid foundation in the physics of medical imaging. It is the intent of this text to provide this foundation. We hope that we have done so in a manner that makes learning enriching and enjoyable.

William R. Hendee, Ph.D.
E. Russell Ritenour, Ph.D.

PREFACE TO THE FIRST EDITION

This text was compiled and edited from tape recordings of lectures in medical radiation physics at the University of Colorado School of Medicine. The lectures are attended by resident physicians in radiology, by radiologic technologists and by students beginning graduate study in medical physics and in radiation biology. The text is intended for a similar audience.

Many of the more recent developments in medical radiation physics are discussed in the text. However, innovations are frequent in radiology, and the reader should supplement the book with perusal of the current literature. References at the end of each chapter may be used as a guide to additional sources of information.

Mathematical prerequisites for understanding the text are minimal. In the few sections where calculus is introduced in the derivation of an equation, a description of symbols and procedures is provided with the hope that the use of the equation is intelligible even if the derivation is obscure.

Problem solving is the most effective way to understand physics in general and medical radiation physics in particular. Problems are included at the end of each chapter, with answers at the end of the book. Students are encouraged to explore, discuss and solve these problems. Example problems with solutions are scattered throughout the text.

Acknowledgments

Few textbooks would be written without the inspiration provided by colleagues, students and friends. I am grateful to all of my associates who have contributed in so many ways toward the completion of this text. The original lectures were recorded by Carlos Garciga, M.D., and typed by Mrs. Marilyn Seckler and Mrs. Carolyn McCain. Parts of the book have been reviewed in unfinished form by: Martin Bischoff, M.D., Winston Boone, B.S., Donald Brown, M.D., Frank Brunstetter, M.D., Duncan Burdick, M.D., Lawrence Coleman, Ph.D., Walter Croft, Ph.D., Marvin Daves, M.D., Neal Goodman, M.D., Albert Hazle, B.S., Donald Herbert, Ph.D., F. Bing Johnson, M.D., Gordon Kenney, M.S., Jack Krohmer, Ph.D., John Pettigrew, M.D., Robert Siek, M.P.H., John Taubman, M.D., Richard Trow, B.S., and Marvin Williams, Ph.D. I appreciate the comments offered by these reviewers. Edward Chaney, Ph.D., reviewed the entire manuscript and furnished many helpful suggestions. Robert Cadigan, B.S., assisted with the proofreading and worked many of the problems. Geoffrey Ibbott, Kenneth Crusha, Lyle Lindsey, R.T., and Charles Ahrens, R.T., obtained much of the experimental data included in the book.

Mrs. Josephine Ibbott prepared most of the line drawings for the book, and I am grateful for her diligence and cooperation. Mrs. Suzan Ibbott and Mr. Billie Wheeler helped with some of the illustrations, and Miss Lynn Wisehart typed the appendixes. Mr. David Kuhner of the John Crerar Library in Chicago located many of the references to early work. Representatives of various instrument companies have helped in many ways. I thank Year Book Medical Publishers for encouragement and patience and Marvin Daves, M.D., for his understanding and support.

I am indebted deeply to Miss Carolyn Yandle for typing each chapter many times, and for contributing in many other ways toward the completion of the book.

Finally, I wish to recognize my former teachers for all they have contributed so unselfishly. In particular, I wish to thank Fred Bonte, M.D., and Jack Krohmer, Ph.D., for their guidance during my years as a graduate student. I wish also to recognize my indebtedness to Elda E. Anderson, Ph.D., and to William Zebrun, Ph.D. I shall not forget their encouragement during my early years of graduate study.

WILLIAM R. HENDEE

ACKNOWLEDGMENTS

One of the greatest pleasures of teaching is the continuing opportunity to work with former students on projects of mutual interest. This text is a good example of such an opportunity. Russ Ritenour received postdoctoral training in medical physics at the University of Colorado while I was chair of the Department of Radiology at that institution. After his NIH postdoctoral fellowship, he stayed on the faculty and we published several papers together, both before and after each of us left Colorado for new adventures. When it came time to write a 3rd edition of *Medical Imaging Physics* about ten years ago, I realized that I needed a co-author to share the workload. Russ was the person I wanted, and I am glad he agreed to be a co-author. He was equally willing to co-author this 4th edition. Future editions will bear his imprint as principal author of *Medical Imaging Physics*.

Several other persons deserve recognition for their support of this project. Foremost are Ms. Terri Komar and Ms. Mary Beth Drapp, both of whom have been instrumental in moving the fourth edition to completion. Terri worked with me as Executive Assistant for almost 10 years before moving to North Carolina. She was succeeded most ably in the position by Mary Beth Drapp. It has been my great fortune to be able to work in my publication efforts with two such competent individuals. Our editor, Ms. Luna Han of John Wiley Publishers, should be recognized for her quiet but firm insistence that we meet our own deadlines. I also am indebted to Jim Youker, M.D., Chair of Radiology at the Medical College of Wisconsin, for his friendship and inspiration over the years and for his enthusiasm for various academic ventures that we have collaborated in.

Most of all, I want to thank my wife Jeannie. Her tolerance of my writing habits, including stacks of books and papers perched on the piano, on the dining table, and, most precariously, in my study, is unfathomable. I certainly don't question it, but I do appreciate it—very much.

William R. Hendee, Ph.D.

I'm delighted to have been able to contribute once again to a new edition of this text and am particularly delighted to work once again with Bill Hendee. There was a lot to do, as so many things have changed and evolved in radiology since the time of the last edition in 1992. But, to me, the change is the fun part.

This was a fun project for another reason as well. Bill and I both enjoy using anecdotes as we teach. I'm referring to historical vignettes, illustrations of radiologic principles through examples in other fields, and, in my case I'm told, terrible jokes. While we felt that the terrible jokes were too informal for a textbook, we have included a number of vignettes and examples from other fields in the hope that it will make the reading more enjoyable and provide the kind of broader framework that leads to a deeper understanding. At least it might keep you awake.

I have to thank more people than there is space to thank. In particular, though, I must thank Pam Hansen, for dealing so patiently with many drafts of a complicated electronic manuscript. Also, I must thank two of my colleagues here at the University of Minnesota, Richard Geise and Bruce Hasselquist, who are endless sources of information and who never hesitate to tell me when I'm wrong. Rolph Gruetter of the Center for Magnetic Resonance Research, was very helpful in reviewing and commenting upon some of the new MR material in this edition. Finally, I want to thank Dr. William M. Thompson, who recently stepped down as chair of radiology here. He has been a tireless supporter of learning at all levels and he will be missed.

Once again, my wife, Julie, and our children, Jason and Karis, have supported me in so many ways during a major project. In this case, they've also been the source of a few of the medical images, although I won't say which ones.

E. Russell Ritenour, Ph.D.

CHAPTER

1

IMAGING IN MEDICINE

OBJECTIVES

After completing this chapter, the reader should be able to:

- Identify the energy sources, tissue properties, and image properties employed in medical imaging.
- Name several factors influencing the increasing role of imaging in healthcare today.
- Define the expression "molecular medicine" and give examples.
- Provide a summary of the history of medical imaging.
- Explain the pivotal role of x-ray computed tomography in the evolution of modern medical imaging.

INTRODUCTION

Natural science is the search for "truth" about the natural world. In this definition, truth is defined by principles and laws that have evolved from observations and measurements about the natural world. The observations and measurements are reproducible through procedures that follow universal rules of scientific experimentation. They reveal properties of objects and processes in the natural world that are assumed to exist independently of the measurement technique and of our sensory perceptions of the natural world. The mission of science is to use observations and measurements to characterize the static and dynamic properties of objects, preferably in quantitative terms, and to integrate these properties into principles and, ultimately, laws and theories that provide a logical framework for understanding the world and our place in it.

Whether the external (natural) world can really be known, and even whether there is a world external to ourselves, has been the subject of philosophical speculation for centuries. It is for this reason that "truth" in the first sentence is offset in quotes.

It is not possible to characterize all properties of an object with exactness. For example, if the location of a particle is exactly known, its velocity is highly uncertain, and vice versa. Similarly, if the energy of a particle is exactly known, the time at which the particle has this energy is highly uncertain, and vice versa. This fundamental tenet of physics is known as the Heisenberg Uncertainty Principle.

As a part of natural science, human medicine is the quest for understanding one particular object, the human body, and its structure and function under all conditions of health, illness, and injury. This quest has yielded models of human health and illness that are immensely useful in preventing disease and disability, detecting and diagnosing illness and injury, and designing therapies to alleviate pain and suffering and to restore the body to a state of wellness or, at least, structural and functional capacity. The success of these efforts depends on (a) our depth of understanding of the human body and (b) the delineation of ways to intervene successfully in the progression of disease and the effects of injuries.

MARGIN FIGURE 1-1
The Human Genome project is a massive undertaking to determine the exact sequence of nucleotides (i.e., the DNA code) on all 24 human chromosomes.

Progress toward these objectives has been so remarkable that the average life span of humans in developed countries is almost twice its expected value a century ago. Greater understanding has occurred at all levels, from the atomic through molecular, cellular, and tissue to the whole body, and includes social and lifestyle influences on disease patterns. At present a massive research effort is focused on acquiring knowledge about genetic coding (the Human Genome Project) and about the role of genetic coding in human health and disease. This effort is progressing at an astounding rate, and it causes many medical scientists to believe that genetics, computational biology (mathematical modeling of biological systems), and bioinformatics (mathematical modeling of biological information, including genetic information) are the major research frontiers of medical science for the next decade or longer.

The number of deaths per 100,000 residents in the United States has declined from more than 400 in 1950 to less than 200 in 1990.

The human body is an incredibly complex system. Acquiring data about its static and dynamic properties results in massive amounts of information. One of the major challenges to researchers and clinicians is the question of how to acquire, process, and display vast quantities of information about the body so that the information can be assimilated, interpreted, and utilized to yield more useful diagnostic methods and therapeutic procedures. In many cases, the presentation of information as images is the most efficient approach to addressing this challenge. As humans we understand this efficiency; from our earliest years we rely more heavily on sight than on any other perceptual skill in relating to the world around us. Physicians increasingly rely

TABLE 1-1 Energy Sources and Tissue Properties Employed in Medical Imaging

Energy Sources	*Tissue Properties*	*Image Properties*
X rays	Mass density	Transmissivity
γ rays	Electron density	Opacity
Visible light	Proton density	Emissivity
Ultraviolet light	Atomic number	Reflectivity
Annihilation Radiation	Velocity	Conductivity
Electric fields	Pharmaceutical Location	Magnetizability
Magnetic fields	Current flow	Resonance
Infrared	Relaxation	Absorption
Ultrasound	Blood volume/flow	
Applied voltage	Oxygenation level of blood	
	Temperature	
	Chemical state	

as well on images to understand the human body and intervene in the processes of human illness and injury. The use of images to manage and interpret information about biological and medical processes is certain to continue its expansion, not only in clinical medicine but also in the biomedical research enterprise that supports it.

Images of a complex object such as the human body reveal characteristics of the object such as its transmissivity, opacity, emissivity, reflectivity, conductivity, and magnetizability, and changes in these characteristics with time. Images that reveal one or more of these characteristics can be analyzed to yield information about underlying properties of the object, as depicted in Table 1-1. For example, images (shadowgraphs) created by x rays transmitted through a region of the body reveal intrinsic properties of the region such as effective atomic number Z, physical density (grams/cm^3), and electron density (electrons/cm^3). Nuclear medicine images, including emission computed tomography (ECT) with pharmaceuticals releasing positrons [positron emission tomography (PET)] and single photons [single-photon emission computed tomography (SPECT)], reveal the spatial and temporal distribution of target-specific pharmaceuticals in the human body. Depending on the application, these data can be interpreted to yield information about physiological processes such as glucose metabolism, blood volume, flow and perfusion, tissue and organ uptake, receptor binding, and oxygen utilization. In ultrasonography, images are produced by capturing energy reflected from interfaces in the body that separate tissues with different acoustic impedances, where the acoustic impedance is the product of the physical density and the velocity of ultrasound in the tissue. Magnetic resonance imaging (MRI) of relaxation characteristics following magnetization of tissues is influenced by the concentration, mobility, and chemical bonding of hydrogen and, less frequently, other elements present in biological tissues. Maps of the electrical field (electroencephalography) and the magnetic field (magnetoencephalography) at the surface of the skull can be analyzed to identify areas of intense neuroelectrical activity in the brain. These and other techniques that use the energy sources listed in Table 1-1 provide an array of imaging methods that are immensely useful for displaying structural and functional information about the body. This information is essential to improving human health through detection and diagnosis of illness and injury.

The intrinsic properties of biological tissues that are accessible through acquisition and interpretation of images vary spatially and temporally in response to structural and functional changes in the body. Analysis of these variations yields information about static and dynamic processes in the human body. These processes may be changed by disease and disability, and identification of the changes through imaging often permits detection and delineation of the disease or disability. Medical images are pictures of tissue characteristics that influence the way energy is emitted, transmitted, reflected, and so on, by the human body. These characteristics are related to, but not the same as, the actual structure (anatomy), composition (biology and

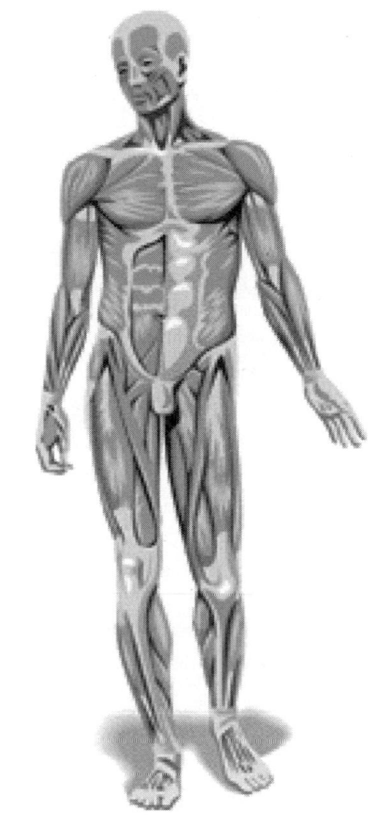

MARGIN FIGURE 1-2
Drawing of the human figure.

The effective atomic number Z_{eff} actually should be used in place of the atomic number Z in this paragraph. Z_{eff} is defined later in the text.

Promising imaging techniques that have not yet found applications in clinical medicine are discussed in the last chapter of the text.

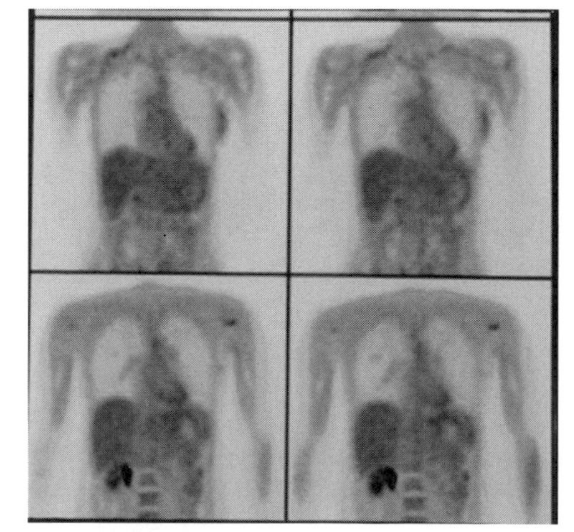

MARGIN FIGURE 1-3
^{18}F-FDG PET scan of breast cancer patient with lymph node involvement in the left axilla.

chemistry), and function (physiology and metabolism) of the body. Part of the art of interpreting medical images is to bridge among image characteristics, tissue properties, human anatomy, biology and chemistry, and physiology and metabolism, as well as to determine how all of these parameters are affected by disease and disability.

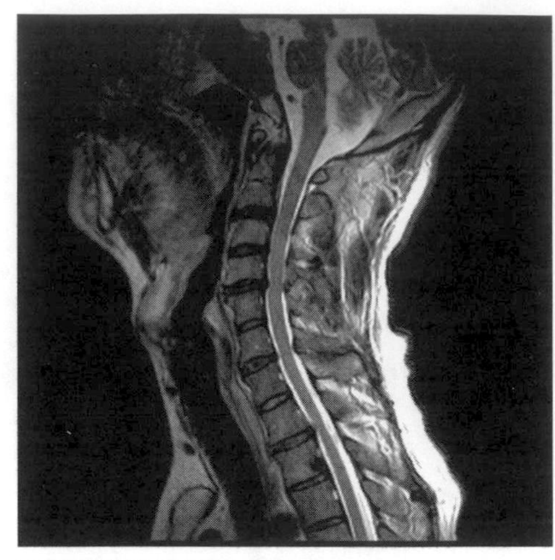

MARGIN FIGURE 1-4
MRI of the human cervical spine.

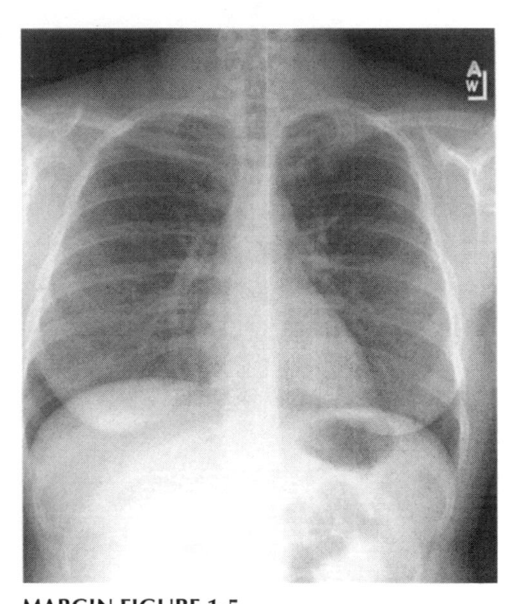

MARGIN FIGURE 1-5
A normal chest radiograph. (Courtesy of Lacey Washington, M.D., Medical College of Wisconsin.)

Advances in Medical Imaging

Advances in medical imaging have been driven historically by the "technology push" principle. Especially influential have been imaging developments in other areas, notably in the defense and military sectors, that have been imported into medicine because of their potential applications to detection and diagnosis of human illness and injury. Examples include ultrasound developed initially for submarine detection (Sonar), scintillation detectors, and reactor-produced isotopes (including ^{131}I, ^{60}Co, and ^{99m}Tc) that emerged from the Manhattan Project, rare-earth fluorescent compounds synthesized initially in defense and space research laboratories, electrical devices for detection of rapid blood loss on the battlefield, and the evolution of microelectronics and computer industries from research funded initially for security, surveillance, defense, and military purposes. Basic research laboratories have also produced several imaging technologies that have migrated successfully into clinical medicine. Examples include (a) reconstruction mathematics for computed tomographic imaging and (b) laboratory techniques in nuclear magnetic resonance that evolved into magnetic resonance imaging, spectroscopy, and other methods useful in clinical medicine. The migration of technologies from other arenas into medicine has not always been successful. For example, infrared detection devices developed for night vision in military operations have so far not proven to be useful in medicine in spite of initial enthusiasm for infrared thermography as an imaging method for early detection of breast cancer.

Technology "push" means that technologies developed for specific applications, or perhaps for their own sake, are driven by financial incentives to find applications in other areas, including healthcare.

Sonar is an acronym for *SO*und *N*avigation *A*nd *R*anging.

The Manhattan Project was the code name for the U.S. project to develop a nuclear weapon during World War II.

Today the emphasis in medical imaging is shifting from a "technology push" approach toward a "biological/clinical pull" emphasis. This shift reflects both (a) a deeper understanding of the biology underlying human health and disease and (b) a growing demand for accountability (proven usefulness) of technologies before they are introduced into clinical medicine. Increasingly, unresolved biological questions important to the diagnosis and treatment of human disease and disability are used to encourage development of new imaging methods, often in association with nonimaging probes. For example, the functions of the human brain, along with the causes and mechanisms of various mental disorders such as dementia, depression, and schizophrenia, are among the greatest biological enigmas confronting biomedical scientists and clinicians. A particularly fruitful method for penetrating this conundrum is the technique of functional imaging employing tools such as ECT and MRI. Functional magnetic resonance imaging (fMRI) is especially promising as an approach to unraveling some of the mysteries related to how the human brain functions in health, disease, and disability. Another example is the use of x-ray computed tomography and MRI as feedback mechanisms to shape, guide, and monitor the surgical and radiation treatment of cancer.

The growing use of imaging techniques in radiation oncology reveals an interesting and rather recent development. Until about three decades ago, the diagnostic and therapeutic applications of ionizing radiation were practiced by a single medical specialty. In the late 1960s these applications began to separate into distinct medical specialties, diagnostic radiology and radiation oncology, with separate training programs and clinical practices. Today, imaging is used extensively in radiation oncology to characterize the cancers to be treated, design the plans of treatment, guide the delivery of radiation, monitor the response of patients to treatment, and follow patients over the long term to assess the success of therapy, occurrence of complications, and frequency of recurrence. The process of accommodating to this development in the training and practice of radiation oncology is encouraging a closer working relationship between radiation oncologists and diagnostic radiologists.

Evolutionary Developments in Imaging

Six major developments are converging today to raise imaging to a more prominent role in biological and medical research and in the clinical practice of medicine. These developments are[1]:

- Ever-increasing sophistication of the biological questions that can be addressed as knowledge expands and understanding grows about the complexity of the human body and its static and dynamic properties.
- Ongoing evolution of imaging technologies and the increasing breadth and depth of the questions that these technologies can address at ever more fundamental levels.
- Accelerating advances in computer technology and information networking that support imaging advances such as three- and four-dimensional representations, superposition of images from different devices, creation of virtual reality environments, and transportation of images to remote sites in real time.
- Growth of massive amounts of information about patients that can best be compressed and expressed through the use of images.
- Entry into research and clinical medicine of young persons who are highly facile with computer technologies and comfortable with images as the principal pathway to information acquisition and display.
- Growing importance of images as effective means to convey information in visually-oriented developed cultures.

Medical Imaging Trends

From	To
Anatomic	Physiobiochemical
Static	Dynamic
Qualitative	Quantitative
Analog	Digital
Nonspecific agents	Tissue-Targeted agents
Diagnosis	Diagnosis/Therapy

Biological/clinical "pull" means that technologies are developed in response to recognized clinical or research needs.

A major challenge confronting medical imaging today is the need to efficiently exploit this convergence of evolutionary developments to accelerate biological and medical imaging toward the realization of its true potential.

Images are our principal sensory pathway to knowledge about the natural world. To convey this knowledge to others, we rely on verbal communication following accepted rules of human language, of which there are thousands of varieties and dialects. In the distant past, the acts of knowing through images and communicating through languages were separate and distinct processes. Every technological advance that brought images and words closer, even to the point of convergence in a single medium, has had a major cultural and educational impact. Examples of such advances include the printing press, photography, motion pictures, television, video games, computers, and information networking. Each of these technologies has enhanced the shift from using words to communicate information toward a more efficient synthesis of images to provide insights and words to explain and enrich insights.[2] Today this synthesis is evolving at a faster rate than ever before, as evidenced, for example, by the popularity of television news programs and documentaries and the growing use of multimedia approaches to education and training.

For purposes of informing and educating individuals, multiple pathways are required for interchanging information. In addition, flexible means are needed for mixing images and words, and their rate and sequence of presentation, in order to capture and retain the attention, interest, and motivation of persons engaged in the educational process. Computers and information networks provide this capability. In medicine, their use in association with imaging technologies greatly enhances the potential contribution of medical imaging to resolution of patient problems in the clinical setting. At the beginning of the twenty-first century, the six evolutionary developments discussed above provide the framework for major advances in medical imaging and its contributions to improvements in the health and well-being of people worldwide.

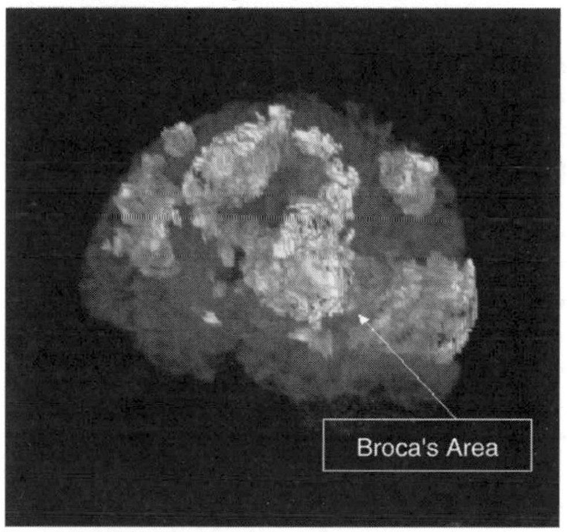

MARGIN FIGURE 1-6
fMRI image of brain tumor in relation to the motor cortex.

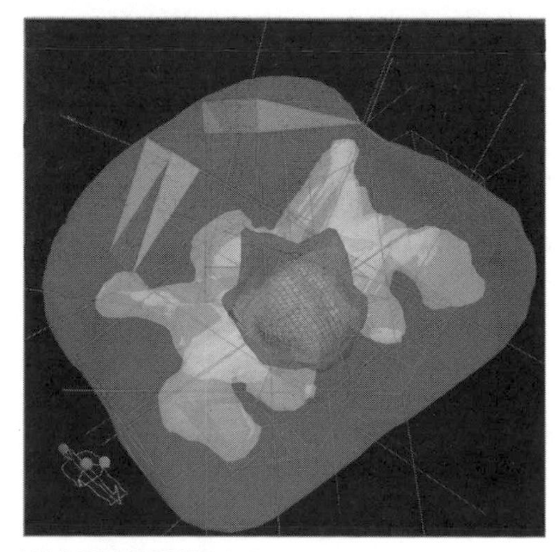

MARGIN FIGURE 1-7
Multifield treatment plan superimposed on a 3-D reconstructed CT image. (From G. Ibbott, Ph.D., MD Anderson Hospital. Used with permission.)

Molecular Medicine

Medical imaging has traditionally focused on the acquisition of structural (anatomic) and functional (physiologic) information about patients at the organ and tissue levels.

If a scientist reads two articles each day from the world's scientific literature published that day, at the end of one year the scientist will be 60 centuries behind in keeping up with the current scientific literature. To keep current with the literature, the scientist would have to read 6000 articles each day.

Each new generation adapts with ease to technologies that were a challenge to the previous generation. Examples of this "generation gap" in today's world include computers, software engineering, and video games.

Imaging technologies useful or potentially useful at the cellular and molecular levels:

- Multiphoton microscopy
- Scanning probe microscopy
- Electron energy-loss spectroscopic imaging
- Transmission electron microscopes with field-emission electron guns
- 3-D reconstruction from electron micrographs
- Fluorescent labels
- Physiologic indicators
- Magnetic resonance imaging microscopy
- Single-copy studies of proteins and oligonucleotides
- Video microscopy
- Laser-scanning confocal microscopy
- Two-photon laser scanning microscopy

Antisense agents (molecules, viruses, etc.) are agents that contain DNA with a nucleotide configuration opposite that of the biological structures for which the agents are targeted.

A major challenge to the use of molecular mechanisms to enhance contrast are limitations on the number of cells that can be altered by various approaches.

This focus has nurtured the correlation of imaging findings with pathological conditions and has led to substantial advances in detection and diagnosis of human disease and injury. All too often, however, detection and diagnosis occur at a stage in the disease or injury where radical intervention is required and the effectiveness of treatment is compromised. In many of these cases, detection and diagnosis at an earlier stage in the progression of disease and injury would improve the effectiveness of treatment and enhance the well-being of patients. This objective demands that medical imaging expand its focus from the organ and tissue levels to the cellular and molecular levels of human disease and injury. Many scientists believe that medical imaging is well-positioned today to experience this expanded focus as a benefit of knowledge gained at the research frontiers of molecular biology and genetics. This benefit is often characterized as the entry of medical imaging into the era of molecular medicine.

Contrast agents are widely employed with x-ray, ultrasound, and magnetic resonance imaging techniques to enhance the visualization of properties correlated with patient anatomy and physiology. Agents in wide use today localize in tissues either by administration into specific anatomic compartments (such as the gastrointestinal or vascular systems) or by reliance on nonspecific changes in tissues (such as increased capillary permeability or alterations in the extracellular fluid space). These localization mechanisms frequently do not provide a sufficient concentration of the agent to reveal subtle tissue differences associated with an abnormal condition. New contrast agents are needed that exploit growing knowledge about biochemical receptor systems, metabolic pathways, and "antisense" molecular technologies to yield concentration differentials sufficient to reveal the presence of pathological conditions.

Another important imaging application of molecular medicine is the use of imaging methods to study molecular and genetic processes. For example, cells may be genetically altered to attract ions that (1) alter the magnetic susceptibility, thereby permitting their identification by magnetic resonance imaging techniques; or (2) are radioactive and therefore can be visualized by nuclear imaging methods. Another possibility is to transect cells with genetic material that causes expression of cell surface receptors that can bind radioactive compounds.[3] Conceivably, this technique could be used to monitor the progress of gene therapy.

Advances in molecular biology and genetics are yielding new knowledge at an astonishing rate about the molecular and genetic infrastructure underlying the static and dynamic processes that comprise human anatomy and physiology. This new knowledge is likely to yield increasingly specific approaches to the use of imaging methods to visualize normal and abnormal tissue structure and function at increasingly fundamental levels. These methods will in all likelihood contribute to continuing advances in molecular medicine.

Historical Approaches to Diagnosis

In the 1800s and before, physicians were extremely limited in their ability to obtain information about the illnesses and injuries of patients. They relied essentially on the five human senses, and what they could not see, hear, feel, smell, or taste usually went undetected. Even these senses could not be exploited fully, because patient modesty and the need to control infectious diseases often prevented full examination of the patient. Frequently, physicians served more to reassure the patient and comfort the family rather than to intercede in the progression of illness or facilitate recovery from injury. More often than not, fate was more instrumental than the physician in determining the course of a disease or injury.

The twentieth century witnessed remarkable changes in the physician's ability to intervene actively on behalf of the patient. These changes dramatically improved the health of humankind around the world. In developed countries, infant mortality decreased substantially, and the average life span increased from 40 years at the beginning of the century to 70+ years at the century's end. Many major diseases, such as smallpox, tuberculosis, poliomyelitis, and pertussis, had been brought under control, and some had been virtually eliminated. Diagnostic medicine has improved dramatically, and therapies have evolved for cure or maintenance of persons with a variety of maladies.

Diagnostic probes to identify and characterize problems in the internal anatomy and physiology of patients have been a major contribution to these improvements. By far, x rays are the most significant of these diagnostic probes. Diagnostic x-ray studies have been instrumental in moving the physician into the role of an active intervener in disease and injury and a major influence on the prognosis for recovery.

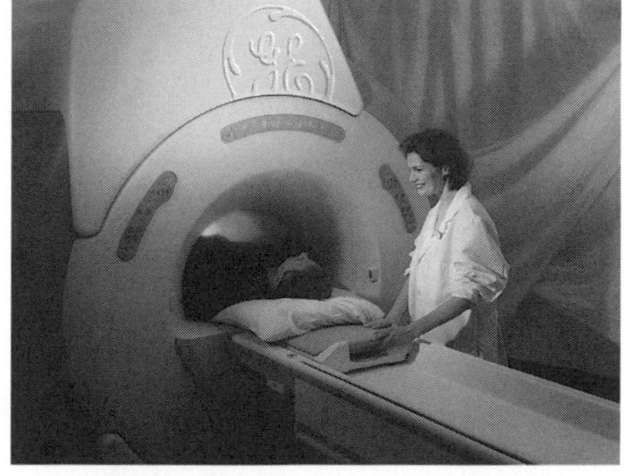

(a)

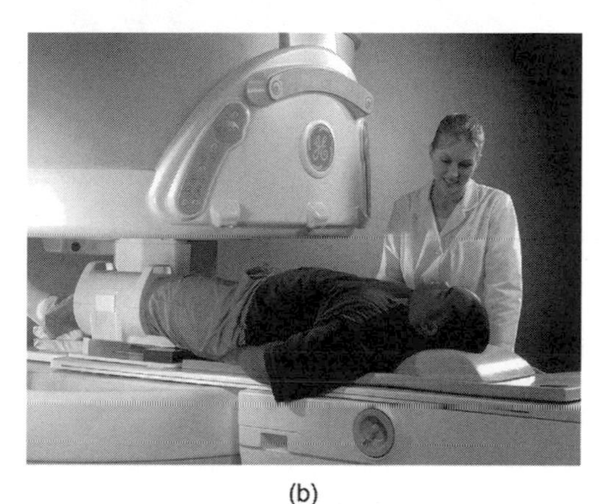

(b)

MARGIN FIGURE 1-8
Closed-bore (**top**) and open-bore (**bottom**) MRI units. (Courtesy of General Electric Medical Systems.)

Capsule History of Medical Imaging

In November 1895 Wilhelm Röntgen, a physicist at the University of Würzburg, was experimenting with cathode rays. These rays were obtained by applying a potential difference across a partially evacuated glass "discharge" tube. Röntgen observed the emission of light from crystals of barium platinocyanide some distance away, and he recognized that the fluorescence had to be caused by radiation produced by his experiments. He called the radiation "x rays" and quickly discovered that the new radiation could penetrate various materials and could be recorded on photographic plates. Among the more dramatic illustrations of these properties was a radiograph of a hand (Figure 1-1) that Röntgen included in early presentations of his findings.[4] This radiograph captured the imagination of both scientists and the public around the world.[5] Within a month of their discovery, x rays were being explored as medical tools in several countries, including Germany, England, France, and the United States.[6]

In 1901, Röntgen was awarded the first Nobel Prize in Physics.

Two months after Röntgen's discovery, Poincare demonstrated to the French Academy of Sciences that x rays were released when cathode rays struck the wall of a gas discharge tube. Shortly thereafter, Becquerel discovered that potassium uranyl sulfate spontaneously emitted a type of radiation that he termed Becquerel rays, now

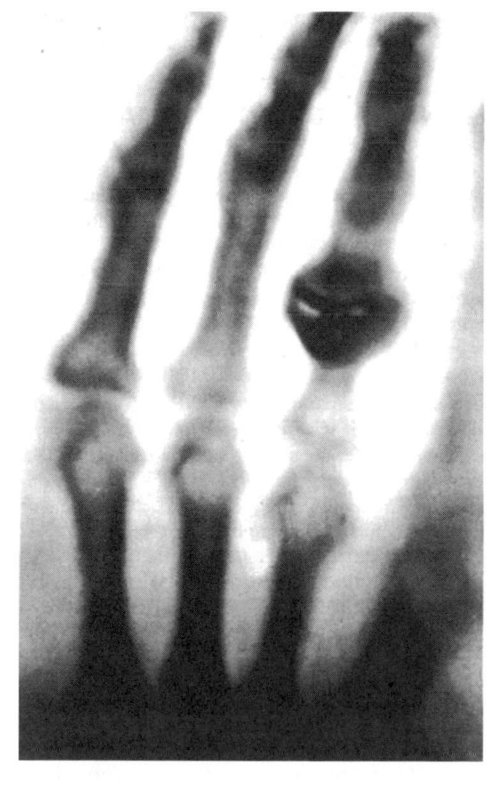

FIGURE 1-1
A radiograph of the hand taken by Röntgen in December 1895. His wife may have been the subject. (From the Deutsches Röntgen Museum, Remscheid-Lennap, Germany. Used with permission.)

popularly known as β-particles.[7] Marie Curie explored Becquerel rays for her doctoral thesis and chemically separated a number of elements. She discovered the radioactive properties of naturally occurring thorium, radium, and polonium, all of which emit α-particles, a new type of radiation.[8] In 1900, γ rays were identified by Villard as a third form of radiation.[9] In the meantime, J. J. Thomson reported in 1897 that the cathode rays used to produce x rays were negatively charged particles (electrons) with about 1/2000 the mass of the hydrogen atom.[10] In a period of 5 years from the discovery of x rays, electrons and natural radioactivity had also been identified, and several sources and properties of the latter had been characterized.

Over the first half of the twentieth century, x-ray imaging advanced with the help of improvements such as intensifying screens, hot-cathode x-ray tubes, rotating anodes, image intensifiers, and contrast agents. These improvements are discussed in subsequent chapters. In addition, x-ray imaging was joined by other imaging techniques that employed radioactive nuclides and ultrasound beams as radiation sources for imaging.

Through the 1950s and 1960s, diagnostic imaging progressed as a coalescence of x-ray imaging with the emerging specialties of nuclear medicine and ultrasonography. This coalescence reflected the intellectual creativity nurtured by the synthesis of basic science, principally physics, with clinical medicine. In a few institutions, the interpretation of clinical images continued to be taught without close attention to its foundation in basic science. In the more progressive teaching departments, however, the dependence of radiology on basic science, especially physics, was never far from the consciousness of teachers and students.

In the early years of CT, an often-heard remark was "why would anyone want a new x-ray technique that when compared with traditional x-ray imaging:

- yields 10 times more coarse spatial resolution
- is 1/100 as fast in collecting image data
- costs 10 times more

Introduction of Computed Tomography

In the early 1970s a major innovation was introduced into diagnostic imaging. This innovation, x-ray computed tomography (CT), is recognized today as the most significant single event in medical imaging since the discovery of x rays.

The importance of CT is related to several of its features, including the following:

1. Provision of cross-sectional images of anatomy
2. Availability of contrast resolution superior to traditional radiology
3. Construction of images from x-ray transmission data by a "black box" mathematical process requiring a computer
4. Creation of clinical images that are no longer direct proof of a satisfactory imaging process so that intermediate control measures from physics and engineering are essential
5. Production of images from digital data that are processed by computer and can be manipulated to yield widely varying appearances.

EMI Ltd., the commercial developer of CT, was the first company to enter CT into the market. They did so as a last resort, only after offering the rights to sell, distribute, and service CT to the major vendors of imaging equipment. The vendors rejected EMI's offer because they believed the market for CT was too small.

Adoption of CT by the medical community was rapid and enthusiastic in the United States and worldwide. A few years after introduction of this technology, more than 350 units had been purchased in the United States alone. Today, CT is an essential feature of most radiology departments of moderate size and larger.

The introduction of CT marked the beginning of a transition in radiology from an analog to a digitally based specialty. The digital revolution in radiology has opened opportunities for image manipulation, storage, transmission, and display in all fields of medicine. The usefulness of CT for brain imaging almost immediately reduced the need for nuclear brain scans and stimulated the development of other applications of nuclear medicine, including qualitative and quantitative studies of the cardiovascular system. Extension of reconstruction mathematics to nuclear medicine yielded the techniques of single-photon emission computed tomography (SPECT) and positron emission tomography (PET), technologies that have considerable potential for revealing new information about tissue physiology and metabolism. Reconstruction mathematics also are utilized in magnetic resonance image (MRI), a technology introduced into clinical medicine in the early 1980s. Today, MRI provides insights into fundamental properties of biologic tissues that were beyond the imagination a few years ago. Digital methods have been incorporated into ultrasonography to provide "real time" gray scale images important to the care of patients in cardiology, obstetrics, and several other specialties. In x-ray imaging, digital methods are slowly but inexorably replacing analog methods for data acquisition and display.

Many believe that the next major frontier of imaging science is at the molecular and genetic levels.

Radiology is a much different field today than it was three decades ago. With the introduction of new imaging methods and digital processing techniques, radiology has become a technologically complex discipline that presents a paradox for physicians. Although images today are much more complicated to produce, they are simultaneously simpler to interpret—and misinterpret—once they are produced. The simplicity of image interpretation is seductive, however. The key to retrieval of essential information in radiology today resides at least as much in the production and presentation of images as in their interpretation.

A physician who can interpret only what is presented as an image suffers a severe handicap. He or she is captive to the talents and labors of others and wholly dependent on their ability to ensure that an image reveals abnormalities in the patient and not in the imaging process. On the other hand, the physician who understands the science and technology of imaging can be integrally involved in the entire imaging process, including the acquisition of patient data and their display as clinical images. Most important, the knowledgeable physician has direct input into the quality of the image on which the diagnosis depends.

A thorough understanding of diagnostic images requires knowledge of the science, principally physics, that underlies the production of images. Radiology and physics have been closely intertwined since x rays were discovered. With the changes that have occurred in imaging over the past few years, the linkage between radiology and physics has grown even stronger. Today a reasonable knowledge of physics, instrumentation, and imaging technology is essential for any physician wishing to perfect the science and art of radiology.

It is wrong to think that the task of physics is to find out what nature is. Physics concerns what we can say about nature." Niels Bohr (as quoted in Pagels, H., *The Cosmic Code*, Simons and Schuster, 1982.)

CONCLUSIONS

- Medical imaging is both a science and a tool to explore human anatomy and to study physiology and biochemistry.
- Medical imaging employs a variety of energy sources and tissue properties to produce useful images.
- Increasingly, clinical pull is the driving force in the development of imaging methods.
- Molecular biology and genetics are new frontiers for imaging technologies.
- Introduction of x-ray computed tomography was a signal event in the evolution of medical imaging.

REFERENCES

1. Hendee, W. R. Physics and applications of medical imaging. *Rev. Mod. Phys.* 1999; **71**(2), Centenary:S444–S450.
2. Beck, R. N. Tying Science and Technology Together in Medical Imaging, in Hendee, W., and Trueblood, J. (eds.), *Digital Imaging*, Madison, WI, Medical Physics Publishing Co., 1993, pp. 643–665.
3. Thrall, J. H. How molecular medicine will impact radiology. *Diagn. Imag.* 1997; **Dec.**:23–27.
4. Röntgen, W. Uber eine neue Art von Strahlen (vorläufige Mitteilung). *Sitzungs-Berichte der Physikalisch-Medicinischen Gesellschaft zu Würzburg* 1895; **9**:132.
5. Glaser, O. Evolution of radiologic physics as applied to isotopes. *Am. J. Roentgenol. Radium Ther.* 1951; **65**:515.
6. Laughlin, J. History of Medical Physics, in Webster, J. (ed.), *Encyclopedia of Medical Devices and Engineering*, Vol. 3. New York, John Wiley & Sons, 1988, p. 1878.
7. Becquerel, H. Sur les radiation émises par phosphorescence. *Compt. Rend.* 1896; **122**:420.
8. Curie, M. *Traité de Radioactivité*. Paris, Gauthier-Villars, 1910.
9. Villard, P. Sur la réflexion et la réfraction des rayons cathodiques et des rayons déviables du radium. *Compt. Rend.* 1900; **130**:1010.
10. Thomson, J. Cathode Rays. *Philos. Mag. 5th Ser.* 1897; **44**:293.

CHAPTER

2

STRUCTURE OF MATTER

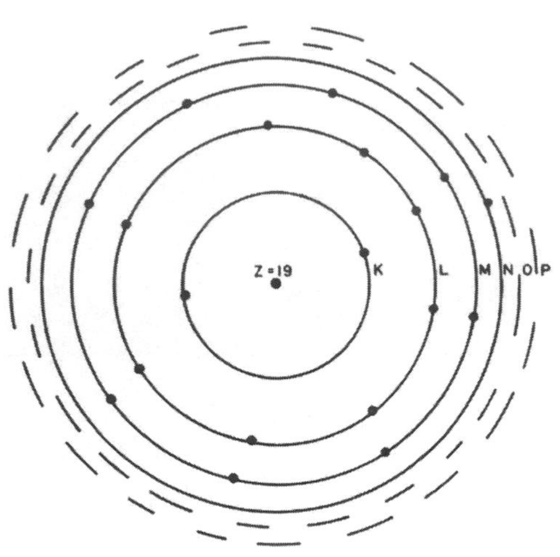

MARGIN FIGURE 2-1
Electron configuration showing electron shells in the Bohr model of the atom for potassium, with 19 electrons ($Z = 19$).

A Brief History of the Development of Evidence for the Existence of Atoms

400–300 B.C.: Demokritos and the Epicurean School in Greece argued for the existence of atoms on philosophical grounds. Aristotle and the Stoics espoused the continuum philosophy of matter.[1]

300 B.C.–1800s: Aristotelian view of matter predominated.

1802: Dalton described the principle of multiple proportions. This principle states that chemical constituents react in specific proportions, suggesting the discreteness of material components.[2]

1809: Gay-Lussac discovered laws that predicted changes in volume of gases.[3]

1811: Avogadro hypothesized the existence of a constant number of atoms in a characteristic mass of an element or compound.[4]

1833: Faraday's law of electrolysis explained specific rates or proportions of elements that would be electroplated onto electrodes from electrolytic solutions.[5]

1858: Cannizaro published data concerning the atomic weights of the elements.[6]

1869–1870: Meyer and Mendeleev constructed the Periodic Table.[7, 8]

1908: Perrin demonstrated that the transfer of energy from atoms to small particles in solution, the cause of a phenomenon known as Brownian motion, leads to a precise derivation of Avogadro's number.[9]

OBJECTIVES

After completing this chapter, the reader should be able to:

- Define the terms: element, atom, molecule, and compound.
- Describe the electron shell structure of an atom.
- Explain the significance of electron and nuclear binding energy.
- List the events that result in characteristic and auger emission.
- Compare electron energy levels in solids that are:
  - Conductors
  - Insulators
  - Semiconductors
- Describe the phenomenon of superconductivity.
- List the four fundamental forces.
- Explain why fission and fusion result in release of energy.
- State the source of the nuclear magnetic moment.
- Define:
  - Isotopes
  - Isotones
  - Isobars
  - Isomers

THE ATOM

All matter is composed of atoms. A sample of a pure element is composed of a single type of atom. Chemical compounds are composed of more than one type of atom. Atoms themselves are complicated entities with a great deal of internal structure. An atom is the smallest unit of matter that retains the chemical properties of a material. In that sense, it is a "fundamental building block" of matter. In the case of a compound, the "fundamental building block" is a molecule consisting of one or more atoms bound together by electrostatic attraction and/or the sharing of electrons by more than one nucleus.

The basic structure of an atom is a positively charged nucleus, containing electrically neutral neutrons and positively charged protons, surrounded by one or more negatively charged electrons. The number and distribution of electrons in the atom determines the chemical properties of the atom. The number and configuration of neutrons and protons in the nucleus determines the stability of the atom and its electron configuration.

Structure of the Atom

One unit of charge is 1.6×10^{-19} coulombs. Each proton and each electron carries one unit of charge, with protons positive and electrons negative. The number of units of positive charge (i.e., the number of protons) in the nucleus is termed the atomic number Z. The atomic number uniquely determines the classification of an atom as one of the elements. Atomic number 1 is hydrogen, 2 is helium, and so on.

Atoms in their normal state are neutral because the number of electrons outside the nucleus (i.e., the negative charge in the atom) equals the number of protons (i.e., the positive charge) of the nucleus. Electrons are positioned in energy levels (i.e., shells) that surround the nucleus. The first ($n = 1$) or K shell contains no more than 2 electrons, the second ($n = 2$) or L shell contains no more than 8 electrons, and the third ($n = 3$) or M shell contains no more than 18 electrons (Margin Figure 2-1). The outermost electron shell of an atom, no matter which shell it is, never contains more than 8 electrons. Electrons in the outermost shell are termed valence electrons and determine to a large degree the chemical properties of the atom. Atoms with an outer shell entirely filled with electrons seldom react chemically. These atoms constitute elements known as the inert gases (helium, neon, argon, krypton, xenon, and radon).

TABLE 2-1 Quantum Numbers for Electrons in Helium, Carbon, and Sodium

Element	n	l	m_l	m_s	Orbital	Shell
Helium($Z = 2$)	1	0	0	$-1/2$	$1s^2$	K
	1	0	0	$+1/2$		
Carbon($Z = 6$)	1	0	0	$-1/2$	$1s^2$	
	1	0	0	$+1/2$		
	2	0	0	$-1/2$	$2s^2$	L
	2	0	0	$+1/2$		
	2	1	−1	$-1/2$	$2p^2$	
	2	1	−1	$+1/2$		
Sodium($Z = 11$)	1	0	0	$-1/2$	$1s^2$	K
	1	0	0	$+1/2$		
	2	0	0	$-1/2$	$2s^2$	L
	2	0	0	$+1/2$		
	2	1	−1	$-1/2$	$2p^6$	
	2	1	−1	$+1/2$		
	2	1	0	$-1/2$		
	2	1	0	$+1/2$		
	2	1	1	$-1/2$		
	2	1	1	$+1/2$		
	3	0	0	$-1/2$	$3s$	M

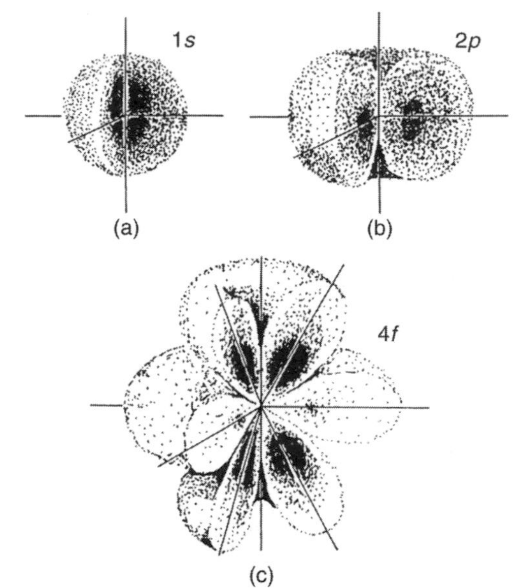

MARGIN FIGURE 2-2
Probability of location of the electron in the hydrogen atom for three different energy states or combinations of principal and angular momentum quantum numbers l. **A**: $n = 1, l = 0$; **B**: $n = 2, l = 1$; **C**: $n = 4, l = 3$.

Energy levels for electrons are divided into sublevels slightly separated from each other. To describe the position of an electron in the extranuclear structure of an atom, the electron is assigned four quantum numbers.

The principal quantum number n defines the main energy level or shell within which the electron resides ($n = 1$ for the K shell, $n = 2$ for the L shell, etc.). The orbital angular-momentum (azimuthal) quantum number l describes the electron's angular momentum ($l = 0, 1, 2, \ldots, n-1$). The orientation of the electron's magnetic moment in a magnetic field is defined by the magnetic quantum number m_l ($m_l = -l, -l+1, \ldots, l-1, l$). The direction of the electron's spin upon its own axis is specified by the spin quantum number m_s ($m_s = +1/2$ or $-1/2$). The Pauli exclusion principle states that no two electrons in the same atomic system may be assigned identical values for all four quantum numbers. Illustrated in Table 2-1 are quantum numbers for electrons in a few atoms with low atomic numbers.

The values of the orbital angular-momentum quantum number, $l = 0$, 1, 2, 3, 4, 5 and 6, are also identified with the symbols, *s, p, d, f, g, h,* and *i*, respectively. This notation is known as "spectroscopic" notation because it is used to describe the separate emission lines observed when light emitted from a heated metallic vapor lamp is passed through a prism. From the 1890s onward, observation of these spectra provided major clues about the binding energies of electrons in atoms of the metals under study. By the 1920s, it was known that the spectral lines above *s* could be split into multiple lines in the presence of a magnetic field. The lines were thought to correspond to "orbitals" or groupings of similar electrons within orbits. The modern view of this phenomenon is that, while the *s* "orbital" is spherically symmetric (Margin Figure 2-2), the other orbitals are not. In the presence of a magnetic field, the *p* "orbital" can be in alignment along any one of three axis of space, x, y, and z. Each of these three orientations has a slightly different energy corresponding to the three possible values of m_l (−1, 0, and 1). According to the Pauli exclusion principle, each orbital may contain two electrons (one with $m_s = +1/2$, the other with $m_s = -1/2$).

The K shell of an atom consists of one orbital, the $1s$, containing two electrons. The L shell consists of the $2s$ subshell, which contains one orbital (two electrons), and the $2p$ subshell, which contains a maximum of three orbitals (six electrons). If an L shell of an atom is filled, its electrons will be noted in spectroscopic notation as $2s^2$, $2p^6$. This notation is summarized for three atoms—helium, carbon, and sodium—in Table 2-1.

Models of the Atom
1907: J. J. Thompson advanced a "plum pudding" model of the atom in which electrons were distributed randomly within a matrix of positive charge, somewhat like raisins in a plum pudding.[10]
1911: Experiments by Rutherford showed the existence of a small, relatively dense core of positive charge in the atom, surrounded by mostly empty space with a relatively small population of electrons.[11]

Why are electron shells identified as K, L, M, and so on, instead of A,B,C, and so on?
Between 1905 and 1911, English physicist Charles Barkla measured the characteristic emission of x rays from metals in an attempt to categorize them according to their penetrating power (energy). Early in his work, he found two and named them B and A. In later years, he renamed them K and L, expecting that he would find more energetic emissions to name B, A, and R. It is thought that he was attempting to use letters from his name. Instead, he discovered lower energy (less penetrating) emissions and decided to continue the alphabet after L with M and N. The naming convention was quickly adopted by other researchers. Thus electron shells (from which characteristic x rays are emitted) are identified as K, L, M, N, and so on. Elements with shell designations up to the letter Q have been identified.[12]

Quantum Mechanical Description of Electrons

Since the late 1920s it has been understood that electrons in an atom do not behave exactly like tiny moons orbiting a planet-like nucleus. Their behavior is described more accurately if, instead of defining them as point particles in orbits with specific velocities and positions, they are defined as entities whose behavior is described by "wave functions." While a wave function itself is not directly observable, calculations may be performed with this function to predict the location of the electron. In contrast to the calculations of "classical mechanics" in which properties such as force, mass, acceleration, and so on, are entered into equations to yield a definite answer for a quantity such as position in space, quantum mechanical calculations yield probabilities. At a particular location in space, for example, the square of the amplitude of a particle's wave function yields the probability that the particle will appear at that location. In Margin Figure 2-2, this probability is predicted for several possible energy levels of a single electron surrounding a hydrogen nucleus (a single proton). In this illustration, a darker shading implies a higher probability of finding the electron at that location. Locations at which the probability is maximum correspond roughly to the "electron shell" model discussed previously. However, it is important to emphasize that the probability of finding the electron at other locations, even in the middle of the nucleus, is not zero. This particular result explains a certain form of radioactive decay in which a nucleus "captures" an electron. This event is not explainable by classical mechanics, but can be explained with quantum mechanics.

Electron Binding Energy and Energy Levels

The extent to which electrons are bound to the nucleus determines several energy absorption and emission phenomena. The binding energy of an electron (E_b) is defined as the energy required to completely separate the electron from the atom. When energy is measured in the macroscopic world of everyday experience, units such as joules and kilowatt-hours are used. In the microscopic world, the electron volt is a more convenient unit of energy. In Margin Figure 2-3 an electron is accelerated between two electrodes. That is, the electron is repelled from the negative electrode and attracted to the positive electrode. The kinetic energy (the "energy of motion") of the electron depends on the potential difference between the electrodes. One electron volt is the kinetic energy imparted to an electron accelerated across a potential difference (i.e., voltage) of 1 V. In Margin Figure 2-3B, each electron achieves a kinetic energy of 10 eV.

The electron volt may be used to express any level of energy. For example, a typical burner on "high" on an electric stove emits heat at an approximate rate of 3×10^{22} eV/sec.

The electron volt can be converted to other units of energy:

$$
\begin{aligned}
1\text{ eV} &= 1.6 \times 10^{-19}\text{ J} \\
&= 1.6 \times 10^{-12}\text{ erg} \\
&= 4.4 \times 10^{-26}\text{ kW-hr}
\end{aligned}
$$

$$
\begin{aligned}
\textit{Note: } 10^3\text{ eV} &= 1\text{ keV} \\
10^6\text{ eV} &= 1\text{ MeV}
\end{aligned}
$$

The electron volt describes potential as well as kinetic energy. The binding energy of an electron in an atom is a form of potential energy.

An electron in an inner shell of an atom is attracted to the nucleus by a force greater than that exerted by the nucleus on an electron farther away. An electron may be moved from one shell to another shell that is farther from the nucleus only if energy is supplied by an external source. Binding energy is negative (i.e., written with a minus sign) because it represents an amount of energy that must be supplied to remove an electron from an atom. The energy that must be imparted to an atom to move an electron from an inner shell to an outer shell is equal to the arithmetic difference in binding energy between the two shells. For example, the binding energy is −13.5 eV for an electron in the K shell of hydrogen and is −3.4 eV for an electron in the L shell. The energy required to move an electron from the K to the L shell in hydrogen is (−3.4 eV) − (−13.5 eV) = 10.1 eV.

Electrons in inner shells of high-Z atoms are near a nucleus with high positive charge. These electrons are bound to the nucleus with a force much greater than that exerted upon the solitary electron in hydrogen. Binding energies of electrons in high- and low-Z atoms are compared in Margin Figure 2-4.

All of the electrons within a particular electron shell do not have exactly the same binding energy. Differences in binding energy among the electrons in a particular shell are described by the orbital, magnetic, and spin quantum numbers, l, m_l, and m_s. The combinations of these quantum numbers allowed by quantum mechanics provide three subshells (L_I to L_{III}) for the L shell (Table 2-1) and five subshells (M_I to M_V) for the M shell (the M subshells occur only if a magnetic field is present). Energy differences between the subshells are small when compared with differences between shells. These differences are important in radiology, however, because they explain certain properties of the emission spectra of x-ray tubes. Table 2-2 gives values for the binding energies of K, and L shell electrons for selected elements.

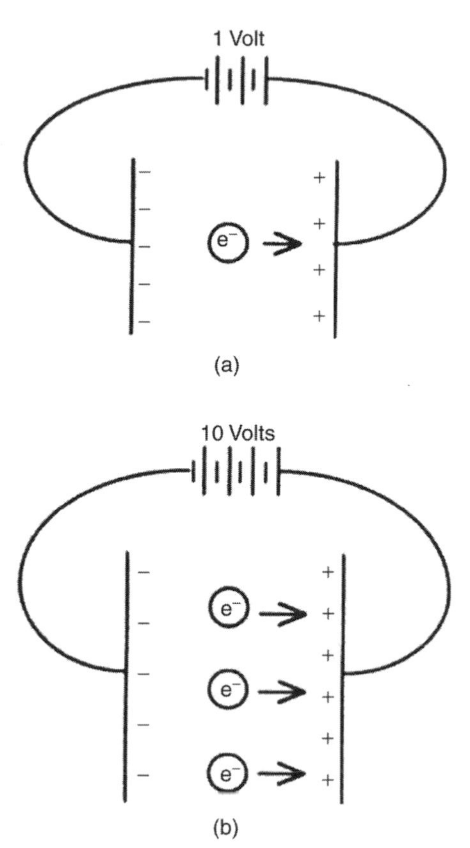

MARGIN FIGURE 2-3
Kinetic energy of electrons specified in electron volts. **A**: The electron has a kinetic energy of 1 eV. **B**: Each electron has a kinetic energy of 10 eV.

Electron Transitions, Characteristic and Auger Emission

Various processes can cause an electron to be ejected from an electron shell. When an electron is removed from a shell, a vacancy or "hole" is left in the shell (i.e., a quantum "address" is left vacant). An electron may move from one shell to another to fill the vacancy. This movement, termed an *electron transition*, involves a

TABLE 2-2 Binding Energies of Electron Shells of Selected Elements[a]

		Binding Energies of Shells (keV)			
Atomic Number	*Element*	*K*	*L_I*	*L_{II}*	*L_{III}*
1	Hydrogen	0.0136			
6	Carbon	0.283			
8	Oxygen	0.531			
11	Sodium	1.08	0.055	0.034	0.034
13	Aluminum	1.559	0.087	0.073	0.072
14	Silicon	1.838	0.118	0.099	0.098
19	Potassium	3.607	0.341	0.297	0.294
20	Calcium	4.038	0.399	0.352	0.349
26	Iron	7.111	0.849	0.721	0.708
29	Copper	8.980	1.100	0.953	0.933
31	Gallium	10.368	1.30	1.134	1.117
32	Germanium	11.103	1.42	1.248	1.217
39	Yttrium	17.037	2.369	2.154	2.079
42	Molybdenum	20.002	2.884	2.627	2.523
47	Silver	25.517	3.810	3.528	3.352
53	Iodine	33.164	5.190	4.856	4.559
54	Xenon	34.570	5.452	5.104	4.782
56	Barium	37.410	5.995	5.623	5.247
57	Lanthanum	38.931	6.283	5.894	5.489
58	Cerium	40.449	6.561	6.165	5.729
74	Tungsten	69.508	12.090	11.535	10.198
79	Gold	80.713	14.353	13.733	11.919
82	Lead	88.001	15.870	15.207	13.044
92	Uranium	115.591	21.753	20.943	17.163

[a]Data from Fine, J., and Hendee, W. *Nucleonics* 1955; **13**:56.

"No rest is granted to the atoms throughout the profound void, but rather driven by incessant and varied motions, some after being pressed together then leap back with wide intervals, some again after the blow are tossed about within a narrow compass. And those being held in combination more closely condensed collide with and leap back through tiny intervals." Lucretius (94–55 B.C.), *On the Nature of Things*

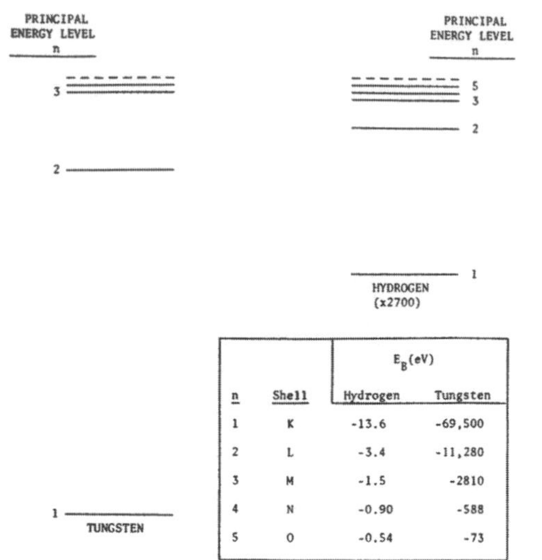

n	Shell	E_B(eV) Hydrogen	Tungsten
1	K	-13.6	-69,500
2	L	-3.4	-11,280
3	M	-1.5	-2810
4	N	-0.90	-588
5	O	-0.54	-73

MARGIN FIGURE 2-4
Average binding energies for electrons in hydrogen ($Z = 1$) and tungsten ($Z = 74$). Note the change of scale required to show both energy diagrams on the same page.

"As the statistical character of quantum theory is so closely linked to the inexactness of all perceptions, one might be led to the presumption that behind the perceived statistical world there still hides a 'real' world in which causality holds. But such speculations seem to us, to say it explicitly, fruitless and senseless."
W. Heisenberg, *The Physical Content of Quantum Kinematics and Mechanics*, 1927.

change in the binding energy of the electron. To move an inner-shell electron to an outer shell, some external source of energy is required. On the other hand, an outer-shell electron may drop spontaneously to fill a vacancy in an inner shell. This spontaneous transition results in the release of energy. Spontaneous transitions of outer-shell electrons falling to inner shells are depicted in Margin Figure 2-5.

The energy released when an outer electron falls to an inner shell equals the difference in binding energy between the two shells involved in the transition. For example, an electron moving from the M to the K shell of tungsten releases (−69,500) − (−2810) eV = −66,690 eV or −66.69 keV. The energy is released in one of two forms. In Margin Figure 2-5B, the transition energy is released as a photon. Because the binding energy of electron shells is a unique characteristic of each element, the emitted photon is called a characteristic photon. The emitted photon may be described as a K, L, or M characteristic photon, denoting the destination of the transition electron. An electron transition creates a vacancy in an outer shell that may be filled by an electron transition from another shell, leaving yet another vacancy, and so on. Thus a vacancy in an inner electron shell produces a cascade of electron transitions that yield a range of characteristic photon energies. Electron shells farther from the nucleus are more closely spaced in terms of binding energy. Therefore, characteristic photons produced by transitions among outer shells have less energy than do those produced by transitions involving inner shells. For transitions to shells beyond the M shell, characteristic photons are no longer energetic enough to be considered x rays.

Margin Figure 2-5C shows an alternative process to photon emission. In this process, the energy released during an electron transition is transferred to another electron. This energy is sufficient to eject the electron from its shell. The ejected electron is referred to as an Auger (pronounced "aw-jay") electron. The kinetic energy of the ejected electron will not equal the total energy released during the transition, because some of the transition energy is used to free the electron from its shell. The Auger electron is usually ejected from the same shell that held the electron that made the transition to an inner shell, as shown in Margin Figure 2-5C. In this case, the kinetic energy of the Auger electron is calculated by twice subtracting the binding energy of the outer-shell electron from the binding energy of the inner-shell electron. The first subtraction yields the transition energy, and the second subtraction accounts for the liberation of the ejected electron.

$$E_{\text{ka}} = E_{\text{bi}} - 2E_{\text{bo}}$$

where E_{ka} is the kinetic energy of the Auger electron, E_{bi} is the binding energy of the inner electron shell (the shell with the vacancy), and E_{bo} is the binding energy of the outer electron shell.

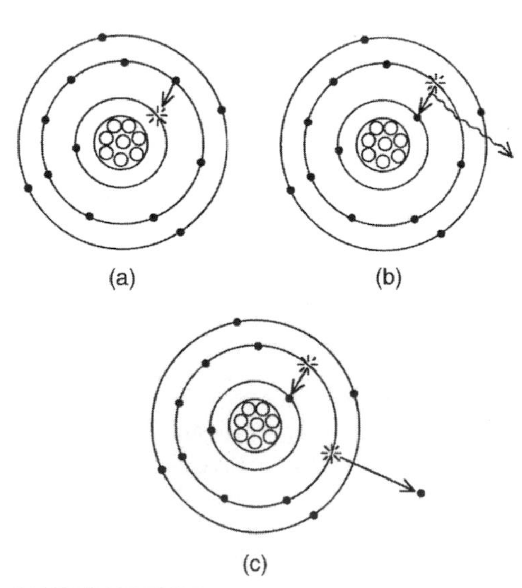

MARGIN FIGURE 2-5
A: Electron transition from an outer shell to an inner shell. **B**: Electron transition accompanied by the release of a characteristic photon. **C**: Electron transition accompanied by the emission of an Auger electron.

Example 2-1

A vacancy in the K shell of molybdenum results in an L to K electron transition accompanied by emission of an Auger electron from the L shell. The binding energies are

$$E_{\text{bK}} = -20{,}000 \text{ eV}$$
$$E_{\text{bL}} = -2521 \text{ eV}$$

What is the kinetic energy of the Auger electron?

$$\begin{aligned} E_{\text{ka}} &= E_{\text{bi}} - 2E_{\text{bo}} \\ &= (-20{,}000 \text{ eV}) - 2(-2521 \text{ eV}) \\ &= -14{,}958 \text{ eV} \\ &= -14.958 \text{ keV} \end{aligned}$$

Fluorescence Yield

Characteristic photon emission and Auger electron emission are alternative processes that release excess energy from an atom during an electron transition. Either process may occur. While it is impossible to predict which process will occur for a specific atom, the probability of characteristic emission can be stated. This probability is termed the fluorescence yield, ω, where

$$\omega = \frac{\text{Number of characteristic photons emitted}}{\text{Number of electron shell vacancies}}$$

The fluorescence yield for K shell vacancies is plotted in Margin Figure 2-6 as a function of atomic number. These data reveal that the fluorescence yield increases with increasing atomic number. For a transition to the K shell of calcium, for example, the probability is 0.19 that a K characteristic photon will be emitted and 0.81 that an Auger electron will be emitted. For every 100 K shell vacancies in calcium, an average of $0.19 \times 100 = 19$ characteristic photons and $0.81 \times 100 = 81$ Auger electrons will be released. The fluorescence yield is one factor to be considered in the selection of radioactive sources for nuclear imaging, because Auger electrons increase the radiation dose to the patient without contributing to the diagnostic quality of the study.

A Brief History of Quantum Mechanics
1913: Bohr advanced a model of the atom in which electrons move in circular orbits at radii that allow their momenta to take on only certain specific values.[13]
1916–1925: Bohr's model was modified by Sommerfeld, Stoner, Pauli, and Uhlenbeck to better explain the emission and absorption spectra of multielectron atoms.[14–18]
1925: de Broglie hypothesized that all matter has wavelike properties. The wavelike nature of electrons allows only integral numbers of "wavelengths" in an electron orbit, thereby explaining the discrete spacing of electron "orbits" in the atom.[19]
1927: Davisson and Germer demonstrated that electrons undergo diffraction, thereby proving that they may behave like "matter waves."[20]
1925–1929: Born, Heisenberg, and Schrödinger describe a new field of physics in which predictions about the behavior of particles may be made from equations governing the behavior of the particle's "wave function."[21–23]

"You believe in the God who plays dice, and I in complete law and order in a world which objectively exists, and which I, in a wildly speculative way, am trying to capture. . . . Even the great initial success of the quantum theory does not make me believe in the fundamental dice game, although I am well aware that our younger colleagues interpret this as a consequence of senility."
Albert Einstein

SOLIDS

Electrons in individual atoms have specific binding energies described by quantum mechanics. When atoms bind together into solids, the energy levels change as the electrons influence each other. Just as each atom has a unique set of quantum energy levels, a solid also has a unique set of energy levels. The energy levels of solids are referred to as energy bands. And just as quantum "vacancies," or "holes," may exist when an allowable energy state in a single atom is not filled, energy bands in a solid may or may not be fully populated with electrons. The energy bands in a solid are determined by the combination of atoms composing the solid and also by bulk properties of the material such as temperature, pressure, and so on.

Two electron energy bands of a solid are depicted in Margin Figure 2-7. The lower energy band, called the valence band, consists of electrons that are tightly bound in the chemical structure of the material. The upper energy band, called the conduction band, consists of electrons that are relatively loosely bound. Conduction band electrons are able to move in the material and may constitute an electrical current under the proper conditions. If no electrons populate the conduction band, then the material cannot support an electrical current under normal circumstances. However, if enough energy is imparted to an electron in the valence band to raise it to the conduction band, then the material can support an electrical current.

Solids can be separated into three categories on the basis of the difference in energy between electrons in the valence and conduction bands. In *conductors* there is little energy difference between the bands. Electrons are continuously promoted from the valence to the conduction band by routine collisions between electrons. In *insulators* the conduction and valence bands are separated by a band gap (also known as the "forbidden zone") of 3 eV or more. Under this condition, application of voltage to the material usually will not provide enough energy to promote electrons from the valence to the conduction band. Therefore, insulators do not support an electrical current under normal circumstances. Of course, there is always a "breakdown voltage" above which an insulator will support a current, although probably not without structural damage.

If the band gap of the material is between 0 and 3 eV, then the material exhibits electrical properties between those of an insulator and a conductor. Such a material, termed a *semiconductor*, will conduct electricity under certain conditions and act as an insulator under others. The conditions may be altered by the addition to the material of trace amounts of impurities that have allowable energy levels that fall within the

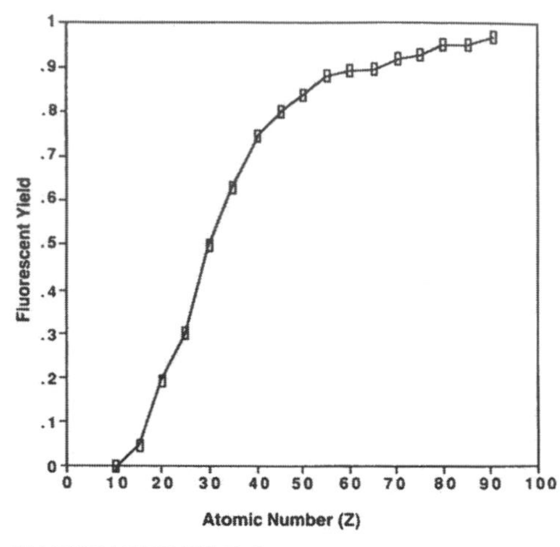

MARGIN FIGURE 2-6
Fluorescence yield for the K shell versus atomic number.

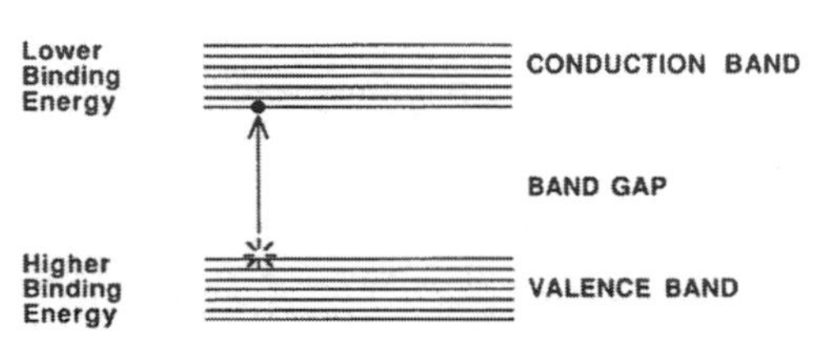

MARGIN FIGURE 2-7
Energy level diagram for solids. An electron promoted from the valence band to the conduction band may move freely in the material to constitute an electric current.

Superconductivity was first discovered in mercury, by Onnes in 1911.[24] However, a satisfactory theoretical explanation of superconductivity, using the formalism of quantum mechanics, did not evolve until 1957 when BCS theory was proposed by Bardeen, Cooper, and Schrieffer.[25, 26]

The 1987 Nobel Prize in Physics was awarded jointly to J. George Bednorz and K. Alexander Müller "for their important breakthrough in the discovery of superconductivity in ceramic materials."

band gap of the solid. Semiconductors have many applications in radiation detection and are discussed further in Chapter 8.

SUPERCONDUCTIVITY

In a conductor, very little energy is required to promote electrons to the conduction band. However, free movement of electrons in the conduction band is not guaranteed. As electrons attempt to move through a solid, they interact with other electrons and with imperfections such as impurities in the solid. At each "encounter" some energy is transferred to atoms and molecules and ultimately appears as heat. This transfer of energy (the basis of electrical "resistance") is usually viewed as energy "loss." Sometimes, however, this "loss" is the primary goal of electrical transmission as, for example, in an electric iron or the filament of an x-ray tube.

There are materials in which, under certain conditions, there is no resistance to the flow of electrons. These materials are called superconductors. In a superconductor the passage of an electron disturbs the structure of the material in such a way as to encourage the passage of another electron arriving after exactly the right interval of time. The passage of the first electron establishes an oscillation in the positive charge of the material that "pulls" the second electron through. This behavior has been compared to "electronic water skiing" where one electron is swept along in another electron's "wake." Thus electrons tend to travel in what are known as "Cooper pairs."[26] Cooper pairs do not (in a probabilistic sense) travel close to each other. In fact, many other electrons can separate a Cooper pair. Cooper pair electrons are also paired with other electrons. It has been shown mathematically that the only possible motion of a set of interleaved pairs of electrons is movement as a unit. In this fashion, the electrons do not collide randomly with each other and "waste" energy. Instead, they travel as a unit with essentially no resistance to flow. Currents have been started in superconducting loops of wire that have continued for several years with no additional input of energy.

Many types of materials exhibit superconductive behavior when cooled to temperatures in the range of a few degrees Kelvin (room temperature is approximately 295°K). Lowering the temperature of some solids promotes superconductivity by decreasing molecular motion, thereby decreasing the kinetic energy of the material. Twenty-six elements and thousands of alloys and compounds exhibit this behavior. Maintenance of materials at very low temperatures requires liquid helium as a cooling agent. Helium liquefies at 23°K, is relatively expensive, and is usually insulated from ambient conditions with another refrigerant such as liquid nitrogen.

On theoretical grounds it had been suspected for many years that superconductivity can exist in some materials at substantially higher temperatures, perhaps even room temperature. In January 1986, Bednorz and Müller discovered a ceramic, an oxide of barium, lanthanum, and copper, that is superconducting at temperatures up to 35°K.[27] With this discovery, superconductivity was achieved for the first time at temperatures above liquid helium. This finding opened new possibilities for cheaper and more convenient refrigeration methods. Subsequently, several other ceramics have exhibited superconductivity at temperatures of up to 135°K. There is, at present, no satisfactory theoretical explanation for superconductivity in ceramics. Some believe that there is a magnetic phenomenon that is analogous to the electronic phenomenon cited in BCS theory. The current record for highest superconducting temperature in a metal is in magnesium diboride at 39°K.[28] While this temperature is not as high as has been achieved in some ceramics, this material would be easier to fashion into a wire for use as a winding in a superconducting magnet. These materials have great potential for yielding "perfect" conductors (i.e., with no electrical resistance) that are suitable for everyday use. Realization of this potential would have a profound impact upon the design of devices such as electrical circuits and motors. It would revolutionize fields as diverse as computer science, transportation, and medicine, including radiology.

THE NUCLEUS

Nuclear Energy Levels

A nucleus consists of two types of particles, referred to collectively as nucleons. The positive charge and roughly half the mass of the nucleus are contributed by protons. Each proton possesses a positive charge of $+1.6 \times 10^{-19}$ coulombs, equal in magnitude but opposite in sign to the charge of an electron. The number of protons (or positive charges) in the nucleus is the atomic number of the atom. The mass of a proton is 1.6734×10^{-27} kg. Neutrons, the second type of nucleon, are uncharged particles with a mass of 1.6747×10^{-27} kg. Outside the nucleus, neutrons are unstable and divide into protons, electrons, and antineutrinos (see Chapter 3). The half-life of this transition is 12.8 minutes. Neutrons are usually stable inside nuclei. The number of neutrons in a nucleus is the neutron number N for the nucleus. The mass number A of the nucleus is the number of nucleons (neutrons and protons) in the nucleus. The mass number $A = Z + N$.

The standard form used to denote the composition of a specific nucleus is

$$^{A}_{Z}\mathrm{X}$$

where X is the chemical symbol (e.g., H, He, Li) and A and Z are as defined above. There is some redundancy in this symbolism. The atomic number, Z, is uniquely associated with the chemical symbol, X. For example, when $Z = 6$, the chemical symbol is always C, for the element carbon.

Expressing the mass of atomic particles in kilograms is unwieldy because it would be a very small number requiring scientific notation. The atomic mass unit (amu) is a more convenient unit for the mass of atomic particles. 1 amu is defined as 1/12 the mass of the carbon atom, which has six protons, six neutrons, and six electrons. Also,

$$1 \text{ amu} = 1.6605 \times 10^{-27} \text{ kg}$$

The shell model of the nucleus was introduced to explain the existence of discrete nuclear energy states. In this model, nucleons are arranged in shells similar to those available to electrons in the extranuclear structure of an atom. Nuclei are extraordinarily stable if they contain 2, 8, 14, 20, 28, 50, 82, or 126 protons or similar numbers of neutrons. These numbers are termed magic numbers and may reflect full occupancy of nuclear shells. Nuclei with odd numbers of neutrons or protons tend to be less stable than nuclei with even numbers of neutrons and protons. The pairing of similar nucleons increases the stability of the nucleus. Data tabulated below support this hypothesis.

Number of Protons	Number of Neutrons	Number of Stable Nuclei
Even	Even	165
Even	Odd	57
Odd	Even	53
Odd	Odd	6

Nuclear Forces and Stability

Protons have "like" charges (each has the same positive charge) and repel each other by the electrostatic force of repulsion. One may then ask the question, How does a nucleus stay together? The answer is that when protons are very close together, an attractive force comes into play. This force, called the "strong nuclear force," is 100 times greater than the electrostatic force of repulsion. However, it acts only over distances of the order of magnitude of the diameter of the nucleus. Therefore, protons can stay together in the nucleus once they are there. Assembling a nucleus by forcing protons together requires the expenditure of energy to overcome the electrostatic repulsion. Neutrons, having no electrostatic charge, do not experience the electrostatic force. Therefore, adding neutrons to a nucleus requires much less energy. Neutrons are, however,

Superconductors have zero resistance to the flow of electricity. They also exhibit another interesting property called the Meisner Effect: If a superconductor is placed in a magnetic field, the magnetic field lines flow around it. That is, the superconductor excludes the magnetic field. This can be used to create a form of magnetic levitation.

As of this writing, magnetic resonance imagers are the only devices using the principles of superconductivity that a typical layperson might encounter. Other applications of superconductivity are found chiefly in research laboratories.

The word "atom" comes from the Greek "atomos" which means "uncuttable."

The half-life for decay of the neutron is 12.8 minutes. This means that in a collection of a large number of neutrons, half would be expected to undergo the transition in 12.8 minutes. Every 12.8 minutes, half the remaining number would be expected to decay. After 7 half-lives (3.7 days), fewer than 1% would be expected to remain as neutrons.

Masses of atomic particles are as follows:

Electron = 0.00055 amu

Proton = 1.00727 amu

Neutron = 1.00866 amu

Note that the proton and neutron have a mass of approximately 1 amu and that the neutron is slightly heavier than the proton.

Quantum Electrodynamics (QED)
Modern quantum mechanics explains a force in terms of the exchange of "messenger particles." These particles pass between (are emitted and then absorbed by) the particles that are affected by the force.
The messenger particles are as follows:

Force	Messenger
Strong nuclear	Gluon
Electromagnetic	Photon
Weak nuclear	"W" and "Z"
Gravity	Graviton

affected by a different "weak nuclear force." The weak nuclear force causes neutrons to change spontaneously into protons plus almost massless virtually noninteracting particles called neutrinos. The opposite transition, protons turning into neutrons plus neutrinos, also occurs. These processes, called beta decay, are described in greater detail in Chapter 3. The fourth of the traditional four "fundamental forces" of nature, gravity, is extremely overshadowed by the other forces within an atom, and thus it plays essentially no role in nuclear stability or instability. The relative strengths of the "four forces" are as follows:

Type of Force	Relative Strength
Nuclear	1
Electrostatic	10^{-2}
Weak	10^{-13}
Gravitational	10^{-39}

Of these forces, the first three have been shown to be manifestations of the same underlying force in a series of "Unification Theories" over recent decades. The addition of the fourth, gravity, would yield what physicists term a "Grand Unified Theory," or GUT.

Nuclear Binding Energy

The mass of an atom is less than the sum of the masses of its neutrons, protons, and electrons. This seeming paradox exists because the binding energy of the nucleus is so significant compared with the masses of the constituent particles of an atom, as expressed through the equivalence of mass and energy described by Einstein's famous equation.[29]

$$E = mc^2$$

The mass difference between the sum of the masses of the atomic constituents and the mass of the assembled atom is termed the *mass defect*. When the nucleons are separate, they have their own individual masses. When they are combined in a nucleus, some of their mass is converted into energy. In Einstein's equation, an energy E is equivalent to mass m multiplied by the speed of light in a vacuum, c (2.998×10^8 m/sec) squared. Because of the large "proportionality constant" c^2 in this equation, one kilogram of mass is equal to a large amount of energy, 9×10^{16} joules, roughly equivalent to the energy released during detonation of 30 megatons of TNT. The energy equivalent of 1 amu is

$$\frac{(1 \text{ amu})(1.660 \times 10^{-27} \text{ kg/amu})(2.998 \times 10^8 \text{ m/sec})^2}{(1.602 \times 10^{-13} \text{ J/Mev})} = 931 \text{ MeV}$$

The binding energy (mass defect) of the carbon atom with six protons and six neutrons (denoted as ${}^{12}_{6}C$) is calculated in Example 2-2.

Example 2-2

$$\begin{aligned}
\text{Mass of 6 protons} &= 6(1.00727 \text{ amu}) = 6.04362 \text{ amu} \\
\text{Mass of 6 neutrons} &= 6(1.00866 \text{ amu}) = 6.05196 \text{ amu} \\
\text{Mass of 6 electrons} &= 6(0.00055 \text{ amu}) = 0.00330 \text{ amu} \\
\text{Mass of components of } {}^{12}_{6}\text{C} &= 12.09888 \text{ amu} \\
\text{Mass of } {}^{12}_{6}\text{C atom} &= 12.00000 \text{ amu} \\
\text{Mass defect} &= 0.09888 \text{ amu} \\
\text{Binding energy of } {}^{12}_{6}\text{C atom} &= (0.09888 \text{ amu})(931 \text{ MeV/amu}) \\
&= 92.0 \text{ MeV}
\end{aligned}$$

Almost all of this binding energy is associated with the $^{12}_{6}C$ nucleus. The average binding energy per nucleon of $^{12}_{6}C$ is 92.0 MeV per 12 nucleons, or 7.67 MeV per nucleon.

In Margin Figure 2-8 the average binding energy per nucleon is plotted as a function of the mass number A.

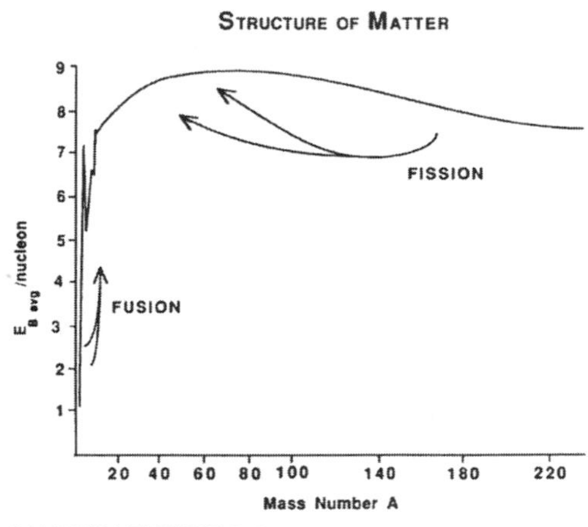

MARGIN FIGURE 2-8
Average binding energy per nucleon versus mass number.

NUCLEAR FISSION AND FUSION

Energy is released if a nucleus with a high mass number separates or fissions into two parts, each with an average binding energy per nucleon greater than that of the original nucleus. The energy release occurs because such a split produces low-Z products with a higher average binding energy per nucleon than the original high-Z nucleus (Margin Figure 2-8). A transition from a state of lower "binding energy per nucleon" to a state of higher "binding energy per nucleon" results in the release of energy. This is reminiscent of the previous discussion of energy release that accompanies an L to K electron transition. However, the energy available from a transition between nuclear energy levels is orders of magnitude greater than the energy released during electron transitions.

Certain high-A nuclei (e.g., ^{235}U, ^{239}Pu, ^{233}U,) fission spontaneously after absorbing a slowly moving neutron. For ^{235}U, a typical fission reaction is

$$^{235}_{92}U + \text{neutron} \rightarrow \,^{236}_{92}U \rightarrow \,^{92}_{36}Kr + \,^{141}_{56}Ba + 3 \text{ neutrons} + Q$$

The energy released is designated as Q and averages more than 200 MeV per fission. The energy is liberated primarily as γ radiation, kinetic energy of fission products and neutrons, and heat and light. Products such as $^{92}_{36}Kr$ and $^{141}_{56}Ba$ are termed fission by-products and are radioactive. Many different by-products are produced during fission. Neutrons released during fission may interact with other ^{235}U nuclei and create the possibility of a chain reaction, provided that sufficient mass of fissionable material (a critical mass) is contained within a small volume. The rate at which a material fissions may be regulated by controlling the number of neutrons available each instant to interact with fissionable nuclei. Fission reactions within a nuclear reactor are controlled in this way. Uncontrolled nuclear fission results in an "atomic explosion."

Nuclear fission was observed first in 1934 during an experiment conducted by Enrico Fermi.[31, 32] However, the process was not described correctly until publication in 1939 of analyses by Hahn and Strassmann[33] and Meitner and Frisch.[34] The first controlled chain reaction was achieved in 1942 at the University of Chicago. The first atomic bomb was exploded in 1945 at Alamogordo, New Mexico.[35]

"The energy produced by the breaking down of the atom is a very poor kind of thing. Anyone who expects a source of power from the transformation of these atoms is talking moonshine."
E. Rutherford, 1933.

The critical mass of ^{235}U is as little as 820 g if in aqueous solution or as much as 48.6 kg if a bare metallic sphere.[30]

Certain low-mass nuclei may be combined to produce a nucleus with an average binding energy per nucleon greater than that for either of the original nuclei. This process is termed nuclear fusion (Margin Figure 2-8) and is accompanied by the release of large amounts of energy. A typical reaction is

$$^{2}_{1}H + \,^{3}_{1}H \rightarrow \,^{4}_{2}He + \text{neutron} + Q$$

In this particular reaction, $Q = 18$ MeV.

To form products with higher average binding energy per nucleon, nuclei must be brought sufficiently near one another that the nuclear force can initiate fusion. In the process, the strong electrostatic force of repulsion must be overcome as the two nuclei approach each other. Nuclei moving at very high velocities possess enough momentum to overcome this repulsive force. Adequate velocities may be attained by heating a sample containing low-Z nuclei to a temperature greater than 12×10^6 °K, roughly equivalent to the temperature in the inner region of the sun. Temperatures this high may be attained in the center of a fission explosion. Consequently, a fusion (hydrogen) bomb is "triggered" with a fission bomb. Controlled nuclear fusion has not

The only nuclear weapons used in warfare were dropped on Japan in 1945. The Hiroshima bomb used fissionable uranium, and the Nagasaki bomb used plutonium. Both destroyed most of the city on which they fell, killing more than 100,000 people. They each released energy equivalent to about 20,000 tons of TNT. Large fusion weapons (H-bombs) release up to 1000-fold more energy.

yet been achieved on a macroscopic scale, although much effort has been expended in the attempt.

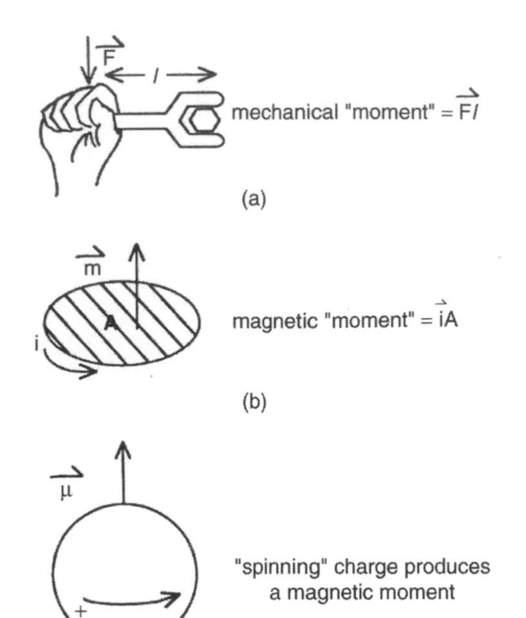

MARGIN FIGURE 2-9
The concept of a "moment" in physics. **A**: A mechanical moment is defined by force **F** and length l. **B**: A magnetic moment is defined by current i and the area A enclosed by the current. **C**: The magnetic moment produced by a spinning charged object.

NUCLEAR SPIN AND NUCLEAR MAGNETIC MOMENTS

Protons and neutrons behave like tiny magnets and are said to have an associated magnetic moment. The term *moment* has a strict meaning in physics. When a force is applied on a wrench to turn a bolt (Margin Figure 2-9A), for example, the mechanical moment is the product of force and length. The mechanical moment can be increased by increasing the length of the wrench, applying more force to the wrench, or a combination of the two. A magnetic moment (Margin Figure 2-9B) is the product of the current in a circuit (a path followed by electrical charges) and the area encompassed by the circuit. The magnetic moment is increased by increasing the current, the area, or a combination of the two. The magnetic moment is a vector, a quantity having both magnitude and direction.

Like electrons, protons have a characteristic called "spin," which can be explained by treating the proton as a small object spinning on its axis. In this model, the spinning charge of the proton produces a magnetic moment (Margin Figure 2-9C).

The "spinning charge" model of the proton has some limitations. First, the mathematical prediction for the value of the magnetic moment is not equal to what has been measured experimentally. From the model, a proton would have a fundamental magnetic moment known as the nuclear magneton, u_n:

$$u_n = \frac{e\hbar}{2m_p} = 3.1525 \times 10^{-12} \text{ eV/gauss}$$

where e is the charge of the proton in coulombs, $\hbar$ is Planck's constant divided by 2π, and m_p is the mass of the proton. The magnetic moment magneton, u_p of the proton, however, is

$$u_p = \text{magnetic moment of the proton} = 2.79u_n$$

The unit of the nuclear magneton, energy (electron volt) per unit magnetic field strength (gauss), expresses the fact that a magnetic moment has a certain (potential) energy in a magnetic field. This observation will be used later to describe the fundamental concepts of magnetic resonance imaging (MRI).

The second difficulty of the spinning proton model is that the neutron, an uncharged particle, also has a magnetic moment. The magnetic moment of the neutron equals $1.91u_n$. The explanation for the "anomalous" magnetic moment of the neutron, as well as the unexplained value of the proton's magnetic moment, is that neutrons and protons are not "fundamental" particles. Instead, they are each composed of three particles called quarks[39] that have fractional charges that add up to a unit charge. Quarks do not exist on their own but are always bound into neutrons, protons, or other particles. The presence of a nonuniform distribution of spinning charges attributable to quarks within the neutron and proton explains the observed magnetic moments.

The magnetic moment of the proton was first observed by Stern and colleagues in 1933.[36,37] The magnetic moment of the neutron was measured by Alvarez and Bloch in 1940.[38] Bloch went on to write the fundamental equations for the "relaxation" of nuclear spins in a material in a static magnetic field that has been perturbed by radiofrequency energy. These equations are the basis of magnetic resonance imaging.

When magnetic moments exist near each other, as in the nucleus, they tend to form pairs with the vectors of the moments pointed in opposite directions. In nuclei with even numbers of neutrons and protons (i.e., even Z, even N), this pairing cancels out the magnetic properties of the nucleus as a whole. Thus an atom such as ${}^{12}_{6}C$ with 6 protons and 6 neutrons has no net magnetic moment because the neutrons and protons are "paired up."

An atom with an odd number of either neutrons or protons will have a net magnetic moment. For example, ${}^{13}_{6}C$ with 6 protons and 7 neutrons has a net magnetic moment because it contains an unpaired neutron. Also, ${}^{14}_{7}N$ with 7 protons and 7 neutrons has a small net magnetic moment because both proton and neutron numbers are odd and the "leftover" neutron and proton do not exactly cancel each other's moments. Table 2-3 lists a number of nuclides with net magnetic moments. The

TABLE 2-3 Nuclides with a Net Magnetic Moment[a]

Nuclide	*Number of Protons*	*Number of Neutrons*	*Magnetic Moment (Multiple of u_n)*
^{1}H	1	0	2.79
^{2}H	1	1	0.86
^{13}C	6	7	0.70
^{14}N	7	7	0.40
^{17}O	8	9	−1.89
^{19}F	9	10	2.63
^{23}Na	11	12	2.22
^{31}P	15	16	1.13
^{39}K	19	20	0.39

[a]Data from Heath, R. L. Table of the Isotopes, in Weast, R. C., et al.(eds.), *Handbook of Chemistry and Physics*, 52nd edition. Cleveland, Chemical Rubber Co., 1972, pp. 245–256.

"Virtual particles" such as quark–antiquark pairs, or "messenger particles" that carry forces between particles, can appear and disappear during a time interval Δt so long as $\Delta t < h/\Delta E$. ΔE is the change in mass/energy of the system that accounts for the appearance of the particles, and h is Planck's constant. This formula is one version of Heisenberg's Uncertainty Principle. In the nucleus, the time interval is the order of magnitude of the time required for a beam of light to cross the diameter of a single proton.

presence of a net magnetic moment for the nucleus is essential to magnetic resonance imaging (MRI). Only nuclides with net magnetic moments are able to interact with the intense magnetic field of a MRI unit to provide a signal to form an image of the body.

NUCLEAR NOMENCLATURE

Isotopes of a particular element are atoms that possess the same number of protons but a varying number of neutrons. For example, ^{1}H (protium), ^{2}H (deuterium), and ^{3}H (tritium) are isotopes of the element hydrogen, and $^{9}_{6}$C, $^{10}_{6}$C, $^{11}_{6}$C, $^{12}_{6}$C, $^{13}_{6}$C, $^{14}_{6}$C, $^{15}_{6}$C, and $^{16}_{6}$C are isotopes of carbon. An isotope is specified by its chemical symbol together with its mass number as a left superscript. The atomic number is sometimes added as a left subscript.

Isotones are atoms that possess the same number of neutrons but a different number of protons. For example, $^{5}_{2}$He, $^{6}_{3}$Li, $^{7}_{4}$Be, $^{8}_{5}$B, and $^{9}_{6}$C are isotones because each isotope contains three neutrons. *Isobars* are atoms with the same number of nucleons but a different number of protons and a different number of neutrons. For example, $^{6}_{2}$He, $^{6}_{3}$Li, and $^{6}_{4}$Be are isobars ($A = 6$). *Isomers* represent different energy states for nuclei with the same number of neutrons and protons. Differences between isotopes, isotones, isobars, and isomers are illustrated below:

	Atomic No. Z	Neutron No. N	Mass No. A
Isotopes	Same	Different	Different
Isotones	Different	Same	Different
Isobars	Different	Different	Same
Isomers	Same	Same	Same (different nuclear energy states)

The general term for any atomic nucleus, regardless of A, Z, or N, is "nuclide."

The full explanation of nuclear spin requires a description of the three main types of quark (up, down, and strange) along with the messenger particles that carry the strong nuclear force between them (gluons) together with the short-lived "virtual" quarks and antiquarks that pop in and out of existence over very short time periods within the nucleus.[39]

PROBLEMS

*2-1. What are the atomic and mass numbers of the oxygen isotope with 16 nucleons? Calculate the mass defect, binding energy, and binding energy per nucleon for this nuclide, with the assumption that the entire mass defect is associated with the nucleus. The mass of the atom is 15.9949 amu.

*2-2. Natural oxygen contains three isotopes with atomic masses of 15.9949, 16.9991, and 17.9992 and relative abundances of 2500:1:5, respectively. Determine to three decimal places the average atomic mass of oxygen.

*For problems marked with an asterisk, answers are provided on page 491.

2-3. Using Table 2-1 as an example, write the quantum numbers for electrons in boron ($Z = 5$), oxygen ($Z = 8$), and phosphorus ($Z = 15$).

*2-4. Calculate the energy required for the transition of an electron from the K shell to the L shell in tungsten (see Margin Figure 2-4). Compare the result with the energy necessary for a similar transition in hydrogen. Explain the difference.

*2-5. What is the energy equivalent to the mass of an electron? Because the mass of a particle increases with velocity, assume that the electron is at rest.

*2-6. The energy released during the atomic explosion at Hiroshima was estimated to be equal to that released by 20,000 tons of TNT. Assume that a total energy of 200 MeV is released during fission of a ^{235}U nucleus and that a total energy of 3.8×10^9 J is released during detonation of 1 ton of TNT. Find the number of fissions that occurred in the Hiroshima explosion, and determine the total decrease in mass.

*2-7. A "4-megaton thermonuclear explosion" means that a nuclear explosion releases as much energy as that liberated during detonation of 4 million tons of TNT. Using 3.8×10^9 J/ton as the heat of detonation for TNT, calculate the total energy in joules and in kilocalories released during the nuclear explosion (1 kcal = 4186 J).

2-8. Group the following atoms as isotopes, isotones,and isobars: $^{131}_{54}Xe$, $^{132}_{54}Xe$, $^{130}_{53}I$, $^{133}_{54}Xe$, $^{131}_{53}I$, $^{129}_{52}Te$, $^{132}_{53}I$, $^{130}_{52}Te$, $^{131}_{52}Te$.

SUMMARY

- Atoms are composed of protons and neutrons in the nucleus, which is surrounded by electrons in shell configurations.
- Electron shells:

Shell	Maximum number of Electrons	Binding energy
K ($N = 1$)	2	Highest
L ($N = 2$)	8	Lower than K
M ($N = 3$)	18	Lower than L

- Transitions of electrons from outer shells to inner shells result in the release of characteristic photons and/or auger electrons
- Materials may be classified in terms of their electrical conductivity as
  - Insulators
  - Semiconductors
  - Conductors
  - Superconductors
- The standard form to denote a nuclide is

$$^{A}_{Z}X$$

where
A is the mass number, the sum of the number of neutrons and protons
Z is the number of protons
X is the chemical symbol (which is uniquely determined by *Z*)
- The four fundamental fources, in order of strength from strongest to weakest are
  - Nuclear
  - Electrostatic
  - Weak
  - Gravitational
- Nuclear fission and fusion both result in product(s) with higher binding energy per nucleon
- Nuclear nomenclature:
  - Isoto*p*es—same number of *p*rotons
  - Isoto*n*es—same number of *n*eutrons
  - Isob*a*rs—same mass number, *A*
  - Isom*e*rs—same *e*verything *e*xcept *e*nergy

REFERENCES

1. Lucretius. *On the Nature of Things* (Lantham translation). Baltimore, Penguin Press, 1951. Bailey, C. *The Greek Atomists and Epicurus*. New York, Oxford University Press, 1928.
2. Dalton, J. Experimental enquiry into the proportions of the several gases or elastic fluids constituting the atmosphere. *Mem Literary Philos. Soc. Manchester* 1805; **1**:244.
3. Gay-Lussac, J. Sur la combinaison des substances gazeuses, les unes avec les autres. *Mem. Soc. d'Arcoeil* 1809; **2**:207.
4. Avogadro, A. D'une manière de déterminer les masses relatives des molécules élémentaires des corps, et les proportions selon lesquelles elles entrent dans ces combinaisons. 1811; *J. Phys.* **73**:58.
5. Faraday, M. Identity of electricities derived from different sources. *Philos. Trans.* 1833; **23**.
6. Cannizzaro, S. An abridgement of a course of chemical philosophy given in the Royal University of Genoa. *Nuovo Cimento* 1858; **7**:321.
7. Meyer, J. Die Natur der chemischen Elemente als Funktion ihrer Atomgewichte. *Ann. Chem. Suppl.* 1870; **7**:354.
8. Mendeleev, D. The relation between the properties and atomic weights of the elements. *J. Russ. Chem. Soc.* 1869; **1**:60.
9. Perrin, J. *Atoms*. London, Constable, 1923.
10. Thomson, J. Papers on positive rays and isotopes. *Philos. Mag. 6th Ser.* 1907; **13**:561.
11. Rutherford, E. The scattering of alpha and beta particles by matter and the structure of the atom. *Philos. Mag.* 1911; **21**:669.
12. Pauling, L. *General Chemistry*. Dover Publications, New York, 1988, p. 129.
13. Bohr, N. On the constitution of atoms and molecules. *Philos. Mag.* 1913; **26**:1, 476, 875.
14. Sommerfeld, A. Zur Quantentheorie der Spektrallinien. *Ann. Phys.* 1916; **1**:125.
15. Wilson, W. The quantum-theory of radiation and line spectra. *Philos. Mag.* 1915; **29**:795.
16. Stoner, E. The distribution of electrons among atomic levels. *Philos. Mag.* 1924; **48**:719.
17. Pauli, W. Über den Zusammenhang des Abschlusses der Elektronengruppen im Atom mit der Komplexstruktur der Spektren. *Z. Phys.* 1925; **31**: 765.
18. Uhlenbeck, G., and Goudsmit, S. Ersetzung der Hypothese vom unmechanischen Zwang durch eine Forderung bezuglich des inneren Verhaltens jedes einzelnen Elektrons. *Naturwissenschaften* 1925; **13**:953.
19. de Broglie, L. Attentative theory of light quanta. *Philos. Mag.* 1926; **47**:446. Recherches sur la theorie des quanta. *Ann. Phys.* 1925; **3**:22.
20. Davisson, C. J., and Germer, L. H. Diffraction of electrons by a crystal of nickel. *Phys. Rev.* 1927; **30**:705–740.
21. Born, M. Quantenmechanik der Stossvorgänge. *Z. Phys.* 1926; **38**:803.
22. Heisenberg, W. *The Physical Principles of the Quantum Theory*. Chicago, University of Chicago Press, 1929.
23. Schrödinger, E. *Collected Papers on Wave Mechanics*. Glasgow, Blackie & Son, Ltd., 1928.
24. Onnes, H. K. *Akad. van Wetenschappen Amsterdam* 1911; **14**:113–118.
25. Bardeen, J., Cooper, L. N., Schrieffer J. R. *Physiol. Rev.* 1957; **108**:1175.
26. Cooper, L. N. *Physiol. Rev.* 1956; **104**:1189.
27. Bednorz, J. G., and Müller, K. A. Z. *Physik B* 1986; **64**:189.
28. Service, R. F. Material sets record for metal compounds. *Science* 23 Feb., 2001; Vol. **291**(No. 5508):1476–1477.
29. Einstein, A. Über einen die Erzeugung and Verwandlung des Lichtes betreffenden heuristischen Geisichtspunkt. *Ann. Phys.* 1905; **17**:132.
30. Cember, H. *Introduction to Health Physics*, 1st edition. New York, Pergamon Press, 1969, pp. 372–374.
31. Fermi, E. Radioactivity induced by neutron bombardment. *Nature* 1934; **133**:757.
32. Fermi, E. Possible production of elements of atomic number higher than 92. *Nature* 1934; **133**:898.
33. Hahn, O., and Strassmann, F. Über den Nachweis und das Verhalten der bei der Bestrahlung des Urans mittels Neutronen entstehenden Erdalkalimetalle. *Naturwissenschaften* 1939; **27**:11, 89.
34. Meitner, L., and Frisch, O. Disintegration of uranium by neutrons: A new type of nuclear reaction. *Nature* 1939; **143**:239.
35. Rhodes, R. *The Making of the Atomic Bomb*. New York, Simon & Schuster, 1986.
36. Friesch, R., and Stern, O. Über die magnetische Ablenkung von Wasserstoff—Molekulen und das magnetische Moment das Protons I. *Z. Phys.* 1933; **85**:4–16.
37. Esterman, I., and Stern, O. Über die magnetische Ablenkung von Wasserstoff—Molekulen und das magnetische Moment das Protons II. *Z. Phys.* 1933; **85**:17–24.
38. Alvarez, L. W., and Bloch, F. *Physiol. Rev.* 1940; **57**:111.
39. Rith, K., and Schafer, A. The mystery of nucleon spin. *Scientific American* July 1999, pp. 58–63.

CHAPTER

3

RADIOACTIVE DECAY

OBJECTIVES

By studying this chapter, the reader should be able to:

- Understand the relationship between nuclear instability and radioactive decay.
- Describe the different modes of radioactive decay and the conditions in which they occur.
- Draw and interpret decay schemes.
- Write balanced reactions for radioactive decay.
- State and use the fundamental equations of radioactive decay.
- Perform elementary computations for sample activities.
- Comprehend the principles of transient and secular equilibrium.
- Discuss the principles of the artificial production of radionuclides.
- Find information about particular radioactive species.

Radioactivity was discovered in 1896 by Henri Becquerel,[1] who observed the emission of radiation (later shown to be beta particles) from uranium salts. A sentence from his 1896 publication reads "We may then conclude from these experiments that the phosphorescent substance in question emits radiations which penetrate paper opaque to light and reduces the salts of silver." Becquerel experienced a skin burn from carrying a radioactive sample in his vest pocket. This is the first known bioeffect of radiation exposure.

This chapter describes *radioactive decay*, a process whereby unstable nuclei become more stable. All nuclei with atomic numbers greater than 82 are unstable (a solitary exception is $^{209}_{83}$Bi). Many lighter nuclei (i.e., with $Z < 82$) are also unstable. These nuclei undergo radioactive decay (they are said to be "radioactive"). Energy is released during the decay of radioactive nuclei. This energy is termed the *transition energy*.

The transition energy released during radioactive decay is also referred to as the "energy of decay."

NUCLEAR STABILITY AND DECAY

Additional models of the nucleus have been proposed to explain other nuclear properties. For example, the "liquid drop" (also known as the "collective") model was proposed by the Danish physicist Niels Bohr[2] to explain nuclear fission. The model uses the analogy of the nucleus as a drop of liquid.

"Bohr's work on the atom was the highest form of musicality in the sphere of thought". A. Einstein as quoted in Moore, R. *Niels Bohr. The Man, His Science and the World They Changed.* New York, Alfred Knopf, 1966.

Neutrons can be transformed to protons, and vice versa, by rearrangement of their constituent quarks.

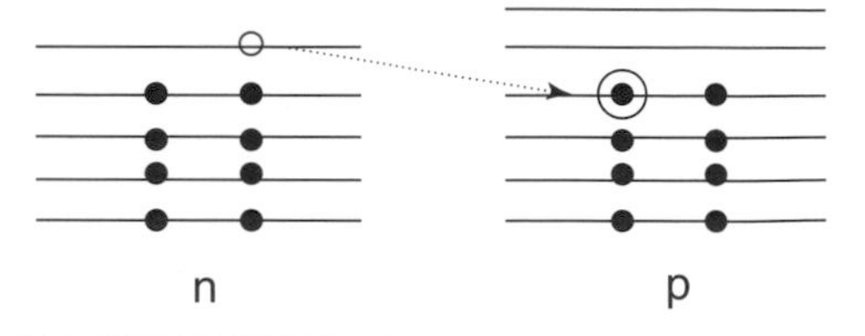

MARGIN FIGURE 3-1
Shell model for the ^{16}N nucleus. A more stable energy state is achieved by an $n \rightarrow p$ transition to form ^{16}O.

The nucleus of an atom consists of neutrons and protons, referred to collectively as nucleons. In a popular model of the nucleus (the "shell model"), the neutrons and protons reside in specific levels with different binding energies. If a vacancy exists at a lower energy level, a neutron or proton in a higher level may fall to fill the vacancy. This transition releases energy and yields a more stable nucleus. The amount of energy released is related to the difference in binding energy between the higher and lower levels. The binding energy is much greater for neutrons and protons inside the nucleus than for electrons outside the nucleus. Hence, energy released during nuclear transitions is much greater than that released during electron transitions.

If a nucleus gains stability by transition of a neutron between neutron energy levels, or a proton between proton energy levels, the process is termed an *isomeric transition*. In an isomeric transition, the nucleus releases energy without a change in its number of protons (Z) or neutrons (N). The initial and final energy states of the nucleus are said to be *isomers*. A common form of isomeric transition is *gamma decay*, in which the energy is released as a packet of energy (a quantum or photon) termed a gamma (γ) ray. An isomeric transition that competes with gamma decay is *internal conversion*, in which an electron from an extranuclear shell carries the energy out of the atom.

It is also possible for a neutron to fall to a lower energy level reserved for protons, in which case the neutron becomes a proton. It is also possible for a proton to fall to a lower energy level reserved for neutrons, in which case the proton becomes a neutron. In these situations, referred to collectively as *beta (β) decay*, the Z and N of the nucleus change, and the nucleus transmutes from one element to another.

In all of the transitions described above, the nucleus loses energy and gains stability. Hence, they are all forms of radioactive decay. In any radioactive process the mass number of the decaying (parent) nucleus equals the sum of the mass numbers of the product (progeny) nucleus and the ejected particle. That is, mass number A is conserved in radioactive decay.

ALPHA DECAY

Some heavy nuclei gain stability by a different form of radioactive decay, termed *alpha (α) decay*. In this mode of decay, an alpha particle (two protons and two neutrons tightly bound as a nucleus of helium ^4_2He) is ejected from the unstable nucleus. The alpha particle is a relatively massive, poorly penetrating type of radiation that can be stopped by a sheet of paper. An example of alpha decay is

$$^{226}_{88}Ra \rightarrow ^{222}_{86}Rn + ^4_2He$$

This example depicts the decay of naturally occurring radium into the inert gas radon by emission of an alpha particle. A decay scheme (see below) for alpha decay is depicted in the right margin.

Parent and progeny nuclei were referred to in the past as "mother" and "daughter." The newer and preferred terminology of parent and progeny are used in this text.

Alpha decay was discovered by Marie and Pierre Curie[3] in 1898 in their efforts to isolate radium, and it was first described by Ernest Rutherford[4] in 1899. Alpha particles were identified as helium nuclei by Boltwood and Rutherford in 1911.[5] The Curies shared the 1902 Nobel Prize in Physics with Henri Becquerel.

DECAY SCHEMES

A decay scheme depicts the decay processes specific for a *nuclide* (nuclide is a generic term for any nuclear form). A decay scheme is essentially a depiction of nuclear mass energy on the *y* axis, plotted against the atomic number of the nuclide on the *x* axis. A decay scheme is depicted in Figure 3-1, where the generic nuclide ^A_ZX has four possible routes of radioactive decay:

1. α decay to the progeny nuclide $^{A-4}_{Z-2}P$ by emission of a ^4_2He nucleus;
2. (a) β^+ (positron) decay to the progeny nuclide $_{Z-1}{}^AQ$ by emission of a positive electron from the nucleus;
   (b) electron capture (ec) decay to the progeny nuclide $_{Z-1}{}^AQ$ by absorption of an extranuclear electron into the nucleus;
3. β^-(negatron) decay to the progeny nuclide $_{Z+1}{}^AR$ by emission of a negative electron from the nucleus.

The processes denoted as 2(a) and 2(b) are competing pathways to the same progeny nuclide. Any of the pathways can yield a nuclide that undergoes an internal shuffling of nucleons to release additional energy. This process, termed an isomeric transition, is shown as pathway 4. No change in *Z* (or *N* or *A*) occurs during an isomeric transition.

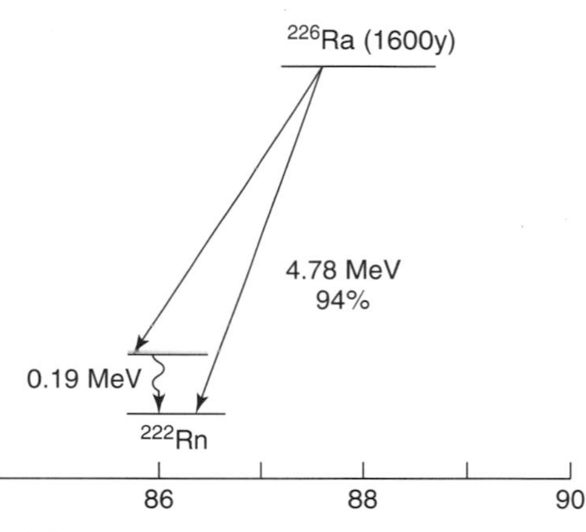

MARGIN FIGURE 3-2
Radioactive decay scheme for α-decay of ^{226}Ra.

Sometimes the symbol *Q* is added to the right side of the decay reaction to symbolize energy released during the decay process.

After a lifetime of scientific productivity and two Nobel prizes, Marie Curie died in Paris at the age of 67 from aplastic anemia, probably the result of years of exposure to ionizing radiation.

A radionuclide is a radioactive form of a nuclide.

Rutherford, revered as a teacher and research mentor, was known as "Papa" to his many students.

A decay scheme is a useful way to depict and assimilate the decay characteristics of a radioactive nuclide.

Negative and positive electrons emitted during beta decay are created at the moment of decay. They do not exist in the nucleus before decay.

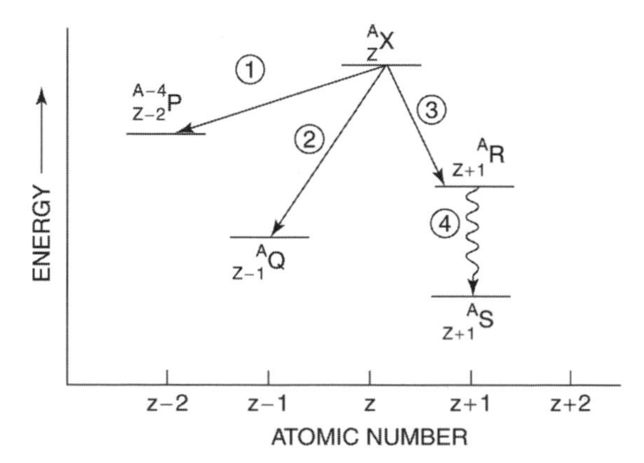

FIGURE 3-1
Radioactive decay scheme.

TABLE 3-1 Nuclear Stability Is Reduced When an Odd Number of Neutrons or Protons Is Present

Number of Protons	*Number of Neutrons*	*Number of Stable Nuclei*
Even	Even	165
Even	Odd	57
Odd	Even	53
Odd	Odd	6

BETA DECAY

Ernest Rutherford first characterized beta decay in 1899.[4]

The *shell model* of the nucleus was introduced by Maria Goeppert in 1942 to explain the magic numbers.

Nuclear Stability

Nuclei tend to be most stable if they contain even numbers of protons and neutrons, and least stable if they contain an odd number of both. This feature is depicted in Table 3-1.

Nuclei are extraordinarily stable if they contain 2, 8, 14, 20, 28, 50, 82, or 126 protons or similar numbers of neutrons. These numbers are termed nuclear *magic numbers* and reflect full occupancy of nuclear shells.

The number of neutrons is about equal to the number of protons in low-Z stable nuclei. As Z increases, the number of neutrons increases more rapidly than the number of protons in stable nuclei, as depicted in Figure 3-2. The shell model of the nucleus accounts for this finding by suggesting that at higher Z, the energy difference is slightly less between neutron levels than between proton levels.

Many nuclei exist that have too many or too few neutrons to reside on or close to the line of stability depicted in Figure 3-2. These nuclei are unstable and undergo radioactive decay. Nuclei above the line of stability (i.e., the n/p ratio is too high for stability) tend to emit negatrons by the process of β^- decay. Nuclei below the line of stability (i.e., the n/p ratio is too low for stability) tend to undergo the competing processes of positron (β^+) decay and electron capture.

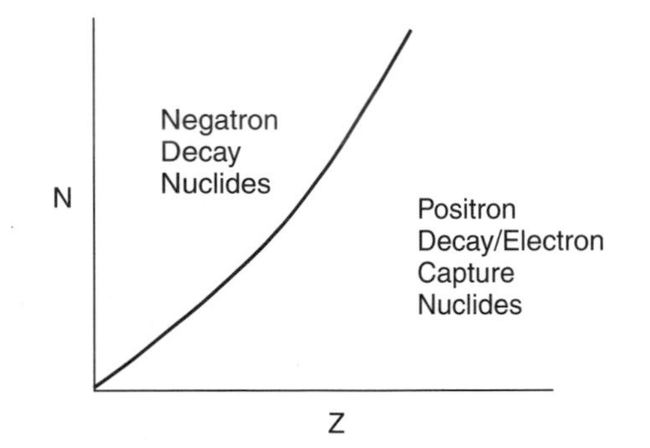

MARGIN FIGURE 3-3
Neutron number N and atomic number Z in stable nuclei.

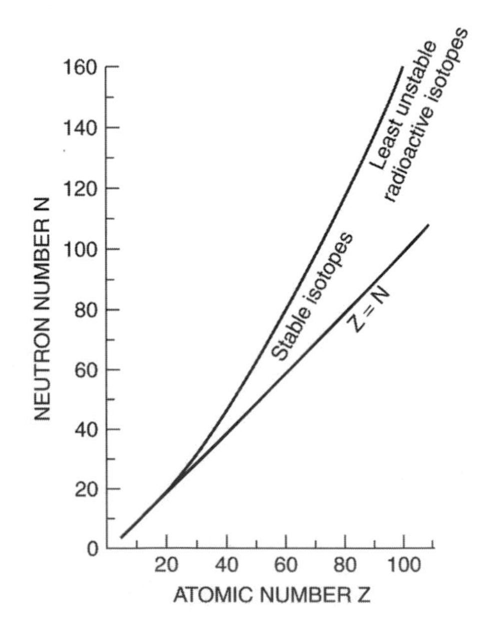

FIGURE 3-2
Number of neutrons (N) for stable (or least unstable) nuclei plotted as a function of the number of protons (Z).

Negatron Decay

In nuclei with an n/p ratio too high for stability, a neutron ${}_0^1n$ may be transformed into a proton ${}_0^1p$:

$$ {}_0^1n \rightarrow {}_1^1p + {}_{-1}^{0}\beta + \tilde{\nu} $$

where ${}_{-1}^{0}\beta$ is a negative electron ejected from the nucleus, and $\tilde{\nu}$ is a massless neutral particle termed an antineutrino (see below). The progeny nucleus has an additional proton and one less neutron than the parent. Therefore, the negatron form of beta decay results in an increase in Z of one, a decrease in N of one, and a constant A. A representative negatron transition is

$$ {}_{55}^{137}\text{Cs} \rightarrow {}_{56}^{137}\text{Ba} + {}_{-1}^{0}\beta + \tilde{\nu} + \text{isomeric transition} $$

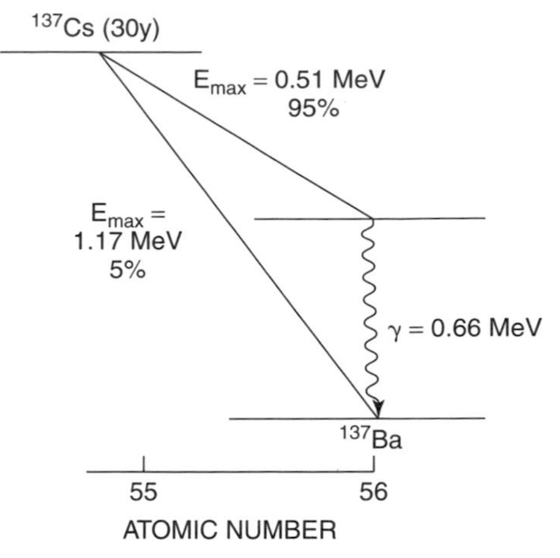

MARGIN FIGURE 3-4
Radioactive decay scheme for negatron decay of ^{137}Cs.

A decay scheme for this transition is shown to the right. A negatron with a maximum energy (E_{max}) of 1.17 MeV is released during 5% of all decays; in the remaining 95%, a negatron with an E_{max} of 0.51 MeV is accompanied by an isomeric transition of 0.66 MeV, where either a γ ray is emitted or an electron is ejected by internal conversion. The transition energy is 1.17 MeV for the decay of ^{137}Cs.

The probability that a nuclide will decay along a particular pathway is known as the branching ratio for the pathway.

Negatron decay pathways are characterized by specific maximum energies; however, most negatrons are ejected with energies lower than these maxima. The average energy of negatrons is about $\frac{1}{3}$ E_{max} along a specific pathway. The energy distribution of negatrons emitted during beta decay of ^{32}P is shown to the right. Spectra of similar shape, but with different values of E_{max} and E_{mean}, exist for the decay pathways of every negatron-emitting radioactive nuclide. In each particular decay, the difference in energy between E_{max} and the specific energy of the negatron is carried away by the antineutrino. That is,

$$ E_{\tilde{\nu}} = E_{max} - E_k $$

where E_{max} is the energy released during the negatron decay process, E_k is the kinetic energy of the negatron, and $E_{\tilde{\nu}}$ is the energy of the antineutrino.

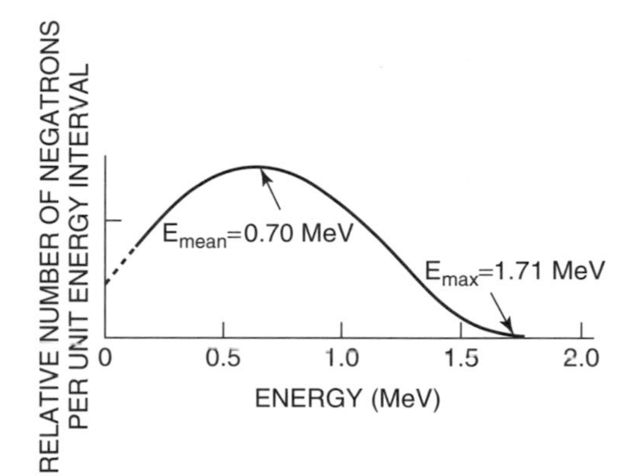

MARGIN FIGURE 3-5
Energy spectrum for negatrons from ^{32}P.

Positron Decay and Electron Capture

Positron decay results from the nuclear transition

$$ {}_1^1p \rightarrow {}_0^1n + {}_{+1}^{0}\beta + \nu $$

where ${}_{+1}^{0}\beta$ represents a positron ejected from the nucleus during decay, and ν is a neutrino that accompanies the positron. The neutrino and antineutrino are similar, except that they have opposite spin and are said to be antiparticles of each other.

In positron decay, the n/p ratio increases; hence, positron-emitting nuclides tend to be positioned below the n/p stability curve shown in Figure 3-2. Positron decay results in a decrease of one in Z, an increase of one in N, and no change in A. The decay of ${}_{15}^{30}$P is representative of positron decay:

$$ {}_{15}^{30}\text{P} \rightarrow {}_{14}^{30}\text{Si} + {}_{+1}^{0}\beta + \nu $$

The n/p ratio of a nuclide may also be increased by electron capture (ec), in which an electron is captured by the nucleus to yield the transition

$$ {}_1^1p + {}_{-1}^{0}e \rightarrow {}_0^1n + \nu $$

Most electrons are captured from the K electron shell, although occasionally an electron may be captured from the L shell or a shell even farther from the nucleus. During electron capture, a hole is created in an electron shell deep within the atom. This vacancy is filled by an electron cascading from a higher shell, resulting in the release of characteristic radiation or one or more Auger electrons.

The difference in the energy released during decay, and that possessed by the negatron, threatened the concept of energy conservation for several years. In 1933 Wolfgang Pauli[6] suggested that a second particle was emitted during each decay that accounted for the energy not carried out by the negatron. This particle was named the *neutrino* (Italian for "little neutral particle") by Enrico Fermi.

Enrico Fermi was a physicist of astounding insight and clarity who directed the first sustained man-made nuclear chain reaction on December 2, 1942 in the squash court of the University of Chicago stadium. He was awarded the 1938 Nobel Prize in Physics.

The antineutrino was detected experimentally by Reines and Cowan[7] in 1953. They used a 10-ton water-filled detector to detect antineutrinos from a nuclear reactor at Savannah River, SC. Reines shared the 1995 Nobel Prize in Physics.

The emission of positrons from radioactive nuclei was discovered in 1934 by Irene Curie[8] (daughter of Marie Curie) and her husband Frederic Joliet. In bombardments of aluminum by α particles, they documented the following transmutation:

$$
{}^{4}_{2}\text{He} + {}^{27}_{13}\text{Al} \rightarrow {}^{30}_{15}\text{P} + {}^{1}_{0}n
$$

$$
{}^{30}_{15}\text{P} \rightarrow {}^{30}_{14}\text{Si} + {}^{0}_{+1}\beta + \nu
$$

Electron capture of K-shell electrons is known as K-capture; electron capture of L-shell electrons is known as L-capture, and so on.

Many nuclei decay by both electron capture and positron emission, as illustrated in Margin Figure 3-6 for ${}^{22}_{11}\text{Na}$.

$$
{}^{22}_{11}\text{Na} \begin{cases} + {}^{0}_{-1}e \rightarrow {}^{22}_{10}\text{Ne} + \nu \\ \rightarrow {}^{0}_{+1}\beta + {}^{22}_{10}\text{Ne} + \nu \end{cases}
$$

The electron capture branching ratio is the probability of electron capture per decay for a particular nuclide. For ${}^{22}\text{Na}$, the branching ratio is 10% for electron capture, and 90% of the nuclei decay by the process of positron emission. Generally, positron decay prevails over electron capture when both decay modes occur.

A decay scheme for ${}^{22}\text{Na}$ is shown in Figure 3-3. The $2m_0c^2$ listed alongside the vertical portion of the positron decay pathway represents the energy equivalent (1.02 MeV or $2m_0c^2$) of the additional mass of the products of positron decay. This additional mass includes the greater mass of the neutron as compared with the proton, together with the mass of the ejected positron. This amount of energy must be supplied by the transition energy during positron decay. Nuclei that cannot furnish at least 1.02 MeV for the transition do not decay by positron emission. These nuclei increase their n/p ratios solely by electron capture.

A few nuclides may decay by negatron emission, positron emission, or electron capture, as shown in Figure 3-3 for ${}^{74}\text{As}$. This nuclide decays

32% of the time by negatron emission
30% of the time by positron emission
38% of the time by electron capture

All modes of decay result in transformation of the highly unstable "odd–odd" ${}^{74}\text{As}$ nucleus ($Z = 33$, $N = 41$) into a progeny nucleus with even Z and even N.

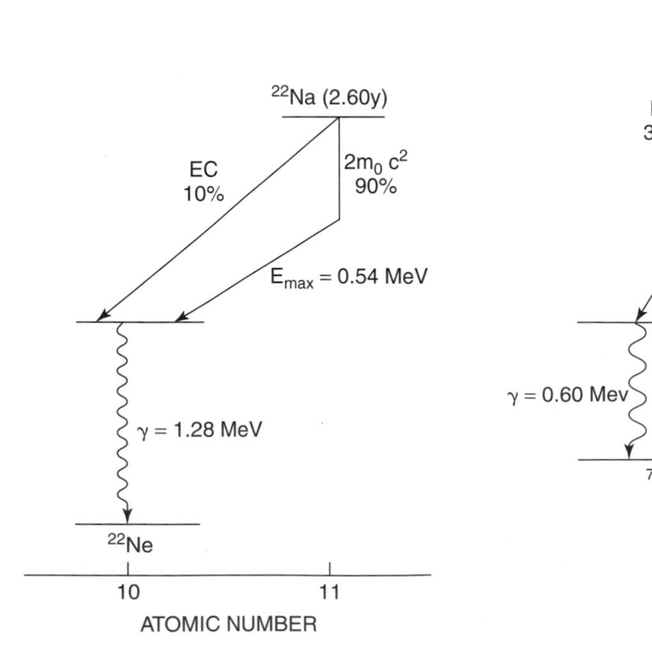

MARGIN FIGURE 3-6
Radioactive decay scheme for positron decay and electron capture of ${}^{22}\text{Na}$.

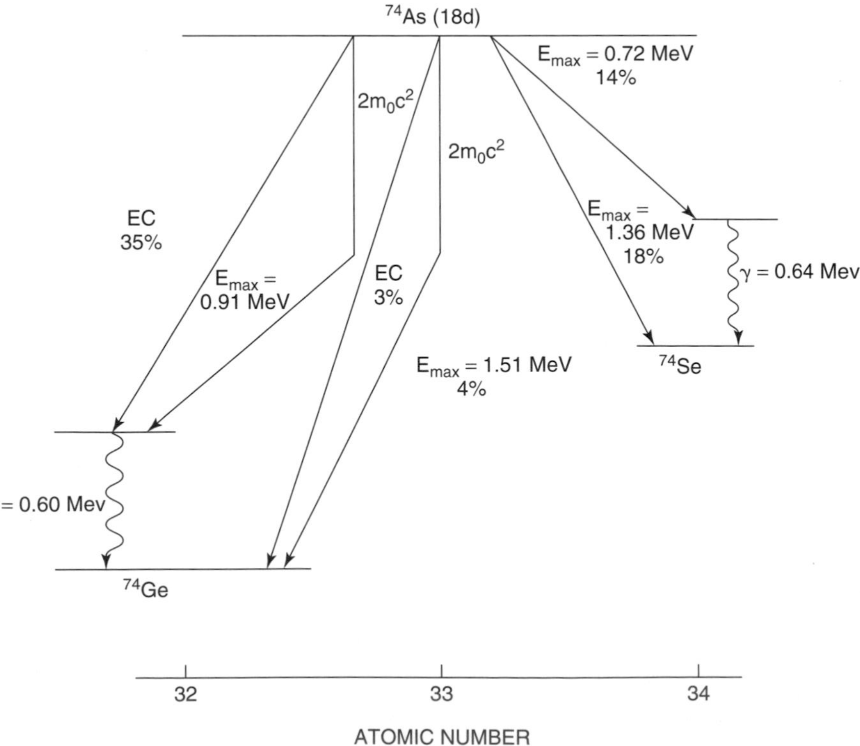

FIGURE 3-3
The decay scheme for ${}^{74}\text{As}$ illustrates competing processes of negatron emission, positron emission, and electron capture.

ISOMERIC TRANSITIONS

As mentioned earlier, radioactive decay often forms a progeny nucleus in an energetic ("excited") state. The nucleus descends from its excited to its most stable ("ground") energy state by one or more isomeric transitions. Often these transitions occur by emission of electromagnetic radiation termed γ rays. γ rays and x rays occupy the same region of the electromagnetic energy spectrum, and they are differentiated only by their origin: x rays result from electron interactions outside the nucleus, whereas γ rays result from nuclear transitions.

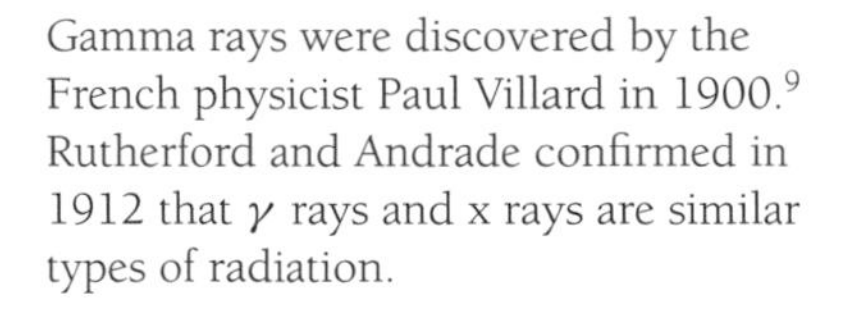

Gamma rays were discovered by the French physicist Paul Villard in 1900.[9] Rutherford and Andrade confirmed in 1912 that γ rays and x rays are similar types of radiation.

No radioactive nuclide decays solely by an isomeric transition. Isomeric transitions are always preceded by either electron capture or emission of an α or $\beta(+ \text{ or } -)$ particle.

Sometimes one or more of the excited states of a progeny nuclide may exist for a finite lifetime. An excited state is termed a *metastable state* if its half-life exceeds 10^{-6} seconds. For example, the decay scheme for ^{99}Mo shown to the right exhibits a metastable energy state, ^{99m}Tc, that has a half-life of 6 hours.

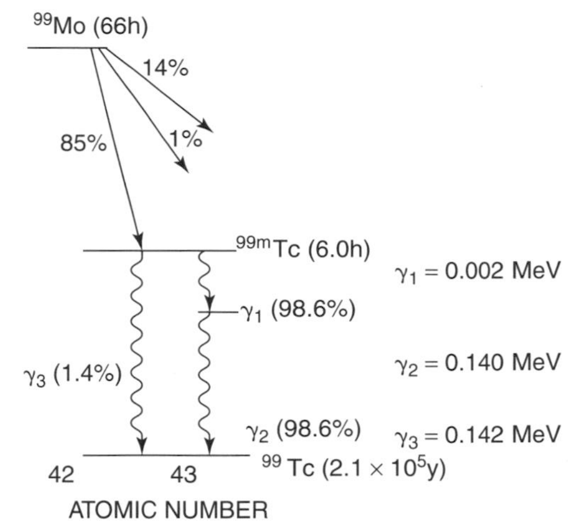

MARGIN FIGURE 3-7
Simplified radioactive decay scheme for ^{99}Mo.

Nuclides emit γ rays with characteristic energies. For example, photons of 142 and 140 keV are emitted by ^{99m}Tc, and photons of 1.17 and 1.33 MeV are released during negatron decay of ^{60}Co. In the latter case, the photons are released during cascade isomeric transitions of progeny ^{60}Ni nuclei from excited states to the ground energy state.

An isomeric transition can also occur by interaction of the nucleus with an electron in one of the electron shells. This process is known as internal conversion (IC). When IC happens, the electron is ejected with kinetic energy E_k equal to the energy E_γ released by the nucleus, reduced by the binding energy E_b of the electron.

$$E_k = E_\gamma - E_b$$

The ejected electron is accompanied by x rays and Auger electrons as the extranuclear structure of the atom resumes a stable configuration.

The internal conversion coefficient for an electron shell is the ratio of the number of conversion electrons from the shell compared with the number of γ rays emitted by the nucleus. The probability of internal conversion increases rapidly with increasing Z and with the lifetime of the excited state of the nucleus.

The types of radioactive decay are summarized in Table 3-2.

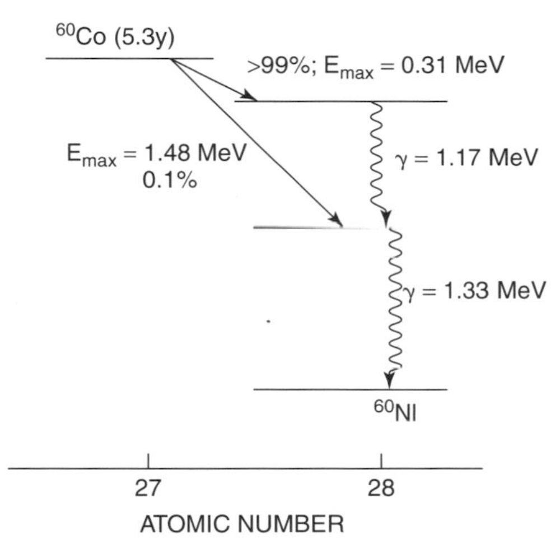

MARGIN FIGURE 3-8
Radioactive decay scheme for ^{60}Co.

MATHEMATICS OF RADIOACTIVE DECAY

Radioactivity can be described mathematically without regard to specific mechanisms of decay. The rate of decay (number of decays per unit time) of a radioactive sample depends on the number N of radioactive atoms in the sample. This concept can be

TABLE 3-2 Summary of Radioactive Decay, ^A_Z (Parent) → ^A_Z (Progeny)

Type of Decay	*A*	*Z*	*N*	*Comments*
Negatron (β^-)	A	$Z+1$	$N-1$	$[E_\beta]_{mean} \cong \frac{[E_\beta]_{max}}{3}$
Positron (β^+)	A	$Z-1$	$N+1$	$[E_\beta]_{mean} \cong \frac{[E_\beta]_{max}}{3}$
Electron capture	A	$Z-1$	$N+1$	Characteristic + auger electrons
Isomeric transition				
Gamma (γ) emission	A	Z	N	Metastable if $T_{1/2} > 10^{-6}$ sec
Internal conversion (IC)	A	Z	N	IC electrons: characteristic + auger electrons
Alpha (α)	$A-4$	$Z-2$	$N-2$	

Equation (3-1) describes the expected decay rate of a radioactive sample. At any moment the actual decay rate may differ somewhat from the expected rate because of statistical fluctuations in the decay rate.

Equation (3-1) depicts a reaction known as a *first-order reaction*.

The decay constant λ is also called the *disintegration constant*.

The decay constant of a nuclide is truly a constant; it is not affected by external influences such as temperature and pressure or by magnetic, electrical, or gravitational fields.

The rutherford (Rf) was once proposed as a unit of activity, where 1 Rf = 10^6 dps. The Rf did not gain acceptance in the scientific community and eventually was abandoned. In so doing, science lost an opportunity to honor one of its pioneers.

The curie was defined in 1910 as the activity of 1 g of radium. Although subsequent measures revealed that 1 g of radium has a decay rate of 3.61×10^{10} dps, the definition of the curie was left as 3.7×10^{10} dps.

^{201}Tl is an accelerator-produced radioactive nuclide that is used in nuclear medicine to study myocardial perfusion.

stated as

$$\Delta N/\Delta t = -\lambda N \tag{3-1}$$

where $\Delta N/\Delta t$ is the rate of decay, and the constant λ is called the *decay constant*. The minus sign indicates that the number of parent atoms in the sample, and therefore the number decaying per unit time, is decreasing.

By rearranging Eq. (3-1) to

$$\frac{-\Delta N/\Delta t}{N} = \lambda$$

the decay constant can be seen to be the fractional rate of decay of the atoms. The decay constant has units of (time)$^{-1}$, such as sec^{-1} or hr^{-1}. It has a characteristic value for each nuclide. It also reflects the nuclide's degree of instability; a larger decay constant connotes a more unstable nuclide (i.e., one that decays more rapidly). The rate of decay is a measure of a sample's *activity*, defined as

$$\text{Activity} = A = -\Delta N/\Delta t = \lambda N$$

The activity of a sample depends on the number of radioactive atoms in the sample and the decay constant of the atoms. A sample may have a high activity because it contains a few highly unstable (large decay constant) atoms, or because it contains many atoms that are only moderately unstable (small decay constant).

The SI unit of activity is the *becquerel (Bq)*, defined as

$$1 \text{ Bq} = 1 \text{disintegration per second (dps)}$$

Multiples of the becquerel such as the kilobecquerel (kBq = 10^3 Bq), megabecquerel (MBq = 10^6 Bq), and gigabecquerel (GBq = 10^9 Bq) are frequently encountered.

An older, less-preferred unit of activity is the curie (Ci), defined as

$$1 \quad \text{Ci} = 3.7 \times 10^{10} \text{ dps}$$

Useful multiples of the curie are the megacurie (MCi = 10^6 Ci), kilocurie (kCi = 10^3 Ci), millicurie (mCi = 10^{-3} Ci), microcurie (μ Ci = 10^{-6} Ci), nanocurie (nCi = 10^{-9} Ci), and femtocurie (fCi = 10^{-12} Ci).

Example 3-1

a. $^{201}_{81}T$ 1has a decay constant of 9.49×10^{-3} hr^{-1}. Find the activity in becquerels of a sample containing 10^{10} atoms. From Eq. (3-1),

$$\begin{aligned} A &= \lambda N \\ &= \frac{(9.49 \times 10^{-3} \text{ hr}^{-1})(10^{10} \text{ atoms})}{3600 \text{ sec/hr}} \\ &= 2.64 \times 10^4 \frac{\text{atoms}}{\text{sec}} \\ &= 2.64 \times 10^4 \text{ Bq} \end{aligned}$$

b. How many atoms of $^{11}_{6}$C with a decay constant of 2.08 hr^{-1} would be required to obtain the same activity as the sample in Part a?

$$\begin{aligned} N &= \frac{A}{\lambda} \\ &= \frac{2.64 \times 10^4 \frac{\text{atoms}}{\text{sec}} \left(3600 \frac{\text{sec}}{\text{hr}}\right)}{(2.08/\text{hr})} \\ &= 4.57 \times 10^7 \text{ atoms} \end{aligned}$$

More atoms of $^{201}_{81}$T1 than of $^{11}_{6}$C are required to obtain the same activity because of the difference in decay constants.

DECAY EQUATIONS AND HALF-LIFE

Equation (3-1) is a differential equation. It can be solved (see Appendix I) to yield an expression for the number N of parent atoms present in the sample at any time t:

$$N = N_0 e^{-\lambda t} \tag{3-2}$$

where N_0 is the number of atoms present at time $t = 0$, and e is the exponential quantity 2.7183. By multiplying both sides of this equation by λ, the expression can be rewritten as

$$A = A_0 e^{-\lambda t} \tag{3-3}$$

where A is the activity remaining after time t, and A_0 is the original activity.

The number of atoms N^* decaying in time t is $N_0 - N$ (see Figure 3-4), or

$$N^* = N_0(1 - e^{-\lambda t}) \tag{3-4}$$

The probability that any particular atom will not decay during time t is N/N_0, or

$$P\,(\text{no decay}) = e^{-\lambda t} \tag{3-5}$$

and the probability that a particular atom will decay during time t is

$$P\,(\text{decay}) = 1 - e^{-\lambda t} \tag{3-6}$$

For small values of λt, the probability of decay can be approximated as

$$P\,(\text{decay}) \approx \lambda t$$

or expressed per unit time, P (decay per unit time) $\approx \lambda$. That is, when λt is very small, the decay constant approximates the probability of decay per unit time. In most circumstances, however, the approximation is too inexact, and the probability of decay must be computed as $1 - e^{-\lambda t}$. The decay constant should almost always be thought of as a fractional rate of decay rather than as a probability of decay.

The *physical half-life* $T_{1/2}$ of a radioactive nuclide is the time required for decay of half of the atoms in a sample of the nuclide. In Eq. (3-2), $N = \frac{1}{2}N_0$ when $t = T_{1/2}$, with the assumption that $N = N_0$ when $t = 0$. By substitution in Eq. (3-2),

$$\frac{1}{2} = \frac{N}{N_0} = e^{-\lambda T_{1/2}}$$

$$\ln\left(\frac{1}{2}\right) = -\lambda T_{1/2}$$

$$T_{1/2} = \frac{0.693}{\lambda}$$

where 0.693 is the ln 2 (natural logarithm of 2).

Equation (3-2) reveals that the number N of parent atoms decreases *exponentially* with time.

The *specific activity* of a radioactive sample is the activity per unit mass of sample.

Radioactive decay must always be described in terms of the probability of decay; whether any particular radioactive nucleus will decay within a specific time period is never certain.

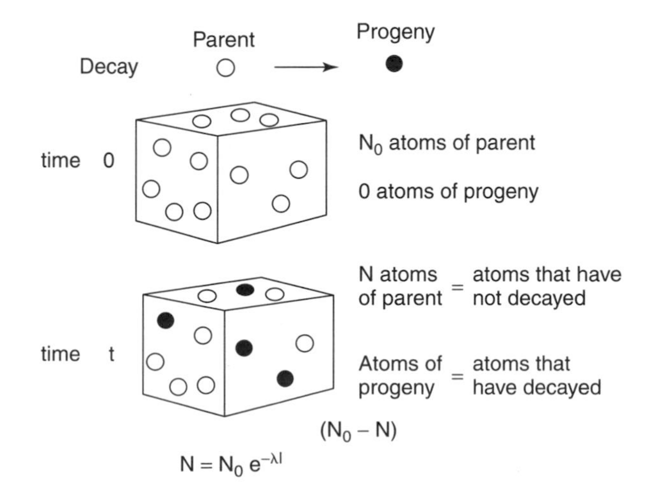

FIGURE 3-4
Mathematics of radioactive decay with the parent decaying to a stable progeny.

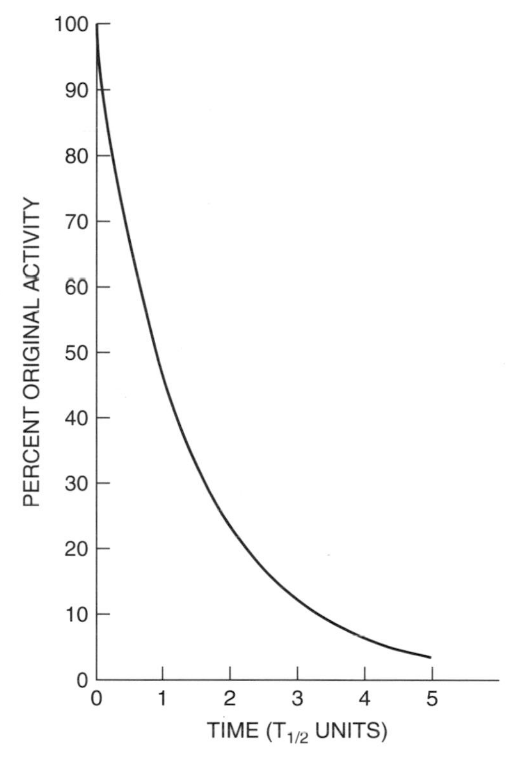

MARGIN FIGURE 3-9
The percent of original activity of a radioactive sample is expressed as a function of time in units of half-life.

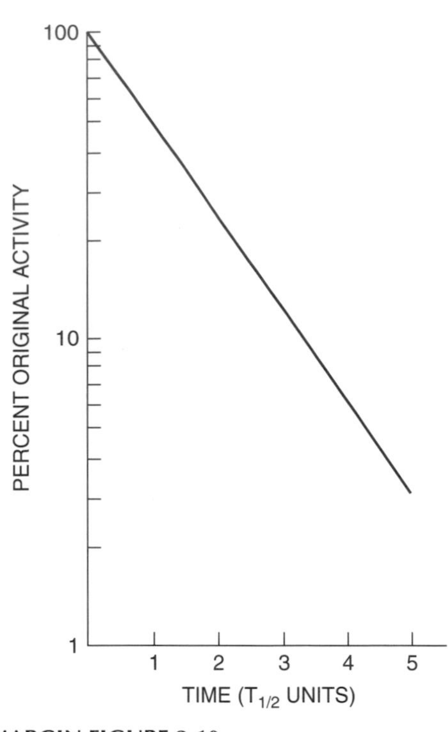

MARGIN FIGURE 3-10
Semilogarithmic graph of the data in Margin Figure 3-9.

^{113}In was used at one time, but no longer, as a blood pool imaging agent.

The mean (average) life is the average expected lifetime for atoms of a radioactive sample. The *mean life* τ is related to the decay constant λ by

$$\tau = \frac{1}{\lambda}$$

Because $T_{1/2} = \frac{(0.693)}{\lambda}$ the mean life may be calculated from the expression

$$\tau = 1.44 T_{1/2}$$

The percent of original activity of a radioactive sample is pictured in Margin Figure 3-9 as a function of time expressed in units of physical half-life. By replotting these data semilogarithmically (activity on a logarithmic axis, time on a linear axis), a straight line can be obtained, as shown in Margin Figure 3-10.

Example 3-2

The physical half-life is 1.7 hours for ^{113m}In.

a. A sample of ^{113m}In has a mass of 2 μg. How many ^{113m}In atoms are present in the sample?

$$\text{Number of atoms } N = \frac{(\text{No. of grams})(\text{No of atoms/gram} - \text{atomic mass})}{\text{No. of grams/gram} - \text{atomic mass}}$$

$$= \frac{(2 \times 10^{-6}\text{ g})(6.02 \times 10^{23}\text{ atoms/gram} - \text{atomic mass})}{113\text{ grams/gram} - \text{atomic mass}}$$

$$= 1.07 \times 10^{16}\text{ atoms}$$

b. How many ^{113m}In atoms remain after 4 hours have elapsed?

$$\text{Number of atoms remaining} = N = N_0 e^{-\lambda t}$$

Since $\lambda = 0.693/T_{1/2}$

$$N = N_0 e^{-(0.693/T_{1/2})t}$$

$$= (1.07 \times 10^{16}\text{ atoms})\, e^{-(0693/1.7\text{ hr})\ 4.0\text{ hr}}$$

$$= (1.07 \times 10^{16}\text{ atoms})(0.196)$$

$$= 2.10 \times 10^{15}\text{ atoms remaining}$$

c. What is the activity of the sample when $t = 4.0$ hours?

$$A = \lambda N = \frac{0.693}{T_{1/2}} N$$

$$= \frac{0.693(2.1 \times 10^{15}\text{ atoms})}{1.7\text{ hr}\,(3600\text{ sec/hr})}$$

$$= 2.4 \times 10^{11}\text{ dps}$$

In units of activity

$$A = 2.4 \times 10^{11}\text{ Bq}$$

Because of the short physical half-life (large decay constant) of ^{113m}In, a very small mass of this nuclide possesses high activity.

d. Specific activity is defined as the activity per unit mass of a radioactive sample. What is the specific activity of the ^{113m}In sample after 4 hours?

$$\text{Specific activity} = \frac{2.4 \times 10^{11}\text{ Bq}}{2 \times 10^{-6}\text{ g}} = 1.2 \times 10^{17}\ \frac{\text{Bq}}{\text{g}}$$

e. Enough ^{113m}In must be obtained at 4 p.m. Thursday to provide 300 kBq at 1 p.m. Friday. How much ^{113m}In should be obtained?

$$A = A_0 e^{-\lambda t}$$

$$A_0 = Ae^{\lambda t} = Ae^{0.693t/T_{1/2}}$$

$$= (300 \text{ kBq})e^{(0.693)(21 \text{ hr})/1.7 \text{ hr}}$$

$$A = 1.57 \times 10^9 \text{ Bq} = 1.57 \text{ GBq} = \text{activity that should be obtained at 4 p.m. Thursday}$$

Example 3-3 may be solved graphically by plotting the activity of each sample as a function of time. The time when activities are equal is indicated by the intersection of the curves for the nuclides.

Equation (3-7) is a Bateman equation. More complex Bateman equations describe progeny activities for sequential phases of multiple radioactive nuclides in transient equilibrium.

Example 3-3

The half-life of ^{99m}Tc is 6.0 hours. How much time t must elapse before a 100-GBq sample of ^{113m}In and a 20-GBq sample of ^{99m}Tc possess equal activities?

$$A^{99m\text{Tc}} = A^{113m} \text{ at time } t$$

$$A_{0(^{99m}\text{Tc})} e^{-(0.693)t/(T_{1/2})^{99m\text{Tc}}} = A_{0(^{113m}\text{In})} e^{-(0.693)t/(T_{1/2})113m\text{In}}$$

$$(20\text{GBq})e^{-(0.693)t/6.0 \text{ hr}} = (100 \text{ GBq})e^{-(0.693)t/1.7 \text{ hr}}$$

$$\frac{100 \text{ GBq}}{20 \text{ GBq}} = e^{(0.408-0.115)t}$$

$$5 = e^{(0.408-0.115)t}$$

$$t = 5.5 \text{ hr before activities are equal}$$

In the ^{132}I–^{132}Te example, transient equilibrium occurs

- at only one instant in time
- when ^{132}Te reaches its maximum activity
- when the activity of ^{132}Te in the sample is neither increasing nor decreasing
- when the activities of ^{132}I and ^{132}Te are equal.

^{99m}Tc is the most widely used radioactive nuclide in nuclear medicine. It is produced in a ^{99}Mo–^{99m}Tc radioactive generator, often called a molybdenum "cow." Eluting ^{99m}Tc from the generator is referred to as "milking the cow." The ^{99}Mo–^{99m}Tc generator was developed by Powell Richards in 1957.

TRANSIENT EQUILIBRIUM

Progeny atoms produced during radioactive decay are sometimes also radioactive. For example, ^{226}Ra decays to ^{222}Rn ($T_{1/2}$ = 3.83 days); and ^{222}Rn decays by α-emission to ^{218}Po, an unstable nuclide that in turn decays with a half-life of about 3 minutes. If the half-life of the parent is significantly longer than the half-life of the progeny (i.e., by a factor of 10 or more), then a condition of *transient equilibrium* may be established. After the time required to attain transient equilibrium has elapsed, the activity of the progeny decreases with an apparent half-life equal to the physical half-life of the parent. The apparent half-life reflects the simultaneous production and decay of the progeny atoms. If no progeny atoms are present initially when $t = 0$, the number N_2 of progeny atoms present at any later time is

$$N_2 = \frac{\lambda_1}{\lambda_2 - \lambda_1} N_0(e^{-\lambda_1 t} - e^{-\lambda_2 t}) \tag{3-7}$$

In Equation (3-7), N_0 is the number of parent atoms present when $t = 0$, λ_1 is the decay constant of the parent, and λ_2 is the decay constant of the progeny. If $(N_2)_0$ progeny atoms are present when $t = 0$, then Eq. (3-7) may be rewritten as

$$N_2 = (N_2)_0 e^{-\lambda_2 t} + \frac{\lambda_1}{\lambda_2 - \lambda_1} N_0(e^{-\lambda_1 t} - e^{-\lambda_2 t})$$

Transient equilibrium for the transition

$$^{132}\text{Te}(T_{1/2} = 78 \text{ hr}) \rightarrow {}^{132}\text{I}(T_{1/2} = 2.3 \text{ hr})$$

is illustrated in the Margin Figure 3-11. The activity of ^{132}I is greatest at the moment of transient equilibrium when parent (^{132}Te) and progeny (^{132}I) activities are equal. At all later times, the progeny activity exceeds the activity of the parent (A_1).

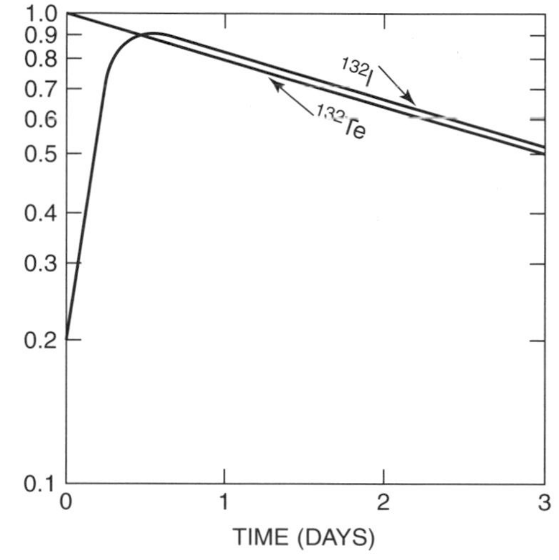

MARGIN FIGURE 3-11
Activities of parent ^{132}Te and progeny ^{132}I as a function of time illustrate the condition of transient equilibrium that may be achieved when the half-life of the parent is not much greater than the half-life of the progeny.

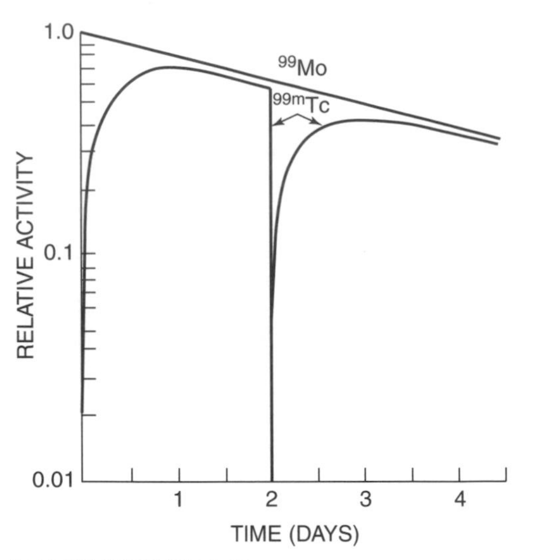

MARGIN FIGURE 3-12
Activities of parent (^{99}Mo) and progeny (^{99m}Tc) in a nuclide generator. About 14% of the parent nuclei decay without the formation of ^{99m}Tc.

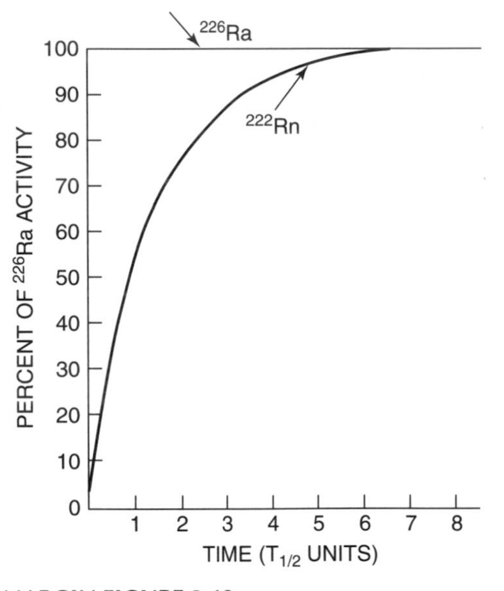

MARGIN FIGURE 3-13
Growth of activity and secular equilibrium of ^{222}Rn produced by decay of ^{226}Ra.

At these times the progeny (A_2) can be shown to equal

$$\frac{A_1}{A_2} = \frac{\lambda_2 - \lambda_1}{\lambda_2}$$

Transient equilibrium is the principle underlying production of short-lived isotopes (e.g., ^{99m}Tc, ^{113m}In, and ^{68}Ga) in generators for radioactive nuclides used in nuclear medicine. For example, the activities of ^{99m}Tc ($T_{1/2} = 6.0$ hours and ^{99}Mo ($T_{1/2} =$ 66 hours) are plotted in Margin Figure 3-12 as a function of time. The ^{99m}Tc activity remains less than that for ^{99}Mo because about 14% of the ^{99}Mo nuclei decay promptly to ^{99}Tc without passing through the isomeric ^{99m}Tc state. The abrupt decrease in activity at 48 hours reflects the removal of ^{99m}Tc from the generator.

Secular Equilibrium

Secular equilibrium may occur when the half-life of the parent is much greater than the half-life of the progeny (e.g., by a factor of 10^4 or more). Since $\lambda_1 \ll \lambda_2$ if $T_{1/2\ \text{parent}} \gg T_{1/2\ \text{progeny}}$, Eq. (3-7) may be simplified by assuming that $\lambda_2 - \lambda_1 = \lambda_2$ and that $e^{-\lambda_1 t} = 1$.

$$N_2 = \frac{\lambda_1}{\lambda_2 - \lambda_1} N_0 \left(e^{-\lambda_1 t} - e^{-\lambda_2 t}\right) \cong \frac{\lambda_1}{\lambda_2} N_0 \left(1 - e^{-\lambda_2 t}\right)$$

After several half-lives of the daughter have elapsed, $e^{-\lambda_2 t} \cong 0$ and

$$N_2 = \frac{\lambda_1}{\lambda_2}(N_0)$$

$$\lambda_2 N_2 = \lambda_1 N_0 \tag{3-8}$$

$$\frac{0.693}{(T_{1/2})_2} N_2 = \frac{0.693}{(T_{1/2})_1} N_0$$

$$\frac{N_2}{(T_{1/2})_2} = \frac{N_0}{(T_{1/2})_1} \tag{3-9}$$

The terms $\lambda_1 N_0$ and $\lambda_2 N_2$ in Eq. (3-8) describe the activities of parent and progeny, respectively. These activities are equal after secular equilibrium has been achieved. Illustrated in Margin Figure 3-13 are the growth of activity and the activity at equilibrium for ^{222}Rn produced in a closed environment by decay of ^{226}Ra. Units for the x axis are multiples of the physical half-life of ^{222}Rn. The growth curve for ^{222}Rn approaches the decay curve for ^{226}Ra asymptotically. Several half-lives of ^{222}Rn must elapse before the activity of the progeny equals 99% of the parent activity.

If the parent half-life is less than that for the progeny ($T_{1/2}$ parent $< T_{1/2}$ progeny, or $\lambda_1 > \lambda_2$), then a constant relationship is not achieved between the activities of the parent and progeny. Instead, the activity of the progeny increases initially, reaches a maximum, and then decreases with a half-life intermediate between the parent and progeny half-lives.

Radioactive nuclides in secular equilibrium are often used in radiation oncology. For example, energetic β-particles from ^{90}Y in secular equilibrium with ^{90}Sr are used to treat intraocular lesions. The activity of the ^{90}Sr–^{90}Y ophthalmic irradiator decays with the physical half-life of ^{90}Sr (28 years), whereas a source of ^{90}Y alone decays with the 64-hour half-life of ^{90}Y. Radium needles and capsules used widely in the past for superficial and intracavitary radiation treatments contain many decay products in secular equilibrium with long-lived ^{226}Ra.

Natural Radioactivity and Decay Series

Almost every radioactive nuclide found in nature is a member of one of three radioactive decay series. Each series consists of sequential transformations that begin with a long-lived parent and end with a stable nuclide. In a closed environment, products

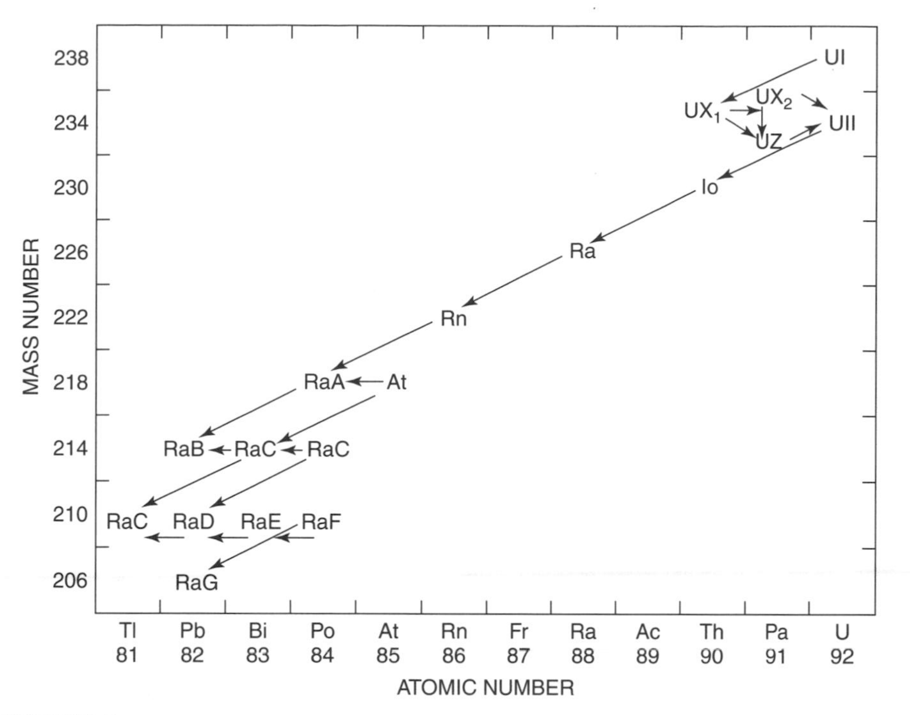

FIGURE 3-5
Uranium ($4n + 2$) radioactive decay series.

of intermediate transformations within each series are in secular equilibrium with the long-lived parent. These products decay with an apparent half-life equal to the half-life of the parent. All radioactive nuclides found in nature decay by emitting either α-particles or negatrons. Consequently, each transformation in a radioactive decay series changes the mass number of the nucleus either by 4 (α-decay) or by 0 (negatron decay).

The inert gas ^{222}Rn produced by decay of naturally occurring ^{226}Ra is also radioactive, decaying with a half-life of 3.83 days. This radioactive gas first called "radium emanation" was characterized initially by Rutherford. Seepage of ^{222}Rn into homes built in areas with significant ^{226}Ra concentrations in the soil is an ongoing concern to homeowners and the Environmental Protection Agency.

The uranium series begins with ^{238}U ($T_{1/2} = 4.5 \times 10^9$ years) and ends with stable ^{206}Pb. The mass number 238 of the parent nuclide is divisible by 4 with a remainder of 2. All members of the uranium series, including stable ^{206}Pb, also possess mass numbers divisible by 4 with remainder of 2. Consequently, the uranium decay series sometimes is referred to as the "$4n + 2$ series," where n represents an integer between 51 and 59 (Figure 3-5). The nuclide ^{226}Ra and its decay products are members of the uranium decay series. A sample of ^{226}Ra decays with a half-life of 1600 years. However, 4.5×10^9 years are required for the earth's supply of ^{226}Ra to decrease to half because ^{226}Ra in nature is in secular equilibrium with its parent ^{238}U.

Other radioactive series are the actinium or "$4n + 3$ series" (^{235}U $\rightarrow\rightarrow$ ^{207}Pb) and the thorium or "$4n$ series" (^{232}Th $\rightarrow\rightarrow$ ^{208}Pb). Members of the hypothetical neptunium or "$4n + 1$ series" (^{241}Am $\rightarrow\rightarrow$ ^{209}Bi) are not found in nature because there is no long-lived parent for this series.

Fourteen naturally occurring radioactive nuclides are not members of a decay series. The 14 nuclides, all with relatively long half-lives, are ^{3}H, ^{14}C, ^{40}K, ^{50}V, ^{87}Rb, ^{115}In, ^{130}Te, ^{138}La, ^{142}Ce, ^{144}Nd, ^{147}Sm, ^{176}Lu, and ^{187}Re, and ^{192}Pt.

ARTIFICIAL PRODUCTION OF RADIONUCLIDES

Nuclides may be produced artificially that have properties desirable for medicine, research, or other purposes. Nuclides with excess neutrons (which will therefore emit negatrons) are created by bombarding nuclei with neutrons from a nuclear reactor,

The first high-energy particle accelerator was the cyclotron developed by Ernest Lawrence in 1931. In 1938, Ernest and his physician brother John used artificially produced ^{32}P to treat their mother, who was afflicted with leukemia.

Nuclear transmutation by particle bombardment was first observed by Rutherford[10] in 1919 in his studies of α-particles traversing an air-filled chamber. The observed transmutation was

$$^{4}_{2}He + ^{14}_{7}N \rightarrow ^{17}_{8}O + ^{1}_{1}H$$

where He represents α-particles and H depicts protons detected during the experiment. This reaction can be written more concisely as

$$^{14}_{7}N(\alpha, p)^{17}_{8}O$$

where $^{14}_{7}N$ represents the bombarded nucleus, $^{17}_{8}O$ the product nucleus, and (α, p) the incident and ejected particles, respectively.

Through their discovery of artificial radioactivity, Irene Curie (the daughter of Marie Curie) and Federic Joliot paved the way to use of radioactive tracers in biomedical research and clinical medicine.

The cross-sectional area presented to a neutron by a target nucleus is usually described in units of barns, with 1 barn equaling 10^{-24} cm^2.

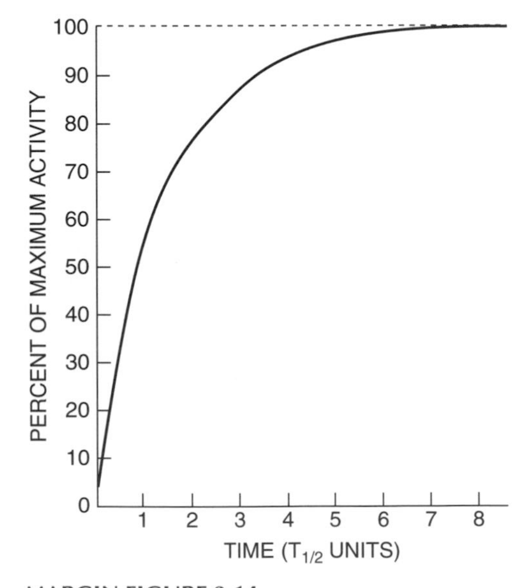

MARGIN FIGURE 3-14
Growth in activity for a sample of ^{59}Co bombarded by slow neutrons in a nuclear reactor.

for example,

$$^{13}_{6}C + ^{1}_{0}n \rightarrow ^{14}_{6}C$$

$$^{31}_{15}P + ^{1}_{0}n \rightarrow ^{32}_{15}P$$

Other isotopes produced by this method include

$$^{3}_{1}H,\ ^{35}_{16}S,\ ^{51}_{24}Cr,\ ^{60}_{27}Co,\ ^{99}_{42}Mo,\ ^{133}_{154}Xe,\ ^{198}_{79}Au$$

Nuclides with excess protons may be produced by bombarding nuclei with protons, α-particles, deuterons, or other charged particles from high-energy particle accelerators. These nuclides then decay by positron decay or electron capture with the emission of positrons, photons, and so on. Representative transitrons include:

$$^{68}_{30}Zn + ^{1}_{1}p \rightarrow ^{67}_{31}Ga + 2n$$

$$^{11}_{5}B + ^{1}_{1}p \rightarrow ^{11}_{6}C + n$$

Useful radioisotopes produced by particle accelerators include

$$^{13}_{7}N,\ ^{15}_{8}O,\ ^{18}_{9}F$$

Other nuclides are obtained as by-products of fission in nuclear reactors; for example,

$$^{235}_{92}U + ^{1}_{0}n \rightarrow ^{A_1}_{Z_1}x + ^{A_2}_{Z_2}y + \text{neutrons}$$

where a range of isotopes X and Y are produced, including

$$^{90}_{38}Sr,\ ^{99}_{42}Mo,\ ^{131}_{53}I,\ ^{137}_{55}Cs$$

Finally, some radionuclides are produced as decay products of isotopes produced by the above methods. For example, $^{201}_{82}Pb$ is produced by bombardment of $^{203}_{81}Tl$ by protons.

$$^{203}_{81}Tl + p \rightarrow ^{201}_{82}Pb + 3n$$

$^{201}_{82}Pb(T_{1/2} = 9.4\text{ hr})$ decays by electron capture to the useful radionuclide $^{201}_{81}Tl$.

MATHEMATICS OF NUCLIDE PRODUCTION BY NEUTRON BOMBARDMENT

The production of artificially radioactive nuclides by neutron bombardment in the core of a nuclear reactor may be described mathematically. The activity A of a sample bombarded for a time t, assuming no radioactivity when $t = 0$, may be written as

$$A = \varphi N\sigma(1 - e)^{-(0.693/T_{1/2})t} \quad \textbf{(3-10)}$$

where

- φ is the neutron flux in neutrons per square centimeter-second
- N is the number of target nuclei in the sample
- σ is the absorption cross section of target nuclei in square centimeters
- $T_{1/2}$ is the half-life of the product nuclide, and
- A is the activity of the sample in becquerels

When the bombardment time t is much greater than $T_{1/2}$, Eq. (3-10) reduces to

$$A_{max} = \varphi N\sigma \quad \textbf{(3-11)}$$

where A_{max} represents the maximum activity in becquerels. Margin Figure 3-14 illustrates the growth of radioactivity in a sample of ^{59}Co bombarded by slow neutrons. This transmutation is written $^{59}_{27}Co\,(n, \gamma)\,^{60}_{27}Co$. The half-life of ^{60}Co is 5.3 years.

Example 3-4

A 20-g sample of ^{59}Co is positioned in the core of a reactor with an average neutron flux of 10^{14} neutrons/cm^2 sec. The absorption cross section of ^{59}Co is 36 barns.

a. What is the activity of the sample after 6.0 years?

$$N = \frac{20\,\text{g}\,(6.02 \times 10^{23}\ \text{atoms/gram} - \text{atomic mass}}{59\ \text{grams/gram} - \text{atomic mass}} = 2.04 \times 10^{23}\ \text{atoms}$$

$$A = \varphi N\sigma(1 - e^{-(0.693)t/T_{1/2}})$$

$$A = (10^{14}\text{neutrons/cm}^2\text{-sec})(2.04 \times 10^{23}\ \text{atoms})(36 \times 10^{-24}\ \text{cm}^2)$$
$$\times\left[1 - e^{-(0.693)6.0\,\text{yrs}/5.3\,\text{yr}}\right]$$
$$= 4.00 \times 10^{14}\ \text{Bq}$$

b. What is the maximum activity for the sample?

$$A_{\text{max}} = \varphi N\sigma$$
$$A = (10^{14}\ \text{neutrons/cm}^2\text{-sec})(2.04 \times 10^{23}\ \text{atoms})(36 \times 10^{-24}\ \text{cm}^2)$$
$$= 7.33 \times 10^{14}\ \text{Bq}$$

c. When does the sample activity reach 90% of its maximum activity?

$$(7.33 \times 10^{14}\ \text{Bq})(0.9) = 6.60 \times 10^{14}\ \text{Bq}$$
$$6.60 \times 10^{14}\ \text{Bq} = 7.33 \times 10^{14}\ \text{Bq}\left[1 - e^{-(0.693)t/5.3\,\text{yr}}\right]$$
$$6.60 \times 10^{14}\ \text{Ci} = 7.33 \times 10^{14} - 7.33 \times 10^{14}\, e^{-0.131t}$$
$$7.33 \times 10^{14}\, e^{-0.131} = 0.73 \times 10^{14}$$
$$e^{-0.131t} = 0.10$$
$$0.131t = 2.30$$
$$t = 17.6\ \text{years to reach } 90\%\ A_{\text{max}}$$

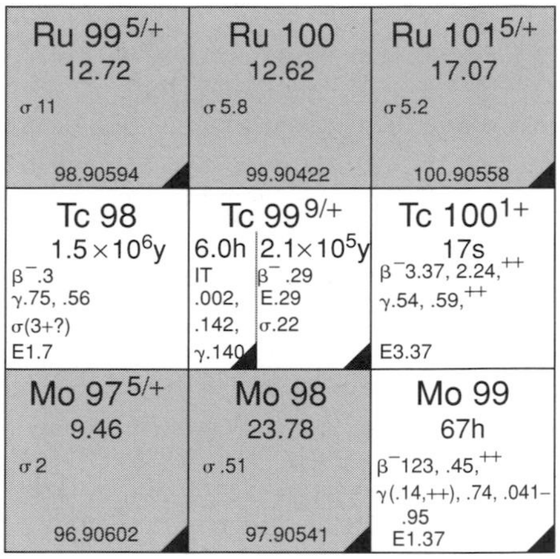

Ru 99 5/+ 12.72 σ 11 98.90594	Ru 100 12.62 σ 5.8 99.90422	Ru 101 5/+ 17.07 σ 5.2 100.90558
Tc 98 1.5×10^6y β^-.3 γ.75, .56 σ(3+?) E1.7	Tc 99 9/+ 6.0h: IT .002, .142, γ.140 2.1×10^5y: β^-.29, E.29, σ.22	Tc 100 1+ 17s β^-3.37, 2.24, ++ γ.54, .59, ++ E3.37
Mo 97 5/+ 9.46 σ 2 96.90602	Mo 98 23.78 σ .51 97.90541	Mo 99 67h β^-123, .45, ++ γ(.14,++), .74, .041–.95 E1.37

MARGIN FIGURE 3-15
Section from a chart on the nuclides.

INFORMATION ABOUT RADIOACTIVE NUCLIDES

Decay schemes for many radioactive nuclei are listed in the *Radiological Health Handbook,*[11] *Table of Isotopes,*[12] and Medical Internal Radiation Dose Committee pamphlets published by the Society of Nuclear Medicine.[13]

Charts of the nuclides (available from General Electric Company, Schenectady, New York, and from Mallinckrodt Chemical Works, St. Louis) contain useful data concerning the decay of radioactive nuclides. A section of a chart is shown to the right. In this chart, isobars are positioned along 45-degree diagonals, isotones along vertical lines, and isotopes along horizonal lines. The energy of radiations emitted by various nuclei is expressed in units of MeV.

The term "isotope" was suggested to the physical chemist Frederick Soddy by the physician and novelist Margaret Todd over dinner in the Glasgow home of Soddy's father-in-law.

PROBLEMS

*3-1. The half-life of ^{132}I is 2.3 hours. What interval of time is required for 3.7×10^9 Bq of ^{132}I to decay to 9.25×10^8 Bq? What interval of time is required for decay of 7/8 of the ^{132}I atoms? What interval of time is required for 3.7×10^9 Bq of ^{132}I to decay to 9.25×10^8 Bq if the ^{132}I is in transient equilibrium with $^{132}Te(T_{1/2} = 78$ hours)?

*3-2. What is the mass in grams of 3.7×10^9 Bq of pure $^{32}P\,(T_{1/2} =$ 14.3 days)? How many ^{32}P atoms constitute 3.7×10^9 Bq? What is the mass in grams of 3.7×10^9 Bq of $Na_3\,^{32}PO_4$ if all the phosphorus in the compound is radioactive?

3-3. Some ^{210}Bi nuclei decay by α-emission, whereas others decay by negatron emission. Write the equation for each mode of decay, and identify the daughter nuclide.

*For problems marked with an asterisk, answers are provided on page 491.

*3-4. If a radioactive nuclide decays for an interval of time equal to its average life, what fraction of the original activity remains?

3-5. From the decay scheme for ^{131}I, determine the branching ratio for the mode of negatron emission that results in the release of γ rays of 364 keV. From the decay scheme for ^{126}I, determine the branching ratios for the emission of negatrons and positrons with different maximum energies. Determine the branching ratios of ^{126}I for positron emission and electron capture.

*3-6. What are the frequency and wavelength of a 100-keV photon?

3-7. ^{126}I nuclei decay may be negatron emission, positron emission, or electron capture. Write the equation for each mode of decay, and identify the daughter nuclide.

*3-8. How much time is required before a 3.7×10^8 Bq sample of ^{99m}Tc ($T_{1/2} = 6.0$ hours) and a 9.25×10^8 Bq sample of ^{113m}In ($T_{1/2} = 1.7$ hours) possess equal activities.

3-9. From a chart of the nuclides determine the following:
a. Whether ^{202}Hg is stable or unstable
b. Whether ^{193}Hg decays by negatron or positron emission
c. The nuclide that decays to ^{198}H by positron emission
d. The nuclide that decays to ^{198}Hg by negatron emission
e. The half-life of ^{203}Hg
f. The percent abundance of ^{198}Hg in naturally occurring mercury
g. The atomic mass of ^{204}Hg

*3-10. How many atoms and grams of ^{90}Y are in secular equilibrium with 1.85×10^9 Bq of ^{90}Sr?

*3-11. How many becquerels of ^{24}Na should be ordered so that the sample activity will be 3.7×10^8 Bq when it arrives 24 hours later?

*3-12. Fifty grams of gold (^{197}Au) are subjected to a neutron flux of 10^{13} neutrons/cm^2-sec in the core of a nuclear reactor. How much time is required for the activity to reach 3.7×10^4 GBq? What is the maximum activity for the sample? What is the sample activity after 20 minutes? The cross section of ^{197}Au is 99 barns.

*3-13. The only stable isotope of arsenic is ^{75}As. What modes of radioactive decay would be expected for ^{74}As and ^{76}As?

*3-14. For a nuclide X with the decay scheme

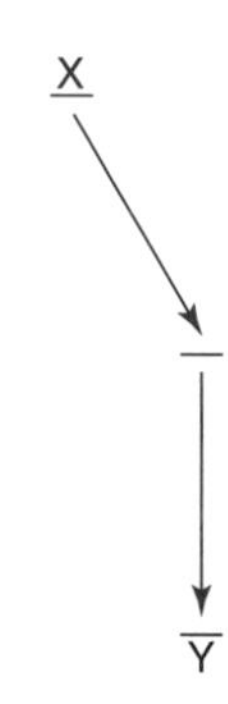

how many γ rays are emitted per 100 disintegrations of X if the coefficient for internal conversion is 0.25?

SUMMARY

- Radioactive decay is the consequence of nuclear instability:
  - Negatron decay occurs in nuclei with a high n/p ratio.
  - Positron decay and electron capture occur in nuclei with a low n/p ratio.
  - Alpha decay occurs with heavy unstable nuclei.
  - Isomeric transitions occur between different energy states of nuclei and result in the emission of γ rays and conversion electrons.
- The activity A of a sample is

$$A = A_0 e^{-\lambda t}$$

where λ is the decay constant (fractional rate of decay).
- The half-life $T_{1/2}$ is

$$T_{1/2} = 0.693/\lambda$$

- The common unit of activity is the becquerel (Bq), with 1 Bq = 1 disintegration/second.
- Transient equilibrium may exist when the progeny nuclide decays with a $T_{1/2} <$ $T_{1/2}$ parent.
- Secular equilibrium may exist when the progeny nuclide decays with a $T_{1/2} \ll$ $T_{1/2}$ parent
- Most radioactive nuclides found in Nature are members of naturally occurring decay series.

REFERENCES

1. Becquerel, H. Sur les radiations émises par phosphorescence. *Compt. Rend.* 1896; **122**:420.
2. Bohr, N. Neutron capture and nuclear constitution. *Nature* 1936; **137**:344.
3. Curie, P., and Curie, S. Sur une substance nouvelle radio-active, contenue dans la pechblende. *C.R. Hebd. Séances Acad. Sci.* 1898; **127**:175–178.
4. Rutherford, E. Uranium radiation and the electrical conduction produced by it. *Philos. Mag.* 1899; **27**:109.
5. Boltwood, B., and Rutherford, E. Production of helium by radium. *Philos. Mag.* 1911; **22**:586.
6. Pauli, W. In *Rapports du Septieme Counseil de Physique Solvay, Bruxelles, 1933.* Paris, Gouthier-Villars & Cie, 1934.
7. Reines, F., and Cowan, C., Jr. Detection of the free neutrino. *Physiol. Rev.* 1953; **92**:830.
8. Curie, I., and Joliot, F. Physique nucléaire: Un nouvean type of radioactivité. *Compt. Rend.* 1934; **198**:254.
9. Villard P: Sur la réfractionn des rayons cathodiques et des rayons déviables du radium. *Compt. Rend.* 1900; **130**:1010.
10. Rutherford, E. Collision of α-particles with light atoms. I. Hydrogen, *Philos. Mag.* 1919; **37**:537.
11. Shleien, B., and Terpilak, M. *The Health Physics and Radiological Health Handbook.* Olney, MD, Nuclear Lectern Associates, 1984.
12. Firestone, R. B., and Shirley, V. S. *Table of Isotopes*, 8th edition. New York, John Wiley and Sons, 1996.
13. Pamphlets of the Medical Internal Radiation Dose Committee of the Society of Nuclear Medicine, 1850 Samuel Morse Drive, Reston, VA 20190-5316.

CHAPTER

4

INTERACTIONS OF RADIATION

OBJECTIVES

After completing this chapter, the reader should be able to:

- Compare directly ionizing and indirectly ionizing radiation and give examples of each.
- Define and state the relationships among linear energy transfer, specific ionization, the *W*-quantity, range, and energy.
- List the interactions that occur for electrons.
- Give examples of elastic and inelastic scattering by nuclei.
- Describe the interactions of neutrons.
- Write an equation for attenuation of photons in a given thickness of a material.
- List the interactions that occur for photons.
- Define atomic, electronic, mass, and linear attenuation coefficients.
- Discuss and compare energy absorption and energy transfer.
- List at least five types of electromagnetic radiation.

Directly and Indirectly Ionizing Radiation

Directly Ionizing Particles (Charged Particles)	Indirectly Ionizing Particles (Uncharged Particles)
Alpha (helium nuclei)	Photons
Any nuclei	Neutrons
Beta (electrons)	
Protons	

The term *interaction* may be used to describe the crash of two automobiles (an example in the macroscopic world) or the collision of an x ray with an atom (an example in the submicroscopic world). This chapter describes radiation interactions on a submicroscopic scale. Interactions in both macroscopic and microscopic scales follow fundamental principles of physics such as (a) the conservation of energy and (b) the conservation of momentum.

CHARACTERISTICS OF INTERACTIONS

In a radiation interaction, the radiation and the material with which it interacts may be considered as a single system. When the system is compared before and after the interaction, certain quantities will be found to be *invariant*. Invariant quantities are exactly the same before and after the interaction. Invariant quantities are said to be *conserved* in the interaction. One quantity that is always conserved in an interaction is the total energy of the system, with the understanding that mass is a form of energy. Other quantities that are conserved include momentum and electric charge.

Some quantities are not always conserved during an interaction. For example, the number of particles may not be conserved because particles may be fragmented, fused, "created" (energy converted to mass), or "destroyed" (mass converted to energy) during an interaction. Interactions may be classified as either *elastic* or *inelastic*. An interaction is elastic if the sum of the *kinetic energies* of the interacting entities is conserved during the interaction. If some energy is used to free an electron or nucleon from a bound state, kinetic energy is not conserved and the interaction is inelastic. Total energy is conserved in all interactions, but kinetic energy is conserved only in interactions designated as elastic.

DIRECTLY IONIZING RADIATION

When an electron is ejected from an atom, the atom is left in an *ionized* state. Hydrogen is the element with the smallest atomic number and requires the least energy (binding energy of 13.6 eV) to eject its K-shell electron. Radiation of energy less than 13.6 eV is termed *nonionizing radiation* because it cannot eject this most easily removed electron. Radiation with energy above 13.6 eV is referred to as *ionizing radiation*. If electrons are not ejected from atoms but merely raised to higher energy levels (outer shells), the process is termed *excitation*, and the atom is said to be "excited." Charged particles such as electrons, protons, and atomic nuclei are directly ionizing radiations because they can eject electrons from atoms through charged-particle interactions. Neutrons and photons (x and γ rays) can set charged particles into motion, but they do not produce significant ionization directly because they are uncharged. These radiations are said to be indirectly ionizing.

Energy transferred to an electron in excess of its binding energy appears as kinetic energy of the ejected electron. An ejected electron and the residual positive ion constitute an *ion pair*, abbreviated IP. An average energy of 33.85 eV, termed the W-quantity or W, is expended by charged particles per ion pair produced in air.[1] The average energy required to remove an electron from nitrogen or oxygen (the most common atoms in air) is much less than 33.85 eV. The W-quantity includes not only the electron's binding energy but also the average kinetic energy of the ejected electron and the average energy lost as incident particles excite atoms, interact with nuclei, and increase the rate of vibration of nearby molecules. On the average, 2.2 atoms are excited per ion pair produced in air.

The specific ionization (SI) is the number of primary and secondary ion pairs produced per unit length of path of the incident radiation. The specific ionization of α-particles in air varies from about 3–7 × 10^6 ion pairs per meter, and the specific ionization of protons and deuterons is slightly less. The linear energy transfer (LET) is the average loss in energy per unit length of path of the incident radiation. The LET is the product of the specific ionization and the W-quantity:

$$\text{LET} = (\text{SI})(W) \tag{4-1}$$

Example 4-1

Assuming that the average specific ionization is 4 × 10^6 ion pairs per meter (IP/m), calculate the average LET of α-particles in air.

$$\begin{aligned}\text{LET} &= (\text{SI})(W)\\ &= 4 \times 10^6\ \frac{\text{IP}}{\text{m}} \times 33.85\ \frac{\text{eV}}{\text{IP}}\\ &= 1.35 \times 10^5\ \frac{\text{keV}}{\text{m}}\end{aligned}$$

The range of ionizing particles in a particular medium is the straight-line distance traversed by the particles before they are completely stopped. For heavy particles with energy E that tend to follow straight-line paths, the range in a particular medium may be estimated from the average LET:

$$\text{Range} = \frac{E}{\text{LET}} \tag{4-2}$$

Example 4-2

Calculate the range in air for 4-MeV α-particles with an average LET equal to that computed in Example 4-1.

$$\begin{aligned}\text{Range} &= \frac{E}{\text{LET}}\\ &= \frac{4\text{MeV}(10^3\text{keV/ MeV})}{1.35 \times 10^5\text{keV/m}}\\ &= \text{approximately } 0.03\text{ m} = 3\text{ cm in air}\end{aligned}$$

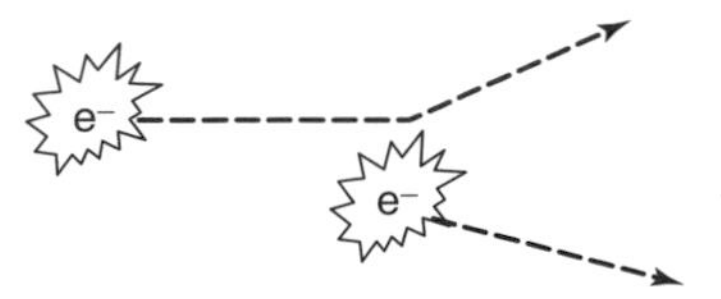

MARGIN FIGURE 4-1
One electron is incident upon a target electron. As the incident electron nears the target electron, the electrostatic fields of the two negatively charged electrons interact, with the result that the stationary electron is set in motion by the repulsive force of the incident electron, and the direction of the incident electron is changed. This classical picture assumes the existence of electrostatic fields for the two electrons that exist simultaneously everywhere in space.

Classical Electrodynamics versus Quantum Electrodynamics (QED)
The classical diagram of particle interactions shows a picture of the interaction in space. The Feynman diagram of quantum electrodynamics (QED) shows the development of the interaction in time. The classical diagram of the interaction between two electrons assumes electrostatic fields for the electrons that exist everywhere in space simultaneously. The Feynman QED diagram of the interaction shows the brief exchange of a virtual photon between the two electrons.

In radiologic physics, we are primarily interested in determining which particles survive an interaction, where they go, and where they deposit their energy. For these purposes, the classical diagrams are adequate.

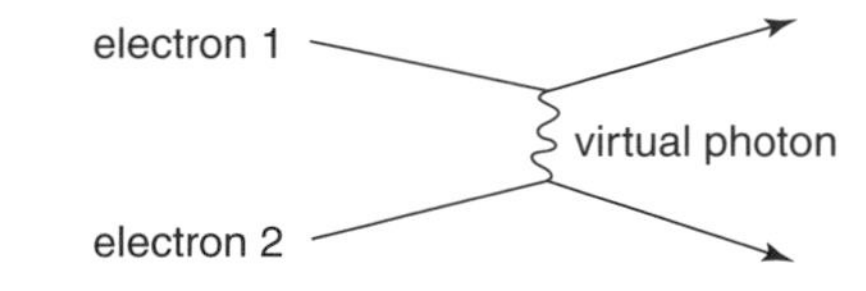

MARGIN FIGURE 4-2
The Feynman diagram of quantum electrodynamics (QED) showing the exchange of a virtual photon between two electrons. When electron 1 emits a virtual photon in a downward direction, conservation of momentum requires the electron to recoil upward. When electron 2 absorbs the virtual photon, it gains momentum and, therefore, must recoil downward. In the QED picture, the electrostatic force between two electrons is explained in terms of exchange of a virtual photon that can travel a specific distance in a specific time.

INTERACTIONS OF ELECTRONS

Interactions of negative and positive electrons may be divided into three categories:

1. Interactions with electrons
2. Elastic interactions with nuclei
3. Inelastic interactions with nuclei

Scattering by Electrons

Negative and positive electrons traversing an absorbing medium transfer energy to electrons of the medium. Impinging electrons lose energy and are deflected at some angle with respect to their original direction. An electron receiving energy may be raised to a shell farther from the nucleus or may be ejected from the atom. The kinetic energy E_k of an ejected electron equals the energy E received minus the binding energy E_B of the electron:

$$E_k = E - E_B \tag{4-3}$$

If the binding energy is negligible compared with the energy received, then the interaction may be considered an elastic collision between "free" particles. If the binding energy must be considered, then the interaction is inelastic.

Incident negatrons and positrons are scattered by electrons with a probability that increases with the atomic number of the absorber and decreases rapidly with increasing kinetic energy of the incident particles. Low-energy negatrons and positrons interact frequently with electrons of an absorber; the frequency of interaction diminishes rapidly as the kinetic energy of the incident particles increases.[2]

Ion pairs are produced by negatrons and positrons during both elastic and inelastic interactions. The specific ionization (ion pairs per meter [IP/m]) in air at STP (standard temperature = 0°C, standard pressure = 760 mm Hg) may be estimated with Eq. (4-4) for negatrons and positrons with kinetic energies between 0 and 10 MeV.

$$\text{SI} = \frac{4500}{(v/c)^2} \tag{4-4}$$

In Eq. (4-4), v represents the velocity of an incident negatron or positron and c represents the speed of light *in vacuo* (3×10^8 m/sec).

Of the many interesting individuals who helped shape our modern view of quantum mechanics, Richard Feynman is remembered as one of the most colorful. Recognized as a math and physics prodigy in college, he came to the attention of the physics community for his work on the Manhattan Project in Los Alamos, New Mexico. In addition to his many contributions to atomic physics and computer science during the project, he was known for playing pranks on military security guards by picking locks on file cabinets, opening safes, and leaving suspicious notes. His late-night bongo sessions in the desert of New Mexico outside the lab were also upsetting to security personnel. After the war, he spent most of his career at Cal Tech, where his lectures were legendary, standing-room-only sessions attended by faculty as well as students.

Example 4-3

Calculate the specific ionization (SI) and linear energy transfer (LET) of 0.1 MeV electrons in air ($v/c = 0.548$). The LET may be computed from the SI or Eq. (4-1) by using an average W-quantity for electrons of 33.85 eV/IP:

$$\begin{aligned}\text{SI} &= \frac{4500}{(v/c)^2} \\ &= \frac{4500}{(0.548)^2} \\ &= 15{,}000 \text{ IP/m}\end{aligned}$$

$$\begin{aligned}\text{LET} &= (\text{SI})\,(W) \\ &= (15{,}000 \text{ IP/m})\,(33.85 \text{ eV/IP})\,(10^{-3} \text{ keV/eV}) \\ &= 508 \text{ keV/m}\end{aligned}$$

After expending its kinetic energy, a positron combines with an electron in the absorbing medium. The particles annihilate each other, and their mass appears as electromagnetic radiation, usually two 0.51-MeV photons moving in opposite directions. These photons are termed *annihilation radiation*, and the interaction is referred to as *pair annihilation*.

Elastic Scattering by Nuclei

Electrons are deflected with reduced energy during elastic interactions with nuclei of an absorbing medium. The probability of elastic interactions with nuclei varies with Z^2 of the absorber and approximately with $1/E_k^2$, where E_k represents the kinetic energy of the incident electrons. The probability for elastic scattering by nuclei is slightly less for positrons than for negatrons with the same kinetic energy. Backscattering of negatrons and positrons in radioactive samples is primarily due to elastic scattering by nuclei.

Probabilities for elastic scattering of electrons by electrons and nuclei of an absorbing medium are about equal if the medium is hydrogen ($Z = 1$). In absorbers with higher atomic number, elastic scattering by nuclei occurs more frequently than electron scattering by electrons because the nuclear scattering cross section varies with Z^2, whereas the cross section for scattering by electrons varies with Z.

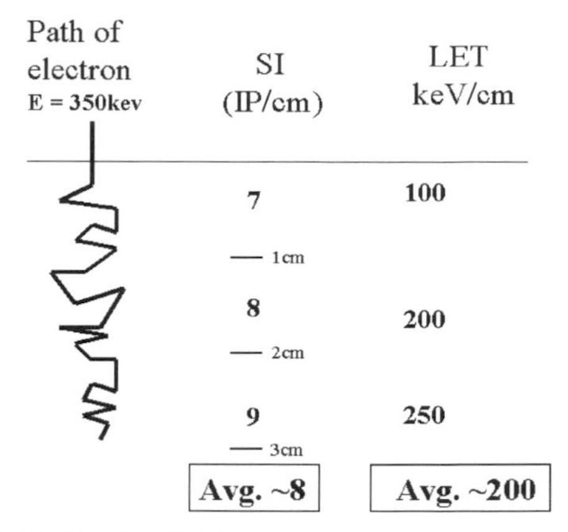

MARGIN FIGURE 4-3
An electron having energy $E = 350$ keV interacts in a tissue-like material. Its actual path is torturous, changing direction a number of times, as the electron interacts with atoms of the material via excitations and ionizations. As interactions reduce the energy of the electron through excitation and ionization, the elctron's energy is transferred to the material. The interactions that take place along the path of the particle may be summarized as specific ionization (SI, ion pairs/cm) or as linear energy transfer (LET, keV/cm) along the straight line continuation of the particle's trajectory beyond its point of entry.

Inelastic Scattering by Nuclei

A negative or positive electron passing near a nucleus may be deflected with reduced velocity. The interaction is inelastic if energy is released as electromagnetic radiation during the encounter. The radiated energy is known as bremsstrahlung (braking radiation). A bremsstrahlung photon may possess any energy up to the entire kinetic energy of the incident particle. For low-energy electrons, bremsstrahlung photons are radiated predominantly at right angles to the motion of the particles. The angle narrows as the kinetic energy of the electrons increases (see Margin Figure 4-6).

The probability of bremsstrahlung production varies with Z^2 of the absorbing medium. A typical bremsstrahlung spectrum is illustrated in Margin Figure 4-4. The relative shape of the spectrum is independent of the atomic number of the absorber.

The ratio of radiation energy loss (the result of inelastic interactions with nuclei) to the energy lost by excitation and ionization (the result of interactions with electrons) is approximately

$$\frac{\text{Radiation energy loss}}{\text{Ionization energy loss}} = \frac{E_k Z}{820} \tag{4-5}$$

where E_k represents the kinetic energy of the incident electrons in MeV and Z is the atomic number of the absorbing medium. For example, excitation-ionization and bremsstrahlung contribute about equally to the energy lost by 10-MeV electrons traversing lead ($Z = 82$). The ratio of energy lost by the production of bremsstrahlung to that lost by ionization and excitation of atoms is important to the design of x-ray tubes in the diagnostic energy range (below 0.1 MeV) where the ratio is much smaller and therefore x ray production is less efficient.

The speed of light in a vacuum (3×10^8 m/sec) is the greatest velocity known to be possible. Particles cannot exceed a velocity of 3×10^8 m/sec under any circumstances. However, light travels through many materials at speeds slower than its speed in a vacuum. In a material, it is possible for the velocity of a particle to exceed the speed of light in the material. When this occurs, visible light known as Cerenkov radiation is emitted. Cerenkov radiation is the cause of the blue glow that emanates from the core of a "swimming pool"-type nuclear reactor. Only a small fraction of the kinetic energy of high-energy electrons is lost through production of Cerenkov radiation.

A cross section is an expression of probability that an interaction will occur between particles. The bigger the cross section, the higher the probability an interaction will occur. This is the same principle as the observation that "the bigger the target, the easier it is to hit." The unit in which atomic cross sections are expressed is the barn, where one barn equals 10^{-28} m^2. Its origin is in the American colloquialism, "as big as a barn" or "such a bad shot, he couldn't hit the broad side of a barn." The unit name was first used by American scientists during the Manhattan Project, the project in which the atomic bomb was developed during World War II. In 1950, the Joint Commission on Standards, Units, and Constants of Radioactivity recommended international acceptance because of its common usage in the United States (Evans, R. D. *The Atomic Nucleus*. Malabar, FL, Krieger Publishing, 1955, p. 9.)

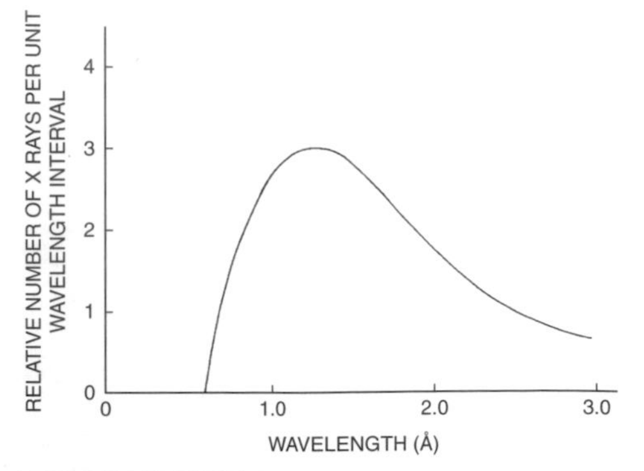

MARGIN FIGURE 4-4
Bremsstrahlung spectrum for a molybdenum target bombarded by electrons accelerated through 20 kV plotted as a function of wavelength in angstroms (Å) = 10^{-10} m. (From Wehr, M., and Richards, J. *Physics of the Atom*. Reading, MA, Addison-Wesley, 1960, p. 159. Used with permission.)

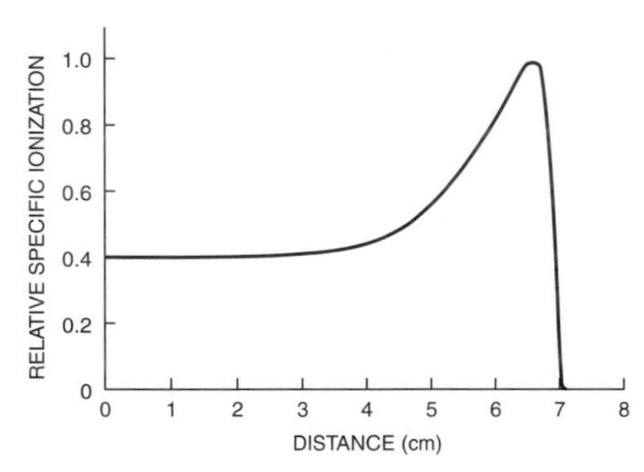

MARGIN FIGURE 4-5
The relative specific ionization of 7.7-MeV α-particles from the decay of ^{214}Po is plotted as a function of the distance traversed in air.

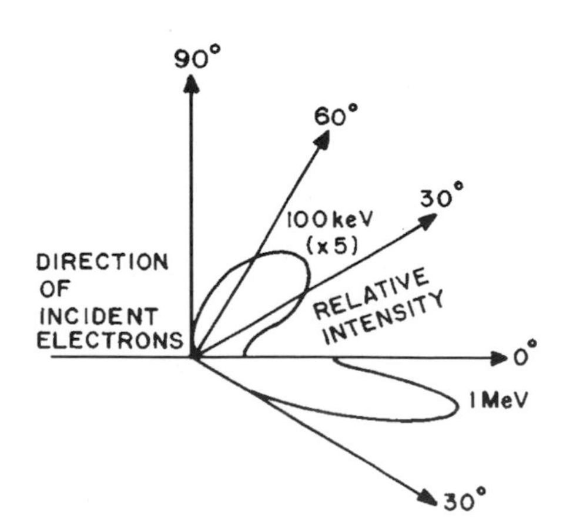

MARGIN FIGURE 4-6
Relative intensity of bremsstrahlung radiated at various angles for electrons with kinetic energies of 100 keV and 1 MeV. (Data from Scherzer, O. *Ann Phys* 1932; **13**:137; and Andrews, H. *Radiation Physics*. Englewood Cliffs, NJ, Prentice-Hall International, 1961.)

INTERACTIONS OF HEAVY, CHARGED PARTICLES

Protons, deuterons, α-particles, and other heavy, charged particles lose kinetic energy rapidly as they penetrate matter. Most of the energy is lost as the particles interact inelastically with electrons of the absorbing medium. The transfer of energy is accomplished by interacting electrical fields, and physical contact is not required between the incident particles and absorber electrons. From Examples 4-1 and 4-2, it is apparent that α-particles produce dense patterns of interaction but have limited range. Deuterons, protons, and other heavy, charged particles also exhibit high specific ionization and a relatively short range. The density of soft tissue (1 g/cm^3) is much greater than the density of air (1.29×10^{-3} g/cm^3). Hence, α-particles of a few MeV or less from radioactive nuclei penetrate soft tissue to depths of only a few microns ($1\ \mu$m $= 10^{-6}$m). For example, α-particles from a radioactive source near or on the body penetrate only the most superficial layers of the skin.

The specific ionization (SI) and LET are not constant along the entire path of monoenergetic charged particles traversing a homogeneous medium. The SI of 7.7-MeV α-particles from ^{214}Po is plotted in Margin Figure 4-5 as a function of the distance traversed in air. The increase of SI near the end of the path of the particles reflects the decreased velocity of the α-particles. As the particles slow down, the SI increases because nearby atoms are influenced for a longer period of time. The region of increased SI is termed the Bragg peak. The rapid decrease in SI beyond the peak is due primarily to the capture of electrons by slowly moving α-particles. Captured electrons reduce the charge of the α-particles and decrease their ability to produce ionization.

INDIRECTLY IONIZING RADIATION

Uncharged particles such as neutrons and photons are said to be indirectly ionizing. Neutrons have no widespread application at the present time in medical imaging. They are discussed here briefly to complete the coverage of the "fundamental" particles that make up the atom.

INTERACTIONS OF NEUTRONS

Neutrons may be produced by a number of sources. The distribution of energies available depends on the method by which the "free" neutrons are produced. Slow, intermediate, and fast neutrons (Table 4-1) are present within the core of a nuclear reactor. Neutrons with various kinetic energies are emitted by ^{252}Cf, a nuclide that fissions spontaneously. This nuclide has been encapsulated into needles and used experimentally for implant therapy. Neutron beams are available from cyclotrons and other accelerators in which low-Z nuclei (e.g., ^{3}H or ^{9}Be) are bombarded by positively charged particles (e.g., nuclei of ^{1}H, ^{2}H, ^{3}H) moving at high velocities. Neutrons are released as a product of this bombardment. The energy distribution of neutrons from these devices depends on the target material and on the type and energy of the bombarding particle.

Neutrons are uncharged particles that interact primarily by "billiard ball" or "knock-on" collisions with absorber nuclei. A knock-on collision is elastic if the kinetic energy of the particles is conserved. The collision is inelastic if part of the kinetic energy is used to excite the nucleus. During an elastic knock-on collision, the energy transferred from a neutron to the nucleus is maximum if the mass of the nucleus equals the neutron mass. If the absorbing medium is tissue, then the energy transferred per collision is greatest for collisions of neutrons with nuclei of hydrogen, because the mass of the hydrogen nucleus (i.e., a proton) is close to the mass of a neutron. Most nuclei in tissue are hydrogen, and the cross section for elastic collision is greater for hydrogen than for other constituents of tissue. For these reasons, elastic collisions with hydrogen nuclei account for most of the energy deposited in tissue by neutrons with kinetic energies less than 10 MeV.

For neutrons with kinetic energy greater than 10 MeV, inelastic scattering also contributes to the energy lost in tissue. Most inelastic interactions occur with nuclei other than hydrogen. Energetic charged particles (e.g., protons or α-particles) are often ejected from nuclei excited by inelastic interactions with neutrons.

ATTENUATION OF X AND γ RADIATION

When an x or γ ray impinges upon a material, there are three possible outcomes. The photon may (1) be absorbed (i.e., transfer its energy to atoms of the target material) during one or more interactions; (2) be scattered during one or more interactions; or (3) traverse the material without interaction. If the photon is absorbed or scattered, it is said to have been attenuated. If 1000 photons impinge on a slab of material, for example, and 200 are scattered and 100 are absorbed, then 300 photons have been attenuated from the beam, and 700 photons have been transmitted without interaction.

Attenuation processes can be complicated. Partial absorption may occur in which only part of the photon's energy is retained in the absorber. Scattering through small angles may not remove photons from the beam, especially if the beam is rather broad.

TABLE 4-1 Classification of Neutrons According to Kinetic Energy

Type	*Energy Range*
Slow	0–0.1 keV
Intermediate	0.1–20 keV
Fast	20 keV–10 Mev
High-energy	>10 MeV

Attenuation of a Beam of X or γ Rays

The number of photons attenuated in a medium depends on the number transversing the medium. If all the photons posses the same energy (i.e., the beam is monoenergetic) and if the photons are attenuated under conditions of good geometry (i.e., the beam is narrow and the transmitted beam contains no scattered photons), then the number I of photons penetrating a thin slab of matter of thickness x is

$$I = I_0 e^{-\mu x} \tag{4-6}$$

where μ is the attenuation coefficient of the medium for the photons and I_0 represents the number of photons in the beam before the thin slab of matter is placed into position. The number I_{at} of photons attenuated (absorbed or scattered) from the beam is

$$\begin{aligned} I_{at} &= I_0 - I \\ &= I_0 - I_0 e^{-\mu x} \\ &= I_0(1 - e^{-\mu x}) \end{aligned} \tag{4-7}$$

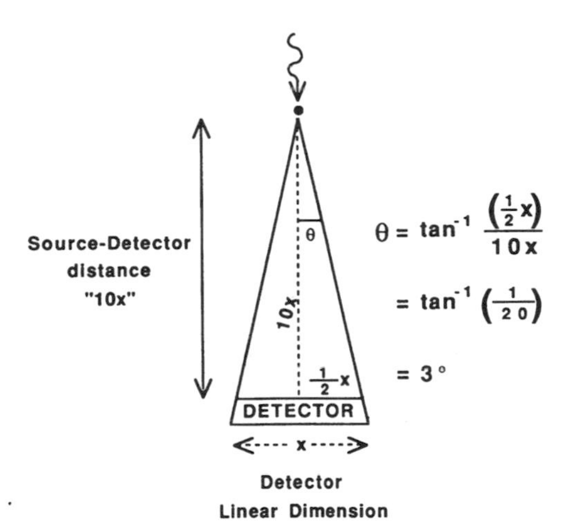

MARGIN FIGURE 4-7
Illustration of "small-angle" scatter. Scatter of greater than 3 degrees generally is accepted as attenuation (removal) from a narrow beam. A generally accepted definition of a broad beam is a beam that is broader than one-tenth the distance to a point source of radiation.

The exponent of e must possess no units. Therefore, the units for μ are 1/cm if the thickness x is expressed in centimeters, 1/in. if x is expressed in inches, and so on. An attenuation coefficient with units of 1/length is called a linear attenuation coefficient. Derivation of Eqs. (4-6), (4-7), and other exponential expressions is given in Appendix I.

The mean path length is the average distance traveled by x or γ rays before interaction in a particular medium. The mean path length is sometimes termed the mean free path or relaxation length and equals $1/\mu$, where μ is the total linear attenuation coefficient. Occasionally the thickness of an attenuating medium may be expressed in multiples of the mean free path of photons of a particular energy in the medium.

The probability is $e^{-\mu x}$ that a photon traverses a slab of thickness x without interacting. This probability is the product of probabilities that the photon does not interact by any of five interaction processes:

$$e^{-\mu x} = (e^{-\omega x})(e^{-\tau x})(e^{-\sigma x})(e^{-\kappa x})(e^{-\pi x}) = e^{-(\omega+\tau+\sigma+\kappa+\pi)x}$$

The coefficients ω, τ, σ, κ, and π represent attenuation by the processes of coherent scattering (ω), photoelectric absorption (τ), Compton scattering (σ), pair production (κ), and photodisintegration (π), as described later. The total linear attenuation coefficient may be written

$$\mu = \omega + \tau + \sigma + \kappa + \pi \tag{4-8}$$

In diagnostic radiography, coherent scattering, photodisintegration, and pair production are usually negligible, and μ is written

$$\mu = \tau + \sigma \tag{4-9}$$

In general, attenuation coefficients vary with the energy of the x or γ rays and with the atomic number of the absorber. Linear attenuation coefficients also depend on the density of the absorber. Mass attenuation coefficients, obtained by dividing linear attenuation coefficients by the density ρ of the attenuating medium, do not vary with the density of the medium:

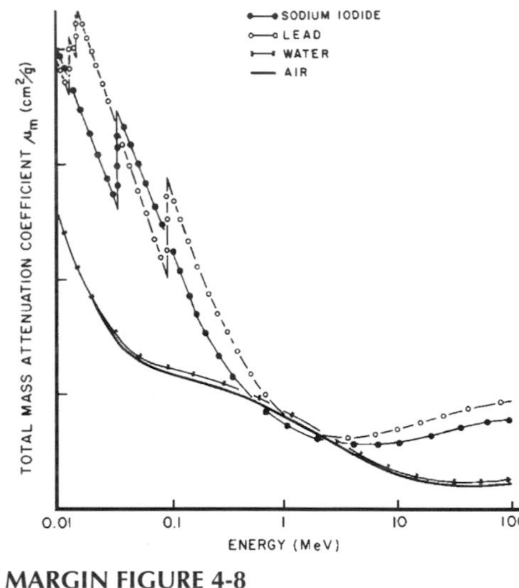

MARGIN FIGURE 4-8
Mass attenuation coefficients for selected materials as a function of photon energy.

$$\mu_m = \frac{\mu}{\rho}, \quad \omega_m = \frac{\omega}{\rho}, \quad \tau_m = \frac{\tau}{\rho}, \quad \sigma_m = \frac{\sigma}{\rho}, \quad \kappa_m = \frac{\kappa}{\rho}, \quad \pi_m = \frac{\pi}{\rho}$$

Mass attenuation coefficients usually have units of square meters per kilogram, although sometimes square centimeters per gram and square centimeters per milligram are used. Total mass attenuation coefficients for air, water, sodium iodide, and lead are plotted in Margin Figure 4-8 as a function of the energy of incident photons.

When mass attenuation coefficients are used, thicknesses x_m are expressed in units such as kilograms per square meter, grams per square centimeter, or milligrams per square centimeter. A unit of thickness such as kilograms per square meter may seem unusual but can be understood by recognizing that this unit describes the mass of a 1-m^2 cross section of an absorber. That is, the amount of material traversed by a beam is indirectly measured by the mass per unit area of the slab. The thickness of attenuating medium may also be expressed as the number of atoms or electrons per unit area. Symbols x_a and x_e denote thicknesses in units of atoms per square meter and electrons per square meter. These thicknesses may be computed from the linear thickness x by

$$\begin{aligned} X_a \frac{\text{atoms}}{\text{m}^2} &= \frac{x(\text{m})\rho(kg/\text{m}^3)N_a(\text{atoms/gram-atomic mass})}{M(\text{kg/gram-atomic mass})} \\ &= \frac{x\rho N_a}{M} \end{aligned} \tag{4-10}$$

$$\begin{aligned} x_e \frac{\text{electrons}}{\text{m}^2} &= \left[x_a\left(\frac{\text{atoms}}{\text{m}^2}\right)\right]\left[Z\left(\frac{\text{electrons}}{\text{atom}}\right)\right] \\ &= x_a Z \end{aligned} \tag{4-11}$$

In these equations, M is the gram-atomic mass of the attenuating medium, Z is the atomic number of the medium, and N_a is Avogadro's number (6.02×10^{23}), the number of atoms per gram-atomic mass. Total atomic and electronic attenuation coefficients μ_a and μ_e, to be used with thicknesses x_a and x_e, may be computed from the linear attenuation coefficient μ by

$$\mu_a\left(\frac{\text{m}^2}{\text{atom}}\right) = \frac{\mu(\text{m}^{-1})M(\text{kg/gram-atomic mass})}{\rho(kg/\text{m}^3)N_a(\text{atoms/gram-atomic mass})} = \frac{\mu M}{\rho N_a} \tag{4-12}$$

$$\mu_e\left(\frac{\text{m}^2}{\text{electron}}\right) = \frac{\mu_a(\text{m}^2/\text{atom})}{Z(\text{electrons/atom})} = \frac{\mu_a}{Z} \tag{4-13}$$

The number I of photons penetrating a thin slab of matter may be computed with any of the following expressions:

$$I = I_0 e^{-\mu x} \qquad I = I_0 e^{-\mu_a x_a}$$

$$I = I_0 e^{-\mu_m x_m} \qquad I = I_0 e^{-\mu_e x_e}$$

Example 4-4

A narrow beam containing 2000 monoenergetic photons is reduced to 1000 photons by a slab of copper 10^{-2} m thick. What is the total linear attenuation coefficient of the copper slab for these photons?

$$I = I_0 e^{-\mu x}$$

$$I/I_0 = e^{-\mu x}$$

$$I_0/I = e^{\mu x}$$

$$\text{In } I_0/I = \mu x$$

$$\text{In}\frac{2000 \text{ photons}}{1000 \text{ photons}} = \mu(10^{-2} \text{ m})$$

$$\text{In } 2 = \mu(10^{-2} \text{ m})$$

$$\mu = \frac{\text{In } 2}{10^{-2} \text{ m}}$$

$$\mu = \frac{0.693}{10^{-2} \text{ m}}$$

$$\mu = 69.3 \text{ m}^{-1}$$

The thickness of a slab of matter required to reduce the intensity (or exposure rate—see Chapter 6) of an x- or γ-ray beam to half is the half-value layer (HVL) or half-value thickness (HVT) for the beam. The HVL describes the "quality" or penetrating ability of the beam. The HVL in Example 4-4 is 10^{-2} m of copper. The HVL of a monoenergetic beam of x or γ rays in any medium is

$$\text{HVL} = \frac{\ln 2}{\mu} \tag{4-14}$$

where $\ln 2 = 0.693$ and μ is the total linear attenuation coefficient of the medium for photons in the beam. The measurement of HVLs for monoenergetic and polyenergetic beams is discussed in Chapter 6.

Example 4-5

a. What are the total mass (μ_m), atomic (μ_a), and electronic (μ_e) attenuation coefficients of the copper slab described in Example 4-4? Copper has a density of 8.9×10^3 kg/m^3, a gram-atomic mass of 63.6, and an atomic number of 29.

$$\mu_m = \mu/\rho$$
$$= \frac{69.3\ \text{m}^{-1}}{8.9 \times 10^3\ \text{kg/m}^3}$$
$$= 0.0078\ \text{m}^2/\text{kg}$$

$$\begin{aligned}\mu_a &= \frac{\mu M}{\rho N} \\ &= \frac{(69.3\ \text{m}^{-1})(63.6\ \text{g/gram-atomic mass})(10^{-3}\ \text{kg/g})}{(8.9 \times 10^3\ \text{kg/m}^3)(6.02 \times 10^{23}\ \text{atoms/gram-atomic mass})} \\ &= 8.2 \times 10^{-28}\text{m}^2/\text{atom}\end{aligned} \tag{4-15}$$

$$\begin{aligned}\mu_e &= \frac{\mu_a}{Z} \\ &= \frac{8.2 \times 10^{-28}\text{m}^2/\text{atom}}{29\ \text{electrons/atom}} \\ &= 2.8 \times 10^{-29}\text{m}^2/\text{electron}\end{aligned} \tag{4-16}$$

b. To the 0.01-m copper slab, 0.02 m of copper is added. How many photons remain in the beam emerging from the slab?

$$\begin{aligned}I &= I_0 e^{-\mu x} \\ &= (2000\ \text{photons})e^{-(69.3\ \text{m}^{-1})(0.03\ \text{m})} \\ &= (2000\ \text{photons})e^{-2.079} \\ &= (2000\ \text{photons})(0.125) \\ &= 250\ \text{photons}\end{aligned} \tag{4-17}$$

A thickness of 0.01 m of copper reduces the number of photons to half. Because the beam is narrow and monoenergetic, adding two more of the same thicknesses of copper reduces the number to one eighth. Only one eighth of the original number of photons remains after the beam has traversed three thicknesses of copper.

c. What is the thickness x_e in electrons per square meter for the 0.03-m slab?

$$\begin{aligned}x_e &= x_a Z \\ &= \frac{x N_a \rho Z}{M} \\ &= \frac{(0.03\ \text{m})(8.9 \times 10^3\ \text{kg/m}^3)(6.02 \times 10^{23}\ \text{atoms/gram-atomic mass})(29\ \text{electrons/atom})}{(63.3\ \text{g/gram-atomic mass})(10^{-3}\ \text{kg/g})} \\ &= 7.3 \times 10^{28}\ \text{electrons/m}^2\end{aligned}$$

d. Repeat the calculation in Part b, but use the electronic attenuation coefficient.

$$\begin{aligned}I &= I_0 e^{-\mu_e x_e} \\ &= (2000\ \text{photons})e^{-(2.8\times10^{-29}\ \text{m}^2/\text{electron})(7.3\times10^{28}\ \text{electrons/m}^2)} \\ &= (2000\ \text{photons})e^{-2.079}\end{aligned}$$

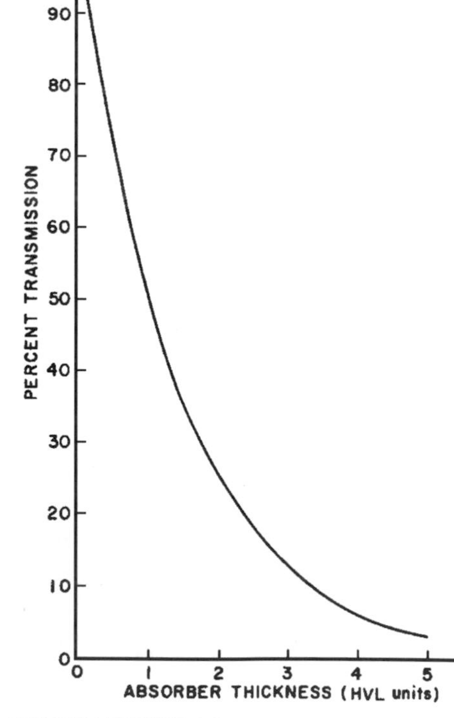

MARGIN FIGURE 4-9
Percent transmission of a narrow beam of monoenergetic photons as a function of the thickness of an attenuating slab in units of half-value layer (HVL). Absorption conditions satisfy requirements for "good geometry."

$$= (2000 \text{ photons})(0.125)$$

$$= 250 \text{ photons}$$

A narrow beam of monoenergetic photons is attenuated exponentially. Exponential attenuation is described by Eq. (4-6). From Example 4-4 we have the equivalent relationship:

$$\ln \frac{I}{I_0} = -\mu x$$

The logarithm of the number I of photons decreases linearly with increasing thickness of the attenuating slab. Hence a straight line is obtained when the logarithm of the number of x or γ rays is plotted as a function of thickness. When the log of the dependent (y-axis) variable is plotted as a function of the linear independent (x-axis) variable, the graph is termed a semilog plot. It should be emphasized that a semilogarithmic plot of the number of photons versus the thickness of the attenuating slab yields a straight line only if all photons possess the same energy and the conditions for attenuation fulfill requirements for narrow-beam geometry.

If broad-beam geometry were used, the falloff of the number of photons as a function of slab thickness would be less rapid. In broad-beam geometry, some scattered photons continue to strike the detector and are therefore not considered to have been "attenuated." A broad-beam measurement is relevant when considering the design of wall shielding for radiation safety near x-ray equipment. Humans on the opposite side of barriers would encounter a broad beam. Narrow-beam geometry is more relevant to situations such as experimental studies of the properties of materials.

When an x-ray beam is polyenergetic (as is the beam emitted from an x-ray tube), a plot of the number of photons remaining in the beam as a function of the thickness of the attenuating material it has traversed does not yield a straight line on a semilogarithmic plot. This is because the attenuation of a polyenergetic beam cannot be represented by a single simple exponential equation with a single attenuation coefficient. It is represented by a weighted average of exponential equations for each of the different photon energies in the beam.

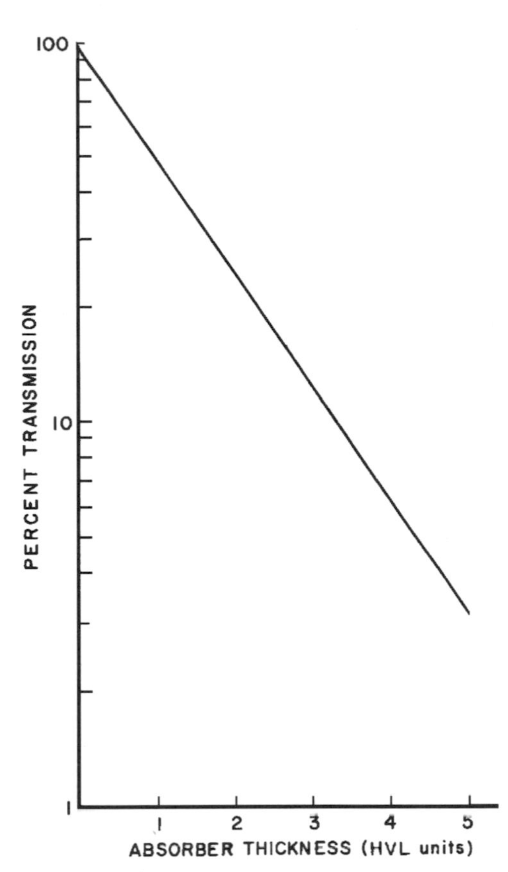

MARGIN FIGURE 4-10
Semilogarithmic plot of data in Margin Figure 4-9.

Filtration

In a polyenergetic beam, lower-energy photons are more likely to be attenuated and higher energy photons are more likely to pass through without interaction. Therefore, after passing through some attenuating material, the distribution of photon energies contained within the emergent beam is different than that of the beam that entered. The emergent beam, although containing fewer total photons, actually has a higher average photon energy. This effect is known as "filtration." A beam that has undergone filtration is said to be "harder" because it has a higher average energy and therefore more penetrating power.

To remove low-energy photons and increase the penetrating ability of an x-ray beam from a diagnostic or therapy x-ray unit, filters of aluminum, copper, tin, or lead may be placed in the beam. For most diagnostic x-ray units, the added filters are usually 1 to 3 mm of aluminum. Because the energy distribution changes as a polyenergetic x-ray beam penetrates an attenuating medium, no single value for the attenuation coefficient may be used in Eq. (4-6) to compute the attenuation of the x-ray beam. However, from the measured half-value layer (HVL) an effective attenuation coefficient may be computed (see Example 4-6):

$$\mu_{\text{eff}} = \frac{\ln 2}{\text{HVL}}$$

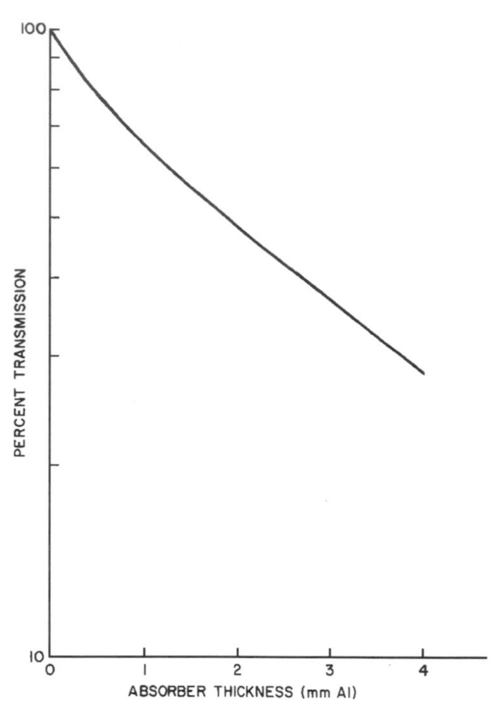

MARGIN FIGURE 4-11
Semilogarithmic plot of the number of x rays in a polyenergetic beam as a function of thickness of an attenuating medium. The penetrating ability (HVL) of the beam increases continuously with thickness because lower-energy photons are selectively removed from the beam. The straight-line relationship illustrated in Margin Figure 4-10 for a narrow beam of monoenergetic photons is not achieved for a polyenergetic x-ray beam.

The effective energy of an x-ray beam is the energy of monoenergetic photons that have an attenuation coefficient in a particular medium equal to the effective attenuation coefficient for the x-ray beam in the same medium.

Example 4-6

An x-ray beam produced at 200 kVp has an HVL of 1.5 mm Cu. The density of copper is 8900 kg/m^3.

a. What are the effective linear and mass attenuation coefficients?

$$\mu_{\text{eff}} = \frac{\ln 2}{1.5 \text{ mm Cu}}$$
$$= 0.46 \ (\text{mm Cu})^{-1}$$
$$(\mu_m)_{\text{eff}} = \frac{\mu_{\text{eff}}}{\rho}$$
$$= \frac{0.46\,(\text{mm Cu})^{-1}(10^3 \text{ mm/m})}{8.9 \times 10^3 \text{ kg/m}^3}$$
$$= 0.052 \text{ m}^2/\text{kg}$$

b. What is the average effective energy of the beam?

Monoenergetic photons of 96 keV possess a total mass attenuation coefficient of 0.052 m^2/kg in copper. Consequently, the average effective energy of the x-ray beam is 96 keV.

Energy Absorption and Energy Transfer

The attenuation coefficient μ (or μ_a, μ_e, μ_m) refers to total attenuation (i.e., absorption plus scatter). Sometimes it is necessary to determine the energy truly absorbed in a material and not simply scattered from it. To express the energy absorbed, the energy absorption coefficient μ_{en} is used, where μ_{en} is given by the expression

$$\mu_{\text{en}} = \mu \frac{E_a}{h\nu} \tag{4-18}$$

In this expression, μ is the attenuation coefficient, E_a is the average energy absorbed in the material per photon interaction, and $h\nu$ is the photon energy. Thus, the energy absorption coefficient is equal to the attenuation coefficient times the fraction of energy truly absorbed. Energy absorption coefficients may also be expressed as mass energy absorption coefficients $(\mu_{\text{en}})_m$, atomic energy absorption coefficients $(\mu_{\text{en}})_a$, or electronic energy absorption coefficients $(\mu_{\text{en}})_e$ by dividing the energy absorption coefficient μ_{en} by the physical density, the number of atoms per cubic meter, or the number of electrons per cubic meter, respectively.

Example 4-7

The attenuation coefficient for 1-MeV photons in water is 7.1 m^{-1}. If the energy absorption coefficient for 1-MeV photons in water is 3.1 m^{-1}, find the average energy absorbed in water per photon interaction.

By rearranging Eq. (4-15), we obtain

$$E_a = \frac{\mu_{\text{en}}}{\mu}(h\nu)$$
$$= \frac{3.1 \text{ m}^{-1}}{7.1 \text{ m}^{-1}}(1 \text{ MeV})$$

$$= 0.44 \text{ MeV}$$
$$= 440 \text{ keV}$$

Example 4-8

An x-ray tube emits 10^{12} photons per second in a highly collimated beam that strikes a 0.1-mm-thick radiographic screen. For purposes of this example, the beam is assumed to consist entirely of 40-keV photons. The attenuation coefficient of the screen is 23 m^{-1}, and the mass energy absorption coefficient of the screen is 5 m^{-1} for 40-keV photons. Find the total energy in keV absorbed by the screen during a 0.5-sec exposure.

The number of photons incident upon the screen is

$$(10^{12} \text{ photons/sec})(0.5 \text{ sec}) = 5 \times 10^{11} \text{ photons}$$

The number of interactions that take place in the screen is [from Eq. (4-7)]

$$\begin{aligned} I_{\text{at}} &= I_0(1 - e^{-\mu x}) \\ &= 5 \times 10^{11}\left[1 - ^{-(23 \text{ m}^{-1})(10^{-3} \text{ m/mm})}\right] \\ &= 1.2 \times 10^9 \text{ interactions} \end{aligned}$$

The average energy absorbed per interaction is [from Eq. (4-15)]

$$\begin{aligned} F_a &= \frac{\mu_{\text{en}}}{\mu}(h\nu) \\ &= \frac{5 \text{ m}^{-1}}{23 \text{ m}^{-1}}(40 \text{ keV}) \\ &= 8.7 \text{ keV} \end{aligned}$$

The total energy absorbed during the 0.5-sec exposure is then

$$(1.2 \times 10^9)(8.7 \text{ keV}) = 1.3 \times 10^{10} \text{ keV}$$

Energy absorption coefficients and attenuation coefficients for air are plotted in Margin Figure 4-12 as a function of photon energy.

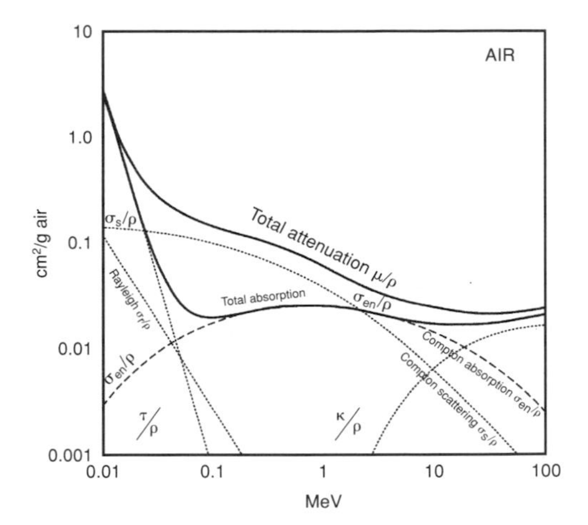

MARGIN FIGURE 4-12
Mass attenuation coefficients for photons in air. The curve marked "total absorption" is $(\mu/\rho) = (\sigma/\rho) + (\tau/\rho) + (\kappa/\rho)$, where σ, τ, and κ are the corresponding linear coefficients for Compton absorption, photoelectric absorption, and pair production. When the Compton mass scattering coefficient σ_s and the Compton absorption coefficient τ_a/ρ are added, the total Compton mass attenuation coefficient σ/ρ is obtained.

Coherent Scattering: Ionizing Radiation

Photons are deflected or scattered with negligible loss of energy by the process of coherent or Rayleigh scattering. Coherent scattering is sometimes referred to as classical scattering because the interaction may be completely described by methods of classical physics. The classical description assumes that a photon interacts with electrons of an atom as a group rather than with a single electron within the atom. Usually the photon is scattered in approximately the same direction as the incident photon. Although photons with energies up to 150 to 200 keV may scatter coherently in media with high atomic number, this interaction is important in tissue only for low-energy photons. The importance of coherent scattering is further reduced because little energy is deposited in the attenuating medium. However, coherent scatter sometimes reduces the resolution of scans obtained with low-energy, γ-emitting nuclides (e.g., ^{125}I) used in nuclear medicine.

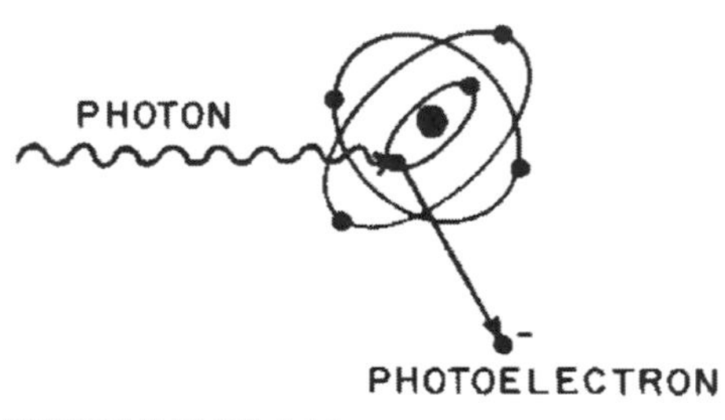

MARGIN FIGURE 4-13
Photoelectric absorption of a photon with energy $h\nu$. The photon disappears and is replaced by an electron ejected from the atom with kinetic energy $E_k = h\nu - E_b$, where E_b is the binding energy of the electron. Characteristic radiation and Auger electrons are emitted as electrons cascade to replace the ejected photoelectron.

Photoelectric Absorption

During a photoelectric interaction, the total energy of an x or γ ray is transferred to an inner electron of an atom. The electron is ejected from the atom with kinetic energy E_κ, where E_κ equals the photon energy $h\nu$ minus the binding energy E_B required

Albert Einstein (1879–1955) was best known to the general public for his theories of relativity. However, he also developed a theoretical explanation of the photoelectric effect—that the energy of photons in a light beam do not add together. It is the energy of each photon that determines whether the beam is capable of removing inner shell electrons from an atom. His Nobel Prize of 1921 mentioned only his explanation of the photoelectric effect and did not cite relativity because that theory was somewhat controversial at that time.

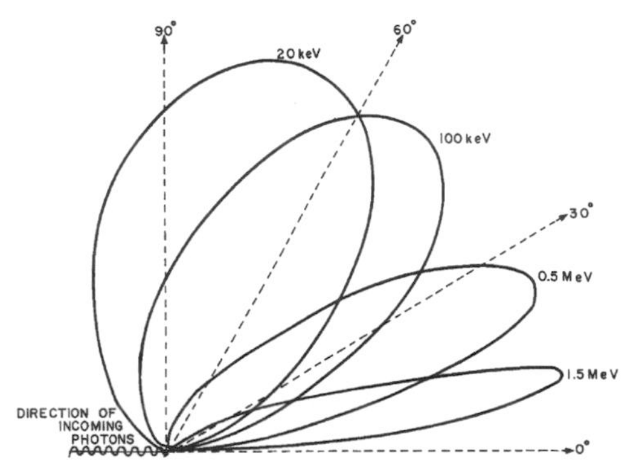

MARGIN FIGURE 4-14
Electrons are ejected approximately at a right angle as low-energy photons interact photoelectrically. As the energy of the photons increases, the angle decreases between incident photons and ejected electrons.

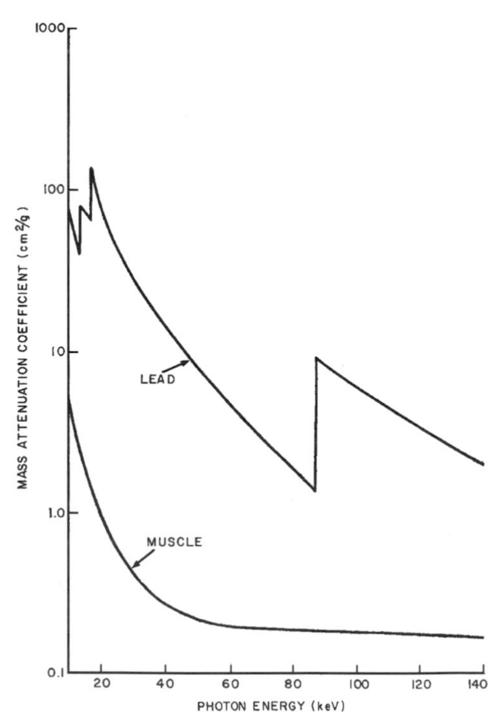

MARGIN FIGURE 4-15
Photoelectric mass attenuation coefficients of lead and soft tissue as a function of photon energy. K- and L-absorption edges are depicted for lead.

to remove the electron from the atom:

$$E_K = h\nu - E_B \tag{4-19}$$

The ejected electron is called a photoelectron.

Example 4-9

What is the kinetic energy of a photoelectron ejected from the K shell of lead (E_B = 88 keV) by photoelectric absorption of a photon of 100 keV?

$$\begin{aligned} E_K &= h\nu - E_B \\ &= 100\text{ keV} - 88\text{ keV} \\ &= 12\text{ keV} \end{aligned}$$

The average binding energy is only 0.5 keV for K electrons in soft tissue. Consequently, a photoelectron ejected from the K shell of an atom in tissue possesses a kinetic energy about 0.5 keV less than the energy of the incident photon.

Photoelectrons resulting from the interaction of low-energy photons are released approximately at a right angle to the motion of the incident photons. As the energy of the photons increases, the average angle decreases between incident photons and released photoelectrons.

An electron ejected from an inner shell leaves a vacancy or hole that is filled immediately by an electron from an energy level farther from the nucleus. Only rarely is a hole filled by an electron from outside the atom. Instead, electrons usually cascade from higher to lower energy levels and produce a number of characteristic photons and Auger electrons with energies that, when added together, equal the binding energy of the ejected photoelectron. Characteristic photons and Auger electrons released during photoelectric interactions in tissue possess an energy less than 0.5 keV. These lower-energy photons and electrons are absorbed rapidly in surrounding tissue.

The probability of photoelectric interaction decreases rapidly as the photon energy increases. In general, the mass attenuation coefficient τ_m for photoelectric absorption varies roughly as $1/(h\nu)^3$, where $h\nu$ is the photon energy. In Margin Figure 4-15, the photoelectric mass attenuation coefficients τ_m of muscle and lead are plotted as a function of the energy of incident photons. Discontinuities in the curve for lead are termed *absorption edges* and occur at photon energies equal to the binding energies of electrons in inner electron shells. Photons with energy less than the binding energy of K-shell electrons interact photoelectrically only with electrons in the L shell or shells farther from the nucleus. Photons with energy equal to or greater than the binding energy of K-shell electrons interact predominantly with K-shell electrons. Similarly, photons with energy less than the binding energy of L-shell electrons interact only with electrons in M and more distant shells. That is, most photons interact photoelectrically with electrons that have a binding energy nearest to but less than the energy of the photons. Hence, the photoelectric attenuation coefficient increases abruptly at photon energies equal to the binding energies of electrons in different shells. Absorption edges for photoelectric attenuation in soft tissue occur at photon energies that are too low to be shown in Margin Figure 4-15. Iodine and barium exhibit K-absorption edges at energies of 33 and 37 keV. Compounds containing these elements are routinely used as contrast agents in diagnostic radiology.

At all photon energies depicted in Margin Figure 4-15, the photoelectric attenuation coefficient for lead ($Z = 82$) is greater than that for soft tissue ($Z_{\text{eff}} = 7.4$) (Z_{eff} represents the effective atomic number of a mixture of elements.) In general, the photoelectric mass attenuation coefficient varies with Z^3. For example, the number of 15-keV photons absorbed primarily by photoelectric interaction in bone ($Z_{\text{eff}} = 11.6$) is approximately four times greater than the number of 15-keV photons absorbed in an equal mass of soft tissue because $(11.6/7.4)^3 = 3.8$. Selective attenuation of photons in media with different atomic numbers and different physical densities is the principal reason for the usefulness of low-energy x rays for producing images in diagnostic radiology.

Effective atomic number (Z_{eff}): A pure element, a material composed of only one type of atom, is characterized by its atomic number (the number of protons in its nucleus). In a composite material, the *effective* atomic number is a weighted average of the atomic numbers of the different elements of which it is composed. The average is weighted according to the relative number of each type of atom.

Compton (Incoherent) Scattering

X and γ rays with energy between 30 keV and 30 MeV interact in soft tissue predominantly by Compton scattering. During a Compton interaction, part of the energy of an incident photon is transferred to a loosely bound or "free" electron within the attenuating medium. The kinetic energy of the recoil (Compton) electron equals the energy lost by the photon, with the assumption that the binding energy of the electron is negligible. Although the photon may be scattered at any angle ϕ with respect to its original direction, the Compton electron is confined to an angle θ, which is 90 degrees or less with respect to the motion of the incident photon. Both θ and ϕ decrease with increasing energy of the incident photon (Margin Figure 4-17).

During a Compton interaction, the change in wavelength [$\Delta\lambda$ in nanometers (nm), 10^{-9} meters] of the x or γ ray is

$$\Delta\lambda = 0.00243\,(1 - \cos\phi) \tag{4-20}$$

where ϕ is the scattering angle of the photon. The wavelength of the scattered photon is

$$\lambda' = \lambda + \Delta\lambda \tag{4-21}$$

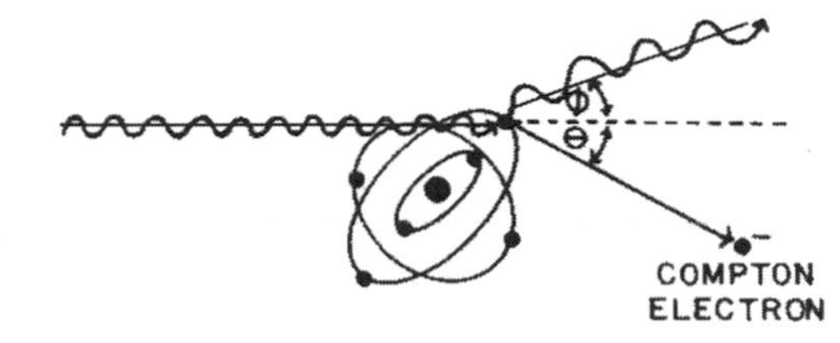

MARGIN FIGURE 4-16
Compton scattering of an incident photon, with the photon scattered at an angle ϕ; the Compton electron is ejected at an angle θ with respect to the direction of the incident photon.

where λ is the wavelength of the incident photon. The energies $h\nu$ and $h\nu'$ of the incident and scattered photons are

$$\begin{aligned} h\nu\ (\text{keV}) &= \frac{hc}{\lambda} \\ &= \frac{(6.62 \times 10^{-34}\ \text{J-sec})(3 \times 10^{8}\ \text{m/sec})}{(\lambda)(10^{-9}\ \text{m/nm})(1.6 \times 10^{-19}\ \text{J/eV})(10^{3}\ \text{eV/keV})} \\ &= \frac{1.24}{\lambda} \end{aligned} \tag{4-22}$$

with λ expressed in nanometers.

Example 4-10

A 210-keV photon is scattered at an angle of 80 degrees during a Compton interaction. What are the energies of the scattered photon and the Compton electron? The wavelength λ of the incident photon is

$$\begin{aligned} \lambda &= \frac{1.24}{h\nu} \\ &= \frac{1.24}{210\ \text{keV}} \\ &= 0.0059\ \text{nm} \end{aligned} \tag{4-23}$$

The change in wavelength $\Delta\lambda$ is

$$\begin{aligned} \Delta\lambda &= 0.00243\,(1 - \cos\phi) \\ &= 0.00243\,(1 - \cos(80\ \text{degrees})) \\ &= 0.00243\,(1 - 0.174) \\ &= 0.0020\ \text{nm} \end{aligned} \tag{4-24}$$

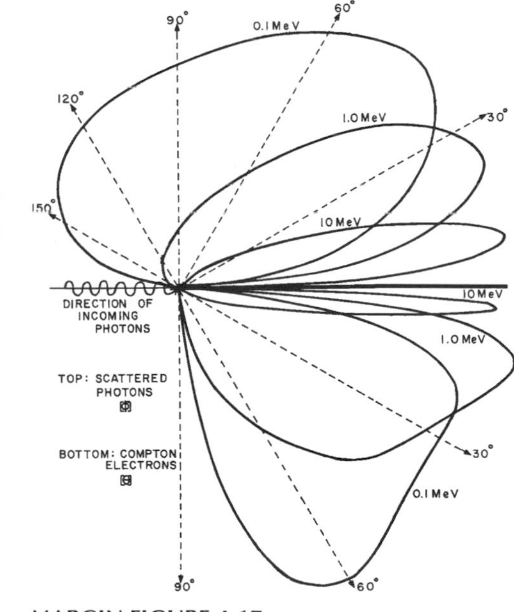

MARGIN FIGURE 4-17
Electron scattering angle θ and photon scattering angle ϕ as a function of the energy of incident photons. Both θ and ϕ decrease as the energy of incident photons increases.

The wavelength of the scattered photon is

$$\begin{aligned} \lambda' &= \lambda + \Delta\lambda \\ &= (0.0059 + 0.0020) \text{ nm} \\ &= 0.0079 \text{ nm} \end{aligned} \tag{4-25}$$

The energy of the scattered photon is

$$\begin{aligned} h\nu' &= \frac{1.24}{\lambda} \\ &= \frac{1.24}{0.0079 \text{ nm}} \\ &= 160 \text{ keV} \end{aligned} \tag{4-26}$$

The energy of the Compton electron is

$$\begin{aligned} E_k &= h\nu - h\nu' \\ &= (210 - 160) \text{ keV} \\ &= 50 \text{ keV} \end{aligned}$$

Example 4-11

A 20-keV photon is scattered by a Compton interaction. What is the maximum energy transferred to the recoil electron?

The energy transferred to the electron is greatest when the change in wavelength of the photon is maximum; $\Delta\lambda$ is maximum when $\phi = 180$ degrees.

$$\begin{aligned} \Delta\lambda_{\text{max}} &= 0.00243 \, [1 - \cos (180)] \\ &= 0.00243 \, [1 - (-1)] \\ &= 0.00486 \text{ nm} \\ &= 0.005 \text{ nm} \end{aligned}$$

The wavelength λ of a 20-keV photon is

$$\begin{aligned} \lambda &= \frac{1.24}{h\nu} \\ &= \frac{1.24}{20 \text{ keV}} \\ &= 0.062 \text{ nm} \end{aligned} \tag{4-27}$$

The wavelength λ' of the photon scattered at 180 degrees is

$$\begin{aligned} \lambda' &= \lambda + \Delta\lambda \\ &= (0.062 + 0.005) \text{ nm} \\ &= 0.067 \text{ nm} \end{aligned} \tag{4-28}$$

The energy $h\nu'$ of the scattered photon is

$$\begin{aligned} h\nu' &= \frac{1.24}{\lambda'} \\ &= \frac{1.24}{0.067 \text{ nm}} \\ &= 18.6 \text{ keV} \end{aligned} \tag{4-29}$$

The energy E_k of the Compton electron is

$$\begin{aligned} E_k &= h\nu - h\nu' \\ &= (20.0 - 18.6) \text{ keV} \\ &= 1.4 \text{ keV} \end{aligned}$$

When a low-energy photon undergoes a Compton interaction, most of the energy of the incident photon is retained by the scattered photon. Only a small fraction of the energy is transferred to the electron.

Example 4-12

A 2-MeV photon is scattered by a Compton interaction. What is the maximum energy transferred to the recoil electron? The wavelength λ of a 2-MeV photon is

$$\begin{aligned} \lambda &= \frac{1.24}{h\nu} \\ &= \frac{1.24}{2000 \text{ keV}} \\ &= 0.00062 \text{ nm} \end{aligned} \tag{4-30}$$

The change in wavelength of a photon scattered at 180 degrees is 0.00486 nm (see Example 4-11). Hence, the wavelength λ' of the photon scattered at 180 degrees is

$$\begin{aligned} \lambda' &= \lambda + \Delta\lambda \\ &= (0.00062 + 0.00486) \text{ nm} \\ &= 0.00548 \text{ nm} \end{aligned} \tag{4-31}$$

The energy $h\nu'$ of the scattered photon is

$$\begin{aligned} h\nu' &= \frac{1.24}{\lambda'} \\ &= \frac{1.24}{0.00548 \text{ nm}} \\ &= 226 \text{ keV} \end{aligned} \tag{4-32}$$

The energy E_k of the Compton electron is

$$\begin{aligned} E_k &= h\nu - h\nu' \\ &= (2000 - 226) \text{ keV} \\ &= 1774 \text{ keV} \end{aligned}$$

When a high-energy photon is scattered by the Compton process, most of the energy is transferred to the Compton electron. Only a small fraction of the energy of the incident photon is retained by the scattered photon.

Example 4-13

Show that, irrespective of the energy of the incident photon, the maximum energy is 255 keV for a photon scattered at 180 degrees and 511 keV for a photon scattered at 90 degrees.

The wavelength λ' of a scattered photon is

$$\lambda' = \lambda + \Delta\lambda \tag{4-33}$$

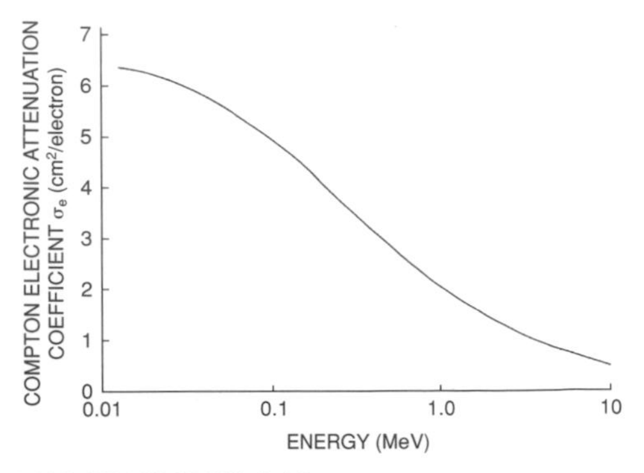

MARGIN FIGURE 4-18
Compton electronic attenuation coefficient as a function of photon energy.

For photons of very high energy, λ is very small and may be neglected relative to $\Delta\lambda$.

For a photon scattered at 180 degrees

$$\begin{aligned}\lambda' \cong \Delta\lambda &= 0.00243\,[1 - \cos(180)]\\ &= 0.00243\,[1 - (-1)]\\ &= 0.00486 \text{ nm}\end{aligned}$$

$$\begin{aligned}h\nu' &= \frac{1.24}{\lambda'} \\ &= \frac{1.24}{0.00486 \text{ nm}} \\ &= 255 \text{ keV}\end{aligned} \tag{4-34}$$

For photons scattered at 90 degrees

$$\begin{aligned}\lambda' \cong \Delta\lambda &= 0.00243\,[1 - \cos(90)]\\ &= 0.00243\,[1 - 0]\\ &= 0.00243 \text{ nm}\end{aligned}$$

$$\begin{aligned}h\nu' &= \frac{1.24}{\lambda'} \\ &= \frac{1.24}{0.00243 \text{ nm}} \\ &= 511 \text{ keV}\end{aligned} \tag{4-35}$$

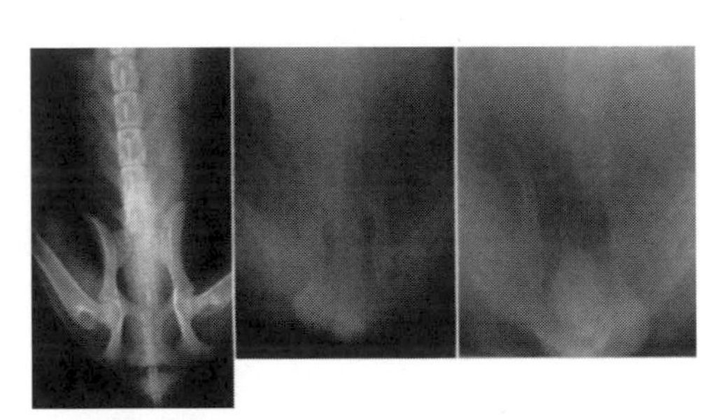

MARGIN FIGURE 4-19
Radiographs taken at 70 kVp, 250 kVp, and 1.25 MeV (^{60}Co). These films illustrate the loss of radiographic contrast as the energy of the incident photons increases.

The Compton electronic attenuation coefficient σ_e is plotted in Margin Figure 4-18 as a function of the energy of incident photons. The coefficient decreases gradually with increasing photon energy. The Compton mass attenuation coefficient varies directly with the electron density (electrons per kilogram) of the absorbing medium because Compton interactions occur primarily with loosely bound electrons. A medium with more unbound electrons will attenuate more photons by Compton scattering than will a medium with fewer electrons.

The Compton mass attenuation coefficient is nearly independent of the atomic number of the attenuating medium. For this reason, radiographs exhibit very poor contrast when exposed to high-energy photons. When most of the photons in a beam of x or γ rays interact by Compton scattering, little selective attenuation occurs in materials with different atomic number. The image in a radiograph obtained by exposing a patient to high-energy photons is not the result of differences in atomic number between different regions of the patient. Instead, the image reflects differences in physical density (kilograms per cubic meter) between the different regions (e.g., bone and soft tissue). The loss of radiographic contrast with increasing energy of incident photons is depicted and discussed more completely in Chapter 7.

Pair Production

An x or γ ray may interact by pair production while near a nucleus in an attenuating medium. A pair of electrons, one negative and one positive, appears in place of the photon. Because the energy equivalent to the mass of an electron is 0.51 MeV, the creation of two electrons requires 1.02 MeV. Consequently, photons with energy less than 1.02 MeV do not interact by pair production. This energy requirement makes pair production irrelevant to conventional radiographic imaging. During pair production, energy in excess of 1.02 MeV is released as kinetic energy of the two electrons:

$$h\nu\,(\text{MeV}) = 1.02 + (E_k)_{e-} + (E_k)_{e+}$$

Although the nucleus recoils slightly during pair production, the small amount of energy transferred to the recoiling nucleus may usually be neglected. Pair production is depicted in Margin Figure 4-20.

Occasionally, pair production occurs near an electron rather than near a nucleus. For 10-MeV photons in soft tissue, for example, about 10% of all pair production interactions occur in the vicinity of an electron. An interaction near an electron is termed *triplet production* because the interacting electron receives energy from the photon and is ejected from the atom. Three ionizing particles, two negative electrons and one positive electron, are released during triplet production. To conserve momentum, the threshold energy for triplet production must be 2.04 MeV. The ratio of triplet to pair production increases with the energy of incident photons and decreases as the atomic number of the medium is increased.

The mass attenuation coefficient k_m for pair production varies almost linearly with the atomic number of the attenuating medium. The coefficient increases slowly with energy of the incident photons. In soft tissue, pair production accounts for only a small fraction of the interactions of x and γ rays with energy between 1.02 and 10 MeV. Positive electrons released during pair production produce annihilation radiation identical to that produced by positrons released from radioactive nuclei.

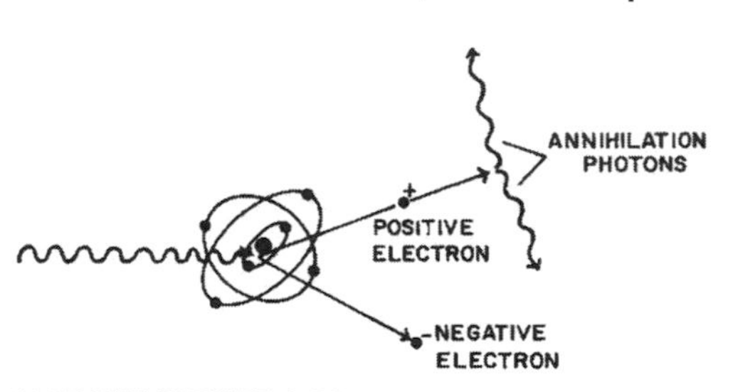

MARGIN FIGURE 4-20
Pair production interaction of a high-energy photon near a nucleus. Annihilation photons are produced when the positron and an electron annihilate each other.

Example 4-14

A 5-MeV photon near a nucleus interacts by pair production. Residual energy is shared equally between the negative and positive electron. What are the kinetic energies of these particles?

$$h\nu\ (\text{MeV}) = 1.02 + (E_k)_{e-} + (E_k)_{e+}$$

$$(E_k)_{e-} = (E_k)_{e+} = \frac{(h\nu - 1.02)\text{MeV}}{2}$$

$$= \frac{(5.00 - 1.02)\ \text{MeV}}{2}$$

$$= (E_k)_{e+} = 1.99\ \text{MeV}$$

Described in Margin Figure 4-21 are the relative importances of photoelectric, Compton, and pair production interactions in different media. In muscle or water ($Z_{\text{eff}} = 7.4$), the probabilities of photoelectric interaction and Compton scattering are equal at a photon energy of 35 keV. However, equal energies are not deposited in tissue at 35 keV by each of these modes of interaction, since all of the photon energy is deposited during a photoelectric interaction, whereas only part of the photon energy is deposited during a Compton interaction. Equal deposition of energy in tissue by photoelectric and Compton interactions occurs for 60-keV photons rather than 35-keV photons.

A summary of the variables that influence the linear attenuation coefficients for photoelectric, Compton, and pair production interactions is given in Table 4-2.

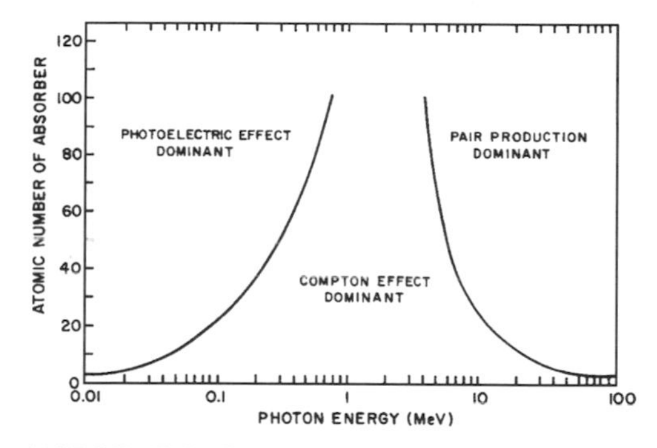

MARGIN FIGURE 4-21
Relative importance of the three principal interactions of x and γ rays. The lines represent energy/atomic number combinations for which the two interactions on either side of the line are equally probable.

TABLE 4-2 Variables that Influence the Principal Modes of Interaction of X and γ Rays

	Dependence of Linear Attenuation Coefficient on			
Mode of Interaction	*Photon Energy $h\nu$*	*Atomic Number Z*	*Electron Density ρ_e*	*Physical Density ρ*
Photoelectric	$\frac{1}{(h\nu)^3}$	Z^3	—	ρ
Compton	$\frac{1}{h\nu}$	—	ρ_e	ρ
Pair production	$h\nu$ (>1.02 MeV)	Z	—	ρ

NONIONIZING RADIATION

As mentioned previously, radiation with an energy less than 13.6 eV is classified as nonionizing radiation. Two types of nonionizing radiation, electromagnetic waves and mechanical vibrations, are of interest. Electromagnetic waves are introduced here, and mechanical vibrations, or sound waves, are discussed in Chapter 19. Because no ionization is produced by the radiation, quantities such as specific ionization, linear energy transfer, and *W*-quantity do not apply. However, the concepts of attenuation and absorption are applicable.

Electromagnetic Radiation

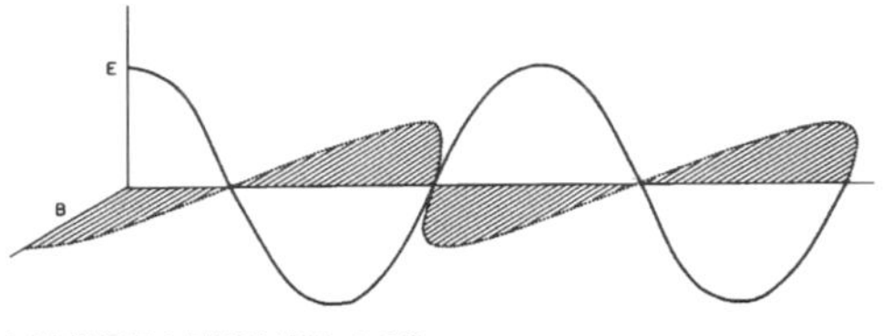

MARGIN FIGURE 4-22
Simplified diagram of an electromagnetic wave.

Electromagnetic radiation consists of oscillating electric and magnetic fields. An electromagnetic wave requires no medium for propagation; that is, it can travel in a vacuum as well as through matter. In the simplified diagram in Margin Figure 4-22, the wavelength of an electromagnetic wave is depicted as the distance between adjacent crests of the oscillating fields. The wave is moving from left to right in the diagram. The constant speed c of electromagnetic radiation in a vacuum is the product of the frequency ν and the wavelength λ of the electromagnetic wave.

$$c = \lambda\nu$$

Often it is convenient to assign wavelike properties to electromagnetic rays. At other times it is useful to regard these radiations as discrete bundles of energy termed photons or quanta. The two interpretations of electromagnetic radiation are united by the equation

$$E = h\nu \tag{4-36}$$

where E represents the energy of a photon and ν represents the frequency of the electromagnetic wave. The symbol h represents Planck's constant, 6.62×10^{-34} J-sec.

The frequency ν is

$$\nu = c/\lambda$$

and the photon energy may be written as

$$E = hc/\lambda \tag{4-37}$$

The energy in keV possessed by a photon of wavelength λ in nanometers may be computed with Eq. (4-38):

$$E = 1.24/\lambda \tag{4-38}$$

Electromagnetic waves ranging in energy from a few nanoelectron volts up to gigaelectron volts make up the electromagnetic spectrum (Table 4-3).

TABLE 4-3 The Electromagnetic Spectrum

Designation	*Frequency [hertz]*	*Wavelength [m]*	*Energy [eV]*
Gamma Rays	1.0×10^{18}–1.0×10^{27}	3.0×10^{-10}–3.0×10^{-19}	4.1×10^{3}–4.1×10^{12}
X-rays	1.0×10^{15}–1.0×10^{25}	3.0×10^{-7}–3.0×10^{-17}	4.1–4.1×10^{10}
Ultraviolet light	7.0×10^{14}–2.4×10^{16}	4.3×10^{-7}–1.2×10^{-8}	2.9–99
Visible light	4.0×10^{14}–7.0×10^{14}	7.5×10^{-7}–4.3×10^{-7}	1.6–2.9
Infra red light	1.0×10^{11}–4.0×10^{14}	3.0×10^{-3}–7.5×10^{-7}	4.1×10^{-4}–1.6
Microwave Radar and Communications	1.0×10^{9}–1.0×10^{12}	3.0×10^{-1}–3.0×10^{-4}	4.1×10^{-6}–4.1×10^{-3}
Television broadcast	5.4×10^{7}–8.0×10^{8}	5.6–0.38	2.2×10^{-7}–3.3×10^{-6}
FM radio broadcast	8.8×10^{7}–1.1×10^{8}	3.4–2.8	3.6×10^{-7}–4.5×10^{-7}
AM radio broadcast	5.4×10^{5}–1.7×10^{6}	5.6×10^{2}–1.8×10^{2}	2.2×10^{-9}–6.6×10^{-9}
Electric power	10–1×10^{3}	3.0×10^{7}–3.0×10^{5}	4.1×10^{-14}–4.1×10^{-12}

Source: Federal Communications Commission: Title 47, Code of Federal Regulations 2.106

Example 4-15

Calculate the energy of a γ ray with a wavelength of 0.001 nm.

$$E = \frac{hc}{\lambda}$$

$$E\ (\text{keV}) = \frac{(6.62 \times 10^{-34}\ \text{J-sec})(3 \times 10^{8}\ \text{m/sec})}{(\lambda\text{nm})(10^{-9}\ \text{m/nm})(1.6 \times 10^{-16}\ \text{J/keV})}$$

$$= \frac{1.24}{\lambda\text{nm}}$$

$$E = \frac{1.24}{0.001\ \text{nm}}$$

$$= 1240\ \text{keV}$$

Example 4-16

Calculate the energy of a radio wave with a frequency of 100 MHz (FM broadcast band), where 100 MHz equals $100 \times 10^{6}\ \text{sec}^{-1}$

The wavelength may be determined from

$$c = \lambda\nu$$

$$\lambda = c/\nu$$

$$\lambda = \frac{(3 \times 10^{8}\ \text{m/sec})(10^{9}\ \text{nm/m})}{100 \times 10^{6}\ \text{sec}^{-1}}$$

$$\lambda = 3 \times 10^{9}\ \text{nm}$$

The energy is then

$$E = \frac{1.24}{\text{nm}}$$

$$= \frac{1.24}{3 \times 10^{9}\ \text{nm}}$$

$$= 4 \times 10^{-10}\ \text{keV}$$

The wavelength of a typical radio wave is many orders of magnitude greater than the wavelength of the γ ray from the previous example. The energy of the radio wave is correspondingly smaller than that of the γ ray.

The interactions of electromagnetic waves vary greatly from one end of the electromagnetic spectrum to the other. Some of the medical uses of parts of the electromagnetic spectrum are outlined below.

Ultraviolet Light

Ultraviolet light is usually characterized as nonionizing, although it can ionize some lighter elements such as hydrogen. Ultraviolet light is used to sterilize medical instruments, destroy cells, produce cosmetic tannings, and treat certain dermatologic conditions.

Visible Light

Because it is the part of the electromagnetic spectrum to which the retina is most sensitive, visible light is used constantly by observers in medical imaging.

Infrared

Infrared is the energy released as heat by materials near room temperature. Infrared-sensitive devices can record the "heat signature" of the surface of the body,[3] and they have been explored for thermographic examinations to detect breast cancer and to identify a variety of neuromuscular conditions. This exploration has not yielded clinically-reliable data to date.

Radar

Electromagnetic waves in this energy range are seldom detected as a pattern of "photons." Rather, they are detected by receiver antennas in which electrons are set into motion by the passage of an electromagnetic field. These electrons constitute an electrical current that can be processed to obtain information about the source of the electromagnetic wave.

AM and FM Broadcast and Television

These electromagnetic waves have resonance absorption properties with nuclei. They may be used as probes for certain nuclei that have magnetic properties in the imaging technique known as magnetic resonance imaging (see Chapter 23).

INTERACTIONS OF NONIONIZING ELECTROMAGNETIC RADIATION

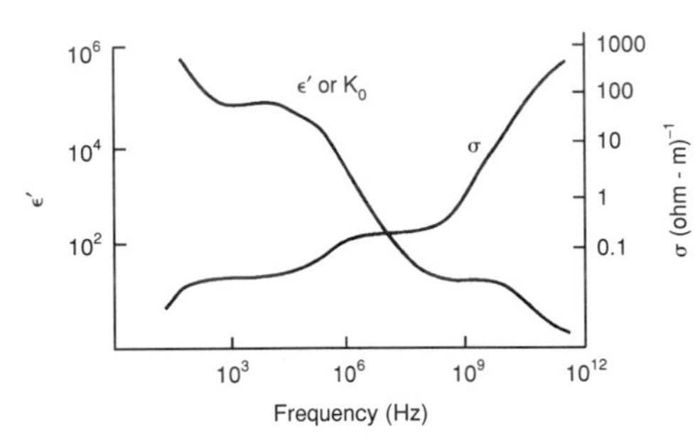

MARGIN FIGURE 4-23
Variation of dielectric constant K_0 and conductivity σ of high water content biological materials as a function of the frequency of nonionizing electromagnetic radiation. (From *Electromagnetic Fields in Biological Media.* HEW Publication No. FDA-78-8068.)

For those parts of the electromagnetic spectrum with energy less than x and γ rays, mechanisms of interaction involve a direct interplay between the electromagnetic field of the wave and molecules of the target material. The energy of the electromagnetic field, and hence the energy of the wave, is diminished when electrons or other charge carriers are set into motion in the target. Energy that electrons absorb directly from the wave is referred to as conduction loss, while energy that produces molecular rotation is referred to as dielectric loss. These properties of a material are described by its conductivity σ and dielectric constant k_e. The degree to which a material is "lossy" to electromagnetic waves (that is, the material absorbs energy and is therefore heated) is a complex function of the frequency of the waves.[4]

Some generalizations about low-energy electromagnetic radiation are possible. For a given material, for example, the increase in conductivity with frequency tends to limit penetration of the radiation. Compared with lower-frequency electromagnetic waves, waves of higher frequency tend to be attenuated more severely in biological materials. Conductivity and dielectric constants are shown in Margin Figure 4-23 for typical "high-water content" biologic tissues as a function of frequency.

The rate at which energy is deposited in a material by nonionizing radiation is described by the specific absorption rate (SAR) of the material, usually stated in units of watts per kilogram. Calculation of the SAR involves consideration of the geometry of the target material as well as energy absorption parameters including conductivity, dielectric constant, and energy loss mechanisms. Specific absorption rates for typical magnetic resonance imaging pulse sequences are given in Chapter 25.

Scattering of Visible Light

Visible light, a nonionizing radiation, may interact with electrons of the atoms and molecules of a target as a group. Coherent scattering results in a change of direction of the incident photons, but essentially no change in energy. The size of the target molecules determines the characteristics of the interaction.

When light from the sun interacts with water molecules in a cloud, a type of coherent scattering called Thomson scattering occurs. Thomson scattering is equally probable for all wavelengths of visible light, and all parts of the visible spectrum are scattered in all directions. The scattered light emanating from a cloud consists of many wavelengths and is perceived as white light.[5]

When visible light from the sun encounters gas molecules of a size commonly present in air, another type of scattering called Rayleigh scattering occurs. Rayleigh scattering is an example of incoherent scattering in which the wavelength (and therefore the energy) of the radiation is changed by the interaction. Rayleigh scattering is much more likely for the shorter-wavelength components of visible light (i.e., the blue end of the spectrum).[6] Therefore, blue components of light are scattered at large angles from molecules in the sky to observers on the ground, and the sky appears blue. At sunset, when light rays reach the earth through relatively straight paths, unscattered rays excluding blue are seen. Sunsets appear red because of the absence of Rayleigh scattering of the longer-wavelength end of the electromagnetic spectrum.

PROBLEMS

*4-1. Electrons with kinetic energy of 1.0 MeV have a specific ionization in air of about 6000 IP/m. What is the LET of these electrons in air?

*4-2. Alpha particles with 2.0 MeV have an LET in air of 0.175 keV/μm. What is the specific ionization of these particles in air?

4-3. The tenth-value layer is the thickness of a slab of matter necessary to attenuate a beam of x or γ rays to one-tenth the intensity with no attenuator present. Assuming good geometry and monoenergetic photons, show that the tenth-value layer equals $2.30/\mu$, where μ is the total linear attenuation coefficient.

*4-4. The mass attenuation coefficient of copper is 0.0589 cm^2/g for 1.0-MeV photons. The number of 1.0-MeV photons in a narrow beam is reduced to what fraction by a slab of copper 1 cm thick? The density of copper is 8.9 g/cm^3.

*4-5. Copper has a density of 8.9 g/cm^3 and a gram-atomic mass of 63.56. The total atomic attenuation coefficient of copper is 8.8×10^{-24} cm^2/atom for 500-keV photons. What thickness (in centimeters) of copper is required to attenuate 500-keV photons to half of the original number?

4-6. Assume that the exponent μx in the equation $I = I_0 e^{-\mu x}$ is equal to or less than 0.1. Show that, with an error less than 1%, the number of photons transmitted is $I_0(1 - \mu x)$ and the number attenuated is $I_0 \mu$. (*Hint*: Expand the term $e^{-\mu x}$ into a series.)

*4-7. K- and L-shell binding energies for cesium are 28 keV and 5 keV, respectively. What are the kinetic energies of photoelectrons released from the K and L shells as 40-keV photons interact in cesium?

4-8. The binding energies of electrons in different shells of an element may be determined by measuring the transmission of a monoenergetic beam of photons through a thin foil of the element as the energy of the beam is varied. Explain why this method works.

*4-9. Compute the energy of a photon scattered at 45 degrees during a Compton interaction, if the energy of the incident photon is 150 keV. What is the kinetic energy of the Compton electron? Is the energy of the scattered photon increased or decreased if the photon scattering angle is increased to more than 45 degrees?

*4-10. A γ ray of 2.75 MeV from ^{24}Na undergoes pair production in a lead shield. The negative and positive electrons possess equal kinetic energy. What is this kinetic energy?

*4-11. Prove that, regardless of the energy of the incident photon, a photon scattered at an angle greater than 60 degrees during a Compton interaction cannot undergo pair production.

*For problems marked with an asterisk, answers are provided on p. 491.

SUMMARY

- Charged particles, such as electrons and alpha particles, are said to be ***directly*** ionizing.
- Uncharged particles, such as photons and neutrons, are said to be ***indirectly*** ionizing.
- LET = (SI) (W).
- Range = $\frac{E}{\text{LET}}$.
- Electrons may interact with other electrons or with nuclei of atoms.
- Inelastic scattering of electrons with a nucleus, bremsstrahlung, is used to produce x rays.

- Neutrons interact primarily by "billiard ball" or "knock on" collisions with nuclei.
- Attenuation of monoenergetic x or γ rays from a narrow beam is exponential.
- Interactions of x rays include coherent, photoelectric, Compton, and pair production.
- Liner attenuation coefficient is the fractional rate of removal of photons from a beam per unit path length.
- Examples of electromagnetic radiation in order of increasing wavelength (decreasing frequency and energy) are x and γ rays, ultraviolet, visible, infrared, and radio waves.

REFERENCES

1. International Commission on Radiation units and Measurements. *Fundamental Quantities and Units for Ionizing Radiation*, ICRU Report No 60. Washington, D.C., 1998.
2. Bichsel, H. Charged-Particle Interactions, in Attix, F., and Roesch, W. (eds.), *Radiation Dosimetry*, Vol. 1. New York, Academic Press, 1968, p. 157.
3. Cameron, J. R., and Skofronick, J. G. *Medical Physics*. New York, John Wiley & Sons, 1978, p. 70.
4. *Electromagnetic Fields in Biological Media*. U.S. Department of Health, Education, and Welfare, HEW Publication (FDA) 78-8068, July 1978, p. 8.
5. Jackson, J. D. *Classical Electrodynamics*, 2nd edition. New York, John Wiley & Sons, 1975, p. 681.
6. Lord Rayleigh. *Philos. Mag.* 1871; **41**:274. 1899; **47**:375. Reprinted in *Scientific Papers*, Vol.1 p. 87; vol. 4, p. 397.

CHAPTER

5

PRODUCTION OF X RAYS

X rays were discovered on November 8, 1895 by Wilhelm C. Röntgen, a physicist at the University of Würzburg in Germany.[1] He named his discovery "x rays" because "x" stands for an unknown quantity. For this work, Röntgen received the first Nobel Prize in Physics in 1901.

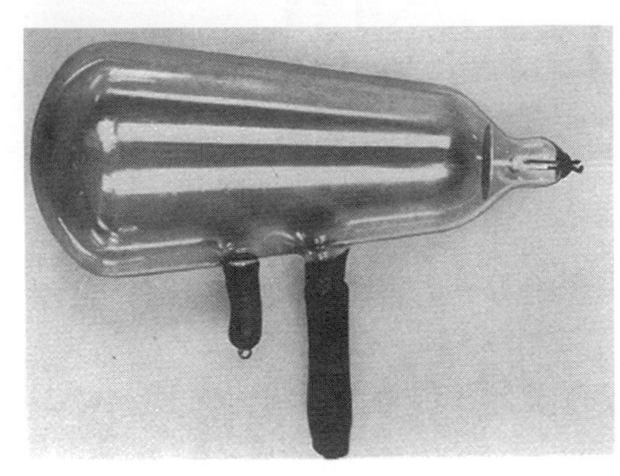

MARGIN FIGURE 5-1
Crookes tube, an early example of a cathode ray tube.

Röntgen was not the first to acquire an x-ray photograph. In 1890 Alexander Goodspeed of the University of Pennsylvania, with the photographer William Jennings, accidentally exposed some photographic plates to x rays. They were able to explain the images on the developed plates only after Röntgen announced his discovery of x rays.

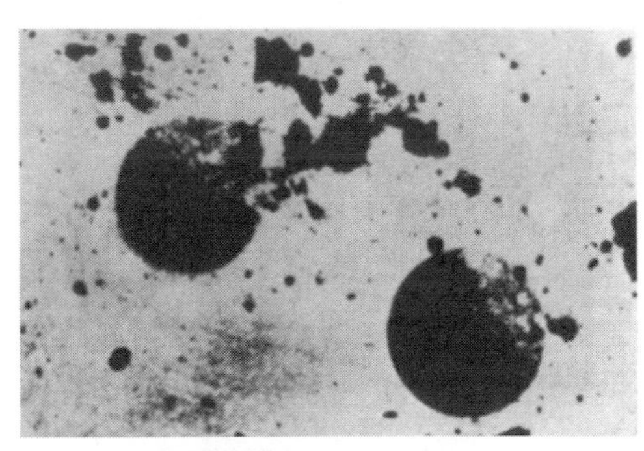

MARGIN FIGURE 5-2
An x-ray picture obtained accidentally in February 1890 by Arthur Goodspeed of the University of Pennsylvania. The significance of this picture, acquired more than 5 years before Röntgen's discovery, was not recognized by Professor Goodspeed.

Heating a filament to release electrons is called thermionic emission or the Edison effect.

Coolidge's contributions to x-ray science, in addition to the heated filament, included the focusing cup, imbedded x-ray target, and various anode cooling devices.

OBJECTIVES

By studying this chapter, the reader should be able to:

- Identify each component of an x-ray tube and explain its function.
- Describe single- and three-phase voltage, and various modes of voltage rectification.
- Explain the shape of the x-ray spectrum, and identify factors that influence it.
- Discuss concepts of x-ray production, including the focal spot, line-focus principle, space charge, power deposition and emission, and x-ray beam hardening.
- Define and apply x-ray tube rating limits and charts.

INTRODUCTION

To produce medical images with x rays, a source is required that:

1. Produces enough x rays in a short time
2. Allows the user to vary the x-ray energy
3. Provides x rays in a reproducible fashion
4. Meets standards of safety and economy of operation

Currently, the only practical sources of x rays are radioactive isotopes, nuclear reactions such as fission and fusion, and particle accelerators. Only special-purpose particle accelerators known as x-ray tubes meet all the requirements mentioned above. In x-ray tubes, bremsstrahlung and characteristic x rays are produced as high-speed electrons interact in a target. While the physical design of x-ray tubes has been altered significantly over a century, the basic principles of operation have not changed.

Early x-ray studies were performed with a cathode ray tube in which electrons liberated from residual gas atoms in the tube were accelerated toward a positive electrode (anode). These electrons produced x rays as they interacted with components of the tube. The cathode ray tube was an unreliable and inefficient method of producing x rays. In 1913, Coolidge[2] improved the x-ray tube by heating a wire filament with an electric current to release electrons. The liberated electrons were repelled by the negative charge of the filament (cathode) and accelerated toward a positive target (the anode). X rays were produced as the electrons struck the target. The Coolidge tube was the prototype for "hot cathode" x-ray tubes in wide use today.

CONVENTIONAL X-RAY TUBES

Figure 5-1 shows the main components of a modern x-ray tube. A heated filament releases electrons that are accelerated across a high voltage onto a target. The stream of accelerated electrons is referred to as the *tube current*. X rays are produced as the electrons interact in the target. The x rays emerge from the target in all directions but are restricted by collimators to form a useful beam of x rays. A vacuum is maintained inside the glass envelope of the x-ray tube to prevent the electrons from interacting with gas molecules.

ELECTRON SOURCE

A metal with a high melting point is required for the filament of an x-ray tube. Tungsten filaments (melting point of tungsten 3370°C) are used in most x-ray tubes. A current of a few amperes heats the filament, and electrons are liberated at a rate that increases with the filament current. The filament is mounted within a negatively charged focusing cup. Collectively, these elements are termed the *cathode assembly*.

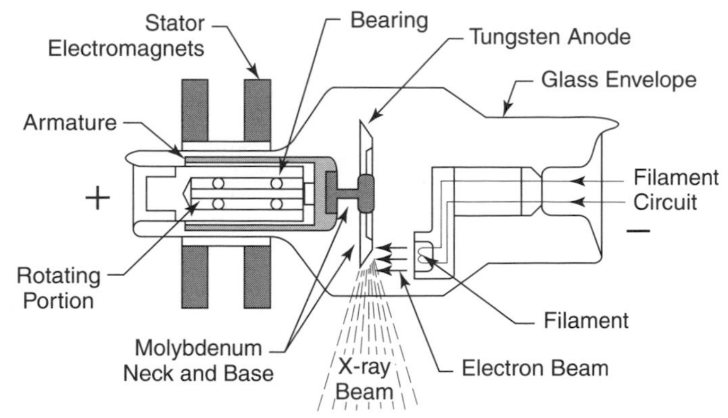

FIGURE 5-1
Simplified x-ray tube with a rotating anode and a heated filament.

Vaporized tungsten from both the filament and anode deposits on the glass envelope of the x-ray tube, giving older tubes a mirrored appearance.

An x-ray tube with two filaments is called a *dual-focus* tube.

The focal spot is the volume of target within which electrons are absorbed and x rays are produced. For radiographs of highest clarity, electrons should be absorbed within a small focal spot. To achieve a small focal spot, the electrons should be emitted from a small or "fine" filament. Radiographic clarity is often reduced by voluntary or involuntary motion of the patient. This effect can be decreased by using x-ray exposures of high intensity and short duration. However, these high-intensity exposures may require an electron emission rate that exceeds the capacity of a small filament. Consequently many x-ray tubes have two filaments. The smaller, fine filament is used when radiographs with high detail are desired and short, high-intensity exposures are not necessary. If high-intensity exposures are needed to limit the blurring effects of motion, the larger, coarse filament is used. The cathode assembly of a dual-focus x-ray tube is illustrated in Margin Figure 5-4.

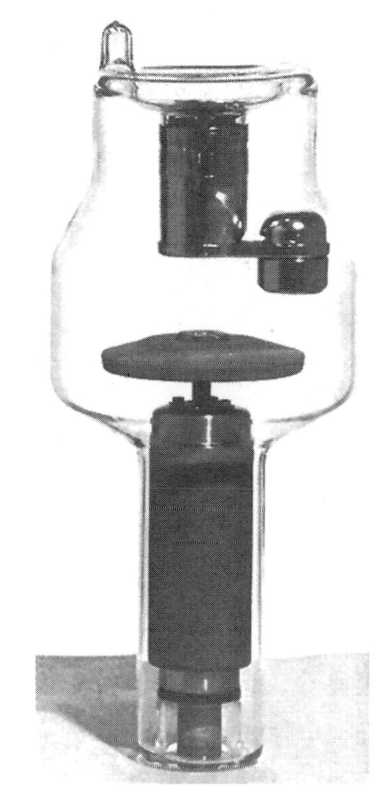

MARGIN FIGURE 5-3
A dual-focus x-ray tube with a rotating anode.

TUBE VOLTAGE AND VOLTAGE WAVEFORMS

The intensity and energy distribution of x rays emerging from an x-ray tube are influenced by the potential difference (voltage) between the filament and target of the tube. The source of electrical power for radiographic equipment is usually alternating current (ac). This type of electricity is by far the most common form available for general use, because it can be transmitted with little energy loss through power lines that span large distances. Figure 5-2 shows a graph of voltage and current in an ac power line. X-ray tubes are designed to operate at a single polarity, with a positive

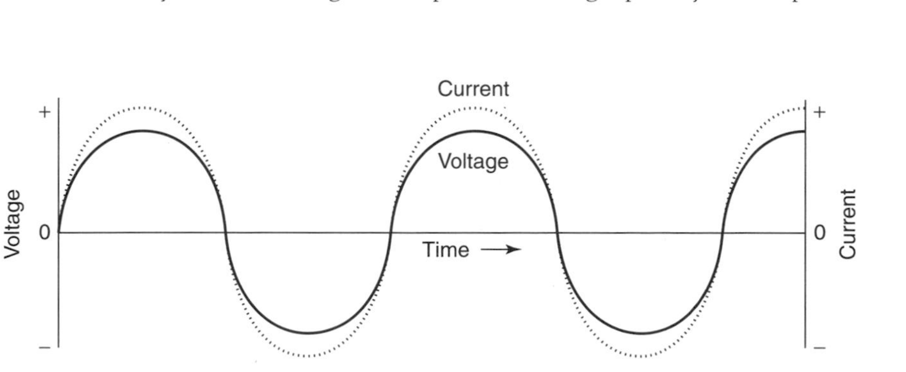

FIGURE 5-2
Voltage and current in an ac power line. Both voltage and current change from positive to negative over time. The relationship between voltage and current (i.e., the relative strength and the times at which they reach their peaks) depends upon a complex quantity called reactance. The positive and negative on the voltage scale refer to polarity, while the positive and negative on the current scale refer to the direction of flow of electrons that constitute an electric current.

MARGIN FIGURE 5-4
Cathode assembly of a dual-focus x-ray tube. The small filament provides a smaller focal spot and a radiograph with greater detail, provided that the patient does not move. The larger filament is used for high-intensity exposures of short duration.

A positive electrode is termed an *anode* because negative ions (anions) are attracted to it. A negative electrode is referred to as a cathode because positive ions (cations) are attracted to it.

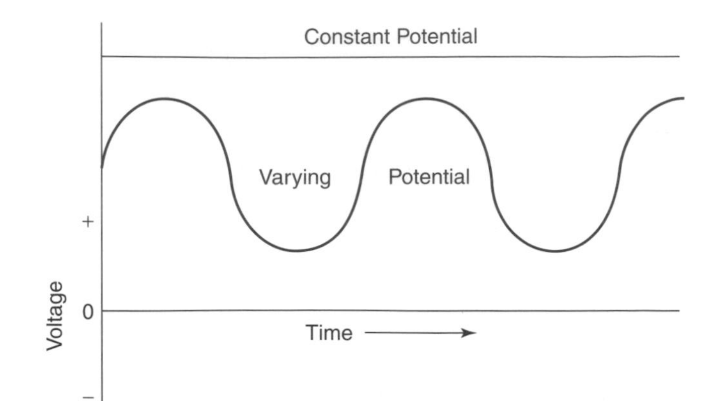

FIGURE 5-3
Voltage in a dc power line. The term *direct current* means that the voltage (and current not shown here) never reverse (change from positive to negative), although they may vary in intensity. X-ray tubes are most efficient when operated at constant potential.

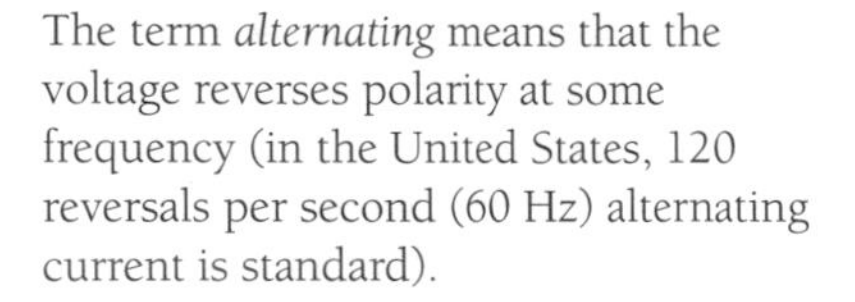

The term *alternating* means that the voltage reverses polarity at some frequency (in the United States, 120 reversals per second (60 Hz) alternating current is standard).

target (anode) and a negative filament (cathode). X-ray production is most efficient (more x rays are produced per unit time) if the potential of the target is always positive and if the voltage between the filament and target is kept at its maximum value. In most x-ray equipment, ac is converted to direct current (dc), and the voltage between filament and target is kept at or near its maximum value (Figure 5-3). The conversion of ac to dc is called *rectification*.

One of the simplest ways to operate an x-ray tube is to use ac power and rely upon the x-ray tube to permit electrons to flow only from the cathode to the anode. The configuration of the filament (a thin wire) is ideal for producing the heat necessary to release electrons when current flows through it. Under normal circumstances the target (a flat disk) is not an efficient source of electrons. When the polarity is reversed (i.e., the filament is positive and the target is negative), current cannot flow in the x-ray tube, because there is no source of electrons. In this condition, the x-ray tube "self-rectifies" the ac power, and the process is referred to as *self-rectification*. At high tube currents, however, the heat generated in the target can be great enough to release electrons from the target surface. In this case, electrons flow across the x-ray tube when the target is negative and the filament is positive. This reverse flow of electrons can destroy the x-ray tube.

A rectified voltage waveform can also be attained by use of circuit components called diodes. Diodes are devices that, like x-ray tubes, allow current to flow in only one direction. A simple circuit containing diodes that produces the same waveform as *self-rectification* is shown in the margin. Rectification in which polarity reversal across the x-ray tube is eliminated is called *half-wave rectification*.

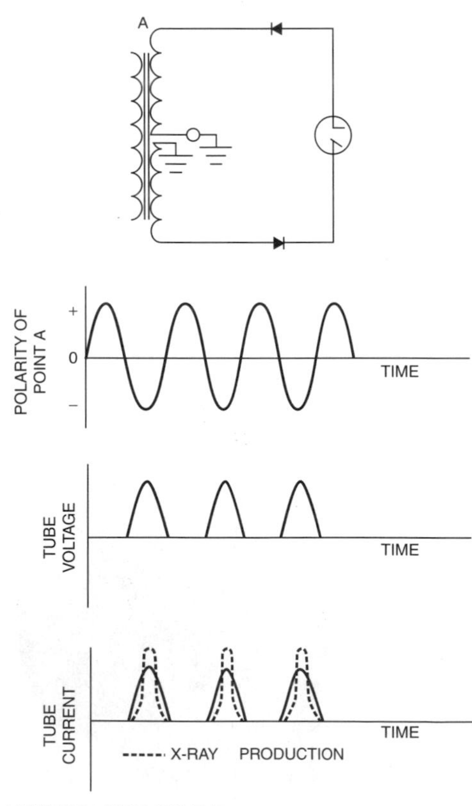

MARGIN FIGURE 5-5
A circuit for half-wave rectification (**top**), with resulting tube voltage, tube current, and efficiency for production of x rays. Rectifiers indicate the direction of conventional current flow, which is opposite to the actual flow of electrons.

A half-wave rectifier converts ac to a dc waveform with 1 pulse per cycle. X-ray production could be made more efficient if the negative half-cycle of the voltage waveform could be used. A more complex circuit called a full-wave rectifier utilizes both half-cycles. In both the positive and negative phases of the voltage waveform, the voltage is impressed across the x-ray tube with the filament (or cathode) at a negative potential and the target (or anode) at a positive potential. This method of rectifying the ac waveform is referred to as *full-wave rectification*. In full-wave rectification the negative pulses in the voltage waveform are in effect "flipped over" so that they can be used by the x-ray tube to produce x rays. Thus a full-wave rectifier converts an ac waveform into a dc waveform having 2 pulses per cycle.

The efficiency of x-ray production could be increased further if the voltage waveform were at high potential most of the time, rather than decreasing to zero at least twice per cycle as it does in full-wave rectification. This goal can be achieved by use of three-phase (3φ) power. Three-phase power is provided through three separate voltage lines connected to the x-ray tube.

The term *phase* refers to the fact that all three voltage lines carry the same voltage waveform, but the voltage peaks at different times in each line. Each phase (line) is rectified separately so that three distinct (but overlapping) full-wave-rectified waveforms are presented to the x-ray tube. The effect of this composite waveform is to supply voltage to the x-ray tube that is always at or near maximum. In a three-phase full-wave-rectified x-ray circuit the voltage across the x-ray tube never drops to zero. With three separate phases of ac, six rectified pulses are provided during each voltage cycle.

A refinement of 3φ circuitry provides a slight phase shift for the waveform presented to the anode compared with that presented to the cathode. This refinement yields 12 pulses per cycle and provides a slight increase in the fraction of time that the x-ray tube operates near peak potential.

Modern solid-state voltage-switching devices are capable of producing "high-frequency" waveforms yielding thousands of x-ray pulses per second. These voltage waveforms are essentially constant potential and provide further improvements in the efficiency of x-ray production.

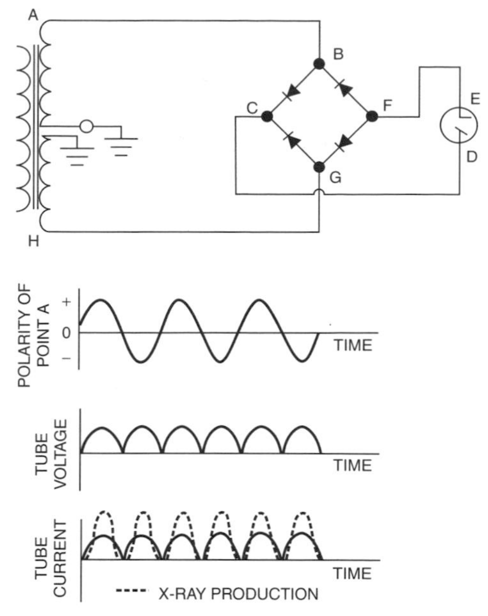

MARGIN FIGURE 5-6
A circuit for full-wave rectification (**top**), with resulting tube voltage, tube current, and efficiency for production of x rays. Electrons follow the path ABFEDCGH when end A of the secondary of the high-voltage transformer is negative. When the voltage across the secondary reverses polarity, the electron path is HGFEDCBA.

RELATIONSHIP BETWEEN FILAMENT CURRENT AND TUBE CURRENT

Two electrical currents flow in an x-ray tube. The *filament current* is the flow of electrons through the filament to raise its temperature and release electrons. The second electrical current is the flow of released electrons from the filament to the anode across the x-ray tube. This current, referred to as the *tube current,* varies from a few to several hundred milliamperes.

The two currents are separate but interrelated. One of the factors that relates them is the concept of "space charge." At low tube voltages, electrons are released from the filament more rapidly than they are accelerated toward the target. A cloud of electrons, termed the *space charge,* accumulates around the filament. This cloud opposes the release of additional electrons from the filament.

The curves in Figure 5-4 illustrate the influence of tube voltage and filament current upon tube current. At low filament currents, a saturation voltage is reached above which the current through the x-ray tube does not vary with increasing voltage. At the saturation voltage, tube current is limited by the rate at which electrons are released from the filament. Above the saturation voltage, tube current can be increased only by raising the filament's temperature in order to increase the rate of electron emission. In this situation, the tube current is said to be *temperature or filament-emission limited.* To obtain high tube currents and x ray energies useful for diagnosis, high filament currents and voltages between 40 and 140 kV must be used. With high filament currents and lower tube voltages, the space charge limits the tube current, and hence the x-ray tube is said to be *space-charge limited.*

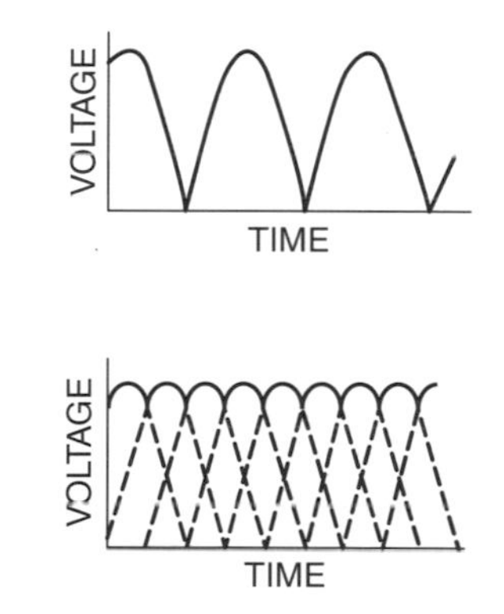

MARGIN FIGURE 5-7
Single-phase (**top**) and three-phase (**bottom**) voltages across an x-ray tube. Both voltages are full-wave rectified. The three-phase voltage is furnished by a six-pulse circuit.

EMISSION SPECTRA

The useful beam of an x-ray tube is composed of photons with an energy distribution that depends on four factors:

- Bremsstrahlung x rays are produced with a range of energies even if electrons of a single energy bombard the target.
- X rays released as characteristic radiation have energies independent of that of the bombarding electrons so long as the energy of the bombarding electrons exceeds the threshold energy for characteristic x ray emission.
- The energy of the bombarding electrons varies with tube voltage, which fluctuates rapidly in some x-ray tubes.

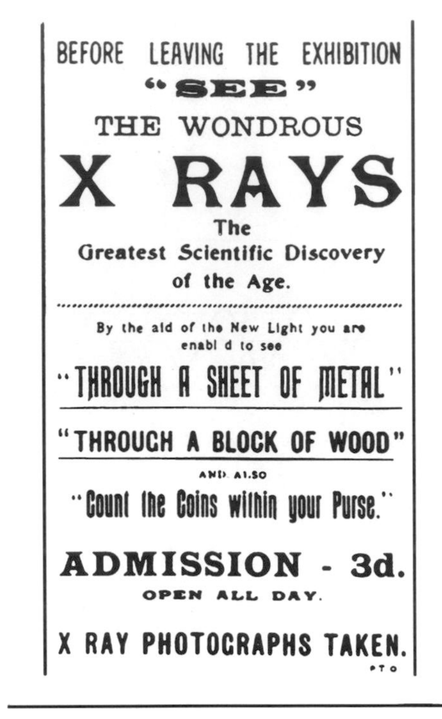

MARGIN FIGURE 5-8
Poster for a public demonstration of x rays, 1896, Crystal Place Exhibition, London.

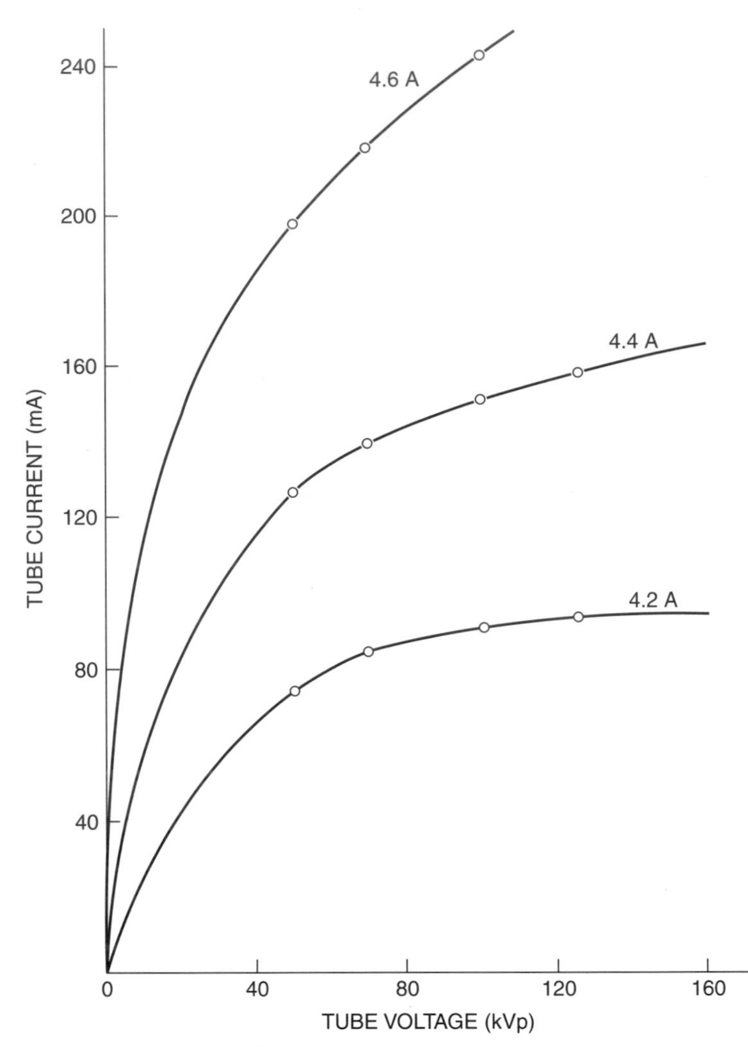

FIGURE 5-4
Influence of tube voltage and filament current upon electron flow in a Machlett Dynamax x-ray tube with a rotating anode, 1-mm apparent focal spot, and full-wave-rectified voltage.

X-ray generators that yield several thousand voltage (and hence x ray) pulses per second are known as *constant potential generators*.

- X rays are produced at a range of depths in the target of the x-ray tube. These x rays travel through different thicknesses of target and may lose energy through one or more interactions.

X-ray tubes can operate in one of two modes:

- filament-emission limited
- space-charge limited

Changes in other variables such as filtration, target material, peak tube voltage, current, and exposure time all may affect the range and intensity of x-ray energies in the useful beam. The distribution of photon energies produced by a typical x-ray tube, referred to as an *emission spectrum*, is shown in Figure 5-5.

As a rough approximation, the rate of production of x rays is proportional to $Z_{\text{target}} \times (\text{kVp})^2 \times \text{mA}$.

FILTRATION

Inherent filtration is also referred to as *intrinsic filtration*.

Mammography x-ray tubes often employ exit windows made of beryllium to allow low-energy x rays to escape from the tube.

An x-ray beam traverses several attenuating materials before it reaches the patient, including the glass envelope of the x-ray tube, the oil surrounding the tube, and the exit window in the tube housing. These attenuators are referred to collectively as the inherent filtration of the x-ray tube (Table 5-1). The aluminum equivalent for each component of inherent filtration is the thickness of aluminum that would reduce the exposure rate by an amount equal to that provided by the component. The inherent filtration is approximately 0.9 mm Al equivalent for the tube described in Table 5-1, with most of the inherent filtration contributed by the glass envelope. The *inherent filtration* of most x-rays tubes is about 1 mm Al.

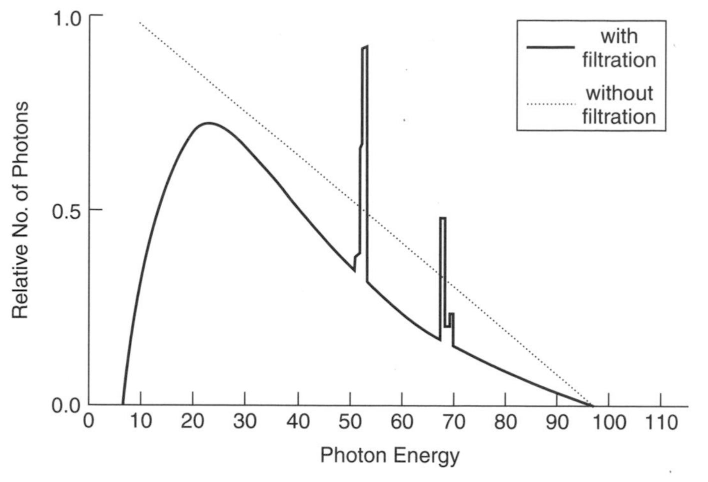

FIGURE 5-5
Emission spectrum for a tungsten target x-ray tube operated at 100 kVp. K-characteristic x-ray emission occurs for tungsten whenever the tube voltage exceeds 69 keV, the K-shell binding energy for tungsten. The *dotted line* represents the theoretical bremsstrahlung emission from a tungsten target. The *solid line* represents the spectrum after self-, inherent, and added filtration. The area under the spectrum represents the total number of x rays.

An x-ray beam of higher average energy is said to be "harder" because it is able to penetrate more dense (i.e., harder) substances such as bone. An x-ray beam of lower average energy is said to be "softer" because it can penetrate only less dense (i.e., softer) substances such as fat and muscle.

In any medium, the probability that incident x rays interact photoelectrically varies roughly as $1/E^3$, where E is the energy of the incident photons (see Chapter 4). That is, low-energy x rays are attenuated to a greater extent than those of high energy. After passing through a material, an x-ray beam has a higher average energy per photon (that is, it is "harder") even though the total number of photons in the beam has been reduced, because more low-energy photons than high-energy photons have been removed from the beam.

The inherent filtration of an x-ray tube "hardens" the x-ray beam. Additional hardening may be achieved by purposefully adding filters of various composition to the beam. The total filtration in the x-ray beam is the sum of the inherent and added filtration as shown in Table 5-1. Usually, additional hardening is desirable because the filter removes low-energy x rays that, if left in the beam, would increase the radiation dose to the patient without contributing substantially to image formation.

Emission spectra for a tungsten-target x-ray tube are shown in Figure 5-6 for various thicknesses of added aluminum filtration. The effect of the added aluminum is to decrease the total number of photons but increase the average energy of photons in the beam. These changes are reflected in a decrease in the overall height of the emission spectrum and a shift of the peak of the spectrum toward higher energy.

Equalization filters are sometimes used in chest and spine imaging to compensate for the large differences in x-ray transmission between the mediastinum and lungs.

Tube Voltage

As the energy of the electrons bombarding the target increases, the high-energy limit of the x-ray spectrum increases correspondingly. The height of the spectrum also

TABLE 5-1 Contributions to Inherent Filtration in Typical Diagnostic X-Ray Tube

Component	*Thickness (mm)*	*Aluminum-Equivalent Thickness (mm)*
Glass envelope	1.4	0.78
Insulating oil	2.36	0.07
Bakelite window	1.02	0.05

[a] Data from Trout, E. *Radiol. Technol.* 1963; **35:**161.

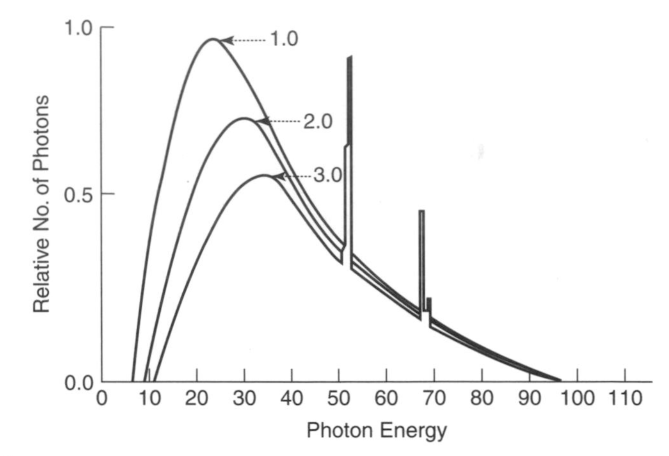

FIGURE 5-6
X-ray emission spectra for a 100-kVp tungsten target x-ray tube with total filtration values of 1.0, 2.0, and 3.0 mm aluminum. kVp and mAs are the same for the three spectra. (Computer simulation courtesy of Todd Steinberg, Colorado Springs.)

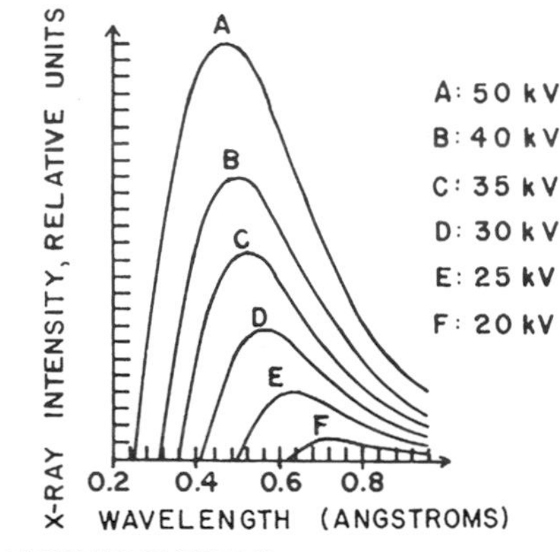

MARGIN FIGURE 5-9
Tungsten-target x-ray spectra generated at different tube voltages with a constant current through the x-ray tube. (From Ulrey, C., *Phys. Rev.* 1918; **1**:401.)

increases with increasing tube voltage because the efficiency of bremsstrahlung production increases with electron energy (see margin).

Tube Current and Time

The product of tube current in milliamperes and exposure time in seconds (mA · sec) describes the total number of electrons bombarding the target.

Example 5-1

Calculate the total number of electrons bombarding the target of an x-ray tube operated at 200 mA for 0.1 sec.

The ampere, the unit of electrical current, equals 1 coulomb/sec. The product of current and time equals the total charge in coulombs. X-ray tube current is measured in milliamperes, where 1 mA = 10^{-3} amp. The charge of the electron is 1.6×10^{-19} coulombs, so

$$1\text{ mA}\cdot\text{s} = \frac{(10^{-3}\text{coulomb/sec})(\text{sec})}{1.6\times 10^{-19}\text{coulomb/electron}} = 6.25\times 10^{15}\text{electrons}$$

For 200 mA and 0.1 sec,

$$\begin{aligned}\text{No. of electrons} &= (200\text{ mA})(0.1\text{ sec})(6.25\times 10^{15}\text{electrons/mA}\cdot\text{sec})\\ &= 1.25\times 10^{17}\text{electrons}\end{aligned}$$

Other factors being equal, more x rays are produced if more electrons bombard the target of an x-ray tube. Hence the number of x rays produced is directly proportional to the product (mA · sec) of tube current in milliamperes and exposure time in seconds. Spectra from the same x-ray tube operated at different values of mA · sec are shown in Figure 5-7. The overall shape of the spectrum (specifically the upper and lower limits of energy and the position of characteristic peaks) remains unchanged. However, the height of the spectrum and the area under it increase with increasing mA · sec. These increases reflect the greater number of x rays produced at higher values of mA · sec.

Target Material

The choice of target material in an x-ray tube affects the efficiency of x-ray production and the energy at which characteristic x rays appear. If technique factors (tube voltage, milliamperage, and time) are fixed, a target material with a higher atomic

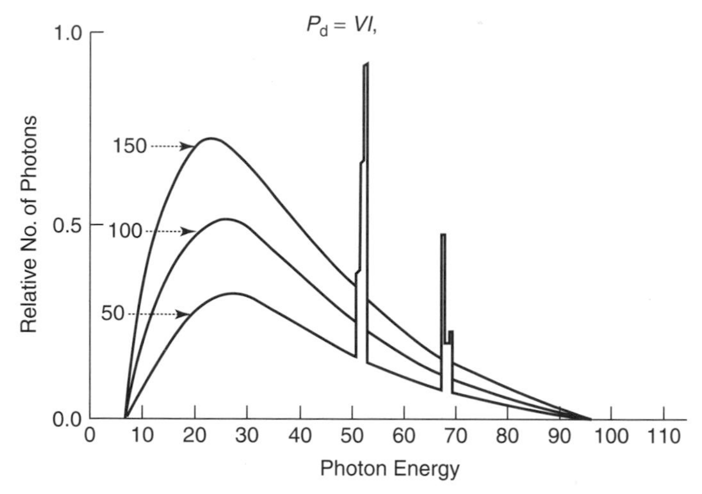

FIGURE 5-7
X-ray emission spectra for a 100-kVp tungsten target x-ray tube operated at 50, 100, and 150 mA. kVp and exposure time are the same for the three spectra.

number (Z) will produce more x rays per unit time by the process of bremsstrahlung.

The efficiency of x-ray production is the ratio of energy emerging as x radiation from the x-ray target divided by the energy deposited by electrons impinging on the target. The rate at which electrons deposit energy in a target is termed the *power deposition* P_d (in watts) and is given by

In the equation $P_d = VI$, the voltage may also be expressed in kilovolts if the current is described in milliamperes.

$$P_d = VI$$

where V is the tube voltage in volts and I is the tube current in amperes. The rate at which energy is released as x radiation,[3] termed the *radiated power* P_r, is

$$P_r = 0.9 \times 10^{-9} Z V^2 I \quad (5\text{-}1)$$

where P_r is the radiated power in watts (W) and Z is the atomic number of the target. Hence, the efficiency of x-ray production is

$$\text{Efficiency} = \frac{P_r}{P_d} = \frac{0.9 \times 10^{-9} Z V^2 I}{VI} \quad (5\text{-}2)$$
$$= 0.9 \times 10^{-9} Z V$$

Equation (5-2) shows that the efficiency of x-ray production increases with the atomic number of the target and the voltage across the x-ray tube.

Example 5-2

In 1 sec, 6.25×10^{17} electrons (100 mA) are accelerated through a constant potential difference of 100 kV. At what rate is energy deposited in the target?

$$P = (10^5 \text{V})(0.1 \text{A})$$
$$= 10^4 \text{W}$$

X-ray production is a very inefficient process, even in targets with high atomic number. For x-ray tubes operated at conventional voltages, less than 1% of the energy deposited in the target appears as x radiation. Almost all of the energy delivered by impinging electrons is degraded to heat within the target.

Efficiency of converting electron energy into x rays as a function of tube voltage.[4]

kV	Heat (%)	X Rays (%)
60	99.5	0.5
200	99	1.0
4000	60	40

The characteristic radiation produced by a target is governed by the binding energies of the K, L, and M shells of the target atoms. Theoretically, any shell could contribute to characteristic radiation. In practice, however, transitions of electrons among shells beyond the M shell produce only low-energy x rays, ultraviolet light,

TABLE 5-2 Electron Shell Binding Energies (keV)

Shell	Molybdenum ($Z = 42$)	Tungsten ($Z = 74$)
K	20	69
L	2.9, 2.6, 2.5[a]	12, 11, 10
M	0.50, 0.41, 0.39, 0.23, 0.22	2.8, 2.6, 2.3, 1.9, 1.8

[a] Multiple binding energies exist within the L and M shells because of the range of discrete values of the quantum number *l*, discussed in Chapter 2.

A characteristic x ray released during transition of an electron between adjacent shells is known as an α x ray. For example, a K_α x ray is one produced during transition of an electron from the L to the K shell. A β x ray is an x ray produced by an electron transition among nonadjacent shells. For example, a K_β x ray reflects a transition of an electron from the M to the K shell.

In the past, x-ray wavelengths were described in units of angstroms (Å), where $Å = 10^{-10}$ m. The minimum wavelength in angstroms would be λ_{min} (Å) $= 12.4/\text{kVp}$. The angstrom is no longer used as a measure of wavelength.

Compared with visible light, x rays have much shorter wavelengths. Wavelengths of visible light range from 400 nm (blue) to 700 nm (red).

and visible light. Low-energy x rays are removed by inherent filtration and do not become part of the useful beam. The characteristic peak for a particular shell occurs only when the tube voltage exceeds the binding energy of that shell. Binding energies in tungsten and molybdenum are shown in Table 5-2.

The characteristic radiation produced by an x-ray target is usually dominated by one or two peaks with specific energies slightly less than the binding energy of the K-shell electrons. The most likely transition involves an L-shell electron dropping to the K shell to fill a vacancy in that shell. This transition yields a photon of energy equal to the difference in electron binding energies of the K and L shells. A characteristic photon with an energy equal to the binding energy of the K shell alone is produced only when a free electron from outside the atom fills the vacancy. The probability of such an occurrence is vanishingly small.

During interaction of an electron with a target nucleus, a bremsstrahlung photon may emerge with energy equal to the total kinetic energy of the bombarding electron. Such a bremsstrahlung photon would have the maximum energy of all photons produced at a given tube voltage. The maximum energy of the photons depicted in Figure 5-5 therefore reflects the peak voltage applied across the x-ray tube, described in units of kilovolts peak (kVp). Photons of maximum energy E_{max} in an x-ray beam possess the maximum frequency and the minimum wavelength.

$$E_{max} = h\nu_{max} = \frac{hc}{\lambda_{min}}$$

The minimum wavelength in nanometers for an x-ray beam may be computed as

$$\lambda_{min} = \frac{hc}{E_{max}} = \frac{hc}{\text{kVp}}$$

where E_{max} is expressed in keV.

$$\lambda_{min} = \frac{(6.62 \times 10^{-34}\ \text{J-sec})(3 \times 10^{8}\ \text{m/sec})(10^{9}\ \text{nm/m})}{\text{kVp}(1.6 \times 10^{-16}\ \text{J/keV})}$$

$$\lambda_{min} = \frac{1.24}{\text{kVp}} \qquad (5\text{-}3)$$

with λ_{min} expressed in units of nanometers (nm).

Example 5-3

Calculate the maximum energy and minimum wavelength for an x-ray beam generated at 100 kVp.

The maximum energy (keV) numerically equals the maximum tube voltage (kVp). Because the maximum tube voltage is 100 kVp, the maximum energy of the photons is 100 keV:

$$\lambda_{min} = \frac{1.24}{100\ \text{kVp}}$$
$$= 0.0124\ \text{nm}$$

TUBE VACUUM

To prevent collisions between air molecules as electrons accelerate between the filament and target, x-ray tubes are evacuated to pressures less than 10^{-5} Hg. Removal of air also reduces deterioration of the hot filament by oxidation. During the manufacture of x-ray tubes, evacuation is accomplished by "outgassing" procedures that employ repeated heating cycles to remove gas occluded in components of the x-ray tube. Tubes still occasionally become "gassy," either after prolonged use or because the vacuum seal is not perfect. Filaments are destroyed rapidly in gassy tubes.

Many x-ray tubes include a "getter circuit" (active ion trap) to remove gas molecules that otherwise might accumulate in the x-ray tube over time.

ENVELOPE AND HOUSING

A vacuum-tight glass envelope (the "x-ray tube") surrounds other components required for the efficient production of x rays. The tube is mounted inside a metal housing that is grounded electrically. Oil surrounds the x-ray tube to (a) insulate the housing from the high voltage applied to the tube and (b) absorb heat radiated from the anode. Shockproof cables that deliver high voltage to the x-ray tube enter the housing through insulated openings. A bellows in the housing permits heated oil to expand when the tube is used. Often the bellows is connected to a switch that interrupts the operation of the x-ray tube if the oil reaches a temperature exceeding the heat storage capacity of the tube housing. A lead sheath inside the metal housing attenuates radiation emerging from the x-ray tube in undesired directions. A cross section of an x-ray tube and its housing is shown in Figure 5-8.

Secondary electrons can be ejected from a target bombarded by high-speed electrons. As these electrons strike the glass envelope or metallic components of the x-ray tube, they interact to produce x rays. These x rays are referred to as off-focus x rays because they are produced away from the target.

The quality of an x-ray image is reduced by off-focus x rays. For example, off-focus radiation contributes as much as 25% of the total amount of radiation emerging from some x-ray tubes with rotating anodes.[5] The effects of off-focus radiation on images may be reduced by placing the beam collimators as close as possible to the x-ray target.

Off-focus radiation can also be reduced by using small auxiliary collimators placed near the output window of the x-ray tube.

For x-ray images of highest quality, the volume of the target from which x rays emerge should be as small as possible. To reduce the "apparent size" of the focal spot,

FIGURE 5-8
Cutaway of a rotating-anode x-ray tube positioned in its housing. (Courtesy of Machlett Laboratories, Inc.)

Electrons accelerated from the filament to the target are absorbed in the first 0.5-mm thickness of target. In this distance, a typical electron will experience 1000 or more interactions with target atoms.

the target of an x-ray tube is mounted at a steep angle with respect to the direction of impinging electrons (see Margin Figure 5-10). With the target at this angle, x rays appear to originate within a focal spot much smaller than the volume of the target absorbing energy from the impinging electrons. This apparent reduction in size of the focal spot is termed the *line-focus principle.* Most diagnostic x-ray tubes use a target angle between 6 and 17 degrees. In the illustration, side a of the projected or apparent focal spot may be calculated by

$$a = A \sin\theta \tag{5-4}$$

where A is the corresponding dimension of the true focal spot and θ is the target angle.

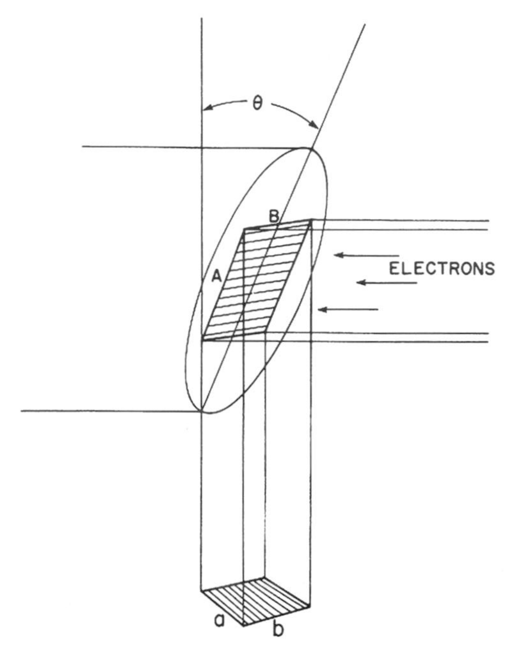

MARGIN FIGURE 5-10
Illustration of the line-focus principle, which reduces the apparent size of the focal spot.

Example 5-4

By using the illustration in the margin and Eq. (5-4), calculate a if $A = 7$ mm and $\theta = 17$ degrees.

$$\begin{aligned} a &= \sin\theta \\ &= (7\text{ mm})(\sin 17\text{ degrees}) \\ &= (7\text{ mm})(0.29) \\ &= 2\text{ mm} \end{aligned}$$

Side b of the apparent focal spot equals side B of the true focal spot because side B is perpendicular to the electron beam. However, side B is shorter than side A of the true focal spot because the width of a filament is always less than its length. When viewed in the center of the field of view, the apparent focal spot usually is approximately square.

As mentioned earlier, dual-focus diagnostic x-ray tubes furnish two apparent focal spots, one for fine-focus radiography (e.g., 0.6 mm^2 or less) produced with a smaller filament, and another for course-focus radiography (e.g., 1.5 mm^2) produced with a larger filament. The apparent focal spot to be used is determined by the tube current desired. The small filament is used when a low tube current (e.g., 100 mA) is satisfactory. The coarse filament is used when a larger tube current (e.g., 200 mA or greater) is required to reduce exposure time. Apparent focal spots of very small dimensions (e.g., $\leq$0.1 mm) are available with certain x-ray tubes used for magnification radiography.

The apparent size of the focal spot of an x-ray tube may be measured with a pinhole x-ray camera.[5,6] A hole with a diameter of a few hundredths of a millimeter is drilled in a plate that is opaque to x rays. The plate is positioned between the x-ray tube and film. The size of the image of the hole is measured on the exposed film. From the dimensions of the image and the position of the pinhole, the size of the apparent focal spot may be computed. For example, the dimension (a) of the apparent focal spot shown in the margin may be computed from the corresponding dimension (a') in the image by

$$a = a'\left(\frac{d_1}{d_2}\right) \tag{5-5}$$

where d_1 is the distance from the target to the pinhole and d_2 is the distance from the pinhole in the film.

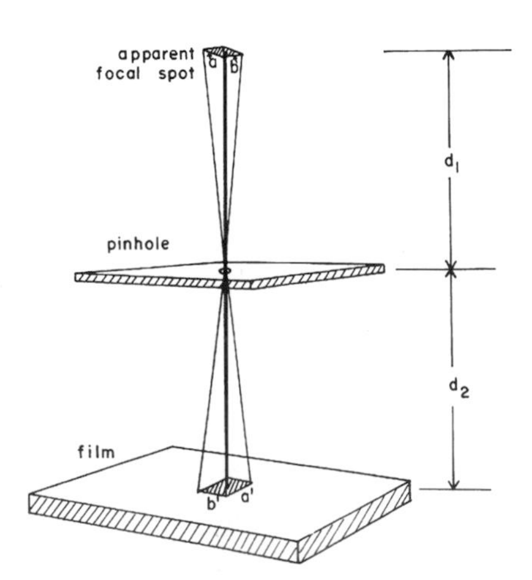

MARGIN FIGURE 5-11
Pinhole method for determining the size *ab* of the apparent focal spot.

Focal spot size also can be measured with a resolution test object such as the star pattern shown in Figure 5-9. The x-ray image of the pattern on the right reveals a blur zone where the spokes of the test pattern are indistinct. From the diameter of the blur zone, the effective size of the focal spot can be computed in any dimension. This effective focal spot size may differ from pinhole camera measurements of the focal spot along the same dimension because the diameter of the blur zone is influenced not only by the actual focal spot dimensions, but also by the distribution of x-ray

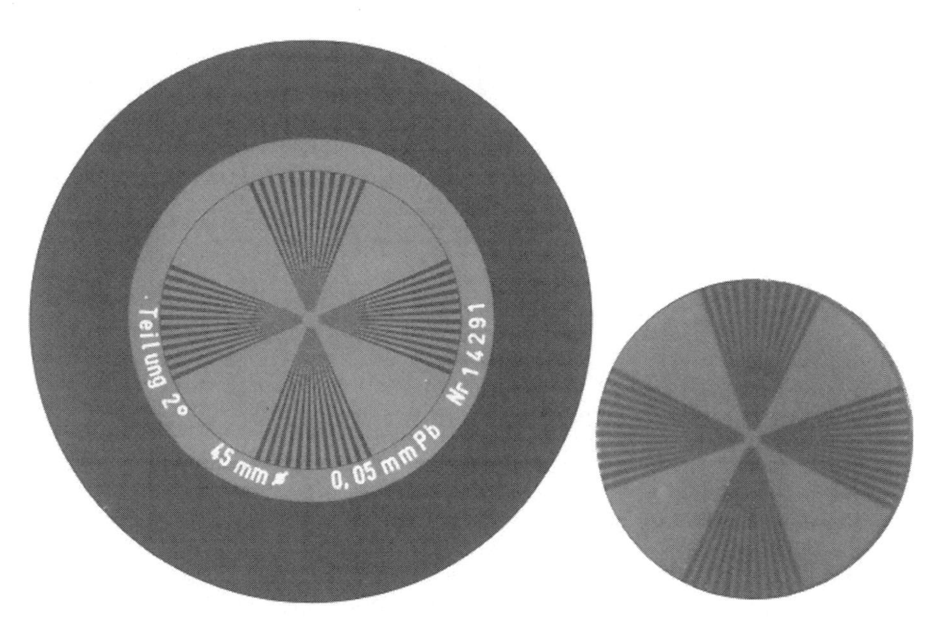

FIGURE 5-9
Contact radiograph (**left**) and x-ray image (**right**) of a star test pattern. The effective size of the focal spot may be computed from the diameter of the blur zone in the x-ray image.

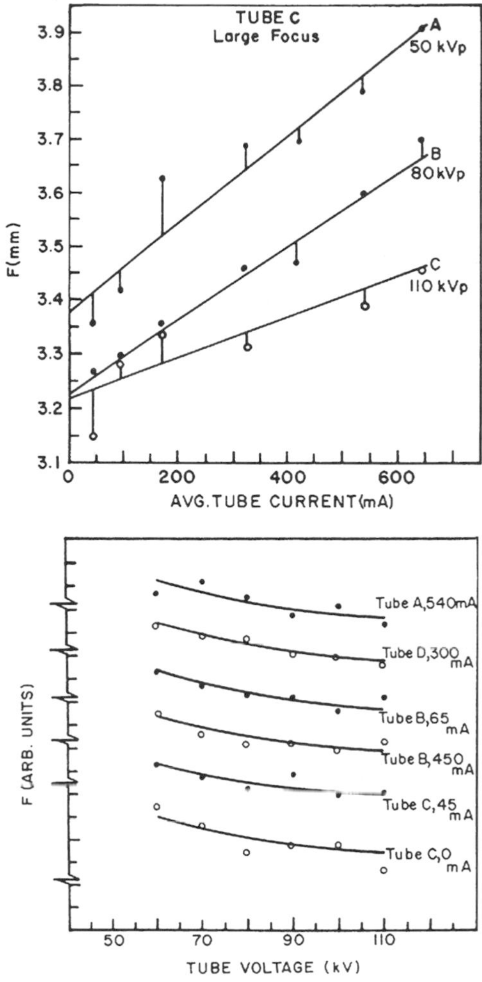

MARGIN FIGURE 5-12
Influence of tube current (**top**) and tube voltage (**bottom**) on the focal spot in a direction parallel to the motion of impinging electrons. (From Chaney, E., and Hendee, W. *Med Phys* 1974; **1**:131.)

intensity across the focal spot. In most diagnostic x-ray tubes, this distribution is not uniform. Instead, the intensity tends to be concentrated at the edges of the focal spot in a direction perpendicular to the electron beam.

For most x ray tubes, the size of the focal spot is not constant. Instead, it varies with both the tube current and the voltage applied to the x-ray tube.[7,8] This influence is shown in the margin for the dimension of the focal spot parallel to the motion of impinging electrons. On the top, the growth or "blooming" of the focal spot with tube current is illustrated. The gradual reduction of the same focal spot dimension with increasing levels of peak voltage is shown on the bottom.

Low-energy x rays generated in a tungsten target are attenuated severely during their escape from the target. For targets mounted at a small angle, the attenuation is greater for x rays emerging along the anode side of the x-ray beam than for those emerging along the side of the beam nearest the cathode. Consequently, the x-ray intensity decreases from the cathode to the anode side of the beam. This variation in intensity across an x-ray beam is termed the *heel effect*. The heel effect is noticeable for x-ray beams used in diagnostic radiology, particularly for x-ray beams generated at low kVp, because the x-ray energy is relatively low and the target angles are steep. To compensate for the heel effect, a filter may be installed in the tube housing near the exit portal of the x-ray beam. The thickness of such a filter increases from the anode to the cathode side of the x-ray beam. Positioning thicker portions of a patient near the cathode side of the x-ray beam also helps to compensate for the heel effect.

The heel effect increases with the steepness of the target angle. This increase limits the maximum useful field size obtainable with a particular target angle. For example, a target angle no steeper than 12 degrees is recommended for x-ray examinations using 14- ×17-in. film at a 40-in. distance from the x-ray tube, whereas targets as steep as 7 degrees may be used if field sizes no larger than 10 × 10 in. are required at the same distance.

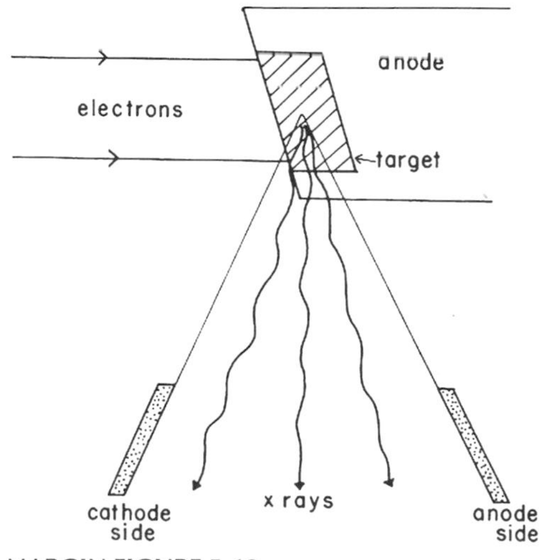

MARGIN FIGURE 5-13
The heel effect is produced by increased attenuation of x rays in the sloping target near the anode side of the x-ray beam.

SPECIAL-PURPOSE X-RAY TUBES

Many x-ray tubes have been designed for special applications. A few of these special-purpose tubes are discussed in this chapter.

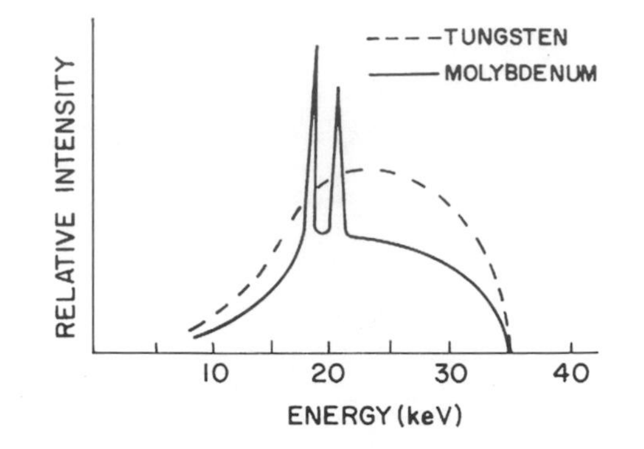

FIGURE 5-10
X-ray spectrum from molybdenum (solid line) and tungsten (dashed line) target x-ray tubes.

Grid-Controlled X-Ray Tubes

Grid-controlled x-ray tubes are also referred to as grid-pulsed or grid-biased x-ray tubes.

In a grid-controlled x-ray tube, the focusing cup in the cathode assembly is maintained a few hundred volts negative with respect to the filament. In this condition, the negative potential of the focusing cap prevents the flow of electrons from filament to target. Only when the negative potential is removed can electrons flow across the x-ray tube. That is, applying and then momentarily removing the potential difference between the focusing cup and the filament provides an off-on switch for the production of x rays. Grid-controlled x-ray tubes are used for very short ("pulsed") exposures such as those required during digital radiography and angiography.

Grid-controlled x-ray tubes cost significantly more than non biased tubes.

Field-Emission X-Ray Tubes

In the field-emission x-ray tube, the cathode is a metal needle with a tip about 1 μm in diameter. Electrons are extracted from the cathode by an intense electrical field rather than by thermionic emission. At diagnostic tube voltages, the rate of electron extraction is too low to provide tube currents adequate for most examinations, and efforts to market field-emission x-ray tubes in clinical radiology have been limited primarily to two applications: (1) pediatric radiography where lower tube currents can be tolerated and (2) high-voltage chest radiography where higher tube voltages can be used (as much as 300 kVp) to enhance the extraction of electrons from the cathode. Neither application has received much acceptance in clinical radiology.

Molybdenum-Target X-Ray Tubes

For low-voltage studies of soft-tissue structures (e.g., mammography), x-ray tubes with molybdenum and rhodium targets are preferred over tubes with tungsten targets. In the voltage range of 25 to 45 kVp, K-characteristic x rays can be produced in molybdenum but not in tungsten. These characteristic molybdenum photons yield a concentration of x rays on the low-energy side of the x-ray spectrum (Figure 5-10), which enhances the visualization of soft-tissue structures. Properties of molybdenum-target x-ray tubes are discussed in greater detail in Chapter 7.

RATINGS FOR X-RAY TUBES

The high rate of energy deposition in the small volume of an x-ray target heats the target to a very high temperature. Hence a target should have high thermal conductivity to transfer heat rapidly to its surroundings. Because of the high energy deposition in the target, rotating anodes are used in almost all diagnostic x-ray tubes. A rotating anode increases the volume of target material that absorbs energy from impinging electrons, thereby reducing the temperature attained by any portion of the anode. The anode is attached to the rotor of a small induction motor by a stem that usually is molybdenum. Anodes are 3 to 5 inches in diameter and rotate at speeds up to 10,000 rpm. The induction motor is energized for about 1 second before high voltage is applied to the x-ray tube. This delay ensures that electrons do not strike the target before the anode reaches its maximum speed of rotation. Energy deposited in the rotating anode is radiated to the oil bath surrounding the glass envelope of the x-ray tube.

Larger diameters and higher rotational speeds are required for applications such as angiography and helical-scan computed tomography that deliver greater amounts of energy to the x-ray target.

The rotating anode, together within the stator and rotor of the induction motor, are known collectively as the *anode assembly*.

Maximum Tube Voltage, Filament Current, and Filament Voltage

The maximum voltage to be applied between filament and target is specified for every x-ray tube. This "voltage rating" depends on the characteristics of the applied voltage (e.g., single phase, three phase, or constant potential) and on the properties of the x-ray tube (e.g., distance between filament and target, shape of the cathode assembly and target, and shape of the glass envelope). Occasional transient surges in voltage may be tolerated by an x-ray tube, provided that they exceed the voltage rating by no more than a few percent.

Limits are placed on the current and voltage delivered to coarse and fine filaments of an x-ray tube. The current rating for the filament is significantly lower for continuous compared with pulsed operation of the x-ray tube, because the temperature of the filament rises steadily as current flows through it.

Maximum Energy

Maximum-energy ratings are provided for the target, anode, and housing of an x-ray tube.[10,11] These ratings are expressed in heat units, where for single phase electrical power

$$\text{Number of heat units (HU)} = \text{(Tube voltage)(Tube current)(Time)}$$
$$= \text{(kVp)(mA)(sec)} \quad \textbf{(5-6)}$$

If the tube voltage and current are constant, then 1 HU = 1J of energy. For three-phase power, the number of heat units is computed as

$$\text{Number of heat units (HU)} = \text{(Tube voltage)(Tube current)(Time)(1.35)}$$
$$= \text{(kVp)(mA)(sec)(1.35)} \quad \textbf{(5-7)}$$

For x-ray tubes supplied with single-phase (1φ), full-wave rectified voltage, the peak current through the x-ray tube is about 1.4 times the average current. The average current nearly equals the peak current in x-ray generators supplied with three-phase (3φ) voltage. For this reason, the number of heat units for an exposure from a 3φ generator is computed with the factor 1.35 in Eq. (5-7). For long exposures or a series of exposures with an x-ray tube supplied with 3φ voltage, more energy is delivered to the target, and the number of exposures in a given interval of time must be reduced. Separate rating charts are usually provided for 1φ and 3φ operation of an x-ray tube.

X-ray tubes are also rated in terms of their maximum power loading in kilowatts. Representative maximum power loadings are shown below[9]:

Focal Spot (mm)	Power Rating (kW)
1.2–1.5	80–125
0.8–1.0	50–80
0.5–0.8	40–60
0.3–0.5	10–30
≤ 0.3	1–10
≤ 0.1	<1

Energy ratings for the anode and the tube housing are expressed in terms of heat storage capacities. The heat storage capacity of a tube component is the total number of heat units that may be absorbed without damage to the component. Anode heat storage capacities for diagnostic x-ray tubes range from several hundred thousand to over a million heat units.

The heat storage capacity of the x-ray tube housing is also important because heat is transferred from the anode to the tube housing. The housing heat storage capacity exceeds the anode capacity and is usually on the order of 1.5 million HU.

To determine whether the target of an x-ray tube might be damaged by a particular combination of tube voltage, tube current, and exposure time, *energy rating charts* furnished with the x-ray tube should be consulted. To use the sample chart shown in Figure 5-11, a horizontal line is drawn from the desired tube current on the y axis to the curve for the desired tube voltage. From the intersection of the line and the curve, a vertical line dropped to the x axis reveals the maximum exposure time

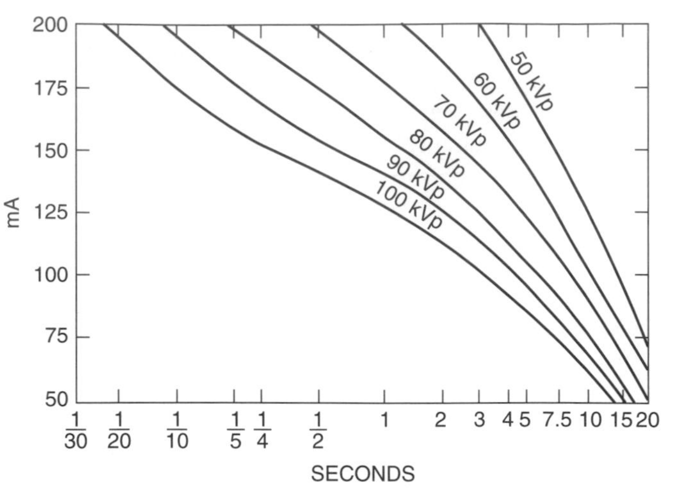

FIGURE 5-11
Energy rating chart for a Machlett Dynamax "25" x-ray tube with a 1-mm focal spot and single-phase, fully rectified voltage.(Courtesy of Machlett Laboratories, Inc.)

that can be used for a single exposure without possible damage to the x-ray target. The area under each voltage curve encompasses combinations of tube current and exposure time that do not exceed the target-loading capacity when the x-ray tube is operated at that voltage. The area above each curve reflects combinations of tube current and exposure time that overload the x-ray tube and might damage the target. Often switches are incorporated into an x-ray circuit to prevent the operator from exceeding the energy rating for the x-ray tube. Shown in Figure 5-12 are a few targets damaged by excess load or improper rotation of the target.

An *anode thermal-characteristics chart* describes the rate at which energy may be delivered to an anode without exceeding its capacity for storing heat (Figure 5-13). Also shown in the chart is the rate at which heat is radiated from the anode to the insulating oil and housing. For example, the delivery of 425 HU per second to the anode of the tube exceeds the anode heat storage capacity after 5.5 minutes. The delivery of 340 HU per second could be continued indefinitely. The cooling curve in Figure 5-13 shows the rate at which the anode cools after storing a certain amount of heat.

The housing-cooling chart in Figure 5-14 depicts the rate at which the tube housing cools after storing a certain amount of heat. The rate of cooling with and without forced circulation of air is shown. Charts similar to those in Figures 5-13 and 5-14 are used to ensure that multiple exposures in rapid succession do not damage an x-ray tube or its housing.

When several exposures are made in rapid succession, a target-heating problem is created that is not directly addressed in any of the charts described above. This target-heating problem is caused by heat deposition that exceeds the rate of heat dissipation in the focal track of the rotating anode. To prevent this buildup of heat from damaging the target, an additional tube-rating chart should be consulted. This chart, termed an angiographic rating chart because the problem of rapid successive exposures occurs frequently in angiography, is illustrated in Figure 5-15. Use of this chart is depicted in Example 5-9.

Example 5-5

From the energy-rating chart in Figure 5-11, is a radiographic technique of 150 mA, 1 second at 100 kVp permissible?

The maximum exposure time is slightly longer than 0.25 seconds for 150 mA at 100 kVp. Therefore, the proposed technique is unacceptable.

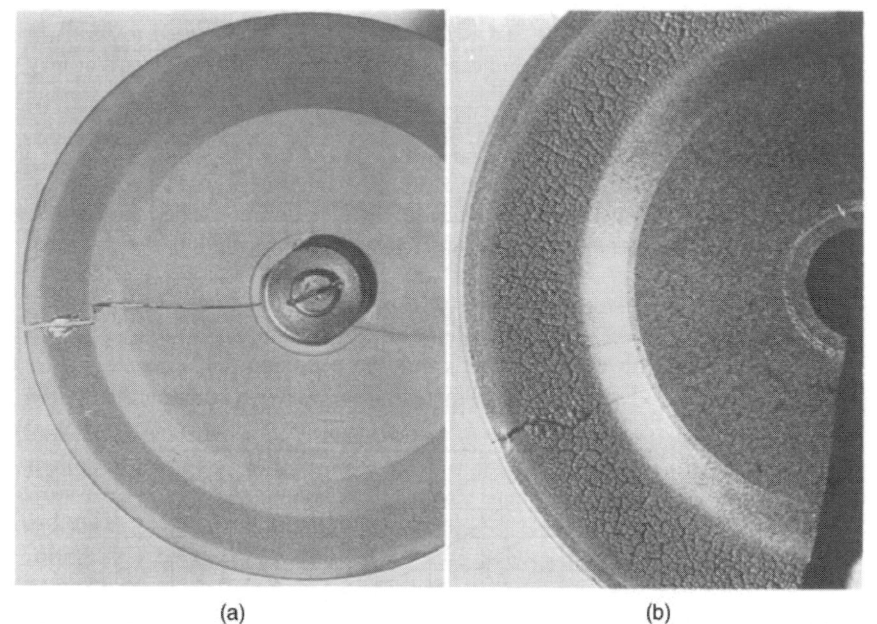

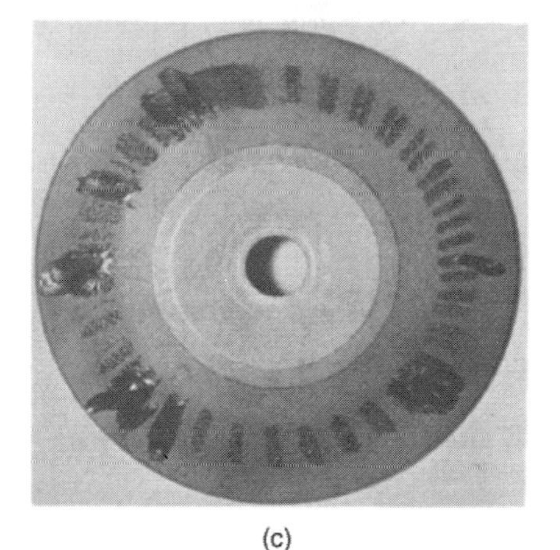

FIGURE 5-12
Rotating targets damaged by excessive loading or improper rotation of the target. **A:** Target cracked by lack of rotation. **B:** Target damaged by slow rotation and excessive loading. **C:** Target damaged by slow rotation.

Is 100 mA at 100 kVp for 1.5 seconds permissible?

The maximum exposure time is 3 seconds for 100 mA at 100 kVp. Therefore the proposed technique is acceptable.

Example 5-6

Five minutes of fluoroscopy at 4 mA and 100 kVp are to be combined with eight 0.5-second spot films at 100 kVp and 100 mA. Is the technique permissible according to Figures 5-11 and 5-13?

The technique is acceptable according to the energy-rating chart in Figure 5-11. By rearranging Eq. (5-6) we calculate that the rate of delivery of energy to the anode during fluoroscopy is (100 kVp) (4 mA) = 400 HU per second. From Figure 5-13, after 5 minutes approximately 60,000 HU have been accumulated by the anode. The eight spot films contribute an additional 40,000 HU [(100 kVp) (100 mA) (0.5 sec) 8 = 40,000 HU]. After all exposures have been made, the total heat stored in the

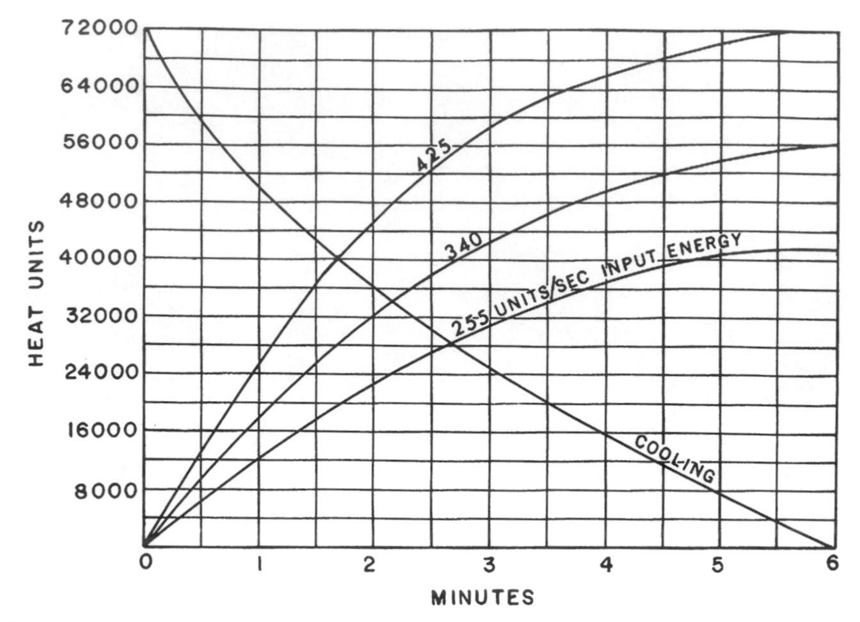

FIGURE 5-13
Anode thermal characteristics chart for a Machlett Dynamax "25" rotating anode x-ray tube. The anode heat-storage capacity is 72,000 HU. (Courtesy of Machlett Laboratories, Inc.)

anode is 40,000 + 60,000 = 100,000 HU. This amount of heat exceeds the anode heat storage capacity of 72,000 HU. Consequently the proposed technique is unacceptable.

Example 5-7

Three minutes of fluoroscopy at 3 mA and 85 kVp are combined with four 0.25-second spot films at 85 kVp and 150 mA. From Figure 5-13, what time must elapse before the procedure may be repeated?

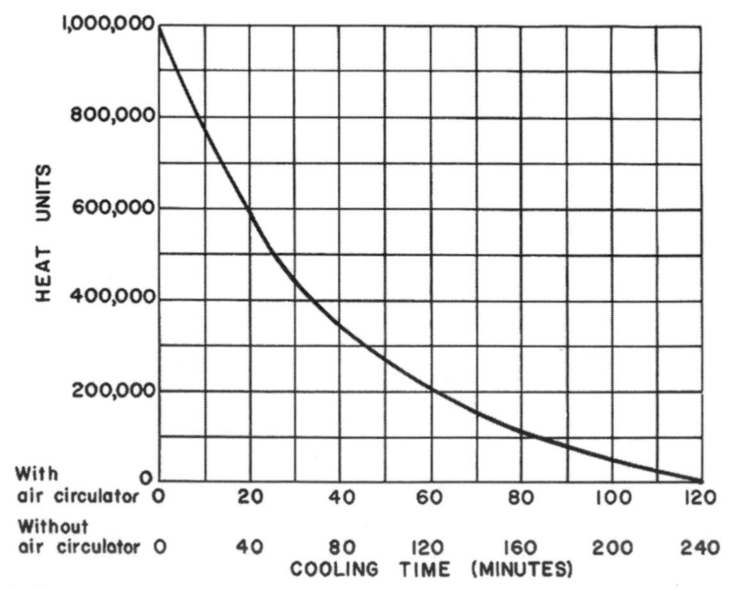

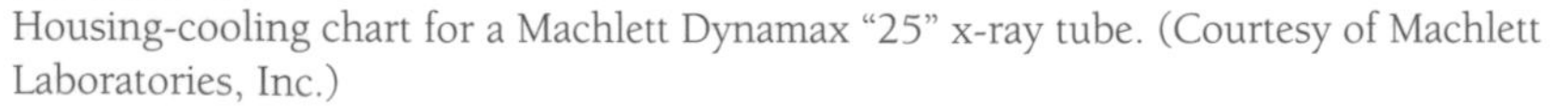

FIGURE 5-14
Housing-cooling chart for a Machlett Dynamax "25" x-ray tube. (Courtesy of Machlett Laboratories, Inc.)

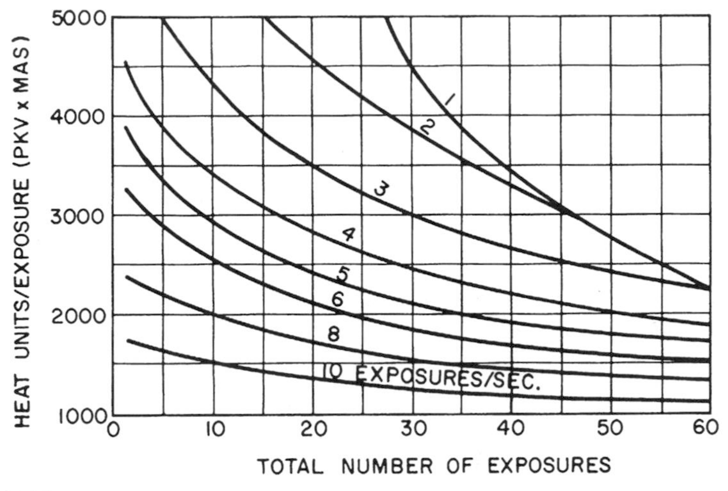

FIGURE 5-15
Angiographic rating chart for a Machlett Super Dynamax x-ray tube, 1.0-mm focal spot, full-wave rectification, single phase.

The rate of delivery of energy to the anode is (85 kVp) (3 mA) = 255 HU per second, resulting in a heat load of 31,000 HU after 3 minutes. To this heat load is added (85 kVp) (150 mA) (0.25 sec) (4) = 12,750 HU for the four spot films to yield a total heat load of 43,750 HU. From the position of the vertical axis corresponding to this heat load, a horizontal line is extended to intersect the anode-cooling curve at 1.4 minutes. From this intersection, the anode must cool until its residual heat load is 72,000 − 43,750 = 28,250 HU, so that when the 43,750 HU from the next procedure is added to the residual heat load, the total heat load does not exceed the 72,000 HU anode heat storage capacity. The time corresponding to a residual heat load of 28,250 HU is 2.6 minutes. Hence the cooling time required between procedures is 2.6 − 1.4 = 1.2 minutes.

Example 5-8

From Figures 5-11 and 5-13, it is apparent that three exposures per minute are acceptable if each 1φ exposure is taken at 0.5 second, 125 mA, and 100 kVp. Could this procedure be repeated each minute for 1 hour? The rate of energy transfer to the housing is

$$= (100\ \text{kVp})\ (125\ \text{mA})\ (0.5\ \text{sec})\ (3\ \text{exposures/min})$$

$$= 18,750\ \text{HU/min}$$

At the end of 1 hour, (18,750 HU/min) (60 min/hr) = 1,125,000 HU will have been delivered to the housing. The heat storage capacity of the housing is only 1 million HU. Without forced circulation of air, the maximum rate of energy dissipation from the housing is estimated to be approximately 12,500 HU per minute (Figure 5-14). With air circulation, the rate of energy dissipation is 25,000 HU per minute. Therefore the procedure is unacceptable if the housing is not air-cooled, but is acceptable if the housing is cooled by forced circulation of air.

Example 5-9

From Figure 5-15, how many consecutive exposures can be made at a rate of six exposures per second if each exposure is taken at 85 kVp, 500 mA, and 0.05 seconds?

Each exposure produced (85 kVp) (500 mA) ().05 sec) = 2125 HU. A horizontal line from this position on the *y* axis intersects the 6 exposures per second curve at a position corresponding to 20 exposures. Hence, no more than 20 exposures should be made at a rate of 6 exposures per second.

PROBLEMS

5-1. Explain why tube current is more likely to be limited by space charge in an x-ray tube operated at low voltage and high tube current than in a tube operated at high voltage and low current. Would you expect the tube current to be space-charge limited or filament-emission limited in an x-ray tube used for mammography? What would you expect for an x-ray tube used for chest radiography at 120 kVp and 100 mA?

*5-2. How many electrons flow from the cathode to the anode each second in an x-ray tube with a tube current of 50 mA? (1A = 1 coulomb/sec.) If the tube voltage is constant and equals 100 kV, at what rate (joules per second) is energy delivered to the anode?

5-3. Explain why off-focus radiation is greatly reduced by shutters placed near the target of the x-ray tube.

*5-4. An apparent focal spot of 1 mm is projected from an x-ray tube. The true focal spot is 5 mm. What is the target angle? Why is the heel effect greater in an x-ray beam from a target with a small target angle?

*5-5. From Figure 5-11, is a radiographic technique of 125 mA for 3 seconds at 90 kVp permissible? Is a 4-second exposure at 90 kVp and 100 mA permissible? Why does the number of milliampere-seconds permitted for a particular voltage increase as the exposure time is increased?

*5-7. From Figure 5-14, how many exposures are permitted each minute over a period of 1 hour if each exposure is made for 1 second at 100 mA and 90 kVp? Is the answer different if the tube is operated with forced circulation of air?

*5-8. What kinetic energy do electrons possess when they reach the target of an x-ray tube operated at 250 kVp? What is the approximate ratio of bremsstrahlung to characteristic radiation produced by these electrons? Calculate the minimum wavelength of x-ray photons generated at 250 kVp.

*5-9. A lead plate is positioned 20 in. from the target of a diagnostic x-ray tube. The plate is 50 in. from a film cassette. The plate has a hole 0.1 mm in diameter. The image of the hole is 5 mm. What is the size of the apparent focal spot?

*5-10. The target slopes at an angle of 12 degrees in a diagnostic x-ray tube. Electrons are focused along a strip of the target 2 mm wide. How long is the strip if the apparent focal spot is 2×2 mm?

*For problems marked with an asterisk, answers are provided on p. 491.

SUMMARY

- Conventional x-ray tubes contain the following components:
  - Heated filament
  - Cathode assembly
  - X-ray target
  - High voltage supply
  - Rectification circuitry
  - Filtration
  - Glass envelope
- The apparent size of the x-ray focal spot depends on
  - Filament size
  - Tube current and voltage
  - Target angle
  - Position along the anode-cathode axis
- X-ray production efficiency is increased by
  - Voltage rectification
  - Use of three-phase and constant-potential tube voltage
  - Increased tube voltage
  - Higher-Z target
- Special-purpose x-ray tubes include
  - Grid-controlled
  - Field emission
  - Mammography
- X-ray tube rating charts include
  - Energy rating charts
  - Anode thermal characteristic charts
  - Housing cooling charts
  - Angiographic rating charts

REFERENCES

1. Röntgen, W. Über eine Art von Strahlen (vorlaufige Mitteilung). *Sitzungs-Berichte der Physikalisch-medicinschen Gesellschaft zu Wurzurg* 1895; **9**:132.
2. Coolidge, W. A powerful roentgen ray tube with a pure electron discharge. *Phys Rev* 1913; **2**:409.
3. Ter-Pogossian, M. *The Physical Aspects of Diagnostic Radiology*. New York, Harper & Row Publishers, 1967, p. 107.
4. Dendy, P. P., Heaton, B. *Physics for Diagnostic Radiology*, 2nd edition. Bristol, Institute of Physics, 1999.
5. International Commission on Radiological Units and Measurements. Methods of Evaluating Radiological Equipment and Materials. Recommendations of the ICRU. *National Bureau of Standards Handbook* 89, 1962.
6. Parrish, W. Improved method of measuring x-ray tube focus. *Rev. Sci. Instrum.* 1967; **38**:1779.
7. Hendee, W., and Chaney, E. X-ray focal spots: Practical considerations. *Appl. Radiol.* 1974; **3**:25.
8. Chaney, E., and Hendee, W. Effects of x-ray tube current and voltage on effective focal-spot size. *Med. Phys.* 1974; **1**:141.
9. Bushberg, J. T., Seibert, J. A., Leidholdt, E. M, Boone JM. *The Essential Physics of Medical Imaging*. Baltimore, Williams & Wilkins, 1994.
10. Hallock, A. A review of methods used to calculate heat loading of x-ray tubes extending the life of rotating anode x-ray tubes. *Cathode Press* 1958; **15**:1.
11. Hallock A. Introduction to three phase rating of rotating anode tubes. *Cathode Press* 1966; **23**:30.

CHAPTER

6

RADIATION QUANTITY AND QUALITY

OBJECTIVES

After completing this chapter, the reader should be able to:

- Define the following terms: intensity, fluence, flux.
- Define and give the SI and traditional units for exposure, dose, RBE dose, dose equivalent, equivalent dose, and effective dose.
- Calculate the photon fluence required to obtain a unit exposure from the photon and energy fluence.
- Name seven methods that are used to measure radiation dose and describe how each method works.
- Explain the concept of electron equilibrium as it applies to ionization chambers.
- Calculate the effective atomic number of a composite material.
- Define the term "half value layer."
- Explain what is meant by the energy spectrum of an x-ray beam.

The specification and measurement of radiation demands assessment of the spatial and temporal distribution of the radiation. For some types of radiation the distribution of energies must also be considered. A complete description of the effects of radiation on a material requires analysis of how the various components of the radiation field interact with atoms and molecules of the material.

To avoid the burden of maintaining a large multidimensional database of detailed information, techniques have been developed that allow multiple interaction parameters to be summarized into singular values. These singular values—the rem, the sievert, the gray, and so on—are the radiation units described in this chapter. While they may be susceptible to the pitfalls of oversimplification, they are all attempts to provide a succinct answer to the question, What will happen when radiation strikes a material?

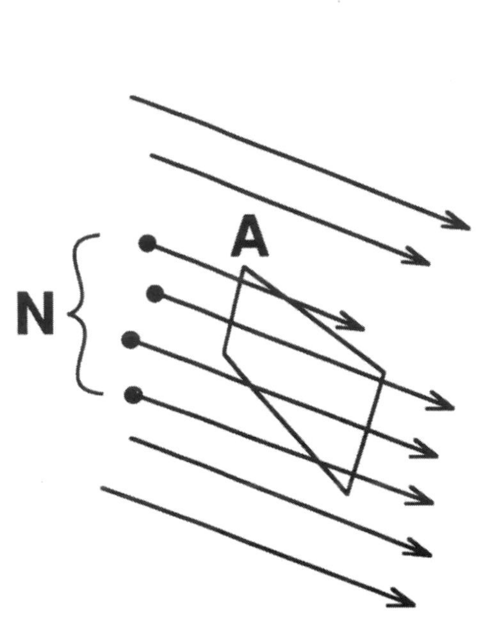

MARGIN FIGURE 6-1
The particle fluence of a beam of radiation is defined as the number of particles (*N*) passing through a unit area (*A*) that is perpendicular to the direction of the beam. If the beam is uniform, the location and size of the area *A* are arbitrary.

INTENSITY

The term *radiation* is defined as energy in transit from one location to another. The term *radiation intensity* is used colloquially to refer to a number of attributes of the output of a radiation source. In physics and engineering, the term "intensity" is defined more specifically in terms of energy per unit area, per unit time. In this text, the term rdiation intensity is given a specific definition.

A beam of radiation is depicted in Margin Figure 6-1. The radiation *fluence* of the beam Φ is defined as the number N of particles or photons per area A:

$$\Phi = \frac{N}{A} \tag{6-1}$$

If the beam is uniform, then the location or size of the area A is irrelevant so long as it is in the beam and perpendicular to the direction of the beam. If the beam is not uniform over its entire area, then the fluence must be averaged over a number of small areas, or specified separately for each area. The time rate of change of fluence, known as the radiation flux ϕ, is

$$\phi = \frac{\Phi}{t} = \frac{N}{A \cdot t} \tag{6-2}$$

If the fluence varies with time, then the flux must be averaged over time or specified at some instant.

If all particles or photons in the radiation beam possess the same energy, the energy fluence Ψ is simply the product of the radiation fluence Φ and the energy E per particle or photon:

$$\Psi = \Phi E = \frac{NE}{A} \tag{6-3}$$

Radiation flux is sometimes termed the *radiation flux density*.

TABLE 6-1 Fluence and Flux (Intensity) of a Beam of Radiation[a]

Quantity	*Symbol*	*Definition*[b]	*Units*
Particle (photon) fluence	Φ	$\frac{N}{A}$	$\frac{\text{Particles (photons)}}{\text{m}^2}$
Particle (photon) flux	ϕ	$\frac{N}{At}$	$\frac{\text{Particles (photons)}}{\text{m}^2 \cdot \text{sec}}$
Energy fluence	Ψ	$\frac{NE}{A}$	$\frac{\text{MeV}}{\text{m}^2}$
Energy flux (intensity)	ψ	$\frac{NE}{At}$	$\frac{\text{MeV}}{\text{m}^2 \cdot \text{sec}}$

[a] *Note:* These expressions assume that the number of particles or photons does not vary over time or over the area A, and that all particles or photons have the same energy.

[b] N = number of particles or photons; E = energy per particle or photon; A = area; t = time.

Similarly, the particle or photon flux ϕ may be converted to the energy flux ψ, also known as the intensity I, by multiplying by the energy E per particle or photon:

$$I = \psi = \phi E = \frac{NE}{At} \tag{6-4}$$

If the radiation beam consists of particles or photons having different energies ($E_1, E_2, \ldots, E_m$), then the intensity (or energy flux) is determined by

$$I = \psi = \sum_{i=1}^{m} f_i \phi E_i \tag{6-5}$$

where f_i represents the fraction of particles having energy E_i, and the symbol

$$\sum_{i=1}^{m}$$

indicates that the intensity is determined by adding the components of the beam at each of m energies. The concepts of radiation fluence and flux are summarized in Table 6-1.

Expressions similar to Eqs. (6-1) to (6-5) may be derived for any type of particle beam such as α, β, neutron, or high-energy nuclei. In the case of electromagnetic radiation such as x or γ rays, the photons have an energy $E = h\nu$. In the case of radiation that is described in terms of waves (such as ultrasound or radio waves), the definition of intensity (Equation 6-4) is modified as

$$I = \frac{E_t}{At} \tag{6-6}$$

where E_t is the total energy delivered by the wave during the time t.

Example 6-1

An abdominal radiograph uses 10^{13} photons to expose a film with an area of 0.15 m^2 (1.5×10^{-1} m^2 or 1500 cm^2) during an exposure time of 0.1 second. All photons have an energy of 40 keV. Find the photon fluence Φ, the photon flux ϕ, the energy fluence Ψ, and the intensity I.

$$\text{Photon fluence } \Phi = \frac{N}{A} = \frac{1 \times 10^{13} \text{ photons}}{1.5 \times 10^{-1} \text{ m}^2} = 6.7 \times 10^{13} \frac{\text{photons}}{\text{m}^2}$$

$$\text{Photon flux } \phi = \frac{N}{At} = \frac{\Phi}{t} = \frac{6.7 \times 10^{13} \frac{\text{photons}}{\text{m}^2}}{0.1 \text{ sec}} = 6.7 \times 10^{14} \frac{\text{photons}}{\text{m}^2 \cdot \text{sec}}$$

$$\text{Energy fluence } \Psi = \frac{NE}{A} = \Phi E = \left(6.7 \times 10^{13}\ \frac{\text{photons}}{\text{m}^2}\right)\left(40 \frac{\text{keV}}{\text{photon}}\right)$$
$$= 2.68 \times 10^{15}\ \frac{\text{keV}}{\text{m}^2}$$
$$= 2.68 \times 10^{12}\ \frac{\text{MeV}}{\text{m}^2}$$

$$\text{Intensity } I = \Psi = \frac{NE}{At} = \frac{\Psi}{t} = \frac{2.68 \times 10^{12}\ \frac{\text{MeV}}{\text{m}^2}}{0.1\ \text{sec}} = 2.68 \times 10^{13}\ \frac{\text{MeV}}{\text{m}^2 \cdot \text{sec}}$$

Note: A beam produced by an x-ray tube actually contains a spectrum of energies (Chapter 5), and a more accurate estimate of intensity would involve a weighted sum of the contributions of the various photon energies to the total intensity. The next example shows a source of radiation that produces two different photon energies and also demonstrates the concept of detector geometry—that is, the effect of area.

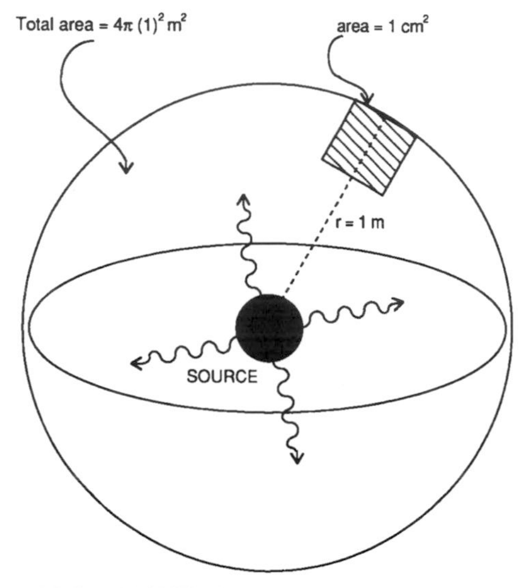

MARGIN FIGURE 6-2
Photons emitted in any direction from an isotopic source eventually cross the surface of a sphere.

Example 6-2

A radionuclide releases 270-keV photons in 90% of its decays and 360-keV photons in approximately 10% of its decays. When 10^6 photons have been released by decay of the radionuclide, what is the photon fluence and energy fluence over a 1-cm^2 area at a distance of 1 m from the source?

Because photons are emitted isotropically (i.e., with equal probability in any direction) from the source, 10^6 photons cross the surface of a sphere with a radius of 1 m (Margin Figure 6-2). The number of photons that cross an area of 1 cm^2 (10^{-4} m^2) on the sphere surface is the fraction of the 10^6 photons that are intercepted by the 1-cm^2 area. The surface area of a sphere is $4\pi r^2$, where r is the sphere's radius.

$$\text{Fraction of total emissions} = \frac{10^{-4}\ \text{m}^2}{4\pi(1\ \text{m})^2} = 7.96 \times 10^{-6}$$

$$\text{Photons crossing the 1-cm}^2\text{ area} = (7.96 \times 10^{-6})10^6 = 7.96\ \text{photons}$$

That is, the photon fluence is approximately 8 photons/cm^2.

The energy fluence must be weighted according to the fraction of photons having each of the two possible energies.

$$\Psi = \sum_{i=1}^{2} f_i \Phi E_i$$
$$= (0.9)\left(8\ \frac{\text{photons}}{\text{cm}^2}\right)(270\ \text{keV}) + (0.1)\left(8\ \frac{\text{photons}}{\text{cm}^2}\right)(360\ \text{keV})$$
$$= (1944 + 288)\ \frac{\text{keV}}{\text{cm}^2}$$
$$= 2.23\ \frac{\text{MeV}}{\text{cm}^2}$$

Although photon and energy flux densities and fluences are important in many computations in radiologic physics, these quantities cannot be measured easily. Hence, units have been defined that are related more closely to common methods for measuring radiation quantity.

TRADITIONAL VERSUS SYSTÈME INTERNATIONAL UNITS

The measurement of x and γ radiation has presented technical challenges from the time of their discovery. In his initial studies, Röntgen used the blackening of a photographic emulsion (i.e., film), as a "dosimeter."[1] Nonlinearity and energy dependence of the film's response caused difficulties. The "erythema dose" (the amount

TABLE 6-2 Traditional (T) and Système International (SI) Quantities

	Unit		
Quantity	*T*	*SI*	*To Convert from T to SI, Multiply by*
Exposure	Roentgen (R)	Coulomb/kg (C/kg)	2.58×10^{-4}
Absorbed dose	rad	Gray (Gy)	0.01
Absorbed dose equivalent	rem	Sievert (Sv)	0.01

of radiation required to produce reddening of the skin) was for many years the major dosimetric method for evaluating the effects of radiation treatment. Biologic variability and the lack of an objective measure of "reddening" were major problems with this technique.

In 1928, the roentgen (R) was defined as the principal unit of radiation quantity for x rays of medium energy. The definition of the roentgen has been revised many times, with each revision reflecting an increased understanding of the interactions of radiation as well as improvements in the equipment used to detect these interactions. Over this same period, several other units of radiation quantity have been proposed. In 1958, the International Commission on Radiation Units and Measurements (ICRU) organized a continuing study of the units of radiation quantity. Results of this continuing study are described in various ICRU reports.[2]

The units described in earlier ICRU reports are considered the "traditional" system (T) of units. Just as the United States is converting to the "metric" system of units, radiology is converting to the system of radiation units known as the Système International (SI).[3] Both the traditional and SI systems are used in this text with preference given to SI units. Table 6-2 gives a summary of units and conversion factors between them.

RADIATION EXPOSURE

Primary ion pairs (electrons and positive ions) are produced as ionizing radiation interacts with atoms of an attenuating medium. Secondary ion pairs are produced as the primary ion pairs dissipate their energy by ionizing nearby atoms. The total number of ion pairs produced is proportional to the energy that the radiation deposits in the medium. The concept of *radiation exposure* is based on the assumption that the absorbing medium is air. If Q is the total charge (negative or positive) liberated as x or γ rays interact in a small volume of air of mass m, then the radiation exposure X at the location of the small volume is

$$X = \frac{Q}{m} \tag{6-7}$$

The total charge reflects the production of both primary and secondary ion pairs, with the secondary ion pairs produced both inside and outside of the small volume of air. The traditional unit of radiation exposure is the *roentgen* (R):

$$1\text{R} = 2.58 \times 10^{-4} \text{ coulomb/kg air}$$

This definition of the roentgen is numerically equivalent to an older definition:

$$\begin{aligned} 1\text{R} &= 1 \text{ electrostatic unit (ESU)}/0.001293 \text{ g air} \\ &= 1 \text{ ESU/cm}^3 \text{ air at STP*} \end{aligned}$$

*STP = standard temperature (0°C) and pressure (1 atm or 760 mm Hg).

The roentgen is applicable only to x and γ radiation and to photon energies less than about 3 MeV. That is, the roentgen cannot be used for particle beams or for beams of high-energy photons. For photon energies above 3 MeV, it becomes increasingly difficult to determine how many secondary ion pairs are produced outside of the measurement volume as a result of interactions occurring in the volume, and vice versa. There is no specially named unit for radiation exposure in SI. Instead, the fundamental units of coulomb and kilogram are used (see Table 6-2). The unit roentgen and the concept of radiation exposure in general are disappearing from use as more fundamental expressions of radiation quantity gain acceptance.

Example 6-3

Find the energy absorbed in air from an exposure of 1 coulomb/kg. The W-quantity, the average energy required to produce ion pairs in a material (Chapter 4), is 33.85 eV per ion pair (IP) for air. The energy absorbed during an exposure X of 1 coulomb per kilogram is

$$E_x = \frac{(1 \text{ coulomb/kg})(33.85 \text{ eV/IP})(1.6 \times 10^{-19} \text{ J/eV})}{(1.6 \times 10^{-19} \text{ coulomb/IP})} \tag{6-8}$$
$$= 33.85 \text{ J/kg}$$

Thus, for every coulomb per kilogram of exposure, air absorbs 33.85 J/kg of energy.

Energy and Photon Fluence per Unit Exposure

From the definitions of energy and photon fluence and some fundamental quantities that have been determined for air, we may calculate the energy or photon fluence required to produce a given amount of exposure. From Eq. (6-8), the energy absorbed in air is 33.85X (J/kg) during an exposure to X coulomb/kg. The absorbed energy can also be stated as

Energy absorbed in air = (Energy fluence) (Total mass energy absorption coefficient)

$$= \Psi\left(\frac{\text{J}}{\text{m}^2}\right)(\mu_{\text{en}})_m\left(\frac{\text{m}^2}{\text{kg}}\right) \tag{6-9}$$
$$= \Psi(\mu_{\text{en}})_m\left(\frac{\text{J}}{\text{kg}}\right)$$

where $(\mu_{\text{en}})_m$ is the total mass energy absorption coefficient for x- or γ-ray photons that contribute to the energy fluence. The coefficient $(\mu_{\text{en}})_m$ is defined as (Chapter 4)

$$(\mu_{\text{en}})_m = \mu_m\left(\frac{E_a}{h\nu}\right),$$

where μ_m is the total mass attenuation coefficient of air for photons of energy $h\nu$ and E_a represents the average energy transformed into kinetic energy of electrons and positive ions per photon absorbed or scattered from the x- or γ-ray beam.[4] The average energy E_a is corrected for characteristic x rays radiated from the attenuating medium, as well as for bremsstrahlung produced as electrons interact with nuclei within the attenuating medium. Mass energy absorption coefficients for a few media, including air, are listed in Table 6-3.

By combining Eqs. (6-8) and (6-9), we obtain

$$\Psi(\mu_{\text{en}})_m = 33.85\ X$$

TABLE 6-3 Mass Energy Absorption Coefficients for Selected Materials and Photon Energies

	Mass Energy Absorption Coefficient $(\mu_{en})_m$ (m^2/kg)			
Photon Energy (MeV)	Air	Water	Compact Bone	Muscle
0.01	0.466	0.489	1.90	0.496
0.02	0.0516	0.0523	0.251	0.0544
0.03	0.0147	0.0147	0.0743	0.0154
0.04	0.00640	0.00647	0.0305	0.00677
0.05	0.00384	0.00394	0.0158	0.00409
0.06	0.00292	0.00304	0.00979	0.00312
0.08	0.00236	0.00253	0.00520	0.00255
0.10	0.00231	0.00252	0.00386	0.00252
0.20	0.00268	0.00300	0.00302	0.00297
0.30	0.00288	0.00320	0.00311	0.00317
0.40	0.00296	0.00329	0.00316	0.00325
0.50	0.00297	0.00330	0.00316	0.00327
0.60	0.00296	0.00329	0.00315	0.00326
0.80	0.00289	0.00321	0.00306	0.00318
1.0	0.00280	0.00311	0.00297	0.00308
2.0	0.00234	0.00260	0.00248	0.00257
3.0	0.00205	0.00227	0.00219	0.00225
4.0	0.00186	0.00205	0.00199	0.00203
5.0	0.00173	0.00190	0.00186	0.00188
6.0	0.00163	0.00180	0.00178	0.00178
8.0	0.00150	0.00165	0.00165	0.00163
10.0	0.00144	0.00155	0.00159	0.00154

Source: *National Bureau of Standards Handbook 85*. Washington, D.C., U.S. Government Printing Office, 1964.

The energy fluence per unit exposure (Ψ/X) is

$$\frac{\Psi}{X} = \frac{33.85}{(\mu_{\text{en}})_m} \tag{6-10}$$

where $(\mu_{\text{en}})_m$ is expressed in units of m^2/kg, Ψ in J/m^2, and X in coulombs/kg.

For monoenergetic photons, the photon fluence per unit exposure, Φ/X, is the quotient of the energy fluence per roentgen divided by the energy per photon:

$$\frac{\Phi}{X} = \frac{\Psi}{h\nu\left(1.6 \times 10^{-13}\ \text{J/MeV}\right)}$$

From Eq. (6-10), we obtain

$$\frac{\Phi}{X} = \frac{2.11 \times 10^{14}}{h\nu\,(\mu_{\text{en}})_m} \tag{6-11}$$

with $h\nu$ expressed in MeV and Φ in units of photons/m^2.

The photon and energy fluence per unit exposure are plotted in Margin Figure 6-3 as a function of photon energy. At lower photon energies, the large influence of photon energy upon the energy absorption coefficient of air is reflected in the rapid change in the energy and photon fluence per unit exposure. Above 100 keV, the energy absorption coefficient is relatively constant, and the energy fluence per unit exposure does not vary greatly.[5] However, the photon fluence per unit exposure decreases steadily as the energy per photon increases.

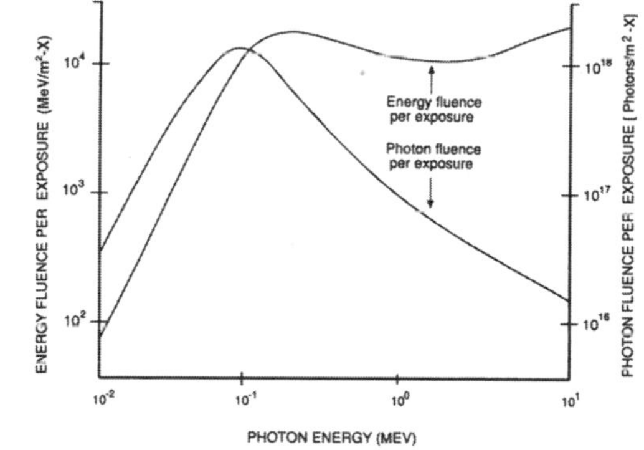

MARGIN FIGURE 6-3
Photon and energy fluence per exposure, plotted as a function of the photon energy in MeV. Exposure is expressed in coulomb/kg. To convert to the traditional unit, roentgen, multiply the vertical scales by 2.58×10^{-4}.

Example 6-4

Compute the energy and photon fluence per unit exposure for ^{60}Co γ rays. The average energy of the photons is 1.25 MeV, and the total energy absorption coefficient is 2.67×10^{-3} m²/kg (Table 6-3).

$$\frac{\Psi}{X} = \frac{33.85}{(\mu_{en})_m}$$

$$= \frac{33.85}{2.67 \times 10^{-3}\ \text{m}^2/\text{kg}}$$

$$= 12{,}600\ \frac{\text{J}}{\text{m}^2} \qquad \textbf{(6-10)}$$

$$\frac{\Phi}{X} = \frac{2.11 \times 10^{14}}{h\nu\,(\mu_{en})_m}$$

$$= \frac{2.11 \times 10^{14}}{(1.25\ \text{MeV/photon})(2.67 \times 10^{-3}\ \text{m}^2/\text{kg})}$$

$$= 6.32 \times 10^{16}\ \frac{\text{photons}}{\text{m}^2} \qquad \textbf{(6-11)}$$

UNITS OF RADIATION DOSE

The Gray

Chemical and biological changes in tissue exposed to ionizing radiation depend upon the energy absorbed in the tissue from the radiation, rather than upon the amount of ionization that the radiation produces in air. To describe the energy absorbed in a medium from any type of ionizing radiation, the quantity of radiation should be described in SI units of gray (or traditional units of rads). The gray (Gy) is a unit of *absorbed dose** and represents the absorption of one joule of energy per kilogram of absorbing material:

The traditional unit of radiation dose, the rad, is an acronym for "radiation absorbed dose."

$$1\ \text{Gy} = 1\ \text{J/kg}$$

The absorbed dose D in gray delivered to a small mass m in kilograms is

$$D(\text{Gy}) = \frac{E/m}{1\ \text{J/kg}} \qquad \textbf{(6-12)}$$

where E, the absorbed energy in joules, is "the difference between the sum of the energies of all the directly and indirectly ionizing particles which have entered the volume, and the sum of the energies of all those which have left it, minus the energy equivalent of any increase in rest mass that took place in nuclear or elementary particle reactions within the volume." [6]

The traditional unit of radiation dose, the rad, represents an absorption of 100 ergs of energy per gram of absorbing material.

This definition means that E is the total energy deposited in a small volume of irradiated medium, corrected for energy removed from the volume in any way.

Example 6-5

If a dose of 0.05 Gy (5 centigray [cGy]) is delivered uniformly to the uterus during a diagnostic x-ray examination, how much energy is absorbed by each gram of the uterus?

The name of the quantity "kerma" is an acronym for the phrase "kinetic energy released in a material" with an "a" added for convenience.

$$D = \frac{E/m}{1\ \text{J/kg-Gy}}$$

*The Gy is also a unit of kerma, defined as the sum of the initial kinetic energies of all charged particles liberated by ionizing radiation in a volume element, divided by the mass of matter in the volume element. Under conditions of charged-particle equilibrium with negligible energy loss by bremsstrahlung, kerma and absorbed dose are identical. The output of x-ray tubes is sometimes described in terms of "air kerma" expressed in terms of energy deposited per unit mass of air.

$$\begin{aligned} E &= (1\ \text{J/kg-Gy})(D)(m) \\ &= (1\ \text{J/kg-Gy})(0.05\ \text{Gy})\left(10^{-3}\ \frac{\text{kg}}{\text{g}}\right)(1\ \text{g}) \\ &= 5 \times 10^{-5}\ \text{J} \end{aligned}$$

During an exposure of 1 C/kg, the energy absorbed per kilogram of air is

$$1\ \text{C/kg} = 33.85\ \text{J/kg in air}$$

or, in traditional units,

$$\begin{aligned} 1\ \text{R} &= 2.58 \times 10^{-4}\ \text{C/kg} \\ &= (2.58 \times 10^{-4}\ \text{C/kg})\left(33.85\ \frac{\text{J/kg}}{\text{C/kg}}\right) \\ &= 86.9 \times 10^{-4}\ \text{J/kg in air} \end{aligned}$$

Because 1 rad = 10^{-2} J/kg, we have

$$\begin{aligned} 1\ \text{C/kg} &= 3385\ \text{rad in air} \\ 1\ \text{R} &= (2.58 \times 10^{-4}\ \text{C/kg})(3385\ \text{rad-kg/C}) \\ &= 0.869\ \text{rad in air} \end{aligned}$$

The Sievert

At the present time, there are five different quantities that use the SI unit of sievert and the traditional unit of rem. In many situations, they are approximately equal in magnitude. The differences lie in the ways that one or more of the four variables that influence the extent of biological damage are managed. These four variables reflect the:

- Damage at the cellular level per unit dose by different types of radiation (alpha, beta, x rays, neutrons, etc.)
- Sensitivity of different body tissues
- Effects that damage to different tissues have upon the overall health of the organism
- Impact upon future generations

RBE Dose

Chemical and biological effects of irradiation depend not only upon the amount of energy absorbed in an irradiated medium, but also upon the distribution of the absorbed energy within the medium. For equal absorbed doses, various types of ionizing radiation often differ in the efficiency with which they elicit a particular chemical or biological response. The *relative biologic effectiveness* (RBE) describes the effectiveness or efficiency with which a particular type of radiation evokes a certain chemical or biological effect. The relative biological effectiveness is computed by comparing results obtained with the radiation in question with those obtained with a reference radiation (e.g., medium-energy x rays or ^{60}Co radiation):

$$\text{RBE} = \frac{\text{Dose of reference radiation required to produce a particular response}}{\text{Dose of radiation in question required to produce the same response}} \tag{6-13}$$

TABLE 6-4 RBE[a] of ^{60}Co Gammas, with Different Biologic Effects Used as a Criterion for Measurement[b]

Effect	*RBE of ^{60}Co Gammas*	*Source*
30-day lethality and testicular atrophy in mice	0.77	Storer et al. (7)
Splenic and thymic atrophy in mice	1	Storer et al. (7)
Inhibition of growth in *Vicia faba*	0.84	Hall (8)
LD_{50}[a] in mice, rat, chick embryo, and yeast	0.82–093	Sinclair (9)
Hatchability of chicken eggs	0.81	Loken et al. (10)
HeLa cell survival	0.90	Krohmer (11)
Lens opacity in mice	0.8	Upton et al. (12)
Cataract induction in rats	1	Focht et al. (13)
L-cell survival	0.76	Till and Cunningham (14)

[a] RBE, relative biological effectiveness; LD_{50}, median lethal dose.

[b] The RBE of 200-kV x rays is taken as 1.

For a particular type of radiation, the RBE may vary from one chemical or biological response to another. Listed in Table 6-4 are the results of investigations of the relative biological effectiveness of ^{60}Co radiation. For these data, the reference radiation is medium-energy x rays. It is apparent that the RBE for ^{60}Co γ rays varies from one biological response to the next.

The RBE dose in sievert (Sv) is the product of the RBE and the dose in gray:

$$\text{RBE dose (Sv)} = \text{Absorbed dose (Gy)} \times \text{RBE} \quad \textbf{(6-14)}$$

The ICRU has suggested that the concept of RBE dose should be limited to descriptions of radiation dose in radiation biology.[6] That is, RBE dose pertains only to an exact set of experimental conditions, and it is not applicable to general radiation protection situations.

DOSE EQUIVALENT

Often the effectiveness with which different types of radiation produce a particular chemical or biological effect varies with the linear energy transfer (LET) of the radiation. The *dose-equivalent* (DE) in sievert is the product of the dose in Gy and a *quality factor* (QF), which varies with the LET of the radiation.

$$\text{DE (Sv)} = D(\text{Gy}) \times \text{QF} \quad \textbf{(6-15)}$$

The dose equivalent reflects a recognition of differences in the effectiveness of different radiations to inflict overall biological damage and is used during computations associated with radiation protection.[15] Quality factors are listed in Table 6-5 as a function of LET, and in Table 6-6 for different types of radiation.

These quality factors should be used to determine shielding requirements and to compute radiation doses to personnel working with or near sources of ionizing radiation.

Example 6-6

A person receives an average whole-body dose of 1 mGy from ^{60}Co γ rays and 0.5 mGy from neutrons with an energy of 10 MeV. What is the dose equivalent to the

TABLE 6-5 Relation Between Specific Ionization, Linear Energy Transfer, and Quality Factor

Average Specific Ionization (IP/μm) in Water	*Average LET (keV/μm) in Water*	*QF*
100 or less	3.5 or less	1
100–200	3.5–7.0	1–2
200–650	7.0–23	2–5
650–1500	23–53	5–10
1500–5000	53–175	10–20

Source: Recommendations of the International Commission on Radiological Protection. *Br J Radiol* 1955; Suppl. 6.

person in millisieverts?

$$\begin{aligned} \text{DE (mSV)} &= [D(\text{mGy}) \times \text{QF}]_{\text{gammas}} + [D(\text{mGy}) \times \text{QF}]_{\text{neutrons}} \\ &= (1\ \text{mGy})(1) + (0.5\ \text{mGy})(20) \\ &= 11\text{mSv} \end{aligned}$$

Example 6-7

A person accidentally ingests a small amount of tritium (β-particle $E_{\max} =$ 0.018 MeV). The average dose to the gastrointestinal tract is estimated to be 5 mGy. What is the dose equivalent in millisieverts?

From Table 6-6, the quality factor is 1.0 from tritium β-particles.

$$\begin{aligned} \text{DE (mSv)} &= D(\text{mGy}) \times \text{QF} \\ &= (5\ \text{mGy})(1.0) \\ &= 5\ \text{mSv} \end{aligned}$$

Effective Dose Equivalent

A further refinement of the prediction of bioeffects can be obtained by multiplying the dose equivalent by a factor to convert the dose to a point within a specific patient tissue to a measure of overall detriment to the patient. In the late 1970s, the available epidemiological data[16] were used to compile weighting factors for many tissues and organs that, when multiplied by the dose equivalent, would yield the effective dose equivalent. The effective dose equivalent, an estimate of overall detriment to the individual, was designed to account for cancer mortality in individuals between the ages of 20 and 60, as well as hereditary effects in the next two generations.[17] These weighting factors have been superseded by factors reformulated in the 1990s.

Equivalent Dose

In the early 1990s, up-to-date dosimetry and epidemiological results from studies of the survivors of the atomic bombs dropped on Hiroshima and Nagasaki in 1945 began to yield statistically robust analyses of the relative radiation sensitivities of different tissues, and the consequences of irradiation of these tissues to the organism as a whole.[18,19] This work also affected the assignment of factors to account for the energy and type of radiation. When the average dose to an organ is known, the risk can be assigned in terms of more recent estimates of fatality from cancer. In addition, the new factors consider the detriment from non fatal cancer and hereditary effects for future generations. In view of these changes, and to differentiate from the older system of dosimetry, the term equivalent dose was adopted to refer to the product

TABLE 6-6 Quality Factors for Different Radiations[a]

Type of Radiation	*QF*
X rays, γ rays, and β particles	1
Thermal neutrons	5
Neutrons and protons	20
α Particles from natural radionuclides	20
Heavy recoil nuclei	20

[a] These data should be used only for purposes of radiation protection. Data from National Council on Radiation Protection and Measurements. *NCRP Report*, no. 91, 1987.

Five quantities are expressed in units of sievert at the present time. It is expected that some of them will become obsolete at some time in the future. Listed below are the five quantities and the factors that are taken into account in specifying the value of those quantities in units of sievert.

1. RBE dose (relative biological effectiveness) - Relative effectiveness of different **types** of radiation in eliciting a **specific biologic effect**, such as reddening of the skin in **specific parts of the body** of animals or humane exposed in experimental conditions. Used in journal articles and textbooks.

2. Dose Equivalent - Relative effectiveness of different **types** of radiation in producing overall bioeffects using data from the 1970's. Still used in federal regulations.

3. Effective Dose Equivalent - Differing biological harm caused by different **types** of radiation and also the **region of the body** irradiated using data from the 1970's. Dose is defined at a point in the patient within each organ or organ system. Still used in federal regulations.

4. Equivalent Dose - Differing biological harm caused by different **types** of radiation using data from the 1990's. Used in current ICRP and NCRP recommendations.

5. Effective Dose - Differing biological harm caused by different **types** of radiation and also the **region of the body** irradiated using data from the 1990's. Used in current ICRP and NCRP recommendations.

Note: There is consistency in that each time that the term "equivalent" is used, the type of radiation is taken into account. Each time the word "effective" is used, both the type of radiation and the region of the body that is irradiated are taken into account.

TABLE 6-7 Radiation Weighting Factors

Source	W_R
X rays, gamma rays, electrons, positrons, muons at all energies	1
Neutrons	
<10 keV	5
10 keV to 100 keV	10
>100 keV to 2 MeV	20
>2 MeV to 20 MeV	10
>20 MeV	5
Protons, >2 MeV	2
Alpha particles, fission fragments, nonrelativistic heavy nuclei	20

Source: National Council on Radiation Protection and Measurements. Limitation of Exposure to Ionizing Radiation, Report No. 116. Bethesda, MD, NCRP, 1993.

of the average organ dose and the newly defined radiation weighting factors, W_R (Table 6-7).

Effective Dose

The newly defined tissue weighting factors, W_T, are given in Table 6-8. They are formulated to yield the effective dose by multiplying by an equivalent dose that represents an average organ dose that has been calculated using the newer radiation weighting factors W_R given in Table 6-7.

Example 6-8

As in Example 6-7, a person ingests a small amount of tritium. The average dose to the gastrointestinal tract is estimated to be 5 mGy. For the purposes of this example, we will assume that a negligible amount appears in the bladder. What is the equivalent dose in millisieverts? And what is the effective dose in millisieverts?

From Table 6-7, the radiation weighting factor is 1 for beta particles.

The equivalent dose is then

$$\begin{aligned}\text{ED (mSv)} &= D(\text{mGy}) \times W_R \\ &= (5\ \text{mGy})(1.0) \\ &= 5\ \text{mSv}\end{aligned}$$

Note that in this situation, 5 mSv is the same numerical value that was calculated for the dose equivalent in Example 6-7. This is because the radiation weighting factor (W_R) for beta particles equals the quality factor (QF) for beta particles. Also, a uniform dose deposition is assumed in this problem so that all point doses are assumed to equal the average organ doses. If the detailed dosimetry were known with greater precision, the values of point doses within organs might have led to a different calculation of dose equivalent.

From Table 6-7, the tissue weighting factor is 0.05 for the esophagus, 0.12 for the stomach, and 0.12 for the colon.

The effective dose is then

$$\begin{aligned}\text{ED (mSv)} &= D(\text{mGy}) \times W_R \times W_T \\ &= (5\ \text{mGy})(0.05) + (5\ \text{mGy})(0.12) + (5\ \text{mGy})(0.12) \\ &= 0.25\ \text{mSv} + 0.60\ \text{mSv} + 0.60\ \text{mSv} \\ &= 1.45\ \text{mSv}\end{aligned}$$

Note that the effective dose is a smaller number than the equivalent dose, because only partial body irradiation has taken place. The value of 1.45 mSv is the dose that, if received uniformly over the whole body, would lead to the same detrimental results (probability of fatal cancers, nonfatal cancers, and hereditary disorders) as the dose actually received by the gastrointestinal tract.

TABLE 6-8 Tissue Weighting Factors

Tissue	W_T
Gonads	0.20
Active bone marrow	0.12
Colon	0.12
Lungs	0.12
Stomach	0.12
Bladder	0.05
Breasts	0.05
Esophagus	0.05
Liver	0.05
Thyroid	0.05
Bone surfaces	0.01
Skin	0.01
Remainder	0.05

Source: Recommendations of the International Commission on Radiological Protection, ICRP Publication 60, Pergamon Press, Elmsford, NY, 1991.

MEASUREMENT OF RADIATION DOSE

The absorbed dose to a medium describes the energy absorbed during exposure of the medium to ionizing radiation. A radiation dosimeter provides a measurable response to the energy absorbed from incident radiation. To be most useful, the dosimeter should absorb an amount of energy equal to what would be absorbed in the medium that the dosimeter displaces. For example, a dosimeter used to measure radiation dose in soft tissue should absorb an amount of energy equal to that absorbed by the same mass of soft tissue. When this requirement is satisfied, the dosimeter is said to be *tissue equivalent*. Few dosimeters are exactly tissue equivalent, and corrections must be applied to most direct measurements of absorbed dose in tissue.

Calorimetric Dosimetry

Almost all the energy absorbed from radiation is eventually degraded to heat. If an absorbing medium is insulated from its environment, then the rise in temperature of the medium is proportional to the energy absorbed. The temperature rise T may be measured with a thermocouple or thermistor.

A thermocouple is a junction of dissimilar metals that yields a voltage that varies with temperature. A thermistor is a solid-state device with an electric resistance that changes rapidly with temperature.

A radiation calorimeter is an instrument used to measure radiation absorbed dose by detecting the increase in temperature of a mass of absorbing material.[20] The absorbed dose in the material is

$$D(\text{Gy}) = \frac{E}{m(1\ \text{J/kg-Gy})} = \frac{4.186\ (\text{J/calorie})s\,\Delta T}{(1\ \text{J/kg-Gy})} \tag{6-16}$$

where E is the energy absorbed in joules, m is the mass of the absorber in kilograms, s is the specific heat of the absorber in cal/kg °C, and ΔT is the temperature rise in Celsius (centigrade) degrees. For Eq. (6-16) to be correct, the absorbing medium must be insulated from its environment. To measure the absorbed dose in a particular medium, the absorber of the calorimeter must possess properties similar to those of the medium. Graphite is often used as the absorbing medium in calorimeters designed to measure the absorbed dose in soft tissue. A calorimeter used to measure radiation absorbed dose is illustrated in Margin Figure 6-4.

If the absorbing medium is thick and dense enough to absorb nearly all of the incident radiation, then the increase in temperature reflects the total energy delivered to the absorber by the radiation beam. Calorimeters that measure the total energy in a beam of radiation usually contain a massive lead absorber.

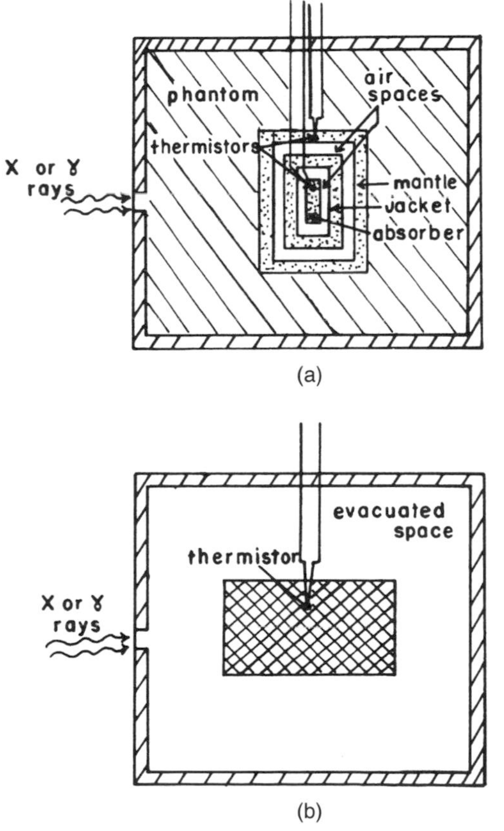

MARGIN FIGURE 6-4
Radiation calorimeters: **A:** For the measurement of absorbed dose in soft tissue. **B:** For the measurement of the total energy in a beam of radiation.

Photographic Dosimetry

The emulsion of a photographic film contains crystals of a silver halide embedded in a gelatin matrix. When the emulsion is developed, metallic silver is deposited in regions that were exposed to radiation. Unaffected crystals of silver halide are removed during fixation. The amount of silver deposited at any location depends upon many factors, including the amount of energy absorbed by the emulsion at that location. The transmission of light through a small region of film varies inversely with the amount of deposited silver. The transmission T is

$$T = \frac{I}{I_0}$$

where I represents the intensity of light transmitted through a small region of film and I_0 represents the light intensity with the film removed. The transmission may be measured with an optical densitometer. Curves relating transmission to radiation exposure, dose, or energy fluence are obtained by measuring the transmission through films receiving known exposures, doses, or fluences. If the problems described below are not insurmountable, the radiation exposure to a region of blackened film may be determined by measuring the transmission of light though the region and referring to a calibration curve.

Accurate dosimetry is difficult with photographic emulsions for reasons including those listed below.[21]

1. The optical density of a film depends not only on the radiation exposure to the emulsion but also on variables such as the energy of the radiation and the conditions under which the film is processed. For example, the optical density of a photographic film exposed to 40-keV photons may be as great as the density of a second film receiving an exposure 30 times greater from photons of several MeV. Photographic emulsions are said to be "energy

dependent" because their response to x and γ rays of various energies differs significantly from the response of air and soft tissue. In most cases, calibration films used to relate optical density to radiation exposure must be exposed to radiation identical to that for which dosimetric measurements are needed. This requirement often is difficult to satisfy, particularly when films are exposed in phantoms or in other situations where a large amount of scattered radiation is present.

2. When compared with air or soft tissue, the high-Z emulsion of a photographic film attenuates radiation rapidly. Errors caused by differences in attenuation may be particularly severe when radiation is directed parallel to the plane of a single film or perpendicular to a stack of films.
3. Differences in the thickness and composition of the photographic emulsion may cause the radiation sensitivity to vary from one film to another.

Chemical Dosimetry

Oxidation or reduction reactions are initiated when certain chemical solutions are exposed to ionizing radiation. The number of molecules affected depends on the energy absorbed by the solution. Consequently, the extent of oxidation or reduction is a reflection of the radiation dose to the solution. These changes are the basis of chemical dosimetry.

The chemical dosimeter used most widely is ferrous sulfate. For photons that interact primarily by Compton scattering, the ratio of the energy absorbed in ferrous sulfate to that absorbed in soft tissue equals 1.024, the ratio of the electron densities (electrons per gram) of the two media. For photons of higher and lower energy, the ratio of energy absorption increases above 1.024 because more energy is absorbed in the higher-Z ferrous sulfate.

The use of solutions of ferrous sulfate to measure radiation dose was described by Fricke and Morse in 1929.[22] A solution of ferrous sulfate used for radiation dosimetry is sometimes referred to as a *Fricke dosimeter*.[23] Although this dosimeter is reliable and accurate (±3%), it is relatively insensitive and absorbed doses of 50 to 500 Gy are required before the oxidation of Fe^{2+} to Fe^{3+} is measurable. The Fricke dosimeter was recommended in earlier years for calibration of high-energy electron beams used in radiation therapy.[24]

The yield of a chemical dosimeter such as ferrous sulfate is expressed by the G value:

$$G = \text{Number of molecules affected per 100 eV absorbed}$$

The G value for the oxidation of Fe^{2+} to Fe^{3+} in the Fricke dosimeter varies with the LET of the radiation. Many investigators have assumed that G is about 15.4 molecules per 100 eV for a solution of ferrous sulfate exposed to high-energy electrons or x and γ rays of medium energy. After exposure to radiation, the amount of Fe^{3+} ion in a ferrous sulfate solution is determined by measuring the transmission of ultraviolet light (305 nm) through the solution. Once the number of affected molecules is known, the energy absorbed by the ferrous sulfate solution may be computed by dividing the number of molecules by the G value.

Example 6-9

A solution of ferrous sulfate is exposed to ^{60}Co γ radiation. Measurement of the transmission of ultraviolet light (305 nm) through the irradiated solution indicates the presence of Fe^{3+} ion at a concentration of 0.0001 mol/L. What was the radiation dose absorbed by the solution?

For a concentration of 0.0001 gram-molecular weight/L and with a density of 10^{-3} kg/mL for the ferrous sulfate solution, the number of Fe^{3+} ions per kilogram is

$$\text{Number of } Fe^{3+} \text{ ions/kg} = \frac{(0.0001 \text{ gram-mol wt}/1000 \text{ mL})(6.02 \times 10^{23} \text{ molecules/gram-mol wt})}{10^{-3} \text{ kg/mL}}$$

$$= 6.02 \times 10^{19} \frac{\text{molecules}}{\text{kg}}$$

For a G value of 15.4 molecules per 100 eV for ferrous sulfate, the absorbed dose in the solution is

$$D(\text{rad}) = \frac{(6.02 \times 10^{19} \text{ molecules/kg})(1.6 \times 10^{-19} \text{ J/eV})}{(15.4 \text{ molecules}/100 \text{ eV})(1 \text{ J/kg-Gy})}$$
$$= 63 \text{ Gy}$$

Scintillation Dosimetry

Certain materials fluoresce or "scintillate" when exposed to ionizing radiation. The rate at which scintillations occur depends upon the rate of absorption of radiation in the scintillator. With a solid scintillation detector (e.g., thallium-activated sodium iodide), a light guide couples the scintillator optically to a photomultiplier tube. In the photomultiplier tube, light pulses from the detector are converted into electronic signals.

Scintillation detectors furnish a measurable response at very low dose rates and respond linearly over a wide range of dose rates. However, most scintillation detectors contain high-Z atoms, and low-energy photons are absorbed more rapidly in the detectors than in soft tissue or air. This *energy dependence* is a major disadvantage of using scintillation detectors to measure radiation exposure or dose to soft tissue. A few organic scintillation detectors have been constructed that are air equivalent or tissue equivalent over a wide range of photon energies.

Thermoluminescence Dosimetry

Diagramed in Margin Figure 6-5 are energy levels for electrons within crystals of a thermoluminescent material such as LiF. Electrons "jump" from the valence band to the conduction band by absorbing energy from ionizing radiation impinging upon the crystals. Some of the electrons return immediately to the valence band; other are "trapped" in intermediate energy levels supplied by impurities in the crystals. The number of electrons trapped in intermediate levels is proportional to the energy absorbed by the LiF phosphor during irradiation. Only rarely do electrons escape from the traps and return directly to the ground state. Unless energy is supplied for their release, almost all of the trapped electrons remain in the intermediate energy levels for months or years after irradiation. If the crystals are heated, energy is supplied to release the trapped electrons. Released electrons return to the conduction band where they fall to the valance band. Light is released as the electrons return to the valence band. This light is directed onto the photocathode of a photomultiplier tube. Because the amount of light striking the photocathode is proportional to the energy absorbed in the LiF during irradiation, the signal from the photomultiplier tube increases with the radiation dose absorbed in the phosphor.

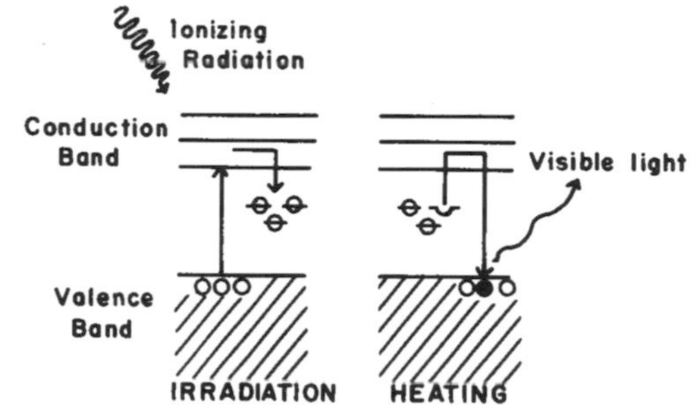

MARGIN FIGURE 6-5
Electron transitions occurring when thermoluminescent LiF is irradiated and heated.

The effective atomic number* of dosimetric LiF (8.18) is close to that for soft tissue (7.4) and for air (7.65).[25] Hence, for identical exposures to radiation, the amount of energy absorbed by LiF is reasonably close to that absorbed by an equal mass of soft tissue or air. Small differences between the energy absorption in LiF and air are reflected in the "energy dependence" curve for LiF shown in Margin Figure 6-6.

Lithium fluoride is used widely for the measurement of radiation dose within patients and phantoms, for personnel dosimetry, and for many other dosimetric measurements. Dosimetric LiF may be purchased as loose crystals, a solid extruded rod, solid chips, pressed pellets, or crystals embedded in a Teflon matrix.

Thermoluminescent dosimeters composed of CaF_2:Mn are frequently used for personnel monitoring and, occasionally, for other measurements of radiation dose. When compared with LiF and $Li_2B_4O_7$, CaF_2:Mn phosphor is more sensitive to ionizing radiation; however, the response of this phosphor varies rapidly with photon energy (Margin Figure 6-6).[26]

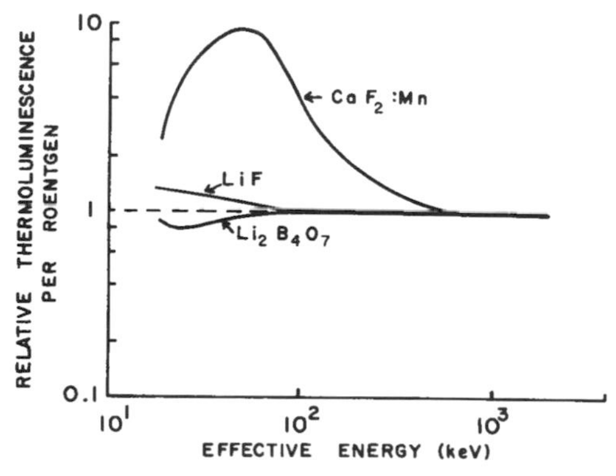

MARGIN FIGURE 6-6
Relative thermoluminescence from various materials per unit exposure is plotted as a function of the effective energy of incident x- or γ-ray photons. Data have been normalized to the response per unit exposure for ^{60}Co γ rays.

*The atomic number of a hypothetical element that attenuates photons at the same rate as the material in question.

Other Solid-State Dosimeters

Photoluminescent dosimeters are similar to thermoluminescent dosimeters except that ultraviolet light rather than heat causes the dosimeters to emit light. Most photoluminescent dosimeters are composed of silver-activated metaphosphate glass of either high-Z or low-Z composition. The response from both types of glass increases rapidly as the energy of incident x or γ rays is reduced. Shields have been designed to reduce the energy dependence of these materials.[27] For exposure rates up to at least 10^8 R/sec, the photoluminescent response of silver-activated metaphosphate glass is independent of exposure rate. Photoluminescent dosimeters are occasionally used for personnel monitoring and other dosimetric measurements.

Radiation-induced changes in the optical density of glasses and plastics have been used occasionally to measure radiation quantity. Changes in the optical density are determined by measuring the transmission of visible or ultraviolet light through the material before and after exposure to radiation. Silver-activated phosphate glass, cobalt-activated borosilicate glass, and transparent plastics have been used as radiation dosimeters.[28] Glass and plastic dosimeters are very insensitive to radiation, and high exposures are required to obtain a measurable response.

To predict the interactions of the vast number of particles produced during nuclear fission, Stanislaw Ulam and John Von Neumann developed a mathematical formalism in 1946 in which the outcomes of individual interactions were determined at random according to the probabilities involved. Large numbers of these "events" were computed, and the results were summed by what was then a relatively new device, the electronic computer. The results were used to predict the overall behavior of the radiations involved in nuclear fission and, later, nuclear fusion. Because of the association of this technique with random outcomes, the developers named it the "Monte Carlo" method, after the famous gaming casino in Monaco.

Ionization Measurements

With designation of the energy absorbed per kilogram of air as E_a, Eq. (6-9) may be rewritten as

$$\Psi = \frac{E_a}{(\mu_{\text{en}})_m}$$

In a unit volume of air, the energy absorbed per kilogram is

$$E_a = \frac{E}{\rho}$$

where E is the energy deposited in the unit volume by impinging x or γ radiation and ρ is the density of air (1.29 kg/m^3 at STP). If J is the number of primary and secondary ion pairs (IPs) produced as a result of this energy deposition, then

$$E = J\,W$$

where $W = 33.85$ eV/IP. The energy fluence Ψ is

$$\Psi = \frac{J\,W}{\rho\,(\mu_{\text{en}})_m} \tag{6-17}$$

Example 6-10

A 1-m^3 volume of air at STP is exposed to a photon fluence of 10^{15} photons per square meter. Each photon possesses 0.1 MeV. The total energy absorption coefficient of air is 2.31×10^{-3} m^2/kg for 0.1-MeV photons. How many ion pairs (IPs) are produced inside and outside of the 1-cm^3 volume? How much charge is measured if all the ion pairs are collected?

$$\begin{aligned}\Psi &= \Phi E \qquad (6\text{-}3')\\ &= \left(10^{15}\,\frac{\text{photons}}{\text{m}^2}\right)\left(0.1\,\frac{\text{MeV}}{\text{photon}}\right)\left(1.6 \times 10^{-13}\,\frac{\text{J}}{\text{MeV}}\right)\\ &= 1.6 \times 10^{1}\,\frac{\text{J}}{\text{m}^2}\\ &= \frac{J\,W}{\rho\,(\mu_{\text{en}})_m}\end{aligned}$$

$$J = \frac{\Psi\rho\,(\mu_{\text{en}})_m}{W} \qquad (6\text{-}18)$$

$$= \frac{(1.6 \times 10^1\ \text{J/m}^2)(1.29\ \text{kg/m}^3)(2.31 \times 10^{-3}\ \text{m}^2/\text{kg})(10^{-6}\ \text{m}^3/\text{cm}^3)(1\ \text{cm}^3)}{(33.8\ \text{eV/IP})(1.6 \times 10^{-19}\ \text{J/eV})}$$

$$= 88 \times 10^8\ \text{ion pairs}$$

The charge Q collected is

$$Q = (88 \times 10^8\ \text{IP})\left(1.6 \times 10^{-19}\ \frac{\text{coulomb}}{\text{IP}}\right)$$

$$= 1.41 \times 10^{-9}\ \text{coulomb}$$

It is not possible to measure all of the ion pairs resulting from the deposition of energy in a small volume of air exposed to x or γ radiation. In particular, secondary ion pairs may escape measurement if they are produced outside the "collecting volume" of air. However, a small volume may be chosen so that energy lost outside the collecting volume by ion pairs created within equals energy deposited inside the collecting volume by ion pairs that originate outside. When this condition of *electron equilibrium* is satisfied, the number of ion pairs collected inside the small volume of air equals the total ionization J. The principle of electron equilibrium is used in the free-air ionization chamber.

Free-Air Ionization Chamber

Free-air ionization chambers are used by standards laboratories to provide a fundamental measurement of ionization in air. These chambers are often used as a reference in the calibration of simpler dosimeters such as thimble chambers described later in this chapter.

X or γ rays incident upon a free-air ionization chamber are collimated by a tungsten or gold diaphragm into a beam with cross-sectional area A (Margin Figure 6-7). Inside the chamber, the beam traverses an electric field between parallel electrodes A and B, with the potential of electrode B highly negative. Electrode A is grounded electrically and is divided into two guard electrodes and a central collecting electrode. The guard electrodes ensure that the electric field is uniform over the collecting volume of air between the electrodes. To measure the total ionization accurately, the range of electrons liberated by the incident radiation must be less than the distance between each electrode and the air volume exposed directly to the x- or γ-ray beam. Furthermore, for electron equilibrium to exist, the photon flux density must remain constant across the chamber, and the distance from the diaphragm to the border of the collecting volume must exceed the electron range. If all of these requirements are satisfied, then the number of ion pairs liberated by the incident photons per unit volume of air is

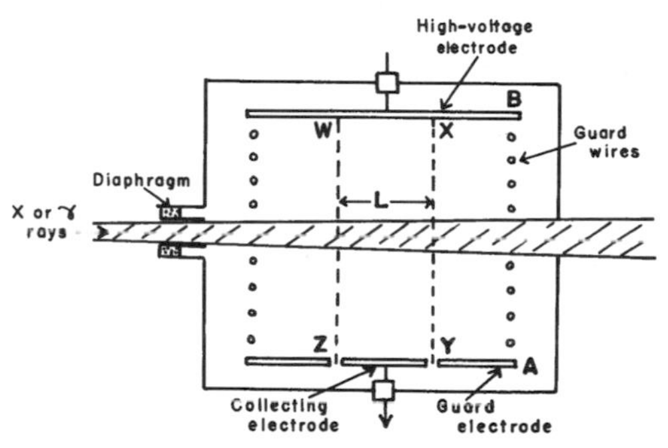

MARGIN FIGURE 6-7
A free-air ionization chamber. The collecting volume of length L is enclosed within the region WXYZ. The air volume exposed directly to the x- or γ-ray beam is depicted by the hatched area.

$$\frac{\text{IP}}{\text{Unit volume}} = \frac{N}{AL}$$

where N is the number of ion pairs collected, A is the cross-sectional area of the beam at the center of the collecting volume, and L is the length of the collecting volume. The charge Q (positive or negative) collected by the chamber is

$$Q = N\left(1.6 \times 10^{-19}\ \frac{\text{coulomb}}{\text{IP}}\right)$$

Because 1 R equals 2.58×10^{-4} coulomb/kg air, the number of roentgens X corresponding to a charge Q in a free-air ionization chamber is

$$X = \left(\frac{1}{2.58 \times 10^{-4}\ \text{coulomb/kg-R}}\right)\frac{Q}{AL\rho} \qquad (6\text{-}19)$$

where ρ is the density of air.

TABLE 6-9 **Range and Percentage of Total Ionization Produced by Photoelectrons and Compton Electrons for X Rays Generated at 100, 200, and 1000 kVp**

X-Ray Tube Voltage (kVp)	*Photoelectrons*		*Compton Electrons*		*Electrode Separation in Free Air Ionization Chamber*
	Range in Air (cm)	*Percentage of Total Ionization*	*Range in Air (cm)*	*Percentage of Total Ionization*	
100	12	10	0.5	90	12 cm
200	37	0.4	4.6	99.6	12 cm
1000	290	0	220	100	4 m

Source: Meredith, W., and Massey, J. *Fundamental Physics of Radiology*. Baltimore, Williams & Wilkins, 1968. Used with permission.

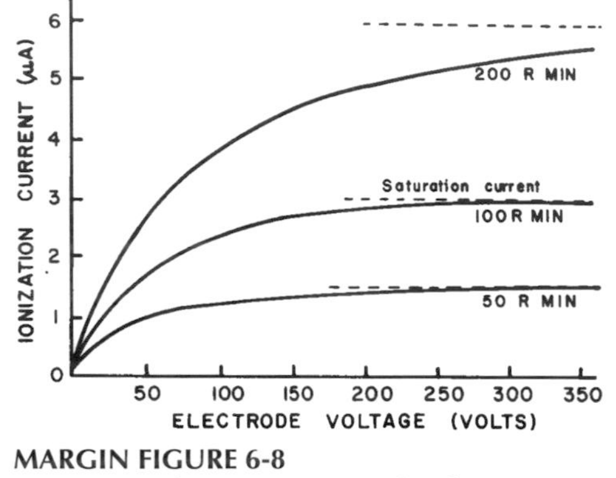

MARGIN FIGURE 6-8
Current in a free-air ionization chamber as a function of the potential difference across the electrodes of the chamber. Saturation currents are shown for different exposure rates. Data were obtained from a chamber with electrodes 1 cm apart. (From Johns, H., and Cunningham, J. *The Physics of Radiology*, 3rd edition. Springfield, IL, Charles C Thomas, 1969. Used with permission.)

To prevent ion pairs from recombining, the potential difference between the electrodes of a free-air ionization chamber must be great enough to attract all ion pairs to the electrodes. A voltage this great is referred to as a *saturation voltage*. Ionization currents in a free-air chamber subjected to different exposure rates are plotted in Margin Figure 6-8 as a function of the potential difference across the electrodes. For any particular voltage, there is an exposure rate above which significant recombination occurs. Unless the potential difference is increased, an exposure rate higher than this limiting value is measured incorrectly because some of the ion pairs recombine and are not collected. Errors due to the recombination of ion pairs may be especially severe during the measurement of exposure rates from a pulsed beam of x rays.[29]

The range of electrons liberated in air increases rapidly with the energy of incident x or γ rays (see Table 6-9). Electrodes would have to be separated by 4 m in a chamber used to measure x rays generated at 1000 kVp. Above 1000 kVp free-air ionization chambers would become very large, and uniform electric fields would be difficult to achieve. Other problems such as a reduction in the efficiency of ion-pair collection are also encountered with chambers designed for high-energy x and γ rays. With some success, chambers containing air at a pressure of several atmospheres have been designed for high-energy photons. Problems associated with the design of free-air ionization chambers for high-energy x and γ rays contribute to the decision to confine the definition of the roentgen to x and γ rays with energy less than about 3 MeV.

A few corrections are usually applied to measurements with a free-air ionization chamber,[30] including the following:

1. Correction for attenuation of the x or γ rays by air between the diaphragm and the collecting volume
2. Correction for the recombination of ion pairs within the chamber
3. Correction for air density and humidity
4. Correction for ionization produced by photons scattered from the primary beam
5. Correction for loss of ionization caused by inadequate separation of the chamber electrodes

With corrections applied, measurements of radiation exposure with a free-air ionization chamber can achieve accuracies to within $\pm$ 0.5%. As mentioned earlier, free-air ionization chambers are used as reference standards for the calibration of x- and γ-ray beams in standards laboratories. Free-air ionization chambers are too fragile and bulky for routine use.

Thimble Chambers

The amount of ionization collected in a small volume of air is independent of the medium surrounding the collecting volume, provided that the medium has an atomic number equal to the effective atomic number of air. Consequently the large distances required for electron equilibrium in a free-air chamber may be replaced by lesser thicknesses of a more dense material, provided that the atomic number is not changed.

The effective atomic number $\overline{Z}$ of a material is the atomic number of a hypothetical element that attenuates photons at the same rate as the material. For photoelectric interactions, the effective atomic number of a mixture of elements is

$$\overline{Z} = \left(a_1 Z_1^{2.94} + a_2 Z_2^{2.94} + \cdots + a_n Z_n^{2.94}\right)^{\frac{1}{2.94}} \qquad \textbf{(6-20)}$$

where, $Z_1, Z_2, \ldots, Z_n$ are the atomic numbers of elements in the mixture and $a_1, a_2, \ldots a_n$ are the fractional contributions of each element to the total number of electrons in the mixture. A reasonable approximation for effective atomic number may be obtained by rounding the 2.94 to 3 in Eq. (6-20).

Example 6-11

Calculate $\overline{Z}$ for air. Air contains 75.5% nitrogen, 23.2% oxygen, and 1.3% argon. Gram-atomic masses are as follows: nitrogen, 14.007; oxygen, 15.999; and argon, 39.948.

Number of electrons contributed by nitrogen to 1 g air

$$= \frac{(1\text{ g})(0.755)(6.02 \times 10^{23}\text{ atoms/gram-atomic mass})(7\text{ electrons/atom})}{14.007\text{ g/gram-atomic mass}}$$

$$= 2.27 \times 10^{23}\text{ electrons}$$

Number of electrons contributed by oxygen to 1 g air

$$= \frac{(1\text{ g})(0.232)(6.02 \times 10^{23}\text{ atoms/gram-atomic mass})(8\text{ electrons/atoms })}{15.999\text{ g/gram-atomic mass}}$$

$$= 0.70 \times 10^{23}\text{ electrons}$$

Number of electrons contributed by argon to 1 g air

$$= \frac{(1\text{ g})(0.013)(6.02 \times 10^{23}\text{ atoms/gram-atomic mass})(18\text{ electrons/atom})}{(39.948\text{ g/gram-atomic mass})}$$

$$= 0.04 \times 10^{23}\text{ electrons}$$

$$\text{Total electrons in 1 g air} = (2.27 + 0.70 + 0.04) \times 10^{23}$$
$$= 3.01 \times 10^{23}\text{ electrons}$$

$$a_1\text{ for nitrogen} = \frac{2.27 \times 10^{23}\text{ electrons}}{3.01 \times 10^{23}\text{ electrons}} = 0.753$$

$$a_2\text{ for oxygen} = \frac{0.70 \times 10^{23}\text{ electrons}}{3.01 \times 10^{23}\text{ electrons}} = 0.233$$

$$a_3\text{ for argon} = \frac{0.04 \times 10^{23}\text{ electrons}}{3.01 \times 10^{23}\text{ electrons}} = 0.013$$

$$\overline{Z}_{\text{air}} = [(0.753)7^{2.94} + (0.233)8^{2.94} + (0.013)18^{2.94}]^{\frac{1}{2.94}}$$
$$= 7.64$$

The large distances in air required for electron equilibrium and collection of total ionization in a free-air chamber may be replaced by smaller thicknesses of "air-equivalent material" with an effective atomic number of 7.64. Chambers with air-equivalent walls are known as *thimble chambers*. Most of the ion pairs collected

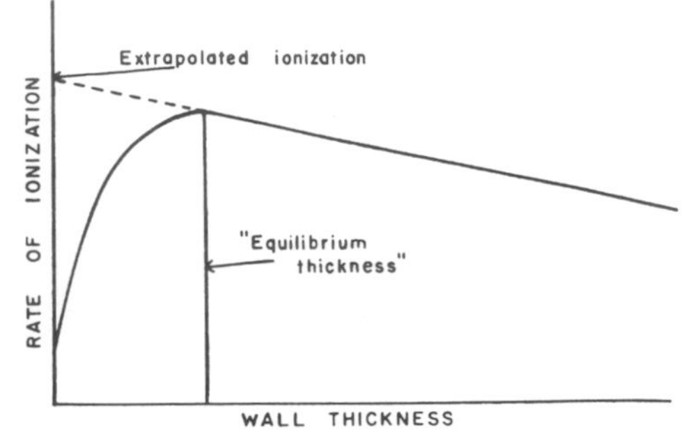

MARGIN FIGURE 6-9
Ionization in an air-filled cavity exposed to ^{60}Co radiation is expressed as a function of the thickness of the air-equivalent wall surrounding the cavity.

within the air volume are produced by electrons released as photons interact with the air-equivalent walls of the chamber. Shown in Margin Figure 6-9 is the ionization in an air-filled cavity exposed to high-energy x rays, expressed as a function of the thickness of the wall surrounding the cavity. Initially, the number of electrons entering the cavity increases with the thickness of the cavity wall. When the wall thickness equals the range of electrons liberated by incident photons, electrons from outer portions of the wall just reach the cavity, and the ionization inside the cavity is maximum. The thickness of the wall should not be much greater than this equilibrium thickness because the attenuation of incident x and γ rays increases with the wall thickness. The equilibrium thickness for the wall increases with the energy of the incident photons. In Margin Figure 6-9, the slow decrease in ionization as the wall thickness is increased beyond the electron range reflects the attenuation of photons in the wall of the chamber. Extrapolation of this portion of the curve to a wall thickness of zero indicates the ionization that would occur within the cavity if photons were not attenuated by the surrounding wall.

A thimble chamber is illustrated schematically in Margin Figure 6-10. Usually the inside of the chamber wall is coated with carbon, and the central positive electrode is aluminum. The response of the chamber may be varied by changing the size of the collecting volume of air, the thickness of the carbon coating, or the length of the aluminum electrode. It is difficult to construct a thimble chamber with a response at all photon energies identical with that for a free-air chamber. Usually the response of a thimble chamber is compared at several photon energies with the response of a free-air ionization chamber. By using the response of the free-air chamber as a standard, calibration factors may be computed for the thimble chamber at different photon energies.

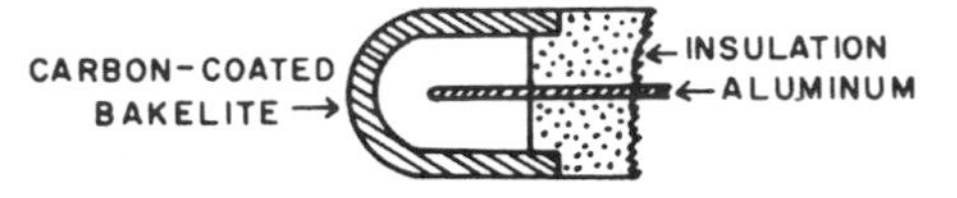

MARGIN FIGURE 6-10
Diagram of a thimble chamber with an air-equivalent wall.

Condenser Chambers

As ion pairs produced by impinging radiation are collected, the potential difference is reduced between the central electrode and the wall of a thimble chamber. If a chamber with volume v is exposed to X coulomb/kg, the charge Q collected is

$$Q(\text{coulomb}) = X\left(\frac{\text{coulomb}}{\text{kg}}\right)\rho\left(\frac{\text{kg}}{\text{m}^3}\right)v(\text{m}^3)$$

If the density of air is 1.29 kg/m^3, we obtain

$$Q = 1.29Xv$$

For a chamber with capacitance C in farads, the voltage reduction V across the chamber is given by

$$V = \frac{Q}{C} = \frac{1.29Xv}{C}$$

The voltage reduction per unit exposure is

$$\frac{V}{X} = \frac{1.29v}{C}$$

where X is the exposure in coulombs/kg and v is the volume of the chamber in cubic meters. The voltage drop per unit exposure, defined as the *sensitivity* of the chamber, may be reduced by decreasing the chamber volume v or by increasing the chamber capacitance C.

If an electrometer with capacitance C_e is used to measure the charge distributed over a condenser chamber with total capacitance C, then the sensitivity of the chamber may be defined as

$$\frac{V}{X} = \frac{1.29v}{C + C_e} \tag{6-21}$$

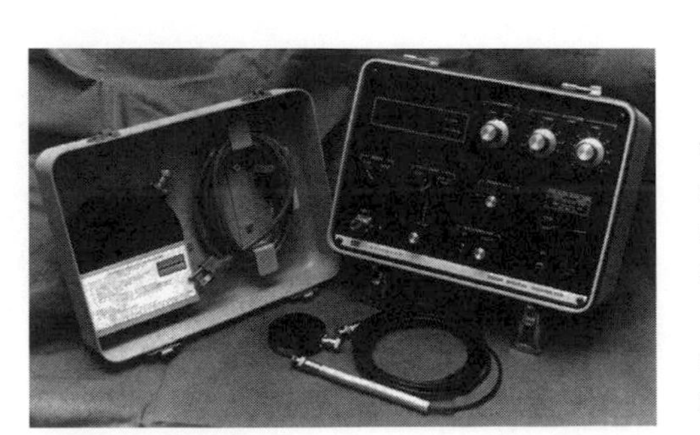

MARGIN FIGURE 6-11
A Keithley model 35020 digital dosimetry system. (Courtesy of Keithley Instruments, Inc.)

The sensitivity of most condenser chamber-electrometer units is defined in this manner.

Example 6-12

The capacitance of a 100-R condenser chamber is 50 picofarads. The capacitance of the charger-reader used with the chamber is 10 picofarads. The volume of the chamber is 0.46 cm^3. What is the sensitivity of the chamber? What reduction in voltage occurs when the chamber is exposed to 1 C/kg?

$$\text{Sensitivity} = \frac{V}{X} = \frac{1.29v}{C + C_e}$$

$$= \frac{1.29(0.046\ \text{cm}^3)(10^{-6}\ \text{m}^3/\text{cm}^3)}{(50 + 10) \times 10^{-12}\ \text{farads}}$$

$$= 989\ \frac{\text{volts}}{\text{C/kg}}$$

$$\text{Sensitivity} \simeq 1000\ \frac{\text{volts}}{\text{C/kg}}$$

Many types of exposure-measuring devices are available in which the chamber remains connected to an electrometer, a device for measuring small electric currents or charge by a shielded cable during exposure. While in the x-ray beam, the chamber may be made sensitive to radiation for a selected interval of time, and the exposure rate or cumulative exposure may be determined. A typical device of this type is illustrated in Margin Figure 6-11.

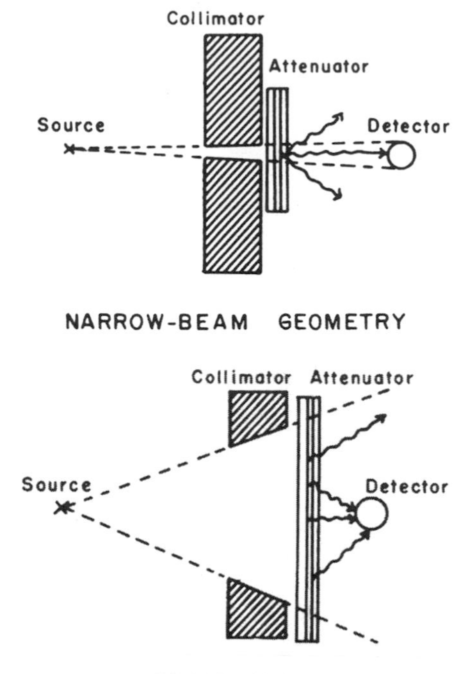

MARGIN FIGURE 6-12
Narrow-beam geometry is achieved by using a narrow beam and a large distance between the attenuator and the chamber used to measure radiation exposure. With narrow-beam geometry, few photons are scattered into the chamber by the attenuator. A significant number of photons are scattered into the chamber when attenuation measurements are made with broad-beam geometry.

Radiation Quality

An x-ray beam is not described completely by stating the exposure or dose it delivers to a small region within an irradiated medium. The penetrating ability of the radiation, often described as the *quality* of the radiation, must also be known before estimates may be made of the following:

1. Exposure or dose rate at other locations within the medium
2. Differences in energy absorption between regions of different composition
3. Biological effectiveness of the radiation

The spectral distribution of an x-ray beam depicts the relative number of photons of different energies in the beam. The quality of an x-ray beam is described explicitly by the spectral distribution. However, spectral distributions are difficult to measure or compute and are used only rarely to describe radiation quality. Usually, the quality of an x-ray beam is described by stating the half-value layer (HVL) of the beam, sometimes together with the potential difference (kilovolts peak) across the x-ray tube. Although other methods have been developed to describe the quality of an x-ray beam, the HVL is adequate for most clinical applications of x radiation.

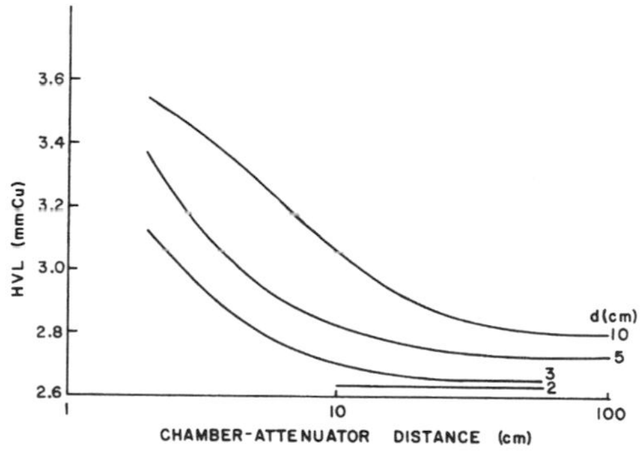

MARGIN FIGURE 6-13
Variation of the HVL with the diameter d of an x-ray beam at the attenuator for various distances between the attenuator and the chamber. The x-ray beam was generated at 250 kVp and filtered by 2 mm Cu. The attenuator was placed 20 cm below the x-ray tube. The geometry improves with decreasing diameter of the x-ray beam and increasing distance between attenuator and chamber. (From International Commission on Radiological Units and Measurements. *Physical Aspects of Irradiation*. National Bureau of Standards Handbook 85, 1964. Used with permission.)

HALF-VALUE LAYER

The HVL of an x-ray beam is the thickness of a material that reduces the exposure rate of the beam to half. Although the HVL alone furnishes a description of radiation quality that is adequate for most clinical situations, the value of a second parameter

Half-value layer (HVL) is sometimes referred to as half-value thickness (HVT).

The homogeneity coefficient of an x-ray beam is the quotient of the thickness of the attenuator required to reduce the exposure rate to half divided by the thickness of attenuator required to reduce the exposure rate from half to one-fourth. The homogeneity coefficient is the ratio of the first and second HVLs.

(e.g., the peak potential difference across the x-ray tube or the homogeneity coefficient of the x-ray beam) is sometimes stated with the HVL. Once the HVL of an x-ray beam is known together with the exposure or dose rate at a particular location, the absorbed dose rate may be computed at other locations within an irradiated medium.

HVLs are measured with solid "absorbers" (more correctly, attenuators) such as thin sheets of aluminum, copper, or lead of uniform thickness. HVLs may be measured by placing attenuators of different thickness but constant composition between the x-ray tube and an ionization chamber designed to measure radiation exposure. The distribution of photon energies changes as attenuators are added to an x-ray beam, and the response of the chamber should be independent of these changes. Measurements of HVL always should be made with narrow-beam geometry. Narrow-beam geometry (sometimes referred to as *"good" geometry*) ensures that the only photons that enter the chamber are primary photons that have been transmitted by the attenuator. Requirements for narrow-beam geometry are satisfied if the chamber is positioned far from the attenuator, and an x-ray beam with a small cross-sectional area is measured. The cross-sectional area of the beam should be just large enough to deliver a uniform exposure over the entire sensitive volume of the chamber. Conditions of narrow-beam and broad-beam "poor" geometry are depicted in Margin Figure 6-12. Measurements of HVL under these conditions are compared in Margin Figure 6-13. With broad-beam geometry, more photons are scattered into the detector as the thickness of the attenuator is increased. Consequently, HVLs measured with broad-beam geometry are greater than those measured under conditions of narrow-beam geometry. HVLs measured under conditions of broad-beam geometry are used for the calculation of shielding requirements for walls surrounding x-ray equipment.

Narrow-beam geometry may be achieved by placing the attenuator midway between the x-ray target and the chamber, with the chamber at least 0.5 m from the attenuator. The area of the field should be no larger than a few square centimeters. Objects that might scatter photons into the chamber should be removed from the vicinity of the x-ray beam. Even greater accuracy can be attained by measuring the HVL for fields of different cross-sectional area and extrapolating the measured HVLs to a field size of zero.[31] However, this procedure is not necessary for routine clinical dosimetry.

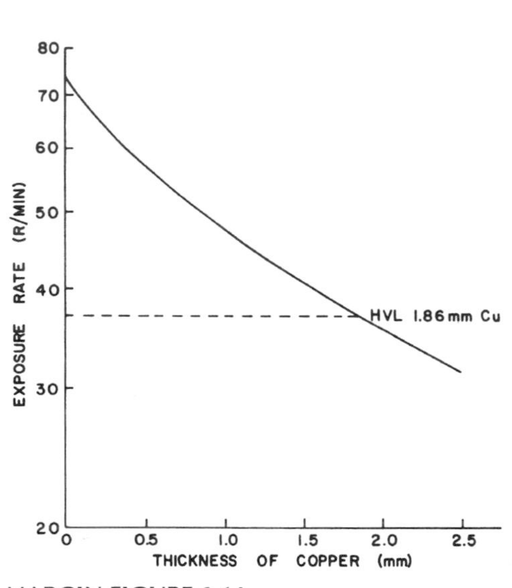

MARGIN FIGURE 6-14
Exposure rate for an x-ray beam as a function of the thickness of copper added to the beam. The HVL of the beam is 1.86 mm Cu.

An attenuation curve for x rays in copper is shown in Margin Figure 6-14. The HVL is 1.86 mm Cu for the x-ray beam described in the figure. A complete attenuation curve is not required for routine dosimetry. Instead, thicknesses of the attenuator are identified that reduce the exposure rate to slightly more than half (50% to 55%) and slightly less than half (45% to 50%). These data are plotted on a semilogarithmic graph and connected by a straight line. The approximate HVL is given by the intersection of this straight line with a horizontal line drawn through the ordinate (y axis) at half the exposure rate measured with no attenuator added to the beam.

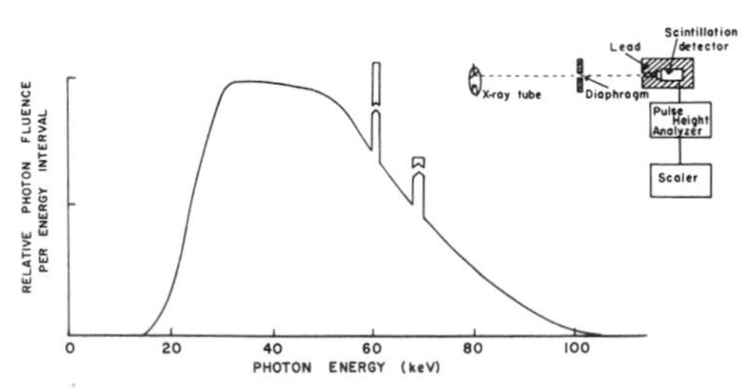

MARGIN FIGURE 6-15
Energy distribution of x-ray photons in a beam generated at 105 kVp and filtered by 2 mm Al. Equipment used to measure the spectra of primary x rays is diagramed in the inset. (From Epp, E., and Weiss, H. *Phys. Med. Biol.* 1966; **11**:225. Used with permission.)

VARIATION IN QUALITY ACROSS AN X-RAY BEAM

The exposure rate on the side of an x-ray beam nearest the anode is less than the exposure rate on the cathode side because x-rays on the anode side are attenuated by a greater thickness of target material before they emerge from the anode. This variation in exposure rate across an x-ray beam is termed the *heel effect* (see Chapter 4). The greater filtration on the anode side increases the HVL of the x-ray beam from the cathode to the anode side. The decrease in exposure rate from cathode to anode across an x-ray beam is less noticeable at depths within a patient because the radiation on the anode side is more penetrating.

SPECTRAL DISTRIBUTION OF AN X-RAY BEAM

The spectral distribution for a beam of x or γ rays may be computed from an attenuation curve measured under conditions of narrow-beam geometry. A variety of curve-fitting techniques may be applied to the attenuation curve to obtain equations that are used to compute the spectral distribution. The accuracy of this method of obtaining a spectral distribution is limited by the accuracy of the curve-fitting technique and the accuracy with which the attenuation curve is measured.

Most measurements of x- and γ-ray spectra are made with a scintillation or semiconductor detector. The height of a voltage pulse from one of these detectors varies with the energy deposited in the detector by an x or γ ray. The pulses are sorted by height in a pulse height analyzer and counted in a scaler. The recorded counts are plotted as a function of pulse height to furnish a *pulse-height distribution*, which reflects the energy distribution of photons impinging upon the detector. To portray the energy distribution accurately, the pulse-height distribution must be corrected for statistical fluctuations in the energy distribution and for incomplete absorption of photons in the detector. Measured spectra for primary and scattered x rays are shown in Margin Figures 6-15 and 6-16.

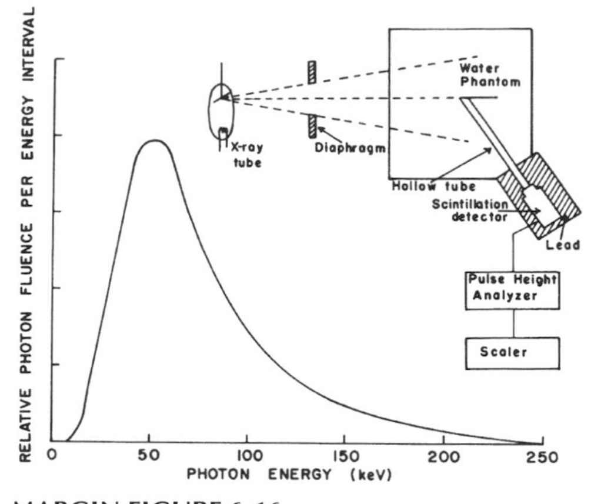

MARGIN FIGURE 6-16
Energy distribution of scattered radiation at 15-cm depth in a water phantom exposed to a 200-cm^2 x-ray beam generated at 250 kVp. Equipment used to measure the spectra of scattered x rays is diagramed in the inset. (From Hettinger, G., and Liden, K. *Acta Radiol.* 1960; **53**:73. Used with permission.)

PROBLEMS

*6-1. The photon flux density is 10^9 photons/(m^2-sec) for a beam of γ rays. One-fourth of the photons have an energy of 500 keV, and three-fourths have an energy of 1.25 MeV. What is the energy flux density of the beam? If the photon flux density is constant, what is the energy fluence over an interval of 10 seconds?

*6-2. During an exposure of 0.02 coulomb/kg, how much energy is absorbed per kilogram of air?

*6-3. The energy absorption coefficient of air is 2.8×10^{-3} m^2/kg for photons of 1.0 MeV. What is the energy fluence required for an exposure of 0.03 coulomb/kg? What is the photon fluence for this exposure?

*6-4. How many ion pairs are produced in 1 cm^3 of air during an exposure of 0.03 coulomb/kg?

*6-5. Water is 89% oxygen (gram-atomic mass, 15.999) and 11% hydrogen (gram-atomic mass, 1.008) by weight. Compute the effective atomic number of water.

*6-6. A thimble chamber with an air-equivalent wall receives an exposure of 0.015 coulomb/kg in 1 minute. The volume of the chamber is 0.46 cm^3. What is the ionization current from the chamber?

*6-7. A condenser ionization chamber has a sensitivity of 7750 V/(C-kg^{-1}). The volume of the chamber is 0.46 cm^3. The capacitance of the chamber is six times the capacitance of the charger-reader. What is the capacitance of the chamber?

*6-8. A miniature ionization chamber has a sensitivity of 3876 V/(C-kg^{-1}). The chamber is discharged by 100 V during an exposure to x radiation. What exposure did the chamber receive?

6-9. A Victoreen condenser chamber receives identical exposures in New Orleans (sea level) and in Denver (5000 ft above sea level). Is the deflection of the platinum wire lower or higher in Denver? Why?

6-10. What is meant by the energy dependence of an air-wall thimble chamber?

*6-11. An organ with a mass of 10 g receives a uniform dose of 10 cGy. How much energy is absorbed in each mass of tissue in the tumor, and how much energy is absorbed in the entire mass?

*6-12. A particular type of lesion recedes satisfactorily after receiving 55 Gy from orthovoltage x rays. When the lesion is treated with ^{60}Co γ radiation, a dose of 65 Gy is required. Relative to orthovoltage x rays, what is the RBE of ^{60}Co radiation for treating this type of lesion?

*6-13. An individual who ingests a radioactive sample receives an estimated dose of 15 mGy to the testes: 11 mGy is delivered by β-particles, and 4 mGy is delivered by γ-radiation. What is the dose to the testes in millisievert?

*For problems marked with an asterisk, answers are provided on p. 491.

*6-14. The specific heat of graphite is 170 cal/kg°C. A uniform dose of 10 Gy is delivered to a graphite block insulated from its environment. What is the rise in temperature of the block?

*6-15. With the assumption that $G = 15.4$, how many Fe^{2+} ions are oxidized if an absorbed dose of 150 Gy is delivered to a 10-mL solution of ferrous sulfate? Assume that the density of ferrous sulfate is 10^{-3} kg/mL.

6-16. Explain clearly what is meant by the energy dependence of a radiation dosimeter.

*6-17. Attenuation measurements for an x-ray beam from a 120-kVp x-ray generator yield the following results:

Added Filtration (mm Al)	Percent Transmission
1.0	60.2
2.0	41.4
3.0	30.0
4.0	22.4
5.0	16.9

Plot the data on semilogarithmic graph paper and determine the following:

a. The first HVL
b. The second HVL
c. The homogeneity coefficient of the x-ray beam.

SUMMARY

- Intensity is energy per unit area per unit time.
- Fluence is particles, photons or energy per unit area.
- Flux is time rate of change of fluence.
- Exposure is a measure of the charge liberated per unit mass of air by ionizing radiation.
- Dose is energy absorbed in any material per unit mass.
- Calculation of RBE dose includes a conversion factor that is based upon experiment.
- Dose equivalent includes a quality factor that accounts for differences in biological harm caused by equal doses of different types of radiation. It assumes that the dose to a point in the organ is known.
- Equivalent dose is similar to dose equivalent, but applies slightly different weighting factors and uses the average dose to an organ or body part.
- Effective dose equivalent used weighting factors to account for biological harm to the individual and genetic effects in the next two generations caused by irradiation of different body regions. It has been superseded by the quantity known as equivalent dose.
- Effective dose converts average organ and body region doses to a measure of overall detriment to the individual and to future generations through genetic effects. It also takes into account the type of radiation received. It is based upon radiation weighting factors W_r and tissue weighting factors W_t that were derived in the early 1990s from analysis of lifetime follow-up of atom bomb survivors.
- Quantities and their units are given below:

Quantity	SI Unit	Traditional Unit
Exposure	coulomb/kg	roentgen
Dose	gray (Gy)	rad
RBE dose	sievert (SV)	rem
Dose equivalent	sievert (SV)	rem
Equivalent dose	sievert (SV)	rem
Effective dose equivalent	sievert (SV)	rem
Effective dose	sievert (SV)	rem

- Methods of measurement of radiation dose include:
  - Calorimetry
  - Photographic methods
  - Chemical
  - Scintillation
  - Thermoluminescence
  - Solid-state detectors
  - Ionization chambers

REFERENCES

1. Glasser, O. *Wilhelm Conrad Roentgen and the Early History of the Roentgen Ray.* Springfield, IL, Charles C Thomas, 1934.
2. International Commission on Radiation Units and Measurements: *The Quality Factor in Radiation Protection.* ICRU Report 40. Bethesda, MD, 1986.
3. Wyckoff, HO, Allisy, A, and Lidén, K. The new special names of SI units in the field of ionizing radiations. *AJR* 1975; **125**:492.
4. Paic, G. Physical Phenomena Linked with the Interactions of Radiation with Matter, in Paic, G. (ed.), *Ionizing Radiation: Protection and Dosimetry.* Boca Raton, FL, CRC Press, 1988, pp. 2–13.
5. Berger, R. The x- or gamma-ray energy absorption or transfer coefficient: Tabulation and discussion. *Radiat. Res.* 1961; **15**:1.
6. International Commission on Radiation Units and Measurements: *Radiation Quantities and Units,* ICRU Report 11. Washington, D.C., U.S. Government Printing Office, 1968.
7. Storer, J., et al. Relative biological effectiveness of various ionizing radiations in mammalian systems. *Radiat. Res.* 1957; **6**:188.
8. Hall, E. Relative biological efficiency of x rays generated at 200 kVp and gamma radiation from cobalt 60 therapy unit. *Br. J. Radiol.* 1961; **34**:313.
9. Sinclair, W. Relative biological effectiveness of 22-MeVp x rays, cobalt 60 gamma rays and 200 kVp rays: 1. General introduction and physical aspects. *Radiat. Res.* 1962; **16**:336.
10. Loken, M., et al. Relative biological effectiveness of ^{60}Co gamma rays and 220 kVp x rays on viability of chicken eggs. *Radiat. Res.* 1960; **12**:202.
11. Krohmer, J. RBE and quality of electromagnetic radiation at depths in water phantom. *Radiat. Res.* 1965; **24**:547.
12. Upton, A., et al. Relative biological effectiveness of neutrons, x rays, and gamma rays for production of lens opacities; observations on mice, rats, guinea pigs and rabbits. *Radiology.* 1956; **67**:686.
13. Focht, E., et al. The relative biological effectiveness of cobalt 60 gamma and 200 kV x radiation for cataract induction. *Am. J. Roentgenol.* 1968; **102**:71.
14. Till, J. and Cunningham, J. unpublished data, cited by Johns, H. and Cunningham, J. (eds). *The Physics of Radiology* (3rd ed., Springfield, IL, Charles C. Thomas, Publisher, 1969), p. 720.
15. Recommendations of the International Commission on Radiological Protection: *Br. J. Radiol.* 1955; Suppl. 6.
16. United Nations Scientific Committee on the Effects of Atomic Radiation. *Sources and Effects of Ionizing Radiation*, Mo. E.77.IX.1, United Nations Publications, New York, 1977.
17. National Council on Radiation Protection and Measurements. *Use of Personal Monitors to Estimate Effective Dose Equivalent and Effective Dose to Workers for External Exposure to Low Level Radiation*, NCRP Report No. 122, National Council on Radiation Protection and Measurements, Bethesda, Maryland, 1995.
18. United Nations Scientific Committee on the Effects of Atomic Radiation. *Sources, Effects and Risks of Ionizing Radiation*, Mo. E.88.IX.7, United Nations Publications, New York, 1988.
19. National Academy of Sciences/National Research Council, Committee on the Biological Effects of Ionizing Radiations. *Health Effects of Exposure to Low Levels of Ionizing Radiations*, BEIR V, National Academy Press, New York, 1990.
20. Laughlin, J., and Genna, S. Calorimetry, in Attix, F., and Roesch, W. (eds.), *Radiation Dosimetry,* Vol 2, 2nd edition. New York, Academic Press, 1968, p. 389.
21. Ehrlich, M. *Photographic Dosimetry of X and Gamma Rays.* National Bureau of Standards Handbook 57, 1954.
22. Fricke, H., and Morse, S. The actions of x rays on ferrous sulfate solutions. *Philos. Mag.* 1929; **7**:129.
23. International Commission on Radiological Units and Measurements: *Physical Aspects of Irradiation, Recommendations of the ICRU.* National Bureau of Standards Handbook 57, 1954.
24. The Subcommittee on Radiation Dosimetry (SCRAD) of the American Association of Physicists in Medicine. Protocol for the dosimetry of high energy electrons. *Phys. Med. Biol.* 1966; **11**:505.
25. Cameron, J., et al. Thermoluminescent radiation dosimetry utilizing LiF. *Health Phys.* 1964; **10**:25.
26. Tochilin, E., and Goldstein, N. Energy and dose rate response of five dosimeter systems. *Health Phys.* 1964; **10**:602.
27. Thornton, W., and Auxier, J. Some x-ray and fast neutron response characteristics of Ag-metaphosphate glass dosimeters, ORNL-2912, 1960.
28. Kreidl, N., and Blair, G. Recent developments in glass dosimetry. *Nucleonics* 1956; **14**:82.
29. Johns, H., and Cunningham, J. *The Physics of Radiology*, 4th edition. Springfield, IL, Charles C Thomas, 1983.
30. Wyckoff, H., and Attix, F. *Design of Free-Air Ionization Chambers.* National Bureau of Standards Handbook 64, 1957.
31. Trout, E., and Kelley, J., and Lucas, A. Determination of half-value layer. *AJR* 1960; **84**:729.

CHAPTER

7

INTERACTION OF X AND γ RAYS IN THE BODY

OBJECTIVES

By studying this chapter, the reader should be able to:

- Explain the origin of the f factor, and determine the dose (Gy) to a medium from knowledge of the exposure (C/kg).
- Describe the attenuation of x rays in different tissues as a function of x-ray energy.
- Discuss the properties of various tissues that permit their distinction in x-ray images.
- Define changes in radiation dose at the interface between bone and soft tissue.
- Identify the applications of high-voltage and low-voltage radiography.
- Characterize the properties and applications of contrast media.

INTRODUCTION

The dominant mode of interaction of x and γ rays in a region of the body varies with the energy of the photons and the effective atomic number and electron density (electrons/kilogram) of the region. For the purpose of discussing interactions, the body may be divided into regions of (1) fat, (2) muscle (or soft tissue excluding fat), (3) bone, and (4) air-filled cavities.

D_{air} was originally defined as $D_{air} = 0.869X$, where D_{air} was expressed in rads and the exposure X was expressed in roentgens. One roentgen was defined as "the amount of x or γ rays that produces 2.58×10^{-4} coulombs of charge of either sign in 1 kg of dry air." The roentgen, the earliest quantitative measure of radiation, is no longer sanctioned by international scientific commissions as an official unit of exposure.

F FACTOR

An exposure X of 1 coulomb/kilogram (C/kg) provides an absorbed dose D_{air} in air of 33.85 gray (Gy) (see Chapter 6):

$$D_{air}(\text{Gy}) = 33.85X \ (\text{C/kg})$$

The dose to a medium such as soft tissue is related to the dose in air at the same location multiplied by the ratio of energy absorption in the medium to that in air:

$$D_{med} = D_{air}\frac{[(\mu_{en})_m]_{med}}{[(\mu_{en})_m]_{air}} \tag{7-1}$$

$$D_{med}(\text{Gy}) = 33.85X\,\frac{[(\mu_{en})_m]_{med}}{[(\mu_{en})_m]_{air}} \tag{7-2}$$

with X in units of C/kg.

In Eq. (7-1), $[(\mu_{en})_m]_{med}$ is the mass energy absorption coefficient of the medium for photons of the energy of interest. This coefficient describes the rate of energy absorption in the medium. The expression $[(\mu_{en})_m]_{air}$ describes the rate of energy absorption in air for photons of the same energy. In Eq. (7-2), the expression may be simplified to

$$D_{med}(\text{Gy}) = (f)(X) \tag{7-3}$$

where

$$f = 33.85\frac{[(\mu_{en})_m]_{med}}{[(\mu_{en})_m]_{air}} \tag{7-4}$$

The expression denoted as f is known as the f factor. This factor varies with the nature of the absorbing medium and the energy of the radiation.

The f factor is used to compute the absorbed dose D in gray in a medium receiving an exposure X in coulombs per kilogram. The f factor is plotted in Figure 7-1 for air, fat, muscle, and compact bone as a function of photon energy.

MARGIN FIGURE 7-1
(Top) Photograph of Wilhelm Conrad Röntgen as a student. **(Bottom)** Röntgen in 1906 while director of the Institute of Physics at the University of Munich.

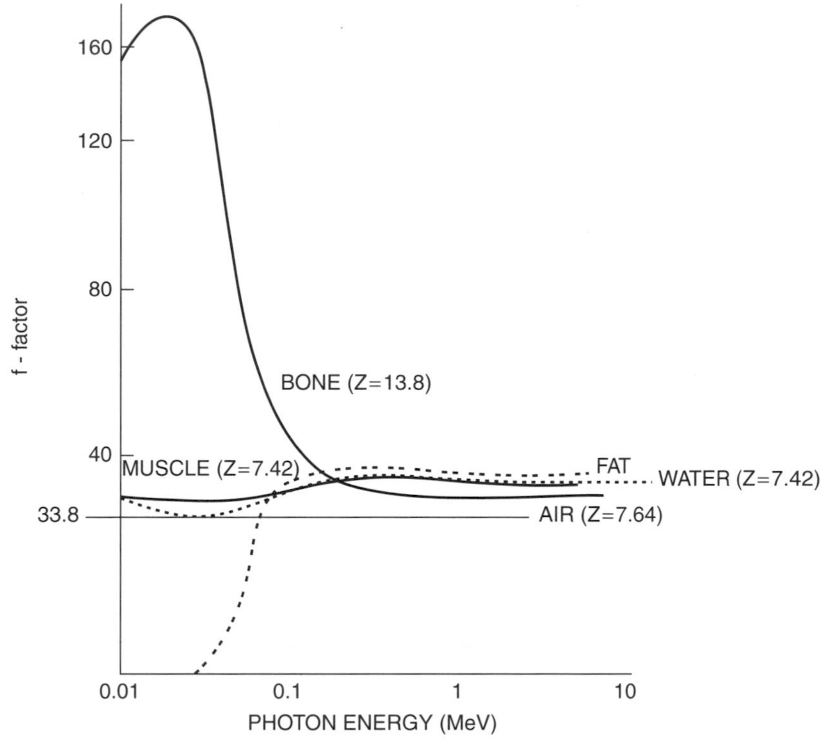

FIGURE 7-1
The f factor for conversion between exposure (C/kg) and absorbed dose (Gy) for air, water, and different constituents of the body, plotted as a function of photon energy. Curves do not extend beyond 3 MeV because radiation exposure (and therefore the f factor) is not applicable to photons of higher energy. Because of the varying composition of fat, the dotted curve for this tissue constituent is only approximately correct.

This plot illustrates why image contrast is reduced at higher photon energies where interactions are primarily Compton scattering rather than photoelectric absorption.

For all photon energies for which exposure (and therefore the f factor) is defined, an exposure of 1 C/kg provides an absorbed dose of 33.85 Gy in air. Consequently, the f factor equals 33.85 for air and is independent of photon energy.

ATTENUATION OF X AND γ RAYS IN TISSUE

A simplified model of the human body consists of three different body tissues: fat, muscle, and bone. Air is also present in the lungs, sinuses and gastrointestinal tract, and a contrast agent may be used to accentuate the attenuation of x rays in a particular region. The elemental composition of the three body tissues, together with their percent mass composition, are shown in Table 7-1. Selected physical properties of the tissues are included in Table 7-2. Mass attenuation coefficients for different tissues as a function of photon energy are shown in Figure 7-2.

In Table 7-2, the data for muscle are also approximately correct for other soft tissues such as collagen, internal organs (e.g., liver and kidney), ligaments, blood, and cerebrospinal fluid. These data are also close to the data for water, because soft tissues, including muscle, are approximately 75% water, and body fluids are 85% to 100% water. The similarity of soft tissues suggests that conventional x-ray imaging yields poor discrimination among them. Sometimes a contrast agent can be used to accentuate the small intrinsic differences in x-ray attenuation among soft tissues.

Compared with other tissue constituents, fat has a greater concentration of low-Z elements, especially hydrogen. Therefore, fat has a lower density and effective atomic number compared with muscle and other soft tissues. Below about 35 keV, x rays interact in fat and other soft tissues predominantly by photoelectric interactions. These

MARGIN FIGURE 7-2
First x-ray "movie" showing 5 views of a frog's leg filmed in 1897.[1]

MARGIN FIGURE 7-3
Advertisement for x-ray studio.[3]

TABLE 7-1 Percent Mass Composition of Tissue Constituents[4-5]

% Composition (by Mass)	*Adipose Tissue*	*Muscle (Striated)*	*Water*	*Bone (Femur)*
Hydrogen	11.2	10.2	11.2	8.4
Carbon	57.3	12.3		27.6
Nitrogen	1.1	3.5		2.7
Oxygen	30.3	72.9	88.8	41.0
Sodium		0.08		
Magnesium		0.02		7.0
Phosphorus		0.2		7.0
Sulfur	0.06	0.5		0.2
Potassium		0.3		
Calcium		0.007		14.7

interactions vary with Z^3 of the tissue. This dependence on Z yields modest image contrast among tissues of slightly different composition (e.g., fat and muscle) when low-energy x rays are used. The contrast disappears with higher-energy x rays that interact primarily by Compton interactions, since these interactions do not vary with Z. Low- energy x rays are used to accentuate subtle differences in soft tissues (e.g., fat and other soft tissues) in applications such as breast imaging (mammography) where the object (the breast) provides little intrinsic contrast. When images are desired of structures with high intrinsic contrast (e.g., the chest where bone, soft tissue, and air are present), higher-energy x rays are used. These x rays suppress x-ray attenuation in bone which otherwise would create shadows in the image that could hide underlying soft-tissue pathology.

When compared with muscle and bone, fat has a higher concentration of hydrogen (~11%) and carbon (~57%) and a lower concentration of nitrogen (~1%), oxygen (30%), and high-Z trace elements (<1%) (Table 7-1).[4,5] Hence, the effective atomic number of fat ($Z_{eff} = 5.9$ to 6.3) is less than that for soft tissue ($Z_{eff} = 7.4$) or bone ($Z_{eff} = 11.6$ to 13.8). Because of its lower Z_{eff}, low-energy photons are attenuated less rapidly in fat than in an equal mass of soft tissue or bone. The reduced attenuation in fat yields a lower f factor for low-energy photons in this body constituent (Figure 7-1). The attenuation of x and γ rays in fat may be estimated from attenuation measurements in mineral oil or polyethylene because the effective atomic numbers, densities, and electron densities of these materials are close to those for fat.[6–8]

As described earlier, the common form of hydrogen (protium) contains no neutrons; hence atoms of this form of hydrogen have one electron for each nucleon (proton). For most other nucleons, including other forms (isotopes) of hydrogen, the ratio of electrons to nucleons (protons and neutrons) is 1:2 or less.

X and γ rays of higher energy interact primarily by Compton scattering with a probability that varies with the electron density of the attenuating medium but not with the atomic number. The electron density of hydrogen is about twice that of other elements. Because more hydrogen is present in fat than in other body constituents, more Compton interactions occur in fat than in an equal mass of muscle or bone. For photons of intermediate energy, therefore, the f factor for fat exceeds that for other body constituents (Figure 7-1).

Hydrogen is absent from air but contributes about 10% of the weight of muscle. Consequently, the electron density is greater for muscle than for air, and the f factor for muscle exceeds that for air (Figure 7-1).

TABLE 7-2 Properties of Tissue Constituents of the Human Body

Material	*Effective Atomic Number*	*Density (kg/m³)*	*Electron Density (electrons/kg)*
Air	7.6	1.29	3.01×10^{26}
Water	7.4	1.00	3.34×10^{26}
Muscle	7.4	1.00	3.36×10^{26}
Fat	5.9–6.3	0.91	$3.34\text{–}3.48 \times 10^{26}$
Bone	11.6–13.8	1.65–1.85	$3.00\text{–}3.10 \times 10^{26}$

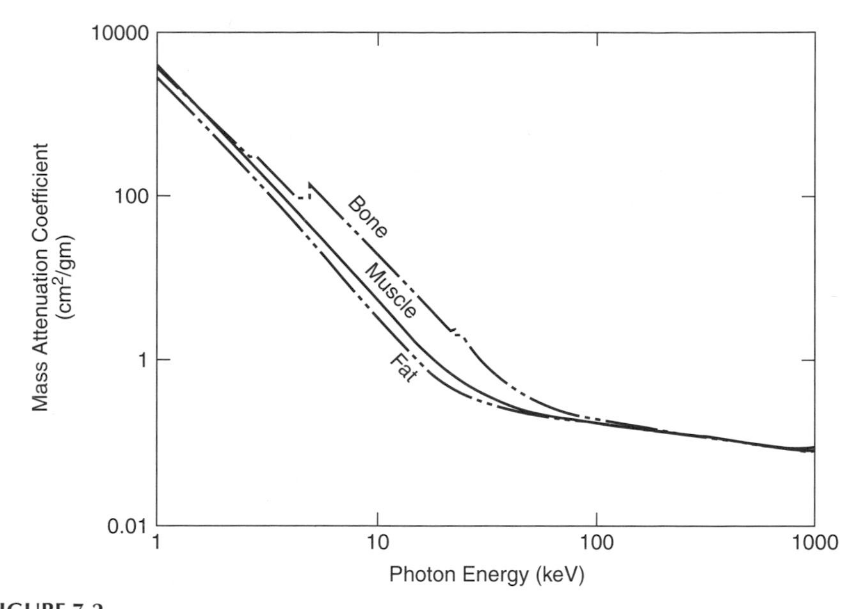

FIGURE 7-2
Mass attenuation coefficients of tissues.[2]

The effective atomic number and physical density are greater for bone than for soft tissue. Hence, x and γ rays are attenuated more rapidly in bone than in an equal volume (not necessarily mass) of soft tissue. This effect reduces the absorbed dose to structures beyond bone. The absorbed dose to soft tissue immediately adjacent to or enclosed within bone may be increased by photoelectrons liberated as photons interact with high-Z atoms (e.g., phosphorus and calcium) in bone.

Compared with muscle and fat, bone contains less hydrogen and therefore its electron density is slightly less as well. For this reason, the *energy absorbed per gram* of bone is slightly less than the energy absorbed per gram of muscle or fat exposed to photons of intermediate energy. As shown in Figure 7-1 the f factor for bone is less than the f-factor for either muscle or fat and intermediate photon energies. However, the physical density of compact bone is almost twice the density of fat or muscle. Therefore, the *energy absorbed per unit volume* of compact bone is almost twice that absorbed in an equal volume of fat or muscle exposed to x and γ rays of intermediate energy.

In a radiograph obtained by exposure of film to high-energy photons (e.g., 4 MV x rays), images of bone are displayed as regions of reduced optical density (i.e., they are more transparent to visible light). Hence the number of photons transmitted by bone is less than the number transmitted by an equal thickness of soft tissue. Although the energy absorbed per unit mass (absorbed dose) is less in bone than in soft tissue exposed to high-energy x rays, the transmission of photons through bone is less because the physical density of bone is greater than that for soft tissue.

DOSE TO SOFT TISSUE BEYOND BONE

The dose delivered to soft tissue by x and γ rays is reduced by bone interposed between the soft tissue and the surface. The reduction in absorbed dose to soft tissue is influenced by:

1. Increased attenuation of primary photons in bone, caused by the higher atomic number and density of this tissue constituent.
2. Changes in the amount of radiation scattered to soft tissue beyond the bone, which depend on many factors, including field size, quality of radiation, and the distance between the bone and the soft tissue.

The effect of bone upon the radiation exposure and absorbed dose at various depths within a patient exposed to diagnostic x rays is illustrated in the margin.[9] The radiation exposure is reduced at locations B, C, and D by bone interposed between the locations and the surface. The reduction primarily reflects the increased attenuation of photons in overlying bone. The dose absorbed in bone is increased at A, B, and C because the attenuation of diagnostic x rays increases with the atomic

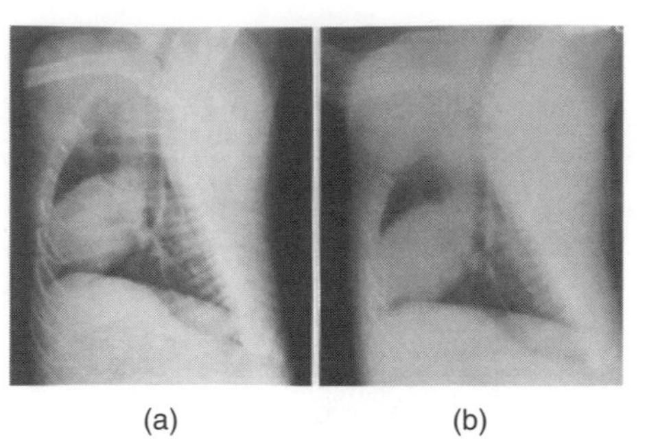

FIGURE 7-3
Radiographs of the chest. **A:** 80 kVp, 1-mm Al filter. **B:** 140 kVp, 1-mm Cu filter.

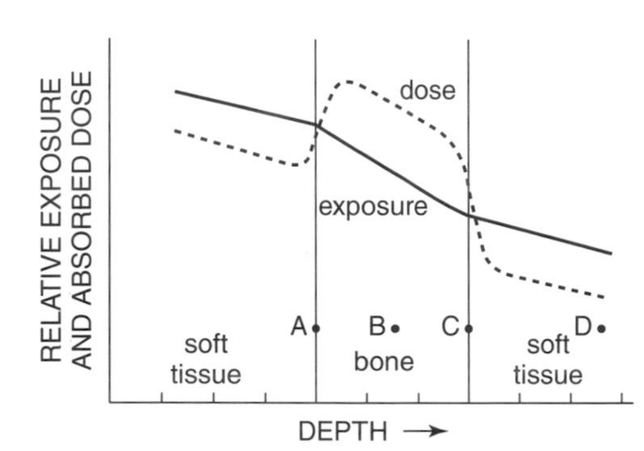

MARGIN FIGURE 7-4
Radiation exposure and absorbed dose versus depth in soft tissue containing bone.

Chest radiography at a lower kVp may be desirable in cases where a broken rib is suspected.

X-ray computed tomography has alleviated the need for pneumoencephalography, a procedure that is so painful and frightening to patients that it is almost barbaric.

number of the attenuating medium. The absorbed dose is reduced to soft tissue beyond bone because more photons are removed from the beam by the overlying bone.

HIGH-VOLTAGE RADIOGRAPHY

Radiographic images obtained with x rays generated below 100 kVp exhibit high contrast between soft tissue and bone. In a radiograph of a chest exposed to 80-kVp x rays (Figure 7-3A), the image of bone obscures the visibility of the trachea. Shadows cast by bone may be reduced by increasing the voltage applied to the x-ray tube and adding filtration to the x-ray beam. The radiograph in Figure 7-3B was obtained with x rays generated at 140 kVp and filtered by 1 mm Cu. The trachea, lung, and retrocardiac markings are displayed more clearly in this radiograph than in Figure 7-3A.

High-voltage x-ray beams are used in radiology primarily for:

- study of air-filled structures such as the chest, larynx, and paranasal sinuses;
- myelography (introduction of air, gas, or other contrast agent into the subarachnoid space of the spinal column);
- pneumoencephalography (introduction of air or inert gas into the subarachnoid space of the spinal column and into the ventricular system of the brain);
- study of the gastrointestinal tract with a contrast agent when retention of some tissue differentiation is desired in structures containing the agent.

Disadvantages of high-voltage radiography include:

- a reduction in contrast between adjacent soft tissues;
- a reduction in radiographic detail caused by an increased amount of scattered radiation;
- reduction in the ability of grids to remove scattered radiation.

LOW-VOLTAGE RADIOGRAPHY

At photon energies below about 35 keV, attenuation is governed mainly by the atomic number of the tissues because most of the interactions are photoelectric. This property is useful in producing radiographs of materials that do not differ greatly in physical density. Figure 7-4 shows a number of such objects imaged with low-energy x rays.

The use of low-energy x-ray beams presents some technical difficulties. At low voltage the efficiency of x-ray production is reduced (see Chapter 5). Low-energy x rays are rapidly attenuated in tissue, and surface exposures must be relatively high

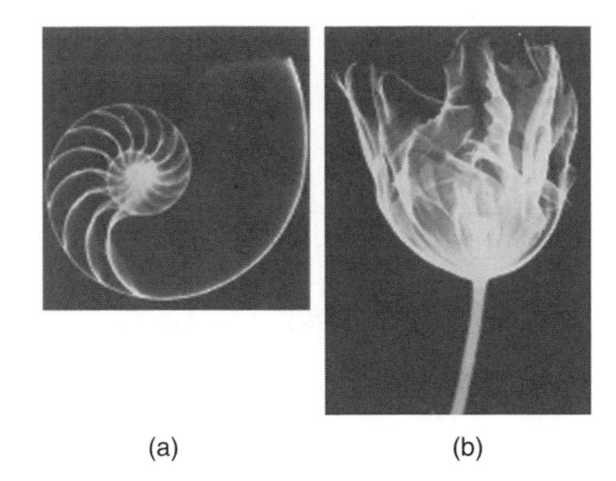

(a) (b)

FIGURE 7-4
Low-voltage radiographs: Parrot tulip at 15 kVp, nautilus shell at 40 kVp. (Courtesy of Mathew J. Ottman, University of Minnesota.)

to transmit enough x rays to the radiation receptor on the exit side of the patient. At very low energies (10–15 keV), attenuation of x rays in air is a significant limitation.

Mammography is one of the principal applications of low-energy x rays. In this technique, breast tissues that are similar in physical density, such as glandular tissue and fat, are distinguishable on the basis of their atomic composition. When compared with glandular tissue, for example, fat has a lower effective atomic number because it contains more hydrogen. The dependence of photoelectric absorption on Z^3 permits delineation of these tissue constituents in the image. Microcalcifications in the breast can be seen because of differences in both atomic number and physical density.

One approach to mammography is to use a tungsten target and relatively low tube voltage (30–45 kVp) to produce a low energy bremsstrahlung x-ray beam. A preferred approach is to use an x-ray target of molybdenum to produce a bremsstrahlung x-ray beam as well as characteristic x rays of approximately 17 and 19 keV. By filtering this x-ray beam with a molybdenum absorber, the very low energy part of the beam is removed so that most of the x rays are above 10 keV. The molybdenum K-absorption edge occurs at 19 keV. X rays with energies above the K-absorption edge are also relatively strongly absorbed. The molybdenum filter is virtually transparent to characteristic x rays from the molybdenum target that fall just below the absorption edge. Hence, most of the characteristic x rays pass through the filter without attenuation (Figure 7-5).

CONTRAST MEDIA

Many anatomic structures may be visualized more clearly with x rays if a material is introduced to increase or decrease the x-ray attenuation. Suitable materials are referred to as *contrast agents* or *contrast media*. A few radiographic examinations that may be improved with contrast media are listed in Table 7-3. Many of the agents contain either iodine or barium because the attenuation coefficient for iodine ($Z = 53$) and barium ($Z = 56$) greatly exceeds that for soft tissue ($Z_{\text{eff}} = 7.4$). Consequently, a structure containing an iodinated or barium-containing compound is clearly distinguishable from adjacent soft tissue. The attenuation coefficients for barium and iodine even exceed the coefficient for lead between the K-absorption edges for iodine (33 keV) and lead (88 keV) (Figure 7-6). Within this range of photon energies, x rays are attenuated more rapidly by iodine or barium than by an equal mass of lead. Because most photons in a diagnostic x-ray beam possess an energy between 33 and 88 keV, iodinated and barium-containing compounds are better contrast agents, gram for gram, than would be compounds containing lead or other high-Z elements, even if the toxicity of these compounds did not prohibit their use. Compounds containing iodine or barium are

TABLE 7-3 Contrast Materials for Radiography

Upper GI	Barium
Lower GI	Barium
Small bowel series	Barium
Angiography	Iodine
Urography	Iodine
Myelography	Iodine, air

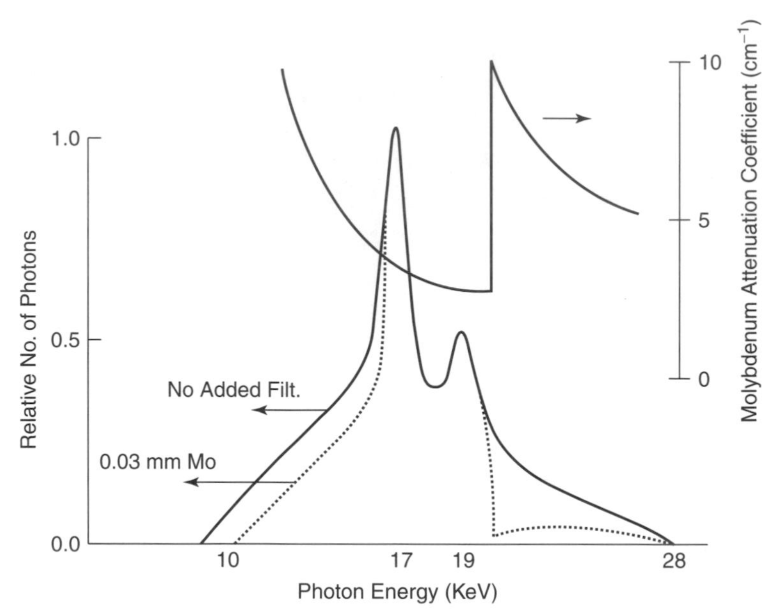

FIGURE 7-5
X-ray spectrum produced by a molybdenum-target x-ray tube operated at 28 kVp (vertical axis on the left) and the attenuation coefficient for molybdenum used to filter the x-ray beam (vertical axis on the right). Larger values of the attenuation coefficient indicate that more photons are attenuated by the filter. The central part of the x-ray spectrum, containing the characteristic peaks at 17 and 19 keV, passes through the filter with little attenuation.

Iodinated contrast agents are water-soluble and, when injected into the circulatory system (angiography), mix with the blood to increase its attenuation relative to surrounding soft tissues. In this manner, blood vessels can be seen that are invisible in x-ray images without a contrast agent.

relatively nontoxic and may be used in a wide variety of radiographic examinations. One advantage of barium over iodine compounds is their miscibility into solutions of higher physical density.

Air may be introduced into certain locations (e.g., the subarachnoid space of the spinal column) to displace tissues and fluids that interfere with visualization of

A thick solution of a barium-containing compound can be introduced into the GI tract by swallowing or enema. The solution outlines the borders of the GI tract to permit visualization of ulcers, polyps, ruptures, and other abnormalities.

Contrast agents have also been developed for use in ultrasound (solutions containing microscopic gas bubbles that reflect sound energy) and magnetic resonance imaging (solutions containing gadolinium that affect the magnetic relaxation of tissues).

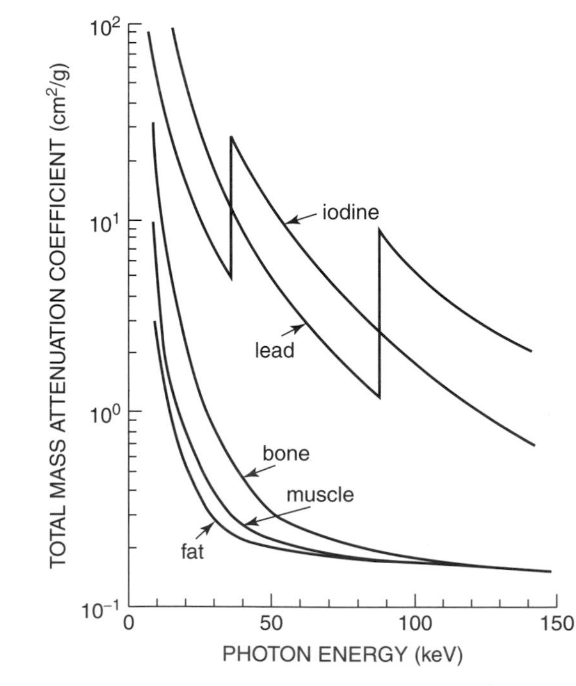

FIGURE 7-6
Mass attenuation coefficients for fat, muscle, bone, iodine, and lead as a function of photon energy.

anatomic structures of interest. The density of air is very low, and x rays are transmitted through the air-filled cavities with little attenuation. Hence, the introduction of air improves the visibility of structures in or adjacent to air-filled cavities.

PROBLEMS

7-1. Referring to Figure 7-1, discuss why x rays generated at low voltage are used to distinguish fat from muscle.

7-2. Referring to Figure 7-6, explain why iodine and barium are used in contrast agents.

7-3. Why is air an effective contrast agent if the effective atomic number of air ($Z_{eff} = 7.65$) is near that for muscle ($Z_{eff} = 7.4$)?

7-4. Discuss the advantages and disadvantages of high-voltage radiography.

7-5. Plot curves of transmission and absorption for a diagnostic x-ray beam penetrating successive layers of muscle, fat, muscle, bone, and muscle.

SUMMARY

- The f factor relates the radiation exposure (C/kg) in air to the absorbed dose (Gy) in a medium.
- Differences in the following properties among tissues permit the differentiation of various tissues in an x-ray image:
  - Atomic number
  - Electron density (electrons/kg)
  - Physical density (kg/m^3)
- Radiography at higher voltages suppresses shadows cast by bone and is helpful in chest radiography and certain GI studies.
- Radiography at lower voltages enhances soft-tissue differentiation in tissues with low subject contrast.
- Contrast media are often used to enhance contrast of accessible tissues and structures such as the gastrointestinal tract and the cardiovascular system.

REFERENCES

1. MacIntyre, J. X-ray records for the cinematograph. *Arch. Skiag.* 1897; **1**:37.
2. Hasegawa, B. H. *The Physics of Medical X-Ray Imaging,* 2nd edition. Madison, WI, Medical Physics Publishing, 1991.
3. Eisenberg, R. L. *Radiology: An Illustrated History*. St. Louis, Mosby-Yearbook, 1992, p. 59.
4. Ter-Pogossian, M. *The Physical Aspects of Diagnostic Radiology*. New York, Harper & Row, 1967.
5. Webb, S. (ed.). *The Physics of Medical Imaging*. Philadelphia, Adam Hilger, 1988.
6. Stanton, L., et al. Physical aspects of breast radiography. *Radiology* 1963; **81**:1.
7. Hendee, W. R. Imaging in Medicine, in Hornack, J. (ed.), *Encyclopedia of Imaging Science and Technology*. New York, John Wiley & Sons, 2002.
8. Bushberg, J. T., Seibert, J. A., Leidholdt, E. M., and Boone, J. M. *The Essential Physics of Medical Imaging*. Baltimore, Williams & Wilkins, 1994.
9. Wingate, C., Gross, W., and Failla, G. Experimental determination of absorbed dose from x-ray near the interface of soft tissue and other material. *Radiology* 1962; **79**:984.

CHAPTER

8

RADIATION DETECTORS FOR QUANTITATIVE MEASUREMENT

OBJECTIVES

After completing this chapter, the reader should be able to:

- Compare and contrast pulse-type and current-type ionization chambers.
- Explain the relationship between ionization current and electrode voltage.
- State four uses of ionization chambers in radiology and nuclear medicine.
- Describe the principle of operation of
  - Proportional counters
  - Geiger–müller tubes
  - Solid scintillation detectors
  - Photomultiplier tubes
  - Liquid scintillation detectors
  - Semiconductor detectors

All image receptors used with ionizing radiation are, in a sense, radiation detectors. The properties of image receptors (image intensifiers, computed radiography plates, flat-panel large area detectors, and film-screen systems) are discussed in chapters where the individual imaging techniques are covered. In this chapter we review the principles of operation of radiation detectors that are used to make quantitative measurements of exposure, dose, dose rate, and count rate.

IONIZATION CHAMBERS

Ion pairs are produced as energy is deposited in a medium by ionizing radiation. If a gas is used as the attenuating medium, the ion pairs may be collected by charged electrodes placed in the medium. The ion pairs migrate toward the charged electrodes with a "drift velocity" that depends on the type and pressure of the gas between the electrodes and on the potential difference and distance between the electrodes. In a gas-filled ionization chamber, the voltage between the electrodes is increased until all ion pairs produced by the impinging radiation are collected. However, the voltage remains below that required to produce additional ion pairs as the ion pairs produced by radiation interactions migrate to the collecting electrodes. Consequently, the electrodes receive only ion pairs that result directly from interactions of ionizing radiation with gas in the chamber.

An ionization chamber designed to detect radiation from radioactive sources of low activity consists of parallel-plate or coaxial electrodes in a volume occupied by a filling gas. A parallel-plate chamber resembles the free-air ionization chamber that serves as a calibration standard for ionization chambers used to calibrate x- and γ-ray beams used in radiation therapy.[1] A coaxial chamber is composed of a central electrode in the form of a straight wire or wire loop that is charged positively with respect to the surrounding cylindrical case (Margin Figure 8-1). The entrance of radiation into the chamber results in an electrical current or voltage pulses produced as ion pairs are collected by the electrodes.

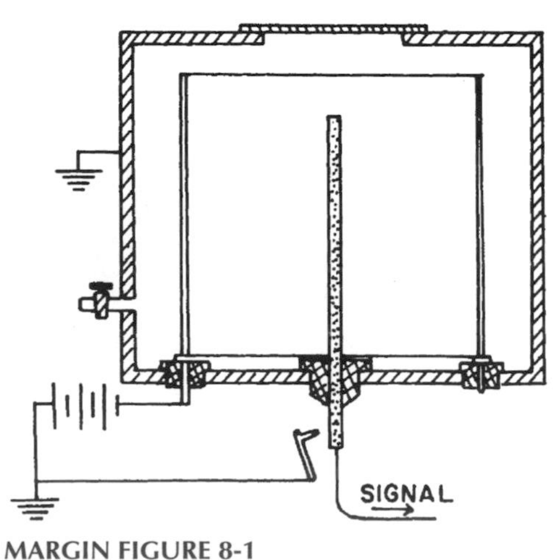

MARGIN FIGURE 8-1
A simple coaxial ionization chamber.

Pulse-Type Ionization Chambers

Consider a 1.75-MeV α-particle traversing the collecting volume of an ionization chamber. If the gas in the collecting volume is air, then the α-particle loses an average energy of 33.85 eV for each ion pair (IP) produced. If the kinetic energy of the α-particle is dissipated completely within the collecting volume, then $(1.75 \times 10^6 \text{ eV})/(33.85 \text{ eV/IP}) = 5 \times 10^4$ ion pairs are produced. Electrons liberated by the radiation migrate rapidly to the central electrode (anode) and reduce the positive charge of this electrode. Usually, electrons are collected within a microsecond after their liberation. The heavier, positively charged ions migrate more slowly toward the negative case (cathode) of the chamber. As the positive ions approach the case, they induce a negative charge on the case that masks the total reduction in voltage between the electrodes. Hence the total reduction in voltage between the electrodes is not attained until all positive ions within the chamber have been neutralized. Usually, a few hundred microseconds are required to neutralize the positive ions. In most pulse-type ionization chambers, to ensure a more rapid response, only that portion of the reduction in voltage that is created by the collection of electrons is utilized in forming a voltage pulse.

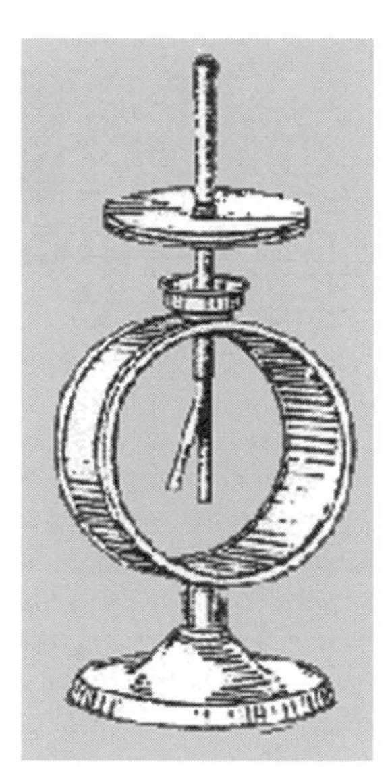

MARGIN FIGURE 8-2
One of the first radiation detectors used was the *gold leaf electroscope*. Two thin strips ("leaves") of gold were suspended at the base of a metal conductor inside of a glass jar. Gold was selected for its electrical conductivity and the ease with which it could be converted into a thin foil. When radiation ionized the air near the outside of the jar, the conductor would acquire an electrostatic charge. This electrostatic charge would distribute itself over the conductor, rendering the two leaves with charge of the same sign. The leaves would then repel each other, causing them to extend outward at some angle. The angle of the leaves was influenced by the amount of electrostatic charge acquired by the conductor, which served as a crude estimate of the amount of ionization in the air outside of the jar.

Example 8-1

What size voltage pulse is produced when a 1.75-MeV α-particle is absorbed totally within an air-filled ionization chamber with a capacitance of 10 picofarads? (10 picofarads = 10×10^{-12} farads.)

A 1.75-MeV α-particle produces 5×10^4 IP = 5×10^4 electrons plus 5×10^4 positive ions.

$Q = (5 \times 10^4 \text{ electrons})(1.6 \times 10^{-19} \text{ coulombs/electron}) = 8 \times 10^{-15}$ coulombs. The voltage pulse produced is

$$
\begin{aligned}
V &= \frac{Q}{C}, \qquad \text{where } C = 10 \times 10^{-12} \text{ farads} \\
&= \frac{8 \times 10^{-15} \text{ coulombs}}{10 \times 10^{-12} \text{ farads}} \\
&= 8 \times 10^{-4} \text{ V} \\
&= 0.8 \text{ mV}
\end{aligned}
$$

In a pulse-type chamber designed to produce voltage pulses by the rapid collection of electrons, no interference with the migration of these electrons can be tolerated. Gases such as oxygen, water vapor, and the halogens have an affinity for electrons. These gases should not be present in the collecting volume of an ionization chamber. Other gases such as helium, neon, argon, hydrogen, nitrogen, carbon dioxide, and methane form negative ions only rarely by combining with electrons. These gases may be used as filling gases in ionization chambers.

The average kinetic energy of β-particles (negatrons and positrons) is less than that of most α-particles. Furthermore, the specific ionization is 1/100 to 1/1000 less for negatrons and positrons than for α-particles. Usually, α-particles expend their entire kinetic energy by interacting with the gas in the collecting volume of an ionization chamber. Negatrons and positrons usually strike the wall of the chamber before dissipating all of their kinetic energy. For these reasons, voltage pulses produced as β-particles traverse an ionization chamber are much smaller than those produced by α-particles. Pulses produced by x and γ rays are even smaller. Ionization chambers are usually operated in the pulse mode for the detection of α-particles; other radiations with lower specific ionization are usually measured by operation of the chamber in the current mode.

Current-Type Ionization Chambers

Electrons collected by the anode of an ionization chamber constitute a direct current that may be amplified and measured with a conventional dc meter. In general, this approach is unsatisfactory because instability and zero drift are introduced by the dc amplifier and accurate measurements are difficult to achieve.

Small currents from an ionization chamber may be measured more accurately with an electrometer. An electrometer converts the signal from the ionization chamber into an alternating current that may be amplified with an ac amplifier. An ac amplifier is not subject to the problems of instability and zero drift encountered with a dc

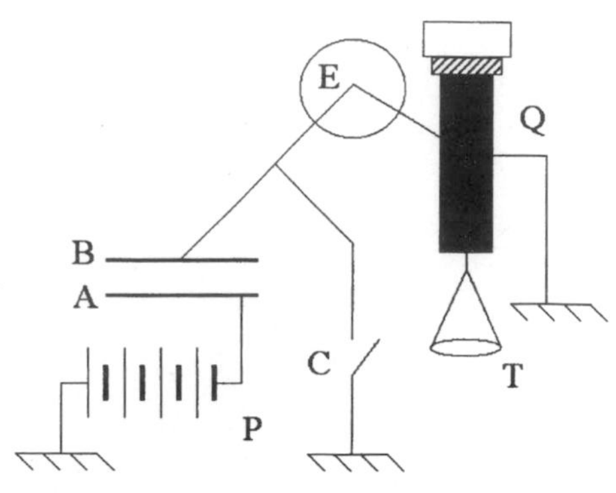

MARGIN FIGURE 8-3
The piezoelectric electrometer/ionization chamber was developed by Pierre Curie in 1897 to measure ionization produced in air due to interactions caused by the emanations from radioactive material. The device was built to aid his wife, Marie Curie, in her efforts to isolate the source of radioactivity from an ore, pitchblende, that was known to fog photographic film and was therefore thought to be "naturally" radioactive. Shown here is a reproduction of a diagram that appeared in Madame Curie's doctoral dissertation. A battery P produces a voltage across parallel plates A and B when switch C is closed. If a powder of the radioactive ore is placed on one of the plates, ions produced in the air between the plates will produce an electrical current. This small "ionization" current may be measured by determining the voltage needed to offset it. The voltage is produced by adding weights to a pan T to apply force to a piezoelectric quartz crystal Q such that the electrometer E records no net current.

amplifier. The amplified alternating current may then be measured to within ± 0.05% precision by one of two methods, referred to as the *voltage-drop method* and the *rate-of-charge method*.

With the voltage-drop method, the amplified alternating current is rectified and directed through a precision resistance. The voltage developed across the resistance is proportional to the current. With the rate-of-charge method, the current is rectified and collected by a precision capacitor. The rate of collection of electrical charge on the plates of the capacitor is proportional to the current. The rate-of-charge method may be used with currents smaller than those measured by the voltage-drop method. The voltage-drop method without current amplification is used in portable survey meters such as the "cutie pie" (Margin Figure 8-4).

Example 8-2*

A 37-Bq sample of $^{14}CO_2$ is contained within a current-type ionization chamber. The ionization current is converted to alternating current and is measured by the voltage-drop method with a precision resistance of 10^{12} Ω. Assuming that all the energy of the β-particles from ^{14}C is deposited in the gas and that the ac signal is not amplified, what voltage is developed across the precision resistance?

$$37 \text{ Bq} = 37 \text{ disintegrations/sec}$$

The average energy of β-particles from ^{14}C is 0.045 MeV = 45×10^3 eV. The average energy dissipated per second in the counting volume is

$$(37 \text{ disintegrations/sec})(45 \times 10^3 \text{ eV/disintegration}) = 16.6 \times 10^5 \text{ eV/sec}$$

The number of electrons released per second is

$$\frac{16.6 \times 10^5 \text{ eV/sec}}{33.85 \text{ eV/IP}} = 4.9 \times 10^4 \text{ IP/sec} = 4.9 \times 10^4 \text{ electrons/sec}$$

The charge liberated per second is

$$(4.9 \times 10^4 \text{ electrons/sec})(1.6 \times 10^{-19} \text{ coulombs/electron})$$
$$= 7.8 \times 10^{-15} \text{ coulombs/sec}$$
$$1 \text{ coulomb/sec} = 1 \text{ A}$$
$$7.8 \times 10^{-15} \text{ coulombs/sec} = 7.8 \times 10^{-15} \text{ A}$$

The voltage drop V across a resistance is the product of the resistance in ohms and the current in amperes:

$$V = IR$$
$$V = (7.8 \times 10^{-15} \text{A})(10^{12}\ \Omega)$$
$$= 7.8 \times 10^{-3} \text{ V}$$
$$= 7.8 \text{ mV}$$

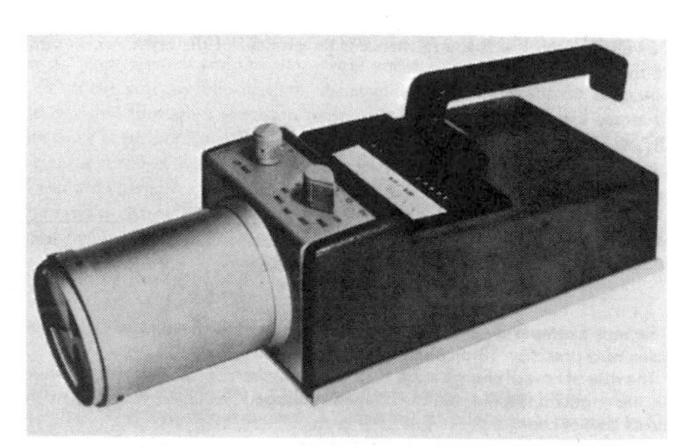

MARGIN FIGURE 8-4
A "cutie pie" portable survey meter. The scale of the meter is calibrated to read directly in units of milliroentgens per hour, with a range switch that decreases the sensitivity of the meter by factors of 10.

Example 8-3†

Repeat Example 8-2, but find the rate of change in voltage across the plates of a precision 10-picofarad capacitor.

The rate of flow of charge is 7.8×10^{-15} coulombs/sec. Because $V = Q/C$ (where C is the capacitance), the rate of change of voltage is given by $V_r = Q_r/C$

*Modified from data of Chase and Rabinowitch.[2]
†Modified from data of Chase and Rabinowitch.[2]

(where) Q_r is the rate of flow of charge):

$$V_r = \frac{7.8 \times 10^{-15} \text{ coulombs/sec}}{10 \times 10^{-12} \text{ farads}}$$
$$= 7.8 \times 10^{-4} \text{ V/sec}$$
$$= (7.8 \times 10^{-4} \text{ V/sec})(60 \text{ sec/min})$$
$$= 46.8 \times 10^{-3} \text{ V/min}$$
$$= 47 \text{ mV/min}$$

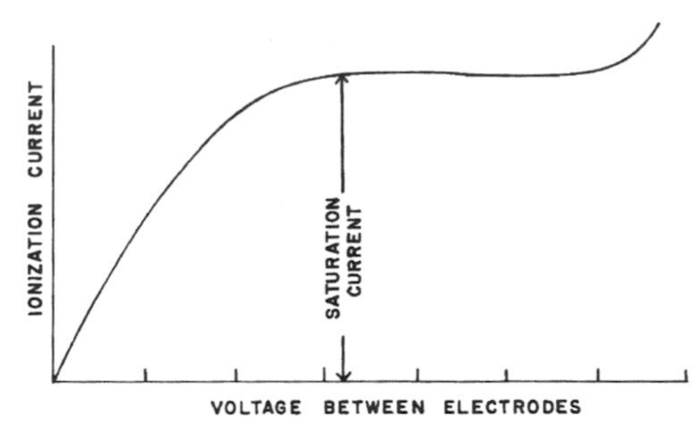

MARGIN FIGURE 8-5
The ionization current from an ionization chamber is plotted as a function of the voltage between electrodes. Recombination of ion pairs occurs at voltages below that furnishing a saturation current. At high voltages, the signal is amplified by ionization produced as electrons are accelerated toward the anode.

In Margin Figure 8-5, the ionization current is plotted as a function of the voltage applied across the electrodes of an ionization chamber. At low voltages, the electrons and positive ions are not strongly attracted to the electrodes, and some of the ion pairs are lost by recombination. The attraction for ion pairs increases with the voltage between the electrodes, and fewer ion pairs recombine. When the voltage between the electrodes exceeds the *saturation voltage*, the electrodes collect all ion pairs produced by the radiation. The saturation voltage for a particular ionization chamber depends upon the design of the chamber, the shape and spacing of the electrodes, and the type and pressure of the gas in the chamber. No increase in ionization current is observed as the electrode voltage is raised a few hundred volts above saturation because all ion pairs produced by the radiation are collected. This region of voltage is referred to as the *ionization chamber plateau*. The ionization current increases abruptly at the end of the plateau. This increase is due to amplification of the signal *caused* by the production of additional ion pairs as electrons liberated by the incident radiation gain energy on their way to the anode. Ionization chambers are operated at a voltage below that which causes signal amplification. At any particular voltage, the ionization current produced by an α-emitting sample is much greater than that produced by a sample that emits β-particles. The reduced signal for the β-emitting sample reflects the reduced ionization produced by β-particles.

Uses of Ionization Chambers

Radiation from solid, liquid, and gaseous samples may be measured with an ionization chamber and electrometer. The activity of liquid samples prepared for administration to patients often is determined by placing the vial or syringe containing the sample into a well-type ionization chamber referred to as an *isotope calibrator* (Margin Figure 8-6). Gaseous samples may be counted by filling an ionization chamber with the radioactive gas. For example, ionization chambers may be used to measure the amount of $^{14}CO_2$ in air expired by patients metabolizing compounds labeled with ^{14}C.

Portable survey instruments such as that depicted in Margin Figure 8-4 are frequently used in nuclear medicine to monitor exposure rates in the vicinity of radioactive sources and patients receiving therapeutic quantities of radioactive material. A neutron detector may be constructed by filling an ionization chamber with BF_3 gas or by coating the wall of an ionization chamber with lithium or boron. Ionization is produced within the chamber by α-particles and recoil nuclei that are liberated during interactions of neutrons with the lithium or boron.

PROPORTIONAL COUNTERS

The small signals from an ionization chamber must be amplified greatly before they are measured. Because of the introduction of electronic noise and instability, amplification by electrical circuits is difficult to achieve without distortion of the signal. This problem may be reduced if the signal is amplified within the counting volume of the chamber. If the potential difference between the electrodes of a chamber is raised beyond a certain voltage, electrons liberated by radiation traversing the chamber are

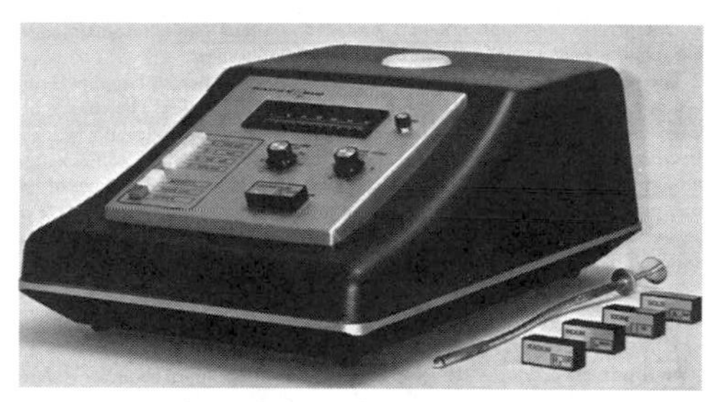

MARGIN FIGURE 8-6
Well-type ionization chamber and electrometer (isotope calibrator) used to measure the activity of a radioactive sample prior to administration of the sample to a patient. The volume of the sample to be administered is computed from the measured activity, often automatically by the isotope calibrator. (Courtesy of RADX Corp.)

Townsend effect
Sir John Sealy Edward Townsend (1868–1957) was a British physicist who was a pioneer in the study of electrical conduction in gases. He did his graduate work in the Cavendish Laboratory under J. J. Thompson, the discoverer of the electron. Townsend made the first direct measurement of the unit electrical charge (e). In 1901 he discovered that gas molecules can be ionized by collision with ions—known today as the "Townsend effect," a principle that is used in the modern proportional counter.

Michael Faraday (1791–1867) was born into the working class of eighteenth-century England. At the age of 14, he was apprenticed to a bookbinder and made a habit of reading the books he was binding. Through reading, he developed an interest in science. In that time, science was done by the wealthy elite of British society. Faraday attended a lecture given by a prominent physicist, Sir Humphry Davy, took careful notes, and presented Sir Davy with a bound set. He convinced Sir Davy to hire him as an assistant and quickly established a reputation as a scientist in his own right. He eventually derived the law of electromagnetic induction—namely, that a changing magnetic field creates an electrical field that can cause current to flow in an electrical conductor. This principle explains the electrical generator and also the origin of the MR signal that is induced in a receiver coil by precessing magnetic moments.

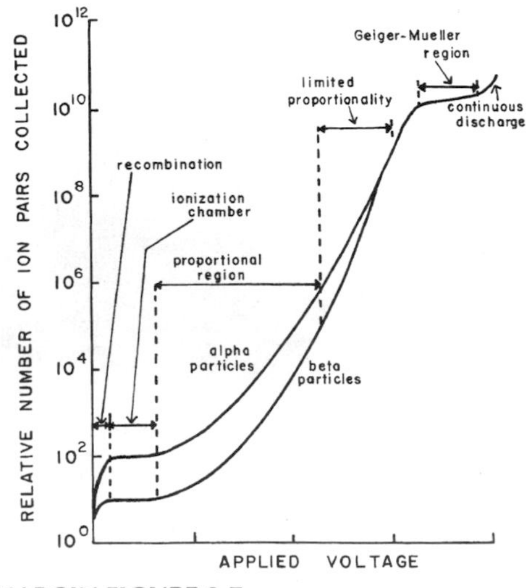

MARGIN FIGURE 8-7
Relationship between the total number of ion pairs produced in a gas-filled detector and the high voltage between electrodes of the detector.

accelerated to a velocity great enough to produce additional ionization. Most of the additional ionization occurs near the anode of the chamber. As a result, many (10^6 to 10^7) electrons and positive ions are collected by the electrodes for a much smaller number (10^3 to 10^5) of ion pairs produced directly by radiation entering the chamber. This process is referred to as *gas amplification* or the *Townsend effect*. The *amplification factor* for the chamber is the ratio of the total number of ion pairs produced within the chamber to the number liberated directly by radiation entering the chamber. The amplification factor depends on the construction of the chamber and the type of gas enclosed within the chamber. The amplification factor varies from 10^2 to 10^4 for most proportional counters providing a signal (approximately 1 mV) that requires only a small amount of external amplification.

The number of ion pairs produced in a gas-filled chamber is plotted in Margin Figure 8-7 as a function of the voltage applied across the electrodes. The voltage between electrodes must be regulated closely because the amplification factor is affected greatly by small changes in voltage. In the *proportional region*, the amount of charge collected by the electrodes increases with the number of ion pairs produced initially by the impinging radiation. Consequently, the size of the signal from a proportional chamber increases with the amount of ionization produced by radiation that traverses the chamber.

Example 8-4*

An α-particle produces 10^5 ion pairs (IP) in a proportional chamber with an amplification factor of 10^3. How many IPs are collected by the electrode?

With an amplification factor of 10^3, 1000 IPs are collected for every ion pair liberated by the incident radiation. The total number of IPs collected is

$$(10^5 \text{ IPs})(10^3) = 10^8 \text{ IPs}$$

Example 8-5*

Repeat the calculation in Example 8-4 for a β-particle that produces 10^3 IPs within the chamber.

For an amplification factor of 10^3, we have

$$(10^3 \text{ IPs})(10^3) = 10^6 \text{ IPs}$$

Because of its rapid response to ionizing events, the multiwire proportional chamber has been investigated as an imaging device for nuclear medicine.[3,4] By raising the pressure of the counting gas to 10 atm or more, the sensitivity of the chamber to γ rays can be improved. The γ-ray sensitivity can also be improved by placing high-Z foils in front of the chamber to convert incoming photons to photoelectrons and Compton electrons. Although the intrinsic resolution of multiwire proportional chambers is excellent, the resolution achievable in practice has been no better than that for other nuclear medicine imaging devices.

At voltages higher than the proportional region, α-particles and other densely ionizing radiations initiate ionization of most of the atoms of gas in the vicinity of the anode. If the chamber is operated at a voltage in this region, then the number of ion pairs collected is not strictly proportional to the ionization produced directly the radiation. Hence this voltage region is referred to as the *region of limited proportionality*. Proportional chambers are not operated routinely in the region of limited proportionality.

GEIGER–MÜLLER TUBES

If the potential difference between the electrodes of a gas-filled detector exceeds the region of limited proportionality (Margin Figure 8-7), then the interaction of a charged particle or x or γ ray within the chamber initiates an *avalanche of ionization*, which represents almost complete ionization of the counting gas in the vicinity of the anode. Because of this avalanche process, the number of ion pairs collected by the electrodes is independent of the amount of ionization produced directly the impinging radiation. Hence the voltage pulses (usually 1 to 10 V) emerging from the detector are similar in size and independent of the type of radiation that initiates the signal. The range of voltage over which signals from the detector are independent of the type of radiation entering the detector is referred to as the *Geiger–Müller (G–M) region*. For detectors operating in this voltage region, the amplification factor is 10^6 to 10^8.

In Margin Figure 8-8, the number of pulses (or counts) recorded per minute is plotted as a function of the voltage across the electrodes of a G–M detector exposed to a radioactive source. No counts are recorded if the voltage is less than the starting voltage because voltage pulses formed by the detector are too small to pass the discriminator and enter the scaler. As the voltage is raised slightly above the starting voltage, some of the pulses are transmitted by the discriminator and recorded. At the *plateau threshold voltage*, all pulses are transmitted to the scaler. Increasing the voltage beyond the plateau threshold does not increase the count rate significantly. Consequently, relatively inexpensive high-voltage supplies that are not exceptionally stable may be used with a G–M detector. G–M detectors usually are operated at a voltage about one-third of the way up the plateau. In Margin Figure 8-8, for example, an operating voltage of 1150 V might be selected for the detector. The voltage range encompassed by the plateau varies with the construction of the G–M detector and with the counting gas used.

Atoms of counting gas near the anode may be ionized spontaneously if the voltage applied to the detector is raised beyond the G–M plateau. This region of voltage is referred to as the *region of spontaneous discharge* or *region of continuous discharge* because the counting gas may be ionized in the absence of radiation. A G–M tube may be permanently damaged by the application of voltages higher than the G–M plateau.

A few G–M detectors are illustrated in Margin Figure 8-9. The anode is a thin wire of tungsten or stainless steel in the center of the detector. The anode is surrounded by a metal or glass cathode that is coated internally with a conducting layer of graphite or evaporated metal. The efficiency of a G–M detector for high-energy x and γ rays may be increased by coating the cathode with a heavy metal such as bismuth or lead. The end window G–M detector (Margin Figure 8-9A) is often used for assay of radioactive samples. The window is usually constructed from split mica with a thickness of a few milligrams per square centimeter. The thin window admits α-particles and low-energy β-particles into the counting volume. Ultrathin Mylar windows between 100 and 200 μg/cm^2 thick are available, but they must be used with a flow counter and a continuous supply of counting gas because windows this thin are permeable to the counting gas.

Detectors with thin walls are used primarily with portable survey meters (Margin Figure 8-9B). The walls are metal or glass and are about 30 mg/cm^2 thick. With the windowless flow detector (Margin Figure 8-9C), the radioactive sample is sealed inside the chamber while it is counted. The thin-wall dipping tube (Margin Figure 8-9D) may be inserted into a radioactive solution. With the needle probe detector (Margin Figure 8-9E), the sensitive volume occupies the tip of a long, thin probe. These detectors may be used to locate radioactive material concentrated in tissues within the body. The chamber with a high-Z wall (Margin Figure 8-9F) is used primarily for the detection of x and γ rays.

When an ionizing event is initiated in a G–M detector, an avalanche of electrons is created along the entire length of the anode. The residual positive ions require 200 μs or longer to migrate to the cathode. During the time required for migration of the positive ions, the detector will not respond fully to additional radiation that enters the counting volume. The curve in Margin Figure 8-10 depicts the response of the detector as a function of time after an ionizing event. During the "dead time," the detector is completely unresponsive to additional radiation. An ionizing event occurring within the "recovery time" produces a voltage pulse that is smaller than

C. T. R. Wilson, a British Scientist, invented the *cloud chamber* in 1912. A cloud chamber is a detector that allows researchers not only to detect the presence of ionizing photons and particulate radiation but also to determine the path taken by the particles through the chamber, even if the path is curvilinear or spiral. He accomplished this feat by preparing a container in which a gas was "supersaturated"—that is, at the appropriate temperature and pressure to condense into a liquid, but still existing in a gaseous state due to the absence of nucleation sites. When ionizing radiation enters a cloud chamber detector, the formation of ion pairs produces nucleation sites that cause a stream of condensation to appear along the path of the particle. A similar phenomenom occurs when a jet plane disturbs supersaturated layers of the upper atmosphere, leaving a "jet stream" pattern of condensation along its path.

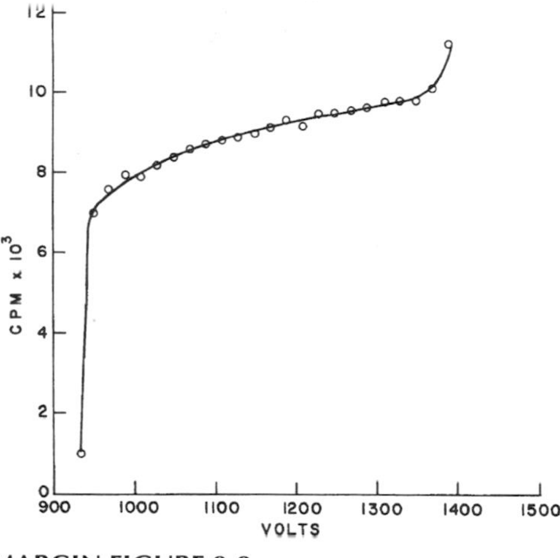

MARGIN FIGURE 8-8
Characteristic curve for a G–M detector.

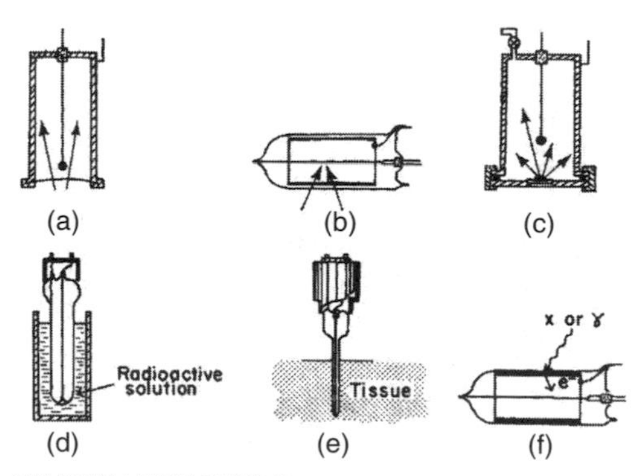

MARGIN FIGURE 8-9
Various G–M detectors. **A:** End window. **B:** Side window. **C:** Windowless flow. **D:** Thin-wall dipping tube. **E:** Needle probe. **F:** Heavy-metal wall.

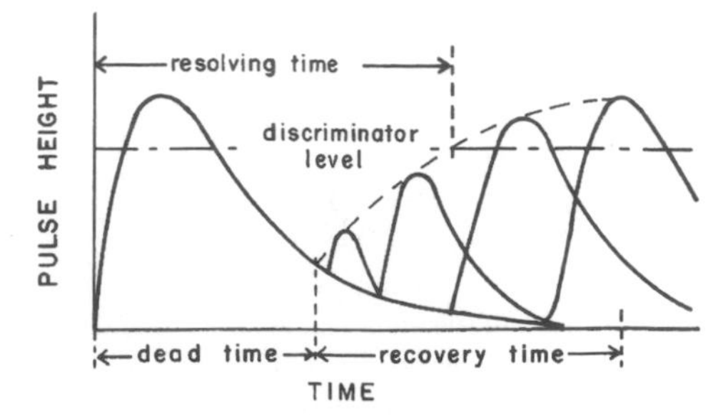

MARGIN FIGURE 8-10
Diagram illustrating the formation of a voltage pulse in a G–M detector as a function of time after an ionizing event.

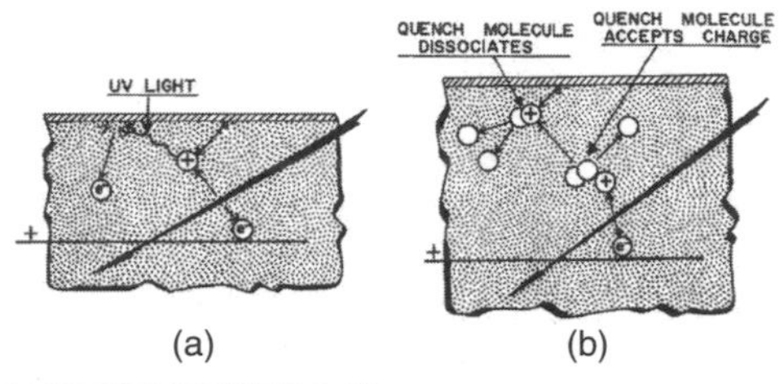

MARGIN FIGURE 8-11
A: Self-perpetuating discharge of a G–M detector caused by bombardment of the cathode by ultraviolet and x-ray photons that are released as positive ions are neutralized. **B:** Molecules of quench gas accept the charge of positive ions and dissociate when neutralized near the cathode. In this way the self-perpetuating discharge is prevented.

The slope of the plateau may be computed as follows:

$$\text{Slope}\,(\%/100\,\text{V}) = \frac{2[(\text{cpm})_2 - (\text{cpm})_1]10^4}{[(\text{cpm})_2 + (\text{cpm})_1](V_2 - V_1)}$$

where $(\text{cpm})_2$ is the count rate at voltage V_2 on the plateau and $(\text{cpm})_1$ is the count rate at voltage V_1 on the plateau.

MARGIN FIGURE 8-12
Portable survey meter equipped with an end window G–M tube.

normal. The "resolving time" is the time between an ionizing event and second event that furnishes a pulse large enough to pass the discriminator and be counted.

Positive ions that approach the cathode of a G–M detector dislodge electrons from the wall of the chamber. As these electrons combine with the positive ions, ultraviolet and x-ray photons are released. Some of these photons strike the chamber wall and release electrons that cause the chamber to remain discharged. If this secondary release of electrons is permitted to occur, the detector will be unresponsive to radiation after the first ionizing event. Self-perpetuating discharge of a G–M detector is diagramed in Margin Figure 8-11A.

Various methods have been devised to "quench" the self-perpetuating discharge of G–M detectors. The most common method is to add a small concentration (about 0.1%) of a selected gas to the counting gas. The gases used most often as "internal quench agents" are polyatomic organic gases (e.g., amyl acetate or ethyl alcohol vapor) or halogens (e.g., Br_2 or Cl_2). Their effect is shown in Margin Figure 8-11B. As the positive ions move toward the cathode, they collide with and transfer charge to molecules of quench gas. The charged molecules of quench gas migrate to the cathode and dislodge electrons from the chamber wall. Energy released as the dislodged electrons combine with the charged molecules causes the dissociation of molecules of the quench gas. The dissociation is irreversible with a polyatomic organic gas, and the useful life is 10^8 to 10^{10} pulses for a G–M detector quenched with one of these agents. Halogen-quenched tubes have an infinite useful life, theoretically, because the molecules recombine after dissociation.

The counting gas used routinely in G–M detectors is an inert gas such as argon, helium, or neon. "Geiger gas" used in flow counters is composed of 99% helium and about 1% butane or isobutane. For G–M tubes quenched with an organic gas, the plateau should be 200 to 300 V long and should have a slope not greater than 1% to 2%/100 V. Halogen-quenched tubes have a shorter plateau (100 to 200 V) and a plateau slope of 3% to 4%/100 V.

The detection efficiency of a G-M counter is about 1% for x and γ rays and nearly 100% for α- and β-particles that enter the counting volume. Of course, many α and low-energy β-particles are absorbed by the window of the detector. Windowless flow counters are often used to detect these particles. Shown in Margin Figure 8-12 is a survey meter equipped with an end-window G–M tube for detecting the presence of radioactive contamination.

SOLID SCINTILLATION DETECTORS

Gas-filled chambers are not efficient detectors for x- and γ-ray photons because these radiations pass through the low-density gas without interacting. The probability of x- and γ-ray interaction is increased if a solid detector with a high density and atomic number is used. Atoms of a solid are immobile, however, and an interaction cannot be registered by the collection of electrons and positive ions. Instead, the interaction must be detected by some alternate method. In a scintillation crystal, light is released as radiation is absorbed. The light impinges upon a photosensitive surface in a photomultiplier tube. Electrons released from this surface constitute an electrical signal. Scintillation detectors may be used to detect particulate radiation as well as x- and γ-ray photons. For example, liquid scintillators often are used to detect low-energy β-particles.

Principles of Scintillation Detection

When an x or γ ray interacts within a scintillation crystal, electrons are raised from one energy state to a state of higher energy. The number of electrons raised to a higher energy level depends upon the energy deposited in the crystal by the incident x or γ ray. Light is released as these electrons return almost instantaneously to the lower energy state. In most scintillation detectors, about 20 to 30 photons of light are released for every keV of energy absorbed. The photons of light are transmitted through the transparent crystal and directed upon the photosensitive cathode (photocathode) of a photomultiplier tube. If the wavelength of light striking the photocathode matches the spectral sensitivity of this photosensitive surface, then electrons are ejected. The number of electrons is multiplied by various stages (dynodes) of the photomultiplier tube, and a signal is provided at the photomultiplier anode that may be amplified electronically and counted. The size of the signal at the anode is proportional to the energy dissipated in the detector by the incident radiation.

MARGIN FIGURE 8-13
The Sudbury Neutrino Observatory (SNO) is located 6800 feet below ground in the Creighton nickel mine near Sudbury, Ontario. SNO captures the elusive neutrino in a tank filled with 10,000 metric tons of heavy water, water in which each ordinary hydrogen atom (hydrogen with a nucleus consisting of a single proton) is replaced by a deuteron, a hydrogen isotope with a nucleus containing a proton, and a neutron. Note the scale of the detector compared to the workers below it. (Photo from the *New York Times*, June 22, 1999.)

Scintillation Crystals

Gamma rays from radioactive samples are often detected with a scintillation crystal. Alkali halide crystals usually are used because the probability of photoelectric interactions is increased by the presence of the high-Z halide component. Sodium iodide is the alkali halide used most frequently, although crystals of cesium iodide and potassium iodide are available at higher cost. Crystals of sodium iodide up to 9 in. in diameter by 9 in. thick or 20 in. in diameter by 0.5 in. thick are available commercially. Smaller crystals (e.g., 2 in. in diameter by 2 in. thick) are used routinely for the assay of γ-emitting samples. The efficiency of a crystal of detecting x and γ rays increases with the size of the crystal.

To be used as a scintillation detector, an alkali halide crystal must be "activated" with an impurity. The impurity is usually thallium iodide at a concentration of about 0.1%, and the crystals are denoted as NaI(T1), CsI(T1), or KI(T1).

Highly purified organic crystals (e.g., anthracene and *trans*-stilbene) are used to detect β-particles. The atomic number of these crystals is relatively low, and the probability is reduced that β-particles will be scattered out of the detector after only part of their energy has been dissipated. The sensitivity of an anthracene or *trans*-stilbene detector to γ rays is low, particularly if the crystal is thin. Consequently, β-particles may be detected with limited interference from γ rays.

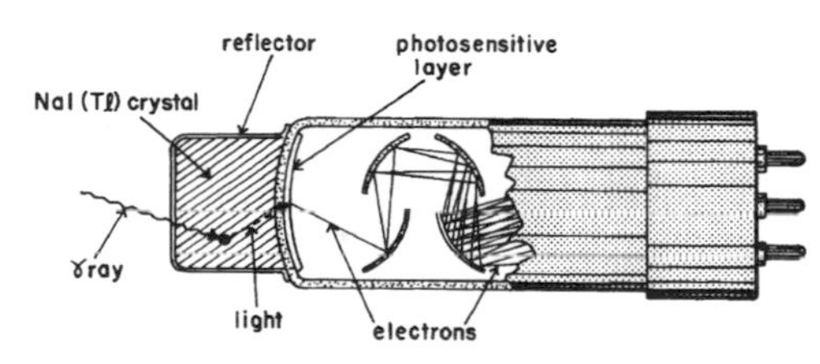

MARGIN FIGURE 8-14
A sodium iodide crystal and photomultiplier tube.

Mounting Scintillation Crystals

Sodium iodide crystals are hygroscopic and must be protected from moisture. If exposed to moisture, NaI(T1) crystal turns yellow and absorbs much of the radiation-induced fluorescence so that less light is emitted (reducing the signal available for detection of radiation). The yellow color reflects the release of free iodine. Crystals are mounted in a dry atmosphere and sealed to prevent the entrance of moisture. A light pipe of Lucite, clear glass, or quartz is sometimes attached to the side of the crystal nearest the photocathode. Other surfaces of the crystal are coated with a light-reflecting material (e.g., Al_2O_3, MgO, or aluminum foil). The crystal may be enclosed within an aluminum canister, perhaps 1/32 in. thick. The canister prevents moisture from reaching the crystal and ambient light from reaching the photocathode. The crystal or light pipe is coupled to the glass face of the photomultiplier tube with a transparent viscous medium such as silicone fluid.

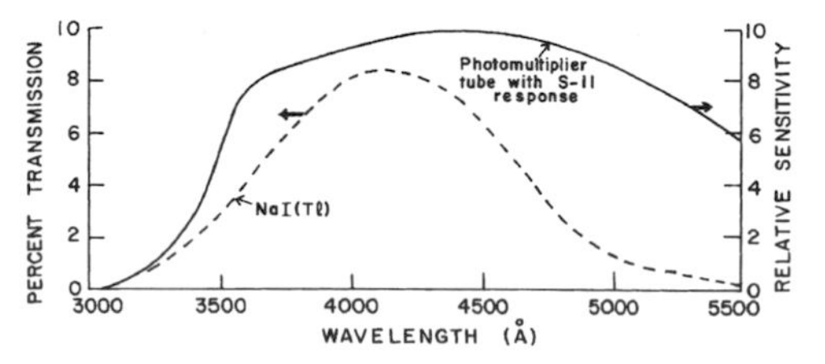

MARGIN FIGURE 8-15
The emission spectrum of NaI(T1) is matched closely to the spectral sensitivity of a photomultiplier tube with an S-11 response.

Photomultiplier Tubes

A photomultiplier tube is diagramed in Margin Figure 8-14. The photocathode is usually an alloy of cesium and antimony, often mixed with sodium and potassium (i.e., a *bialkali* photocathode), from which an acceptable number of electrons are released per light photon absorbed. The spectral sensitivity of the alloy must match the wavelength of light emerging from the crystal. The spectral sensitivity of a photocathode with an "S-11 response" is compared in Margin Figure 8-15 with the emission spectrum of light from irradiated NaI(T1). Only 10% to 30% of the light photons that strike the photocathode cause the ejection of electrons. These electrons are

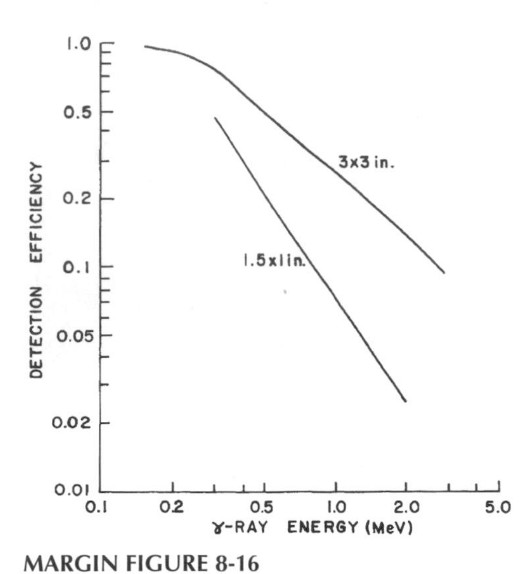

MARGIN FIGURE 8-16
The detection efficiency of 3 × 3-in. and 1.5 × 1-in. NaI(T1) crystals is plotted as a function of the energy of incident γ-rays. The radioactive sources were positioned 7 cm from the 1.5 × 1-in. crystal and 9.3 cm from the 3 × 3-in. crystal. (From Lazar, N., Davis, R., and Bell, P. *Nucleonics* 1956; **14**:52. Used with permission.)

accelerated to the first dynode, a positively charged electrode positioned a short distance from the photocathode. For each electron absorbed by the first dynode, three or four electrons are ejected and accelerated to the second dynode, where more electrons are released. Because photomultiplier tubes contain 6 to 14 dynodes with a potential difference of 100 to 500 V between successive dynodes, 10^6 to 10^8 electrons reach the anode for each electron liberated from the photocathode. The amplification of the signal is very dependent upon the potential difference between dynodes, and the high-voltage supply for dynodes must be very stable.

Electrons collected by the anode are converted to a voltage pulse with an amplitude of a few millivolts to a few volts. This voltage pulse is delivered to the preamplifier, which often is mounted on the photomultiplier tube.

Energy Dependence of NaI(T1) Detectors

The detection efficiency of a NaI(T1) scintillation detector decreases with increasing energy of impinging γ rays (Margin Figure 8-16). The efficiency of a scintillation detector for detection of γ rays may be improved by using a larger crystal and by improving the "counting geometry." For example, a crystal into which the radioactive sample may be inserted (i.e., a well detector) may furnish a detection efficiency greater than that provided by a crystal that receives at best no more than half the γ rays from a radioactive source.

LIQUID SCINTILLATION DETECTORS

With a solid scintillation detector such as a NaI(T1) crystal, the radioactive sample is positioned outside the detector. In liquid scintillation counting, the radioactive sample is mixed intimately with the scintillating material, and attenuation by materials between the radioactive sample and the scintillating material is reduced to a minimum.[5,6] Consequently, the detection efficiency is high for radiations with very short range, including weak β-particles such as those from ^{3}H ($E_{max} = 0.018$ MeV), ^{14}C ($E_{max} = 0.156$ MeV), and ^{35}S ($E_{max} = 0.168$ MeV). Usually light from the mixture of scintillator and radioactive sample is directed toward at least two photomultiplier tubes. The signal from each photomultplier tube is transmitted by a preamplifier and amplifier to a coincidence circuit (Margin Figure 8-17). The coincidence circuit transmits a voltage pulse to the scaler only if a pulse is received simultaneously from both photomultiplier tubes.

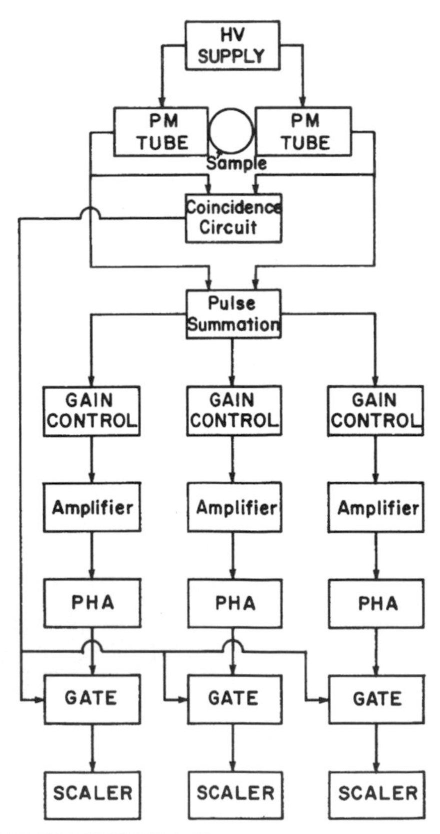

MARGIN FIGURE 8-17
A liquid scintillation counter with two photomultiplier tubes and a coincidence circuit to reduce the background count rate.

Except for very low energy particles and photons, radiation emitted by the sample usually produces a signal in each photomuliplier tube, and a pulse passes to the scaler for most disintegrations of the sample. However, spurious pulses generated by "thermal noise" in the photomultiplier tubes or preamplifiers are received by the coincidence circuit from one direction only, and a pulse is not passed to the scaler. In this manner, the coincidence circuit reduces the background count rate. Without this circuit, the background count rate would be intolerably high. In older liquid scintillation counters, the scintillation mixture, photomultiplier tubes, and preamplifiers were cooled to a few degrees above zero to help eliminate spurious counts. Newer photomultiplier tubes do not require refrigeration.

The scintillating solution, or "cocktail," consists of the radioactive sample, solvent, primary fluor, or solute and, if necessary, a secondary fluor. Solvents dissolve the sample and transfer energy from the sites of interaction of the radiation to the molecules of fluor present at a low concentration (1% or less) in the scintillation mixture.

Molecules of the primary fluor release light upon receipt of energy from the solvent molecules. The wavelengths of light emitted by the scintillation cocktail must correspond to the spectral sensitivity of the photocathodes of the photomultiplier tubes. Some photomultiplier tubes with quartz windows are sensitive to the light emitted by the primary fluor. With other photomultiplier tubes, however, a secondary fluor must be added to the scintillation solution. The secondary fluor is termed a *wavelength shifter* because light of longer wavelength is emitted by the scintillation solution when the secondary fluor is present.

In a liquid scintillation counter, the size of the voltage pulse depends on the energy dissipated in the scintillation cocktail by a photon or particle emitted by the radioactive sample. The number of pulses of different sizes is shown in Margin Figure 8-18 for β-emitting samples of ^{3}H and ^{14}C. By adjusting upper and lower discriminators, one isotope may be counted in the presence of the other. For a number of reasons, these spectra do not exactly correspond to the energy distribution of the emitted particles.

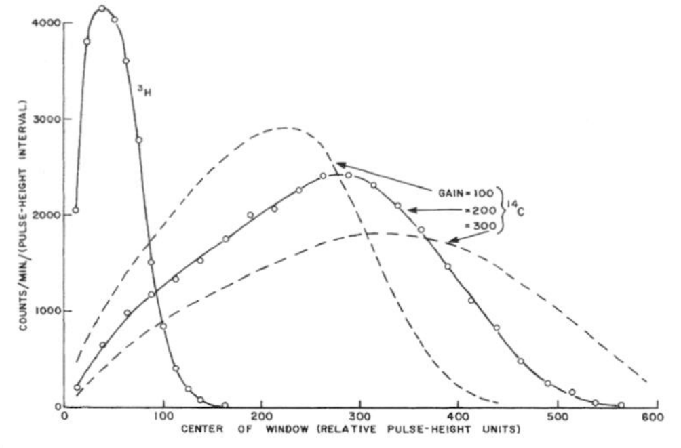

MARGIN FIGURE 8-18

Liquid scintillation spectra for β-particles from ^{3}H ($E_{max} = 0.018$ MeV) and ^{14}C ($E_{max} = 0.156$ MeV). The dashed curves illustrate the effects of the amplifier gain upon the ^{14}C spectrum.

Interference with the production or transmission of light in a liquid scintillation solution is termed *quenching*. Quenching is always present in liquid scintillation counting and is caused by the following:

1. Interference with the mechanism of energy transfer contributed by the sample or other components of the cocktail: *chemical quenching*.
2. Absorption of light by colored materials in the sample: *color quenching*.
3. Passive interference with the mechanism of energy transfer resulting from dilution of the scintillation mixture by the sample or other material: *dilution quenching*.
4. Absorption of light by the scintillation vial, fingerprints on the vial, and so on: *optical quenching*.

Quenching shifts the spectrum for any isotope toward pulses of smaller size. Shown in Margin Figure 8-19 are spectra for a ^{14}C-labeled sample dissolved in a scintillation cocktail and quenched with different amounts of carbon tetrachloride, a chemical quench agent.

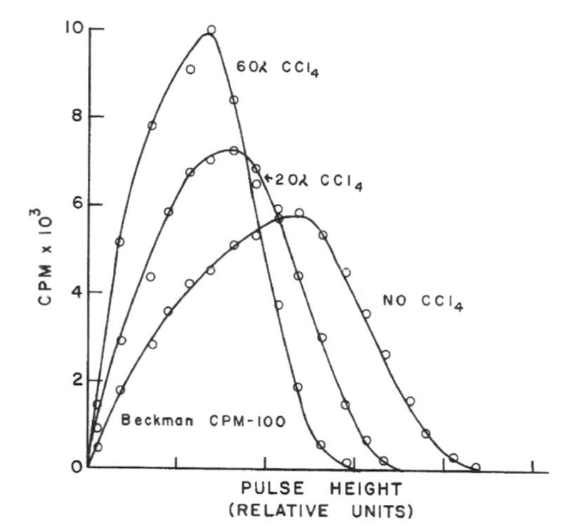

MARGIN FIGURE 8-19

Effects of quenching on the liquid scintillation spectrum for ^{14}C. The quenching agent is CCl_4.

The count rate for a particular sample must be corrected for quenching before the disintegration rate of the sample can be determined. Three methods for quench correction have been developed.[7] With the *internal spike method* (internal standard method), a cocktail is counted before and after a small quantity of "unquenched" material has been added to the sample. The unquenched material and the sample are labeled with the same radioactive isotope. The count rate without the unquenched material is subtracted from the count rate with the unquenched material. The counting efficiency is the difference in count rate divided by the disintegration rate for the unquenched material. With the *channels ratio method* for quench correction, a sample is counted in two separate "counting windows," or "channels," which are defined by upper, lower, and intermediate discriminators. The ratio of the count rates in the two channels varies with the amount of quenching in the cocktail. By reference to a calibration curve of counting efficiency versus the ratio of count rates, the counting efficiency for a particular sample may be determined. With the *external standard method* for quench correction, the sample is counted before and after a γ-emitting source (e.g., ^{137}Cs, radium, or ^{133}Ba) has been positioned adjacent to the scintillation vial. The increase in count rate obtained with the source near the vial varies with the amount of quenching in the sample. The counting efficiency is determined by referring to a calibration curve of counting efficiency versus the ratio of count rates before and after the γ-emitting source has been positioned adjacent to the scintillation vial.

Samples that are not soluble in a liquid scintillation cocktail may be counted by *suspension counting*. Gelling agents such as aluminum stearate, Cab-O-Sil, and Thixcin furnish suspensions of radioactive samples in various counting solutions. Techniques have been developed for counting insoluble samples such as filter paper, paper chromatograms, and Millipore filter disks. Scintillating beads are sometimes used when liquid or gaseous samples are counted by liquid scintillation.

MARGIN FIGURE 8-20
ALEPH is a particle detector that was built and is operated by a collaboration of several hundred physicists and engineers from 32 universities and national laboratories from around the world. Its purpose is to explore the Standard Model of particle physics and search for manifestations of new physics through observation of particles produced by the large electron–positron collider at the CERN laboratory in Geneva, Switzerland. The detector includes eight separate subsystems, making use of calorimetry, solid state, and other types of detectors. It is surrounded by one of the largest superconducting magnets in the world. Shown here is a photo of the front of the detector.

SEMICONDUCTOR RADIATION DETECTORS

Semiconductor detectors are often used for the detection of charged particles and photons emitted by radioactive nuclei. Semiconductor detectors exhibit many desirable properties, including (1) a response that varies linearly with the energy deposited in the detector and does not depend upon the type of radiation that deposits the energy, (2) a negligible absorption of energy in the entrance window of the detector, (3) excellent energy resolution, (4) the formation of pulses with fast rise times, and (5) small detector size.

The mechanism of response of a semiconductor detector resembles that for an ionization chamber. Ionization produced within the sensitive volume of the detector is converted to a voltage pulse that is amplified and counted. The size of the voltage pulse is proportional to the energy expended in the detector by the incident radiation. When compared with an ionization chamber, the voltage pulse is larger and more accurately reflects the energy deposited in the detector. Also, the rise time of the pulse is shorter because the ionization is collected more rapidly.

In most gases, an average energy of 30 to 40 eV is expended per ion pair produced. An ion pair is produced in a silicon semiconductor detector for each 3.5 eV deposited by incident radiation; in a germanium detector, only 2.9 eV is required to produce an ion pair. When compared with an ionization chamber, therefore, many more ion pairs are produced in a semiconductor detector for a given amount of energy absorbed.

Response of Semiconductor Detectors

A semiconductor radiation detector is similar to a transistor and is diagramed in Margin Figure 8-21. The p-type region is composed of a semiconducting element (e.g., germanium or silicon) "doped" with an *electron acceptor impurity* with fewer valence electrons. For example, a p-type semiconductor may be obtained by doping tetravalent germanium with trivalent boron, indium, or gallium. The n-type region is composed of germanium doped with an *electron donor impurity* such as antimony or lithium. Electrons flow from the n-type region to the p-type region and establish an electric field across the junction between the two regions. The region in the vicinity of the junction is termed the *depletion region* and may be increased in width by applying a *reverse bias* across the junction (positive potential to the n-type region, negative potential to the p-type region). The width of the depletion region may also be increased by a process known as *lithium drifting* to produce a Ge(Li) or Si(Li) detector.

If a charged particle or an x or γ ray loses energy within the depletion region, electrons are raised from the valence band to the conduction band, where they can migrate to the positive electrode (the n-type region). In the valence band, electrons move closer to the positive terminal by jumping to holes left by the released electrons. Other electrons fill the holes left by the jumping electrons. In this manner, holes migrate toward the negative terminal as if they were positively charged particles. The migration of positive holes in the valence band constitutes a current similar to that provided by electrons moving in the conduction band to the positive terminal. In fact, electrons released from the valence band, together with the positive holes left behind, constitute the ion pairs for a semiconductor detector.

In an intrinsic semiconductor, such as germanium, electrons are bound to atoms via covalent bonds in specific positions within a crystal structure. If an electron is freed from an atom, it can participate in current conduction by moving through the material. The atom from which the conduction electron originated has a deficiency that may be filled by neighboring electrons, causing the site of the deficiency to move. This moving positively charged "deficiency" region may be regarded as a fictitious particle called a "hole." At room temperature, intrinsic semiconductors rarely develop electron–hole pairs.

If a potential difference is applied across a pure semiconductor, a current is produced even if the semiconductor is not exposed to ionizing radiation. This current is the sum of (1) a *bulk current* that is dependent on the resistance of the semiconductor and the number of electron–hole pairs produced by thermal excitation and (2) a current caused by *charge leakage* at the surface of the semiconductor. These currents interfere with the identification of signals produced as radiation interacts within the detector. The bulk current is reduced with the p–n junction described above, and this barrier to current flow is required in a semiconductor radiation detector. The p–n

junction reduces the bulk current in silicon to an acceptable level at room temperature. Even with a p–n junction, however, the bulk current in a germanium detector is too great at room temperature. Consequently, germanium semiconductor detectors must be operated at reduced temperature. Germanium detectors are usually mounted in a cryostat and maintained at the temperature of liquid nitrogen (−190°C). The surface-leakage current is reduced in germanium and silicon by special techniques for constructing the detectors.

The size V of a voltage pulse from a semiconductor detector equals the charge Q collected by the electrodes divided by the capacitance C of the depletion region. The charge Q equals Ne, where N is the number of electron–hole pairs produced and e is the charge of the electron. That is, the size of the voltage pulse is proportional to the energy lost in the detector by the incident radiation. The size of the pulse is not dependent on the specific ionization of the radiation, because the ion pairs are swept away immediately and cannot recombine. Consequently, the response of the detector depends on the energy deposited in the detector but not on the type of radiation that deposits the energy.

The energy required to produce an ion pair in a semiconductor detector is only one-tenth of the energy required in a gas, and the voltage pulse from a semiconductor detector is about 10 times larger than the pulse from a gas-filled ionization detector. For example, a 1-MeV α-particle absorbed completely in the depletion region of a silicon semiconductor detector produces about $[10^6 \text{ eV}/(3.5 \text{ eV/IP})] = 3 \times 10^5$ ion pairs. The same particle produces only about $[10^6 \text{ eV}/(33.85 \text{ eV/IP})] = 3 \times 10^4$ ion pairs in an air-filled ionization chamber. The estimated percent standard deviation $\%\sigma/N$ (Chapter 11) for the pulse is

$$\frac{\%\sigma}{N} = \frac{100}{\sqrt{\text{Pulse size}}}$$

For the semiconductor detector, $\%\sigma/N$ is 0.18 for the pulse produced by a 1-MeV α particle. For the gas-filled detector, $\%\sigma/N$ is 0.58. Consequently, the range of pulse heights produced by the absorption of a given amount of energy is much narrower for a semiconductor detector than for a gas-filled ionization chamber, and the energy resolution of the semiconductor detector is much better. A similar analysis is applicable to the comparison of a semiconductor detector to a scintillation detector. Often, the maximum resolution obtainable with a semiconductor detector is limited by the preamplifier rather than by the detector.

However, if impurities, called doping agents, are introduced, a higher level of conduction is produced. In the doped state, light, heat, or ionizing radiation may trigger formation of electron–hole pairs. For example, if an atom of arsenic, having five outer electrons, is introduced into the crystal structure of germanium, which has four outer electrons, four of the arsenic electrons are taken up in covalent bonds while the fifth is free to move about as a negative charge carrier. Therefore, germanium doped with arsenic is an n-type semiconductor. Less than one arsenic atom per million germanium atoms is required to achieve this effect.

Similarly, if boron atoms, which have three outer-shell electrons, are used as a doping agent in germanium, holes appear, yielding a p-type semiconductor.

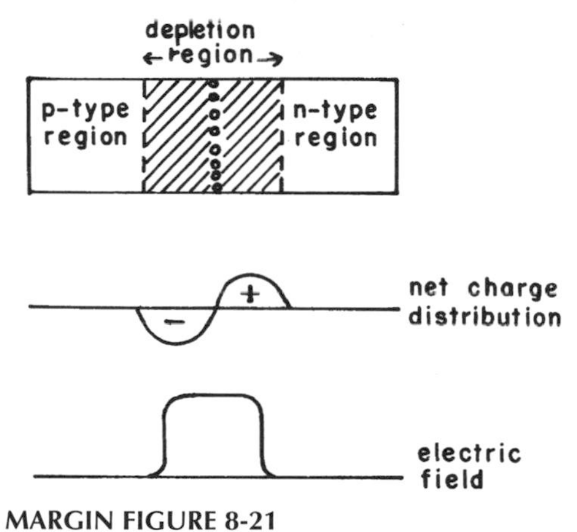

MARGIN FIGURE 8-21
Diagram of a p–n junction semiconductor detector.

Properties of Semiconductor Detectors

The efficiency of semiconductor detectors is nearly 100% for particulate radiations and relatively high for low-energy x and γ rays. The efficiency for detecting high-energy x and γ rays is lower because the depletion regions of the detectors are small. The atomic number is greater for germanium ($Z = 32$) than for silicon ($Z = 14$), and the γ-detection efficiency is higher.

Applications of Semiconductor Detectors

Because of their excellent energy resolution, semiconductor detectors are used widely for x- and γ-ray spectrometry and for similar laboratory measurements. These detectors have also been used for fluorescence scanning, where quantitative estimates of iodine in the thyroid are obtained by measurement of iodine x rays released as the thyroid is exposed to low-energy x or γ radiation.[8] Some effort has been directed toward extension of semiconductor detectors to imaging applications in nuclear medicine. Although not in widespread use at the present time, one vendor offers a photodiode coupled to each of 4096 CsI(Tl) scintillation crystals, thereby eliminating the light guide and photomultiplier tubes (see Chapter 12) and reducing the size and weight of the scanner.[9]

The capacitance C of a semiconductor detector may be computed with the expression:

$$C = 1.44\,\frac{A}{W}$$

where A is the area of the junction and W is the width of the depletion region.

PROBLEMS

*8-1. For a G–M detector operated with a voltage on the G–M plateau:
   a. Is the pulse produced by an α-particle larger than a pulse produced by a β-particle?
   b. Do the size and shape of the pulse vary with the length of the anode? With the diameter of the detector?

*8-2. Should a G–M detector be used to measure the exposure rate in the vicinity of a storage safe for radioactive nuclides?

8-3. What type of radiation detector would you recommend for the following:
   a. Detection of γ rays from ^{131}I in a patient's thyroid?
   b. Detector of β-particles from [^{3}H]thymidine dissolved in water?
   c. Detection of β-particles from ^{14}C in a gaseous sample of CO_2?
   d. Detection of α-particles from a plated source of ^{210}Po?
   e. Detection of radioactive contamination on a workbench in a laboratory where ^{32}P is used? In a laboratory where ^{3}H is used?
   f. Measurement of the exposure rate in the vicinity of a brachytherapy patient?
   g. ^{131}I in aqueous solution with ^{32}P?
   h. A mixture of ^{51}Cr and ^{131}I in blood?

*8-4. A γ ray from ^{241}Am (60 keV) is absorbed completely in an NaI(T1) crystal. The photomultiplier tube has 10 dynodes, with each dynode providing an electron multiplication factor of 3. About 80% of the light from the crystal is absorbed by the photocathode, which has a photocathode efficiency (number of electrons emitted per light photon absorbed) of 0.05. Assuming that 30 photons of light are produced in the NaI(T1) crystal per kiloelectron volt of energy absorbed, compute the number of electrons received at the anode of the photomultiplier tube.

*8-5. An α-particle from ^{210}Po (5.30 MeV) is absorbed completely in an air-filled pulse-type ionization chamber. Assuming a capacitance of 40 picofarads, compute the size of the voltage pulse.

8-6. Compare the efficiency and resolution of a semiconductor detector and a scintillation detector exposed to γ rays. Why are the efficiency and resolution different?

*For problems marked with an asterisk, answers are provided on p. 491.

SUMMARY

- An ionization chamber consists of parallel plate or coaxial electrodes in a volume occupied by a filling gas. The gas is sometimes air at ambient temperature and pressure.
- Quantities measured with an ionization chamber include:
  - Size of voltage pulse
  - Total charge collected
  - Magnitude of electrical current
- Uses of ionization chambers in radiology and nuclear medicine include:
  - Isotope calibrators
  - Gaseous sample counting
  - Portable survey meters
  - Neutron detection (with BaF_3 gas or lithium, or boron electrode coating)
- Proportional counters employ the principle of gas amplification to increase the magnitude of the signal.
- Geiger counters use further increases in voltage to obtain complete discharge of the gas in the counting volume with each counting event. Because all detection events elicit the same response, energy resolution is sacrificed to achieve a greater pulse size.
- Solid scintillation detectors produce light in response to absorption of ionizing radiation. Scintillation crystals are often coupled with photomultiplier tubes to achieve a measurable electrical signal, but may also be coupled with diode/thin film transistor arrays.
- Liquid scintillation employs the same principles of solid scintillation, but can be used to study dissolved samples of low-energy beta emitters.
- Semiconductor detectors record ionization events within their depletion regions as bulk current or as a voltage pulse.

REFERENCES

1. Hendee, W., Ibbott, G., and Hendee, E. *Radiation Therapy Physics*. 3rd edition. New York, John Wiley, 2002.
2. Chase, G., and Rabinowitch, J. *Principles of Radioisotope Methodology*, 3rd edition. Minneapolis, Burgess Publishing Co., 1967.
3. Reynolds, R., Snyder, R., and Overton, T. A multiwire proportional chamber positron camera: Initial results. *Phys. Med. Biol.* 1975; **20**:136.
4. Lim, C., et al. A multiwire proportional chamber positron camera: Preliminary imaging device. *J. Nucl. Med.* 1975; **16**:546.
5. Rapkin, E. Samples for Liquid Scintillation Counting, in Hine, G. (ed.). *Instrumentation in Nuclear Medicine*, Vol 1. New York, Academic Press, 1967, p. 182.
6. Knoll, G. F. *Radiation Detection and Measurements*. New York, John Wiley & Sons, 1989.
7. Hendee, W. *Radioactive Isotopes in Biological Research*. New York, John Wiley & Sons, 1973.
8. Esser, P., and Liter, D. A new apparatus for fluorescent scanning: A moving x-ray tube. *J. Nucl. Med.* 1977; **18**:640.
9. Digirad, San Diego, CA.

CHAPTER

9

ACCUMULATION AND ANALYSIS OF NUCLEAR DATA

OBJECTIVES

By studying this chapter, the reader should be able to:

- Identify the major components of a nuclear counting system and describe their properties and functions.
- Distinguish between differential and integral sorting in a pulse height analyzer.
- Define the advantages of a multichannel analyzer over a single-channel analyzer.
- Identify the factors contributing to determinate errors in radioactivity measurements.
- Discuss the property of resolving time of radiation detectors, and differentiate between nonparalyzable and paralyzable systems.
- Explain how a pulse-height spectrum is acquired, and identify the mechanisms that give rise to various peaks on a spectrum.
- Delineate the properties of a photopeak on a pulse-height spectrum, and explain the mechanism of photopeak counting.
- Describe the various properties of a radioactive pharmaceutical that are important to applications in nuclear imaging.

INTRODUCTION

Signals from a radiation detector used to detect and measure radioactivity are usually voltage pulses formed by collecting electrical charge or light flashes in the detector. Ideally, each voltage pulse corresponds to the emission of a single ionizing particle or γ-ray photon from the radioactive sample. The pulses may be transmitted from the detector to a succession of electronic circuits for analysis and display. With some detectors, the amplitude of the pulses depicts the energy deposited in the detector during interactions of the radiation. The pulses can be selectively registered if a discriminator or pulse height analyzer is present in the electronic circuitry. Data displayed by output devices must be interpreted in terms of their statistical significance and their relationship to the activity and modes of decay of the radioactive sample. Some of the difficulties encountered during this interpretation are considered in this chapter.

COUNTING SYSTEMS

Counting systems are assembled by combining various electronic circuits and display devices. A general-purpose counting system is outlined in Figure 9-1. Each component of this system is discussed separately in the following sections.

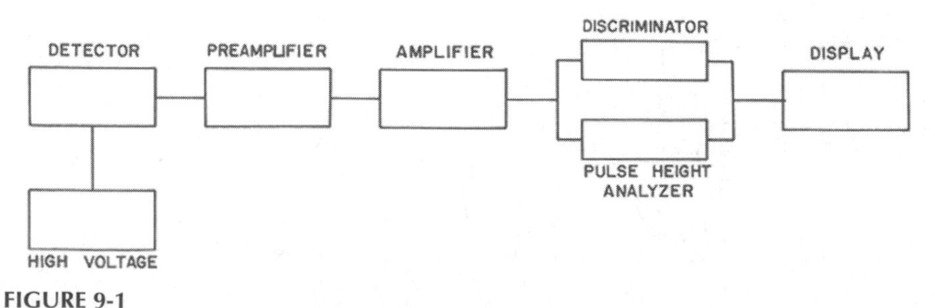

FIGURE 9-1

Components of a general-purpose counting system.

Preamplifiers

Almost all radiation detectors exhibit low capacitance and high impedance. Signals from these detectors are distorted and severely attenuated if they are transmitted by coaxial cable directly from the detector to an amplifier some distance away. To reduce these effects, a preamplifier may be inserted near the detector. The purpose of the preamplifier is to match the impedance of the detector with that of the amplifier. When this is done, the cable to the amplifier may be several feet long. Preamplifiers are used also to "clip" and "shape" the voltage pulse from the detector to meet the specifications of the amplifier.

A high impedance means that a detector has a high internal resistance to the flow of electrons.

The term *preamplifier* is misleading when applied to impedance matching circuits, because the voltage pulse from the detector usually is not amplified by the circuit. In fact, the gain of some preamplifiers is less than unity. Other preamplifiers may have a gain as high as 10^4 or 10^5.

Impedance matching means that the preamplifier presents a high impedance input to the detector and a low impedance output to succeeding electronic components. With this arrangement, signals from the detector are transmitted to the counting electronics in an undistorted fashion.

Amplifiers

An amplifier is used to increase the size and vary the shape of signals from the detector and preamplifier. An amplifier may be either voltage sensitive or charge sensitive, depending on the type of signal received at the input terminal. The increase in signal size is described as the *amplifier gain*, which is the ratio of the height of the voltage pulse leaving the amplifier to the size of the signal received at the input terminal of the amplifier. Depending on the type of detector and the characteristics of circuits in the counting system, an amplifier gain of 1000 to 10,000 may be required.

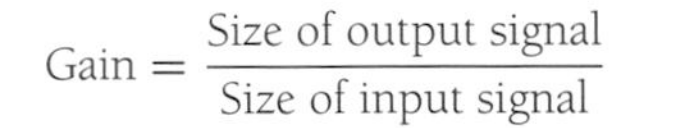

$$\text{Gain} = \frac{\text{Size of output signal}}{\text{Size of input signal}}$$

The pulse furnished by a typical amplifier is shown in the margin. The *pulse rise time* is the time required for the pulse amplitude to increase from 10% to 90% of its maximum amplitude. The *pulse decay time* is the time required for the pulse to decrease from maximum amplitude to 10% of maximum. The rise time of an amplifier should be less than the time required to collect the ion pairs or light produced during interaction of a particle or photon in the detector. The voltage pulse should be terminated rapidly to prevent the amplifier from summing successive pulses from the detector. The *integration time* of an amplifier is the time required to form an output pulse. The integration time is a compromise between the time required for complete collection of a signal from the detector and the time that causes a significant number of successive pulses to combine. The integration time is significantly less than 1 μs for amplifiers used with NaI(Tl) scintillation detectors.

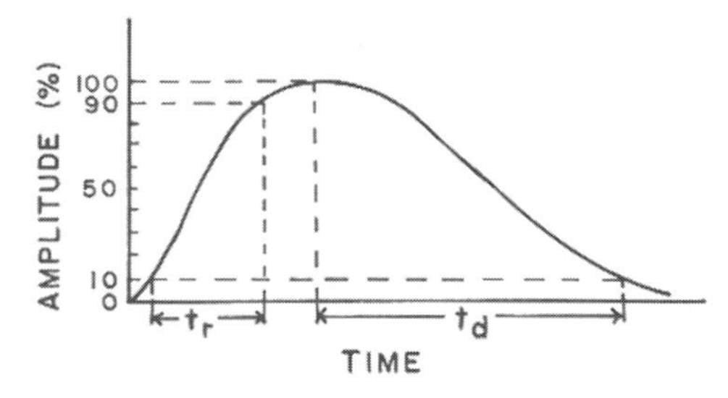

MARGIN FIGURE 9-1
Pulse delivered by an amplifier. Amplitude may be either voltage or instantaneous current. Pulse rise time t_r and pulse decay time t_d are shown.

Large input pulses or high rates of reception of input pulses may cause the characteristics of an amplifier to change. For example, a few output pulses may be distorted after a very large pulse (e.g., a pulse produced by interaction of a cosmic ray) has been received by the amplifier. This distortion of output pulses is termed *pulse-amplitude overloading*. *Count-rate overloading* refers to pulse-shape distortion caused by delivery of pulses to the amplifier at too high an input rate. Pulse-amplitude overloading and count-rate overloading may severely distort the data displayed by the output device.[1]

Pulses are amplified linearly in most amplifiers used in counting circuits. Linear amplification may be a handicap if pulses from the detector are widely variable in size. For these applications, an amplifier with logarithmic gain may be useful. With a logarithmic amplifier, the size of the output pulse is proportional to the logarithm of the size of the input pulse. In such an amplifier, a wide range of input pulses may be amplified without pulse-amplitude overloading and without rejection of very small pulses.

Performance criteria of an amplifier in a nuclear counting circuit include:

- Signal amplification
- Accurate measurement of the detector signal
- Acceptable performance at high count rates
- Maximum signal-to-noise ratio
- Rejection of pulse pile-up at high count rates

These criteria are not independent, and choices among them are required for any particular counting system.

Pulse Height Analyzers

With a detector such as a scintillation or semiconductor detector, the height of a voltage pulse from the detector (and amplifier) is proportional to the energy expended in the detector by a charged particle or an x or γ ray. A typical train of pulses from

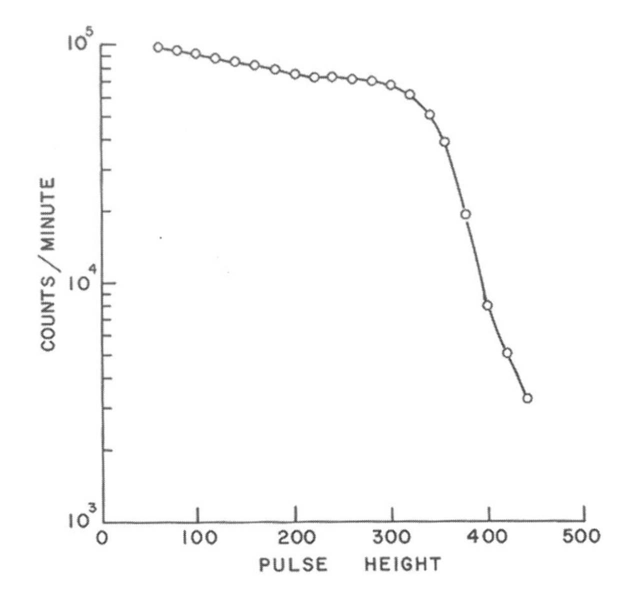

FIGURE 9-2
Integral pulse height spectrum for ^{131}I measured with a 2 × 2-in. NaI(Tl) well crystal.

A single-channel pulse-height analyzer is also known as a differential discriminator.

an amplifier connected to a scintillation or semiconductor detector is depicted in the margin. These pulses may be sorted by a single-channel pulse height analyzer (PHA) to yield a pulse height spectrum. This spectrum reflects the distribution of energies deposited in the detector during interactions of incident photons or particles. Two techniques, *differential sorting and integral sorting*, are used for pulse height analysis.

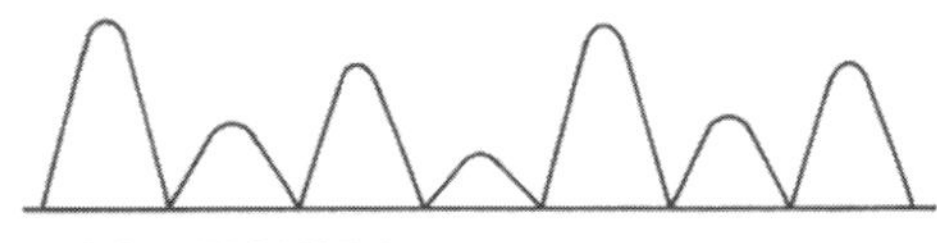

MARGIN FIGURE 9-2
Train of voltage pulses emerging from an amplifier that follows a preamplifier and a scintillation or semiconductor detector. The height of each pulse reflects the energy expended in the detector during interaction of an incident photon or particle.

For integral sorting, a single discriminator in the pulse height analyzer is varied from a position where all pulses are transmitted to the display device, to a position where all pulses are rejected. Shown in Figure 9-2 is an integral spectrum for ^{131}I. At any value of pulse height on the x axis, the height of the curve denotes the number of pulses that are large enough to pass the input discriminator and reach the display device.

A single-channel pulse height analyzer is composed basically of two discriminators connected to an anticoincidence circuit. The discriminators transmit pulses above a certain minimum size. In Figure 9-3, for example, the lower discriminator transmits pulses larger than size V_1, and the upper discriminator transmits pulses larger than size V_2. Pulses from the amplifier are applied simultaneously to both discriminators. Pulses too small ($<V_1$) to be transmitted by either discriminator are rejected and not transmitted to the anticoincidence circuit. Pulses of size between V_1 and V_2 are transmitted by the lower discriminator only and are delivered to one input terminal of the anticoincidence circuit. Pulses large enough ($>V_2$) to be transmitted by both discriminators are delivered simultaneously to both input terminals of the anticoincidence circuit. The anticoincidence circuit transmits a pulse to the display device when it receives a pulse at only one input terminal. A pulse is not transmitted to the display device when signals are received simultaneously at both input terminals.

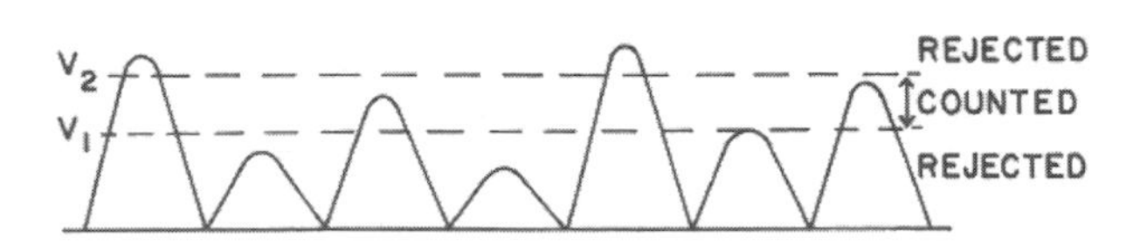

FIGURE 9-3
Diagram of a series of voltage pulses impinging on discriminators V_1 and V_2 of a single-channel pulse height analyzer. Pulses with height between V_1 and V_2 are counted. Pulses with height less than V_1 are rejected by both discriminators, and pulses with height greater than V_2 are rejected by an anticoincidence circuit.

Consequently, the display device registers only the number of pulses of size between V_1 and V_2. The range of pulse sizes registered by the display device may be varied by changing the settings V_1 and V_2 of the lower and upper discriminators. These settings may be labeled on the pulse height analyzer as "lower discriminator" and "upper discriminator," "E_1" and "E_2," or "lower level" and "upper level." Occasionally (e.g., in some scintillation cameras), the lower and upper discriminators of a pulse height analyzer are not adjustable. With this type of analyzer, differential pulse height analysis may be achieved by changing the range of pulse sizes emerging from the detector or by varying the amplification of the pulses in the amplifier. These changes may be accomplished by varying the high voltage of the photomultiplier tube in a scintillation detector, or by changing the gain of the amplifier.

The range of pulse sizes displayed on the output device may be varied by changing the following:

- Detector high voltage
- Amplifier gain
- Lower and upper discriminators

In a pulse height analyzer with discriminators that can be adjusted independently of each other, the range of pulse sizes transmitted to the anticoincidence circuit may be affected severely by small fluctuations in voltage applied to the discriminators. To reduce this dependence on voltage stability, the upper discriminator may be arranged to "ride" on the lower discriminator. In this manner, a constant difference in pulse size may be maintained between the discriminators. The lower discriminator may be termed "window level," "threshold," "E," "baseline," or "peak voltage," and the difference in pulse size between the two discriminators may be referred to as "window width," "window," "slit width," "ΔE," or "percent window." The position of the lower discriminator determines the minimum size of pulses transmitted to the display device, and the width of the window determines the increment of pulse sizes transmitted. As the window width is reduced, fewer pulses are transmitted to the display device, but the pulse-size (energy) resolution of the pulse height spectrum is increased. A differential pulse height spectrum for ^{131}I is shown in Figure 9-4.

Most pulse height analyzers may be operated in either the integral or differential mode. In the differential mode, some analyzers may be operated with independent lower and upper discriminators or with a variable lower discriminator and a dependent window.

The *linearity* of a pulse height analyzer describes the relationship between the position of the lower discriminator (or middle of the window) and the size of pulses transmitted to the display device. For a counting system with linear amplification, a straight line should be obtained if the average size of pulses admitted to the display

In 1927, Niels Bohr described the Principle of Complementarity for quantum systems. As a companion to Heisenberg's Uncertainty Principle, the Principle of Complementarity states that for every measurable quantity there exists a "complementary" measurable quantity. Increasing the accuracy of one quantity reduces the accuracy with which the complementary quantity can be known.

At the end of the lecture in which Bohr presented the Principle of Complementarity, a member of the audience asked "What, then, is complementary to Truth?" Without hesitation, Bohr responded "Clarity!"

Reported by L. Litt, M.D.
New York Times, May 2000

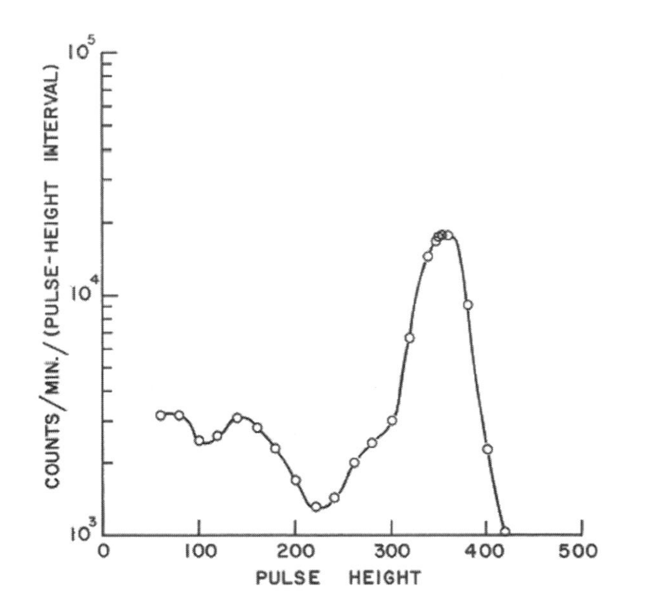

FIGURE 9-4
Differential pulse height spectrum for ^{131}I measured with a 2 × 2-in. NaI(Tl) well crystal. The peak in the spectrum represents pulses produced by total absorption of 364-keV γ rays in the scintillation crystal.

Pulses selected by a pulse-height analyzer for measurement or image display are referred to as "Z pulses."

The multichannel analyzer began to replace single-channel analyzers in most counting systems in the late 1960s.

In a scintillation camera, pulses transmitted by the pulse-height analyzer may be counted in a scaler, displayed on a cathode-ray tube, recorded on film, transmitted to a computer, stored on magnetic tape or laser disk, or any combination thereof.

Determinate (also termed systematic) errors are caused by inadequate design of the measurement system. Determinate errors are distinguishable from indeterminate (random) errors that are intrinsic to measurements of radioactivity.

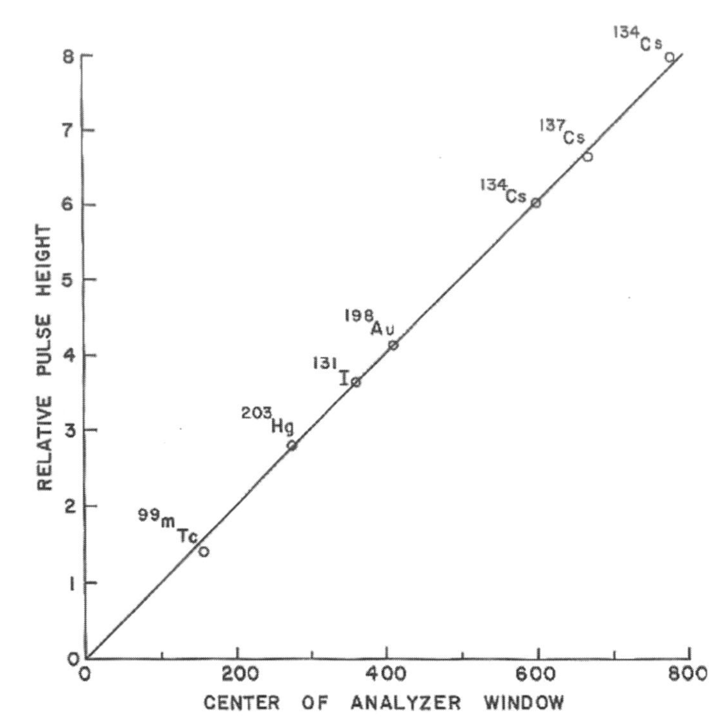

FIGURE 9-5
The average size of pulses transmitted by a pulse height analyzer as a function of the position of the center of the analyzer window.

device is plotted as a function of the position of the lower discriminator or center of the analyzer window (Figure 9-5). The maximum departure from a straight line is termed the *integral nonlinearity* and is less than 1% in a satisfactory counting system. Zero offset describes the positive or negative displacement from the origin of a curve such as that in Figure 9-5.

Multichannel Analyzers

To determine the number of pulses of different sizes impinging upon a single-channel pulse height analyzer, counts must be recorded for selected intervals of time while the window of the analyzer is moved incrementally from the smallest to the largest pulse size encountered. This procedure is tedious and imprecise because each pulse height "channel" is sampled independently and for only a short time. Also, pulse height spectra for isotopes with short half-lives are difficult to measure with a single-channel analyzer. With a single-channel analyzer and a counting time of 1 minute per channel, more than 100 minutes are required to sample a pulse height distribution that is divided into 100 parts. With a 100-channel analyzer, the same data may be collected in 1 minute. Multichannel analyzers are available commercially with thousands of channels.

In the multichannel analyzer, a pulse from the amplifier is fed to an analog-to-digital converter (ADC). Within the ADC, charge is stored in a capacitor in proportion to the amplitude of the incoming pulse. After receipt of the pulse, the capacitor is discharged while an oscillator emits pulses at a constant rate. The number of pulses emitted by the oscillator reflects the amplitude of the original pulse fed to the ADC. The number of oscillator pulses determines the location in the computer memory where a binary number is increased by 1 to reflect receipt by the ADC of a pulse of specified magnitude. Each storage location in the computer memory corresponds to a specific pulse amplitude, and the binary number stored at each location reflects the number of pulses of a specific amplitude received by the ADC during the counting period. These numbers can be displayed graphically on a display device to portray

Detectors such as whole-body counters used to measure small amounts of radioactivity must be well shielded to reduce the background count rate to very low levels. Pre–World War II steel from decommissioned battleships has been employed as a detector shield because it is not contaminated with radioactive fallout.

the pulse height spectrum for the radionuclide. By arithmetic manipulation of data stored in the multichannel analyzer, the pulse height spectrum can be corrected for background counts, the presence of more than one radionuclide in the sample, and other influences.

Scalers and Timers

A scaler is used to record and display the number of pulses received from an amplifier or pulse height analyzer. Many scalers are equipped with an electronic timer to indicate elapsed counting time or to stop the accumulation of counts by the scaler after a preset counting time. Timers consist of a scaler pulsed by a constant-frequency oscillator. Elapsed or preset time may be displayed visually or printed automatically.

High-Voltage Supplies

The specifications for a high-voltage supply for a counting system vary with the detector used. A high-voltage supply for a G–M tube or a semiconductor detector need not be as stable as that for a scintillation detector, for example, because the signal from a G–M or semiconductor detector is affected little by fluctuations in applied voltage. On the other hand, the signal from a scintillation detector may vary by as much as 10% for a 1% change in high voltage. The stability of a high-voltage supply may be affected by changes in temperature, fluctuations in line voltage, or variations in the impedance (load) across the output terminals.

DETERMINATE ERRORS IN RADIOACTIVITY MEASUREMENTS

The count rate measured for a radioactive sample reflects the rate of decay of atoms in the sample. However, the influence of several determinate errors must be known before the activity of the sample can be determined from the measured count rate. This influence must be included in an expression for the relationship between sample activity A and measured count rate c_{s+b} [Eq. (9-1)]:

$$A = \frac{c_{s+b}}{\{[1-(c)_{s+b}]\tau\}\, EfGBf_o f_w f_s} - (c)_b \quad \text{(9-1)}$$

where corrections to the background count rate are assumed to be negligible.

- $(c)_{s+b}$ = Sample-plus-background count rate in counts per minute (cpm)
- $(c)_b$ = Background count rate (cpm)
- τ = Resolving time in minutes
- E = Detector efficiency
- f = Fractional emission of source
- G = Geometry correction
- B = Backscatter correction
- f_0 = Sidescatter correction
- f_w = Correction for attenuation in detector window, air, sample covering, etc.
- f_s = Correction for self-absorption

These corrections are discussed in the following sections.

Background Count Rate

A counting system used to measure radiation from a radioactive sample almost always indicates the presence of small amounts of radiation when the radioactive sample is removed. The residual radiation (background) originates from a number of sources,

Background radiation contributes an annual whole-body radiation dose to each person on earth. At sea level, the background dose-equivalent rate is approximately 3.6 mSv/yr (360 mrem/yr), including about 2 mSv/yr (200 mrem/yr) from radon. The latter value is highly variable from one location to another.

When compared with other common radiation detectors, the resolving time of a G–M detector is rather long (100–300 μs), and the resolving time of a NaI(Tl) scintillation detector is rather short (0.5–10 μs).

The intense radiation belts (Van Allen belts) encircling the earth were discovered because of paralysis of radiation detectors on an early unmanned space craft.

In some texts the resolving time is called the dead time, and the coincidence loss is termed the *dead-time loss*.

The detector efficiency is also termed the *intrinsic efficiency*.

When a pulse-height window is used, the number of counts is further reduced by the fraction of counts actually transmitted by the pulse-height analyzer.

Well counters, in which the source is almost entirely surrounded by the detector, approach the condition of 4π geometry.

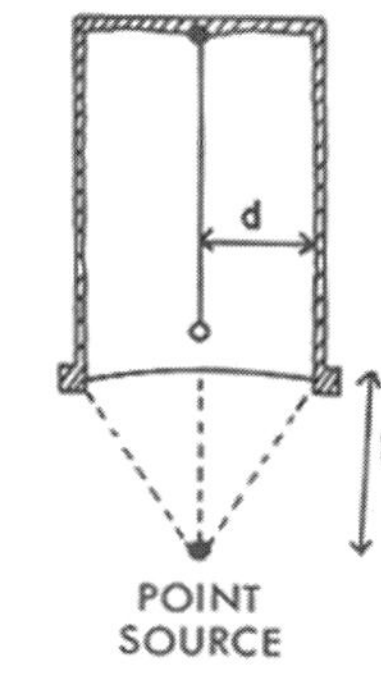

MARGIN FIGURE 9-3
Geometric relationship between a detector and a point source of radioactive material.

including (1) cosmic radiation, (2) radioactive materials such as ^{226}Ra, ^{14}C, and ^{40}K in the earth, human body, walls of buildings, etc., (3) radioactive materials stored or used nearby, (4) radioactive materials in devices such as watches and instrument dials, (5) radioactive contamination of the counting equipment, laboratory benches, etc., and (6) radioactive fallout (this source contributes no more than 1% to 2% to the total background radiation). Usually radiation detectors are surrounded by lead or other shielding material to reduce the background count rate. Other methods to reduce background include the use of pulse height analyzers and coincidence or anticoincidence circuits.

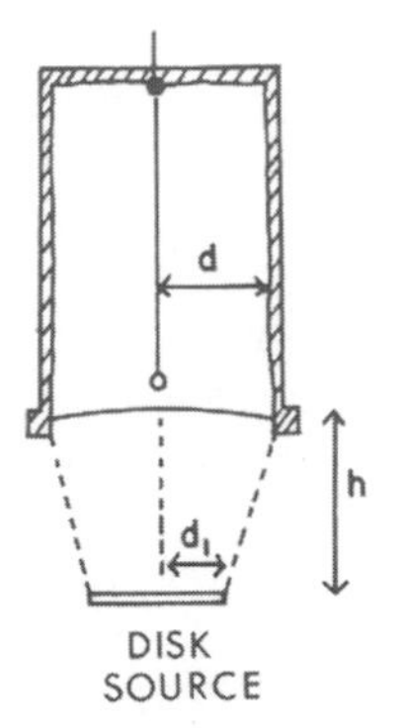

MARGIN FIGURE 9-4
Geometric relationship between a detector and a disk source of radioactive material.

Resolving Time

The resolving time of a radiation detector is the interval of time required between successive interactions in the detector for the interactions to be recorded as independent events. The resolving time of a detector may be measured with an oscilloscope that displays pulses from the detector. Sequential measurements of the count rate for a short-lived radioactive nuclide may also be used to determine resolving time.[2]

Detectors, and the counting systems in which they are used, may be classified as one of two types:

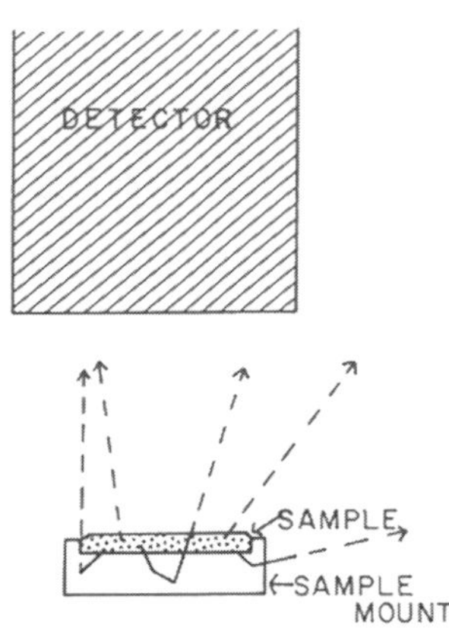

MARGIN FIGURE 9-5
Backscattering of radiation.

- *Nonparalyzable systems* in which, if an interaction occurs within the resolving time of a preceding event, the second interaction is not registered but has no effect on subsequent events. Most electronic components, with the exception of radiation detectors, exhibit nonparalyzable behavior.
- *Paralyzable systems* in which the second interaction, even though it is not registered in computer memory, extends the resolving time beyond that for the first event. Most radiation detectors exhibit paralyzable behavior, a phenomenon referred to as "pulse pile-up."

Observed versus true count rates are depicted in Figure 9-6 for nonparalyzable and paralyzable counting systems.

For nonparalyzable systems, the count rate c_0 corrected for the loss of counts caused by resolving time is

$$c_0 = \frac{c}{1 - c\tau} \tag{9-2}$$

where c is a measured count rate in counts per minute and τ is the resolving time in minutes. The coincidence loss is the difference between the uncorrected count rate c and the corrected count rate c_0.

Detector Efficiency

The efficiency of a radiation detector is the quotient of the number of particles or photons that interact in the detector divided by the number of particles or photons

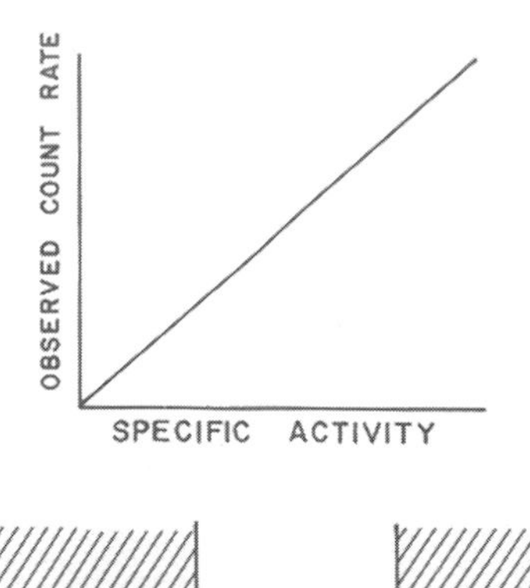

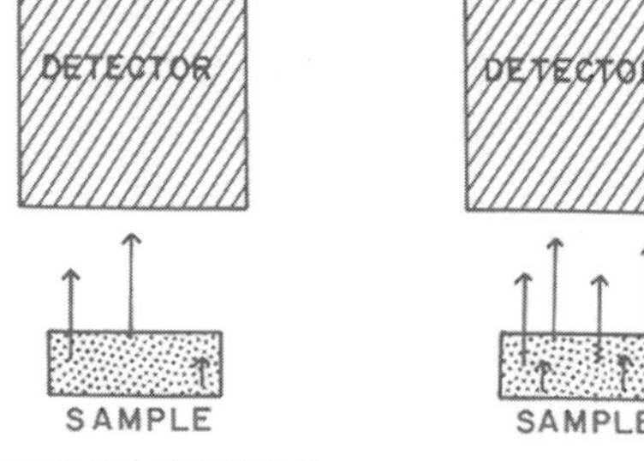

MARGIN FIGURE 9-6
Count rate for a sample with constant volume but increasing specific activity.

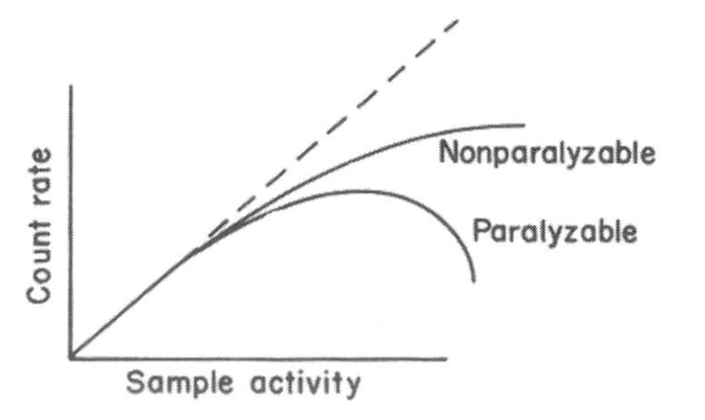

FIGURE 9-6
Count rate characteristics of paralyzable and nonparalyzable counting systems.

that enter the detector. The efficiency of most gas-filled detectors is close to 1 for α- and β-particles and about 0.01 for x and γ rays. The efficiency of a NaI(Tl) crystal or semiconductor detector for the detection of x and γ rays varies with the thickness of the crystal and energy of the x and γ rays.

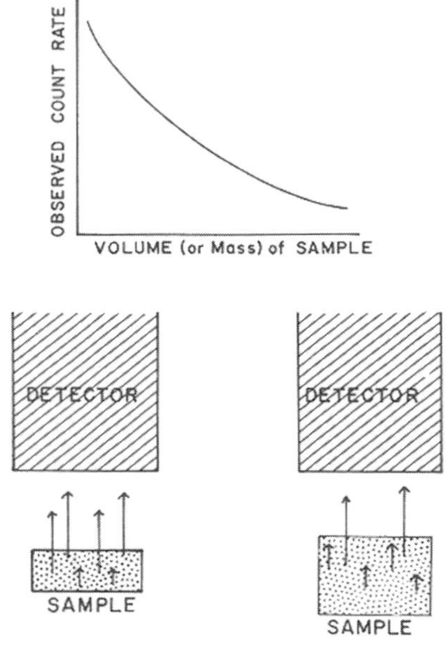

MARGIN FIGURE 9-7
Count rate for a sample with constant activity but increasing sample volume.

Fractional Emission

The fractional emission of a radioactive sample is the fractional number of decays that yield the radiation to be detected. For example, suppose that γ rays (0.393 MeV) from ^{113m}In are detected. Because the coefficient of internal conversion is 0.55 for ^{113m}In, the number of decays that furnish a γ ray of 0.393 MeV is $1/(1 + 0.55) = 0.645$. Hence the fractional emission is 0.645 for the 0.393 MeV γ rays from ^{113m}In.

Detector Geometry

The *geometry correction* is the ratio of the number of particles or photons emitted in the direction of a detector to the total number of particles or photons emitted by the sample. Usually, the detector is said to "subtend" a certain solid angle Ω measured in steradians. A sphere subtends a solid angle of 4π steradians, and the geometry correction G is

$$G = \frac{\Omega}{4\pi}$$

The expressions *2π geometry* and *4π geometry* are used to describe counting conditions where half of all of the radiation emitted by the source is intercepted by the detector.

The geometry correction varies with the radius d of the detector and with the distance b between the detector and the source. The correction is described in Eq. (9-3) for point and disk sources of radioactive material:
Point source:

$$G = 0.5\left(1 - \frac{h}{\sqrt{h^2 + d^2}}\right) \tag{9-3}$$

Disk source (radius of disk = d_1):

$$G = 0.5\left(1 - \frac{h}{\sqrt{h^2 + d^2}}\right) - \frac{3}{16}\left(\frac{dd_1}{h^2}\right)^2\left(\frac{h}{\sqrt{h^2 + d^2}}\right)$$

Scattering

Radiation emitted from a radioactive source in a direction outside the solid angle subtended by a detector may be scattered toward the detector as the radiation interacts with the sample container or with shielding around the sample and detector. The scattered particles or photons are backscattered if they were originally emitted in a direction away from the detector (see margin illustration). If the backscattered radiation originates entirely within the mount for the sample, the percent backscatter (%B) and backscatter factor (B) are

$$\%B = \left[\frac{\text{Counts with mount present} - \text{Counts with mount removed}}{\text{Counts with mount removed}}\right] \times 100 \tag{9-4}$$

$$B = \frac{\text{Counts with mount present}}{\text{Counts with mount removed}} \tag{9-5}$$

The backscatter factor is related to the percent backscatter by

$$\%B = 100(B - 1)$$

For β-particles, the backscatter factor increases with the atomic number of the sample mount and, initially, with the thickness of the mount. *Saturation thickness* is

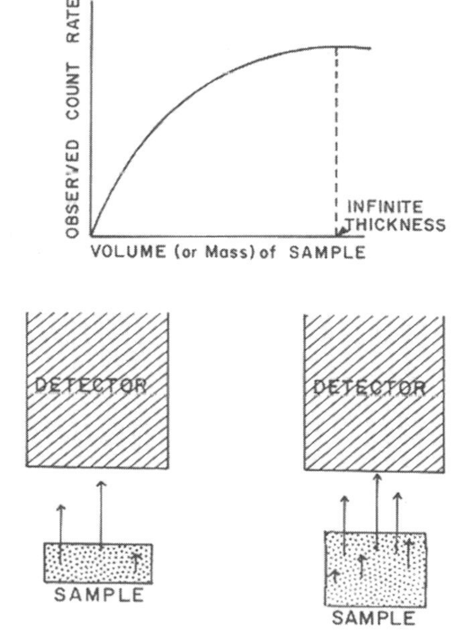

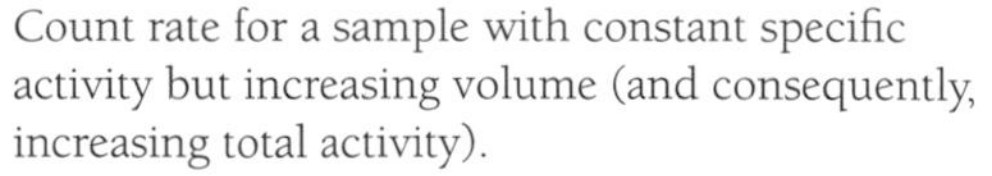

MARGIN FIGURE 9-8
Count rate for a sample with constant specific activity but increasing volume (and consequently, increasing total activity).

Scintillations created in solid substances by ionizing radiation were first observed by Sir William Crookes in 1903.

The study of pulse height spectra is termed *pulse height spectroscopy*, sometimes referred to as *scintillation spectroscopy*.

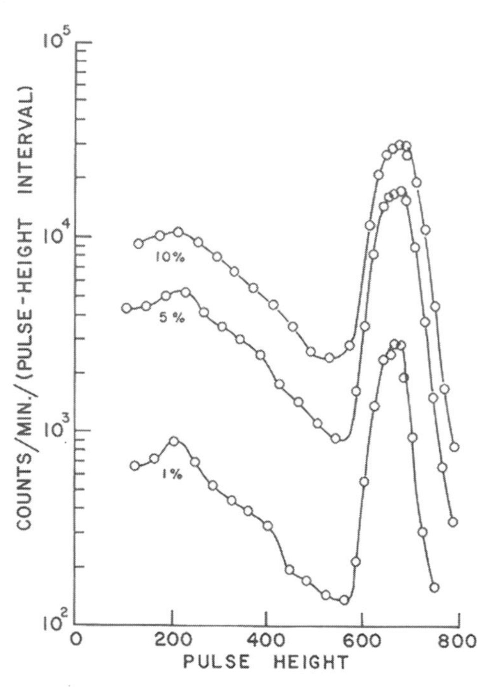

MARGIN FIGURE 9-9
Increased count and decreased resolution (depicted by a widening of the photopeak) are achieved as the window width is increased. Window widths are expressed as a percentage of the highest setting of the pulse height scale. Spectra were measured for a ^{137}Cs source in a 2 × 2-in. NaI(Tl) well detector.

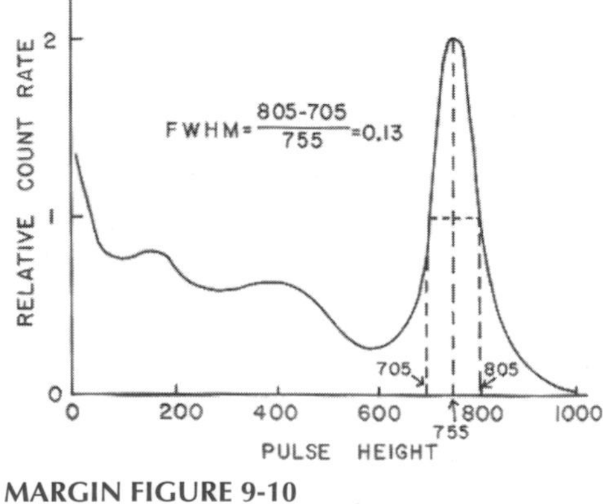

MARGIN FIGURE 9-10
"Full width at half-maximum."

achieved when an increase in thickness does not increase the backscattered radiation. For β-particles, saturation thickness equals about three-tenths of the range of the particles in the mount. Backscattering may be reduced by using a sample mount of low atomic number. For β counting, for example, aluminum planchets are often preferred over copper or steel planchets because the backscatter factor is lower with aluminum. If the sample mount provides saturation thickness, then the backscatter factor for β-particles is virtually independent of the energy of the β-particles.

Sidescatter may be reduced by moving all scattering material (e.g., shielding around the source and detector) far from the path between the source and the detector. Sidescatter may also be reduced with a sleeve of plastic, aluminum, or other low-Z material inside the high-Z shield around the source and detector.

Air and Window Absorption

Radiation is attenuated by material between the radioactive source and the sensitive volume of the detector. Usually, at least three different attenuators are encountered by radiation moving toward the detector. These attenuators are (1) the covering over the radioactive source, (2) air between the source and the detector, and (3) the entrance window of the detector.

Self-Absorption

The term *self-absorption* refers to the attenuation of radiation within the radioactive sample itself. Shown in the margin is the increase in count rate for a radioactive sample achieved if the activity of the sample is increased with no change in sample volume. The data describe a straight line because the fraction of radiation absorbed by the sample remains unchanged. However, if the volume of the sample increases with no change in activity, then the count rate for the sample decreases because an increasing fraction of the radiation is absorbed. For a sample with increasing volume and constant specific activity (specific activity equals the activity of the sample divided by the mass of the sample), the count rate approaches a constant value at large thickness ("infinite thickness") because the increased absorption of radiation by the sample compensates for the increased activity.

GAMMA-RAY SPECTROMETRY

Attenuation coefficients for photoelectric, Compton and pair-production interactions of γ rays in NaI(Tl) are plotted in Figure 9-7 as a function of γ-ray energy. For γ rays with energy less than about 200 keV, photoelectric absorption is the most probable interaction in NaI(Tl). Compton scattering is the dominant interaction between about 200 keV and 7 MeV, and pair production is most important at energies greater than 7 MeV. A γ ray that interacts in a NaI(Tl) crystal may deposit part or all of its energy in the crystal. The size of the resulting voltage pulse from the photomultiplier tube coupled to the crystal is proportional to the energy deposited in the crystal by the γ ray. Pulses from a scintillation detector may be sorted with a scintillation spectrometer to yield a pulse height spectrum that reflects the energy distribution of the interactions of incident photons.[3,4]

PULSE HEIGHT SPECTRA

The pulse height spectrum for a radioactive source consists of peaks and valleys that reveal the energy of the γ rays from the source. The spectrum also reflects interactions that occur before the photons reach the scintillation detector, interactions in the detector, and the operating characteristics of components of the counting system. If the amplifier gain or the high voltage across the photomultiplier tube is increased,

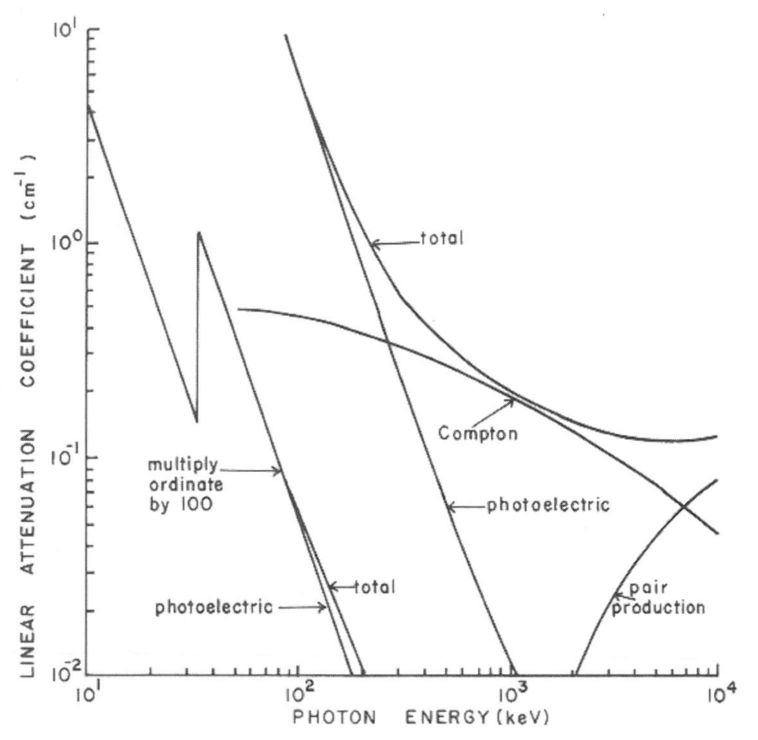

FIGURE 9-7
Linear attenuation coefficients of NaI(T1) for photoelectric absorption, Compton scattering, and pair production as a function of γ-ray energy.

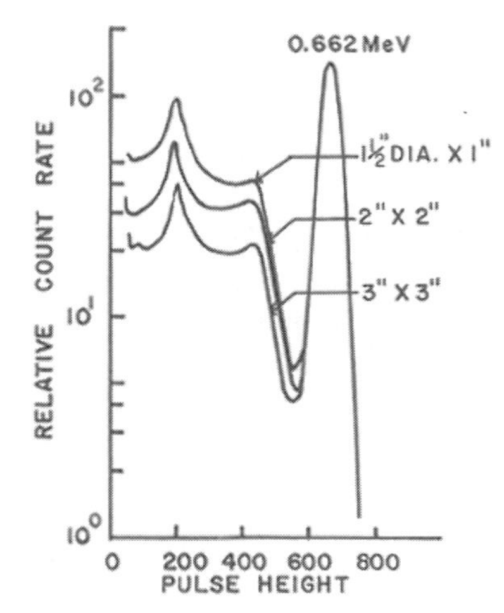

MARGIN FIGURE 9-11
Pulse height spectra for ^{137}Cs measured with NaI(Tl) crystals of different sizes. The photopeaks are normalized to the same height. (From Heath, R.[3] Used with permission.)

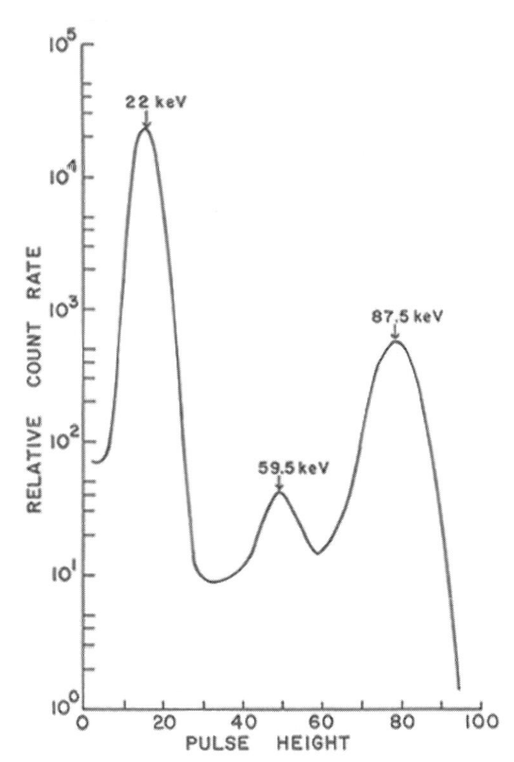

MARGIN FIGURE 9-12
Pulse height spectrum for ^{109}Cd. An iodine x-ray escape peak is present at a position 28 keV below the photopeak for the 87.5-keV γ rays from ^{109}Cd. The large photopeak at 22 keV is produced by absorption of characteristic x rays released during decay of ^{109}Cd by electron capture. (From O'Kelley, G.[4] Used with permission.)

then each pulse is amplified proportionally and displayed at a position farther along the pulse height scale. Hence an increase in high voltage or gain expands the pulse height spectrum over a greater range of pulse heights (Figure 9-8). The height of the spectrum decreases because the voltage pulses encompass a wider range of pulse heights and each pulse height interval contains fewer pulses.

As the window of the pulse height analyzer is widened, more pulses are transmitted to the display device for each position of the window along the pulse height scale. Consequently, the count rate at each window position increases with window width. With a wider window, however, peaks and valleys in the spectrum are defined less precisely, and the resolution of the spectrum is reduced. Pulse height spectra for ^{137}Cs measured with analyzer windows of different widths are shown in Margin Figure 9-9.

Photopeaks

A *photopeak* on a pulse height spectrum encloses pulses produced by total absorption of γ rays of a particular energy in the scintillation crystal. The energy of the γ ray may have been deposited in the crystal during one interaction or a series of interactions. The *photofraction* or *peak-to-total ratio* for a photopeak is the area enclosed within the photopeak divided by the total area enclosed within the pulse height spectrum. The intrinsic peak efficiency is the fraction of the γ rays of a particular energy impinging upon a detector that produces pulses enclosed within the photopeak. The *resolution* of a pulse height spectrum is sometimes described as the *full width at half-maximum* (FWHM), where

$$\text{FWHM} = \frac{\text{Width of the photopeak at half its maximum height}}{\text{Position of the photopeak along the pulse height scale}} \quad (9\text{-}6)$$

The full width at half-maximum is illustrated in Margin Figure 9-10.

The x-ray escape peak is occasionally referred to as an "iodine escape peak" because the peak reflects escape of an x ray produced by an L→K electron transition in iodine.

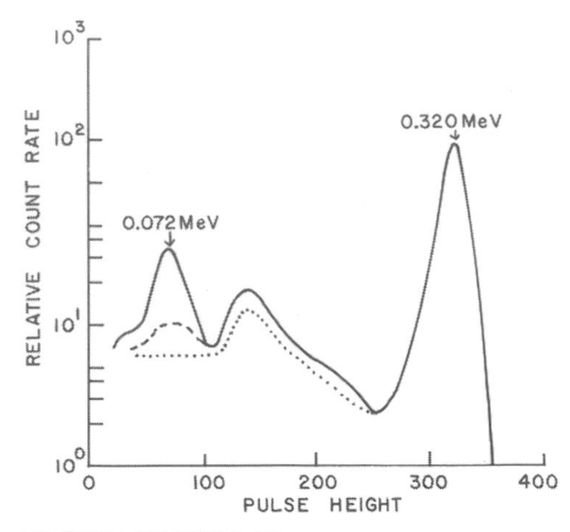

MARGIN FIGURE 9-13
Characteristic x-ray peak at 72 keV produced by x-rays from a 6 x 6-in. lead shield surrounding a scintillation detector and a ^{51}Cr source. The x-ray peak is reduced by increasing the inner dimensions of the shield to 32 × 32 in. (dotted curve) or by lining the shield with 0.03 in. of cadmium (dashed curve). (From Heath, R.[3] Used with permission.)

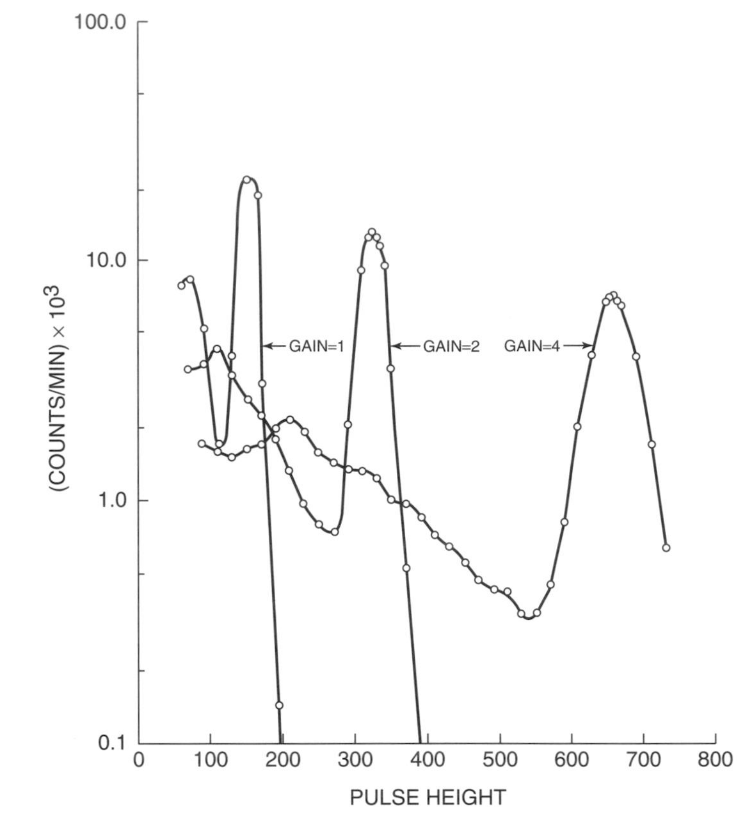

FIGURE 9-8
Pulse height spectra for ^{137}Cs measured with different settings of the amplifier gain.

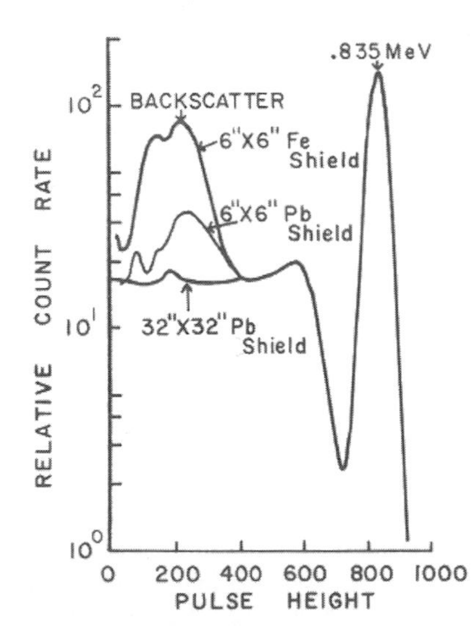

MARGIN FIGURE 9-14
The backscatter peak at about 200 keV is produced by absorption in the detector of photons scattered at wide angles during Compton interactions of primary γ rays in the detector-source shield. The backscatter peak is greatest for the 6 × 6-in. iron shield because the relative probability of Compton interaction of the 0.835-MeV γ rays from ^{54}Mn is higher for iron than for lead. The backscatter peak is reduced if the distance between the shield and the detector is increased.

Compton Valley, Edge, and Plateau

Compton interactions of incident γ rays in a NaI(Tl) crystal, with escape of the scattered photons from the crystal, result in the formation of voltage pulses smaller than those enclosed within a photopeak. The valley just below a photopeak is referred to as the *Compton valley*. The *Compton edge* is the portion of the pulse height spectrum just below the Compton valley where the height of the spectrum increases rapidly. The portion of the pulse height spectrum below the Compton edge is termed the *Compton plateau*.

The relative heights of the photopeak and the Compton plateau vary with the size of the NaI(Tl) crystal. In a larger crystal, more scattered photons interact before they escape from the crystal. Hence the height of the photopeak is increased, and the height of the Compton plateau is reduced (see Margin Figure 9-11).

X-Ray Escape Peaks

A γ ray that interacts photoelectrically in a NaI(Tl) crystal usually ejects an electron from the K shell of iodine. As the vacancy left by the photoelectron is filled by an electron from the L shell, a characteristic x ray is released with an energy equal to the difference (28 keV) in binding energy of L and K electrons in iodine. If the 28-keV x ray interacts in the scintillation crystal, then the light released during this interaction contributes to formation of the voltage pulse for the primary γ ray, and this voltage pulse falls within the photopeak. However, if the x ray escapes from the crystal, then the pulse for the primary γ ray is smaller than those enclosed within the photopeak. If many x rays escape during γ ray interactions, then a peak occurs on the pulse height spectrum at a location about 28 keV less than the photopeak (Margin Figure 9-12). This peak is referred to as an *x-ray escape peak*. X-ray escape peaks are not observed for

γ rays with energy greater than about 200 keV because the lateral spread of the photopeak for photons of higher energy obscures the presence of the x-ray escape peak. Also, characteristic x rays released during interactions of higher-energy γ rays have a lower probability of escape from the crystal because more energetic γ rays tend to penetrate farther into the crystal before interacting.

Characteristic X-Ray Peak

Characteristic x rays are released as γ rays from a radioactive source undergo photoelectric interaction in the lead shield surrounding the source and detector. Some of these x rays escape from the shield and strike the NaI(Tl) crystal to produce peaks in the pulse height spectrum. In Margin Figure 9-13, the peak at 0.072 MeV reflects the absorption of characteristic x rays from a lead shield surrounding a scintillation detector and a ^{51}Cr source. The number of characteristic x rays reaching the detector may be reduced by increasing the distance between the detector and shield or by lining the shield with a material with atomic number less than that of lead.

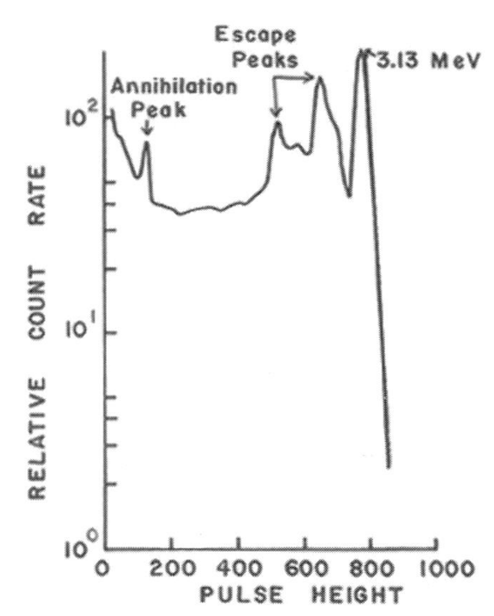

MARGIN FIGURE 9-15
The pulse height spectrum for ^{37}S illustrates single and double escape peaks and an annihilation peak at 0.511 MeV.

Backscatter Peak

If a photon with energy greater than 200 keV is scattered at a wide angle during a Compton interaction, then the scattered photon has an energy of about 200 keV irrespective of the energy of the primary γ ray. For this reason, photons scattered at wide angles during Compton interactions in the detector-source shield produce a peak at about 200 keV on the pulse height spectrum (see Margin Figure 9-14). This peak is referred to as the *backscatter peak*. The number of scattered photons received by the detector may be reduced by increasing the distance between the shield and the detector or by choosing a material for the shield in which the number of Compton interactions is reduced.

Annihilation Peak

Pair-production interaction of γ rays with an energy greater than 1.02 MeV is accompanied by release of 511-keV annihilation photons. If a primary γ ray interacts by pair production in the detector-source shield, then one of the annihilation photons may escape from the shield and interact in the crystal. This process results in the production of an annihilation peak of 511 keV in the pulse height spectrum (see margin). An annihilation peak at 511 keV is also present in pulse height spectra for positron-emitting nuclides because annihilation radiation is released as the positron combines with an electron.

Annihilation Escape Peaks

Pair production of high-energy (>1.02 MeV) photons in the scintillation crystal results in the emission of two 511-keV annihilation photons, one or both of which may escape from the crystal. Pulses that reflect the loss of one annihilation photon contribute to the *single-escape peak* that occurs at a location 0.511 MeV below the photopeak for the primary γ ray. Pulses that reflect the loss of both annihilation photons contribute to the *double-escape peak* that occurs at a location 1.02 MeV below the photopeak. Single- and double-escape peaks, together with an annihilation peak at 0.511 MeV, are present in the pulse height spectrum for ^{37}S, a nuclide that decays by the emission of a β-particle and a γ ray with an energy of 3.13 MeV (Margin Figure 9-15).

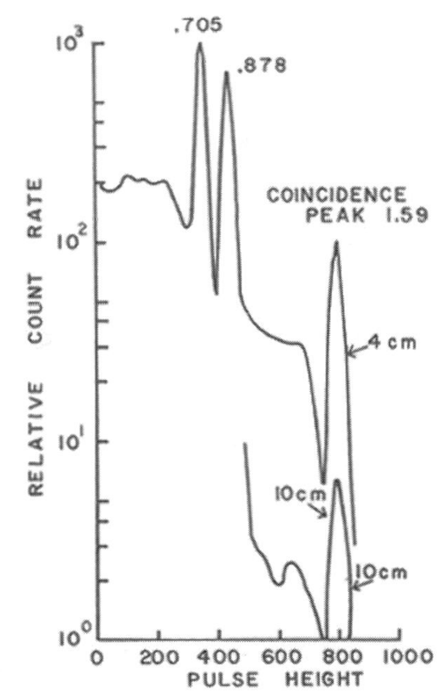

MARGIN FIGURE 9-16
The coincidence peak reflects the simultaneous absorption of 0.705-MeV and 0.878-MeV γ rays from ^{94}Nb. The relative height of the coincidence peak is reduced if the distance between source and detector is increased from 4 cm to 10 cm.

Coincidence Peak

Many radioactive sources decay with the emission of two or more γ rays in cascade. When γ rays from these sources are detected, more than one of the cascading γ rays may be completely absorbed in the crystal. Coincidence peaks in the pulse height

Coincidence peaks are often termed "sum peaks."

spectrum reflect the simultaneous absorption of more than one γ ray in the crystal (Margin Figure 9-16). The relative height of a coincidence peak is reduced if the distance is increased between source and detector, because the probability is reduced that two or more photons will strike the crystal simultaneously. For high-activity samples, coincidence peaks may be produced by simultaneous absorption of γ rays released during the decay of different atoms of the sample.

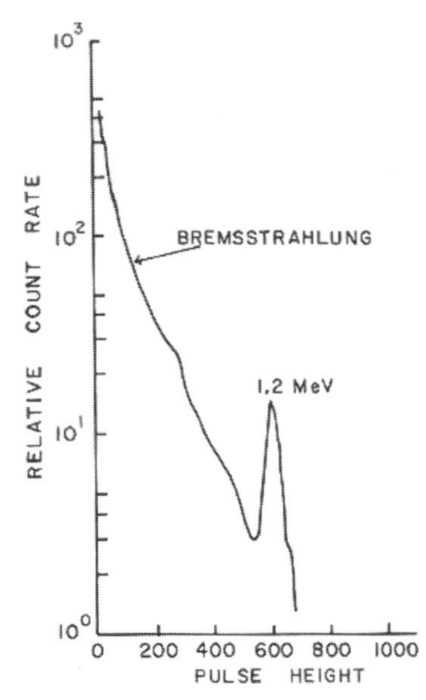

MARGIN FIGURE 9-17
The pulse height spectrum of ^{91}Y illustrates the bremsstrahlung contribution characteristic of a radioactive source for which the ratio of γ rays to β-particles is low.

Bremsstrahlung

Bremsstrahlung is released as β-particles from a radioactive source are absorbed in the source or source-detector shield. If the ratio of γ rays to β-particles emitted by the source is low, then the contribution of bremsstrahlung to the pulse height distribution may be noticeable, particularly at the lower end of the pulse height scale (Margin Figure 9-17).

PHOTOPEAK COUNTING

Radioactive sources that emit γ rays are often counted by sending to the display device only those pulses that are enclosed within a pulse height window centered over the photopeak. Other pulses are rejected because they are larger or smaller than the pulses transmitted by the window. To center the window over the photopeak, the window is positioned at a desired location along the pulse height scale. Then the amplifier gain or the high voltage is increased until a maximum count rate is obtained, indicating that the photopeak is centered in the window. For example, if the photopeak for ^{137}Cs is desired at 662 on the pulse height scale, then the window is centered at 662, and the gain or high voltage is increased until a maximum count rate is obtained. If the size of the voltage pulse (i.e., the pulse height) varies linearly with energy deposited in the crystal, then this procedure furnishes a pulse height scale calibrated from 0 to 1000 keV, with each division on the pulse height scale representing 1 keV. The pulse height scale may be calibrated from 0 to 2000 keV by centering the ^{137}Cs photopeak at 331 and from 0 to 500 keV by centering the photopeak for a lower-energy γ ray at an appropriate position on the pulse height scale. A pulse height scale may be calibrated from 0 to 500 keV, for example, by centering the photopeak for ^{99m}Tc (primary γ ray of 140 keV) at 280 on the pulse height scale.

The fraction of detected γ rays that contribute to the photopeak is known as the *photofraction* or *photopeak efficiency*.

The width of the analyzer window most desirable for photopeak counting varies with the relative heights of the photopeak and background at the position of the window. For many counting situations, the background pulse-height spectrum has relatively little structure and is much lower than the photopeak for the sample. For these situations, a relatively narrow (e.g., 2%) window should be centered over the photopeak. If the photopeak is not much higher than background, then the value of the sample net count rate squared divided by the background count rate may be determined as a function of window width. When the ratio (known as the square of the signal divided by the background, or S^2/B ratio) is maximum, the fractional standard deviation is minimum for the net count rate of the sample.[5] A plot of S^2/B as a function of window width is shown in Figure 9-9 for a ^{137}Cs source and a 2 × 2-in. NaI(Tl) well detector. The optimum window is 70 keV.

The width of the photopeak is a reflection of statistical variations that occur in the process of forming voltage pulses in the detector. These variations include:

- Variations in the number of light photons released in the crystal per unit energy absorbed
- Variations in the number of light photons that must impinge on the photocathode to eject a photoelectron
- Variations in the number of electrons released at each successive dynode in the pm tube.

Ross and Harris[6] have shown that an acceptable window usually is achieved if lower and upper discriminators are positioned on either side of the photopeak at a location where the count rate is equal to one-third of the count rate at the center of the photopeak. About 85% of all pulses enclosed within the photopeak are transmitted by a window defined by discriminators in these positions, and the window is termed an *85% window*. The window width should be increased if the window or amplifier gain drifts or if the width of the window fluctuates. Under these circumstances the lower and upper discriminators should intersect the pulse height spectrum at equal heights near the base of the photopeak. If the gain of the amplifier is unstable, then the photopeak should be intercepted by the lower and upper discriminator at relative

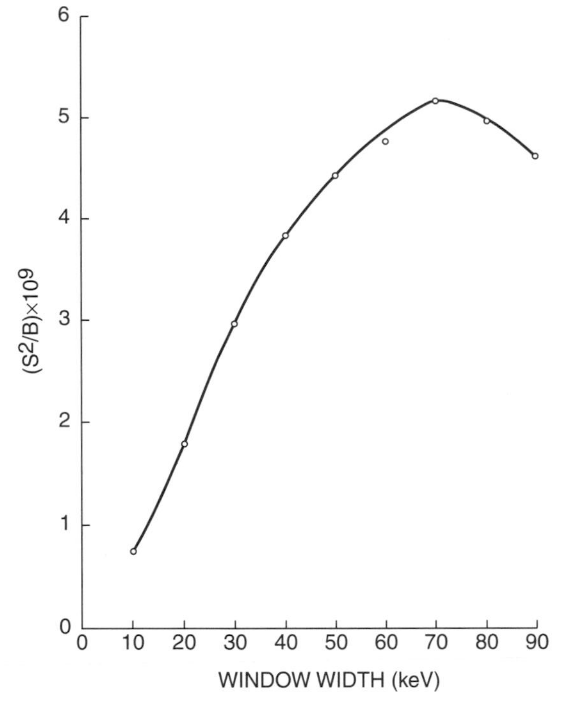

FIGURE 9-9
S^2/B versus window width for a ^{137}Cs source and a 2 × 2-in. NaI(Tl) well detector.

Optimum characteristics of a radionuclide for nuclear medicine imaging include:

- A suitable physical half-life
- Emission of γ rays of suitable energy
  - High enough to penetrate the body
  - Low enough to be absorbed by the detector collimator
- Minimum particulate radiation
- Availability in a convenient, economic and uncontaminated form

count rates R_1 and R_2, respectively, where R_1 and R_2 are determined by the expression $R_1/R_2 = V_2/V_1$ and V_1 and V_2 are the positions of the lower and upper discriminators along the pulse height scale.

An exception to the half-life rule are the radioactive nuclides (^{11}C, ^{13}N, and ^{15}O) with extremely short half-lives (on the order of seconds or minutes) used for positron emission tomography (PET). Use of these nuclides requires a source such as a cyclotron adjacent to the PET imaging area.

RADIOACTIVE AGENTS FOR CLINICAL STUDIES

Many radioactive nuclides have been administered to patients for the diagnosis and treatment of a variety of conditions. For a nuclide to be acceptable for most diagnostic applications, the half-life for radioactive decay should be between 1 hour and a few days. Nuclides with half-lives shorter than 1 hour decay too rapidly for the labeling and sterilizing of radioactive compounds, and nuclides with long half-lives often deliver high radiation doses to patients.

The uptake ratio of a pharmaceutical in an organ is a measure of the difference in the concentration of activity in the organ (target) compared with surrounding tissues (nontarget).

Most radioactive agents administered to patients are injected intravenously and must be sterile. Agents are sterilized by autoclaving (e.g., heating for 15 to 20 minutes at a temperature of about 125°C and a pressure of roughly 20 psi) or by filtration through fine cellulose-ester membranes. The injected material should have a pH near 7 and should either dissolve in an aqueous medium or form a colloidal suspension or a suspension of larger particles. An agent injected intravenously must be free of *pyrogens*, particles that have a diameter of 0.05 to 1 μm and do not decompose at elevated temperature.[7] The presence of pyrogens in an agent administered intravenously to a patient causes the rapid appearance of characteristic symptoms that have been documented by Miller and Tisdal.[8] Solutions are free from pyrogens only if they are prepared by meticulous chemical techniques. Procedures for testing for the presence of pyrogens are described in the *United States Pharmacopoeia*[9] and other publications.[10]

Radioiodinated human serum albumin (RISA) is 99.9% radiochemically pure when freshly prepared. Over time, some of the radioactive iodine detaches from the RISA molecule, and the radiochemical purity declines. The rate of decline depends strongly on storage conditions.

An example of a problem with radionuclidic purity is ^{123}I which invariably is contaminated with ^{124}I. The ^{124}I does not contribute to a study with ^{123}I, but it does increase the radiation dose to the patient and degrades the quality of the ^{123}I image.

Other important considerations for a radioactive pharmaceutical to be used clinically are:

1. *Radiochemical purity*, the ratio of the quantity of the desired isotope (or nuclide) present in the desired chemical form to the total quantity of the isotope (or nuclide) in the pharmaceutical.

A carrier is a nonradioactive isotope of a particular element that is present in the pharmaceutical.

Radiolysis is the alteration of chemical structure caused by exposure to radiation. Compounds tagged with long-live radioactive nuclides and stored for significant periods of time can experience substantial radiolysis. This problem can be reduced by storing the compound as follows:

- In a diluted or dispersed form
- At a low temperature
- With free-radical scavengers added to the solution
- At reduced oxygen concentration
- Under conditions that provide maximum chemical stability

Mechanisms of Radiopharmaceutical Localization[11]

Mechanism	*Example*
Active transport	Thyroid uptake and scanning with iodine
Compartmental localization	Blood pool scanning with human serum albumin, plasma, or red blood cell volume determinations
Simple exchange or diffusion	Bone scanning with ^{99m}Tc-labeled phosphate compounds
Phagocytosis	Liver, spleen, and bone marrow scanning with radiocolloids
Capillary blockade	Lung scanning with macroaggregates (8 to 75 μm); organ perfusion studies with intraarterial injection of macroaggregates
Cell sequestration	Spleen scanning with damaged red blood cells
Receptor binding/ antibody–antigen	Tumor imaging with somatosin, ^{111}In-penetreotide, ^{111}In-OncoScint for tumors

2. *Radioisotopic purity* (or *radionuclidic purity*), the ratio of the activity of the desired radioactive isotope (or nuclide) to the total activity of radioactive material in the pharmaceutical. Progeny nuclides are usually excluded from the description of radionuclidic purity.
3. Presence of a *carrier* in the pharmaceutical. A pharmaceutical is *carrier free* if all isotopes of a particular element in the pharmaceutical are radioactive.
4. Presence of an *isotope effect*, which is an expression used to describe changes in chemical behavior of the pharmaceutical that are caused by differences in isotopic mass between the radioactive nuclide and its stable counterpart. An isotope effect is frequently observed with compounds labeled with tritium. Occurrence of the isotope effect decreases rapidly with increasing atomic number of the radioactive nuclide and is rare with compounds labeled with radioactive nuclides with atomic numbers greater than about 15.
5. Presence of *isotope exchange reactions* during which the radioactive nuclide detaches from the pharmaceutical and adheres to another compound in the body.
6. Radiation dose delivered to critical organs and to the whole body of the patient.
7. Presence of degradation products caused by chemical instability and by self-irradiation of the pharmaceutical.

Localization of a radioactive pharmaceutical administered intravenously, as well as the rate and mode of excretion of the pharmaceutical from the body, is determined by a number of properties, including[12,13]:

1. The oxidation (valence) state of the pharmaceutical labeled with the nuclide at the pH (7.4) of blood.
2. The solubility of the labeled pharmaceutical in an aqueous medium. If the compound is insoluble, then the localization and excretion of the compound are determined by the size of the particles formed.
3. The tendency of the pharmaceutical to incorporate into or bind to organic compounds.

During most clinical applications of radioactive nuclides, radiation emitted by a nuclide distributed within the body is measured with a detector positioned outside the body. For these applications, the nuclide must emit a γ ray with an energy between about 20 and 700 keV. Photons with energy less than 20 keV are severely attenuated by the patient, and photons with energy greater than 700 keV are attenuated inadequately by the collimator attached to the detector. Nuclides (e.g., ^{99m}Tc) that decay by isomeric transition are preferred for *in vivo* studies, because no particulate radiation is emitted from the nucleus during the decay of these nuclides and, consequently, the radiation dose delivered to the patient is reduced.

PROBLEMS

*9-1. A G–M detector that is 60% efficient for counting β-particles subtends a solid angle of 0.2 steradians around a sample of ^{204}Tl ($T_{1/2} = 3.8$ years). The counter registers 2 cps. Assuming that corrections for backscatter, sidescatter, air, window, and self-absorption all equal 1.0, compute the number of ^{204}Tl atoms in the sample.

*9-2. A G–M detector has a resolving time of 300 μs. What is the count rate corrected for coincidence loss if the measured count rate is 10,000 cpm? What is the maximum count rate if the coincidence loss must not exceed 1%?

*9-3. The following data were recorded for identical samples:

"Weightless" Mylar mount	2038 cmp
Silver backing	3258 cpm

Compute the backscatter factor and the percent backscatter.

9-4. Explain why there is a saturation or infinite thickness for curves of backscattering (count rate versus thickness of backing material) and self-absorption (count rate versus weight of sample of constant specific activity).

9-5. The following data were obtained by counting a series of weighed fractions of a radioactive sample:

Mass (mg)	2	4	6	8	10	15	20	25	30	53
cpm (net)	85	160	235	305	335	360	380	398	408	514

Plot a self-absorption curve and determine the specific count rate (cpm/mg) for the sample corrected for self-absorption.

*9-6. A ^{131}I standard solution was obtained from the National Bureau of Standards (NBS). The following information was supplied concerning the solution: assay of ^{131}I at 8 A.M., January 1, 2.09×10^4 dps/mL $\pm$ 2%. On January 15 at 8 A.M., a sample of ^{131}I was counted. This sample had been obtained from a shipment of ^{131}I in the following manner: 20 λ ($1\ \lambda = 1\ \mu L$) of the shipment was diluted in a volumetric flask to 25 mL; 100 λ of this solution was diluted to 10 mL; and 50 λ of this solution was withdrawn, mounted on Mylar, dried, and counted. The dried sample provided a count rate of 4280 cpm corrected for background. At the same time (8 A.M., January 15), a 50-λ sample of the NBS standard ^{131}I solution was mounted, dried, and counted in the same manner. The count rate for the standard sample was 1213 cpm corrected for background.

a. What was the specific activity (millicuries per milliliter) of the ^{131}I shipment at the time the sample from the shipment was counted?

b. What was the total correction factor between the count rate and the activity of the system used to count ^{131}I mounted on Mylar?

9-7. A series of radioactive samples of short half-life will be counted one per day over several days. Explain how variations in the observed counts caused by fluctuations in the sensitivity of the counting system can be compensated.

9-8. A pulse height spectrum for ^{131}I (γ ray of 0.393 MeV) was obtained with a single-channel pulse height analyzer calibrated from 0 to 500 keV. The window width was 2%. Describe the changes in the spectrum if

a. A 5% window were used

b. The amplifier gain were twice as great

c. The amplifier gain were half as great

d. The upper discriminator of the window were removed

*9-9. ^{24}Na decays to ^{24}Mg by β decay with a $T_{1/2}$ of 15 hours. Gamma rays of 4.14 MeV (<1%), 2.76 MeV (100%), and 1.38 MeV (100%) are emitted during the decay process. A pulse height spectrum for this nuclide exhibits peaks at 2.76 MeV, 2.25 MeV, 1.74 MeV, 1.38 MeV, 0.87 MeV, 0.51 MeV, and about 0.20 MeV. Explain the origin of each of the peaks.

9-10. Describe the changes in a pulse height spectrum for ^{51}Cr if

a. A larger NaI(Tl) crystal were used

b. A 1-in. slab of Lucite were interposed between the ^{51}Cr source and the scintillation detector

c. A 1-in. slab of Lucite were placed behind the ^{51}Cr source

*9-11. The position of the photopeak for ^{137}Cs γ rays (0.662 MeV) changes with high voltage as shown in the following table when the gain is set at 4.

Position of photopeak on pulse height scale	520	580	650	730	820
High voltage across photomultiplier tube	730	750	770	790	810

If the gain is reduced to 2 and the high voltage is set at 750, where is the photopeak for the 1.33-MeV γ ray from ^{60}Co on the pulse height scale?

9-12. Describe the decay characteristics desired for a radioactive nuclide used as a scanning agent.

9-13. Describe the characteristics desired for a chemical compound used as a scanning agent.

*9-14. A shipment is labeled "30% ^{14}C-enriched $Ba^{14}CO_3$—weight 2285 mg." This label is interpreted to mean "30% by weight $Ba^{14}CO_3$, the rest being $Ba^{12}CO_3$ and $Ba^{13}CO_3$." What is the activity of ^{14}C in the sample?

*9-15. Radioactive ^{14}C is produced in the upper atmosphere by neutrons released during cosmic ray bombardment:

$$^1_0n + ^{14}_7N \rightarrow ^1_1H + ^{14}_6C$$

The radioactive carbon is distributed over the earth as $^{14}CO_2$ in the air and $NaH^{14}Co_3$ in the sea. Plants and animals incorporate ^{14}C while alive. After a plant or animal dies, the incorporated ^{14}C decays with a half-life of 5600 years. The radioactivity in the remains of the plant or animal indicates the time elapsed since the death of the organism. Determination of the age of objects by measurement of their ^{14}C content is termed *carbon dating*. For example, wood from an old mummy case provides a specific count rate of 10.7 cpm/g. New wood furnishes 16.4 cpm/g. How old is the mummy case?

*9-16. What is the specific activity (curies per gram) of a pure sample of ^{32}P?

*For problems marked with an asterisk, answers are provided on p. 492.

SUMMARY

- Components of a nuclear counting system include:
  - Detector
  - High voltage supply
  - Preamplifier

- Amplifier
- Pulse height analyzer
- Scaler
- Timer
- Factors contributing to determinate error in nuclear counting include:
  - Background count rate
  - Resolving time
  - Detector efficiency
  - Fractional emission of source
  - Geometry
  - Backscatter
  - Sidescatter
  - Attenuation
  - Self-absorption
- Characteristics of a pulse-height spectrum include:
  - Photopeak
  - Compton valley, edge, and plateau
  - X-ray escape peak
  - Backscatter peak
  - Characteristic x-ray peak
  - Annihilation peak
  - Annihilation escape peak
  - Coincidence peak
  - Bremsstrahlung
- Properties of radioactive pharmaceuticals include:
  - Sterility
  - Pyrogen-free
  - Radiochemical purity
  - Radionuclidic purity
  - Presence of a carrier
  - Presence of an isotopic effect
  - Presence of isotope exchange reactions
  - Radiation dose
  - Presence of degradation products
- Localization and excretion of a radioactive pharmaceutical depend on:
  - The oxidation state of the pharmaceutical
  - The solubility of the pharmaceutical
  - The incorporation or binding capacity of the pharmaceutical

REFERENCES

1. Knoll, G. F. *Radiation Detection and Measurement*, 2nd edition. New York, John Wiley & Sons, 1989.
2. Hendee, W. *Radioactive Isotopes in Biological Research*. New York, John Wiley & Sons, 1973.
3. Heath, R. *Scintillation Spectrometry Gamma-Ray Spectrum Catalogue*, 1st edition. AEC report IDO-16408, 1957; 2nd edition, AEC report IOD-16880, 1964.
4. O'Kelley, G. *Detection and Measurement of Nuclear Radiation*, NAS-NS3105. Washington, D.C., Office of Technical Services, Department of Commerce, 1962.
5. Sorenson, J., and Phelps, M. *Physics in Nuclear Medicine*, 2nd edition. New York, Grune & Stratton, 1987.
6. Ross, D., and Harris, C. *Measurement of Clinical Radioactivity*, ORNL-4153. Washington, D.C., Office of Technical Information, Atomic Energy Commission, 1968.
7. Bennett, I., and Beeson, P. Properties and biological effects of bacterial pyrogens. *Medicine (Baltimore)* 1950; **29**:365.
8. Miller, E., and Tisdal, L. Reactions to 10,00 pooled liquid human plasma transfusions. *JAMA* 1945; **123**:863.
9. *United State Pharmacopoeia*, re 20. Rockville, MD, U.S. Pharmacopoeial Convention, 1979.
10. Moore, M., and Hendee, W. *Workshop Manual for Quality Assurance of Radiopharmaceuticals and Radionuclide Handling*. Rockville, MD, Bureau of Radiological Health, 1978.
11. Chandra, R. *Nuclear Medicine Physics: The Basics*, 5th edition. Baltimore, Williams & Wilkins, 1998, p. 39.
12. Durbin, D. Metabolic characteristics within a chemical family. *Health Phys.* 1960; **2**:224.
13. Owunwanue, A., Patel, M., and Sadek, S. *The Handbook of Radiopharmaceuticals*. London, Chapman & Hall, 1995.

CHAPTER

10

COMPUTERS AND IMAGE NETWORKING

Babbage once remarked that his calculating machine "could do everything but compose country dances."

Moore's Law: In 1965, Gordon Moore, co-founder of Intel Corporation, mentioned in a speech that the amount of information that can be stored in a silicon computer chip (related to the number of transistors per chip) doubles every 18–24 months. Moore's Law, as it came to be known, held through the rest of that century. It is not expected to hold for more than another decade or so into the twenty-first century, because fundamental limitations such as the quantum uncertainty of how individual electrons behave in a circuit may be reached.

A

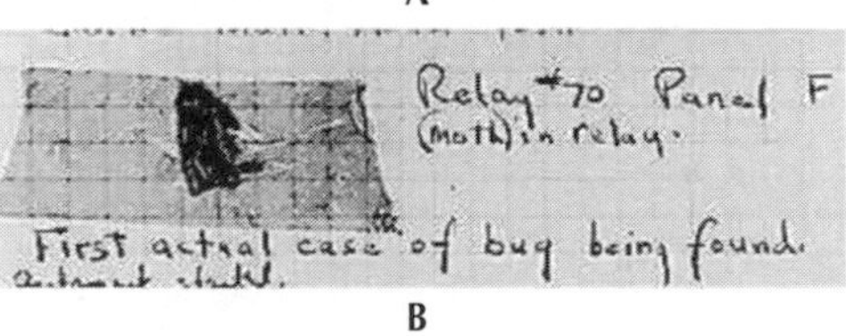

B

MARGIN FIGURE 10-1
Grace Hopper, Ph.D., a pioneer of electronic computing and computer language development, worked on most of the "mainframe" computers in existence from the 1940s to the 1980s. She developed many of the approaches to programming and data base management that are taken for granted today. Much of this effort was during military service, in which she rose to the rank of Admiral in the U.S. Navy. In 1951, Grace Hopper coined the term "bug" for the cause of any computer malfunction, after finding an actual bug in an electronic relay of an early computer.

OBJECTIVES

After completing this chapter, the reader should be able to:

- Convert numbers from base 10 to other bases.
- Explain how electrical circuits mimic logical operations and how that allows mathematical operations to be performed by a computer.
- Define the terms bit, byte, and word.
- Describe how letters, numbers, and images are stored by a computer.
- Compute image size and transmission time over a network.
- Calculate the error in a digitized value.
- Name the main hardware components of a computer.
- List some standard input and output devices.
- Discuss the difference between serial and parallel operations.
- Give several examples of software.
- Name the typical hardware components of a network.
- Give examples of the speed of transmission commonly found in
  - Copper wire
  - Wireless
  - Fiber optic
- State the requirements for monitor performance and matrix size found in the American College of Radiology standard.

HISTORY

A computer is any device other than the human brain that records and operates with numbers and symbols. By this definition, computers were invented by primitive humans when they counted on their fingers or piled up sticks and stones to record changes in some quantity. The modern electronic computer is fundamentally no different, except that it replaces sticks and stones with a series of on–off switches. These switches can be set and changed automatically, which makes the modern computer much more useful than a simple counting device. Moreover, multiple meanings and interrelationships have been developed among the on–off switches. This is the property that sets the modern computer apart from any other computational device or technique.

One of the earliest calculating devices was the abacus, which in its most common form consists of beads on a wire frame. There is evidence of early forms of the abacus in use by the Babylonians as early as 3000 B.C.[1]

One of the first mechanical calculating devices was invented by Pascal in the 1600s.[2,3] It was an adding machine that freed accountants from summing long chains of numbers by hand. It performed only the single operation of addition. In 1801, the automated loom was invented by Jacquard. This device used punched cards to control the operation of the loom. It could be reprogrammed to produce different weaves by changing the pattern of holes in the cards. The automated loom did not actually perform calculations; however, it introduced the concept of programming, which is essential to modern computers.

In 1833, Babbage, an English mathematician, designed an "analytical engine" to produce navigational tables for commercial shipping.[4] The Babbage invention used decks of punched cards to program different functions. It represented the zenith of mechanical calculating devices. Although the working model of the engine was not produced during the inventor's lifetime, advances in the technology of precision machine tooling in this century have led to construction of working models that verify the soundness of the original design.

In 1890, a new type of tabulating machine was developed by Hollerith for the U.S. Census Bureau.[5] This device received data from punched cards, kept track of separate categories of data, and kept an updated sum of numbers as more information was entered. It represented the first system for automated entry of information into a "data base" for the management of large amounts of information.

The first true automatic, electronic, digital computer was built in 1939 at Iowa State University by Atanasoff and Berry. This device was designed to help solve equations in quantum mechanics.[6] Additional electronic computers that were stimulated by the technologic demands of World War II evolved a few years later. The Colossus was built in 1943 in Betchley, England, by Turing and others to penetrate German codes during World War II.[7] The Electronic Numerical Integrator and Calculator (ENIAC) (Margin Figure 10-2) was constructed by Mauchly and Eckert at the University of Pennsylvania in 1945 to prepare ballistics tables.[8]

In the 1930s the word "computer" meant a person (typically female) whose job was doing computations (Davis, M. *The Universal Computer. The Road from Leibniz to Turing*. New York, W. H. Norton, 2000.)

MARGIN FIGURE 10-2
The Electronic Numerical Integrator and Computer (ENIAC) was constructed in 1945 under military contract at the University of Pennsylvania for approximately $500,000. It contained 17,468 vacuum tubes, weighed 60,000 pounds, occupied 16,200 cubic feet of space, and used 174 kilowatts of power. It operated for nine years, during which time it was the fastest computer on earth. It could add 5000 numbers/second, whereas a typical PC today can add 70 million/second. As part of ENIAC's 50th Anniversary Celebration, a team of graduate students at the University of Pennsylvania duplicated its functionality on a $7.44 \times 5.29\text{-mm}^2$ chip that contains about 174,569 transistors. A typical PC central processing chip used today contains over three million transistors.

MACHINE REPRESENTATION OF DATA

The most fundamental unit of information is a simple yes–no decision. Components in a computer can record and store this information as a simple "current will pass–current will not pass" or "voltage is high–voltage is low" condition. Symbolically, we refer to this polarized situation with the digits 1 and 0. The binary "base 2" number system describes the use of ones and zeros to describe quantities numerically and to perform mathematical operations. The binary number system is a natural way to describe the operation of a computer or to "communicate" with one. For the same reason, number systems based upon powers of 2 such as octal (base 8) and hexadecimal (base 16) are also used in computer applications.

Numbers and Bases

The number system with which we are most familiar is the decimal, or "base 10," system. It consists of the 10 digits 0 through 9, and it denotes a power of 10 through the position of the digit in a number. If a certain power of 10 is not needed to express a quantity, its position is occupied by a zero. These characteristics, a series of digits in a "place holder" configuration that connotes different powers of a base, are shared by most number systems including binary, octal, base 10, and hexadecimal.

The binary, or base 2, system is illustrated in Margin Figures 10-3 and 10-4. The numbers 0 and 1 are the same in both base 2 and base 10 systems. However, the decimal number 2 has no symbol in binary since only 0 and 1 are available as digits. The problem of binary expression of the number 2 is analogous to the problem of decimal expression of the number 10 (Margin Figure 10-5). Because there is no single digit in the decimal system to express the number 10, a place is marked in the rightmost column with the digit zero, and a 1 is placed in the next column to the left, the "ten's" place, to express the decimal number "10." Similarly, to express the number 2 in binary, a zero is positioned in the rightmost column, and a 1 is placed in the next column to the left, the "two's" place.

Octal and hexadecimal number systems are based upon the numbers 8 and 16, respectively. The properties of these number systems are illustrated in Margin Figure 10-3. Conversion of number representations among the decimal, binary, octal, and hexadecimal systems is illustrated in Margin Figure 10-6. When working with numbers in more than one base, it is important to remember that the quantity itself—for example, "twenty things"—is the same in any number system. It is only the notation that changes when moving from one number base to another.

Binary (Base 2)	Decimal (Base 10)
0	0
1	1
10	2
11	3
100	4
101	5
110	6
111	7
1000	8
1001	9
1010	10

MARGIN FIGURE 10-3
Binary and decimal notation for numbers.

Numerical and Logical Operations

One may pose the question, "How do computers think?" One should then counter with the question, "How do humans think?" These questions are topics of current research interest, and it is tempting to presume that the best, or at least the most

Claude Shannon (1917–2001) founded what is known today as "Information Science." In a landmark 1940 paper, he wrote down the equations that specify how much information communication networks can carry, described how signals could be conveyed by a series of ones and zeros (which we now call bits), and introduced the concept of data compression.

1	2^{10}	one thousand twenty fours	1024
0	2^{9}	five hundred twelves	512
0	2^{8}	two hundred fifty sixes	256
1	2^{7}	one hundred twenty eights	128
0	2^{6}	sixty fours	64
1	2^{5}	thirty twos	32
1	2^{4}	sixteens	16
0	2^{3}	eights	8
0	2^{2}	fours	4
1	2^{1}	twos	2
1_2	2^{0}	ones	1
=	↑	↑	↑
1203_{10}	power of two	place	decimal equivalent

MARGIN FIGURE 10-4
Placeholder system, powers of 2 for binary. The example binary number shown is equivalent to the decimal number 1203.

interesting, computers would think the way humans do. It is plausible that a computer could go beyond its specifically programmed instructions to "adapt" to new situations and "learn" from its mistakes. Providing these and other capabilities to computers is the goal of a branch of computer science called artificial intelligence (AI).[9,10] Computers in operation at the present time, however, are far more simple than the human mind. They basically do exactly what they are programmed to do, no more and no less. Hence the tasks that computers are currently asked to perform must be dissectable into simple, completely defined steps.

Computers are able to perform numerical operations such as addition, subtraction, multiplication, and division through the use of electronic components called "gates." The key to understanding how gates work is to examine the way that electronic signals are modified in passing through them. At the input end of a gate the two impinging signals are either "high" (e.g., 5 V), representing a binary number 1, or "low" (e.g., 0 V), representing a zero. The function of the gate is to alter the voltage in a fashion that mimics a mathematical or logical operation. The alteration is illustrated by the "truth table" for the gate showing the output values achieved for all possible combinations of input values. For example, Margin Figure 10-8 shows two gates, an AND gate and an exclusive-OR (X OR) gate. The inputs are labeled a and b. The truth table for the AND gate shows that the output will be 1 (high voltage) if and only if both of the inputs are 1. The truth table for the X OR gate shows that the output will be 1 if and only if the two inputs are different (either $a = 1, b = 0$ or $a = 0, b = 1$).

Gates and combinations of gates function as mathematical and logical operators for binary numbers. For example, the X OR and the AND gate, when connected as shown in Margin Figure 10-9, produce a truth table that corresponds to the addition operation for binary numbers. When the inputs are added, the output is a 2-bit representation of the result. Adding 1 and 0 in either order produces the binary number 01. When both inputs are 1, the output is the binary number 10 (decimal number 2). Other combinations of gates carry out operations such as subtraction and multiplication, as well as logical operations such as "If a is true (i.e., high voltage), then the result (output of the gate) is true." A large assembly of gates is capable of attaining a high degree of complexity.

It may be difficult to understand how a problem such as the reconstruction of an image from input signals is related to simple gates. Nevertheless, this operation can be traced to individual gates that are actuated in sequence to arrive at an appropriate output in binary form.

Bits, Bytes, and Words

Information in a computer is handled in chunks of various sizes. The fundamental unit of information is the binary digit, contracted to the term "bit," which refers to the individual ones and zeros that make up a binary number. In this context a one or a zero is considered as a single bit of information. Eight bits make up a "byte" of information, and 2 or more bytes make up a "word" of information. If in a particular type of computer the "word length" is 16 bits, then a word is composed of 2 bytes (Margin Figure 10-10). Common word lengths are 16 bits (a 2-byte word), 32 bits (a 4-byte word), and 64 bits (an 8-byte word). The size of the chunk of information that a particular computer can handle at one time determines the precision of numerical operations and the amount of storage the computer can maintain.

Certain prefixes have become standard for specifying large numbers of bits, bytes, and words. Because the information computers deal with is routinely divided into powers of 2, the prefixes kilo and mega in computer terminology invoke the powers of 2 that are nearest to the quantities 1000 and 1,000,000, respectively. The number 2 raised to the tenth power equals 1024 and is referred to as one "K" or kilo. Thus, 5 Kbytes is actually $5 \times 1024 = 5120$ bytes rather than 5000 bytes. Similarly, 2 raised to the 20th power equals 1,048,576 and is referred to as one "M" or mega. So, 10 Mwords equals 10,485,760 words, or 485,750 words more than one might expect from the metric meaning of "mega." This terminology is used consistently in computers, as there is little cause for confusion in practice. A computer with 2 Mbyte of internal memory has twice as much memory as one with 1 Mbyte of memory.

The expression "memory" to describe the storage capacity of the computer was suggested by John von Neumann in 1945. The first all-purpose computers marketed by IBM were referred to as "johnniacs." This acknowledgment reflected von Neumann's many contributions to computer designs, including the concept of a program (a code that directs computer operations).

Storage of Letters and Symbols

The storage of binary values is accomplished directly by influencing the status (high or low) of voltages in various circuits in the computer. A row of switches that are "high, low, low, high" in voltage might represent the binary number 1001 (decimal 9). Letters and other symbols have no such "natural" representation. Instead, they are recorded in code as specific patterns of binary digits that may be stored and manipulated by the computer. When the computer "communicates" with the outside world (e.g., with a printer), these patterns of bits are converted to the shapes humans recognize as letters and symbols.

The 26 letters of the alphabet, the digits 0 to 9, and a number of special symbols (such as $, #, &, . . .) are referred to collectively as the "alphanumeric character set" of the computer. Depending on the number of symbols and whether both uppercase and lowercase letters are allowed, the alphanumeric character set may consist of a many as 128 elements. Table 10-1 shows, as an example, one of the most widely used schemes of encoding alphanumeric information. This scheme is the American National Standard Code for Information Interchange (ASCII).

(a)

Ten's place	One's place
	0
	1
	2
	3
	4
	5
	6
	7
	8
	9
1	0

DECIMAL (BASE 10)

(b)

Two's place	One's place
	0
	1
1	0

BINARY (BASE 2)

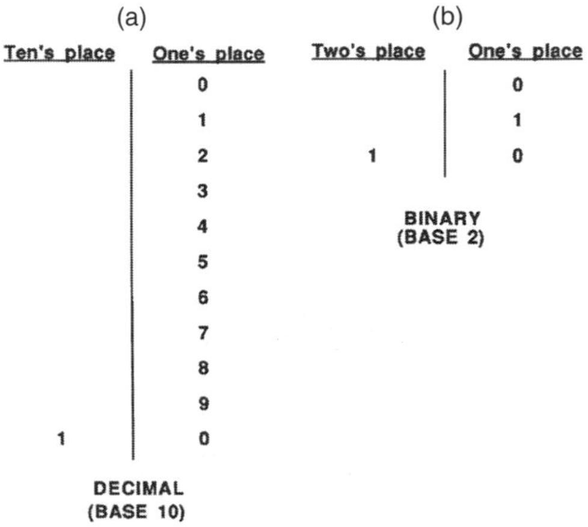

MARGIN FIGURE 10-5
A: Decimal expression of the number 10 involves the use of zero as a placeholder in the one's place and the digit "1" in the ten's place. **B:** Binary expression of the number 2 involves the use of zero as a placeholder in the one's place and the digit "1" in the two's place.

Image Storage: Pixels

To store images in a computer, the image must be divided into smaller sections called picture elements, or "pixels" (Margin Figure 10-12). Each pixel is assigned a single numerical value that denotes the color, if a color image is stored, or the shade of gray (referred to as the gray "level"), if a black and white image is stored. The image is like a jigsaw puzzle, except that each piece is a square and has a uniform color or shade

TABLE 10-1 American National Standard Code for Information Interchange (ASCII)

Character	*Binary Code*	*Character*	*Binary Code*
A	100 0001	0	011 0000
B	100 0010	1	011 0001
C	100 0011	2	011 0010
D	100 0100	3	011 0011
E	100 0101	4	011 0100
F	100 0110	5	011 0101
G	100 0111	6	011 0110
H	100 1000	7	011 0111
I	100 1001	8	011 1000
J	100 1010	9	011 1001
K	100 1011		
L	100 1100		
M	100 1101	Blank	010 0000
N	100 1110	.	010 1110
O	100 1111	(	010 1000
P	101 0000	+	010 1011
Q	101 0001	$	010 0100
R	101 0010	*	010 1010
S	101 0011	)	010 1001
T	101 0100	-	010 1101
U	101 0101	/	010 1111
V	101 0110	,	010 1100
W	101 0111	=	011 1101
X	101 1000		
Y	101 1001		
Z	101 1010		

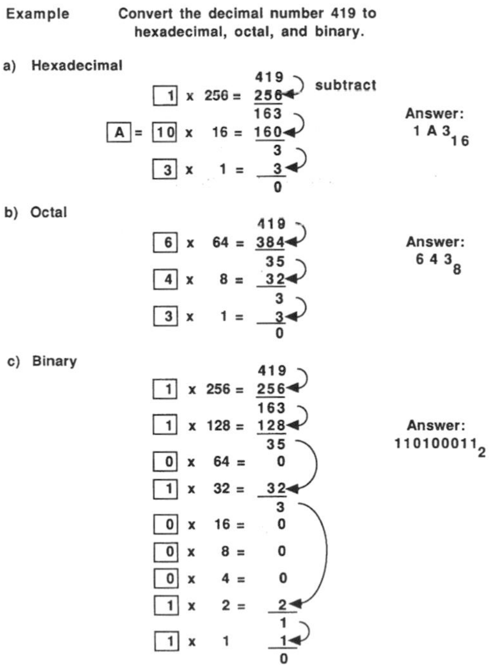

Example — Convert the decimal number 419 to hexadecimal, octal, and binary.

a) Hexadecimal

1 x 256 = 256 (419 − 256 = 163; subtract)
A = 10 x 16 = 160 (163 − 160 = 3)
3 x 1 = 3 (3 − 3 = 0)
Answer: $1A3_{16}$

b) Octal

6 x 64 = 384 (419 − 384 = 35)
4 x 8 = 32 (35 − 32 = 3)
3 x 1 = 3 (3 − 3 = 0)
Answer: 643_8

c) Binary

1 x 256 = 256 (419 − 256 = 163)
1 x 128 = 128 (163 − 128 = 35)
0 x 64 = 0
1 x 32 = 32 (35 − 32 = 3)
0 x 16 = 0
0 x 8 = 0
0 x 4 = 0
1 x 2 = 2 (3 − 2 = 1)
1 x 1 = 1 (1 − 1 = 0)
Answer: 110100011_2

MARGIN FIGURE 10-6
Conversion of decimal numbers to hexadecimal, octal, and binary.

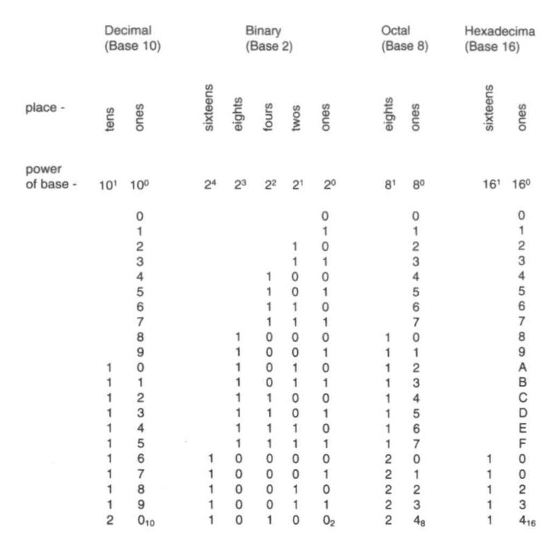

	Decimal (Base 10)		Binary (Base 2)					Octal (Base 8)		Hexadecima (Base 16)	
place -	tens	ones	sixteens	eights	fours	twos	ones	eights	ones	sixteens	ones
power of base -	10^1	10^0	2^4	2^3	2^2	2^1	2^0	8^1	8^0	16^1	16^0
		0					0		0		0
		1					1		1		1
		2				1	0		2		2
		3				1	1		3		3
		4			1	0	0		4		4
		5			1	0	1		5		5
		6			1	1	0		6		6
		7			1	1	1		7		7
		8		1	0	0	0	1	0		8
		9		1	0	0	1	1	1		9
	1	0		1	0	1	0	1	2		A
	1	1		1	0	1	1	1	3		B
	1	2		1	1	0	0	1	4		C
	1	3		1	1	0	1	1	5		D
	1	4		1	1	1	0	1	6		E
	1	5		1	1	1	1	1	7		F
	1	6	1	0	0	0	0	2	0	1	0
	1	7	1	0	0	0	1	2	1	1	0
	1	8	1	0	0	1	0	2	2	1	2
	1	9	1	0	0	1	1	2	3	1	3
	2	0_{10}	1	0	1	0	0_2	2	4_8	1	4_{16}

MARGIN FIGURE 10-7
Representation of the decimal numbers 1–20 in other bases.

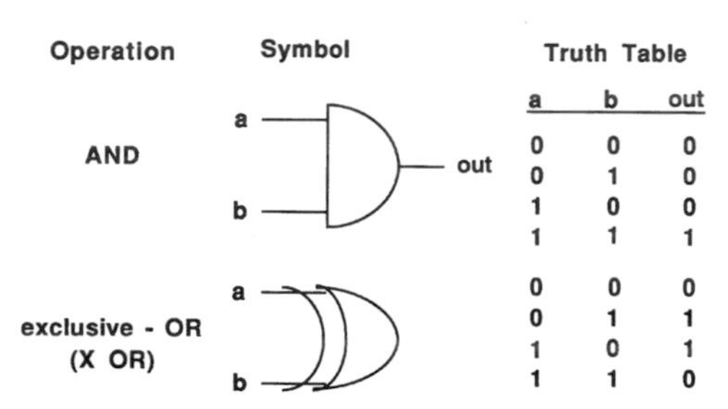

MARGIN FIGURE 10-8
Logic gates convert the input signals a and b to an output signal.

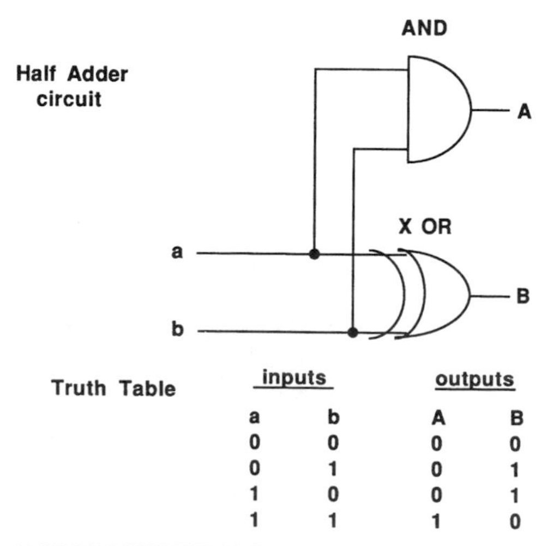

MARGIN FIGURE 10-9
The AND gate and the exclusive-OR (XOR) gate, when connected as shown here, produce a 2-bit output that is identical to the binary addition of the a and b signals.

of gray. The numerical values are represented in binary (or octal or hexadecimal) notation, and the image is said to be "digital."

The set of pixels composing the image is referred to as the image "matrix." That is, a digital image consists of a matrix of pixels.

Some images are computer generated directly in pixel or digital form. Computed tomography (CT), magnetic resonance, and computed radiography are examples of technologies that produce digital images directly. An image such as a chest film that was not originally "digital" can be entered and stored in a computer by dividing it into a matrix of squares and assigning the average color (or shade of gray) within each square as a single numerical value for the pixel (Margin Figure 10-13). An image modified in this fashion is said to have been "digitized."

If a few pixels are used in digitizing an image, the result is a coarse or "blocky" appearance. If a larger number of pixels is used, the observer may not even notice that the image is digital (Margin Figure 10-14). It is not necessarily better to break an image into the largest matrix size (smallest pixels). If the computer has limited storage or if the processing and transmission time for the images is restricted, then a smaller matrix size (larger pixels) may be desirable. Matrix sizes are typically powers of 2, reflecting the binary nature of computer storage (e.g., 64, 128, 256, 512, 1024, . . .).

Example 10-1

A digital image is constituted as a 128 × 128 matrix of pixels. Transmission of the image from one computer workstation to another requires 13.7 sec. If the matrix size is changed to 256 × 256, how long will it take to transmit the image?

- The 128 × 128 image consists of $128^2 = 16{,}384$ pixels.
- The 256 × 256 image consists of $256^2 = 65{,}536$ pixels.
- There are $65{,}536/16{,}384 = 4$ times as many pixels in the image with the larger matrix size.

Note: Increasing the length of one side of a matrix by a factor of 2 increases the total number of pixels by a factor of 4 since image matrices are usually square.

Therefore the new transmission time is

- 4×13.7 sec $= 54.8$ sec

In this example, the transmission time depends strictly upon the number of pixels. We will see in another section that some data compression techniques can decrease the transmission time.

There is no universal standard code for pixel values as there is for alphanumeric symbols. On the contrary, the ability to reassign numerical values that signify gray levels or colors is one of the most useful principles underlying image manipulation with a computer. A range of numbers can be assigned to any image, for example, with lower numerical values representing darker and higher numerical values representing lighter shades of a gray-scale image. The range of values affects the appearance of the image as shown in Margin Figure 10-15.

The number of bits used to describe the gray level is referred to as the "bit depth" of the image. This terminology comes from thinking of the digital image not as a two-dimensional object but as a three-dimensional block with the gray level as the third dimension (Table 10-2; see also Margin Figure 10-12).

Example 10-2

How many bits are required to store a 512 × 512 digital image with a bit depth of 8 bits?

- $512 \times 512 \times 8 = 2{,}097{,}152$ bits
- Recall that $2^{20} = 1{,}048{,}576$ bits $= 1$ Mbit
- Therefore $\frac{2{,}097{,}152}{1{,}048{,}576} = 2$ Mbit are required to store the image

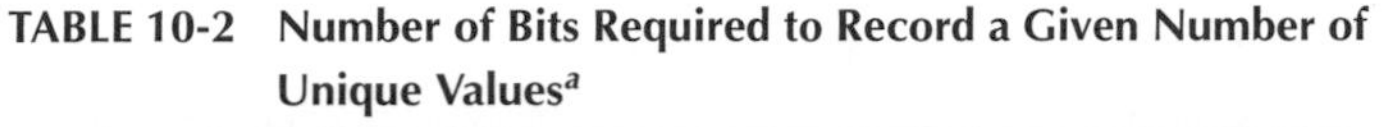

TABLE 10-2 Number of Bits Required to Record a Given Number of Unique Values[a]

Number of Bits	*Possible Values*				*Number of Values*
1	0	1			2
2	00	01	10	11	4
3	000	001	010	011	8
	100	101	110	111	
4	0000	0001	0010	0011	16
	0100	0101	0110	0111	
	1000	1001	1010	1011	
	1100	1101	1110	1111	

[a] The number of unique values that may be recorded equals 2 raised to the number of bits used to record the values.

The problem may be solved without a calculator by noting that

$$512 = 2^9 \quad \text{and} \quad 8 = 2^3$$

Therefore, the total number of bits is

$$2^9 \times 2^9 \times 2^3 = 2^{21} \text{ bits}$$

(superscripts add when multiplying powers of the same base number), but

$$\frac{2^{21} \text{ bits}}{2^{20} \text{ bits/Mbit}} = 2^1 = 2 \text{ Mbits}$$

(superscripts subtract when dividing powers of the same base number).

Example 10-3

How many bytes are required to store the above image?

$$\frac{512 \times 512 \times 8 \text{ bits}}{8 \text{ bits/byte}} = 512 \times 512$$

$$= 2^9 \times 2^9$$

$$= 2^{18} \text{ bytes}$$

In this example, each pixel is 1 byte "deep," so the number of pixels is the same as the number of bytes.

Example 10-4

How many bytes are required to store the above image if the bit depth is reduced to 6? Increased to 9?

A byte takes up 8 bits. By using only 6 bits we are simply not filling each byte. But unless some special memory configuration is used, the basic unit of information is the byte. So, each pixel cannot take up less than 1 byte. Therefore, even though the bit depth may be less than 8, 1 byte per pixel is still required. So for a bit depth of 6, the number of bytes required is

$$512 \times 512 \times 1 \text{ byte} = 2^{18} \text{ bytes}$$

If the bit depth is any number between 8 and 16, 2 bytes per pixel are required. Therefore, if the bit depth is increased to 9, the number of bytes required is

$$512 \times 512 \times 2 \text{ bytes} = 2^{19} \text{ bytes}$$

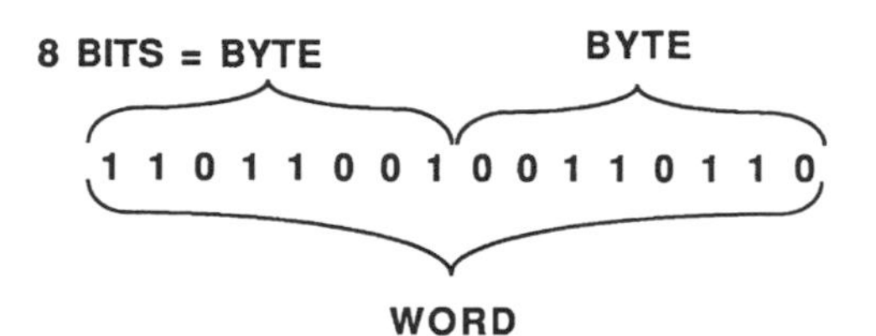

MARGIN FIGURE 10-10
A byte is composed of 8 binary digits, or bits. A word is composed of an integral number of bytes. The word shown here consists of two bytes, or 16 bits. However, word length is not universally recognized, so there are 16-, 32-, 64-, and so on, bit words.

Power of 10 Nearest Power of 2 (prefixes or "units")	Examples of files that would occupy 1-1000 of these units (in bytes)
$1{,}000 = 10^3$	
$1{,}024 = 2^{10}$ = 1 K ("key" or "kilo")	text files
$1{,}000{,}000 = 10^6$	
$1{,}048{,}576 = 2^{20}$ = 1 M ("Meg")	a medical image or a book
$1{,}000{,}000{,}000 = 10^9$	
$1{,}073{,}741{,}824 = 2^{10}$ = 1 G ("gig")	short term, on-site storage of medical images
$1{,}000{,}000{,}000{,}000 = 10^{12}$	
$1{,}099{,}511{,}627{,}776 = 2^{40}$ = 1 T ("tera")	One year of image storage in an "all digital" hospital
$1{,}000{,}000{,}000{,}000{,}000 = 10^{15}$	
$1{,}125{,}899{,}906{,}842{,}624 = 2^{50}$ = 1 P ("peta")	10 year medical image storage for an "all digital" hospital"
$1{,}000{,}000{,}000{,}000{,}000{,}000 = 10^{18}$	
$1{,}152{,}921{,}504{,}606{,}846{,}976 = 2^{60}$ = 1 E ("exa")	All of the angiographic studies ever performed
***Really* big numbers:**	
1, . . . (80 zeros) . . . $= 10^{80}$	
$\sim 1.186 \times 10^{80} = 2^{266} = 1?$	Estimated number of protons in the Universe
1, . . . (100 zeros) . . . $= 10^{100}$	
$\sim 1.750 \times 10^{100} = 2^{333}$ = 1 gogol	The biggest number formally named. (except for the gogolplex, which is 10^{gogol}) (and except for 2 gogolplex, 3 gogolplex, etc.)

MARGIN FIGURE 10-11
Powers of 10 and prefixes.

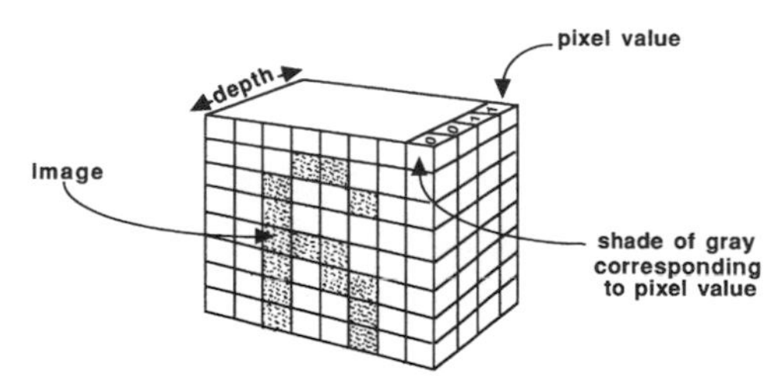

MARGIN FIGURE 10-12
A digital image may be thought of as a three-dimensional object composed of a number of small cubes, each of which contains a binary digit (bit). The image appears on the front surface of the block. The "depth" of the block is the number of bits required to describe the gray level for each pixel. The total numbers of cubes that make up the large block is the number of bits required to store the image.

Two mathematicians, Alan Turing and Kurt Gödel, were among those who made substantial contributions to the logic underlying computer design. Both were included in *Time* magazine's (March 29, 1999) list of Greatest Scientists and Thinkers in the twentieth century. Turing set standards for artificial intelligence and Gödel set limits on what could be computed within a given logical framework.

MARGIN FIGURE 10-13
The photograph on the *left* has been digitized for the purpose of manipulation and storage in a computer.

MARGIN FIGURE 10-14
As the matrix size of a digital image is increased, the relative size of each pixel decreases and the image appears less "blocky." The images shown here, beginning at *top left*, consist of 8 × 8, 16 × 16, 64 × 64, 128 × 128, 256 × 256 pixels. Each image has a bit depth of 8, corresponding to 256 shades of gray.

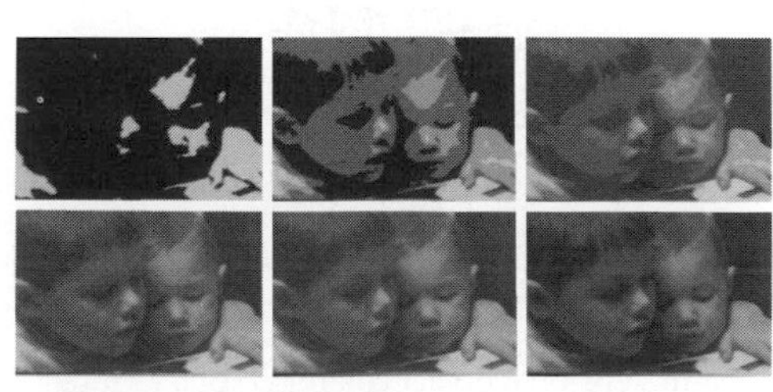

MARGIN FIGURE 10-15
As the bit depth of a digital image is increased, the number of shades of gray that are stored for each pixel is increased. The images shown here are, *top left*, 1 bit (2 shades), 2 bit (4 shades), 3 bit (8 shades), 4 bit (16 shades), 6 bit (64 shades), 8 bit (256 shades). Each image has a matrix size of 768 × 512 pixels.

Analog-to-Digital and Digital-to-Analog Converters

Measurable quantities may be separated into two categories, analog and digital. Analog quantities are continuously variable and can take on any value between two limits. The list of possible values is limited only by the tools of measurement. Digital quantities are discrete and can assume only specific values separated by gaps. For example, the height of an individual is an analog quantity, but the number of people in a classroom is a digital quantity. Time can be measured in analog fashion with an analog watch (a watch with hands) or in a digital fashion with a digital watch (a watch with a display showing several decimal digits).

Computers operate with digital quantities, but much of the information we routinely use is in analog form. Therefore, conversion from one form of information to the other often is necessary. For example, electrical signals from some radiation detectors (such as photomultiplier tubes in a gamma camera) are analog. If a computer is to use these signals, an analog-to-digital conversion (ADC) must be performed. A computer used to control devices in a laboratory must produce signals that at least appear to the equipment to be analog. The production of a digital image is a representation of digital information as an analog picture. That is, the presentation of an image to an observer is a form of digital-to-analog conversion (DAC).

When an analog quantity is translated into a digital value, the issue of the precision of the translation arises. For example, suppose that a radiation detector produces a signal of 3.675 V out of a possible range of 0 to 10 V. A computer that uses 3 bits to record the signal can register, at most, 8 unique values (they are 000, 001, 010, 011, 100, 101, 110, and 111). For a range of 0 to 10 V, these eight values differ by increments of 10/8, or 1.25 (Margin Figure 10-10). An analog quantity between 0 and 1.25 might be recorded as 0, a quantity between 1.25 and 2.50 might be recorded as 1.25, and so on. The value of 3.675 is between 2.50 and 3.75 and would be recorded as 2.5, with an absolute error of 1.175. In this example, the maximum digitization error is the size of the increment (i.e., 1.25 V) because a value of 1.249999 V would be recorded as 0. The maximum error is therefore 1.25 V. If the analog values were randomly distributed, we would expect to see all errors from zero to slightly less than 1.25. Thus, the average digitization error is half the digitization increment, or 0.625 V. We can generalize to the following formulas:

$$\mathrm{err}_{\max} = \frac{R_a}{N} \tag{10-1}$$

where $\mathrm{err}_{\max}$ is the maximum digitization error, R_a is the analog range, and N is the number of distinct values available to the digitizer. Also,

$$N = 2^n, \tag{10-2}$$

where n is the number of bits. Combining Eqs. (10-1) and (10-2) yields

$$\mathrm{err}_{\max} = R_a 2^{-n} \tag{10-3}$$

Then

$$\mathrm{err}_{\mathrm{avg}} = \frac{\mathrm{err}_{\max}}{2} = \frac{R_a}{2^{n+1}} \tag{10-4}$$

where $\mathrm{err}_{\mathrm{avg}}$ is the average error.

COMPUTER SYSTEM HARDWARE

A computer system needs both hardware and software to function. Hardware is any physical component of the computer; it has mass and takes up space. Software is any set of instructions required to operate the computer or perform mathematical operations.

There are many types of computers. The purpose of this section is to introduce the features that are common to most computers and to illustrate some operating principles of computers used in medical imaging. A diagram of a computer is shown in Margin Figure 10-17.

Input and Output

A device that allows information to be entered into the computer is an input device, and a device that transfers information from the computer to the outside world is an output device. Many devices can perform both functions. For example, a disk drive may be used to read information from disks as an input function. The same disk drive may also be used to write information from the computer to a disk for storage, an output function.

A list of devices and their usual roles as input or output devices is shown in Table 10-3. Many of the devices that perform input or output functions also perform storage, interfacing, and networking functions that are discussed in a later section. All of the devices in Table 10-3 may be used in diagnostic imaging. The output devices most used are the display screen and the hard-copy device. Commonly used input devices in diagnostic radiology include digital radiographic equipment, CT gantries, magnetic resonance imaging receiver coils, ultrasound transducers, and gamma cameras. The computer can be considered as the actual imaging device, with the equipment mentioned above serving to gather and transmit data about the patient to the computer.

Central Processing Unit

The central processing unit (CPU) acts as the "traffic cop" of the computer. Its function is so central to the operation that if any one component can be said to be a computer in and of itself, it is the CPU. It determines when different components of the computer are needed to perform required tasks. For example, a simple task such as "add two numbers" requires the following steps: (1) Locate the first number from an input

TABLE 10-3 Sampling of Input/Output Devices

Device	*Purpose*	*Input*	*Output*
Keyboard	Manual entry of alphanumeric data	X	
Light pen	Entry of data by a hand-held wand	X	
Joystick	Manual entry of directional information	X	
Mouse	Manual entry of directional information	X	
Modem	Communication through phone lines	X	X
Hard copy	Conversion of a video signal to a permanent image		X
Display screen	Visual display of information		X
Printer	Provides paper copy of alphanumeric information, graphs, and charts		X
X/Y plotter	Provides continuous lines, graphs		X
Touchscreen	User-interactive display	X	X
Disk/tape drive	External storage	X	X
Digitizer	Used to enter graphs or images directly	X	
Network interface	Communications	X	X
Equipment interface	Communications	X	X

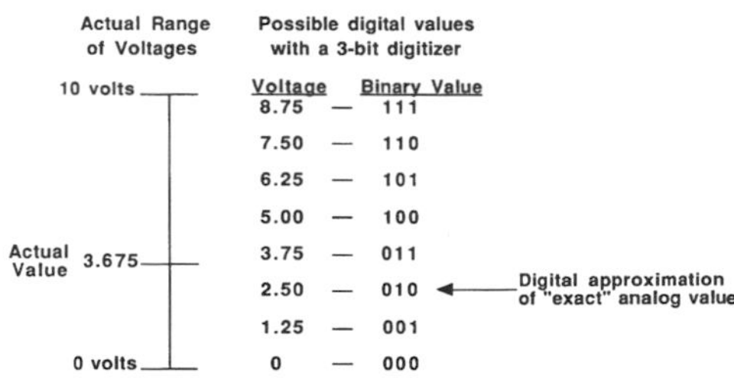

MARGIN FIGURE 10-16
A 3-bit digitizer is used to record voltage signals over the range of 0 to 10 V. Because only eight digital values are possible for a 3-bit digitizer, one scheme, illustrated here, would assign all voltages between 0 and 1.25 V to the value 0 V (binary 000), all voltages between 1.23 and 2.50 V to the value 1.25 V (binary 001), and so on.

Quantum Computers: In the quest for greater computing power, one area of research involves using some of the properties of objects, such as small numbers of photons, electrons, or nuclei, whose behavior is dominated by quantum mechanical effects.

One such phenomenon is "quantum teleportation" in which a state that could represent one bit of information (such as the direction of electromagnetic fields in a light beam) may be changed instantaneously over large distances. The implications for high-speed communications are obvious.

Another potentially useful property of a quantum mechanical system is that it exists in all possible combinations of states up to the moment of measurement. This property suggests that the various states could represent ones and zeros in computations and that, until the moment of measurement, the "quantum computer" could explore all possible combinations of solutions to a problem. One way to construct such a system of quantum mechanical spins involves the same nuclear magnetic resonance phenomenon in liquid samples that is used in magnetic resonance imaging of the human body.

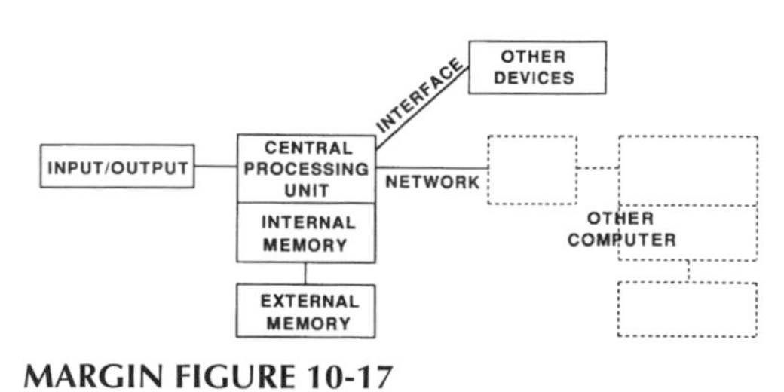

MARGIN FIGURE 10-17
Computer system hardware.

In February 1996, the chess-playing computer Deep Blue defeated World Champion Garry Kasparov. The significance of this victory has been debated ever since in the artificial intelligence community.

Molecular Computers: One approach to the "next generation" of computers is molecular computing. The nucleotide sequence of each DNA molecule could be used to encode a potential solution to a problem, such as the pixel values of a computed tomography image that might correspond to the x rays that were detected by the system. Various techniques in molecular biology could, in principle, be used to recombine the molecules in all possible combinations and select specific sequences that correspond to the best solution to the overall problem. A liter of a solution of DNA molecules contains as many as 10^{20} molecules. If used as active components of a "molecular computer," it would represent a vast improvement over today's chips, which typically contain "only" billions of components.

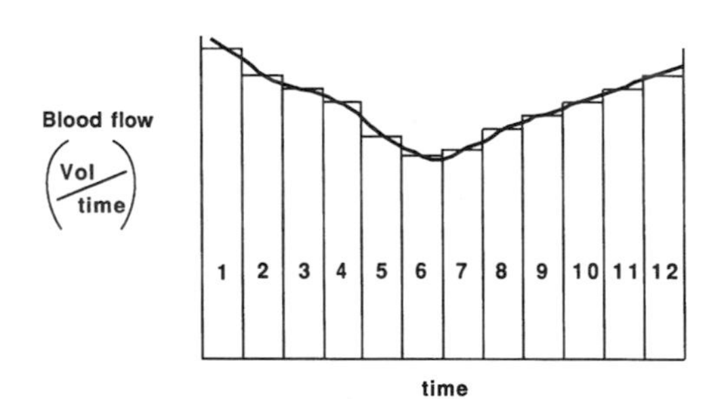

MARGIN FIGURE 10-18
Total area under a blood flow curve represents the total volume of blood. The area may be approximated. The computer estimates the blood volume by finding the area of each of the 12 rectangles and summing the results.

device; (2) store the number in a temporary storage location, and keep track of where it is; (3) locate the second number from an input device; (4) perform the addition function (itself a multistep process that may be "hard-wired" as a circuit such as the adder shown in Margin Figure 10-9 and route the result of the calculation to a third temporary storage location; (5) identify the temporary storage location where the sum is placed; and (6) report the result back to the main program, and look for the next instruction.

A CPU operates on data by assigning memory locations or "addresses" as indicated in the example above. Each address can be a binary (octal or hexadecimal) number no longer than the maximum word length of the computer. As shown earlier, the number of bits determines the number of unique values that a computer may recognize. Thus, the number of unique items that can be referenced by the computer is determined by the number of addresses it can hold at one time. This number is determined by the word length recognized by the computer. The word length is referred to as the "bit depth" of the CPU. If a certain processor has bit depth of 32, for example, it can process 2 to the 32nd power, or 4,294,967,296 instructions.

A 64-bit processor is inherently more powerful than a 32-bit CPU. A processor with greater bit depth is able to handle more storage locations and its numerical operations have a higher precision compared with those performed by a CPU with a smaller bit depth.

Most computers currently in operation use serial processing. This approach requires that one task be completed before another is begun. In many problems, the result of one task is the input for the next, and serial processing is necessary. In some problems, however, the completion of one task is independent of at least some other tasks. In these cases it may be possible for one processor to work on one task while a separate processor works on another. These tasks are said to be performed "in parallel." A computer that can perform tasks in parallel is termed a "parallel processor."

Serial and parallel processing are illustrated by the problem of determining the area under the curve in Margin Figure 10-18. This curve could represent the relative volume of blood in the heart at different times as measured by a gated blood pool nuclear medicine study. The area under the curve represents the volume of blood flowing through the heart during a cardiac cycle. In the serial approach to the problem, the computer might "draw" a series of rectangles under the curve and keep a running sum of the area of the rectangles. At the end of the curve, the summed area of the rectangles approximates the total area under the curve. In the parallel approach the computer recognizes that it has the data for the entire curve in memory. Because the areas of the rectangles are independent, the computer could "draw" and compute the area under each of the rectangles simultaneously, provided that it had enough independent processors. The problem could then be solved more rapidly.

At the present time, medical imaging systems do not make extensive use of parallel processors because they are costly and complex. Problems such as the blood flow example above can be solved in less than a second in serial fashion. Unless a large number of such problems must be solved quickly, there is little reason to use parallel processors. These processors are used for problems in other fields that cannot be solved by serial processing in a reasonable time. Some examples include large-scale problems in aerodynamics, meteorology, cosmology, and nuclear physics.[11] The use of parallel processors in radiology could increase in the future as imaging applications become more complex.

Special Processors

The array processor is a special-purpose processor that can be considered a small-scale parallel processor. An array processor uses a single instruction to perform the same computation on all elements of an array of data. Tremendous savings in computer time are achieved by using array processors to manipulate digital images. Such processors are routinely used in CT and magnetic resonance imaging units and are becoming standard in all types of digital imaging systems.

Another special processor that is used routinely in radiology is the "arithmetic processor." Many mathematical functions such as exponential values, square roots, and trigonometric functions can be calculated in clever but time-consuming fashion by repeated application of procedures such as additions that are "hard-wired" in gate circuits. The CPU of a typical computer is designed to perform alphanumeric and logical operations as well as handle mathematical operations. A faster (and usually more accurate) alternative involves the use of an arithmetic processor, a device that executes fewer steps in performing certain mathematical equations. The arithmetic processor is optimized to perform mathematical functions only. It is called upon by the CPU to provide these functions and return the results to memory.

Speed: MHz, MIPS, and MFLOPS

The speed of a CPU may be specified in a number of ways. One of the most fundamental specifications is the "clock rate." As the CPU executes the various steps required to complete a task, it keeps synchrony among the various components (input-output devices, memory, etc.) with a "master clock." The minimum time required to execute an instruction is the clock's "cycle time." The inverse of this cycle time, the theoretical upper limit of the number of instructions the CPU can execute per second, is the clock rate. Because the clock rate is a frequency in cycles per second, it is commonly expressed in multiples of hertz, such as megahertz (MHz) or gigahertz (GHz). For example, a CPU having a cycle time of 1.25 nanoseconds has a clock rate of

$$\frac{1 \text{ instruction}}{1.25 \times 10^{-9} \text{ seconds}} = 800 \times 10^{6} \text{ instructions per second}$$

$$= 800 \times 10^{6} \text{ hertz}$$

$$= 800 \text{ MHz}$$

Currently, CPUs in personal computers have a maximum clock rate of 1.5 gigahertz. Because of economies of scale in sale and manufacture, CPUs in computers associated with medical devices are the same CPUs that are sold commercially in quantity to PC users.

The CPU may take several clock cycles to execute an instruction. Therefore, a more realistic measure of speed of a CPU is the number of instructions per second that it can execute. For modern computers, this is often specified in millions of instructions per second (MIPS). Today's PCs can perform up to 100 MIPS. Some special purpose machines that have been built to test the limits of design can perform up to 10^{12} (prefix "tera") instructions per second (TIPS).

In numerical calculations it is often convenient to specify numbers in scientific notation (e.g., 1.25×10^{-9}), with a mantissa (e.g., 1.25) specifying a numerical value and an exponent (e.g.,−9) specifying the location of the decimal point. This specification facilitates operations on numbers that cover a wide range of values. In computer jargon, scientific notation is referred to as "floating point." Several bits of a floating-point word are used for the mantissa and several others for the exponent. The exponent is stored as a power of two, but the principle is the same. A few more bits are required for the signs of the mantissa and exponents. The instructions for floating-point arithmetic are more complicated (and therefore time-consuming) than for integer arithmetic, but the use of floating point is often necessary for scientific calculations. One million floating-point operations per second is abbreviated as 1 MFLOPS.

Supercomputers with parallel processing capability can attain processor speeds of 1000 MIPS and floating-point speeds of over 300 MFLOPS. So-called "massively parallel" systems with several hundred microprocessors achieve giga-FLOP speeds.[12]

The speed with which the entire computer can complete a task is usually limited by things other than the CPU. The rate at which the device can transfer data, access disk drives, or transfer information to peripheral devices may play a major role in determining how quickly it can accomplish complex tasks such as digital image processing and display. For this reason, overall computing speed is often measured by the time required to run standard programs successfully. These standard programs, called "benchmark" programs, are selected on the basis of their representation of the particular tasks that the computer in question would be expected to perform.

Memory

Memory in a computer can be divided into internal memory and external memory (or external storage). Internal memory is actively used by the computer during execution of a program, and it is directly accessible during an instruction cycle. External memory requires that the computer read information from an external storage device such as magnetic tape or disk. An analogy can be made with the student taking notes in a course. If the student knows the material without looking at the notes, the formation is stored in "internal memory." If the student must refer to the notes, then the information is in "external memory," or storage.

There are many types of internal memory. Random-access memory (RAM) is used as a scratch pad during calculations. Data can be stored there, read out, and replaced with new data. RAM is available when the computer is on. It contains information placed there for the problem that the computer is working on at the moment. Most RAM is "volatile," meaning that the information is lost when the computer is turned off. Read-only memory (ROM) is not volatile. It is used to store sets of instructions that the computer needs on occasion. The information in ROM is "hard-wired" (i.e., it is a physical part of the electronic components) and cannot be changed by the computer (hence the term "read only"). A variation of ROM, called programmable read-only memory (PROM), can be changed by exposure to an external electrical or optical device such as a laser so that reprogramming of the PROM is possible but not routine.

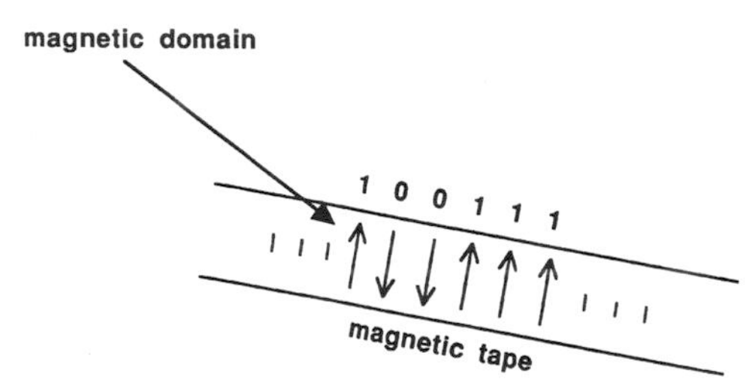

MARGIN FIGURE 10-19
Magnetic storage media such as magnetic tapes utilize magnetic "domains," regions of magnetization that are allowed two possible configurations to store binary digits.

In a digital imaging device the internal memory must be large enough to contain at least one digital image. A $512 \times 512 \times 8$-bit image requires $\frac{1}{4}$ Mbyte. Therefore, an imaging device that manipulates several such images needs several Mbytes of internal memory.

External memory devices are used to store images that are not needed for an immediate imaging task. There are various levels of separation from the computer. Images for which the diagnosis has been made are often stored for several years for legal purposes. These images are said to be "archived" (e.g., on magnetic tape), and they have to be remounted on the computer to retrieve the information. Images that are on hard disk drives are physically mounted on the computer and can be accessed within milliseconds. In magnetic storage devices the individual bits are recorded as magnetic "domains" where "north pole up" might mean a 1 and "north pole down" might mean a 0 (Margin Figure 10-19).

The most commonly used external memory devices store information in one of two ways. Magnetic tape, a mainstay of computer storage into the 1970s in the form of large reel-to-reel devices, now plays a role as a relatively cheap "off-line" storage option in the form of smaller closed cassettes. Although it may take a few seconds or even minutes to load the modern tape cassette and forward it to the location where the data of interest are stored, the long-term cost of buying and maintaining this storage medium are low.

Various optical storage devices are also available. The time required to access a particular string of data on an optical storage device may be nearly as short as the time required to access data on a (magnetic) hard disk. Optical disks contain tiny imperfections in an otherwise smooth mirror-like surface. A hole burned into the surface of the disk may be used to indicate a binary 1, while the absence of a hole indicates a 0. Optical disks that are created by burning are examples of "write once, read many" times (WORM) devices. This feature makes them suitable for long-term archival storage while still retaining a greater access speed. Disks 12 cm in diameter have become standard for musical recording and for routine PC use. These disks are known as "compact disks" or CD ROMs. Higher-capacity disks of the same size as

TABLE 10-4 Characteristics of Storage Media

Media	*Capacity*	*Access Time*
Magnetic tape	Gbyte	sec-min
Floppy disks	2 Mbyte	msec
Hard disks	10 Gbyte	msec
Optical disks	5 Gbyte	msec

CD ROMs, but capable of storing enough digital information to hold several hours of high-quality video, are known as digital video disks (DVDs). DVDs are becoming more prevalent. Mechanical devices are sometimes used to fetch and mount disks on demand from a large collection of optical disks. These devices are known as "jukeboxes." Table 10-4 shows some of the characteristics of commonly used external storage media.

SOFTWARE

The term *software* refers to any set of instructions the computer uses to function effectively. Specific sets of instructions are necessary for any computer. The "operating system" consists of instructions that govern the use of input and output devices, set aside internal storage locations, direct the computer in tasks such as copying, naming, and erasing files from external storage, and perform other necessary tasks. The operating system always must be executed before any other programs are run. In many computers the operating system program is executed automatically when the power is turned on. Some of the operating system instructions are usually resident in ROM or PROM. Running the operating system programs is referred to as "booting" the computer.

The term "booting", meaning the startup process for a computer, comes from the phrase "to pull oneself up by the bootstraps" (i.e., to be a self-starter).

Software that has been written to perform a specific task, such as to calculate ejection fractions for a nuclear cardiology study, is called a "program."

Several levels of software are usually present when a computer is operating, and instructions may be written in various "languages." High-level languages are most accessible to the typical programmer. They are written in statements that use everyday language such as "go to, return, and continue." In a high-level language, mathematical formulas are expressed just as they would be written on paper. Assembly language is a step closer to direct communication with the computer. Statements in assembly language direct the computer to perform specific tasks such as to store data in certain locations. High-level languages such as Visual BASIC, C++, or SQL tell the computer what to do in a general way. Assembly languages give the computer specific instructions.

Jukebox: Juke is a term from Gullah, a language spoken by descendants of West African slaves living on islands off the coast of Georgia and South Carolina. Juke, one of the few terms from Gullah to be adopted into common English usage, means bad, wicked, or disorderly. It was used to refer to roadside drinking establishments in the Southeastern United States beginning in the 1920s that came to be known as "juke joints." When mechanical devices capable of selecting and playing individual selections of music from a collection of vinyl disks became available, the devices, often located in "juke joints," became known as "jukeboxes." Today, this term is used to refer to mechanical devices for selecting and interfacing multiple digital storage disks (CD-ROM, Optical, Magneto-Optic, etc.). The earlier connotation of wickedness is not generally associated with the information management specialists who use these devices.

Both high-level and assembly languages must ultimately be translated into machine language. Machine language is represented simply as strings of ones and zeros. It directs the fundamental setting and resetting of various electronic components that yield signals sent to an output device. Assembly language programs are converted into machine language by a translation program called an "assembler." Higher-level languages may be converted into machine language in one of two ways. They may be translated all at once into machine language by a program called a "compiler," or they may be translated line by line each time the program is executed by a program called an "interpreter." The interpreter is used when the programmer needs to make changes in the program. Once the high-level language is compiled, the program cannot be changed, because individual lines in the program are no longer accessible. The compiler is used when further changes are not required, and speed of execution is important. Compared with interpreted programs, compiled programs usually require much less time to execute. Different types of software and their interrelationships are compared in Margin Figure 10-20.

NETWORKING

The increased use of computers in all areas of medicine, including diagnostic imaging, creates the desire to link the computers together, thereby making information present in one computer available to all computers in an institution or community. This linking of computers, called "networking," is a topic of current interest.[13–16]

In the late 1960s, the U.S. Department of Defense wanted to develop a computer network that would survive a nuclear attack. The network would continue to deliver

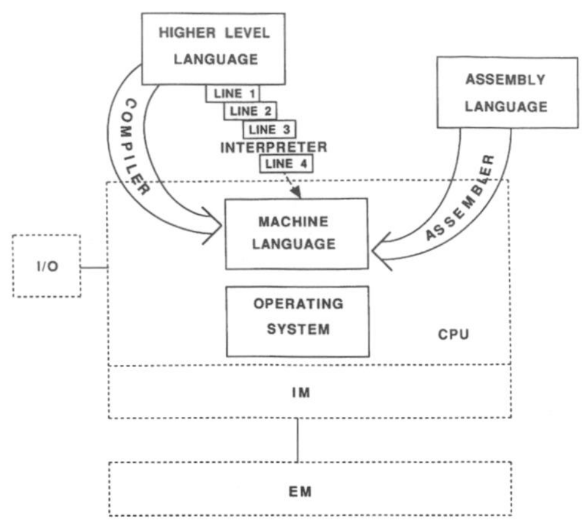

MARGIN FIGURE 10-20
Relationships among software and hardware in a typical computer. Hardware is shown inside *dashed lines*. I/O = input/output; CPU = central processing unit; IM = internal memory; EM = external memory.

History of the Internet: The World Wide Web was created in March 1989, at CERN, a high-energy particle physics laboratory on the Franco-Swiss border. Tim Berners-Lee, a physicist at CERN, proposed the idea of using a hypertext system that would link data from diverse information sources and different computer platforms.

streams of digital data, even if a number of its components or linkages were destroyed. The solution was "packet switching" a system in which data streams are broken into smaller pieces, called cells or frames (see Margin Figure 10-18). Each cell contains not only a piece of the data, but also information such as the place of the cell in the original sequence, the priority of this stream of data compared with other data streams, and so forth. One key feature of packet switching is that each cell also contains the address of the component to which it is being sent. Another key feature of packet switching is that the network consists of interconnected routers. Each router is a computer whose purpose is to maintain information about the addresses of surrounding routers. When a packet arrives at a router, it is automatically sent on to another router that is "closer" to its destination. Thus, if part of the network is disabled, the routers update their information and simply send cells via different routes. This earliest wide-area network was known as Arpanet, after the Department of Defense's Advanced Research Projects Administration.

When electronic mail (e-mail) was developed, the usefulness of Arpanet became apparent to a wider community of users in academics and in government, and the number of network users continued to grow. Arpanet was officially decommissioned in 1989. By that time, a wide community of users required that the network be continued. The administration of the network was turned over the National Science Foundation. The network of routers, file servers, and other devices that have become the foundation of modern communications has been known as the Internet since 1989.

In 1989, the World Wide Web (WWW) was first proposed at CERN, a high-energy physics laboratory on the border between France and Switzerland. This proposal embodied all of the modern components of the WWW, including file servers that maintain "pages" of data in standard format with "hypertext" links to other sources of data or other servers. In 1991 there were only 10 file servers on the WWW, at various physics laboratories. Today, the number of servers is well into the millions.

Network Components and Structure

Computer networks for medical imaging are known by many names, such as information management, archiving, and communications systems (IMACS), picture archiving and communications systems (PACS), digital imaging networks (DIN), and local area networks (LAN). These networks face some fundamental problems, including (1) how to transmit images quickly, (2) how to avoid bottlenecks or "data collision" when the network is experiencing heavy use, (3) how to organize and maintain the data base or "log" that records the existence of images and their locations on the system, and (4) how to retain as many images as possible on the network for as long as necessary.

The main components of a computer network and its typical organization are shown in Margin Figure 10-21. Components include image acquisition (CT, ultrasound, etc.), archiving (tape, disk, etc.), central controller, data base manager, and display station. Components that allow the network to communicate with the outside world through display and archiving techniques are referred to as "nodes." All components are not necessarily nodes. For example, several ultrasound units might be connected to a single formatting device that translates the digital images into a standard format recognizable by the network. The formatting device communicates directly with the network and is the node for the ultrasound units.

Networks can be divided into two categories on the basis of their overall structure. Centralized networks use a single computer (a central controller) to monitor and control access to information for all parts of the network. A distributed network contains components that are connected together with no central controller. Tasks are handled as requested until a conflict arises such as more demands being placed on a component than it can handle. While it may appear chaotic, a distributed network has the advantage that other components are not affected by slowdown or stoppage of one or more components of the system. In particular, there is no central controller that would shut the whole system down if it were to malfunction.

Interfaces

The interface of an imaging device such as a CT scanner with a computer network is usually a more complex matter than simply connecting a few wires. Transfer of images and other data requires that the network must be ready to receive, transmit, and store information from the device. These operations could interfere with other activities taking place on the network. Therefore, the imaging device may have to send an "interrupt" signal that indicates that a transmission is ready. The data may have to be transmitted in blocks of specified size (e.g., 256-Kbyte transmission packets) that are recognizable to the network as part of a larger file of information.

The above-mentioned problems require both hardware and software solutions. Programs that control transmission of data to and from various components are called "device drivers." The physical connectors must also be compatible with (i.e., can be attached and transmit signals to) other network components. Both hardware and software are implied in the term "interface." Connecting a component to a computer or computer network in such a way that information may be transmitted is described as interfacing the component. There have been a number of attempts to standardize interfaces for imaging devices, including the American College of Radiology/National Electronics Manufacturers Association (ACR/NEMA) standards.[17]

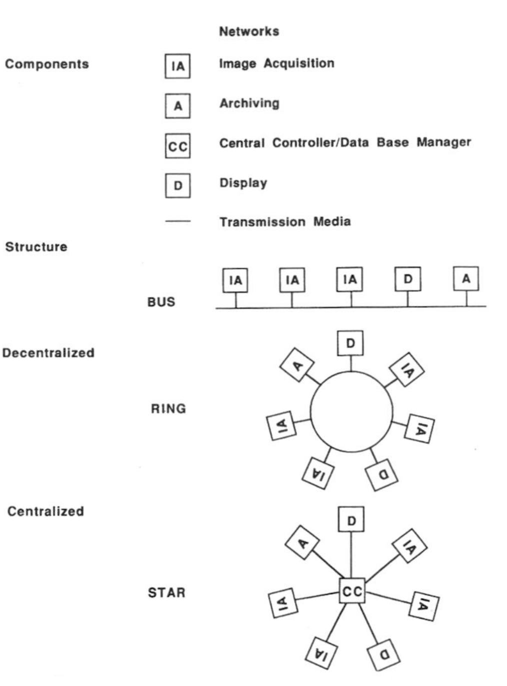

MARGIN FIGURE 10-21
Components and structures commonly used in image networks.

Transmission Media

Components of a network may be physically separated by distances varying from a few feet to several hundred miles or more. The transmission of digital information requires a transmission medium that is appropriate to the demands of the particular network. One of the most important factors to consider is the rate at which data are transmitted, measured in bits per second (bps).

Example 10-5

A network is capable of transmitting data at a rate of 1 megabit per second (Mbps). If each pixel has a bit depth of 8 bits, how long will it take to transmit 50 512 × 512 images?

Because each pixel consists of 8 bits, the total number of bits is

$$\begin{aligned} 50[8 \text{ bits/pixel} \times (512 \times 512 \text{ pixels})] &= 50(2^3 \times 2^9 \times 2^9) \text{ bits} \\ &= 50(2^{21}) \text{ bits} \\ &= (50)(2 \text{ Mbits}) = 100 \text{ Mbits} \end{aligned}$$

The transmission time is then

$$\frac{100 \text{ Mbits}}{1 \text{ Mbps}} = 100 \text{ sec}$$

One of the least expensive transmission media is telephone wire (sometimes called "twisted pairs"). It is inexpensive and easy to install and maintain. However, the transmission rate does not usually exceed a few Mbps. Higher rates up to hundreds of Mbps are achievable with coaxial cable. However, coaxial cable is more expensive, needs inline amplifiers, and is subject to electrical interference problems in some installations. The highest transmission rates are achievable with fiber-optic cable. This transmission medium consists of glass fibers that transmit light pulses, thereby eliminating electrical interference problems. Transmission rates for fiber-optic cables are currently just below a terabit per second (Tbps). Telephone, coaxial cable, and fiber-optic cable are shown in Margin Figure 10-22.

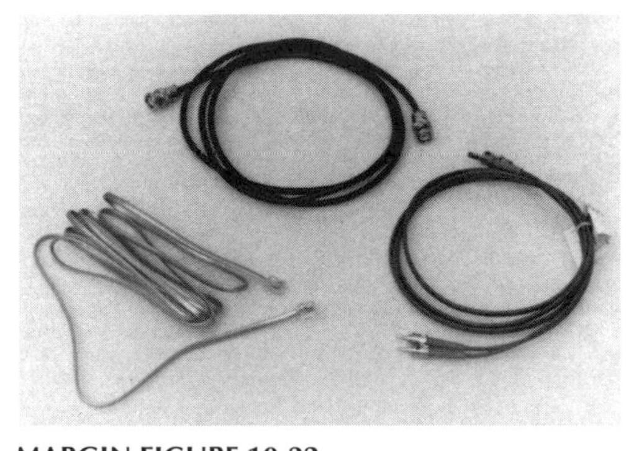

MARGIN FIGURE 10-22
Transmission media used in imaging networks. Clockwise from the top: coaxial cable, fiber-optic cable, unshielded "twisted pair" (telephone cable).

Digital Communications Options

Copper Wire:	
Modulator/Demodulator (modem)	56 kbps
Integrated Services Digital Network (ISDN)	64 or 128 kbps
Digital Subscriber Line (DSL)	Several Versions up to 4 Mbps Asynchronous DSL 3-4 Mbps to the home <1 Mbps from the home
Hybrid-Fiber/Coax:	
Cable modem	40 MBps to the home, However, multiple users slow it down Usually, ≤10 MBps from cable box to your computer
Wireless (high frequency GHz radio waves):	
Local Multi-Point Distributed Services (LMDS)	data rates of up to 155 Mbps problems: "rain fade", trees, etc.
Cellular Telephone	Now: 10 to 50 kbps Future: 1 Mbps
Satellites	data rates of up to 210 Mbps problem: cost
Direct Fiber:	
fiber-to-the-home	Now: <10 Mbps Ethernet for most PCs <100 Mbps FDDI for some devices 10's of Gbps Asynchronous Transfer Mode Future: Tbps

Data Compression

Images are transmitted faster and require less storage space if they are composed of fewer bits. Decreasing the number of pixels reduces spatial resolution, however, and decreasing the bit depth decreases contrast sensitivity. It is possible, however, to "compress" image data in such a way that fewer bits of information are needed without significant loss of spatial or contrast resolution.

One way to reduce the number of bits is to encode pixel values in some sequence (e.g., row by row) to indicate the value of each pixel and the number of succeeding pixels with the same value. This decreases the total number of bits needed to describe the image, because most images have several contiguous pixels with identical values (e.g., the black border surrounding a typical CT image). Other techniques for data compression include (a) analysis of the probability of occurrence of each pixel value and (b) assignment of a code to translate each pixel value to its corresponding probability. Such a "probability mapping" uses fewer bits for the more probable pixel values.[19]

In the examples above, the full information content of the image is preserved. When the image is expanded, it is exactly the same as it was before compression. These approaches to data compression are known as lossless (bit-preserving, or nondestructive) techniques. They may yield a reduction in the number of required bits by a factor of 3 or 4. When a greater reduction is needed, methods of data compression may be used that do not preserve the exact bit structure of an image, but still maintain acceptable diagnostic quality for the particular application. These techniques, known as irreversible (non–bit-preserving or destructive) data compression methods, can reduce the number of bits by an arbitrarily high factor.[12]

An example of an irreversible data compression method involves the use of the Fourier transform to describe blocks of pixel values in the image. Some of the high- or low-frequency components of the image are then eliminated to reduce the number of bits required to store the image. When an inverse transform is used to restore the image, the loss of high or low spatial frequencies may not significantly detract from the diagnostic usefulness of the image.

Display Stations and Standards

The part of a computer network that is most accessible to the observer is the digital display or monitor, usually referred to as the image workstation. Some display stations are capable of displaying more data than are presented on the screen at one time. They may, initially, show images at reduced resolution (e.g., 1024 × 1024) while preserving the full "high-resolution" data set (e.g., 2048 × 2048) in memory. The stored data can be recalled in sections through user-selectable windows (analogous to using a magnifying glass on a part of the image). Alternately, only part of the image may be presented, but the part that is presented may be "panned" or moved around over the full image.

Standards for digital matrix size and remote display of medical images that are acceptable for primary diagnosis from computer monitors have been published by the Radiological Society of North America.[18] These standards will continue to evolve as equipment performance (particularly display monitors) continues to improve. Currently, two classes of images are recognized: small matrix systems and large matrix systems. Small matrix systems (CT, MRI, ultrasound, nuclear medicine, and digital fluoroscopy) must have a format of at least 5 k × 5 k × 8 bits. The display must be capable of displaying at least 5 k × 0.48 k × 8 bits. Large matrix systems (digitized radiographs, computed radiography) are held to a standard based upon required spatial and contrast resolution. For those imaging methods, the digital data must provide a resolution of at least 2.5 line pairs per millimeter and a 10-bit gray scale. The display must be capable of resolving 2.5 line pairs per millimeter with a minimum 8-bit gray scale.

The industry standard format for transferring images between components of a network is the Digital Imaging and Communications in Medicine or DICOM standard. The standard consists of specifications for various data "fields" that must occur in the image header. In addition to pixel values, certain patient data should accompany each image. For example, each image should be preceded by "header" information that gives patient demographic data as well as details of the structure of data in the image file. These fields describe attributes such as the matrix size of the image, whether it is part of a series (e.g., one slice of a multislice CT series), and patient demographic data.

The development of the DICOM standard was initiated by the American College of Radiology (ACR) and the National Electrical Manufacturer's Association (NEMA) in 1985, and it continues to evolve as equipment capabilities change.

Although the capabilities of "high-end" workstations are far from being standardized, some general features have been. A monitor should be able to display enough written information concerning the patient to obviate the need for transport of paper documents. Obviously, it must be able to display text and perform at least rudimentary editing and file organization functions. The display station should be able to run image processing software as well as provide simple display functions such as variable window level, window width, and magnification. Dimensional reconstruction and tissue segmentation (display of bones only or vessels only) are becoming more common.

PROBLEMS

*10-1. How many bits of information are contained in 5 kilobits (kbits)?

*10-2. How many bits of information are contained in 60 megabytes (Mbytes)? How many bytes?

*10-3. How many bits of information are contained in 0.5 Mwords if a 32-bit word length is used? How many bytes?

*10-4. A 512 × 512 × 12-bit image is transmitted along fiber-optic lines at a rate of 1 gigabit/sec (1 Gbit = 2^{30} bit). How many of these images can be transmitted per minute at such a rate? (*Note*: Demographic and other textual information would accompany each image, reducing the actual rate of transmission.)

*10-5. A 4-bit A to D converter is used to record signals that range from 0 to 1 V. What is the average digitization error in millivolts?

*10-6. Define the following acronyms: RAM, ROM, PROM. Which is the most appropriate term to describe a computer chip that is altered during an on-site equipment upgrade?

*10-7. A digitized radiograph is viewed on a 12-in.-wide monitor. If the actual width of the patient's chest is 14 in. and if the monitor resolution is set to 1024 × 1280 (height × width), does this monitor meet the ACR standard for use for primary diagnosis?

*10-8. A 2048 × 2048 × 12-bit image is compressed at 10 to 1 compression ratio. How many bits are present in the compressed image?

*For problems marked with an asterisk, answers are provided on p. 492.

SUMMARY

- The number of **distinct values N** that may be recorded with ***n*** **bits** is given by the formula $N = 2^n$.
- **Bandwidth** is the maximum number of bits per second that may be transmitted over a communications channel.
  - Bandwidth in medical communications varies from thousands of bits per second (10^3 or **kbps**) in relatively inexpensive systems to billions of bits per second (10^9 or **Gbps**) in more expensive and specialized systems.
- Images may be thought of as three-dimensional data blocks, where a **matrix of pixels** is the "face" of the block and where the number of bits allocated for representation of the pixel value is the "depth" of the block.
- Medical images may be generally divided into **small matrix** and **large matrix** images.
  - **Small matrix images** (CT, MRI, ultrasound, nuclear medicine, digital flurography) are 512 × 512 or smaller.
  - **Large matrix images** (CR, digitized film, direct digital capture x-ray images) are 1024 × 1024 or greater.

- In images, edges are generally associated with **higher spatial frequencies,** and contrast between large regions is generally associated with **lower spatial frequencies.**
- The Radiological Society of North America standard for official diagnostic interpretation of soft-copy images requires:
  - For **small matrix**:
    Acquisition: 0.5 k × 0.5 k × 8 bits or better
    Display: 0.5 k × 0.48 k × 8 bits or better
  - For **large matrix**:
    Acquisition: 2.5 line pairs/mm and 1024 gray scale or better
    Display: 2.5 line pairs/mm and 256 gray scale or better
- **Local image processing** involves movement of a small (e.g., 5 × 5 pixel) **kernel** over the image.
- **Global image processing** involves the application of a **filter function** or **operator** to the entire image.
- **High-pass filters** remove lower spatial frequencies while allowing higher spatial frequencies to remain.
- **Low-pass filters** remove higher spatial frequencies while allowing lower spatial frequencies to remain.
- Images may be **compressed** so that fewer bits are required for transmission or storage of the images.
  - In **loss-less compression,** the original image data block may be reconstructed to its original matrix size and depth with no error.
  - In **lossy compression**, some error is allowed in exchange for a reduction in the number of bits required to transmit or store the image.
- Picture Archiving and Communication Systems **(PACS)** consist of modules for acquisition, image processing, transmission, display and storage of images.
  - The internationally recognized image and data format for PACS systems is the Digital Imaging and Communications in Medicine **(DICOM)** standard.
- **Transmission media** include: copper wire ("twisted pair"), coaxial cable, optical fiber, and microwave carrier.
- **Transmission protocols** include: point-to-point presence **(PPP)**, transmission control protocol/internet protocol **(TCP/IP)**, and asynchronous transfer mode **(ATM)**.
- The set of **file servers** and **routers** that are generally accessible throughout the world using a TCP/IP protocol is known as the **internet**.

REFERENCES

1. Considine, D. M. (ed.). *Van Nostrand's Scientific Encyclopedia,* 5th edition. New York, Van Nostrand Reinhold, 1976.
2. Spencer, D. D. *Introduction to Information Processing*. Columbus, OH, Charles E. Merrill, 1981.
3. Kuni, C. C. *Introduction to Computers and Digital Processing in Medical Imaging*. St. Louis, Mosby–Year Book, 1988.
4. Ritenour, E. R. *Computer Applications in Diagnostic Radiology, Instructor's Manual*. St. Louis, Mosby–Year Book, 1983.
5. Stubbe, J. W. *Computers and Information Systems*. New York, McGraw-Hill, 1984.
6. Mackintosh, A. R. The first electronic computer. *Phys. Today* 1987; **March**:25–32.
7. Singh, S. J. *Fermat's Enigma*. New York, Walker and Co., 1997, pp. 147–157.
8. Mauchly, K. R. *Annals in the History of Computing,* Vol. 6, 1984, p. 116.
9. Waldrop, M. M. The necessity of knowledge. *Science* 1984; **223**:1279–1282.
10. Hofstadter, D. R. *Metamagical Themas*. New York, Basic Books, 1985.
11. Doing physics with computers. Special issue of *Physics Today*, Vol. 36, Woodbury, NY, American Institute of Physics, 1983.
12. Glantz, J. Microprocessors deliver teraflops. *Science* 1996; **271:**598.
13. Johnson, N. D., Garofolo, G., and Geers, W. Demystifying the hospital information system/radiology information system integration process. *J. Digital Imaging* 2000; **13**(2 Suppl. 1):175–179.
14. Langer, S. G. Architecture of an image capable, Web-based, electronic medical record. *J. Digital Imaging* 2000; **13**(2):82–89.
15. Abbing, H. R. Medical confidentiality and electronic patient files. *Medicine & Law* 2000; **19**(1):107–112.
16. Staggers, N. The vision for the Department of Defense's computer-based patient record. [Review] [19 refs]. *Military Medicine* 2000; **165**(3):180–185.
17. Wang, Y., Best, D. E., Hoffman, J. G., et al. ACR/NEMA digital imaging and communications standards: Minimum requirements. *Radiology* 1988; **166**:529–532.
18. Radiological Society of North America. *Handbook of Teleradiology Applications*. Oakbrook, IL, Radiological Society of North America, 1997.
19. Huang, H. K. *PACS: Basic Principles and Applications*. New York, John Wiley and Sons, 1999, Chapter 6.

CHAPTER

11

PROBABILITY AND STATISTICS

OBJECTIVES

After studying this chapter, the reader will be able to:

- Distinguish between determinate and indeterminate errors, and give examples of each.
- Explain the concepts of

Accuracy	Mean
Precision	Median
Bias	Mode
True value	Standard deviation
Randomness	Standard error
Noise	Confidence limits

- Describe the properties of Poisson and normal (Gaussian) distributions.
- Delineate the rules for error propagation.
- Apply selected statistical tests:
  - Student's *t*-test
  - Chi-squared test

INTRODUCTION

A random event is one in which the outcome is uncertain and can only be predicted by applicable rules of probability.

Statistics is the study of probability rules and their applications.

Analysis of error is a cornerstone of all research and clinical service. The presentation of patient data in the form of images involves uncertainty and requires statistical analysis. An appreciation of error allows the interpreter of images to judge the value of visual data and to develop more informed decisions about techniques of image acquisition. The choice of technique factors (e.g., kVp, mA, and time in radiography) influences (a) the appearance of statistical fluctuations (visual "noise") in film images and (b) the random variations in pixel values in digital systems. Digital images are affected by the choice of pixel size, gray scale, and acquisition time. Trade-offs among these and other variables are dealt with continuously in diagnostic imaging. Nuclear medicine and ultrasonographic data are influenced by uncertainties that arise at all stages in the image-forming process. Data acquisition techniques in magnetic resonance imaging provide a wide variety of choices among variables such as pulse sequences, field strengths, and receiver coils, all of which affect the statistical quality of the images. For all these reasons, the imaging specialist must acquire a reasonable command of rudimentary statistical concepts.

NATURE OF ERROR

For every human problem, there is a solution that is simple, neat . . . and wrong. H. L. Mencken

The error of a measurement is the difference between the measured value and the "true value" for the variable being measured. Two categories of error exist, *determinate* and *indeterminate* errors. Determinate errors result from factors such as inadequate experimental design, malfunctioning equipment, and incomplete correction for extraneous influences. The influence of determinate errors upon experimental results can often be reduced by better instrumentation and thorough planning and execution of an experiment. For imaging systems, some examples of determinate errors include overexposure or underexposure to films in radiography, incorrect time-gain compensation (TGC) settings in ultrasonography, and a falloff in surface coil sensitivity in magnetic resonance images.

Determinate errors are sometimes called *systematic errors*; indeterminate errors are often termed *random errors*.

Indeterminate errors are errors that cannot be reduced by eliminating or correcting extraneous factors. They are caused by fundamental uncertainties in the variables being measured and in the processes of obtaining and displaying the measurements. In nuclear medicine, for example, the random nature of radioactive decay and the statistical uncertainties in detecting ionizing radiations give rise to significant indeterminate errors.

The precision of a series of measurements describes the range or spread of the individual measurements from the average value of the series. Precision improves as the influence of indeterminate error is reduced. Precision does not describe the accuracy of a series of measurement. Accuracy is achieved only if the measured values agree with the true value determined by some method other than the measurement technique under consideration.

Usually, the "true value" for a measurement cannot be determined (if it could, the need for the measurement would disappear). However, a technique may be available that yields results that are generally accepted as being as close as possible to the true value. This technique and the result that it provides are referred to as a "gold standard." An example of a gold standard is the use of biopsy results against which interpretations of diagnostic images are evaluated.

To increase the accuracy of experimental results, the influence of both indeterminate and determinate errors must be reduced. The contribution of determinate errors to a reduction in accuracy of a set of measurements is termed the bias of the measurements. The distinctions among precision, accuracy, and bias are illustrated in the margin.

Precision: Is the consistency of repeated measurements of the same thing. Precision is sometimes referred to as *reliability*.

Accuracy is sometimes termed *validity*.

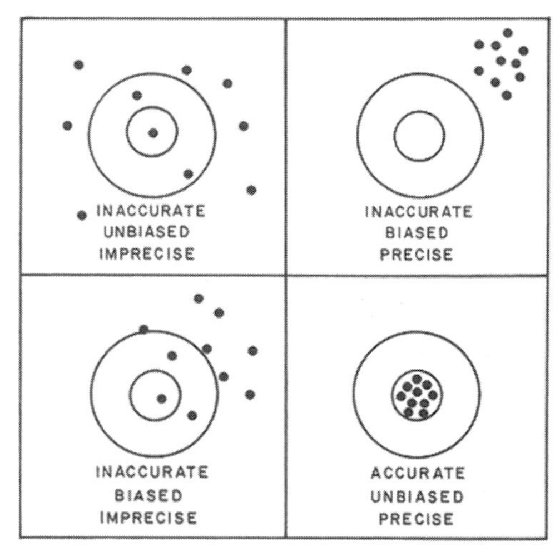

MARGIN FIGURE 11-1
Precision, bias, and accuracy.

PROBABILITY DISTRIBUTIONS

Indeterminate error implies that the results of a measurement may be different if the measurement is repeated. After many measurements have been recorded, a histogram can be constructed to show the number of observations for each measurement value (Figure 11-1). A histogram plotted as the frequency or fraction of times that each measurement is observed gives the probability that each specific value will be measured if the experiment is repeated. An accurate assessment of the probability of occurrence of each value would require an infinite number of measurements. If the sources of indeterminate error are known, however, a choice can be made among different theoretical probability distributions whose properties are depicted by mathematical analysis.

Bias is a systematic (i.e., not random) difference between the true value of a property and individual measurements of the property.

Bias is sometimes used incorrectly as a synonym for *prejudice*. Prejudice is a state of mind of an individual that creates a desire for a particular outcome.

A probability distribution reveals the likelihood of each of all possible outcomes of a set of measurements. The sum of all likelihoods (probabilities) (i.e., the area under a probability curve) is unity.

Poisson Distribution

Radioactive decay is a probabilistic (random) process. When many atoms are treated as a group, we may make statements such as "the number of atoms will be reduced to half of the initial number after a certain time (half-life) has elapsed." With respect to a

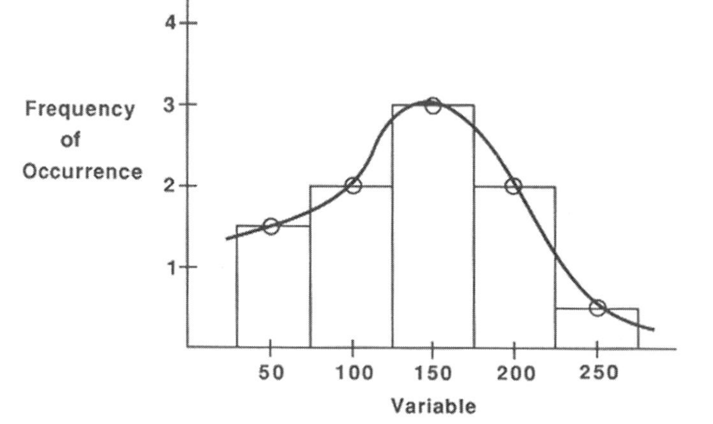

FIGURE 11-1
A histogram showing the frequency of occurrence of some variable. A histogram may be presented as bars, points, or a continuous line.

Simeon Denis Poisson (1781–1840) was a French mathematician who began studies as a physician but switched to mathematics after one of his patients died.

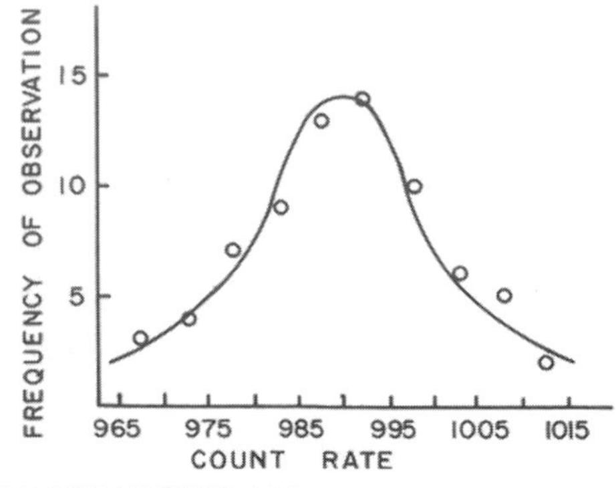

MARGIN FIGURE 11-2
Data for ^{137}Cs source that provided a count rate of about 1000 cpm. A count of 10,000 was accumulated for each measurement. The results of 100 measurements were plotted as a function of the number of times that the count rate fell within successive increments of 5 (i.e., between 975 and 980, 980, and 985, etc.). The Poisson curve of best fit is drawn through the data points (From F. Low.[2] Used with permission.)

A Poisson distribution is said to be a "discrete" distribution because it may contain only 0 or positive integers.

single atom, however, we cannot say with certainty when it will decay. The occurrence of radioactive decay (detection of "counts" in a radioactive sample) follows the Poisson probability distribution.[1] The shape of the Poisson probability distribution is given by the formula

$$p_n = \frac{r^n e^{-r}}{n!}$$

where p_n is the probability of obtaining a count n and r is the true average count for the sample. The term $n!$ [n-factorial] equals $[(n)\ (n-1)(n-2)\ \ldots\ (2)(1)]$. The probability p_n is termed the Poisson probability density, and radioactive decay is said to follow a Poisson probability law. Because the true average count r (often called the true mean) for the sample cannot be determined, it is assumed to equal the average measured count (estimated mean), often referred to as the sample mean. Shown in the margin are the results of an experiment in which the count rate (number of counts measured over a given counting time) was measured repeatedly.

Example 11-1

What is the probability of obtaining a count of 12 when the average count is 15?

$$\begin{aligned} p_n &= \frac{r^n e^{-r}}{n!} \\ p_{12} &= \frac{(15)^{12} e^{-15}}{12!} \\ &= \frac{(129.7 \times 10^{12})(30.6 \times 10^{-8})}{(4.79 \times 10^8)} \\ &= 0.0829 \end{aligned} \quad \text{(11-1)}$$

The probability is 0.0829 (or 8.29%) that a count of 12 will be obtained when the average count is 15.

Normal Distribution

A Gaussian (normal) distribution is a bell-shaped symmetrical distribution characterized by two values: the mean of a set of measurements and the standard deviation of the measurements.

If a large number of counts is accumulated for a radioactive sample, then the Poisson distribution curve of frequency of occurrence versus counts is closely approximated by a Gaussian distribution curve. In this case the data are described reasonably well by the equation for a Gaussian (normal) probability density function. Illustrated in Figure 11-2 is the approach of a Poisson probability distribution to a Gaussian distribution. The equation for the Gaussian probability density g_n is

$$g_n = \frac{1}{\sqrt{2\pi r}} e^{-(n-r)^2/2r} \quad \text{(11-2)}$$

The Poisson distribution can be written with σ^2 in place of r.

where g_n is the probability of observing a count n when the true count (estimated as the average count) is r.

Properties of a Gaussian distribution include the following:

- The distribution is continuous and symmetrical, with both tails extending to infinity.
- The arithmetic mean, median, and mode are identical.
- The likelihood of occurrence of a particular value depends on its distance from the mean value.
- The standard deviation is represented by the distance from the mean value to the inflection point of the distribution curve.

Example 11-2

By using Eq. (11-2), estimate the probability of obtaining a count of 12 when the true or average count is 15.

$$\begin{aligned} g_n &= \frac{1}{\sqrt{2\pi r}} e^{-(n-r)^2/2r} \\ &= \frac{1}{\sqrt{2\pi(15)}} e^{-(12-15)^2/2(15)} \\ &= 0.103 e^{-0.3} \\ &= 0.0764 \end{aligned} \quad \text{(11-3)}$$

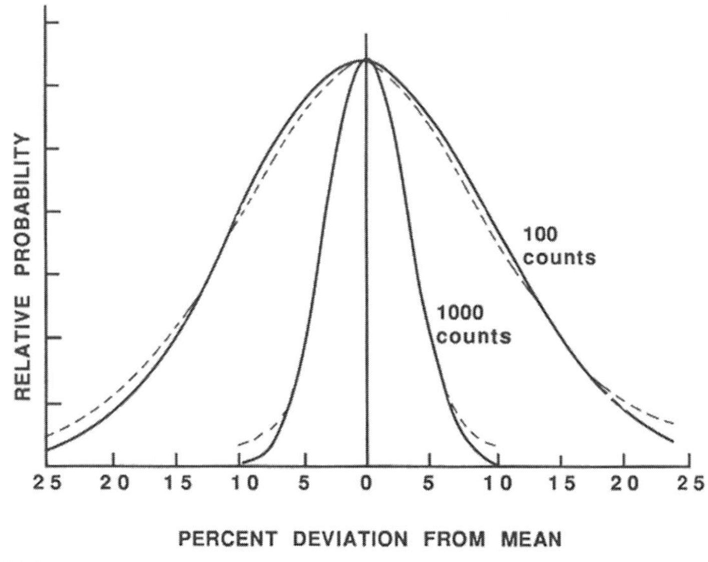

FIGURE 11-2
Probability of observing a particular count is plotted for 100 measurements and 1000 measurements. *Dashed curves* illustrate Gaussian (normal) probability density functions, and *solid curves* illustrate Poisson probability density functions. Curves have been normalized to equal heights at the mean.

The probability is 0.0764 (or 7.64%) that a count of 12 will be obtained when the average count is 15, estimated by assuming that radioactive decay follows a Gaussian rather than a Poisson probability distribution.

SIGNAL AND NOISE

Many of the challenges of imaging systems may be reduced to the simple problem of detecting a signal against a background of noise. The principle of signal to noise can be illustrated by examining the electrical signal in a wire. The current in a wire plotted as a function of time yields a graph such as shown in Figure 11-3 where there is a fluctuation, or noise σ, at all times. The current i is measured at three different times t_1, t_2, and t_3. The current i_0 is the average "background level." The problem of signal detection may then be stated as a question: Are the currents i_1, i_2, and i_3 measured at times t_1, t_2, and t_3 significantly different from the background level i_0? If so, then they can be interpreted as signals occurring at times t_1, t_2, and t_3. If not, they should be interpreted as statistical fluctuations in the background current and ignored. Unfortunately, there are no definitive answers to these questions. If the noise σ is truly random, we cannot say with absolute certainty that the currents i_1, i_2, and i_3 are signals rather than simply fluctuations in the background current, no matter how different the levels may be.

A series of events is said to be *random* if every event has an equal probability of occurring. "White noise" refers to the presence of background signals that are random in nature.

There are two approaches to this dilemma. The first approach is to state the probability that currents i_1, i_2, and i_3 are different from the background i_0. This leaves the final decision up to the reader. However, there are levels of probability that are commonly accepted as sufficient to say that two signals are significantly different. For example, one might say that there is a 95% chance that a peak such as i_2 is not a random fluctuation in the background current but instead represents a true signal superimposed on the background current. This approach is the foundation of the statistical t-test discussed later in this chapter.

The background signals that constitute white noise are "independent events" which have the property that each event has no influence on the likelihood or value of a previous or subsequent event.

The second approach to the separation of signal and background is to determine beforehand that a multiple of noise or background will be accepted as a signal. For

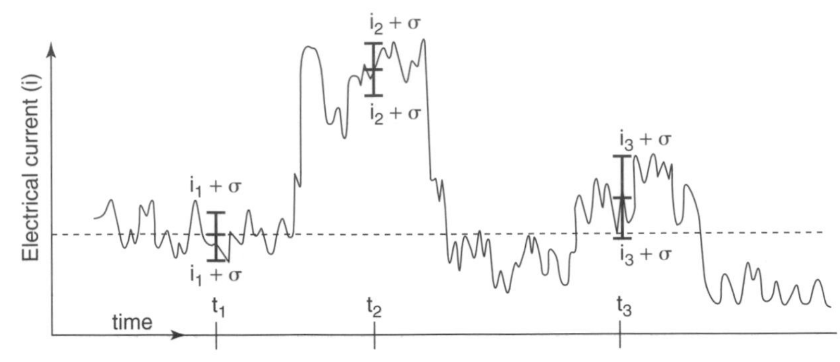

FIGURE 11-3
The problem of signal detection against background noise. Electrical currents i_1, i_2, and i_3 are measured at three times, t_1, t_2, and t_3. For each measurement there is a mean value i and a standard deviation σ. A current i_0 is assumed to be the background value. That is, i_0 was measured at a time when, through prior information, it was known to be unlikely that there would be a signal (a current significantly above background level and noise) resulting in our estimate of background level (i_0) and noise (σ). The problem is to determine if there is a "signal" at time t_1, t_2 or t_3. The central problem of signal detection is to determine the current i that is required to be accepted as a signal that is distinct from background noise. In the example shown here, it appears that there is a clear signal (i.e., distinct from the noise) at t_2, while the signals at t_1 and t_3 are not strong enough to "rise above the noise." Statistical techniques, such as specialized versions of the t-test, may be used to determine the probability that it was a "real" signal.

signal i_2 we define (with reference to Figure 11-3)

$$\text{Signal-to-background ratio} = \frac{i_2}{i_0}$$

$$\text{Signal-to-noise ratio} = \frac{i_2}{\sigma}$$

where the noise σ is a measure of the fluctuation around the background level i_0. In diagnostic imaging, a value of the ratio of 3 to 5 often is accepted as indicative of a signal.

The problem of signal detection against background noise can be extended to three dimensions, as shown in Figure 11-4 (in this image, spatial coordinates are plotted for two dimensions, vertical and horizontal, and gray scale is displayed as the third dimension). In a Poisson distribution, noise varies with the average value of a series of measurements. This relationship is assumed in many applications of digital imaging where the noise (standard deviation) associated with a pixel value is assumed to equal the square root of the pixel value.

METHODS TO DESCRIBE PROBABILITY DISTRIBUTIONS

Mean, Median, and Mode

When describing the results of several measurements, a single value that summarizes the results is useful. For example, consider the data in Table 11-1. A typical approach to the problem is to average the data. However, the term "average value" is nonspecific. Exact reporting requires use of one of the following three terms: mean, median, or mode.

Mean

The mean is the arithmetic average of measured values. Usually, the mean is intended when the term "average" or "average value" is used.

The mean is the result obtained by summing (adding together) all measured values and dividing by the number of values. The sum of the nine values in Table 11-1 is 39. Therefore, the sample mean is $39/9 = 4.3$.

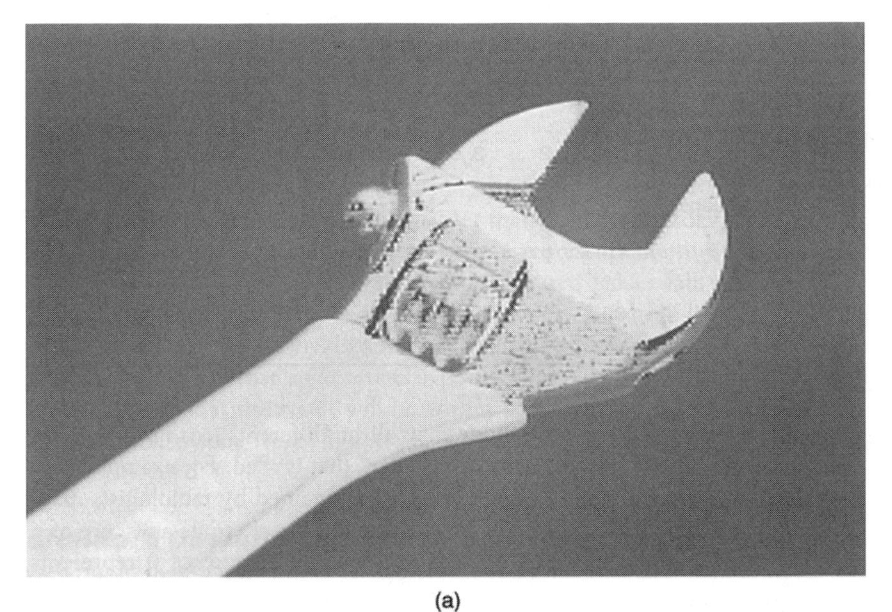

(a)

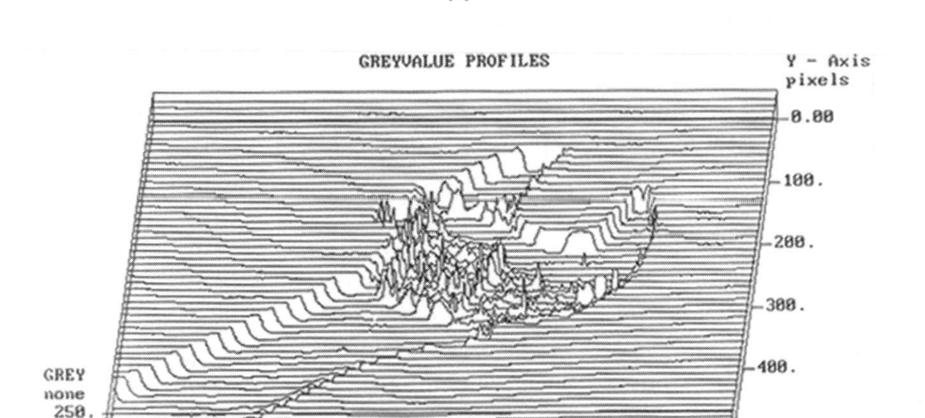

(b)

FIGURE 11-4
A: An image may be analyzed as a three-dimensional graph. **B:** Each spatial coordinate (x, y) has a gray level represented by amplitude in the z dimension.

Median

When values are placed in order of magnitude (from smallest to largest, for example), the median is the center value. Half of the values are greater than the median, and half are less. The data in Table 11-1, when placed in rank order, have the value 4 as median. Because there is an odd number of values, the median is an actual data value.

TABLE 11-1 Data for Sample Calculation of Mean, Median, and Mode

Random order ("as measured")	6 4 6 6 3 2 6 4 2
Ranked by order of magnitude	**Median** 2 2 3 4 4 6 6 6 6
Ranked by frequency	**Mode**
Value	3 2 4 6
Frequency	1 2 2 4
	Mean $\frac{\text{Sum of values}}{\text{Total number of values}} = \frac{39}{9} = 4.3$

The median is also known as the 50th percentile.

The mode is the most common value in a set of measurements.

If there were an even number of values, the median would be the average of the two values on either side of a hypothetical center value.

Mode

The mode is the value that appears most often in a set of data. When the data in Table 11-1 are placed in order of frequency of occurrence, the value 6 occurs most often. Hence, 6 is the mode for this series of measurements.

Because the mean, median, and mode may all be different, it is important to specify the type of "average" that is used. For example, suppose that the data in Table 11-1 are the scores obtained by radiologists diagnosing a disorder with a new type of imaging system. In these data a score of 6 represents perfect agreement with biopsy results, and a score of 0 represents complete disagreement. We may compare the mean scores for other imaging systems against the mean score of 4.3 for the new system. The mode, or most frequently reported score, is the high score of 6. Perhaps the experimental setup should be examined to see whether the viewing conditions, training, experience, or other factors were more favorable for the four radiologists who scored 6. The 50th percentile score (median) was 4, and the next highest score was the maximum. This suggests that the radiologists can be separated into two groups, one whose performance with the new system is excellent and another whose performance is inferior. This conclusion suggests that any differences between the two groups (e.g., training, experience, visual acuity, viewing conditions, etc.) might explain the difference in performance. Of course, one possibility is that the sample size (nine radiologists) is too small to represent radiologists in general.

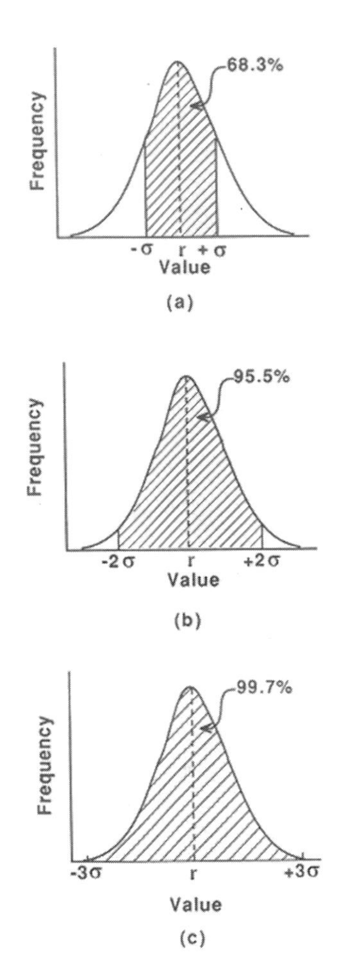

MARGIN FIGURE 11-3
The three Gaussian probability distributions shown above are centered about the mean value r. **A:** The area defined by ±1 SD is 68.3% of the total area under the Gaussian. **B:** Two standard deviations on either side of the mean encompass 95.5% of the total area. **C:** Three standard deviations encompass 99.7%.

Standard Deviation

The standard deviation for a series of measurements describes the precision or reproducibility of the measurements. If a series of measurements is distributed according to the Poisson probability distribution, the standard deviation σ is given by

$$\sigma = \sqrt{r_t}$$

where r_t is the true mean (value) for the series. If the true mean is estimated as the mean value r for a series of measurements, then the estimated standard deviation for the measurements is

$$\sigma = \sqrt{r}$$

If only one measured value, n, is available, then the standard deviation may be estimated by assuming that the value n represents the true mean.

In a normal distribution, 68.3% of all measured values fall within 1 σ on either side of the mean, 95.5% within 2 σ, and 99.7% within limits of 3 σ. If an experiment yields results that follow a normal distribution, there is a 68.3% likelihood that any particular result will fall within 1 σ of the mean, a 95.5% chance that the result will be within 2 σ, and a 99.7% chance that the result will be within 3 σ of the mean.

Three "types" of standard deviations are commonly cited. These types are described briefly below.

Standard Deviation of True Distribution

This is the value that we would like to report for an experiment but can never actually achieve because an infinite number of measurements would be required.

In a Gaussian distribution, one σ on either side of the mean define the inflection points of the distribution.

Standard Deviation Estimated from a Sample Mean

If a series of measurements is obtained and the mean value determined, the standard deviation can be estimated by taking the square root of the estimated mean. An extreme case of this procedure occurs when a single measurement is obtained of a variable assumed to follow a Poisson distribution. In this case, we might assume that the measured value is the mean of the distribution. We then report the single measurement as the mean, and we report the square root of the single measurement as the standard deviation. The number of samples actually taken should always be reported when a standard deviation is estimated from a mean. The rather inexact technique of estimating the standard deviation from a single measurement is often necessary in digital imaging techniques where the "noise" of a pixel value is estimated as the square root of the pixel value. When a value is measured only once, this inexact estimate is the only choice available.

Standard Deviation of a Series of Measurements

From a series of measurements the quantity known as the sample standard deviation can be constructed. The sample standard deviation σ_s is

$$\sigma_s = \sqrt{\frac{\sum_{i=1}^{N} (x_i - r)^2}{N - 1}} \tag{11-4}$$

where N measurements of a variable are made. The individual measurements yield the values x_1, where i takes on the values $1, 2, 3, \ldots, N$. The value x_1 is the first measurement, x_2 is the second measurement, and so on. The mean value of the sample is r. The quantity

$$\frac{\sum\limits_{i=1}^{N} (x_i - r)^2}{N - 1}$$

is known as the variance of the measurements. That is, the variance is σ_s^2, the square of the sample standard deviation.

If the formula for the sample standard deviation were applied to a "true" Poisson population (this can only be done mathematically since a "true" population or data set requires an infinite amount of data), it would reveal that the standard deviation equals the square root of the mean value. For any actual series of measurements, the sample standard deviation may yield a number that is "close" to the square root of the sample mean. The closer the sample distribution approximates a true Poisson distribution, the more nearly the sample standard deviation equals the square root of the sample mean.

A Poisson distribution has the following properties:

- It is composed of discontinuous discrete data;
- A large number of measurements is performed;
- The likelihood of any particular result is small;
- The chance of any particular measurement is independent of measurements that preceded it.

Example 11-3

Determine the standard deviation for the set of measurements given in Table 11-2:

$$\text{Mean} = r = 43{,}950$$

$$\sum (x_i - r)^2 = 1{,}302{,}400$$

$$\sigma_s = \sqrt{\frac{\sum (x_i - r)^2}{N - 1}}$$

$$= \sqrt{\frac{1{,}302{,}400}{10 - 1}}$$

$$\sigma_s = 380$$

The mean may be expressed as 43,950 ± 380, where 380 is understood to be 1 σ.

Assume that the data in Table 11-2 are from a population having a Poisson distribution, and estimate the standard deviation from the mean value alone.

$$\sigma \sqrt{r} = \sqrt{43{,}950}$$

$$= 210$$

TABLE 11-2 Data for Computation of Mean and Standard Deviation

x_1 *(Measurement)*	$x_i - r$ *(Deviation)*	$(x_1 - r)^2$
43,440	−510	260,100
43,720	−230	52,900
44,130	180	32,400
43,760	−190	36,100
44,390	440	193,600
43,810	−140	19,600
44,740	790	624,100
43,750	−200	40,000
44,010	60	3,600
43,750	−200	40,000
mean $= \frac{\sum x_i}{N} = 43{,}950$		$1{,}302{,}400 = \sum (x_i - r)^2$

Note that this value differs from the sample standard deviation of 380 derived above. This result reveals that the data are distributed more widely than would be predicted for a Poisson distribution having the same mean value.

PROPAGATION OF ERROR

Variables should be managed according to appropriate statistical models. When variables are modified by operations such as scaling (multiplication or division by a constant), statistical parameters such as the standard deviation must be treated appropriately. For example, counts measured in a nuclear decay experiment have a Poisson distribution, and it may be assumed that the standard deviation is equal to the square root of the mean. However, a count rate obtained by dividing the number of counts by the counting time must be handled differently. If all of the data values x_i in Eq. (11-4) are divided by the counting time t_1, then the sample mean r is also divided by t, and the standard deviation of a count rate σ_c is

In estimating the precision of a count rate, the count is assumed to be the product of a random process whereas the time over which the count is measured is assumed to be known exactly.

$$\sigma_c = \sqrt{\frac{\sum_{i=1}^{N} \left(\frac{x_i}{t} - \frac{r}{t}\right)^2}{N-1}}$$

$$= \sqrt{\frac{1}{t^2} \frac{\sum_{i=1}^{N} (x_i - r)^2}{N-1}}$$

$$= \frac{1}{t} \sqrt{\frac{\sum_{i=1}^{N} (x_i - r)^2}{N-1}}$$

$$= \frac{1}{t} \sigma_s$$

The estimated standard deviation σ_c of a count rate is then

$$\sigma_c = \frac{\sigma_s}{t}$$

where σ_s is the estimated standard deviation of a count n, and the average count rate is determined by dividing the count n by the counting time t. Because

$$\sigma_s = \sqrt{n}$$

we have

$$\sigma_c = \frac{\sqrt{n}}{t}$$

TABLE 11-3 Results of Arithmetic Operations with Numbers A and B[a]

Arithmetic Operation	*First Number*	*Second Number*	*Result ± Standard Deviation*
Addition	$(A \pm \sigma_A)$	$+(B \pm \sigma_B)$	$(A + B) \pm \sqrt{\sigma_A^2 + \sigma_B^2}$
Subtraction	$(A \pm \sigma_A)$	$-(B \pm \sigma_B)$	$(A - B) \pm \sqrt{\sigma_A^2 + \sigma_B^2}$
Multiplication	$(A \pm \sigma_A)$	$\times(B \pm \sigma_B)$	$(AB)[1 \pm \sqrt{(\sigma_A/A)^2 + (\sigma_B/B)^2}]$
Division	$(A \pm \sigma_A)$	$\div(B \pm \sigma_B)$	$(A/B)[1 \pm \sqrt{(\sigma_A/A)^2 + (\sigma_B/B)^2}]$
Scaling	$C(A \pm \sigma_A)$		$CA \pm C\sigma_A$

[a] The precision of A and B is described by the standard deviations σ_A and σ_B. C is a constant.

The count n is the average count rate c multiplied by the counting time t. Therefore

$$\sigma_c = \frac{\sqrt{ct}}{t} = \sqrt{\frac{c}{t}} \tag{11-5}$$

Hence the estimated standard deviation of a count rate (average or instantaneous) is determined by first dividing the count rate by the counting time and then taking the square root of the result.

A frequent mistake is to assume that the standard deviation of a count rate is the square root of the count rate c, that is, $\sigma_c \neq \sqrt{c}$.

If the precision of numbers A and B is denoted by standard deviations σ_A and σ_B, then the standard deviations of the results of arithmetic operations involving A and B may be computed with the expressions in Table 11-3.

Example 11-4

What is the net count and standard deviation for a sample from a nuclear counting experiment if the count is 400 ± 22 for the sample plus background and the background count is 64 ± 10?

$$\begin{aligned}\text{Sample net count} &= (A - B) \pm \sqrt{\sigma_A^2 + \sigma_B^2} \\ &= (400 - 64) \pm \sqrt{(22)^2 + (10)^2} \\ &= 336 \pm 24 \text{ counts}\end{aligned}$$

Example 11-5

From repeated counts of a radioactive sample, the mean sample plus background count was 232 ± 16. The counting time for each count was 2.0 ± 0.1 minute. What are the mean (sample plus background) count rate and the estimated standard deviation of the mean count rate?

$$\begin{aligned}\text{Mean sample plus background count rate} &= \frac{\text{Mean (sample plus background)}}{\text{Time}} \\ &= \left(\frac{A}{B}\right)\left[1 \pm \sqrt{\left(\frac{\sigma_A}{A}\right)^2 + \left(\frac{\sigma_b}{B}\right)^2}\right] \\ &= \left(\frac{232}{2.0}\right)\left[1 \pm \sqrt{\left(\frac{16}{232}\right)^2 + \left(\frac{0.1}{2.0}\right)^2}\right] \\ &= 116 \pm 10 \text{ cpm}\end{aligned}$$

Example 11-6

A 10-minute count of sample plus background yields 10,000 counts, and a 6-minute count of background alone yields 1296 counts. What are the net sample count rate and its standard deviation?

$$\text{Sample plus background count rate} = \frac{10{,}000 \text{ counts}}{10 \text{ min}} = 1000 \text{ cpm}$$

$$\text{Background count rate} = \frac{1296 \text{ counts}}{6 \text{ min}} = 216 \text{ cpm}$$

$$\sigma_C \text{ for sample plus background count rate} = \sqrt{\frac{C}{t}} = \sqrt{\frac{1000 \text{ cpm}}{10 \text{ min}}} = \sqrt{100}$$
$$= 10 \text{ cpm}$$

$$\sigma_C \text{ for background count rate} = \sqrt{\frac{C}{t}} = \sqrt{\frac{216 \text{ cpm}}{6 \text{ min}}} = \sqrt{36} = 6 \text{ cpm}$$

$$\text{Net sample count rate} = (A - B) \pm \sqrt{\sigma_A^2 + \sigma_B^2}$$
$$= (1000 - 216) \pm \sqrt{(10)^2 + (6)^2}$$
$$= 784 \pm \sqrt{136}$$
$$= 784 \pm 12$$

OTHER METHODS FOR DESCRIBING PRECISION

In addition to the standard deviation, a number of other methods can be used to describe the precision of data. Only a brief explanation of these methods is provided here, and the reader is referred to standard texts on probability and statistics for a more complete explanation.[3–5]

Probable error (PE) may be used in place of the standard deviation to describe the precision of data. The probability is 0.5 that any particular measurement will differ from the true mean by an amount greater than the PE. The probable error PE $= 0.6745\sigma$.

In place of the standard deviation, 9/10 error may be used to describe the precision of data. The probability is 0.1 that any particular measurement will differ from the true mean by an amount greater than the 9/10 error.

A 95/100 error is similar to the 9/10 error, except that the probability is 0.05 that the difference between a particular measurement and the true mean is greater than the 95/100 error.

Confidence limits are also referred to as a confidence interval, with the limits serving as the lower and upper bounds of the interval. Although 95% and 99% confidence intervals are frequently encountered, an interval of any magnitude could be used.

Confidence limits may be used to estimate the precision of a number when repeated measurements are not made. For example, if a result is stated as $A \pm a$, where $\pm a$ defines the limits of the 95% confidence internal, then it may be said with "95% confidence" that a second measurement of A will fall between $A - a$ and $A + a$. For a normal distribution, the 95% confidence interval is approximately $A \pm 2\sigma$.

Fractional Standard Deviation

The fractional standard deviation (σ/r) may be used in place of the standard deviation to describe the precision of data. If r represents the mean, then $\sigma = \sqrt{r}$, and the fractional standard deviation (σ/r) is $1/\sqrt{r}$. For example, σ for the number 10,000 is 100, and the fractional standard deviation is $100/10{,}000 = 0.01$. The percent standard deviation (%σ) is the fractional standard deviation multiplied by 100.

In the example above, $\%\sigma = 1$. The percent standard deviation of a count rate ($\%\sigma_c$) is

$$\%\sigma_c = 100\sqrt{\frac{1}{ct}} \tag{11-6}$$

where c is the average count rate measured over a time t.

The fractional standard deviation is sometimes (and misleadingly) referred to as the fractional error.

Example 11-7

For a count rate of 1000 cpm, what counting time is necessary to achieve a percent standard deviation of 1%?

$$\%\,\sigma_c = 100\sqrt{\frac{1}{ct}}$$

$$1\% = 100\sqrt{\frac{1}{(1000\text{ cpm})t}}$$

$$1000t = (100)^2 = 10{,}000$$

$$t = 10\text{ min}$$

Estimated percent standard deviations for different values are as follows:

Value	% σ
10	32
100	10
1000	3.2
10,000	1
100,000	0.32
1,000,000	0.1

Standard Deviation of the Mean

The standard deviation of the mean σ_m yields an estimate of the precision of the mean of a set of measurements. For N individual measurements constituting a mean, $\sigma_M = \sigma_s/\sqrt{N}$, where σ_s is the standard deviation of the individual measurements.

The standard deviation of the mean is sometimes referred to as the *standard error.*

Example 11-8

Determine the standard deviation of the mean σ_M for the data in Table 11-2.

For these data,

$$\sigma_s = 380 \text{ (Example 11-3)}$$

$$N = 10$$

Therefore

$$\sigma_M = \frac{\sigma}{\sqrt{N}} = \frac{380}{3.16} = 120$$

The standard deviation of the mean (120) always is smaller than the sample standard deviation.

The standard deviation of the mean is a measure of the spread of mean values expected when sets of data (such as the 10 measurements in Table 11-2) are repeated several times. If the experiment were repeated once, there is a 68% chance that the mean value of the next data set would fall within ± 1 standard deviation of the mean of the first set.

The standard deviation of the mean should not be confused with the standard deviation of a set of data. It is misleading to report the standard deviation of the mean instead of the standard deviation of individual values of a set of data unless it has been made clear that the spread of the *mean* values is the parameter of interest. It is usually the range of individual measurements that is of interest. Under such circumstances, reporting the standard deviation of the mean simply makes the data look more reproducible than they actually are.

Two factors influence the value of σ_M: (1) the variability of the original set of measurements, as represented by σ; and (2) the size of the sample N.

SELECTED STATISTICAL TESTS

A number of statistical tests may be applied to a set of measurements. A few of these tests are described briefly in this section.

Student's t-Test

The t-test was published in 1908 by W. S. Gosset, a brewer who worked for the Guinness Company. Gosset used the pseudonym "Student" in the publication, and the test became known as Student's t-test.

Student's t-test is a method for testing for the significance of the difference between two measurements, or two sets of measurements. The t-value for the difference between two measurements n_1 and n_2 is

$$t\text{-value} = \frac{|n_1 - n_2|}{\sqrt{\sigma_1^2 + \sigma_2^2}} \quad (11\text{-}7)$$

where σ_1 and σ_2 are the standard deviations for the measurements n_1 and n_2, and the vertical bars enclosing n_1 and n_2 indicate that the absolute or positive value of the difference is used for the computation. With the computed t-value and Table 11-4, the probability p may be determined that the difference in the numbers is simply statistical in nature and is not a real difference between dissimilar samples.

The probability p is the probability of a deviation from a true value equal to or greater than the measured value, if the Null Hypothesis is true (i.e., the deviation is simply statistical and not real).

Example 11-9

Activity in livers excised from two rabbits given ^{99m}TcS intravenously was counted for 1 minute. The net count rate was 1524 ± 47 cpm for the first liver and

TABLE 11-4 Cumulative Normal Frequency Distribution

t-Value	*p*	*t*-Value	*p*
0.0	1.000	2.5	0.0124
0.1	0.920	2.6	0.0093
0.2	0.841	2.7	0.0069
0.3	0.764	2.8	0.0051
0.4	0.689	2.9	0.0037
0.5	0.617	3.0	0.00270
0.6	0.548	3.1	0.00194
0.7	0.483	3.2	0.00136
0.8	0.423	3.3	0.00096
0.9	0.368	3.4	0.00068
1.0	0.317	3.5	0.00046
1.1	0.272	3.6	0.00032
1.2	0.230	3.7	0.00022
1.3	0.194	3.8	0.00014
1.4	0.162	3.9	0.00010
1.5	0.134	4.0	0.0000634
1.6	0.110	4.1	0.0000414
1.7	0.090	4.2	0.0000266
1.8	0.072	4.3	0.0000170
1.9	0.060	4.4	0.0000108
2.0	0.046	4.5	0.0000068
2.1	0.036	4.6	0.0000042
2.2	0.028	4.7	0.0000026
2.3	0.022	4.8	0.0000016
2.4	0.016	4.9	0.0000010

1601 ± 49 cpm for the second. Is the difference significant?

$$\begin{aligned} t\text{-value} &= \frac{|n_1 - n_2|}{\sqrt{\sigma_1^2 + \sigma_2^2}} \\ &= \frac{|1{,}524 - 1{,}601|}{\sqrt{(47)^2 + (49)^2}} \\ &= \frac{77}{68} \\ &= 1.13 \end{aligned}$$

From Table 11-4, the probability p is 0.259 (or 25.9%) that the difference is attributable to random variation of the count rate for similar samples. The probability is $1 - 0.259 = 0.741$ that the difference between the samples is significant.

Values of $p > 0.01$ seldom are considered indicative of a significant difference in values between measurements.

The Null Hypothesis is the assumption that a measured value represents the true value and that any deviation between the two is purely statistical and not real. The Null Hypothesis gives rise to two types of error:

- **Type I Error:** The Null Hypothesis is rejected as false even though it is true;
- **Type II Error:** The Null Hypothesis is accepted as true even though it is false.

Chi-Square (χ^2) Test

The chi-square test is used to determine the goodness of fit of measured data to a probability density function. From a series of repeated measures, the value of χ^2 may be computed:

$$\chi^2 = \frac{1}{r}\sum_{i-1}^{N}(x_i - r)^2 \tag{11-8}$$

where x_i represents each of N individual measurements and r is the sample mean. The formula for computing χ^2 from a sample resembles the formula for calculating the sample standard deviation [Eq. (11-4)]. A little algebra shows that the two are related by the equation

$$\chi^2 = \frac{(N-1)\sigma_s^2}{r} \tag{11-9}$$

If the variable we are measuring has a Poisson distribution, then the sample standard deviation σ_s should be close to the square root of the mean r of the sample. That is, the square of the sample standard deviation (the variance) should equal the sample mean. If that is true, then Eq. (11-9) shows that χ^2 is equal to $N - 1$, or one less than the number of data points.

χ^2 may be used as a measure of how closely the sample data correspond to a Poisson distribution. From the computed value of χ^2, the number of degrees of freedom $(N - 1)$, and data in Table 11-5, a probability p may be determined. Values of p less than 0.05 suggest that the data are distributed over a wider range of values than would be expected for a Poisson distribution. Values of p greater than 0.95 suggest that the data are confined to a smaller range of values than that predicted for a Poisson distribution.

The number of degrees of freedom is the number of observations that are free to vary without restriction. For example, there are $(N - 1)$ degrees of freedom in a box of N chocolates, because after $(N - 1)$ chocolates have been selected, there is no freedom in the selection of the last chocolate.

Example 11-10

A series of 25 measurements on a radioactive sample provides a mean of 950 and a value of 17,526 for

$$\sum_{i=1}^{N}(x_i - r)^2$$

TABLE 11-5 Chi-Square Values

Degrees of Freedom (N − 1)	*There Is a Probability of*						
	0.99	*0.95*	*0.90*	*0.50*	*0.10*	*0.05*	*0.01*
	That the Calculated Value of Chi-Square Will Be Equal to or Greater Than						
2	0.020	0.103	0.211	1.386	4.605	5.991	9.210
3	0.115	0.352	0.584	2.366	6.251	7.815	11.345
4	0.297	0.711	1.064	3.357	7.779	9.488	13.277
5	0.554	1.145	1.610	4.351	9.236	11.070	15.086
6	0.872	1.635	2.204	5.348	10.645	12.592	16.812
7	1.239	2.167	2.833	6.346	12.017	14.067	18.475
8	1.646	2.733	3.490	7.344	13.362	15.507	20.090
9	2.088	3.325	4.168	8.343	14.684	16.919	21.666
10	2.558	3.940	4.865	9.342	15.987	18.307	23.209
11	3.053	4.575	5.578	10.341	17.275	19.675	24.725
12	3.571	5.226	6.304	11.340	18.549	21.026	26.217
	4.107	5.892	7.042	12.340	19.812	22.362	27.688
	4.660	6.571	7.790	13.339	21.064	23.685	29.141
	5.229	7.261	8.547	14.339	22.307	24.996	30.578
	5.812	7.962	9.312	15.338	23.542	26.296	32.000
	6.408	8.672	10.085	16.338	24.769	27.587	33.409
	7.015	9.390	10.865	17.338	25.989	28.869	34.805
	7.633	10.117	11.651	18.338	27.204	30.144	36.191
	8.260	10.851	12.443	19.337	28.412	31.410	37.566
	8.897	11.591	13.240	20.337	29.615	32.671	38.932
	9.542	12.338	14.041	21.337	30.813	33.924	40.289
	10.196	13.091	14.848	22.337	32.007	35.172	41.638
	10.856	13.848	15.659	23.337	33.196	36.415	42.980
	11.534	14.611	16.473	24.337	34.382	37.382	44.314
	12.198	15.379	17.292	25.336	35.563	38.885	45.642
	12.879	16.151	18.114	26.336	36.741	40.113	46.963
	13.565	16.928	18.939	27.336	37.916	41.337	48.278
	14.256	17.708	19.768	28.336	39.087	42.557	49.588

Do the data appear to follow a normal distribution?

$$\chi^2 = \frac{1}{r}\sum_{i=1}^{N}(x_i - r)^2$$

$$= \frac{1}{950}(17{,}526)$$

$$= 18.5$$

With $N - 1$, or 24 degrees of freedom, this value for χ^2 falls between probabilities 0.9 and 0.5 and suggests that the data are distributed as a Poisson data set.

The χ^2 test is often used in the classroom to demonstrate that the data from Gregor Mendel's famous experiments with peas, which formed the basis of his theory of inheritance, were probably falsified.[6,7]

Example 11-11

Determine whether the data in Table 11-2 are distributed according to a Poisson distribution.

From Table 11-2,

$$\sum_{i=1}^{N}(x_i - r)^2 = 1{,}302{,}400$$

$$\chi^2 = \frac{1}{r}\sum_{i=1}^{N}(x_i - r)^2 = \frac{1{,}302{,}400}{43{,}950} = 29.6$$

With $N - 1$, or 9 degrees of freedom, this value of χ^2 falls below a p value of 0.01 and suggests that the data are not distributed according to a Poisson distribution.

SUMMARY

- Errors can be categorized as determinate (systematic) and indeterminate (random).
- A series of measurements may be characterized in terms of its accuracy, precision, and bias.
- Common probability distributions are the
  - Poisson distribution
  - Normal (Gaussian) distribution
- The mean, median, and mode are single values that can be determined for any probability distribution; the values are identical for a normal distribution.
- The precision of a series of measurements is described by the standard deviation (sometimes expressed as the fractional standard deviation).
- Mathematical manipulations of imprecise data lead to the propagation of errors that can be described analytically for a normal distribution.
- Student's t-test is a method for determining the statistical significance of a difference between sets of measurements.
- The chi-square test is used to determine the "goodness of fit" of measured data to an expected probability distribution.

PROBLEMS

*11-1. Suppose 3600 counts were accumulated for a sample counted for 10 minutes. Background contributed 784 counts over a counting interval of 6 minutes.
a. Find the estimated standard deviation and the percent estimated standard deviation of the sample count rate uncorrected for background.
b. What are the estimated standard deviation and the percent estimated standard deviation for the background count rate?
c. What are the estimated standard deviation and the percent estimated standard deviation of the sample count rate corrected for background?

11-2. A medical researcher carefully measures the length of the fetal humerus in 50 ultrasound studies by using electronically generated distance-measuring "calipers" on the display. The researcher obtains a mean value of 5.5 cm with a standard deviation of 2 mm. After the measurements have been obtained, the researcher discovers through studies of excised bone that the electronically generated caliper measurements are consistently 2% larger than the true length.
a. Describe the researcher's original results in terms of bias, precision, and accuracy.
b. How can the mean and standard deviation of the original data be corrected?

*11-3. A 2-minute count of 3310 was measured for a radioactive sample. A second sample furnished 4709 counts in 3 minutes. Is the difference in count rate between the two samples significant?

11-4. Show that the formula for [Eq. (11-8)] can be rewritten in terms of the formula for the sample standard deviation [Eq. (11-4)] to yield Eq. (11-9).

11-5. A series of radioactive sample of short half-life will be counted one per day over a period of several days. Explain how variations in the observed counts caused by fluctuations in the sensitivity of the counting system can be compensated.

*For those problems marked with an asterisk, answers are provided on p. 492.

REFERENCES

1. Knoll, GF. *Radiation Detection and Measurement,* 2nd edition. New York, John Wiley & Sons, 1989.
2. Low, F. Basic considerations in nuclear instrumentation, in Hine, G. (ed.). *Instrumentation in Nuclear Medicine,* Vol. 1. New York, Academic Press, 1967.
3. Glantz, S. A. *Primer of Biostatistics*. New York, McGraw-Hill, 1981, pp. 129–148.
4. Ingelfinger, J. A., Mostller, F., and Thibodeau, L. A., et al. *Biostatistics in Clinical Medicine*. New York, Macmillan, 1983.
5. Bohm, A. K. *Basic Medical Statistics*. New York, Grune & Stratton, 1972.
6. Mould, R. F. *Introductory Medical Statistics,* 3rd edition. Philadelphia, Institute of Physics, 1998.
7. Ingelfinger, J. A., Mosteller, F., Thibodeau, L. A., and Ware, J. H. *Biostatistics in Clinical Medicine*. New York, Macmillan, 1983.

CHAPTER

12

INSTRUMENTATION FOR NUCLEAR IMAGING

OBJECTIVES

By studying this chapter, the reader should be able to:

- Describe the principles and application of rate measurements with radioactive nuclides.
- Compute volumes using the dilution method.
- Identify the function of each component of a scintillation camera.
- Explain the properties and uses of different collimators.
- List the advantages of a single-crystal scintillation camera.
- Trace the progression of imaging data in a single-crystal camera from release of a γ ray in the patient to display of a bright spot on the display.
- Explain the origin and correction of the resolving time of a scintillation camera.
- Discuss the acquisition, presentation and properties of images for single-photon emission tomography.
- Discuss the acquisition, presentation and properties of images for positron emission tomography.
- Document recent advances in positron emission tomography (PET).

Nuclear diagnostic studies can be categorized into four general groups:

- Localization
- Dilution
- Flow or diffusion
- Biochemical and metabolic properties

Many of these properties were first demonstrated in the 1920s by George de Hevesy, a Hungarian physicist working in Denmark. One of his more well-known experiments was use of a radioactive tracer to prove that the hash served at his boarding house contained meat he had left on his plate the evening before.

In accepting the Nobel Prize in Chemistry in 1943, de Hevesy stated: "The most remarkable result obtained in the study of the application of isotope indicators is perhaps the discovery of the dynamic state of the body constituents. The molecules building up the plant or animal organism are incessantly renewed."

Before the development of the scintillation detector, thyroid uptakes were measured with a collimated Geiger-Müller counter.

In the 1930s and 1940s, thyroid function was studied with a variety of radioactive isotopes of iodine, including ^{126}I, ^{128}I, ^{130}I and ^{131}I. After World War II, ^{131}I became the nuclide of choice because it could be produced in copious amounts in nuclear reactors. More recently, ^{123}I has replaced ^{131}I as the preferred nuclide for thyroid uptake measurements.

INTRODUCTION

Most nuclear medicine studies require the use of a detector outside the body to measure the rate of accumulation, release, or distribution of radioactivity in a particular region inside the body. The rate of accumulation or release of radioactivity may be measured with one or more detectors, usually NaI(Tl) scintillation crystals, positioned at fixed locations outside the body. Images of the distribution of radioactivity usually are obtained with a stationary imaging device known as a scintillation camera.

MEASUREMENT OF ACCUMULATION AND EXCRETION RATES

Measurement of the accumulation of radioactive iodine in the thyroid gland was the first routine clinical use of a radioactive nuclide. Since early tests of thyroid function with radioactive iodine, many diagnostic applications of radioactive nuclides have been developed. Many of these applications require measurement of the change in radioactivity in a selected region of the body as a function of time. Some of the applications are discussed briefly in this section.

Thyroid Uptake

The rate of accumulation (uptake) of iodine in the thyroid gland may be estimated from measurements of the radioactivity in the thyroid at specified times (usually 6 and 24 hours) after a selected amount of radioactive iodine (e.g., ~10 MBq of ^{123}I) has been administered orally. Radioactivity in the thyroid is usually measured with a NaI(Tl) scintillation crystal at least 3 cm in diameter by 3 cm thick that is positioned 20 to 30 cm in front of the patient and in line with the isthmus of the thyroid. The *field of view* of the detector is the area perpendicular to the extended axis of the detector at the surface of the patient over which the count rate remains above 90% of the count rate at the center of the field. The detector should be collimated to restrict the field of view to a diameter not greater than 15 cm for adults and proportionately less for children.

To provide a standard sample for determination of thyroid uptake, a vessel resembling the average thyroid (e.g., a 30-ml polyethylene bottle 3 cm in diameter) is filled with an amount of ^{123}I equal to that delivered orally to the patient. This standard sample is placed at a depth of 0.5 cm behind the surface of a neck phantom composed of a cylinder of Lucite 15 cm in diameter and 15 cm tall. The percent uptake of iodine

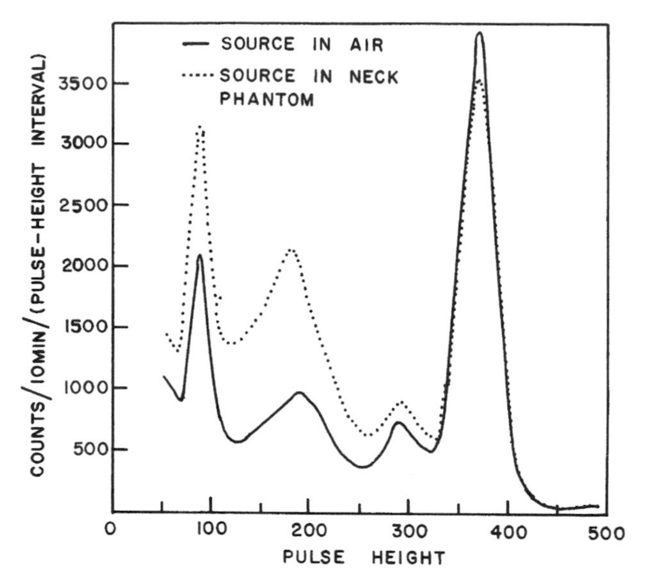

FIGURE 12-1
Pulse height spectra for a 1-ml, 5-μCi ^{131}I source in air and in a neck phantom at a distance of 30 cm from a 1.5-in.-diameter, 1-in.-thick NaI(Tl) crystal. Radiation scattered by material around the radioactive source increases the height of the Compton plateau.

in the thyroid is the ratio of the net count rate measured in front of the thyroid to that measured under identical conditions with the detector in front of the neck phantom:

$$\text{Uptake (\%)} = (100)\frac{(c)_t}{(c)_s} \tag{12-1}$$

Sometimes a standard count rate is obtained by counting the activity of the capsule of ^{123}I before it is administered to the patient.

where c_t is the count rate for the thyroid and c_s is the count rate for the standard.

Radiation scattered from extrathyroidal tissue should be considered in the computation of percent uptake (Figure 12-1). With one method of correction for extrathyroid activity, count rates for the patient and the phantom are measured before and after a lead block (B-block) about 1/2 in. thick has been placed over the thyroid and the radioactive source in the phantom.[1] The percent uptake is

$$\text{Uptake (\%)} = (100)\frac{c_t - c_{tL}}{c_s - c_{sL}} \tag{12-2}$$

where c_t is the count rate for the thyroid, c_{tL} is the count rate for the thyroid shielded by a lead block, c_s is the count rate for a standard, and c_{sL} is the count rate for a standard shielded by a lead block.

Example 12-1

The following count rates were obtained 24 hours after administration of 0.5 MBq of ^{123}I to a patient:

For the patient's thyroid	16,110 cpm
For the patient's thyroid shielded by a lead block	5,790 cpm
For a standard	96,450 cpm
For a standard shielded by a lead block	5,460 cpm

What is the percent uptake?

$$\begin{aligned}\text{Uptake (\%)} &= (100)\frac{c_t - c_{tL}}{c_s - c_{sL}}\\ &= (100)\frac{16{,}110 - 5{,}790}{96{,}450 - 5{,}460}\\ &= 11\%\end{aligned} \tag{12-2}$$

de Hevesy was the "father" of the tracer method of using radioactive isotopes to map biological processes. For this method, he needed relatively long-lived radioactive isotopes of elements that occur naturally in the body. His work received a significant boost in 1934 with arrival of artificially produced ^{32}P ($T_{1/2}$ = 14.3 days) from Enrico Fermi's laboratory in Rome.

Dilution Studies

Dilution measurements with radioactive nuclides are used for a variety of diagnostic tests in clinical medicine. If a solution of volume v_i that contains a known concentration $(\text{cpm/ml})_i$ of radioactive material is thoroughly mixed with a large volume of nonradioactive solution, then the volume v_f of the final solution may be estimated by measuring the specific count rate $(\text{cpm/ml})_f$ of the final solution:

$$v_f = \frac{(\text{cpm/ml})_i}{(\text{cpm/ml})_f} v_i \qquad \textbf{(12-3)}$$

Because $(\text{cpm/ml})_i\ v_i$ equals the total counts per minute ($[\text{CPM}]_i$) of the injection solution, Eq. (12-3) may be written as

$$v_f = \frac{(\text{cpm})_i}{(\text{cpm/ml})_f} \qquad \textbf{(12-4)}$$

Among the earliest dilution experiments were those of de Hevesy. In 1934, for example, de Hevesy measured his total body water by drinking "heavy water" (water containing deuterium in place of ordinary hydrogen) and measuring the concentration of deuterium in his urine.

The specific count rates $(\text{cpm/ml})_i$ and $(\text{cpm/ml})_f$ must be corrected for background. The difference $v_f - v_i$ is the volume of the nonradioactive solution with which the solution of volume v_i is mixed.

During some applications of the dilution technique, the specific count rate $(\text{cpm/ml})_f$ of the final solution must be corrected for activity that escapes from the volume to be measured. For example, estimates of blood volume with serum albumin labeled with ^{131}I sometimes are made at intervals over the first hour after intravenous injection of the labeled compound. To correct for the escape of serum albumin from the circulation, the specific count rate may be plotted as a function of time and extrapolated back to the moment of injection. Alternately, the blood volume may be measured at a specific time (e.g., 10 minutes after injection of the labeled albumin), and a predetermined correction factor may be applied to the measured volume.

Example 12-2

0.5 MBq of serum albumin tagged with ^{131}I (radioiodinated human serum albumin [RISA]) are injected intravenously into a patient. After the labeled albumin has thoroughly mixed in the patient's circulatory system, 5 ml of the patient's blood is withdrawn. The 5-ml sample yields a count rate of 2675 cpm. Another 0.5 MBq sample of RISA is diluted to 2000 ml. A 5-ml aliquot from this standard solution yields a count rate of 6343 cpm. The background count rate is 193 cpm. What is the total blood volume?

$$\begin{aligned}
\text{Sample net specific count rate} &= (2675 - 193)\ \text{cpm/5 ml} \\
&= 496\ \text{cpm/ml} \\
\text{Standard net specific count rate} &= (6343 - 193)\ \text{cpm/5 ml} \\
&= 1230\ \text{cpm/ml} \\
\text{Count rate for a 0.5 MBq sample} &= (1230\ \text{cpm/ml})\ (2000\ \text{ml}) \\
&= 246 \times 10^4\ \text{cpm}
\end{aligned}$$

If the volume of the injected RISA is ignored, the blood volume v_f is

$$\begin{aligned}
v_f &= \frac{(\text{cpm})_i}{(\text{cpm/ml})_f} \\
&= \frac{246 \times 10^4\ \text{cpm}}{496\ \text{cpm/ml}} \\
&\approx 5000\ \text{ml} \qquad \textbf{(12-4)}
\end{aligned}$$

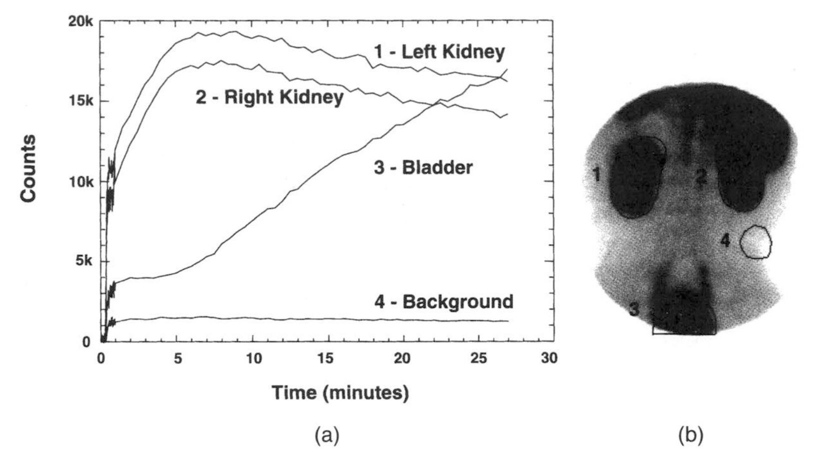

FIGURE 12-2
A: Dynamic renal study showing excretion (changes in count rate as a function of time) of technetium ^{99m}Tc-MAG$_3$. **B:** Regions of interest for left and right kidneys, bladder, and background are drawn from the static image. (Courtesy of Bruce Hasselquist, Ph.D.)

The first application of radioactive tracers to experimental medicine were studies by Blumgart, Weiss, and Yens in 1926 that used a decay product of radium to measure the circulation time of blood.

Time–Activity (Dynamic) Studies

Static images display the spatial distribution of radioactivity in an organ in an effort to demonstrate structural abnormalities, reveal physiologic defects, or estimate relative mass. Some nuclear medicine studies have been devised to measure the radioactivity in an organ over time after administration of a radioactive compound. These studies are termed *dynamic studies*. Examples are renal studies to evaluate blood flow and functional renograms to determine kidney function. Dynamic and static studies often involve different radiopharmaceuticals. Static renal studies are performed with ^{99m}Tc-dimercaptosuccinic acid (DMSA) or ^{99m}Tc-glucoheptonate to image the renal parenchyma, whereas effective renal plasma flow is evaluated with ^{131}I-Hippuran or ^{99m}Tc mertiatide (^{99m}Tc-MAG$_3$). Dynamic studies may include the summation of data over time to provide a "static" image. For example, both static and dynamic renal studies can be acquired with a single injection of ^{99m}Tc-glucoheptonate.

A dynamic renal study obtained with ^{99m}Tc-MAG$_3$ is shown in Figure 12-2. Regions of interest are drawn for the left and right kidneys, the bladder, and background. The activity of ^{99m}Tc is shown as it is excreted by the kidneys via active tubular secretion and glomerular filtration.

SINGLE-CRYSTAL SCINTILLATION CAMERA

Modern cameras can be used for dynamic as well as static studies.

The first single-crystal scintillation camera was assembled in 1956 by Anger at the University of California.[2] This development was a major advance in static imaging of radioactivity distributions, because it allowed simultaneous collection of γ rays from all regions of small organs. The camera contained a 6-mm-thick sodium iodide [NaI(Tl)] crystal only 10 cm in diameter. It was viewed by seven photomultiplier (PM) tubes, each 3.8 cm in diameter. Six PM tubes were hexagonally arranged about the seventh tube at the center. In 1960, work began at Nuclear-Chicago Inc. on an improved version of the Anger camera for commercial marketing. In 1963, a larger scintillation camera was built with a 28-cm crystal and 19 PM tubes arranged in hexagonal fashion. In this camera, as in earlier versions, the PM tubes were optically coupled to the scintillation crystal with mineral oil, a technique that furnished efficient transmission of light from the crystal but restricted the position of the detector to an overhead view of the patient. A major improvement in camera design was replacement of the mineral oil by a solid plastic light guide and optical coupling grease. This change eliminated the oil seals required for earlier cameras and permitted rotation of the detector head to any desired position.

The first commercial scintillation camera was installed at Ohio State University in 1962.

Scintillation cameras are sometimes referred to as gamma cameras or Anger cameras.

Although the basic design of the scintillation camera has not changed much since 1963, many improvements have been added that have increased its speed and versatility and enhanced its usefulness in clinical medicine. Among these improvements are replacement of vacuum tubes by integrated circuits and the use of precision electronic components to obtain a more rapid and reliable signal that is less sensitive to environmental and power fluctuations. Currently, scintillation cameras are available with crystals up to 0.5 m in diameter coupled to as many as 91 PM tubes. This expansion has improved the quality of scintillation camera images. Another development that has greatly enhanced the usefulness of the scintillation camera is interfacing of the camera to data storage and processing systems. Other modifications include the use of more efficient bialkali PM tubes, utilization of multichannel analyzers for window selection and adjustment, and incorporation of digital computers into the image acquisition and display network.

The element technetium has no stable isotopes, and consequently does not occur in nature. It was first discovered in 1937 by the Italian physicist E. Segrè in molybdenum heat shields and deflectors discarded from E. O. Lawrence's first cyclotron at Berkeley. Segrè was awarded the 1959 Nobel Prize in Physics for the discovery of the antiproton.

The widespread availability of ^{99m}Tc is one factor that has contributed greatly to use of the scintillation camera and to nuclear medicine procedures in general. This radionuclide emits a γ ray of 140 keV, an ideal energy for scintillation camera imaging, with only slight contamination by particles and photons of undesired energies. In addition, ^{99m}Tc has a short half-life (6 hours) and can be incorporated into a variety of compounds and drugs. The short half-life of ^{99m}Tc permits administration of large amounts of radioactive material without excessive radiation doses to patients. Increased administration of radioactivity means that more γ rays are emitted over a shorter period of time, permitting the scintillation camera to be used to its full advantage.

Advantages of ^{99m}Tc for nuclear medicine studies include:

- Short high life (6 hr)
- Ideal γ-ray energy (140 keV)
- Little particulate radiation
- Ease of labeling
- Administration of relatively high activities

PRINCIPLES OF SCINTILLATION CAMERA OPERATION

A single-crystal scintillation camera can be separated into four sections: the collimator, the detector, the data processing section, and the display unit.

Collimators

To produce an image of the distribution of radioactivity in a patient, the sites of absorption of γ rays in the scintillation camera crystal must be related clearly to the origin of the γ rays within the patient. This relationship can be achieved by placing a collimator between the crystal and the patient. The most commonly used collimator is the parallel multihole collimator. This collimator is composed of small, straight cylinders separated by lead or tungsten septa (Margin Figure 12-1). Relatively thin septa are used in low-energy collimators designed for use with radionuclides emitting low-energy photons. Thicker septa are required for collimators used with nuclides emitting higher-energy photons. The ability of high-energy photons to traverse even relatively thick septa establishes an upper limit of about 600 keV on the photon energy range useful for scintillation camera imaging. As a general rule, the minimum path length through septa of the collimator should be at least five mean free paths (<1% transmission) for the photons with which the collimator is to be used.

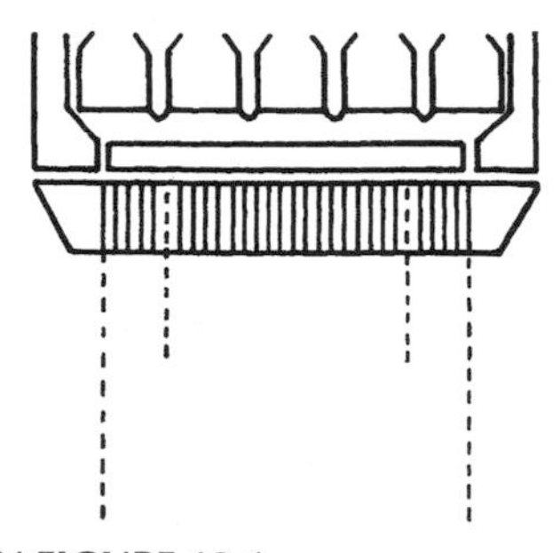

MARGIN FIGURE 12-1
Parallel multihole collimator.

A γ ray moving from the patient toward the crystal in a direction in line with one of the collimator cylinders can easily pass through the cylinder and interact in the crystal. However, a γ ray emerging obliquely from the patient must traverse one or more of the lead septa in the collimator before reaching the crystal (Figure 12-3). Hence, γ rays emitted obliquely have a much lower probability of reaching the detector. By this mechanism, γ rays interacting in the crystal can be related spatially to their origin in small discrete volumes of tissue directly in front of the collimator.

A collimator with many cylinders of small diameter is termed *a high-resolution collimator*. A collimator with fewer cylinders of larger diameter, or with shorter cylinders of the same diameter, allows more photons to reach the crystal and is termed a *high-sensitivity collimator*. In the selection of parallel multihole collimators, a gain in sensitivity is accompanied by a loss of resolution and vice versa.

A high-resolution collimator provides relatively low sensitivity, whereas a high-sensitivity collimator yields relatively low spatial resolution.

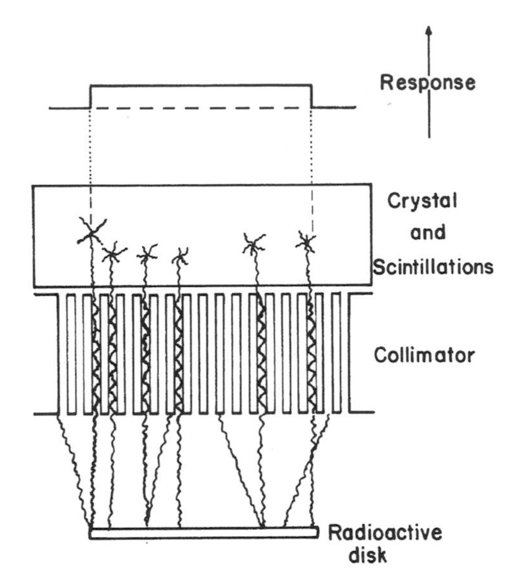

FIGURE 12-3
With a parallel multihold collimator, γ rays emitted in line with the collimator holes are transmitted to the crystal, whereas γ rays emitted obliquely are absorbed by the collimator septa.

As the diameter of the collimator cylinders is reduced, fewer photons are admitted through each cylinder into the scintillation crystal. Hence, the statistical influence of variations in photon emission (i.e., quantum noise or mottle) is increased. The only way to prevent an increase in image noise is to increase the imaging time or the emission from the source. Either of these changes ensures that the number of photons passing through each cylinder of the collimator remains constant. As an example, suppose that the resolution of a nuclear medicine imaging system is improved by a factor of 2 by decreasing the diameter of the collimator cylinders. Also suppose that this improvement is accompanied by a twofold reduction in the number of photons reaching the crystal. To keep the quantum noise of the image at the original level, the number of photons incident upon the collimator must be increased by a factor of 2. To reduce the quantum noise even further to a level corresponding to the finer resolution of the new collimator, however, an additional fourfold increase would be required in the number of photons incident upon the collimator. That is, a twofold improvement in the spatial resolution of an imaging system achieved by reducing the diameter of the collimator cylinders should be accompanied by an eightfold increase in photon emission or imaging time to obtain a parallel improvement in image noise.

The spatial resolution of a scintillation camera is termed the *intrinsic resolution*. Its value is typically 4–5 mm. This resolution is degraded significantly in practice by the presence of a collimator. The spatial resolution with a collimator in place is referred to as the *extrinsic resolution*.

As the distance is increased between the face of a parallel multihole collimator and the patient, the resolution decreases because each open cylinder in the collimator views a larger cross-sectional area of the object. For the same reason, the sensitivity does not change significantly with the distance between a multihole collimator and the patient. Although the number of photons collected by each collimator cylinder from a unit area of the patient decreases with the square of the collimator–object distance, the area viewed by each cylinder increases as the square of the distance. These two influences cancel each other so that the sensitivity does not change appreciably with alterations in the collimator–patient distance.

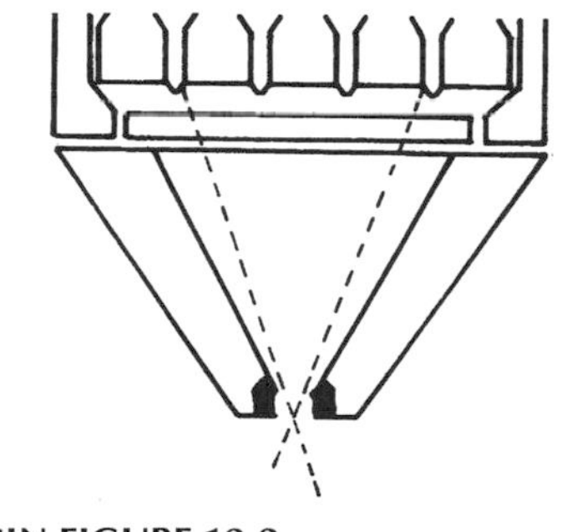

MARGIN FIGURE 12-2
Pinhole collimator.

Other types of collimators used occasionally with scintillation cameras include the pinhole collimator, the diverging collimator, and the converging collimator.

The pinhole collimator consists of a pinhole aperture in a lead, tungsten, or platinum absorber. It is used primarily to obtain high-resolution images of small organs (e.g., the thyroid) that can be positioned near the pinhole. The major disadvantage of the pinhole collimator is its low sensitivity because only a few of the photons emitted by the object are collected to form an image. Occasionally, a pinhole collimator with three apertures is used to provide three different views of a small organ simultaneously.

In the early years of single-photon emission computed tomography (SPECT), a multihole collimator was used for nuclear tomographic imaging. This approach has been superseded by other tomographic approaches.

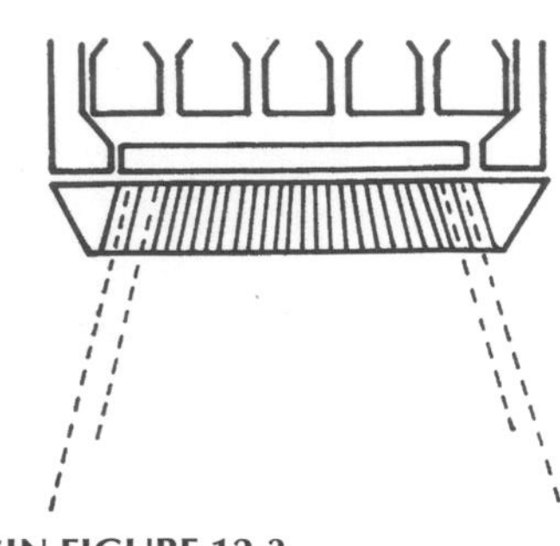

MARGIN FIGURE 12-3
Diverging collimator.

The diverging collimator permits the scintillation camera to image anatomic regions larger than the dimensions of the crystal. The large crystals of up to 20 in. in diameter available with modern scintillation cameras have greatly alleviated the need for diverging collimators. When compared with a parallel multihole collimator, there is some reduction in resolution with a diverging collimator.

With the converging collimator, images with improved resolution can be obtained for organs smaller than the area of the crystal. The resolution is improved because the object is projected over the entire crystal face rather than onto a limited region of the crystal. Also, the photon collection efficiency is greater for a converging collimator than for a parallel multihole collimator used to image a small object. Hence, the influence of quantum noise is often less noticeable in images obtained with a converging collimator.

Coded Aperture Substitutes for Collimators

Collimators that can be used in both converging and diverging modes are referred to as *div/con* collimators.

Conventional collimators provide the necessary positional correspondence between the interaction of a γ ray in the crystal and the origin of the γ ray in the patient. However, they are inefficient collectors of γ rays because they reject all γ rays except those arising within a relatively small solid angle. Also, conventional collimators are often the major limitation in the spatial resolution of a scintillation camera image. As substitutes for conventional collimators, various coded apertures have been proposed. One of the simplest of these devices is the Fresnel zone plate. With this device, each point of radioactivity in the patient casts a unique shadow of the zone plate onto the detector (Figure 12-4). That is, points of radioactivity at the same distance from the zone plate cast shadows on the detector that are laterally displaced from each other, and points at different distances from the zone plate cast shadows of different sizes. Because of these effects, information is projected onto the detector about the three-dimensional distribution of radioactivity in the patient.

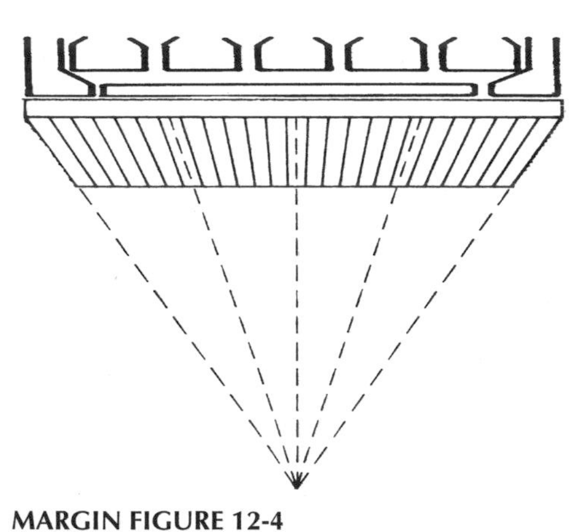

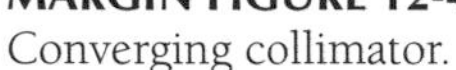

MARGIN FIGURE 12-4
Converging collimator.

To retrieve this information, the transmission data projected onto the detector may be recorded in coded form on film. When this coded image is placed in a coherent light beam, diffraction of the light produces an interpretable image that can be viewed telescopically (Figure 12-5). This image is tomographic, and movement of the eyepiece of the telescope allows one to visualize the distribution of radioactivity at different depths in the patient.

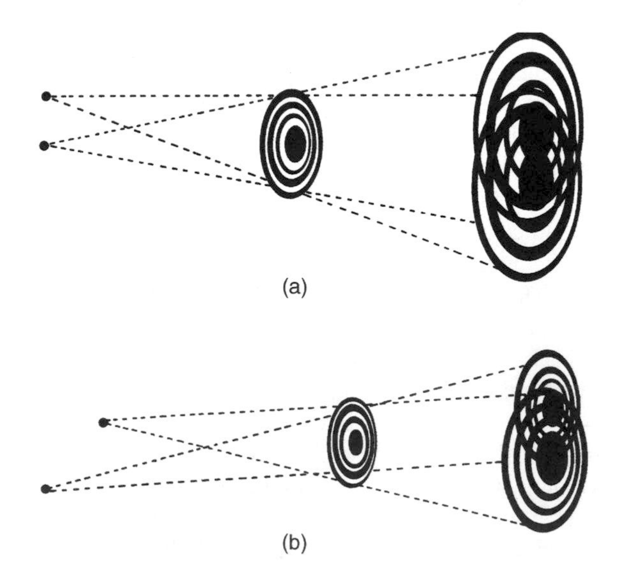

FIGURE 12-4
In zone plate imaging, each point of radioactivity casts a unique shadow onto the detector. (From Farmelant, M., et al.[3] Used with permission.)

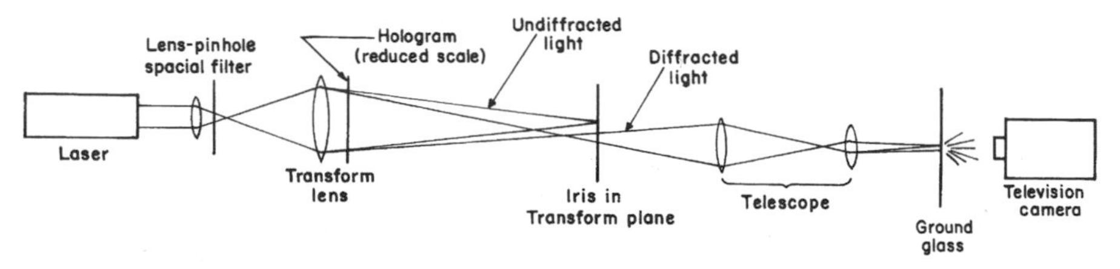

FIGURE 12-5
Optical reconstruction system for coded aperture imaging. (From Farmelant, M., et al.[3] Used with permission.)

Many coded apertures have been studied, including some that change continuously over the time in which radiation from the patient is detected. All coded apertures developed so far have one major limitation, however; they exhibit very low sensitivity to large areas of radioactivity such as those usually encountered in nuclear medicine. Consequently, coded apertures have not found useful applications in clinical nuclear medicine.

Scintillation Crystal

Although NaI(Tl) crystals up to 0.5 m in diameter are used in single-crystal scintillation cameras, the periphery of the crystal is always masked and does not contribute to image formation. With a 28-cm-diameter crystal, for example, only the center 25 cm is used to create an image. Masking is necessary to prevent distortion in the image caused by inefficient collection of light at the periphery of the crystal. To prevent physical damage to the crystal and to keep it free of moisture, the crystal is sealed between a thin (<1 mm) aluminum canister on the exterior surface and a glass or plastic plate on the interior surface.

The photopeak detection efficiency of a 1.25-cm-thick NaI(Tl) detector varies from about 80% for ^{99m}Tc (140 keV) to about 10% at 511 keV. Increased detection efficiency could be achieved with thicker crystals. However, multiple scattering within the crystal would also increase and cause a reduction in the photon-positioning accuracy of the camera.

Although NaI(T1) is the detector used today in almost all clinical scintillation cameras, other detectors have been or are being explored. Among these detectors are bismuth germanate [$Bi_4Ge_3O_{12}$, often called BGO], barium fluoride [BaF_2], lutetium orthosilicate [$Lu_2[SiO_4]O(Ce)$, often called LSO], yttrium orthosilicate [YSO], and lutetium orthophosphate (LOP).

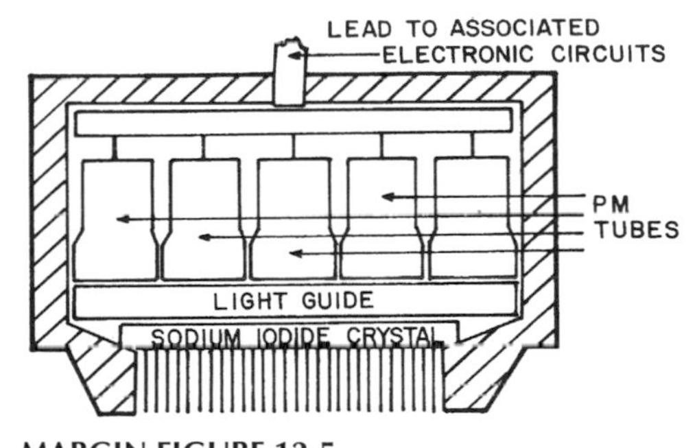

MARGIN FIGURE 12-5
Detector assembly of a scintillation camera.

Light Guide

In addition to the protective glass plate, camera detectors may have a light guide (usually composed of Plexiglas) that transmits and distributes the light from the primary interaction site in the crystal to the surrounding PM tubes. The distribution of light within a light guide of properly chosen thickness (usually about 4 cm from the central plane of the crystal to the face of the PM tubes) results in increased counting efficiency and positioning accuracy. The light is transmitted with almost no loss because the sides of the light guide are coated with a reflective covering. Because the light guide is solid, the detector assembly can be positioned in different orientations without difficulty. A light guide is not required in digital scintillation cameras.

Photomultiplier Tubes

The light transmitted to a particular PM tube strikes the photocathode element of the tube. In response, electrons are ejected from the photocathode and focused upon the first of 8 to 14 positively charged electrodes referred to as *dynodes*. The number of electrons is multiplied in the remaining dynodes to yield 10^6 to 10^8 electrons collected at the final electrode or anode for each electron released at the photocathode. The actual number of electrons finally collected at the anode (i.e., the size of the output signal) varies in direct proportion to the amount of light striking the photocathode.

The photocathode of a PM tube is usually an alloy of cesium (Cs) and antimony (Sb) that has a relatively low *work function* (i.e., electrons are easily ejected by impinging light photons).

This lower limit of 70 keV, together with the upper limit of 600 keV, define the range of useful γ-ray energies for nuclear imaging.

With a single-crystal scintillation camera, one of the restrictions on image quality is statistical fluctuation in the number of electrons ejected from the photocathode for a given amount of incident light. The statistical fluctuation is determined primarily by the average number of electrons ejected per unit of incident light. It has been improved markedly by replacement of older S11-type PM tubes with tubes having photocathodes that release more electrons per unit of incident light. Another improvement in noise reduction is the use of threshold preamplifiers that reject small signals from PM tubes distant from the interaction site in the crystal. These signals are rejected because they add significant noise without contributing substantially to image formation. Even with the newer tubes, however, statistical fluctuations in electron ejection from PM tube photocathodes establish a lower limit of about 70 keV for the energy range of γ rays amenable to single-crystal scintillation camera imaging. Also contributing to this lower energy limit is the difficulty in using pulse height analysis to distinguish primary from scatter photons when the primary photon energies are low. This difficulty arises because γ rays scattered during Compton interactions have about the same energy as the primary γ rays. Finally, low-energy γ rays are strongly absorbed in tissue, and few can reach the external detector.

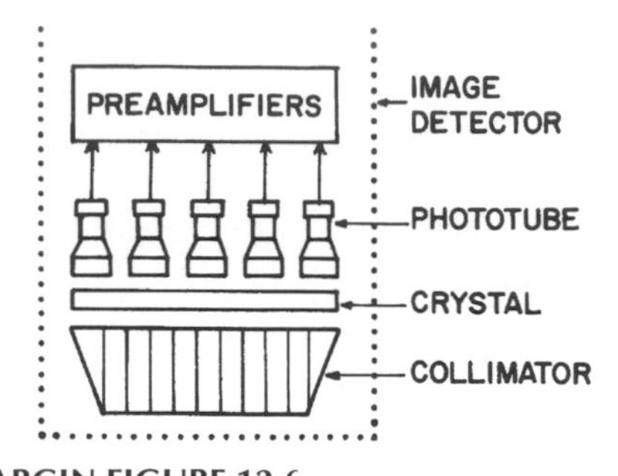

MARGIN FIGURE 12-6
Detector assembly and preamplifiers of a scintillation camera.

Photomultiplier Tube Preamplifiers

Each PM tube transmits its output electrical signal to its own preamplier. This circuit is used primarily to match the PM tube impedance to that of the succeeding electronic circuits. Preamplifiers can be "tuned" so that signals from the PM tubes are identical for a given amount of energy absorbed in corresponding regions of the detector. Many scintillation cameras use this approach to tuning the PM tube array.

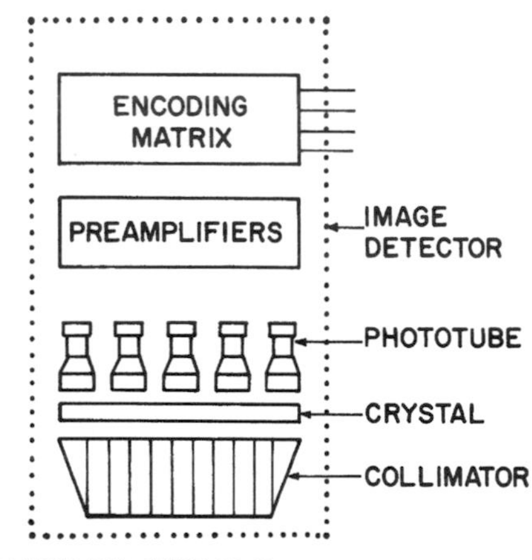

MARGIN FIGURE 12-7
Preamplifiers and position-encoding matrix of the scintillation camera.

Position-Encoding Matrix

The PM tubes are positioned in a hexagonal array above the scintillation crystal, with the spacing and separation from the crystal selected to provide an optimum combination of spatial uniformity and resolution. For the distribution of electrical signals emerging from the preamplifiers of these PM tubes, some operation must be performed to determine the site of energy absorption in the crystal. This operation is the function of the position-encoding matrix. The signal from each preamplifier enters the matrix and is applied across four precision resisters to produce four separate electrical signals for each PM tube. These signals are labeled $+x$, $-x$, $+y$, and $-y$.

For the PM tube in the center of the array, all four resistors are equal, and all four signals are identical. On the other hand, a PM tube in the lower left quadrant in the array is represented in the matrix by $+x$ and $+y$ resistors that are significantly different from the $-x$ and $-y$ resistors. A signal from the preamplifier of this PM tube furnishes $-x$ and $-y$ signals that are larger than the $+x$ and $+y$ signals. The actual magnitude of the signals from each four-resistor component of the encoding matrix reflects the proximity of the corresponding PM tube to the origin of the light flash within the crystal.

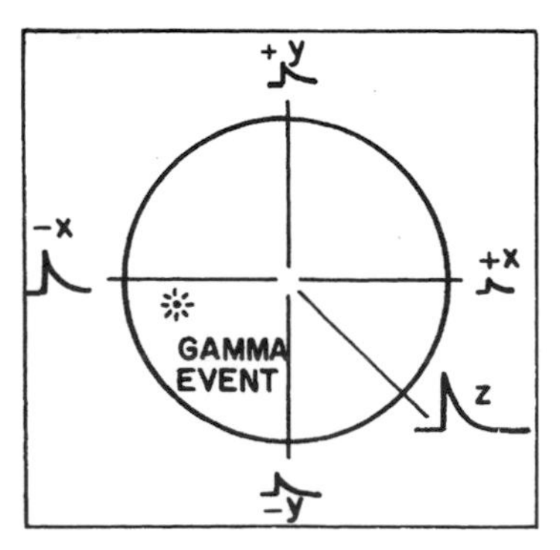

MARGIN FIGURE 12-8
In the position-encoding matrix for each PM tube, the PM signal is separated into four signals of relative sizes that reflect the position of the particular PM tube within the PM tube array.

Summation Amplifiers

From the encoding matrix, the position signals for each PM tube are transmitted to four summation amplifiers. The $+x$ signals for all of the PM tubes are added in one amplifier, the $-x$ signals for all of the tubes are added in a second summation amplifier, and so forth. The product of the summation operation is four electrical signals, one for each of the four deflection plates in the cathode-ray tube (CRT).

Deflection Amplifiers

Before the signals are applied to the deflection plates, they are transmitted to four deflection amplifiers where they are amplified, differentiated, and shaped (Figure 12-6).

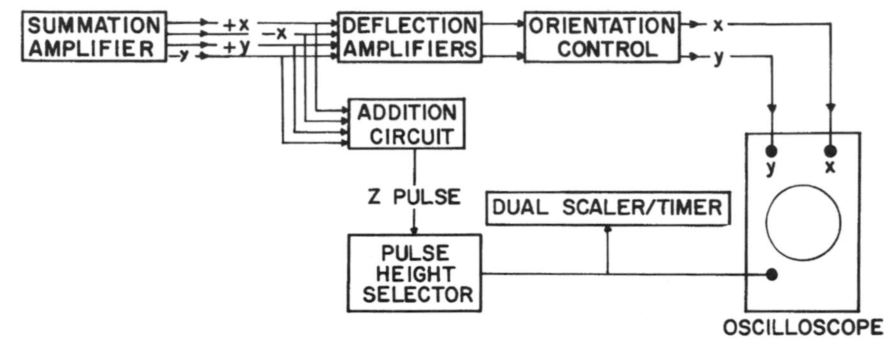

FIGURE 12-6
Position and pulse height control circuitry of a scintillation camera.

Here the signals also are divided by the Z pulse (see below) to normalize the positioning signals to the summation signal representing the particular interaction. This normalization process permits the use of a relatively wide pulse-size acceptance range (i.e., a relatively wide window) without an unacceptable reduction in position accuracy. Among other adjustments in the deflection amplifiers, the signals are modified to furnish the exact size of image desired on the CRT. Malfunctions of the deflection amplifiers often are responsible for distortions of the image such as the production of elliptical images for circular distributions of radioactivity. Following the deflection amplifiers, the positioning signals are fed to the orientation control where rotation switches on the camera console can be manipulated to alter the left–right and top–bottom presentation of the image on the camera console.

Pulse Summation (Addition) Circuit

In a circuit parallel to the deflection amplifiers, the four position signals are added to form the Z pulse. The Z pulse is sent to the pulse height analyzer for analysis of pulse height. In many cameras the pulse height analyzer is a fixed bias circuit, and the pulse height selection window is nonvariable. With a fixed-bias analyzer, pulses must be attenuated with an isotope range switch so that pulses representing total absorption of photons of a desired energy will be of the proper size to be accepted by the analyzer. Usually the isotope range switch, with different attenuation settings for different photon energy ranges, is positioned in the camera circuitry before the positioning signals branch to the addition circuit and deflection amplifiers.

If a particular pulse triggers the lower, but not the upper, discriminator of the analyzer, a negative pulse is transmitted to one input of an AND gate. The second input to the AND gate is furnished as soon as any previous pulse has dissipated. When the AND gate is activated, a microsecond unblanking pulse is furnished to the display CRT, where it activates the CRT electron beam to produce a short burst of electrons. This burst of electrons produces a tiny light flash on the cathode-ray screen. The number of electrons in the burst, and hence the brightness of the light flash on the cathode-ray screen, is adjusted with the intensity control on the console of the scintillation camera. The unblanking pulse is also transmitted to a scaler that records the counts forming the image.

Cathode-Ray Tube

The CRT consists essentially of an electron gun, four deflection plates, and a fluorescent screen that furnishes a flash of light when struck by the electron beam from the electron gun (Figure 12-7). Normally, the electron beam is prevented from flowing across the CRT. Only during the brief interval when the signal from the AND gate is received at the CRT is the electron beam permitted to flow and produce a light flash on the screen. During this brief interval, the position signals are applied to the

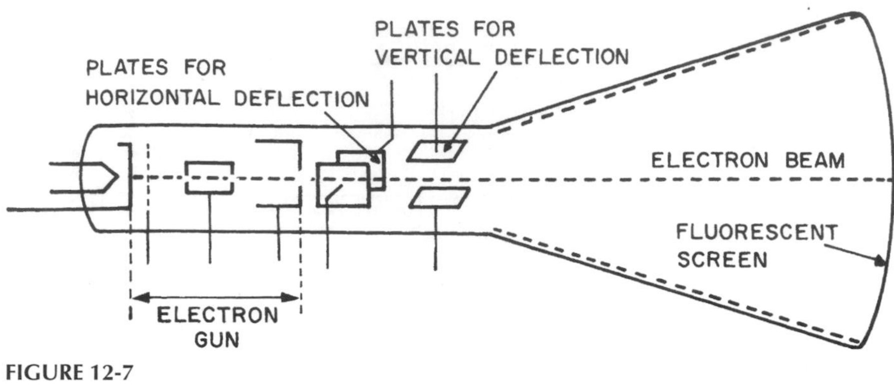

FIGURE 12-7
Cathode-ray image display device.

deflection plates so that the light flashes on the screen at a position corresponding to the site of γ-ray absorption in the crystal.

Scintillation cameras that digitize the signals after pulse-height and positioning analysis are called *hybrid analog-digital cameras*. Scintillation cameras that digitize the signals before pulse-height and positioning analysis are called *all-digital cameras*. Virtually all scintillation cameras marketed today are all-digital cameras. These cameras furnish more precise pulse-height and positioning analysis, and yield images with superior spatial resolution.

Processing, Display, and Recording

The image of a radionuclide distribution created in the scintillation camera may be digitized, stored in computer memory, and processed to enhance its features. The analog x and y signals from the camera electronics are converted to digital signals with analog-to-digital (ADC) converters. The digitized x and y signals determine the location of a pixel, and the number of counts recorded at that position determine the pixel value. Thus, images are stored as matrices in the computer in a special image display memory. The display system contains a video processor that reads the computer image display memory and performs a digital-to-analog conversion (DAC) to send the information to the monitor at video frame rates. A permanent record of the image may be made by exposing Polaroid or x-ray film to the light on the monitor. Recording devices with scanning laser beams may also be used to print images directly onto film.

Resolving Time Characteristics of Scintillation Cameras

At relatively low sample activities, the count rate from a single-crystal scintillation camera varies linearly with the sample activity. At higher sample activities, however, this relationship is not maintained, and the count rate increases less rapidly than expected as the activity of the sample is increased. The resulting loss of counts is termed the coincidence loss, and its magnitude depends on the resolving time of the camera. As described in Chapter 9, the resolving time is the minimum time required between successive interactions in the scintillation camera crystal for the interactions to be recorded as independent events.

The loss of counts at high count rates is said to be caused by "pulse pile-up." This effect is produced by two or more γ-ray interactions occurring within the resolving time of the scintillation crystal. Pulse pile-up can also cause image distortion if the energy deposited during Compton interactions occurring in different locations is summed to yield an energy equivalent to that of total absorption of a primary γ ray. In this case the events will be registered as a single primary γ-ray interaction at a location between the sites of the Compton interactions. The net effect of this positional distortion is a loss of contrast in the image.

When scintillation cameras are used for dynamic function studies, the loss of counts at high count rates may be a significant problem. In studies of the heart, for example, a bolus of activity on the order of 500 MBq or more may be present in the camera's field of view. This high activity may yield as many as 80,000 scintillations per second in the detector, and possibly even more if a converging collimator is used. At count rates this high, significant count rate losses occur, and quantitative interpretation of count rate data can be misleading.

For purposes of resolving time considerations, a scintillation camera may be considered a two-component system, with a paralyzable component followed by a nonparalyzable component.[4] In a paralyzable counting system, the minimum time required for the independent detection of two separate interactions is extended if an interaction occurs within the recovery interval. With such a system, the count rate may decrease dramatically (i.e., the system may be paralyzed) for samples of exceptionally high activity (Figure 9-6). In a nonparalyzable system, a certain minimum time is required to process each interaction, but this time is not extended by an interaction occurring within the processing interval. With a nonparalyzable system and high sample activity, the count rate may reach a maximum plateau but will not decrease as the activity is increased. Many scintillation cameras exhibit a behavior intermediate between paralyzable and nonparalyzable. In general, the paralyzable component is the part of the camera that precedes the pulse height analyzer and reflects primarily the fluorescence decay time of the scintillation crystal. The nonparalyzable component is the part of the camera, including scalers and data processing equipment, that follows the pulse height analyzer.

A typical NaI(Tl) crystal exhibits a fluorescence decay time of 0.25 μsec; 98% of all light is released within 0.8 μsec.

Several methods have been proposed to correct scintillation camera data for coincidence count rate loss. If the resolving time is known for a particular camera, for example, true counts can be estimated from observed counts, provided that the resolving time is independent of factors such as activity position across the detector face and that the summation of small coincident pulses to produce pulses transmitted by the PHA can be estimated. One widely used method to correct for coincidence count rate loss utilizes the count rate from a small marker source of known activity positioned near the periphery of the camera's field of view. In place of the radioactive marker source, an electronic marker such as a pulse generator may be used.

MULTIPLE-CRYSTAL SCINTILLATION CAMERA

The multiple-crystal approach to scintillation camera design was pioneered by Bender and Blau, who introduced the concept of the autofluoroscope in 1963.[5] In this device, 294 discrete NaI(Tl) crystals are arranged in a matrix of 21 columns and 14 rows, with each crystal 0.8 cm by 0.8 cm in cross section and about 4 cm thick.

Later versions of the autofluoroscope were marketed as "System 70."

In a single-crystal scintillation camera, the detection of γ-ray interactions includes determination of the spatial coordinates of the interaction. In a multiple-crystal camera, the detection of an interaction does not require the simultaneous collection of position information, and much higher count rates can be processed without significant count rate loss or positioning errors. In the multiple-crystal camera, the resolving time is determined essentially by the speed with which events can be processed electronically rather than by the fluorescence decay time of the crystal. Most multiple-crystal cameras in clinical use have been reserved for studies such as cardiac first-pass studies where the viewing area is limited and the count rates are very high.

The resolution of the multiple-crystal camera is limited intrinsically by the size of each detector element. However, the resolution can be improved markedly by moving the patient incrementally in the x and y directions so that each resolution element is divided into 16 subelements. With this motion of the scanning bed supporting the patient, the 294 resolution elements are subdivided essentially into $294 \times 16 = 4704$ subelements, with the final image reassembled by computer. For anatomic regions larger than the 15-cm × 23-cm viewing area of the detector, two successive adjacent 16-position images can be reassembled by computer to furnish a single image containing 9408 picture elements.

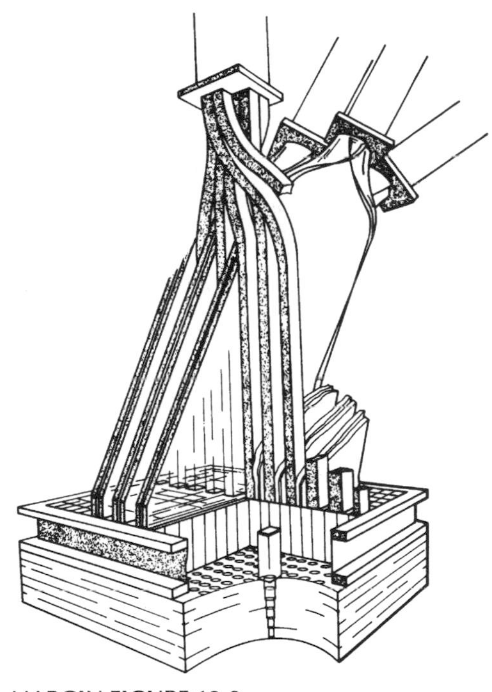

MARGIN FIGURE 12-9
Cutaway view of the detector assembly and light pipes for the multicrystal scintillation camera.

SOLID-STATE CAMERA

Nuclear cameras that employ direct digital detectors have been in development for several years. One digital camera introduced commercially just recently employs 4096 3-mm × 3-mm cadmium–zinc–tellerium [CdZnTe] detector elements. This detector maps γ-ray interactions in the elements directly into an image, so no PM tubes or x- and y-positioning circuits are needed. The detector is 21.6 × 21.6 cm square, so the field of view is rather limited. For small organs, this version of the direct digital camera provides excellent spatial and contrast resolution.

The commercial version of the direct digital camera described in the text is referred to as Digital Corporation's 2020tc Imager.

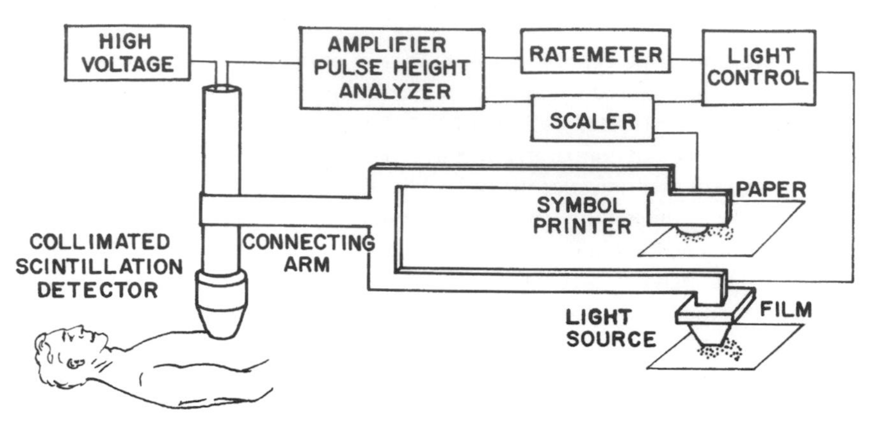

FIGURE 12-8
Rectilinear scanner with mechanical linkage.

RECTILINEAR SCANNER

A rectilinear scanner consists of a radiation detector that moves in a plane above or below the patient, and a recording device that moves in synchrony with the detector (Figure 12-8). In earlier scanners, the motion of the detector was synchronized with the recording device by a mechanical linkage. In later versions, electronic rather than mechanical linkage has been used, and the recording device is usually physically separated from the detector.

The first practical rectilinear scanner was developed in 1951 by Cassen and colleagues.[6] In this scanner, a small crystal of calcium tungstate was used as the radiation detector. However, the superior properties of sodium iodide (NaI) for detecting γ rays were quickly recognized, and essentially all scanners now in use employ this type of detector. In earlier scanners, the NaI crystal was typically 7.5 cm in diameter and 5 cm thick. In more recent scanners, the crystal diameter has increased to 13 or 20 cm to improve the efficiency of γ-ray detection. With a larger crystal, images may be obtained more quickly; or alternately, less "noisy" images may be obtained in comparable imaging times. Although the larger crystals collect more γ rays because of their larger field of view, they exhibit a more limited depth of focus, and radioactivity above and below the focal plane of the collimator is imaged less sharply.

Cassen's first rectilinear scanner employed a moving $CaWo_4$ crystal coupled to a PM tube and equipped with a single-hole lead collimator. His first studies were ^{131}I scans of the thyroid.

As the detector moves across the patient, it samples activity from a small region of the patient. This region is defined by a focused collimator and is referred to as the focal plane. The distance between the focal plane and the collimator face is the focal length. Pulse height analysis is used to record the signal. An energy "window" is set to accept only those pulses that correspond to total absorption of the γ rays of interest. Lower-energy pulses correspond to events such as interactions in the crystal of photons that were scattered originally in the patient. Such events would reduce contrast in the final image and are rejected by the PHA.

Advances in stationary scintillation cameras, coupled with the excessive time required to complete a rectilinear scan, have greatly reduced the value and use of rectilinear scanners in recent years.

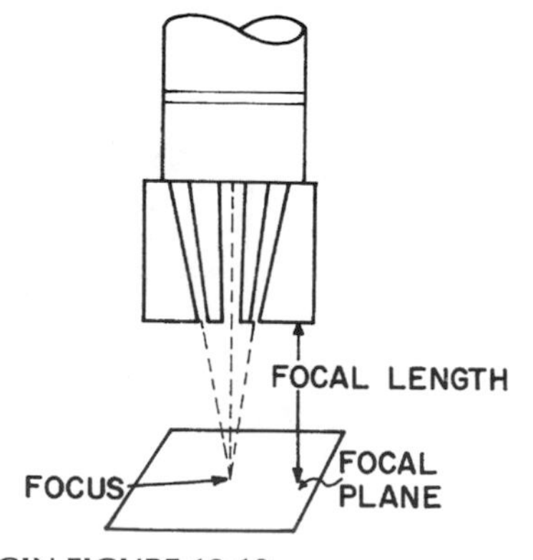

MARGIN FIGURE 12-10
Multihole focused collimator illustrating the focal length, focal plane, and focus of the collimator.

EMISSION COMPUTED TOMOGRAPHY

Tomography is the process of collecting data about an object from multiple views (projections), and using these data to construct an image of a "slice" through the object. In traditional x-ray tomography, multiple views of an object are built up on a single image receptor in such a way that uninteresting anatomy is blurred while a selected plan or region of the patient is presented as a sharp image. In computed tomography (CT), a computer is used to reconstruct an image of the patient from multiple views. CT using x rays transmitted through the patient is termed *transmission CT*

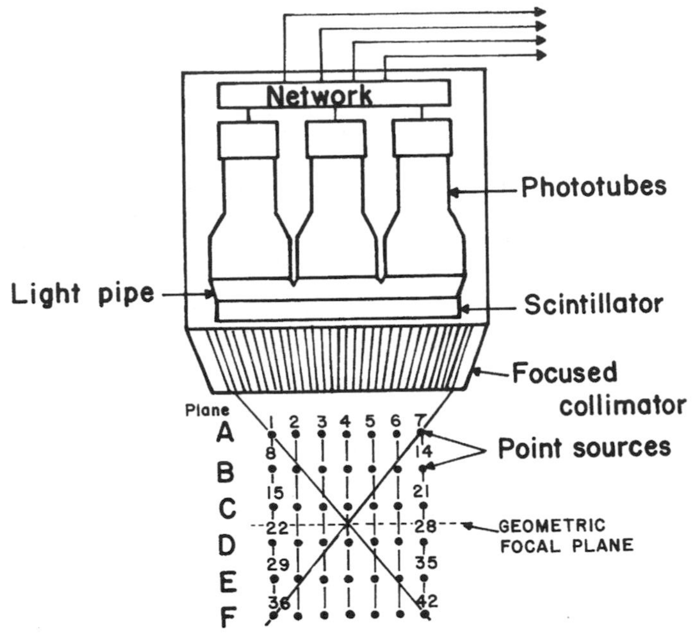

FIGURE 12-9
Detector assembly for multiplane tomographic scanner (From Anger, H.[7] Used with permission.)

(x-ray CT). A similar type of study can be performed in nuclear medicine by detecting photons emitted from a radiopharmaceutical distributed within the body. The term *emission computed tomography* (ECT) refers to this procedure. Production of tomographic images by detection of annihilation photons released during positron decay is referred to as *positron emission tomography* (PET). Tomographic images computed from the registration of interactions of individual γ rays in a crystal is referred to as *single-photon emission computed tomography* (SPECT).

Single-Photon Emission Computed Tomography

Longitudinal-Section SPECT

Also known as focal-plane SPECT, this general category of techniques involves the use of multiple views, and possibly motion, to provide a coronal, sagittal, or oblique slice of information within the patient. A diagram of a longitudinal-section SPECT unit is shown in Figure 12-9. In this unit, two or more views of the object are obtained from different angles. Super-position of data from the multiple views leads to multiple images and therefore blurring of all objects except those at the plane defined by the junction of the viewing areas of the detector. Images may be compiled by simple addition of counts from the multiple views. With the aid of computers, several ingenious techniques are available to reconstruct images from any plane or arbitrary surface within the object. If the fixed spacings between views and the angle between the views at the imaging device are known, the origin of each detected "event" within the crystal may be matched to the "event" seen by another view. In this manner, a source of radiation can be located in three dimensions within the patient.

The first longitudinal-section SPECT unit was built by H. Anger in 1965.[7]

Various methods are used to obtain multiple views for focal-plane SPECT. Special collimators have been designed to acquire the multiple views, and dedicated software has been developed to process the information. Examples of these systems include multiple pinhole techniques[8] sometimes combined with collimator or detector motion.[9–10] Although longitudinal-section SPECT systems continue to have research applications, they have been replaced in clinical nuclear medicine by tránsverse-section SPECT systems.

Single-Photon Emission Computed Tomography

Transverse-Section SPECT

Development of SPECT can be traced to the work of Kuhl and Edwards in the early 1960s.[11]

This nuclear imaging technique has several features in common with x-ray CT. Multiple views are obtained to compile an image of one or more transverse slices through the patient. In fourth-generation x-ray CT, the x-ray tube rotates and the detectors are stationary. Most transverse-section SPECT systems employ a rotating detector (a scintillation camera) and a stationary source of γ rays (the patient). One, two, or three scintillation cameras are mounted on a rotating gantry. Each scintillation "head" encompasses enough anatomy in the axial direction to permit acquisition of multiple slices in a single scan.

In earlier SPECT units, the detectors rotated in a circular orbit around the body. In newer units, the detectors follow an elliptical path so that they remain close to the skin surface during the entire scan.

One major difference between SPECT and x-ray CT is that a SPECT image represents the distribution of radioactivity across a slice through the patient, whereas an x-ray CT image reflects the attenuation of x rays across a tissue slice. In SPECT, photon attenuation interferes with the imaging process because fewer photons are recorded from voxels at greater depths in the patient. A correction must be made for this property. Methods for attenuation correction usually involve estimating the body contour, sampling the patient from both sides (i.e., 360° rather than 180° rotation), and constructing a correction matrix to adjust scan data for attenuation.

Transverse-section SPECT units can also provide coronal and sagittal images by interpolation of data across multiple transverse planes.

Transverse-section SPECT is useful in several clinical applications including cardiac imaging,[12,13] liver/spleen studies,[14,15] and chest–thorax procedures.[16] It is also important in the imaging of receptor-specific pharmaceuticals such as monoclonal antibodies, because it provides improved detection and localization of small sites of radioactivity along the axial dimension.[17]

For 360° SPECT imaging, a single-head unit must rotate 360°; a dual-head unit must rotate 180°; and a triple-head unit must rotate 120°. Hence, data acquisition is three times faster with a triple-head unit than with a single-head unit.

Quality control for SPECT systems is similar to that for stationary scintillation cameras. In general, uniformity requirements are rather stringent for SPECT, and variations in count rate sensitivity over the image should not exceed 2%. The uniformity of count rate sensitivity should be verified in several orientations of each detector because distortions may be introduced by the influence of the earth's magnetic field on the PM tubes. Also, coupling of the PM tubes to the crystal may change as a detector changes position. Certain parameters are critical, such as assignment of the center of rotation for transverse-section SPECT. If a source of radioactivity is positioned incorrectly in the image matrix for different views during transverse SPECT, it will appear as a ring in the final image. To verify the center of rotation, a point source of activity is placed at the center of rotation and imaged from multiple views. In each view, the source will appear in an individual pixel. The pixel in which the source appears should be the same in views taken from different angles. Another approach to verifying alignment of the center of rotation is to position a point source slightly away from the center of rotation. The point source should appear as a symmetrical donut in the image.

Reconstruction of a SPECT image is performed with the same class of algorithms used in x-ray CT. The most common algorithm is convolution (filtered) backprojection. However, a shift is occurring from convolution algorithms to algebraic iterative approaches for image reconstruction. Iterative algorithms allow easy incorporation of noise-reduction techniques and *a priori* information to improve spatial and temporal information.

Quality control is critical in transverse-section SPECT. Additional information about SPECT quality-control procedures is available in texts on nuclear imaging.[18–20]

Positron Emission Tomography

The principle of transverse SPECT is applicable to positron emission tomography (PET). This principle is that radiation emitted from a radiopharmaceutical distributed in the patient is registered by external detectors positioned at different orientations. Then, the distribution of radioactivity is estimated by backprojection methods. In PET, however, the radiation detected is annihilation radiation released as positrons interact with electrons. The directionality of the annihilation photons (two 511-keV annihilation photons emitted in opposite directions) provides a mechanism for localizing the origin of the photons and hence the radioactive decay process that resulted in their emission.

In transverse SPECT, data acquisition can either be continuous [i.e., the detector(s) move continuously] or discontinuous [i.e., the detector(s) stop rotating during data acquisition]. The latter method is referred to as "step and shoot."

In a typical PET system (Figure 12-10) the patient is surrounded by a ring of detectors. When detectors on opposite sides of the patient register annihilation photons simultaneously (e.g., within 10^{-9} seconds), the positron decay process that created the photons is assumed to have occurred along a line between the detectors.

Because the directionality of annihilation photons provides information about the origin of the photons, collimators are not needed in PET imaging. This approach, referred to as *electronic collimation*, greatly increases the detection efficiency of PET imaging.

The PET scanner was developed in the early 1950s by Brownell, Sweet, and Aronow at Massachusetts General Hospital.[21,22]

(a)

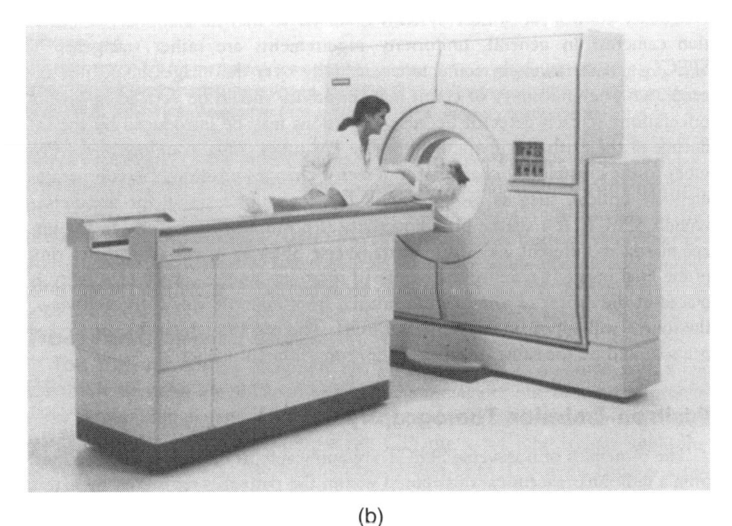

(b)

FIGURE 12-10
A: In positron emission tomography (PET), two detectors D_1 and D_2 record the interaction of 0.511-MeV photons in coincidence, indicating that a positron decay event has taken place somewhere along a line connecting the two detectors within the patient P. **B:** PET scanner (Siemens ECAT953B brain scanner) at the MRC cyclotron Unit, London. (Part B, courtesy of Siemens Medical Systems, Inc.)

The PET image reveals the number of decays (counts) occurring in each of the voxels represented by pixels in the image.

Radionuclides suitable for PET are the short-lived (half-lives on the order of seconds, minutes, or a few days) positron-emitting nuclides ^{11}C, ^{13}N, ^{15}O, ^{18}F, ^{62}Cu, ^{68}Ga, and ^{82}Rb. To date, all clinical applications of PET have employed ^{18}F-labeled fluorodeoxyglucose [^{18}FDG]. The positron-emitting isotopes ^{11}C, ^{13}N, and ^{15}O have significant physiologic potential because they can replace atoms in molecules that are essential for metabolism. Thus direct studies of metabolism in tissue are possible (Figure 12-11).

The first multicrystal PET unit was developed in the mid-1970s by TerPogossian, Phelps, and Hoffman[23] of Washington University in St. Louis.

Positron emitters are produced in a cyclotron. For very short-lived nuclides such as ^{11}C, ^{13}N, and ^{15}O, the cyclotron must be located near the scanner so the nuclides can be used before they decay. Thus the facilities and support personnel required for PET imaging with these radionuclides can be a major financial investment. ^{18}F has a half-life of 1.7 hours and can often be obtained from a nearby supplier. PET imaging confined to use of ^{18}F is less costly but somewhat limited in its universality. There is great interest in further development of PET. The availability of positron sources such as ^{68}Ga that are produced in radionuclide generators, along with advances in miniaturization and simplification of cyclotrons, decreases site planning and facilities

Replacement of an atom in a molecule with its radioactive counterpart is known as *isotopic labeling*.

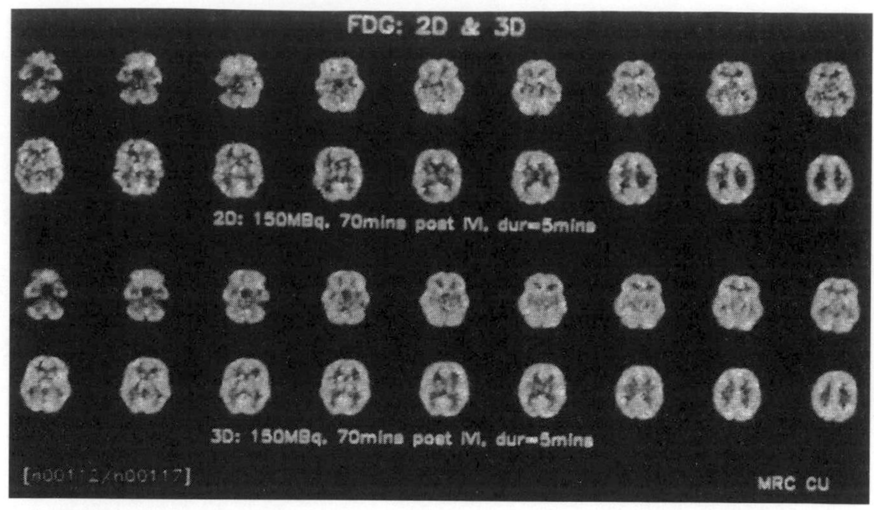

FIGURE 12-11
PET images from a fluorodeoxyglucose study of normal brain glucose metabolism. Signal-to-noise comparison of 2D (with collimators) versus 3D (without collimators) coincidence plane collection of annihilation events. Courtesy of Siemens Medical Systems, Inc.

Self-shielded, computer-controlled, miniaturized cyclotrons with automated chemical synthesis are providing ^{18}F-deoxyglucose (CT-DG)-labeled pharmaceuticals for a variety of PET applications.

support requirements and is stimulating the development of more clinical sites for PET imaging.

It is more challenging to obtain multiple image slices in PET than in SPECT. Most PET scanners employ several detector rings for this purpose. However, attenuation corrections are inherently more accurate in PET than in SPECT because the pair of annihilation photons defines a line through the patient such that the total path length through the patient is always traversed when a decay is detected.

PET Radionuclides

Nuclide	$T_{1/2}$ (minutes)
^{11}C	20.4
^{13}N	10
^{15}O	2
^{18}F	110
^{62}Cu	9.8
^{68}Ga	68
^{82}Rb	1.25

Detector sensitivity is a major design consideration in PET. Bismuth germanate (BGO) crystals are used in place of the NaI(Tl) detectors employed in SPECT, because BGO has a greater intrinsic efficiency for the higher-energy (511 keV) photons. Resolution of PET systems is limited to 5 mm or so by the number of detectors used, distance traveled by the positron from its origin before annihilation, and deviation from exact 180°-annihiliation photon emission. Although PET resolution is poorer than that of x-ray CT or magnetic resonance imaging (MRI), interest in the development of PET is substantial because the physiologic and metabolic information available in a PET study is unique.

A relatively recent development is the integration of PET and SPECT into a single imaging unit. With this integrated unit, SPECT and PET imaging can be performed on the same patient. Also, the merger of PET and x-ray CT into the same imaging unit provides PET images fused with CT images. This approach is particularly useful for tumor identification and localization in oncology, which at this time constitutes the major clinical application of PET. In addition, x-ray CT furnishes a rapid and accurate method of correcting PET images for photon attenuation.

PROBLEMS

*12-1. The uptake of ^{131}I was measured for a patient treated for hyperthyroidism with the same radioactive nuclide 2 months previously. The count rates were as follows:

For a ^{131}I standard on day 0	13,260 cpm
For a ^{131}I standard 24 hour later	12,150 cpm
For the patient on day 0 before ^{131}I administered	6,140 cpm
For the patient 24 hours later	12,840 cpm

Compute the percent uptake of ^{131}I in the thyroid gland. Ignore the contribution of background radiation to the count rates, but include the contribution from residual ^{131}I.

*12-2. A patient was given 10×10^4 Bq of ^{131}I-RISA. A second 10×10^4 Bq sample was diluted to 3000 ml. A 5-ml aliquot of this solution provided a count rate of 3050 cpm. A 5-ml sample of the patient's blood provided a count rate of 1660 cpm. What is the patient's blood volume?

12-3. Describe a method with a radioactive nuclide to determine the volume of a large container.

*12-4. The following count rates were recorded 24 hours after administration of 0.4 MBq of ^{131}I to the patient:

For the patient's thyroid	1831 cpm
For the patient's thyroid shielded by a lead block	246 cpm
For a standard source in a thyroid phantom	3942 cpm
For a standard shielded by a lead block	214 cpm

What is the percent uptake?

*12-5. Line-source response functions were measured for a NaI(Tl) scintillation detector and a multihole focused collimator. Which functions resemble a normal probability curve more closely:

a. A response function measured with a ^{198}Au source (γ ray of 412 keV) or a response function measured with a ^{241}Am source (γ ray of 60 keV)?

b. A response function measured with a ^{51}Cr source (γ ray of 320 keV) or a response function measured with a ^{197}Hg source (γ and x rays of about 70 keV)?

12-6. Explain the use with a scintillation camera of (a) a pinhole collimator, (b) a parallel multihole collimator, (c) a diverging collimator, and (d) a converging collimator. Explain why the sensitivity does not change greatly as the distance varies between the patient and the face of a parallel multihole collimator.

12-7. Explain why the minimum resolution distance of a multiple-crystal camera is about equal to the width of a crystal in the detector mosaic.

12-8. Describe the reasons for limiting the useful range of a scintillation camera from about 70 keV to about 600 keV.

12-9. Explain why, when compared with a parallel multihole collimator, a converging collimator yields superior images for small distributions of radioactivity.

*For those problems marked with an asterisk, answers are provided on p. 492.

SUMMARY

- Radioactive nuclides are widely used to measure rate and volumes in clinical medicine. Examples include:
  - Thyroid uptakes
  - Blood volumes
  - Kidney flow studies
- The single-crystal scintillation camera, together with generator-produced ^{99m}Tc, are the principal factors responsible for the widespread use of nuclear medicine procedures in medicine.
- Collimators provide a one-to-one relationship between the origin of a γ ray in the patient and the site of interaction of the γ ray in the crystal. Types of collimators include:
  - Parallel multihole
  - Pinhole
  - Converging
  - Diverging
- Basic components of a single-crystal scintillation camera include:
  - Detector (crystal, PM tube, preamplifier)
  - Position-encoding matrix
  - Summation and deflection amplifiers
  - Pulse summation circuit
  - Cathode-ray display
- The resolving time of a scintillation camera is the shortest time between individually measurable events in the camera.
  - Emission computed tomography includes:
  - Longitudinal-section SPECT
  - Transverse-section SPECT
  - Positron emission tomography (PET)

REFERENCES

1. Saha, G. *Physics and Radiobiology of Nuclear Medicine, 2nd Edition*. New York, Springer, 2001.
2. Anger, H. Scintillation camera. *Rev. Sci. Instrum.* 1958; **29**:27.
3. Farmelant, M., DeMeester, G., Wilson, D., et al. *J. Nucl. Med.* 1975; **16**:183.
4. Sorensen, J. Deadtime characteristics of Anger cameras. *J Nucl. Med.* 1975; **16**:284.

5. Bender, M., and Blau, M. The autofluoroscope. *Nucleonics* 1963; **21**:52.
6. Cassen, B., et al. Instrumentation for ^{131}I use in medical studies. *Nucleonics* 1951; **9**:46.
7. Anger, H. Tomography and other depth-discrimination techniques, in Hine, G., and Sorenson, J. (eds.), *Instrumentation in Nuclear Medicine*, Vol 2. New York, Academic Press, 1974.
8. Budinger, T. Physical attributes of single-photon tomography. *J. Nucl. Med.* 1980; **21**:579–592.
9. Muehllehner, G. A tomographic scintillation camera. *Phys. Med. Biol.* 1971; **16**:87.
10. Anger, H. *Tomographic Gamma-Ray Scanner with Simultaneous Readout of Several Planes.* UCRL Report 16899. Berkeley, University of California, Press, 1966.
11. Kuhl, D., and Edwards, R. Image separation radioisotope scanning. *Radiology* 1963; **80**:653–662.
12. Garcia, E., et al. Quantification of rotational thallium-201 myocardial tomography. *J. Nucl. Med.* 1985; **26**:17–26.
13. Depasquale, E., et al. Quantitative rotational thallium-201 tomography for identifying and localizing coronary artery disease. *Circulation* 1988; **77**:316–327.
14. Van Heertum, R. Current advances in hepatic SPECT imaging, in *Clinical SPECT Symposium*. American College of Nuclear Physicians, Washington, D.C., 1986, pp. 58–64.
15. Friman, L., and Soderberg, B. Spleen–liver ratio in RES scintigraphy: A comparison between posterior registration and emission computed tomography. *Acta Radiol.* 1987; **28**:439–441.
16. Khan, B., Ell, P., Jarritti, P., et al. Radionuclide section scanning of the lungs in pulmonary embolism. *Br. J. Radiol.* 1981; **54**:586–591.
17. Chatal, J., et al. Comparison of SPECT imaging using monoclonal antibodies with computed tomography and ultrasonography for detection of recurrences of colorectal carcinoma: A prospective clinical study. *J. Nucl. Med.* 1985; **26**:15.
18. Peters, A., and Myers, M. *Physiological Measurements with Radionuclides in Clinical Practice*. New York, Oxford Univ. Press, 1998.
19. Chandra, R. *Nuclear Medicine Physics: The Basics,* 5th edition. Baltimore, Williams & Wilkins, 1998.
20. Powsner, R., and Powsner, E. *Essentials of Nuclear Medicine Physics.* Malden, MA, Blackwell, 1998.
21. Brownell, G., and Sweet, W. Localization of brain tumors. *Nucleonics* 1953; **11**:40–45.
22. Aronow, S., Brownell, G., Lova, S., and Sweet, W. Statistical analysis of eight years' experience in positron scanning for brain tumor localization. *J. Nucl. Med.* 19562; **3**:198.
23. TerPogossian, M., Phelps, M., and Hoffman, E. A positron-emission transaxial tomograph for nuclear imaging (PETT). *Radiology* 1975; **114**:89–98.

CHAPTER

13

RADIOGRAPHY

OBJECTIVES

After completing this chapter, the reader should be able to:

- Describe the construction of radiographic film.
- Define the following terms concerning radiographic film:
  - Transmittance
  - Optical density
  - H and D curve
  - Latitude
  - Average gradient
  - Speed
  - Reciprocity
- State the effect of the following upon the H and D curve of film:
  - X-ray tube voltage
  - Development time
  - Development temperature
- State the effect of changes in development time and temperature upon:
  - Average gradient
  - Relative speed
  - Fog
- Describe the principle of operation of the radiographic screen.
- Name five different types of radiographic grids.
- Define the following terms concerning radiographic grids:
  - Grid ratio
  - Contrast improvement factor
  - Selectivity
  - Cutoff
- Explain how the air gap technique may be used to exclude scattered photons from a radiographic image.
- Discuss direct and indirect energy conversion in large-area digital detectors.

The expression "radiography" refers to procedures for recording, displaying, and using information carried to a film by an x-ray beam. Satisfactory images are recorded only if enough information is transmitted to the film by x-rays emerging from the patient. Also, unnecessary information and background "noise" must not interfere with extraction of the information desired. Procedures for recording the information desired and for reducing extraneous information and noise are described in this chapter.

X-RAY FILM

X-ray film is available with an emulsion on one side (single-emulsion film) or both sides (double-emulsion film) of a transparent film base that is about 0.2 mm thick (Margin Figure 13-1). The base is either cellulose acetate or a polyester resin. Single-emulsion film is less sensitive to radiation and is used primarily when exceptionally fine detail is required in the image (e.g., laser printing and mammography). The emulsion is composed of silver halide granules, usually silver bromide, that are suspended in a gelatin matrix. The emulsion is covered with a protective coating (T coat) and is sensitive to visible and ultraviolet light and to ionizing radiation. Film used with intensifying screens is most sensitive to the wavelengths of light emitted by the screens. Nonscreen film is designed for direct exposure to x rays and is less sensitive to visible light. Virtually all film used in diagnostic radiology is designed to be used with intensifying screens.

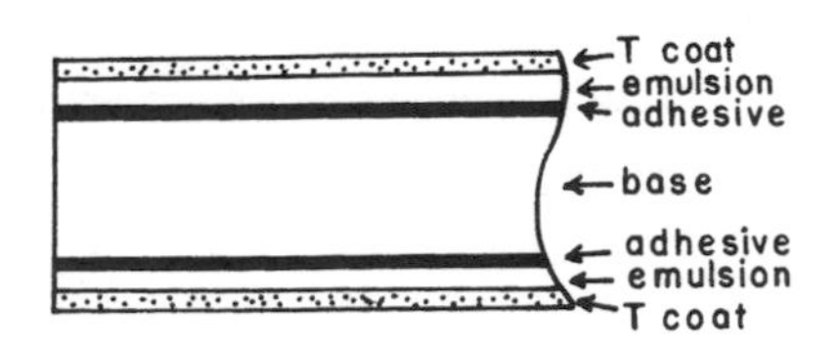

MARGIN FIGURE 13-1
Cross section of an x-ray film. The base is cellulose acetate or a polyester resin, and the emulsion is usually silver bromide suspended in a gelatin matrix.

Photographic Process

The granules of silver bromide in the emulsion of an x-ray film are affected when the film is exposed to ionizing radiation or to visible light. Electrons are released as energy is absorbed from the incident radiation. These electrons are trapped at "sensitivity centers" in the crystal lattice of the silver bromide granules. The trapped electrons attract and neutralize mobile silver ions (Ag^{+}) in the lattice. Hence small quantities of metallic silver are deposited in the emulsion, primarily along the surface of the silver bromide granules. Although these changes in the granules are not visible, the deposition of metallic silver across a film exposed to an x-ray beam is a reflection of the information transmitted to the film by the x radiation. This information is captured and stored as a *latent image* in the photographic emulsion.

When the film is placed in a developing solution, additional silver is deposited at the sensitivity centers. Hence the latent image induced by the radiation serves as a catalyst for the deposition of metallic silver on the film base. No silver is deposited along granules that are unaffected during exposure of the film to radiation, and these granules are removed by the sodium thiosulfate or ammonium thiosulfate present in the fixing solution. The fixing solution also contains potassium alum to harden the emulsion and acetic acid to neutralize residual developer present on the film. The degree of blackening of a region of the processed film depends on the amount of free silver deposited in the region and, consequently, on the number of x rays absorbed in the region.

Until the 1960s, the material used for the base of radiographic film was the same material that was used for motion picture film, cellulose nitrate. The same drawback was recognized in both fields; cellulose nitrate turns brown and disintegrates over periods as short as a few years, unless stored under special environmental conditions. This disintegration causes it to emit toxic fumes. Cellulose nitrate radiographic film was the cause of a series of explosions that resulted in number of fatalities at the Cleveland Clinic in May 15, 1929. Today, more stable plastics such as mylar are used for film base material.

Optical Density and Film Gamma

The amount of light transmitted by a region of processed film is described by the *transmittance T*, where

$$T = \frac{\text{Amount light transmitted by a region of film}}{\text{Amount light received at the same location with film removed}} \tag{13-1}$$

The degree of blackening of a region of film is described as the *optical density* (OD) of the region:

$$\text{OD} = \log\left(\frac{1}{T}\right) \tag{13-2}$$

Often, the optical density is referred to as the "density" of the film.

Example 13-1

A region of processed film transmits 10% ($T = 0.1$) of the incident light. What is the optical density of the region?

$$\begin{aligned} T &= 0.1 \\ \text{OD} &= \log\left(\frac{1}{T}\right) = \log\left(\frac{1}{0.1}\right) = \log(10) \\ &= 1 \end{aligned}$$

Example 13-2

Two processed films each have optical densities of 1 ($T = 0.1$). What is the resulting optical density if one is placed over the other?

$$T = (0.1)(0.1) = 0.01$$

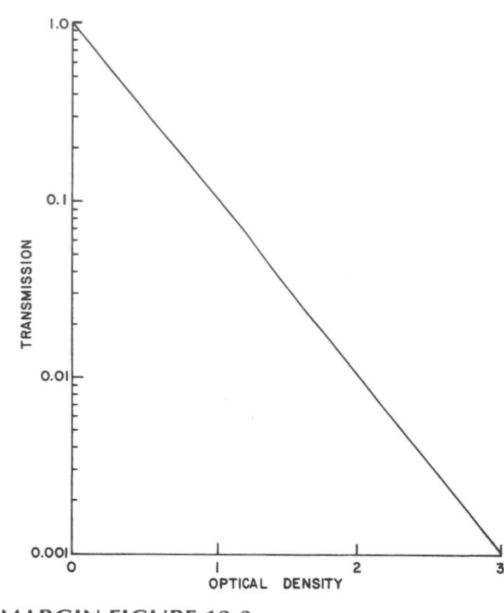

MARGIN FIGURE 13-2
Relationship between optical density and transmittance of x-ray film.

The optical density is

$$OD = \log\left(\frac{1}{T}\right) = \log\left(\frac{1}{0.01}\right) = \log(100)$$

$$OD = 2$$

The relationship between transmittance and optical density is depicted graphically in Margin Figure 13-2. The optical density across most radiographs varies from 0.3 to 2, which corresponds to a range of transmittance from 50% to 1%. Radiographs with an optical density greater than 2 must usually be viewed with a light source more intense than the ordinary viewbox (a "hot light").

The optical density of a particular film or combination of film and intensifying screens may be plotted as a function of the exposure or [log (exposure)] to the film. The resulting curve is termed the *characteristic curve, sensitometric curve*, or *H–D curve* for the particular film or film-screen combination (Margin Figure 13-3). The expression "H–D curve" is derived from the names of Hurter and Driffield, who in 1890 first used characteristic curves to describe the response of photographic film to light. The region below the essentially straight-line portion of the characteristic curve is referred to as the *toe* of the curve. The *shoulder* is the region of the curve above the straight-line portion. A radiograph with optical densities in the region of the toe or the shoulder furnishes an image with inferior contrast (Margin Figure 13-4). The exposure range over which acceptable optical densities are produced is known as the *latitude* of the film. The shape of the characteristic curve for a particular film is affected by the quality of the x-ray beam used for the exposure (Margin Figure 13-5) and by the conditions encountered during development (Margin Figures 13-6 and 13-7). Consequently, the tube voltage used to generate the x-ray beam and the temperature, time, and solutions used for processing the film should be stated when a characteristic curve for a particular film or film–screen combination is displayed.

The radiation exposure to an x-ray film should be sufficient to place the range of optical densities exhibited by the processed film along the essentially straight-line portion of the characteristic curve. The average slope of this portion of the curve is referred to as the *average gradient* of the film:

$$\text{Average gradient} = \frac{D_2 - D_1}{\log X_2 - \log X_1} \quad \textbf{(13-3)}$$

In Eq. (13-3), D_2 is the optical density resulting from an exposure X_2 and D_1 is the optical density produced by an exposure X_1 (see Margin Figure 13-3). Often, D_1 and D_2 are taken as optical densities of 0.3 and 2.0, respectively. Films with higher average gradients tend to furnish images with higher contrast (more blacks and whites and fewer shades of gray) than do films with lower values of the average gradient. Because contrast and latitude are reciprocally related, films with high average gradients also provide relatively short latitudes and vice versa. The maximum value of the slope of the characteristic curve is termed the *gamma* (γ) of the film. The gamma of a film may be defined explicitly as the slope at the inflection point of the characteristic curve.

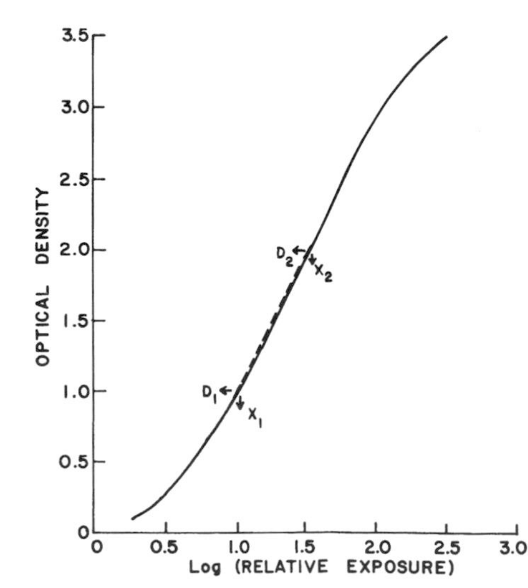

MARGIN FIGURE 13-3
Characteristic curve for an x-ray film. The average gradient for the film is 1.9 over the density range 1.0 to 2.0.

Example 13-3

In Margin Figure 13-3, $D_2 = 2.0$, $\log X_2 = 1.52$, $D_1 = 1.0$, and $\log X_1 = 1.00$. What is the average gradient for the film in this region of optical density?

$$\begin{aligned}\text{Average gradient} &= \frac{D_2 - D_1}{\log X_2 - \log X_1} \\ &= \frac{2.0 - 1.0}{1.52 - 1.00} \\ &= 1.9 \end{aligned} \quad \textbf{(13-3)}$$

Film Speed

Film speed or film sensitivity is a measure of the position of the characteristic curve at a specific value of the exposure axis.

$$\text{Film speed} = \frac{1}{\text{Exposure in R required for an OD of 1.0 above base density}} \quad (13\text{-}4)$$

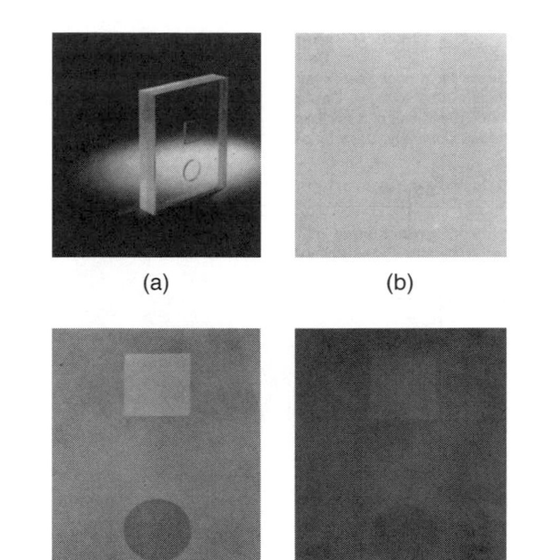

MARGIN FIGURE 13-4
Photograph of a plastic test object **(A)** and radiographs of the object obtained with screen-type films and an exposure on the toe **(B)**, middle **(C)**, and shoulder **(D)** of the characteristic curve. Contrast of the image is highest for the exposure on the straight-line portion of the curve. (From *Sensitometric Properties of X-Ray Films*. Rochester, NY, Radiography Markets Division, Eastman Kodak Company. Used with permission.)

Example 13-4

What is the speed of the x-ray film in Margin Figure 13-3 if 20 mR is required to produce an optical density of 1.0 above base density?

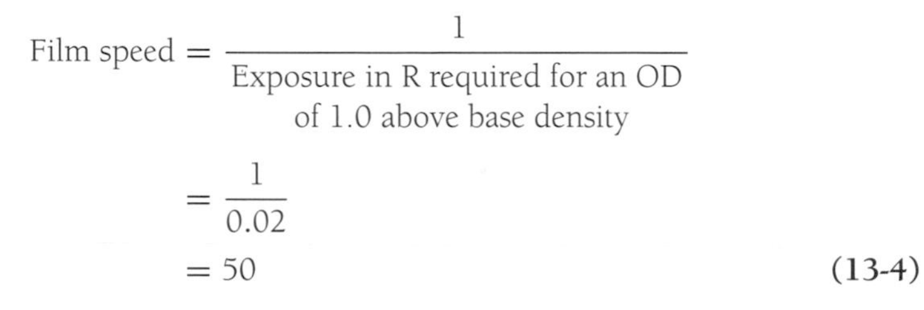

$$\text{Film speed} = \frac{1}{\text{Exposure in R required for an OD of 1.0 above base density}}$$

$$= \frac{1}{0.02}$$

$$= 50 \quad (13\text{-}4)$$

The toe of the H&D curve is short for high-speed film or film-screen combinations and longer for film with lower speed. High-speed film is referred to as "fast" film, and film with lower speed is said to be "slow." The speed of a film depends primarily on the size of the silver bromide granules in the emulsion. A film with large granules is faster than a film with smaller granules. Speed is gained by requiring fewer x-ray or light photons to form an image. Hence, a fast film furnishes a "noisier" image in which fine detail in the object may be less visible (see Chapter 16).

Film speed varies with the energy of the x rays used to expose the film (Margin Figure 13–8). For nonscreen film, the film speed is greatest when the spectral distribution of the x-ray beam is centered over the K-absorption edge (25 keV) of silver.

Film Reciprocity

For most exposure rates encountered in radiography, the shape of the characteristic curve is unaffected by changes in the rate of exposure of the film to x radiation. The independence of optical density and exposure rate is referred to as the *reciprocity law*. The reciprocity law sometimes fails if a film is exposed at either a very high or a very low exposure rate. In these situations, a particular exposure delivered at a very high or very low rate provides an optical density less than that furnished by an exposure delivered at a rate nearer optimum. However, a decrease in optical density is usually not noticeable unless the film is exposed with calcium tungstate intensifying screens and unless the exposure rate differs from the optimum by a factor of at least 8.[1] For film exposed with rare earth intensifying screens, failure of the reciprocity law has been observed over a more restricted range of exposure rates.[2]

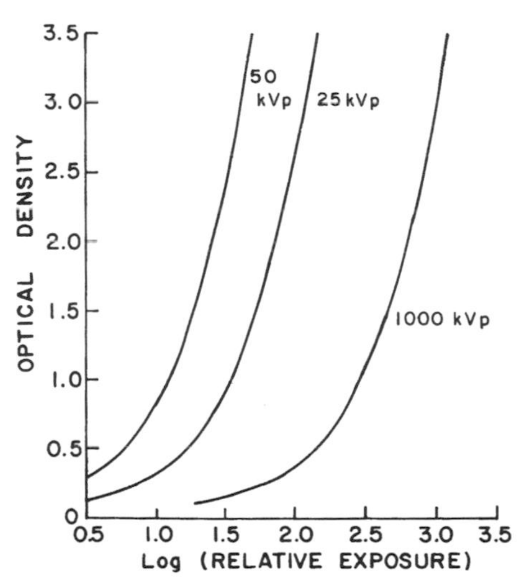

MARGIN FIGURE 13-5
Characteristic curves for a nonscreen x-ray film exposed to heavily filtered x-ray beams generated at different tube voltages. Developing conditions were constant for all curves. (From *Sensitometric Properties of X-Ray Films*. Rochester, NY, Radiography Markets Division, Eastman Kodak Company. Used with permission.)

Inherent Optical Density and Film Fogging

The optical density of film processed without exposure to radiation is referred to as the *inherent optical density (base density)* of the film. The inherent optical density is a reflection primarily of the absorption of light in the film base, which may be tinted slightly. Most x-ray film has an inherent optical density of about 0.07.

Fogging of an x-ray film is caused by the deposition of metallic silver without exposure to radiation and by the undesired exposure of a film to radiation (usually background radiation). For example, exposure as small as 1 mR to a high-speed film may produce a significant amount of fogging. The optical density attributable

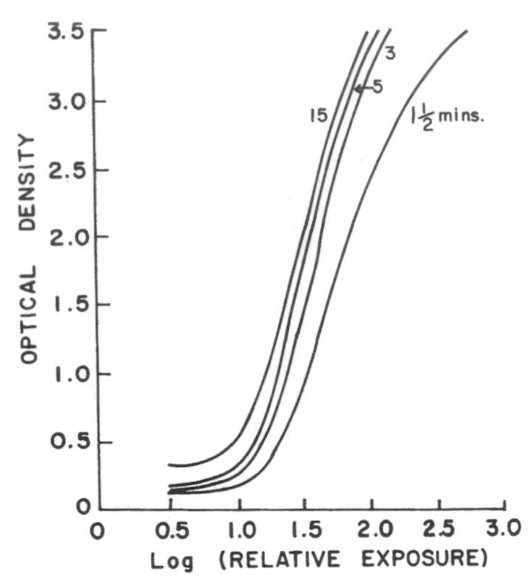

MARGIN FIGURE 13-6
Characteristic curves for a screen-type x-ray film developed for a series of times at 68°F. (From *Sensitometric Properties of X-Ray Films*. Rochester, NY, Radiography Markets Division, Eastman Kodak Company. Used with permission.)

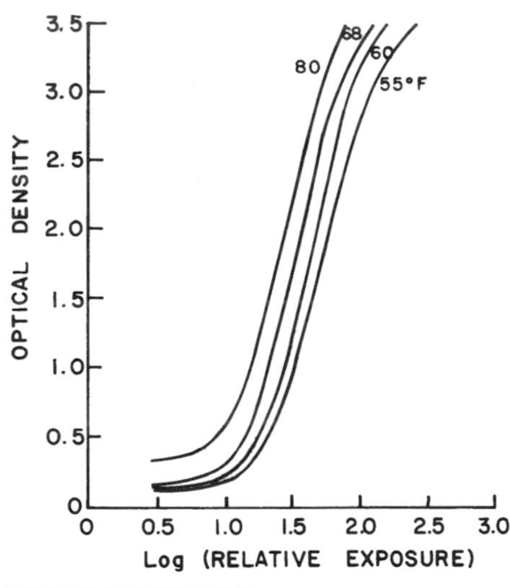

MARGIN FIGURE 13-7
Characteristic curves for a screen-type x-ray film developed for 5 minutes at a series of developing temperatures. (From *Sensitometric Properties of X-Ray Films*. Rochester, NY, Radiography Markets Division, Eastman Kodak Company. Used with permission.)

Some film-screen manufacturers use their own method of specifying relative speed. A "par" or average system is identified, given a value of 100, and the speeds of other systems are normalized to this reference value. For example, very fast film-screen systems might have relative speeds of 800–1000.

to fog may be as low as 0.05 for fresh film but increases with the storage time of the film, particularly if the temperature of the storage facility is not below room temperature. An inherent-plus-fog optical density greater than 0.2 is considered excessive.[3]

Processing X-Ray Film

Properties of x-ray film that affect the quality of a radiographic image include contrast, speed, and the ability of the film to record fine detail. Any two of these properties may be improved at the expense of the third, and a particular type of film is chosen only after a compromise has been reached among the various properties.

The properties of x-ray film are affected to some extent by the conditions encountered during processing. The speed of an x-ray film increases initially with development time. With a long development time, however, film speed reaches a plateau and even may decrease (Margin Figure 13-9). Fogging of an x-ray film increases with time spent in the developing solution. The average gradient of a film increases initially with development time and then decreases. The development time recommended by the supplier of a particular type of x-ray film is selected to provide acceptable film contrast and speed with minimum fogging of the film.

The temperature of the developing solution also affects film speed, fog, and contrast (Margin Figure 13-10). For films processed manually, a temperature of 68°F is usually recommended for the developing solution, and the development time is chosen to furnish satisfactory contrast and speed with minimum fogging. With automatic, high-speed processors, the temperature of the developing solution is elevated greatly (85°F to 95°F), and the development time is reduced. The relationship between development time and temperature of the developing solution is illustrated in Table 13-1.

Contrast, speed, and fog are not affected greatly by changes in the fixation time or the temperature of the fixing solution. However, a decrease in film speed and contrast and an increase in film fog may be noticeable if the fixation time is prolonged greatly.

Film Resolution

The *resolution* (detail, definition, sharpness) of radiographic and fluoroscopic images is discussed in detail in Chapter 16. In general, resolution describes the ability of an x-ray film or a fluoroscopic screen to furnish an image that depicts differences in the transmission of radiation by adjacent small regions of a patient. A detector with "good" or "high" resolution provides an image with good resolution of fine structure in the patient. A detector with "poor" or "coarse" resolution (unsharpness) furnishes an image that depicts only relatively large anatomic structures in the patient.

TABLE 13-1 Relationship Between Development Time and Temperature of Developing Solution for Particular X-Ray Film

Developer Temperature (°F)	*Development Time (min)*
60	9
65	6
68	5
70	4 1/4
75	3

Source: Seemann, H. *Physical and Photographic Principles of Medical Radiography*. New York, John Wiley & Sons, 1968, p. 69. Used with permission.

INTENSIFYING SCREENS

For an x-ray beam of diagnostic quality, only about 2% to 6% of the total energy in the beam is absorbed in the emulsion of an x-ray film exposed directly to the beam. The amount of energy absorbed is even smaller for x rays of greater energy. Consequently, direct exposure of film to x rays is a very inefficient utilization of energy available in the x-ray beam. This procedure is used only when images with very fine detail are required. For most radiographic examinations, the x-ray film is sandwiched between *intensifying screens*. The intensifying screens furnish a light image that reflects the variation in exposure across the x-ray beam. This light image is recorded by film that is sensitive to the wavelengths of light emitted by the screen. The mechanism for the radiation-induced fluorescence of an x-ray intensifying screen is similar to that discussed in Chapter 12 for an NaI(Tl) scintillation crystal.

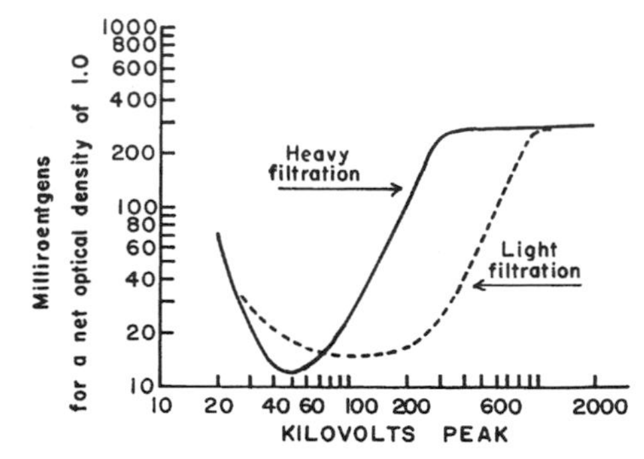

MARGIN FIGURE 13-8
Exposure for an optical density of 1.0 above base density, expressed as a function of the quality of the x-ray beam used to expose the film. The speed of the film is the reciprocal of the exposure in roentgens required for a density of 1.0 above the base density. (From *Sensitometric Properties of X-Ray Films*. Rochester, NY, Radiography Markets Division, Eastman Kodak Company. Used with permission.)

Composition and Properties

The construction of an intensifying screen is diagramed in Margin Figure 13-11. The backing is cardboard, plastic, or, less frequently, metal. The backing is coated with a white pigment that reflects light. The reflecting layer is covered by an active layer that is composed of small granules of fluorescent material embedded in a plastic matrix. The granule diameters range from about 4 to about 8 μm. The thickness of the active layer ranges from perhaps 50 μm for a detail screen to about 300 μm for a fast screen. The active layer is protected by a coating about 0.001 in. thick that is transparent to the light produced in the active layer.

Desirable properties of an x-ray intensifying screen include the following:

1. A high attenuation coefficient for x rays of diagnostic quality.
2. A high efficiency for the conversion of energy in the x-ray beam to light.
3. A high transparency to light released by the fluorescent granules.
4. A low refractive index so that light from the granules will be released from the screen and will not be reflected internally.
5. An insensitivity to handling.
6. An emission spectrum for the radiation-induced fluorescence that matches the spectral sensitivity of the film used.
7. A reasonably short time for fluorescence decay.
8. A minimum loss of light by lateral diffusion through the fluorescent layer. To reduce this loss of light, the fluorescent layer is composed of granules and is not constructed as a single sheet of fluorescent material.
9. Low cost.

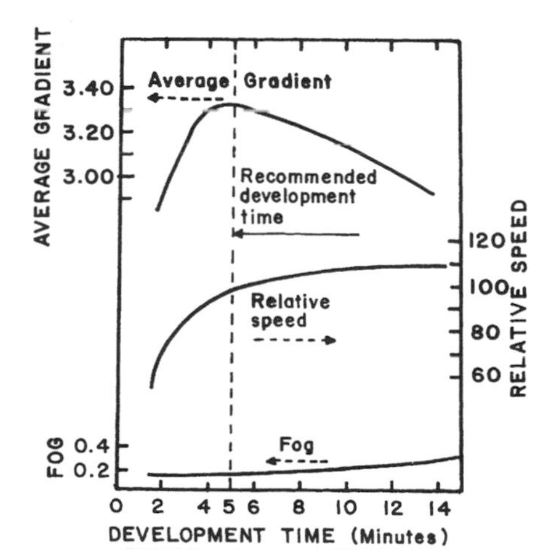

MARGIN FIGURE 13-9
Effect of development time on film speed, film gradient, and fog. (From *Sensitometric Properties of X-Ray Films*. Rochester, NY, Radiography Markets Division, Eastman Kodak Company. Used with permission.)

Crystalline calcium tungstate, with an emission spectrum peaked at 400 nm, was used in most x-ray intensifying screens until the early 1970s. For most calcium tungstate screens, the efficiency is about 20% to 50% for the conversion of energy in the x-ray beam to light.[4,5] Barium lead sulfate, with an emission spectrum peaked in the ultraviolet region (360 nm), is also used as a fluorescent material in intensifying screens. This material is particularly useful for radiography at higher tube voltages.

The K-absorption edge (69.5 keV) of tungsten, the principal absorbing element in calcium tungstate screens, is above the energy of most photons in a diagnostic x-ray beam. Increased absorption of diagnostic x rays would occur if tungsten were replaced by an absorbing element of reduced Z with a K-absorption edge in the range of 35 to 50 keV. Elements with K-absorption edges in this energy range include the elements gadolinium, lanthanum, and yttrium. These elements, sometimes referred to as "rare-earth elements" and complexed in oxysulfide or oxybromide crystals embedded in a plastic matrix, have some advantages over calcium tungstate for use in intensifying screens. Rare-earth screens exhibit not only an increased absorption of diagnostic x rays but also an increased efficiency in the conversion of absorbed x-ray energy to

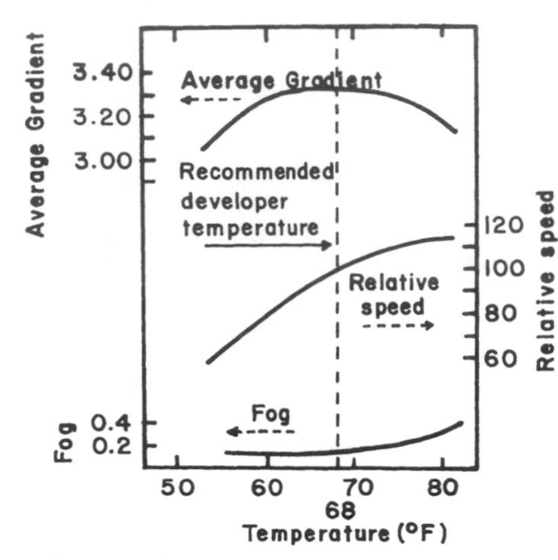

MARGIN FIGURE 13-10
Effect of temperature of the developing solution on film speed, fog, and inherent contrast. (From *Sensitometric Properties of X-Ray Films*. Rochester, NY, Radiography Markets Division, Eastman Kodak Company. Used with permission.)

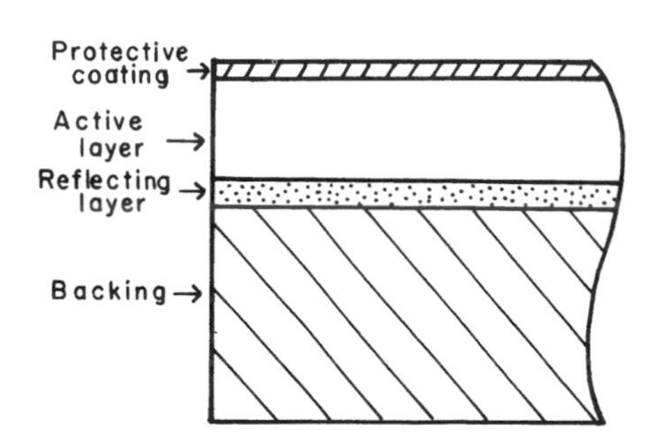

MARGIN FIGURE 13-11
Construction of a typical intensifying screen.

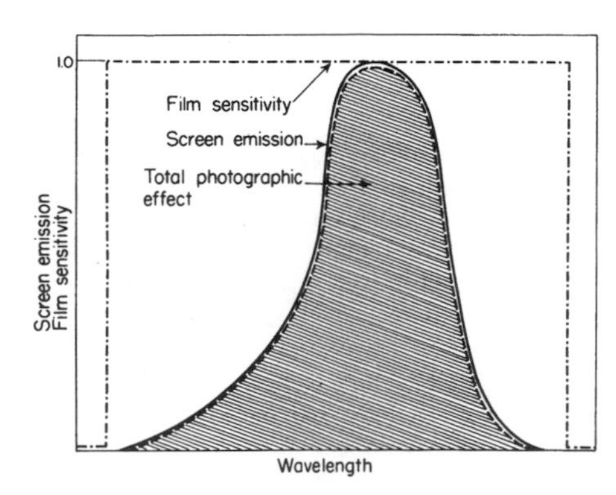

MARGIN FIGURE 13-12
Total photographic effect obtained by combining an intensifying screen and a photographic film with a wavelength sensitivity that extends above and below the emission spectrum for the screen.

light. Hence rare-earth screens are faster than their calcium tungstate counterparts. This increased speed facilitates one or more of the following changes in radiographic technique:

1. Reduced exposure time and decreased motion unsharpness in the image.
2. Reduced tube current allowing more frequent use of small focal spots.
3. Reduced tube voltage and improved contrast in the image.
4. Reduced production of heat in the x-ray tube.
5. Reduced patient exposure.

Some rare earth screens emit blue light and can be used with conventional x-ray film; others emit yellow-green light and must be used with special yellow-green-sensitive film.[6]

The efficiency with which energy in an x-ray beam is converted to light increases with the thickness of the active (fluorescent) layer. Hence the efficiency of energy conversion is greater for a screen with a thick active layer (i.e., a fast screen) than for a screen of medium speed (a par-speed screen) or for a slow screen (a detail screen). However, the resolution of the radiographic image decreases as the thickness of the active layer increases. For example, for typical calcium tungstate screens the maximum resolution is 6 to 9 lines per millimeter for a fast (thick active layer) screen. Using the same screen material, resolution increases to 8 to 10 lines per millimeter for a par-speed screen (thinner), and 10 to 15 lines per millimeter for a detail screen (thinnest) when measured under laboratory conditions. The resolution obtained clinically with these screens is usually considerably less.

The wavelength of light emitted by an intensifying screen should correspond closely with the spectral sensitivity of the film used. Usually, the sensitivity of the film extends to wavelengths both above and below those emitted by the screen (Margin Figure 13-12). The total photographic effect is decreased if the spectral sensitivity of the film is matched too closely to the emission spectrum for the screen (Margin Figure 13-13).

The *intensification factor* of a particular screen-film combination is the ratio of exposures required to produce an optical density of 1.0 without and with the screen in position.

$$\text{Intensification factor} = \frac{\begin{array}{c}\text{Exposure required to produce}\\ OD\ 1.0 \text{ without screen}\end{array}}{\begin{array}{c}\text{Exposure required to produce}\\ OD\ 1.0 \text{ with screen}\end{array}} \qquad \textbf{(13-5)}$$

The intensification factor for a particular screen-film combination varies greatly with the energy of the x-rays used for the radiation exposure, but a range of 50 to 100 is common. Therefore patient exposure is reduced substantially through the use of screens.

Usually, a double-emulsion film is sandwiched in a cassette between two intensifying screens. In some cassettes, the fluorescent layer behind the film is thicker than the layer in front. The resolution of the radiographic image is severely reduced if both screens are not in firm contact with the entire surface of the film.

The major advantage of x-ray intensifying screens is a reduction of the exposure required to form an image of acceptable quality. The radiation exposure of patients is similarly reduced. Because exposure depends linearly upon the product of tube current (milliamperes [mA]) and exposure time (seconds [s]), any reduction of either current or time that leads to a lower product (mA·s) may be used. Shorter exposures (reduced time) may lead to improved resolution by reducing voluntary and involuntary patient motion. Thus, the use of faster screens is generally associated with shorter exposure times.

To summarize factors that describe the speed of fluorescent screens, we may compare conventional calcium tungstate screens with rare-earth screens. Absorption of x-ray photons in calcium tungstate screens varies from 20% to 40% (for par-speed to high-speed screens), while the absorption is approximately 40% to 60% for rare-earth screens. The percentage of absorbed energy converted to light (the conversion efficiency) is approximately 5% for calcium tungstate and 20% for the rare earths. Therefore, rare-earth screens not only absorb more x-ray photons but also convert more of the absorbed energy into visible light.

Example 13-5

Compare the number of visible light photons emitted by a calcium tungstate screen with the number emitted by a gadolinium oxysulfide screen. Assume that 100 x-ray photons, each having an energy of 30 keV, strike the screen. The following data are given:

	Calcium Tungstate	Gadolinium Oxysulfide
Absorption (%)	40	60
Conversion (%)	5	20
Spectral emission peak (nm)	420	550

The energy per visible photon emitted (here we make the simplifying assumption that all photons are emitted at the spectral peak energy) is

$$\text{keV} = \frac{1.24}{\lambda(\text{nm})} = \frac{1.24}{420} = 3 \times 10^{-3}\ \text{keV} = 3\ \text{eV for calcium tungstate}$$

$$= \frac{1.24}{550}$$

$$= 2.25 \times 10^{-3}\ \text{keV} = 2\ \text{eV for gadolinium oxysulfide}$$

The energy converted to visible light photons emitted from each screen is

$$\text{Energy emitted} = (100)(30 \times 10^{3}\ \text{eV})(0.40)(0.05)$$

$$= 6 \times 10^{4}\ \text{eV for calcium tungstate}$$

$$= (100)(30 \times 10^{3}\ \text{eV})(0.60)(0.20)$$

$$= 36 \times 10^{4}\ \text{eV for gadolinium oxysulfide}$$

The number of visible photons emitted from each screen is

$$\text{No. of photons emitted } \frac{6 \times 10^{4}\ \text{eV}}{3\ \text{eV/photon}} = 2 \times 10^{4}\ \text{photons for calcium tungstate}$$

$$\frac{36 \times 10^{4}\ \text{eV}}{2\ \text{eV/photon}} = 18 \times 10^{4}\ \text{photons for gadolinium oxysulfide}$$

Thus the lower energy per visible photon and the higher absorption and conversion efficiencies all act to produce more visible photons per incident photon for the gadolinium oxysulfide screen than for the calcium tungstate screen.

A complete evaluation of screen performance requires analysis of not only speed but also detail and noise (quantum mottle, discussed in Chapter 16). The advantages of higher speed screens (i.e., reduced motion unsharpness, reduced patient dose) may be offset in some clinical situations by the disadvantages of increased noise or loss of detail. Shown in Margin Figure 13-15 are radiographs obtained with and without a fast calcium tungstate screen. The film exposed without a screen required an exposure of 125 mA·s. An exposure of 7 mA·s was required for the film exposed with the intensifying screen.

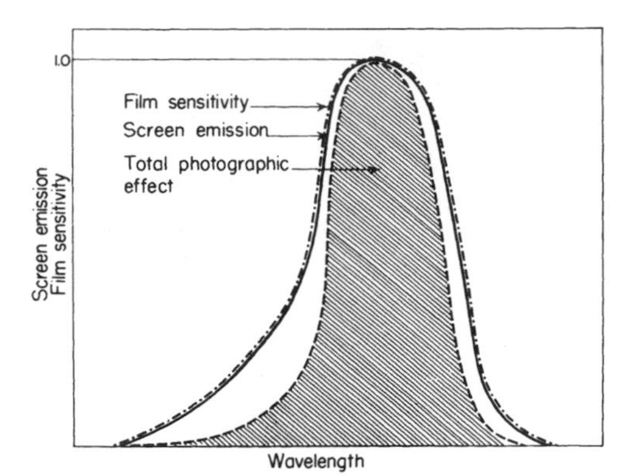

MARGIN FIGURE 13-13
The total photographic effect is reduced if the spectral sensitivity of the film is matched too closely with the emission spectrum of the intensifying screen.

Industrial radiography is used to determine the structural integrity of steel and other materials. When thicknesses of steel greater than 0.25 inch are radiographed, voltages above 130 kVp are used and the intensifying screen is usually constructed from lead foil. The lead converts the photon energy to kinetic energy of electrons, which go on to deposit energy in the film. For photon energies below 2 MeV, the lead foil need be only 0.004 to 0.006 inch thick.

Neutron Radiography:
Radiographs obtained by exposing objects to slow neutrons (kinetic energy = 0.025 eV) are remarkably dissimilar from those obtained by exposing the same objects to x rays (Margin Figure 13-14). Slow neutrons are attenuated primarily during interactions with low-Z nuclei. In tissue, neutrons are attenuated more rapidly in muscle and fat than in bone because the concentration of low-Z nuclei (primarily hydrogen) is greater in muscle and fat. Neutron radiographs display excellent contrast between air and soft tissue and between soft tissue and structures that contain a contrast agent (e.g., Gd_2O_3) for slow neutrons. The contrast is poor between fat and muscle but may be improved by exposing the specimen to heavy water (water with an increased concentration of deuterium). Bone transmits slow neutrons readily, and neutron radiographs of regions containing bone furnish excellent detail of soft tissue without the distracting image of surrounding bone.

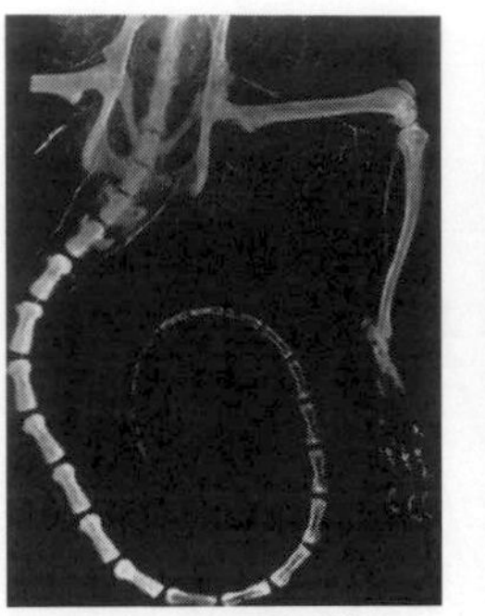
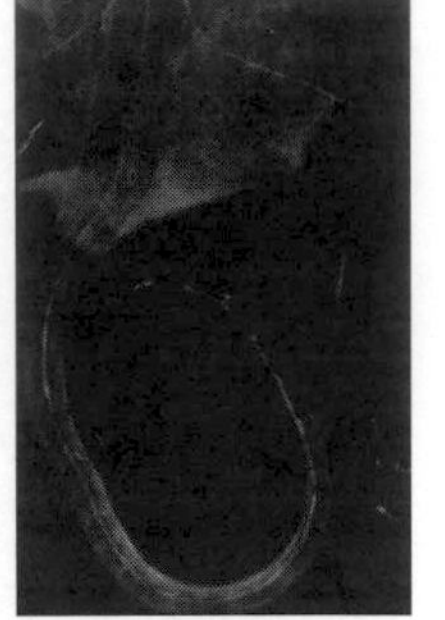

(a) (b)

MARGIN FIGURE 13-14
Arteriograms of a rat leg, paw, and tail obtained by using Gd_2O_3 as a contrast medium.
A: Radiograph obtained with 20-kVp x-rays.
B: Slow neutron radiograph exposed with an antiscatter grid. (From Brown, M., Allen, J., and Parks, P. *Biomed. Sci. Instrum.* 1969; **6**. p. 100 & p. 102. Used with permission.)

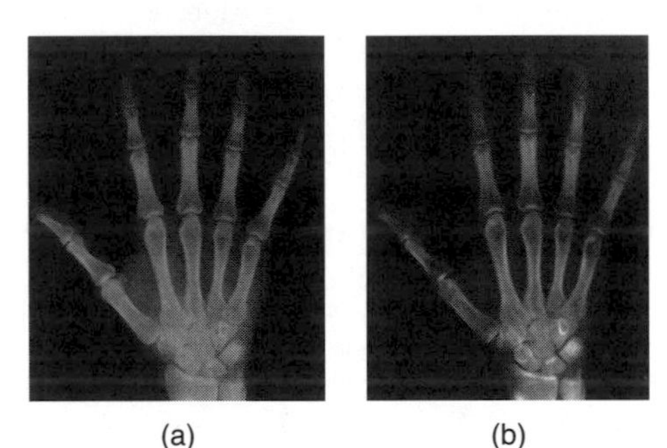

(a) (b)

MARGIN FIGURE 13-15
Effect of intensifying screens on exposure time and image detail. **A:** Radiograph exposed without a screen and requiring 125 mA·s. **B:** Radiograph exposed with a fast calcium tungstate screen and requiring 7 ma·s.

The first American photographic plate (before screens were used) was manufactured in 1896 by John Carbutt and Arthur W. Goodspeed of Philadelphia. Called the Roentgen x-ray plate, it used a thicker and heavier silver emulsion than was used for ordinary photography and thereby permitted a dramatic reduction in time of exposure. The time required to obtain an image of extremities was reduced from an hour to "only" 20 minutes.

The influence of scattered radiation on the radiographic image is reduced somewhat when intensifying screens are used because scattered x rays with reduced energy are absorbed preferentially in the upper portions of the phosphor layer of the screen. Many of the light photons produced by these scattered x rays are absorbed before they reach the film.

A few precautions are necessary to prevent damage to an intensifying screen. For example, cassettes with intensifying screens should be loaded carefully with film to prevent scratching of the screen surface and accumulation of electrical charge, which can produce images of static discharges on the radiograph. Moisture on the screen may cause the film to adhere to the screen. When the film and screen are separated, part of the screen may be removed. A screen may be stained permanently by liquids such as developing solution, and care must be taken to ensure that liquids are not splashed onto the screen. The surface of the screen must be kept clean and free from lint and other particulate matter. Although screens may be washed with a soap solution or a weak wetting agent, organic solvents should be avoided because the active layer may be softened or otherwise damaged by these materials.

RADIOGRAPHIC GRIDS

Information is transmitted to an x-ray film by unattenuated primary radiation emerging from a patient. Radiation scattered within the patient and impinging on the film tends to conceal this information by producing a general photographic fog on the film. The amount of radiation scattered to a film increases with the volume of tissue exposed to the x-ray beam. Hence a significant reduction in scattered radiation may be achieved by confining the x-ray beam to just the region of interest within the patient. In fact, proper collimation of an x-ray beam is essential to the production of radiographs of highest quality.

Much of the scattered radiation that would reduce the quality of the radiographic image may be removed by a radiographic grid between the patient and the x-ray film or screen-film combination. A radiographic grid is composed of strips of a dense, high-Z material separated by a material that is relatively transparent to x rays. The first radiographic grid was designed by Bucky in 1913. The ability of a grid to remove scattered radiation is depicted in Margin Figure 13-15.

Construction of Grids

Ideally, the thickness of the strips in a radiographic grid should be reduced until images of the strips are not visible in the radiographic image. Additionally, the strips should be completely opaque to scattered radiation and should not release characteristic x-rays ("grid fluorescence") as scattered x-ray photons are absorbed. These requirements are satisfied rather well by lead foil about 0.05 mm thick, and this material is used for the strips in most radiographic grids.

The interspace material between the grid strips may be aluminum, fiber, or plastic. Although fiber and plastic transmit primary photons with almost no attenuation, grids with aluminum interspaces are sturdier. Also, grids with aluminum interspaces may furnish a slightly greater reduction of the scattered radiation because scattered x-ray photons that escape the grid strips may be absorbed in the aluminum interspaces.[7]

Types of Grids

Radiographic grids are available commercially with "parallel" or "focused" grid strips in either linear or crossed grid configurations (Margin Figures 13-17 and 13-18). When a focused grid is positioned at the correct distance from the target of an x-ray tube, lines through the grid strips are directed toward a point or "focus" on the target. The focus-grid distance approaches infinity for a grid with parallel strips. With a parallel grid positioned at a finite distance from an x-ray tube, more primary x rays are attenuated along the edge of the radiograph than at the center. Consequently, the optical density decreases slightly from the center to the edge of a radiograph exposed with a parallel grid. The uniformity of optical density is improved in a radiograph exposed with a focused grid, provided that the grid is positioned correctly.

A linear grid is constructed with all parallel or focused grid strips in line (Margin Figure 13-18). A crossed grid is made by placing one linear grid on top of another, with the strips in one grid perpendicular to those in the other. In most situations, a crossed grid removes more scattered radiation than does a linear grid with the same grid ratio (see below) because a linear grid does not absorb photons scattered parallel to the grid strips. However, a linear grid is easier to use in situations where proper alignment of the x-ray tube, grid, and film cassette is difficult.

Describing Radiographic Grids

A grid usually is described in terms of its *grid ratio* (Margin Figure 13-19):

$$\text{Grid ratio} = \frac{\text{Height of grid strips}}{\text{Distance between grid strips}} \qquad (13\text{-}6)$$

$$r = \frac{h}{d}$$

The grid ratio of a crossed grid is $r_1 + r_2$, where r_1 and r_2 are the grid ratios of the linear grids used to form the crossed grid. Radiographic grids are available commercially with grid ratios as high as 16. However, grids with ratios of 8 to 12 are used most frequently because the removal of scattered radiation is increased only slightly with grids with a higher ratio. Also, grids with a high grid ratio are difficult to align and require a greater exposure of the patient to radiation. The effectiveness of a radiographic grid for removing scattered radiation is illustrated in Margin Figure 13-20 as a function of grid ratio.

The improvement in radiographic contrast provided by grids with different grid ratios is depicted in Table 13-2. From data in this table, it is apparent that (1) all grids improve radiographic contrast significantly, (2) the effectiveness of a grid for improving radiographic contrast increases with the grid ratio, (3) the increase in radiographic contrast provided by a grid is less for an x-ray beam generated at a higher voltage, (4) a crossed grid removes scattered radiation more effectively than a linear grid with an equal grid ratio, and (5) the radiation exposure of the patient increases with the grid ratio. The exposure must be increased because the film is exposed to fewer scattered x-ray photons. Also, more primary x-ray photons are absorbed by the grid strips.

TABLE 13-2 Relative Increase in Radiographic Contrast and Radiation Exposure for Grids with Different Ratios[a]

Type of Grid and Grid Ratio	*Improvement in Contrast*			*Increase in Exposure*		
	70 kVp	*95 kVp*	*120 kVp*	*70 kVp*	*95 kVp*	*120 kVp*
None	1	1	1	1	1	1
5 linear	3.5	2.5	2	3	3	3
8 linear	4.75	3.25	2.5	3.5	3.75	4
12 linear	5.25	3.75	3	4	4.25	5
16 linear	5.75	4	3.25	4.5	5	6
5 cross	5.75	3.5	2.75	4.5	5	5.5
8 cross	6.75	4.25	3.25	5	6	7

[a]The thickness of the grid strips and interspaces were identical for all grids.

Source: Characteristics and Applications of X-Ray Grids. Cincinnati, Liebel-Flarsheim Co., 1968. Used with permission.

Michael Pupin of Columbia University was the first person to use an intensifying screen in medical radiography. In early 1896 a well-known lawyer had sustained a shotgun blast to the hand, and he was sent to Pupin by a surgeon to obtain a radiograph that would aid in localization and removal of the shotgun pellets. The first two attempts at radiography were unsuccessful because the patient could not endure the hour-long process of exposure. Fortunately, Pupin's friend, Thomas Edison, had recently sent him a calcium tungstate fluoroscopic screen. Pupin placed the fluoroscopic screen upon the photographic plate and obtained a clear image of the pellets and the bones of the hand with an exposure that lasted only a few seconds. The lawyer offered to establish a fellowship for Pupin at a local establishment that would entitle him to two free drinks per day for life. (*Source:* Pupin, M. *From Immigrant to Inventor*. New York, Scribners, 1923.)

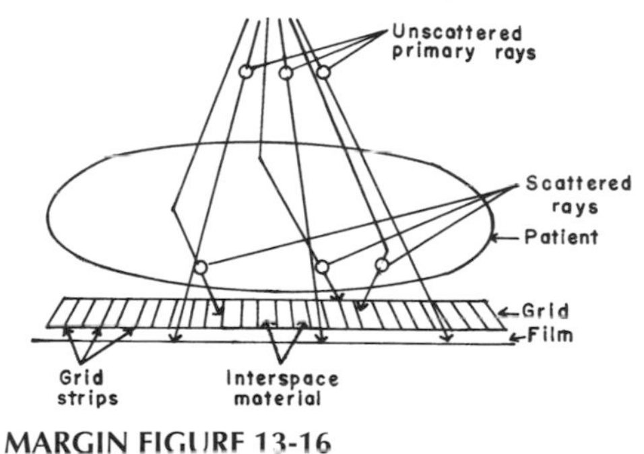

MARGIN FIGURE 13-16
A radiographic grid is used to remove scattered radiation emerging from the patient. Most primary photons are transmitted through the grid without attenuation.

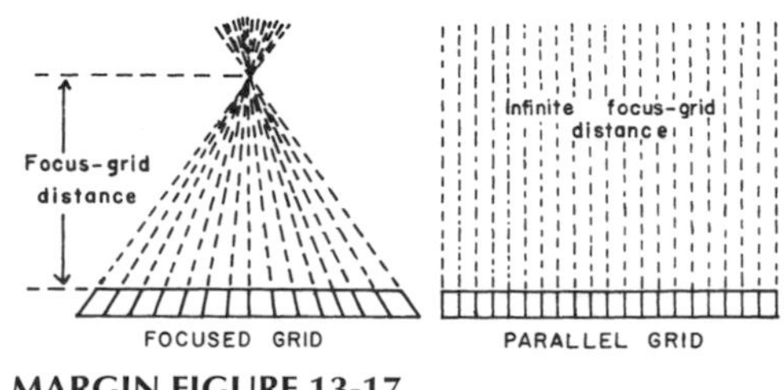

MARGIN FIGURE 13-17
Focused and parallel grids.

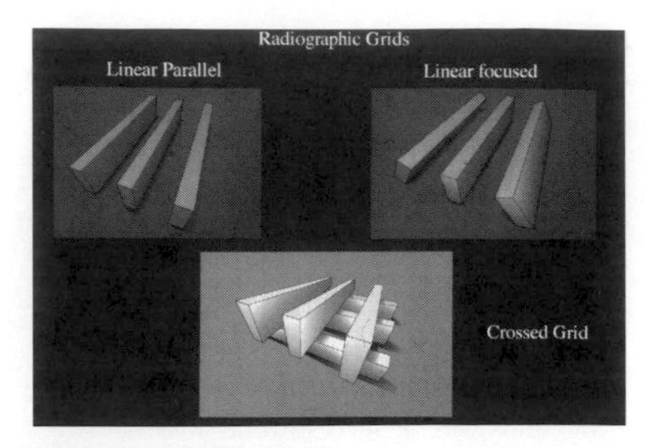

MARGIN FIGURE 13-18
Types of focused and parallel grids.

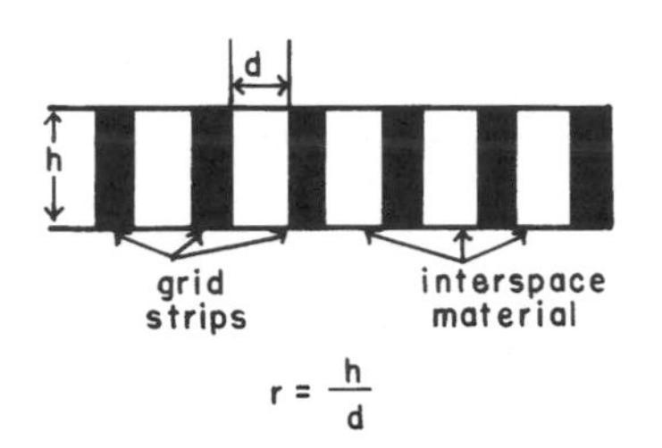

MARGIN FIGURE 13-19
The grid ratio r is the height h of the grid strips divided by the distance between the strips.

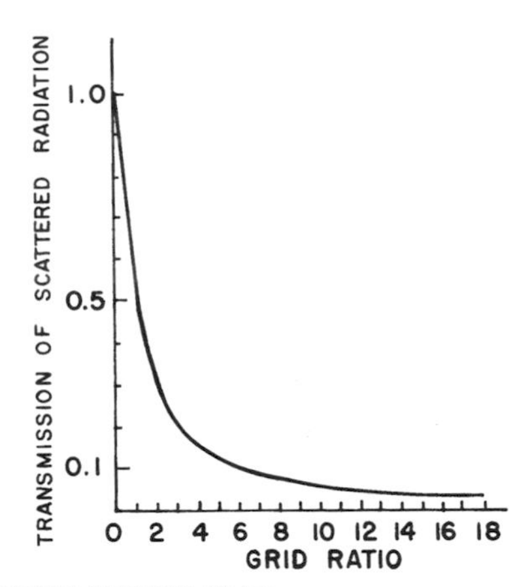

MARGIN FIGURE 13-20
The effectiveness of a radiographic grid for removing scattered radiation is pronounced for small grid ratios but increases only gradually with grid ratios above 8. (Courtesy of S. Ledin and the Elema Shönander Co.)

In 1915, G. Bucky developed the moving grid, which, after it was perfected by H. E. Potter in 1919, became known as the Potter–Bucky diaphragm.

A radiographic grid is not described explicitly by the grid ratio alone because the grid ratio may be increased by either increasing the height of the grid strips or reducing the width of the interspaces. Consequently, the number of *grid strips per centimeter* (or per inch) usually is stated with the grid ratio. Grids with many strips ("lines") per inch produce shadows in the radiographic image that may be less distracting than those produced by thicker strips in a grid with fewer strips per inch. Grids with as many as 110 strips per inch are available commercially. The *lead content* of the grid in units of grams per square centimeter sometimes is stated with the grid ratio and the number of strips per inch.

The *contrast improvement factor* for a grid is the quotient of the maximum radiographic contrast obtainable with the grid divided by the maximum contrast obtainable with the grid removed. The contrast improvement factor may be used to compare the effectiveness of different grids for removing scattered radiation. The contrast improvement factor for a particular grid varies with the thickness of the patient and with the cross-sectional area and energy of the x-ray beam. Usually the contrast improvement factor is measured with a water phantom 20 cm thick irradiated by an x-ray beam generated at 100 kVp.[9]

The *selectivity* of a grid is the ratio of primary to scattered radiation transmitted by the grid. The efficiency of a grid for removing scattered radiation is described occasionally as *grid cleanup*. Grids may be described as "heavy" or "light," depending upon their lead content. One popular but rather ambiguous description of grid effectiveness is the *Bucky factor*, defined as the exposure to the film without the grid divided by the exposure to the film with the grid in place and exposed to a wide x-ray field emerging from a thick patient.

Moving Grids

The image of grid strips in a radiograph may be distracting to the observer. Also, grid strip shadows sometimes interfere with the identification of small structures such as blood vessels and trabeculae of bone. In the early 1900s, Bucky and Potter developed the moving grid, referred to as the *Potter–Bucky diaphragm*, which removes the distracting image of grid strips by blurring their image across the film. Early Potter–Bucky diaphragms moved in one direction only. Modern moving grids use a reciprocating motion, with the grid making several transits back and forth during an exposure. The linear distance over which the grid moves is small (1 to 5 cm) and permits the use of a focused grid. The motion of the grid must not be parallel to the grid strips and must be rapid enough to move the image of a number of strips across each location in the film during exposure. The motion of the grid must be adjusted to prevent synchronization between the position of the grid strips and the rate of pulsation of the x-ray beam. The direction of motion of the grid changes very rapidly at the limits of grid travel, and the "dwell time" of the grid at these limits is insignificant.

Orthogonal crossed grids are not often used as moving grids. Rhombic crossed grids used in "super-speed recipromatic diaphragms" provide excellent removal of scattered radiation with no image of the grid strips in the radiograph. The travel of the grid is very short and does not cause significant off-center cutoff (see below). Most stationary radiographic tables contain a recipromatic Potter–Bucky diaphragm. However, the development of grids with many strips per inch has reduced the need for moving grids. The maintenance cost for reciprocating grids and their poor efficiency for removing the image of grid strips during short exposure times enhance the attractiveness of stationary grids that contain many strips per inch.

The choice of a radiographic grid for a particular examination depends on factors such as the amount of primary and scattered radiation emerging from the patient, the energy of x rays in the x-ray beam, and the variety of radiographic techniques provided by the x-ray generator.[8]

Grid Cutoff

The expression "grid cutoff" is used to describe the loss of primary radiation caused by improper alignment of a radiographic grid. With a parallel grid, cutoff occurs near the edges of a large field because grid strips intercept many primary photons along the edges of the x-ray beam. The width of the shadow of grid strips in a parallel grid increases with distance from the center of the grid (Margin Figure 13-21).

The use of a focused grid at an incorrect target–grid distance also causes grid cutoff. This effect is termed *axial decentering* or *off-distance cutoff* and is depicted in Margin Figure 13-22. The optical density of a radiograph with off-distance cutoff decreases from the center of the radiograph outward. The variation in optical density increases with displacement of the grid from the correct target–grid distance. However, the effect is not objectionable until the displacement exceeds the target–grid distance limits established for the grid. The limits for the target–grid distance are narrow for grids with a high grid ratio and wider for grids with a smaller ratio. The effects of off-distance cutoff are more severe when the target–grid distance is shorter than that recommended and less severe when the target–grid distance is greater.

Lateral decentering or *off-center cutoff* occurs when x rays parallel to the strips of a focused grid converge at a location that is displaced laterally from the target of the x-ray tube (Margin Figure 13-23). *Off-level cutoff* results from tilting of the grid (Margin Figure 13-24). Both off-center cutoff and off-level cutoff cause an overall reduction in optical density across the radiograph. The importance of correct alignment of the x-ray tube, grid, and film increases with the grid ratio.

To x-ray photons traveling from an x-ray tube through the patient to an image receptor, a moving grid does not appear to be moving very much at all. Therefore, the function of the grid, to allow photons that are traveling in a straight line from the focal spot to reach the image receptor, is preserved, even though the final image of the grid is blurred. A typical moving grid travels at a maximum speed of 15 cm/sec. It takes x-ray photons only 3 nanoseconds to travel the 100-cm distance from the x-ray tube to the image receptor. During this time, a grid moving at 15 cm/sec will have traveled only 0.5 nanometers—the width of 5 atoms.

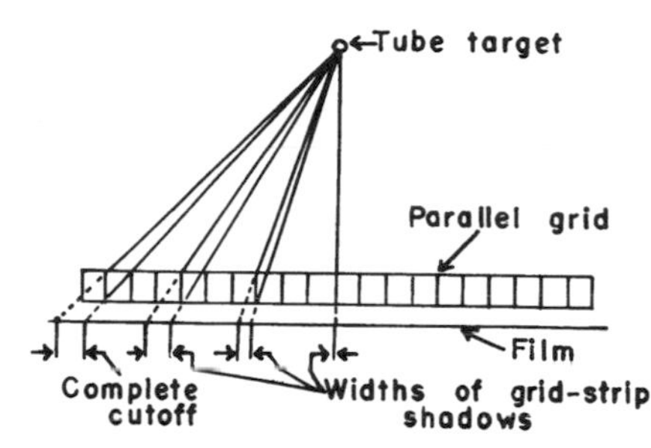

MARGIN FIGURE 13-21
Cutoff of primary x-ray photons with a parallel grid.

Air Gap

The amount of scattered radiation reaching an x-ray film or film-screen combination may be reduced somewhat by increasing the distance between the patient and the film. This procedure is referred to as an air grid or air gap. The use of an air gap is accompanied by magnification of the image and an increase in geometric unsharpness (see Chapter 16).

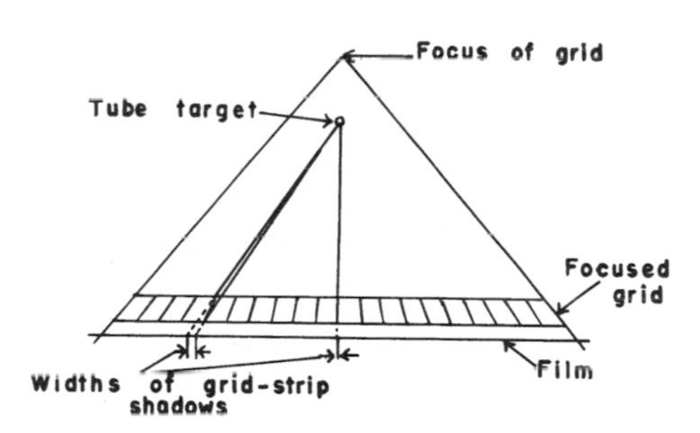

MARGIN FIGURE 13-22
Off-distance cutoff with a focused grid.

Moving-Slit Radiography

More efficient techniques for removal of scattered radiation have been sought for many years. One technique that has been proposed periodically is moving-slit radiography, in which one or more slits in an otherwise x-ray-opaque shield move above the patient in synchrony with an equal number of slits in a shield between the patient and the film (Margin Figure 13-25). The long, narrow x-ray beams emerging through the upper slits are scanned across the patient and transmitted to the film through the lower slits below the patient. Radiation scattered by the patient is intercepted by the opaque shield below the patient and does not reach the film. The principal disadvantage of moving-slit radiography is the possibility of image distortion due to motion during the time required for the x-ray beam(s) to scan across the patient. This disadvantage is less cumbersome with newer, "fast" x-ray systems using high-milliamperage generators and rare-earth intensifying screens.[10]

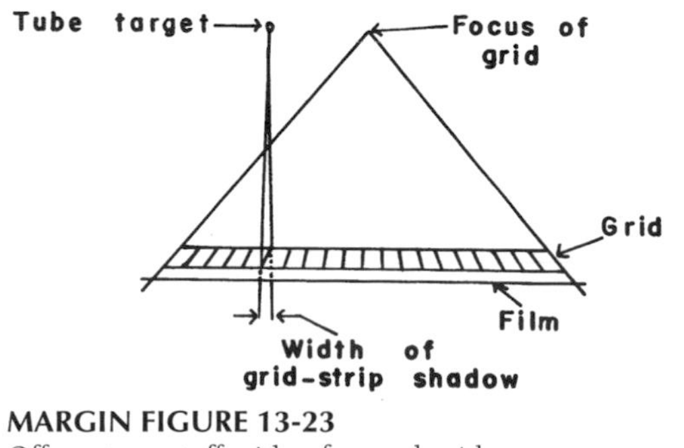

MARGIN FIGURE 13-23
Off-center cutoff with a focused grid.

MAGNIFICATION RADIOGRAPHY

If x rays are assumed to originate from a single point on the target of the x-ray tube, then the magnification of a radiographic image is

$$\text{Magnification} = \frac{\text{Image size}}{\text{Object size}} = \frac{\text{Target} - \text{film distance}}{\text{Target} - \text{object distance}}$$

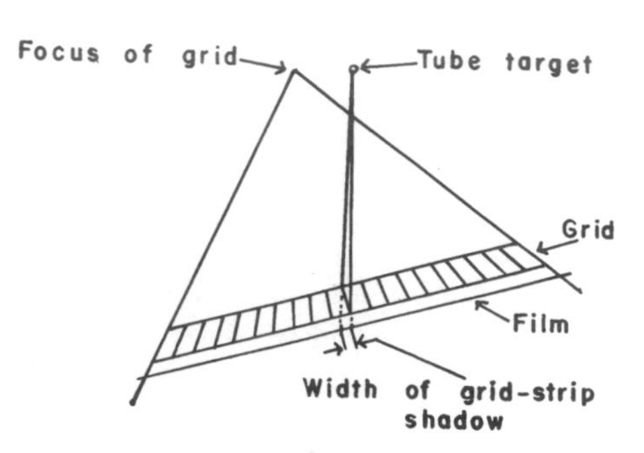

MARGIN FIGURE 13-24
Off-level cutoff with a focused grid.

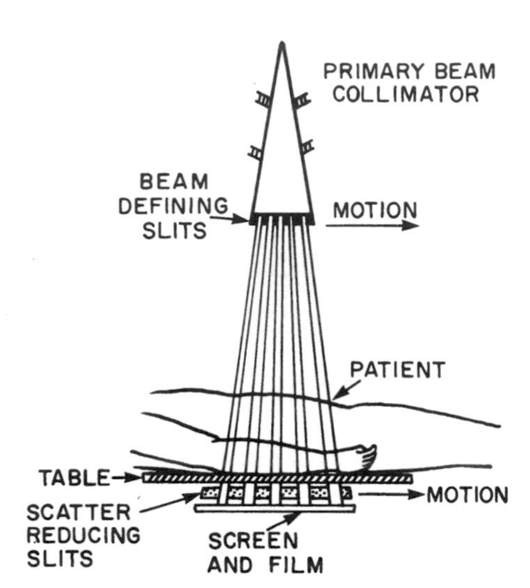

MARGIN FIGURE 13-25
Moving-slit approach to rejection of scattered radiation. (From Hendee, W., Chaney, E., and Rossi, R. *Radiologic Physics, Equipment and Quality Control*. St. Louis, Mosby-Year Book, 1977. Used with permission.)

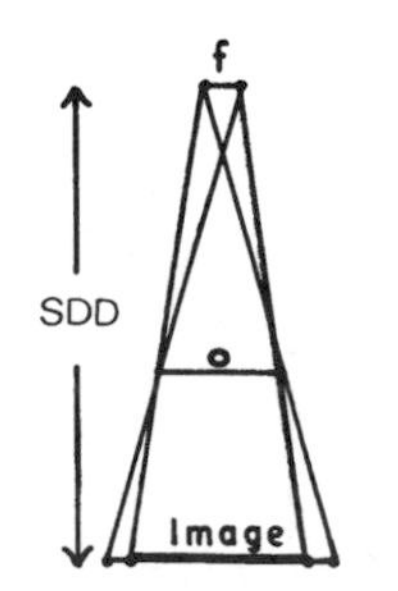

MARGIN FIGURE 13-26
Principle of enlargement radiography: f is the apparent focal spot of the x-ray tube, o is the object size, and SDD represents the distance between the image receptor and the target of the x-ray tube.

If the distance between the target of the x-ray tube and the x-ray film is constant, then the ratio of image to object size may be increased by moving the object toward the x-ray tube. This method of image enlargement is referred to as *object-shift enlargement* (Margin Figure 13-26). If the distance between the target and object is constant, then the ratio of image size to object size may be increased by moving the film farther from the object. This technique of image enlargement is referred to as *image-shift enlargement*.

The amount of enlargement possible without significant loss of image detail should be determined when radiologic images are magnified. With the image receptor close to the patient, the visibility of image detail is usually determined primarily by the unsharpness of the intensifying screen. As the distance between the screen and the patient is increased by object-shift or image-shift enlargement, no change occurs in the contribution of screen unsharpness to the visibility of image detail. However, the contribution of geometric unsharpness increases steadily with increasing distance between the patient and image receptor. At some patient–receptor distance, geometric unsharpness equals screen unsharpness. Beyond this distance, the visibility of image detail deteriorates steadily as geometric unsharpness increasingly dominates image unsharpness. As a general rule, the image should not be magnified beyond the point at which geometric and screen unsharpness are about equal.

An image also may be magnified by optical enlargement of a processed radiograph. This technique may enhance the visibility of detail but offers no improvement in intrinsic resolution of the image. For most procedures, magnification radiographs obtained directly by geometric enlargement of the image are preferred over those obtained by optical enlargement.[11,12]

DIGITAL RADIOGRAPHY

A number of techniques exist to acquire radiographic images digitally or to convert images acquired in analog fashion into digital format. Of course, it may be argued that even film screen systems are not truly analog because the grain size in film and crystal size in fluorescent screens are examples of the fundamental discreetness of all image receptors and indeed all matter. However, structures such as film grain do not limit current imaging technology, whereas the resolution of currently available display systems such as cathode-ray tubes and the limitations of transmission time and data storage for the massive amounts of information present in images are limitations. Furthermore, the discrete components of "analog" image acquisition devices (e.g., film grains) cannot be manipulated or adjusted individually. A number of interesting techniques for accessing discrete components of images have recently moved from research into clinical usage.

Discrete Digital Detectors

The use of an array of detectors is routine for ultrasound and computed tomography (CT). The exquisite information density of film obviated the need for such devices in radiography until imaging demands that required digital techniques were identified. These demands include increased dynamic range, simultaneous viewing at multiple locations, and the use of digital image processing and enhancement. These newer demands have recently begun to be matched by technologic capabilities. A number of discrete detectors that produce electronic signals that are easily digitized have been used, including fluorescent crystals such as sodium iodide and bismuth germanate, photodiodes, and various semiconductor devices. The idea of individual detectors or groups of detectors intercepting narrowly collimated x-ray beams for the purpose of reducing the influence of scattered photons on image formation has been exploited in the form of line (slit) scanning and point scanning systems. The problem of dynamic range has been addressed in the technique of scanning equalization radiography.[13,14] This system incorporates a feedback mechanism in which the magnitude of signals from the detector causes automatic adjustment of the output of the x-ray tube. The adjustment of tube output in a scanning equalization radiography system reduces the range of exposure rates over which the detector must respond.

Storage Phosphor

Another approach to digital imaging is acquisition of the image on a continuous medium that is specially designed to be digitized. In storage phosphor technology (also known as computed radiography [CR]) the image is acquired on a plate containing crystals of a photostimulatable phosphor. A material such as barium fluorobromide (BaFBr) is capable of storing the energy from an x-ray exposure. When exposed to a strong light source of the appropriate wavelength, the photostimulatable phosphor re-emits the energy as visible light that can be detected by a photomultiplier tube. Thus the storage phosphor plate records a latent image that may be read out some time after x-ray exposure. The readout may be accomplished with a well-collimated intense light source, such as a helium–neon laser beam, so that the size of the stimulated region of the plate remains small to yield good resolution. The electrical signals from the photomultiplier or other light-sensing device are digitized with an analog-to-digital converter. Once the image is stored in digital form, it may be viewed on a high-resolution monitor or printed out on film.

One of the advantages of a stimulatable phosphor plate system over conventional film is an improvement in dynamic range. Radiographic film operates over possible exposures ranging from the exposure required to reach the shoulder of the characteristic curve to that required to rise above the toe. This range typically causes exposure differences of a factor of approximately 100. The dynamic range of a storage phosphor is on the order of 10,000. Thus the storage phosphor has greater latitude (i.e., is more "forgiving") if an incorrect exposure is used. To preserve the information contained within such a broad dynamic range, the analog-to-digital converter must have a sufficient bit depth (see Chapter 10). A typical storage phosphor readout system is capable of digitizing a 2000 × 2000 pixel matrix or greater, with each pixel taking one of $2^{10} = 1024$ possible values. The images may then be processed according to a number of possible gray-scale mapping functions or simply windowed and leveled in a fashion similar to that used for magnetic resonance and CT images. Image processing techniques may also be used to enhance the appearance of edges of structures such as tubes in chest radiographs taken in intensive care units.[15]

Some of the advantages of storage phosphor or CR systems over conventional film-screen approaches include the potential for reduction of patient exposure and the virtual elimination of retakes due to improper technique. There is some advantage of the storage phosphor system over other digital radiographic techniques in that the storage phosphor plate simply replaces the screen/film cassette and does not represent a significant change in patient positioning or other imaging procedures.

Example 13-6

Calculate the spatial resolution of a CR system in which a 14 × 17-in. storage phosphor plate is digitized at a matrix size of 2000 × 2510.

$$\text{Resolution element or pixel size} = r_{14} = \frac{14 \text{ in.}}{2000} = 0.007 \text{ in.} = 0.18 \text{ mm}$$

$$= r_{17} = \frac{17 \text{ in.}}{2510} = 0.007 \text{ in.} = 0.18 \text{ mm}$$

Thus the resolution element size is the same in both the 14- and 17-in. dimensions. The number of line pairs per millimeter that could be displayed equals the number of pixel pairs per millimeter.

$$\frac{\text{Line pairs}}{\text{millimeter}} = \frac{1}{2(r)} = \frac{1}{2(0.18 \text{ mm})} = 2.8 \frac{\text{Line pairs}}{\text{millimeter}}$$

Note that the spatial resolution of the current generation of storage phosphor systems is less than the six to ten line pairs per millimeter available with film-screen systems.

Film Scanning

Radiographic films may be digitized after they are acquired with conventional film/screen systems. In a typical film digitizer, a laser beam is scanned across a film. The pattern of optical densities on the film modulates the transmitted light. A light detector on the opposite side of the film converts the transmitted laser light to an electrical signal that is digitized by an analog-to-digital converter.[16] Spatial resolution of a film scanner is determined by spot size, defined as the size of the laser beam as it strikes the film. Spot sizes down to 50 μm (50×10^{-6} m) are available, but most clinical systems use 100 μm or larger. Because the laser spot must be scanned across the entire area of the image, a decrease in the diameter of the spot size by a factor of 2 requires a scan time that is 4 × longer, if no other adjustments are made in the system. Commercially available laser film scanners have matrix sizes of at least 2000 × 2000 × 10 or 12 bits.

Example 13-7

Find the total number of bytes stored for a film scanning system with a 2048 × 2048 matrix if an 8-bit analog-to-digital converter is used.

$$\begin{aligned}\text{Matrix size} &= 2048 \times 2048 \times 8 \text{ bits}\\ &\simeq 33{,}570{,}000 \text{ bits}\end{aligned}$$

However, 8 bits = 1 byte, and

$$\begin{aligned}\text{Matrix size} &= 33{,}570{,}000 \text{ bits} \div 8 \text{ bits/byte}\\ &\simeq 4{,}200{,}000 \text{ bytes}\\ &\simeq 4.2 \text{ Mbytes}\end{aligned}$$

Each image requires 4 Mbytes.

A film scanner has the obvious advantage over other digital radiographic techniques in that it does not disrupt routine imaging procedures. All procedures (e.g., technique, patient positioning, film development) are the same up to the point of appearance of the developed radiograph. Compared with other digital techniques, disadvantages of a film scanner include its inability to correct for large variations in film density or gray-scale mapping, the need for handling individual sheets of film, and the time required for scanning.

Large-Area Digital Image Receptors for Radiography

Large-area (e.g., 35 cm × 43 cm) digital detectors have been under development for over 30 years, beginning with the first charge-coupled devices (CCDs) developed at Bell Laboratories in 1969.[17] CCDs convert visible light to electrical signals. Digital detection has revolutionized light and video camera technology, and various digital technologies are ubiquitous in photography and video recording today. However, the extreme demands of resolution, low noise, and high dynamic range required in medical radiography present major technological challenges to application of digital detectors in radiography. Over the past two decades, the promise of digital detectors has remained just out of reach in terms of both performance and cost effectiveness. There has been great incentive to continue development, however, because using detectors that can be read electronically in a matter of seconds eliminates the need for trips to a processor (as is required with film and/or a charged-plate reader used for computed radiography). Also, the possibility of rapid image processing—even just window and level manipulation—yields an overall increase in the quality of the final images.

Only recently have advances in the technology of fabrication and manufacture allowed the production of large-area flat panel detectors that equal, or even surpass (at least in theory), the performance of film-screen and computed radiography systems. The most important feature of flat panel detectors in this comparison is the higher quantum efficiency for detection and conversion of x rays into digital signals with acceptably low noise over a large exposure range.[18] The two basic types of large-area flat panel x-ray detection systems, *direct conversion* and *indirect conversion,* are shown in Margin Figure 13-27.

The objective of a digital detector in radiography is to convert the energy of x-ray photons into electrical signals that signify the magnitude of the exposure in different pixels. In direct capture systems, the energy of the x rays is converted to an electrical signal in a single layer of a material called a *photoconductor*. Various materials can be used to construct photoconductors, including amorphous selenium, cadmium zinc telluride, and lead iodide.[19] When x rays strike a photoconductor conversion plate, electron–hole pairs are created. Electrical fields are applied between the front and back surfaces of the photoconductor to force separation of the electron–hole pairs and their transfer to one of the charged surfaces. Each surface is a thin-film transistor (TFT) array, a grid of transistors that can be interrogated electronically ("read out") to determine the amount of electrical charge present above each transistor. The TFT array corresponds to an array of pixels in the final image, with the number of transistors equaling the number of pixels. Flat-panel detectors having pixel sizes on the order of 200 micrometers (2.5 lp/mm resolution) are available for chest imaging, and a new generation of detectors with spatial resolution in the 50- to 100-micrometer (5- to 10-lp/mm) range is becoming available for mammography.

In indirect conversion systems, a scintillation material converts the energy of the x-ray photons to visible light. Scintillation materials used for this purpose include amorphous silicon and cesium iodide. The pattern of visible light produced by the scintillation conversion plate can be read out by a photoconductor fabricated to convert visible light to electrical signals. Photoconductors used in indirect conversion systems include CCDs and materials such as amorphous silicon coupled with a TFT array.

An important feature of the newer generation of flat-panel detectors is their higher detective quantum efficiency (DQE). DQE is a measure of the detection efficiency that includes the effects of noise as well as resolution at various spatial frequencies. The DQE of current flat-panel digital detectors is over 65%,[20] while that of computed radiography storage phosphor systems is approximately 35%,[21] and that of screen-film systems is approximately 25%.[22] Because of the higher DQE, flat-panel detectors are expected to use radiation dose more efficiently, and they could theoretically provide lower noise images over the range of spatial frequencies encountered in medical imaging.

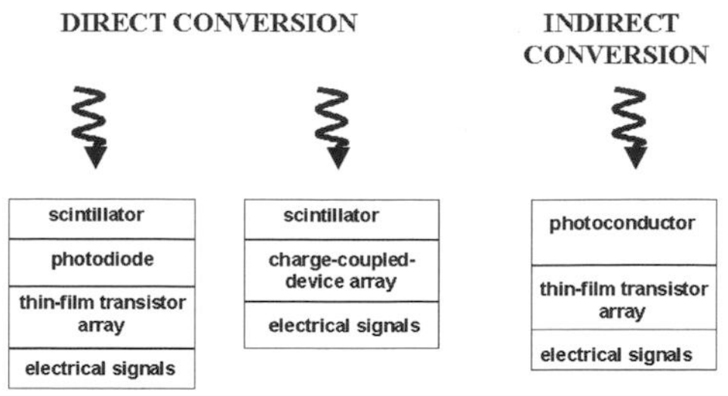

DIRECT CONVERSION | INDIRECT CONVERSION

scintillator / photodiode / thin-film transistor array / electrical signals

scintillator / charge-coupled-device array / electrical signals

photoconductor / thin-film transistor array / electrical signals

MARGIN FIGURE 13-27
Some large-area detectors provide indirect conversion of x-ray energy to electrical charge through intermediate steps involving photodiodes or charge-coupled-devices. Other large-area detectors provide direct conversion of x-ray energy to electrical charge through the use of a photoconductor.

PROBLEMS

*13-1. Adjacent regions of a radiograph have optical densities of 1.0 and 1.5. What is the difference in the transmission of light through the two regions?

*13–2. Optical densities of 1.0 and 1.5 are measured for adjacent regions of a radiograph obtained with x-ray film with a film gamma of 1.0. If the exposure was 40 mR to the region with an optical density of 1.0, what was the exposure to the region with an optical density of 1.5?

13–3. Explain why poor contact between an intensifying screen and an x-ray film reduces the resolution of the radiographic image.

13–4. The effective atomic number is greater for barium lead sulfate than for calcium sulfate or zinc sulfide. Explain why intensifying screens of barium lead sulfate are particularly useful for radiography at higher tube voltages.

13–5. Explain the meaning of the expressions "grid cutoff," "off-distance cutoff," "off-center cutoff," and "off-level cutoff."

13–6. Explain why a crossed grid provides a greater contrast improvement factor than a linear grid with the same grid ratio.

13–7. Discuss the relationships between average gradient, film contrast, and latitude.

13–8. Define film speed and discuss its significance.

13–9. Describe the principle of moving-slit radiography.

13–10. Describe two reasons why rare-earth screens are faster than conventional calcium tungstate screens.

13–11. Discuss the principle of xeroradiography and ionic radiography.

*For those problems marked with an asterisk, answers are provided on page 492.

SUMMARY

- Radiographic film consists of a silver halide emulsion on a support base with a protective coating.
- OD = $\log \frac{1}{T}$
- Latitude is the exposure range over which acceptable optical densities are obtained.
- Average gradient is the slope of the H and D curve in the region where it is linear.
- Gamma is the maximum slope of the H and D curve.
- Speed is the inverse of the exposure required to achieve an optical density of 1.0 above base + fog.
- Grid ratio is the ratio of the height of the grid strips to the width of the spacing between strips.
- The contrast improvement factor of a grid is the ratio of the contrast with the grid to the contrast without the grid.
- Selectivity of a grid is the ratio of primary radiation to scattered radiation that is transmitted by the grid.
- Storage (photostimulable) phosphor image receptors record an image and are then interrogated with a laser beam to "read out" the image.
- Large-area digital detectors may be classified as direct conversion or indirect conversion. Indirect conversion detectors first convert the energy of photons to visible light before recording a signal with a light sensitive digital detector.

REFERENCES

1. Ter-Pogossian, M. *The Physical Aspects of Diagnostic Radiology*. New York, Harper & Row, 1967.
2. Arnold, B., Eisenberg, H., and Bjärngard, B. Measurements of reciprocity law failure in green-sensitive x-ray films. *Radiology* 1978; **126**:493.
3. Seemann, H. *Physical and Photographic Principles of Medical Radiography*. New York, John Wiley & Sons, 1968, pp. 69.
4. Coltman, J., Ebbighauser, E., and Altar, W. Physical properties of calcium tungstate x-ray screens. *J. Appl. Phys.* 1947; **18**:530.
5. Towers, S. X-ray intensifying screens: Part I. *X-Ray Focus* 1967; **8**:2.
6. Rossi, R., Hendee, W., and Ahrens, C. An evaluation of rare earth screen/film combinations. *Radiology* 1976; **121**:465.
7. Hondius-Boldingh, W. *Grids to Reduce Scattered X-Rays in Medical Radiography,* Philips Research Reports Supplement 1. Eindhoven, Netherlands, Philips Research Laboratories, 1964.
8. *Characteristics and Applications of X-Ray Grids*. Cincinnati, Liebel-Flarsheim, 1968.
9. International Commission on Radiological Units and Measurements. *Methods of Evaluating Radiological Equipment and Materials*. Recommendations of the ICRU, Report 10f. National Bureau of Standards Handbook 89, 1963.
10. Barnes, G., Cleare, H., and Brezovich, I. Reduction of scatter in diagnostic radiology by means of a scanning multiple slit assembly. *Radiology* 1976; **120**:691.
11. Enlargement radiography, a review of 0.3 mm techniques with bibliography. *Cathode Press* 1967; **24**:32.
12. Isard, H., et al. *An Atlas of Serial Magnification Roentgenography,* Cathode Press Supplement. Stamford, CT, Machlett Laboratories, 1968.
13. Wandtke, J. C., and Plewes, D. B. Improved chest disease detection with scanning equalization radiography. *AJR* 1985; **145**:979–983.
14. Korhola, O., Riihimaki, E., and Savikurki, S. Reduction of radiation scatter with a multiple pencil-beam imaging device. *Radiology* 1987; **165**:257–259.
15. Merritt, C. R. B., Tutton, R. H., Bell, K. A., et al. Clinical application of digital radiography: Computed radiographic imaging. *Radiographics* 1985; **5**:397–414.
16. Anderson, W. F. A Laser multiformat imager for medical applications, in Schneider, R. H., Dwyer, S. J. II (eds.), *Medical Imaging. Proceedings of the Society of Photo Optical Instrumentation Engineers (SPIE),* Vol. 767, 1987, pp. 516–523.
17. Boyle, W., and Smith, G. Charge coupled semiconductor devices. *Bell Systems Tech. J.* April 1970; **49**:587.
18. Chotas, M. S., Dobbins, J. T., Ravin, C. E. Principles of digital radiography with large-area, electronically readable detectors: A review of the basics. *Radiology* 1999, **210**:595–599.
19. Pisano, E. D., et al. Current status of full-field digital mammography. *Acta Radiol.* 2000; **7**:266–280.
20. Floyd, C. E., et al. Imaging characteristics of an amorphous silicon flat-panel detector for digital chest radiography. *Radiology* 2001; 218(3):683–688.
21. Dobbins, J. T., et al. DQE(f) of four generations of computed radiography acquisition devices. *Med. Phys.* 1995; **22**:1581–1593.
22. Bunch, P. C. Performance characteristics of asymmetric zero-crossover screen-film systems. *Proc. SPIE* 1992; **1653**:46–65.

CHAPTER

14

FLUOROSCOPY

OBJECTIVES

After completing this chapter, the reader should be able to:

- Describe the principle of operation of the image intensifier.
- State how exposure rate varies when the field of view is changed on an image intensifier.
- Explain the principle of operation of television.
- Determine the vertical and horizontal resolution of a television system.
- Explain how image processing can be used to improve signal to noise ratio in digital fluoroscopy.
- Name the events that take place when digital subtraction angiography is performed.
- Give two methods that are used to remove motion artifacts from digital subtraction angiography images.
- Describe how photospot images are obtained.
- Trace the chain of events that take place when an automatic brightness control system varies exposure rate.

Information about moving structures within a patient may be obtained with radiographs exposed in rapid succession. However, the interval of time between serially exposed films may be too great to provide complete information about dynamic processes within the body. Consequently, a technique is required to furnish images that reflect near instantaneous changes occurring in the patient. This technique is referred to as *fluoroscopy*.

FLUOROSCOPY AND IMAGE INTENSIFICATION

In early fluoroscopic techniques, x rays emerging from the patient impinged directly on a fluoroscopic screen. Light was emitted from each region of the screen in response to the rate at which energy was deposited by the incident x rays. The light image on the fluoroscopic screen was viewed by the radiologist from a distance of 10 or 15 in. A thin plate of lead glass on the back of the fluoroscopic screen shielded the radiologist from x radiation transmitted by the screen.

Using this fluoroscopic technique, the radiologist perceived a very dim image with poor visibility of detail. Radiologists had to "dark adapt" their eyes by remaining in the dark for extended periods in order to view the images. Radiologists recognized in the 1940s that the poor visibility of image detail in fluoroscopy was related to the dim image presented by early fluoroscopes.[1,2] They emphasized the need for brighter fluoroscopic images and encouraged the development of the *image intensifier*. Image intensifiers increase the brightness of the fluoroscopic image and permit the observer to use photopic (cone) vision in place of the scotopic (rod) vision required with earlier fluoroscopes. Because of the brighter images, dark adaptation is not required for fluoroscopy with image intensification. Although the image intensifier has increased the cost and complexity of fluoroscopic systems, the use of non-image-intensified fluoroscopy has been outmoded for some time.

X-Ray Image Intensifier Tubes

The x-ray image intensifier "intensifies" or increases the brightness of an image through two processes: (1) minification, in which a given number of light photons emanates from a smaller area, and (2) flux gain, where electrons accelerated by high voltages produce more light as they strike a fluorescent screen. This same principle is employed in "night vision" devices used by the military to observe objects in low light conditions.

An image intensifier tube is diagramed in Margin Figure 14-1. X rays impinge upon a fluorescent screen (input screen) that is from 4 inches to greater than 16 inches in diameter and slightly convex in shape. The fluorescent emulsion is a thin layer of CsI (cesium iodide). Older image intensifier input screens were composed of ZnS:CdS (zinc cadmium sulfide). The principal advantage of CsI over ZnS:CdS:Ag is the increased absorption of x rays because of the presence of higher-Z components in the CsI phosphor and because of the increased packing density of CsI molecules in the phosphor granules.

For each x-ray photon absorbed, 2000 to 3000 photons of light are emitted by the screen. These light photons are not observed directly. Instead, the light falls on a photocathode containing the element antimony (Sb), such as antimony cesium oxide (Sb-CsO).[3] Light photons released in a direction away from the photocathode are reflected toward the photocathode by a mirrored aluminum support on the outside surface of the input screen. If the spectral sensitivity of the photocathode is matched to the wavelength of light emitted by the screen, then 15 to 20 electrons are ejected from the photocathode for every 100 photons of light received. The number of electrons released from any region of the photocathode depends upon the number of light photons incident upon the region. The electrons are accelerated through a potential difference of 25 to 35 kV between the photocathode and the anode of the image intensifier tube. The electrons pass through a large hole in the anode and strike a small fluorescent screen (output screen) mounted on a flat glass support. The emulsion on the output screen resembles that for the input screen, except that the fluorescent granules are much smaller. Diameters of most output screens range from ½ in. to 1 in. Intensifiers with small output screens are used for television fluoroscopy because the diameter of the input screen of a television camera is small also. A coating of metal, usually aluminum, is deposited on the output screen to prevent the entrance of light from outside the intensifier. The metallic layer also removes electrons accumulated by the output screen.

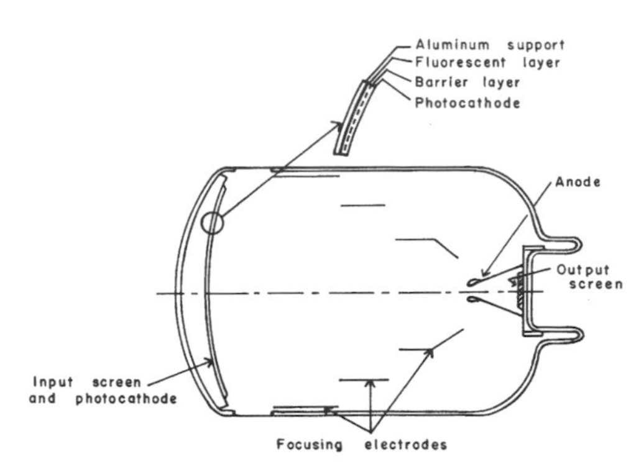

MARGIN FIGURE 14-1
Cross section of a typical x-ray image intensifier tube.

Electrons from the photocathode are focused on the output screen by cylindrical electrodes positioned between the photocathode and the anode. Usually, three focusing electrodes are used (Margin Figure 14-2). The glass envelope is contained within a housing of Mumetal, an alloy containing iron. The housing attenuates magnetic fields that originate outside the intensifier and prevents these fields from distorting the motion of electrons inside. The motion of electrons and, therefore, the image on the output phosphor may still be distorted by a strong magnet field around the intensifier. Also, an intense magnetic field in the vicinity of the image intensifier may magnetize the Mumetal shield and focusing electrodes and cause permanent distortion of the fluoroscopic image. Consequently, image intensifiers should not be located near permanent or transient magnetic fields of high intensity as may be found near magnetic resonance imaging systems (see Chapter 25).

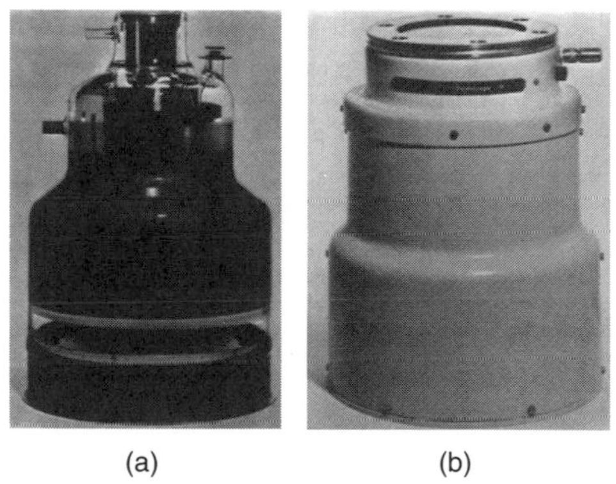

MARGIN FIGURE 14-2
A: Image intensifier tube with an input screen 9 in. in diameter. **B:** Tube encased in its housing.

With an x-ray image intensifier, four different information carriers transmit information about the patient to the radiologist. The x-ray beam transmits information from the patient to the input screen of the image intensifier. At the input screen, the information carrier is changed from x rays to photons of visible light. As the light photons are absorbed by the photocathode, the information is transferred to an electron beam, which is directed upon the output screen of the intensifier. The information is transmitted as a light image from the output screen to the observer's retina.

Gain and Conversion Efficiency of Image Intensifiers

The brightness of the image on the output screen of an image intensifier may be compared with the brightness of the image provided by a standard non-image-intensified fluoroscopic screen. The image-intensifier and the fluoroscopic screen receive identical exposures to radiation, and the ratio of the brightness of the two images is termed the *brightness gain* of the image intensifier.

$$\text{Brightness gain} = \frac{\text{Brightness of the output screen of the image intensifier}}{\text{Brightness of a standard screen}} \quad (14\text{-}1)$$

The brightness gain of image intensifiers varies from 1000 to over 6000, depending on the particular image intensifier used and the fluorescent screen with which it is compared. The gain in brightness results from two independent processes that occur within the intensifier. These processes are *image minification* and *flux gain*.

The luminance or "brightness" of an object is described in units of lamberts. One lambert (L) is the luminance of a perfectly diffusing surface that is emitting or reflecting 1 lumen/cm^2; 1 mL = 0.001 L.

The light image produced as x rays are absorbed in the input screen of an image intensifier is reproduced as a minified image on the output screen of the intensifier. Because the output screen is much smaller than the input screen, the amount of light per unit area from the output screen is greater than that from the input screen. The increase in image brightness furnished by minification of the image is referred to as the *minification gain* g_m and is equal to the ratio of the areas of the input and output screens:

$$\begin{aligned} g_m &= \frac{\text{Area of input screen}}{\text{Area of output screen}} \\ &= \frac{\pi[(\text{diameter of input screen})/2]^2}{\pi[(\text{diameter of output screen})/2]^2} \\ &= \frac{(\text{Diameter of input screen})^2}{(\text{Diameter of output screen})^2} \end{aligned} \quad (14\text{-}2)$$

For example, the minification gain is 81 for an image intensifier with an input screen 9 in. in diameter and an output screen 1 in. in diameter:

$$g_m = \frac{(9\text{ in.})^2}{(1\text{ in.})^2}$$

The brightness of the image on the output screen is also increased because electrons from the photocathode are accelerated as they travel toward the output screen. As these electrons are stopped in the output screen, the number of light photons released varies with the energy of the electrons. The gain in brightness due to electron acceleration is termed the *flux gain* g_f of the image intensifier. A typical image intensifier has a flux gain of at least 50.

The *total gain* g in luminance of an image intensifier is the product of the minification gain g_m and the flux gain g_f:

$$g = (g_m)(g_f) \quad (14\text{-}3)$$

The candela is a unit of luminance and equals 1/60 of the luminance of a square centimeter of a blackbody heated to the temperature of solidifying platinum (1773.5° C). One candela is equivalent to 0.3 millilamberts.

For example, the luminance gain g is 4050 for an image intensifier with a minification gain of 81 and a flux gain of 50:

$$\begin{aligned} g &= (81)(50) \\ &= 4050 \end{aligned}$$

Two image intensifiers may be compared by describing the conversion factor for each intensifier.[4] The conversion factor G_r is the quotient of the luminance of the output screen of the image intensifier divided by the exposure rate at the input screen. The luminance is expressed in units of candela per square meter and the exposure rate in milliroentgens per second:

$$G_r = \frac{\text{Luminance of output screen (candela/m}^2)}{\text{Exposure rate at input screen (mR/sec)}} \quad (14\text{-}4)$$

The conversion factor for an image intensifier is dependent on the energy of the radiation and should be measured with x radiation from a full-wave-rectified or constant-potential x-ray generator operated at about 85 kVp. The conversion factor of most image intensifiers ranges from 50 to100 (candela-sec)/(mR-m^2).

Resolution and Image Distortion with Image Intensifiers

The resolution of an image intensifier is limited by the resolution of the input and output fluorescent screens and by the ability of the focusing electrodes to preserve the image as it is transferred from the input screen to the output screen. The resolution of image intensifiers averages about 4 lines per millimeter for intensifiers with CsI input screens. Contributions to resolution loss that originate outside the image intensifier include the presence of scattered radiation in the x-ray beam received by the input screen and unsharpness in the image contributed by patient motion and the finite size of the focal spot. In addition, the quality of the fluoroscopic image is affected by statistical fluctuations in the number of x rays impinging on the input screen.

The resolution, brightness, and contrast of an image provided by an image intensifier are greatest in the center of the image and reduced toward the periphery. The decrease in brightness along the periphery is usually less than 25%. The reduction in brightness and image quality along the border of the fluoroscopic image is referred to as *vignetting*. Vignetting is a reflection of (a) the reduced exposure rate along the periphery of the input screen and (b) the reduced precision with which electrons from the periphery of the photocathode strike the output screen. Also, the center of the output phosphor receives some scattered light from surrounding regions of the output phosphor, whereas the periphery receives contributions only from the center. Therefore, the lack of a contribution of light from the area beyond the output screen also contributes to vignetting.

Straight lines in an object often appear to curve outward in the fluoroscopic image. This effect is referred to as *pincushion distortion,* caused by the curvature of the input screen and by the reduced precision with which electrons from the periphery of the photocathode are focused upon the output screen. Image lag of the input and output screens of an image intensifier may occur if the phosphor is slow to respond to changes in image brightness. This effect may be of concern during procedures such as high-frame-rate cinefluorography and fast digital imaging.

Size of Image Intensifier

The diameter of the input screen of an image intensifier ranges from 4 to more than 16 in. Intensifiers with small input screens are more maneuverable and less expensive. A small intensifier furnishes a slight improvement in resolution because electrons from the photocathode strike the output screen with greater precision. However, the region of the patient encompassed by the input screen is restricted with a small intensifier. A larger intensifier is more expensive and less maneuverable but furnishes a larger field-of-view and greater opportunity for image magnification.

The diameter of the input screen of an image intensifier is larger than the diameter of the region studied within the patient. In Margin Figure 14-3, for example, an object of length s has a length s' in the image on the input screen. If d is the distance from the target of the x-ray tube to the object and d' is the distance from the target to the input screen, then the apparent length s' of an object of actual length s is

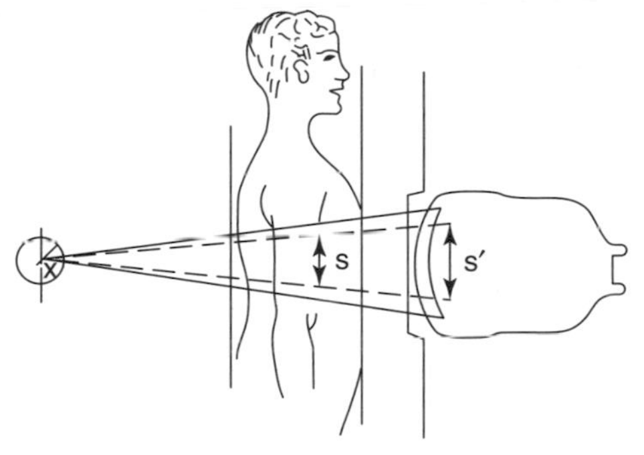

MARGIN FIGURE 14-3
An object of length s has an apparent length s' on the input screen of an image intensifier. The field of view of an image intensifier is smaller than the size of the input screen.

$$s' = s\left(\frac{d'}{d}\right) \tag{14-5}$$

The magnification M of the image is given by

$$M = \frac{s'}{s} = \frac{d'}{d} \tag{14-6}$$

Example 14-1

An x-ray tube is positioned 45 cm below a fluoroscopic table. The input screen of an image intensifier is 15 cm in diameter and 30 cm above the table. What is the maximum length s of an object that is included completely on the screen? What is the magnification of the image? The object is 10 cm above the table.

$$s' = s\left(\frac{d'}{d}\right) \tag{14-5}$$

$$\begin{aligned} s &= s'\left(\frac{d}{d'}\right) \\ &= (15\text{ cm})\left(\frac{10\text{ cm} + 45\text{ cm}}{30\text{ cm} + 45\text{ cm}}\right) \\ &= 11\text{cm} \\ M &= \frac{s'}{s} \\ &= \frac{15\text{ cm}}{11\text{ cm}} \\ &= 1.36 \end{aligned} \tag{14-6}$$

Dual- and Triple-Field Intensifiers

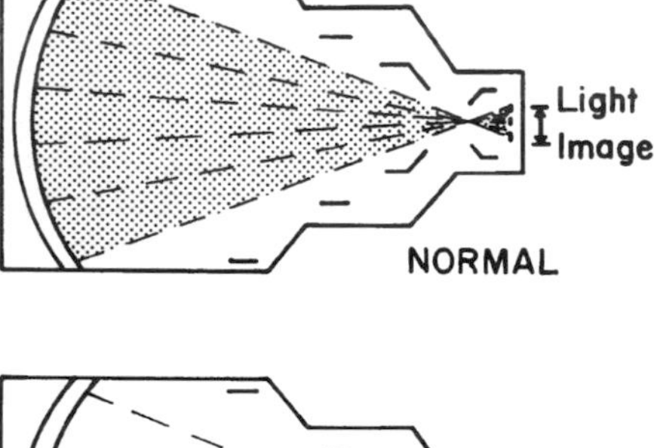

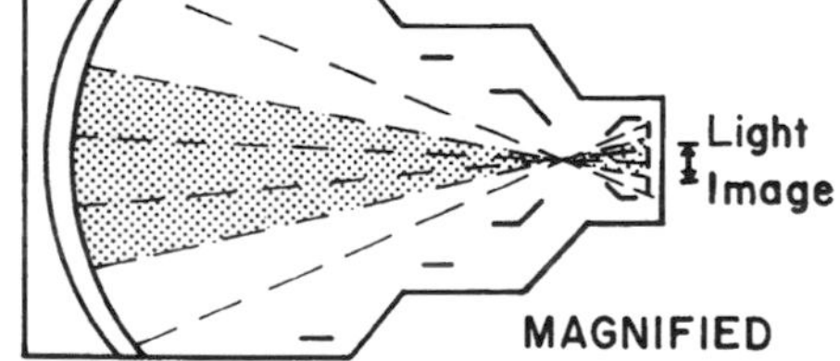

MARGIN FIGURE 14-4
Dual-field image intensifier. By changing the voltage on the focusing electrodes, two different electron crossover points are produced that correspond to normal and magnified modes of viewing.

Many image intensifiers permit magnified viewing of the central region of the input screen. These intensifiers are termed *dual-field* intensifiers if one mode of magnified viewing is provided, and they are called *triple-field* intensifiers if two magnified viewing modes are offered. The operation of a dual-field intensifier is diagramed in Margin Figure 14-4. In the normal viewing mode, electrons from the photocathode converge upon the electron crossover point at the location nearest the output screen and strike only the region of the output screen visible to the TV camera. In this manner, the observer views the entire input screen of the intensifier. When the intensifier is operated in magnified mode, the voltage on the focusing electrodes is changed, and electrons converge upon a crossover point farther from the output screen. Under this condition, electrons originating from the periphery of the input screen strike the output screen outside the region viewed by the TV camera. Only the central region of the input screen is viewed, and this region is seen as a magnified image. In a 22.5-cm (9-in.) dual-field image intensifier, only the central 15 cm (6 in.) of the input screen are seen in the magnified viewing mode.

If the voltage on the focusing electrodes is changed again, electrons from the photocathode can be forced to converge upon a crossover point even farther from the output screen. In this manner, a second mode of magnified viewing is presented to the observer. An intensifier with two magnified viewing modes is termed a *triple-field image intensifier*. Many 22.5-cm triple-field image intensifiers provide an image of the entire input screen in the normal viewing mode and an image of the central 15 cm (6 in.) and 11 cm (4.5 in.) of the input screen during magnified viewing.

In the magnified viewing mode, the image on the output screen is produced by electrons from only the central portion of the input screen. Because fewer electrons are used to produce the image, the brightness of the image would be reduced unless the exposure rate to the input screen is increased. This increase in exposure rate is accomplished automatically as the viewing mode is switched from normal to magnified. The exposure rate increases with decreasing area of the active input phosphor.

Example 14-2

Estimate the increase in exposure rate required to maintain image brightness when an image intensifier field of view is decreased from 9 in. to 6 in. in diameter.

Because the brightness of the image depends on the area of the image intensifier, the increase is determined by the ratio of the square of the two diameters.

$$\frac{\text{Exposure rate at 6 in.}}{\text{Exposure rate at 9 in.}} = \frac{(9\text{ in.})^2}{(6\text{ in.})^2} = 2.25$$

Thus the exposure is increased by a factor of 2.25 when the field of view is decreased from 9 to 6 in. if no other adjustments are made.

TELEVISION DISPLAY OF THE FLUOROSCOPIC IMAGE

The image furnished by an image intensifier may be televised by viewing the intensifier output screen with television camera. The signal from the camera is transmitted by cable to a television monitor. This method for transmission and display of an image is referred to as *closed-circuit television* because the signal is transmitted from the television camera to the monitor by a coaxial cable rather than through air (Margin Figure 14-5).

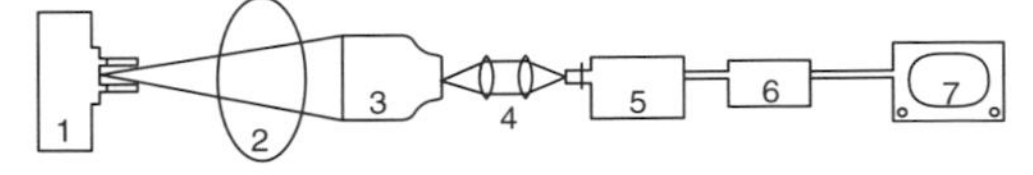

MARGIN FIGURE 14-5
Typical system for television fluoroscopy: 1, x-ray tube; 2, patient; 3, image intensifier; 4, optical system; 5, television camera; 6, camera control unit; 7, television monitor.

Television Cameras

Television cameras, sometimes referred to as television pickup tubes, average about 15 cm in length and 3 cm in diameter (Margin Figure 14-6). In a typical analog camera a thin layer of photoconducting material (usually antimony trisulfide [SbS_3] in the vidicon camera) is plated on the inner surface of the face of a glass envelope and is separated from the glass by a thin, transparent conducting layer. The side of the photoconducting layer nearest to the face of the glass envelope is 20 to 60 V positive with respect to the cathode of the electron gun. The photoconducting layer is an electrical insulator when not exposed to light. When light from the output screen of the image intensifier falls upon the photoconducting layer, the resistance across the layer decreases. That is, the resistance across any region of an illuminated photoconducting layer depends on the amount of light incident on the region.

The scanning electron beam from the electron gun deposits electrons on the photoconducting layer. Some of the electrons migrate toward the positive side of the layer and are collected by the conducting layer. In any region, the number of electrons that migrate to the positive side depends on the resistance and, consequently, the illumination of the region. Electrons collected by the conducting layer furnish an electrical signal that fluctuates in response to variations in the illumination of the photoconducting layer. This signal is amplified and used to control the intensity of the electron beam that is scanned across the fluorescent screen of the television monitor in synchrony with the scanning electron beam in the camera. Hence, the image on the television monitor corresponds closely to the image on the output screen of the image intensifier.[5]

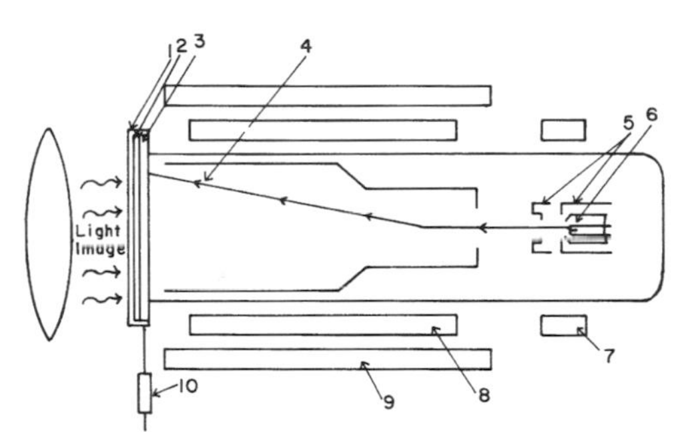

MARGIN FIGURE 14-6
Vidicon television pickup tube: 1, glass envelope; 2, transparent conducting layer; 3, photoconducting layer; 4, scanning electron beam; 5, grids; 6, electron gun; 7, alignment coil for the scanning beam; 8, horizontal and vertical deflection coils; 9, focusing coils; 10, load resistor.

In charge coupled device (CCD) cameras (see Chapter 8 for information on CCD detectors), the photoconduction and scanning are replaced with arrays of CCDs that produce electrical signals in response to the amount of visible light falling upon them. These electrical signals may be stored as arrays of digital values or may be converted directly to standard television signals. As in the consumer market, CCD cameras are replacing analog cameras in newer fluoroscopic systems.

Television Scanning of a Fluoroscopic Image

As the scanning electron beam moves horizontally across the photoconducting layer of the television camera (or as CCD elements are struck by photons), the signal from the conducting layer is modulated in response to the illumination of each photoconducting element in the path of the scanning electron beam. This motion of the scanning electron beam is termed an *active sweep*. As the electron beam is scanned horizontally in the opposite direction *(retrace sweep)*, no information is transmitted to the television monitor. The retrace sweep is about five times faster than the active sweep. As the next active sweep is initiated, it is displaced slightly below the preceding sweep. The number of active sweeps of the scanning electron beam across the image may be 525, 625, 837, 875, 945, or 1024. Hence the image is divided into 525, 625, 837, 875, 945 or 1024 horizontal lines on the screen of the television monitor. (The actual number of lines that appear is slightly smaller because of the vertical retrace; see below.) Images with a large number of horizontal scan lines provide better vertical resolution than those formed with a smaller number of lines.[6]

To prevent brightness flicker in the image, the electron beam of the television camera first scans the photoconducting layer along even-numbered sweeps (2, 4, 6, 8, . . .). This operation requires 1/60 second in standard television systems. The resulting set of horizontal scan lines or *field* is then interlaced with sweeps of the electron beam along odd-numbered lines (1, 3, 5, 7, . . .). The final image, produced by two interlaced fields, is referred to as a *frame*. A frame is completed in 1/30 second. If the two "half images" or fields were not interlaced (if the electron beam swept sequentially along all lines (1, 2, 3, 4, . . .), the image would appear at a rate of 30 times per second and a sensation of flickering light would be perceived. An alternative is the use of special high frame rate monitors that present full frames 60–85 times per second.

After the entire photoconducting layer has been scanned once to compose a television field, the electron beam is returned to the top of the image to begin the next set of interlaced scan lines. The return of the scanning beam requires about 10^{-4} seconds and is referred to as the *vertical retrace*. Because of the time required for the vertical retrace, not all of the scan lines theoretically available are realized in practice. In a television system with 525 possible lines, for example, the image is actually formed with about 490 horizontal lines.[7] The remaining 35 lines are lost during the time required for the vertical retrace.

The vertical resolution of a television image is determined primarily by the number of active sweeps of the electron beam across the anode or photoconducting layer and by the *Kell factor*. The Kell factor is the fraction of the active sweeps that are actually effective in preserving detail in the image. The Kell factor is about 0.7 for most television systems.[8]

$$\text{Vertical resolution} = (\text{Number of active sweeps/frame})(\text{Kell factor}) \qquad \textbf{(14-7)}$$

If the number of active sweeps is increased, then the visibility of detail may be improved in the vertical direction across the television image. Of course, a camera with more active sweeps will not improve the vertical resolution if the detail is not available initially in the image on the output screen of the image intensifier.

The horizontal resolution of a television image is determined primarily by the number of resolution elements available across a single horizontal scan line. The number of resolution elements available depends primarily upon the number of scan lines per frame and upon the range of frequencies transmitted by the electronic circuitry between the television camera and monitor. The highest frequency of signal modulation that is transmitted undistorted from the television camera to the television monitor is referred to as the *frequency bandpass* or *bandwidth* of the television system. Bandwidths for closed-circuit television systems used in radiology range from 3.5 MHz (3.5 megacycles per second) to 12 MHz. A bandwidth of 3.5 MHz is used in broadcast television. The influence of bandwidth in the resolution of the television image is illustrated in the simplified problem in Example 14-3.

Example 14-3

A 525-line television system provides about 490 active sweeps per frame and a vertical resolution of

$$\begin{aligned}\text{Vertical resolution} &= (\text{Number of active sweeps/frame})(\text{Kell factor})\\ &= (490 \text{ active sweeps/frame})(0.7)\\ &= 343 \text{ lines} \qquad \textbf{(14-7)}\end{aligned}$$

What minimum bandwidth is required to provide equal horizontal (343 resolution elements per line) and vertical (343 lines per frame) resolution in the image?

$$\begin{aligned}\text{Number of resolution elements transmitted per second} &= (343 \text{ resolution elements/line})(343 \text{ lines/frame}) \times (30 \text{ frames/sec}) \\ &= 3.53 \times 10^6 \text{ resolution elements/sec}\end{aligned}$$

To transmit this number of resolution elements per second, a television system is required with a bandwidth of at least 3.53×10^6 Hz. That is, a system bandwidth of 3.5 MHz is barely adequate for this application and may sacrifice some image detail for very closely spaced structures in the object.

A more exact expression for the horizontal resolution in resolution elements per line is[7]

$$\text{Horizontal resolution} = \frac{2\,[\text{bandwidth}]\begin{bmatrix}\text{horizontal}\\ \text{fraction}\end{bmatrix}\begin{bmatrix}\text{aspect}\\ \text{ratio}\end{bmatrix}}{(\text{frames/sec})(\text{lines/frame})} \tag{14-8}$$

In Equation 14-8, (1) the bandwidth is expressed in hertz; (2) the horizontal fraction is the quotient of the time required to complete one active sweep of the scanning electron beam divided by the time required for both the active and the retrace sweeps; (3) the number of frames per second is usually 30 but higher frame rate systems are sometimes used; (4) the lines per frame is the number of horizontal scanning lines per frame; (5) the aspect ratio is the ratio of the width of a television frame to its height; and (6) the expression is multiplied by 2 because resolution is described in terms of the total number of resolution elements, both light and dark, that are visible in the television image. If the number of horizontal scanning lines per frame is increased, as in a high-resolution television system, then the visibility of detail in the image may be increased in the vertical direction. However, an increase in bandwidth is required if the image detail is to be maintained or improved in the horizontal direction. Most television systems provide an image with about equal horizontal and vertical resolution.

Example 14-4

An image of light and dark vertical and horizontal lines is presented on the output screen of an image intensifier. This image is transmitted to a television monitor by a 525-line television camera. The bandwidth is 3.5 MHz for the television chain. The framing time of the camera is 1/30 second. What is the maximum number of light and dark lines on the output screen that are duplicated without distortion on the television monitor?

With an aspect ratio of 1 and a horizontal fraction of 0.83 (since the retrace sweep is five times as rapid as the active sweep), the horizontal resolution may be computed with Eq. (14-8):

$$\begin{aligned}\text{Horizontal resolution} &= \frac{2[\text{bandwidth}][\text{horizontal fraction}][\text{aspect ratio}]}{(\text{frames/sec})(\text{lines/frame})} \\ &= \frac{2(3.5 \times 10^6 \text{ Hz})(0.83)(1.0)}{(30 \text{ frames/sec})(50 \text{ lines/frame})} \\ &= 370 \text{ lines}\end{aligned}$$

The image will be transmitted undistorted if it is composed of no more than 185 dark lines separated by 185 light lines. As described in Example 14-3, the vertical resolution for a 525-line camera is 343 lines. Hence the vertical and horizontal resolutions are about equal.

TABLE 14-1 Bandwidth Required for Various Television Systems to Provide Equal Resolution in Horizontal and Vertical Directions Across the Image

No. of Scan Lines/Frame	*Frames/sec*	*Vertical and Horizontal Resolution (Lines)*	*Bandwidth (MHz)*
525	30	343	3.3
625	25	408	3.8
837	30	547	8.3
875	30	571	9.1
945	30	617	10.5
1024	16.7	670	6.9

Data in Table 14-1 show the bandwidths required for different television systems to furnish equal resolution in the vertical and horizontal directions. These data were computed for an aspect ratio of 1.0 and a horizontal fraction of 0.83.

DIGITAL FLUOROSCOPY

One technique for digital fluoroscopy is digitization of the signal from the video camera. Other techniques, in which the video camera is replaced by a digital detector such as a CCD camera or a flat-panel detector (see Chapter 13), are becoming increasingly popular and are expected ultimately to replace digitization of the video camera signal.

There are a number of advantages to digital fluoroscopic techniques. First, the digital still-frame image (digital photospot) serves as an alternative to the spot film (taken from a cassette on the input side of the image intensifier) or the analog photospot image (taken on 105-mm film on the output side of the image intensifier). Digital photospots have the advantages of lower patient dose and the possibility of image processing to enhance structures of interest. Second, the digital image may be used for various purposes in real-time fluoroscopy. For example, the most efficient use of radiation is to produce a bright image on the output phosphor with a pulse of radiation, and to use the TV camera to scan every line of the image to produce the full frame as quickly as possible. The digital "scan converter" may be used to store the full digitized frame. The full frame data are then read out and transmitted as two interlaced fields to a standard television monitor. This "pulsed progressive" technique reduces the loss of resolution that occurs in "continuous interlaced" scanning due to motion. Such loss of resolution is particularly troublesome in applications involving rapidly changing structures, such as cardiac catheterization. "Pulsed progressive" scanning results in a substantial (30% to 50%) dose reduction when compared with the standard video method[9] of scanning every other line to produce a field and repeating the process to produce the interlaced frame, all during continuous x-ray exposure.

Another possible advantage of digital techniques in fluoroscopy is the ability to perform "road mapping," a digital frame store-and-display mode that retains the most recent fluoroscopic image on the screen as an aid to the placement of catheters and guidewires for interventional procedures. Road mapping employing frame-hold procedures reduces the need for continuous x-ray exposure of the patient. Once multiple digital images are available, image processing techniques such as frame averaging and edge enhancement may be used to improve the data presentation.

Various image processing techniques are used to enhance image quality in digital fluoroscopy and to acquire information that would be difficult to obtain if the images were analog (on cut film or cine film). One example of digital image processing in fluoroscopy is frame averaging. In digital fluoroscopy, images are being acquired rapidly, and the level of quantum noise may be quite high in any one frame. When viewed as multiple frames per second, the eye and brain of the observer do their own "averaging" of quantum noise over several frames so that the noise is suppressed. When viewed as individual frames, as in road mapping, a reduction in quantum noise is desirable. This may be accomplished by averaging corresponding pixel values over several frames. Over a period of less than a second, the image features may not change significantly, but the quantum fluctuation (quantum noise) does change. By frame averaging, the effects of noise in each pixel may be reduced without significantly altering the pixel value (signal) that represents the attenuation of x rays in the patient.[10]

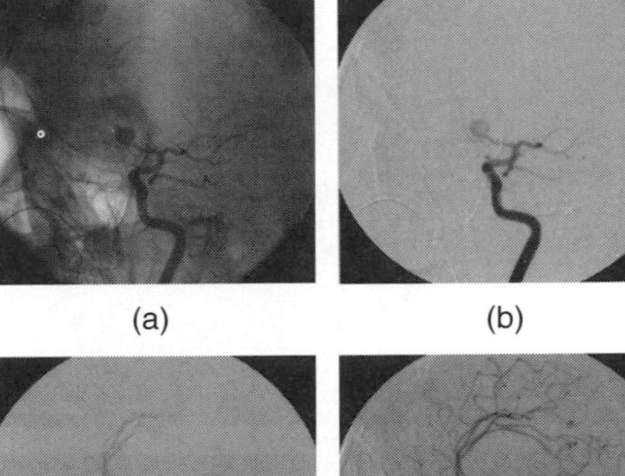

(a) (b)

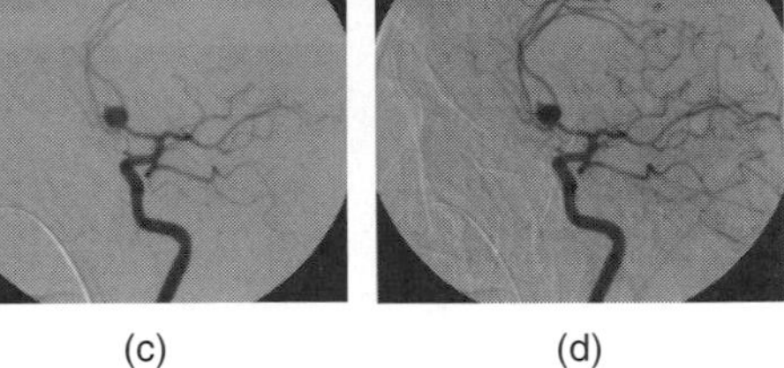

(c) (d)

MARGIN FIGURE 14-7
Digital subtraction angiography. Cerebral arteriogram (oblique transorbital view) shows an aneurysm of the anterior communicating artery (junction of the A_1 and A_2 segments of the anterior cerebral artery). **A:** Unsubtracted original digital fluoroscopic image obtained midway through the contrast material injection. **B–D:** Subtracted DSA images obtained at three progressive time points during contrast material injection. (from ref 10 with permission)

Example 14-5

Find the theoretical improvement in image quality (reduction of noise) if 3, 5, and 7 digital images are averaged.

The digital signal is directly proportional to the number of photons striking the image receptor. Digital noise in a quantum limited system varies as the square root of the number of photons. Therefore, the signal-to-noise ratio (SNR) is

$$\text{SNR} = \frac{\text{Signal}}{\text{Noise}} \propto \frac{N}{\sqrt{N}} = \sqrt{N}$$

and the following table may be constructed:

Images Averaged	SNR
3	1.73
5	2.23
7	2.65

The table shows that averaging 3 frames per second yields an increase in SNR by a factor of 1.73.

Note that the relative improvement diminishes as more frames are averaged. Seven frames is more than twice as many as three frames, but the SNR is not twice its relative value at three frames. This observation, coupled with the fact that averaging more frames is likely to produce more motion blur places a limit, fewer than 10, upon the maximum number of frames that are usually averaged.

Digital Subtraction Angiography

Images (acquired pre- and postinjection of contrast material) may be subtracted digitally to yield images of the contrast-laden vessels without the distraction of other structures in the patient (Margin Figure 14-7). Typically, a "mask" image is selected before contrast injection. The mask is subtracted from succeeding images after contrast injection (Margin Figure 14-8). If patient motion occurs after the mask has been selected, then anatomical structures in succeeding images will not coincide and artifacts will occur. If this occurs, possible solutions include:

- **Remasking:** If other precontrast images occur after motion has ceased but before contrast injection, a new mask may be selected.
- **Pixel shifting:** If motion is not too complex, the mask digital image may be shifted one or more pixels in all directions until the smoothest image is obtained. An algorithm to accomplish this task without human intervention is available in most digital subtraction angiography systems.

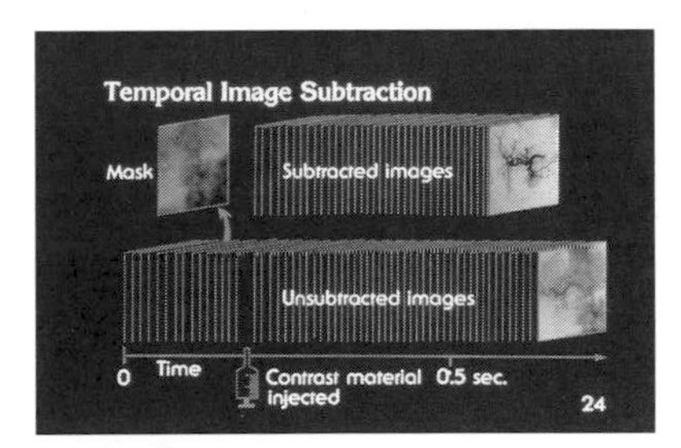

MARGIN FIGURE 14-8
The timing of events in digital subtraction angiography. During serial acquisition, a mask image is selected from images taken prior to injection of contrast material. The mask image is subtracted from images taken after injection of contrast material to obtain the "subtracted" images. From Ritenour ER Computer Applications in Diagnostic Radiology, Mosby, Chicago, 1983.

AUTOMATIC BRIGHTNESS CONTROL

In fluoroscopy, an automatic brightness control (ABC) unit is used to maintain an image of constant brightness on the output screen of the intensifier, irrespective of changes in the attenuation of the x-ray beam in the patient. An image of constant brightness is achieved by automatic adjustment of exposure technique variables (e.g., peak kilovoltage, milliamperes, and x-ray pulse width) in response to fluctuations in brightness of the output screen. For purposes of discussion, an ABC unit may be separated into two components: (1) the brightness-sensing element, with an output used to control (2) the machine's technique variables, which are adjusted automatically to maintain constant image brightness.[11,12]

Brightness-Sensing Elements

The earliest technique to monitor image brightness was continuous measurement of the image intensifier photocathode–anode current. With this approach there is no simple way to compensate for bright-edge effects at the periphery of the image that are caused by x rays unattenuated by the patient. Compensation for variable collimation usually is provided by a shutter compensation control, which reduces the ABC's response as field size is decreased.

In systems that offer television viewing of the fluoroscopic image, a second approach is available to monitor the brightness of the entire image on the output screen. This approach uses the intensity of the television signal as an indication of output screen brightness. The advantage of this technique is that the portion of the image actually monitored can be selected electronically. If the center of the image is monitored, bright-edge and collimation effects are eliminated.

The most common method of brightness sensing uses a photomultiplier tube to monitor the brightness of the output screen of the intensifier. The area monitored can be defined optically with lenses and diaphragms, thereby eliminating edge and collimation effects. This approach has the advantage of less sensitivity to noise and image intensifier aging effects and, along with the television method, is not influenced greatly by x-ray energy.

Technique Variables Controlled by Automatic Brightness Control

Three variables that can be controlled by an ABC unit are the peak kilovoltage, the tube current, and the width of the x-ray pulses in x-ray units with variable pulse width.

Peak Kilovoltage Variability

With systems of this type, the current (in milliamperes) is selected by a continuous or fixed-step current control, and the peak kilovoltage is selected automatically to provide an image of desired brightness. In the automatic mode, the selected current exerts some influence on the peak kilovoltage required to furnish a satisfactory image, that is, high milliamperage lowers the peak kilovoltage and low milliamperage raises the peak kilovoltage. On some machines, the peak kilovoltage control establishes the maximum peak kilovoltage that the ABC can select. On units with an automatic/manual control, the manual setting disables the ABC unit and permits the operator to select both the milliamperage and the peak kilovoltage. X-ray systems with peak-kilovoltage-variable ABC units usually provide a wider dynamic range of brightness control.

Current Variability

A current-variable ABC control offers the advantage of operator control of peak kilovoltage during ABC operation. At the peak kilovoltage selected, however, the tube current must be sufficiently variable to furnish an image of adequate brightness over a wide range of x-ray beam attenuation. Usually, the dynamic range of a current-variable unit is less than that of a peak-kilovoltage-variable unit. Units with variable current have a relatively long response time because they depend on the thermal characteristics of the x-ray tube filament.

Peak-Kilovoltage-Current Variability

Some ABC units vary both the peak kilovotage and the current, with both technique variables changed in the same direction (i.e., either increased or decreased). This approach offers a wide dynamic range but provides little control over the peak kilovoltage selected for the examination. For this reason, some peak kilovoltage/current-variable systems offer the option of manual selection of either peak kilovoltage or current. With these options, operation is identical to that described above for current-variable or peak-kilovoltage-variable units.

Pulse Width Variability

Machines equipped with digital or cine cameras sometimes use grid-controlled x-ray tubes to vary the width (time) of the voltage pulse across the x-ray tube. In this manner, the brightness of the image can be kept constant during fluoroscopy. In most units of this type, the milliamperage is set at a fixed value, and the pulse width of the x-ray beam is varied up to some maximum limit. A peak kilovoltage override circuit increases the kilovoltage above that selected if an image of adequate brightness is not achieved at maximum pulse width. The pulse-width-variable type of control has the distinct advantages of instantaneous response and, with the peak kilovoltage override circuit, a very wide dynamic range.

Combination Circuits

Any combination of the above approaches to ABC operation is possible, but the one most commonly used, other than peak voltage–current variability, is the peak kilovoltage override capability. This method may be applied to either milliamperage-variable or pulse-width-variable machines. The peak kilovoltage override approach provides automatic compensation if the selected peak kilovoltage is not adequate to furnish an image of desired brightness.

CINEFLUOROGRAPHY

The image on the output screen of an image intensifier may be photographed with a 35-mm movie camera. Although this technique has largely been replaced by digital techniques, it is still performed in a few centers.

Grid-controlled x-ray tubes are used to synchronize the x-ray exposure with the cinefluorographic film. With these x-ray tubes, the bias voltage on the focusing cup surrounding the filament is regulated by a timing circuit that is synchronized with the film-supply mechanism of the cinefluorographic camera. Electrons flow across the x-ray tube only when the cinefluorographic film is in position for exposure.

"Flickering" of the image in the viewing mirror is noticeable during cinefluorography with a synchronized cinefluorographic unit operated at frame rates slower than about 30 frames per second. The rate at which the image flickers depends on the number of frames of film exposed per second.

The photographic film used for cinefluorography must be sensitive to the wavelength of light from the output screen of the image intensifier. Orthochromatic emulsions (sensitive to all visible light except red light) and panchromatic emulsions (sensitive to all wavelengths of visible light) are used for cinefluorography. A relatively high-speed film with high inherent contrast should be used for cinefluorography. The film should be able to record fine detail. No single film satisfies all these requirements, and a compromise must be reached when a particular film is chosen for a cinefluorographic examination.

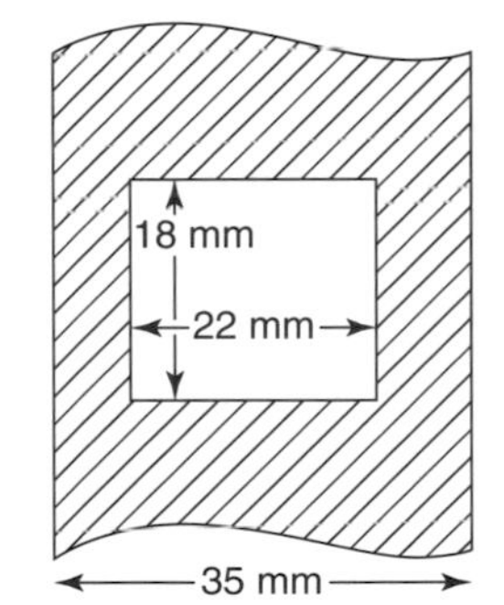

MARGIN FIGURE 14-9
Dimensions of single frames of 35-mm cinefluorographic film.

Each frame of 35-mm film is 22 mm wide and 18 mm long and furnishes 16 frames per linear foot. Each frame of 35-mm film is rectangular rather than square (Margin Figure 14-9). If the size of the recorded image is restricted to the smaller dimension of a frame, then the frame is not exposed at the borders along the larger dimension. Expanding the image to fill the larger dimension of the film is referred to as *overframing*. Overframing causes the loss of part of the image at the borders along the shorter dimension of the frame. If the lost portion of the image is not important clinically, then overframing is useful because it expands the image over a larger area of film and may increase the image detail that is recorded.

Example 14-6

An image intensifier provides a resolution of 1.5 line pairs per millimeter on an input screen 9 in. in diameter. With overframing, the minification of the image on 35-mm cinefluorographic film is

$$\frac{(9 \text{ in.})(2.54 \text{ cm/in.})}{2.2 \text{ cm}} = 10.4$$

Therefore, the film should provide a resolution of

$$(1.5 \text{ line pairs/mm})(10.4) = 15 \text{ line pairs/mm}$$

Cinefluorographic film is capable of this level of resolution.

During cinefluorography, the current through the x-ray tube is increased many times over that required for routine fluoroscopy with an image intensifier. The increased current is necessary to provide an image on the output screen that is bright enough to expose each frame of cinefluorographic film. This exposure requires an exposure rate of approximately 10 μR per frame at the input screen of the image intensifier.[13] The exposure rate at the entrance surface of the patient will be 100 to 1000 times this value. Because exposure rate increases linearly with current through the x-ray tube, exposure of the patient to radiation is also increased many times during a cinefluorographic examination.[14–16] A similar increase in patient dose occurs in digital fluoroscopy, depending upon the number of frames per second and the tube current used.

Photospot Images

Spot photographs of the image on the output screen of an image intensifier may be acquired with 100-mm C-4 film or 105-mm roll film. Photospot images require less exposure of the patient to radiation when compared with spot films.[17] Although the image quality of a photospot image may be slightly less than that furnished by a conventional spot film, photospot images are satisfactory for most examinations.[18]

PROBLEMS

*14-1. An image intensifier has an input screen 9 in. in diameter and an output screen 1 in. in diameter. The flux gain of the intensifier is 40. What is the total gain in luminance?

14-2. In an image intensification system, trace the information and its carrier from the x-ray beam emerging from the patient to the visual image on the observer's retina. Describe the factors that contribute to resolution loss at each stage of this transformation.

*14-3. A patient 12 in. thick is positioned on a table 15 in. from the target of an x-ray tube. The region of interest in the patient is 6 in. in diameter and 3 in. above the table. Can the entire region be displayed at once with an image intensifier with an input screen 9 in. in diameter?

*14-4. A television system with 875 scan lines per frame provides 815 active sweeps per frame. If the Kell factor is 0.7, the horizontal fraction is 0.83, the aspect ratio is 1.0, and the frame rate is 30 frames per second, what bandwidth is required to provide equal resolution horizontally and vertically?

14-5. What is meant by vignetting in an image furnished by image intensification fluoroscopy? What is meant by "pincushion distortion"?

*14-6. The exposure rate at tabletop for an abdominal fluoroscopic study of a particular patient is 1×10^{-3} C/(kg-min) in the 16-in. mode. What exposure rate would be expected on the same equipment if switched to the 12-in. mode? 6-in. mode?

*For those problems marked with an asterisk, answers are provided on p. 492.

SUMMARY

- An image intensifier employs both minification gain and electronic (flux) gain.
- Reducing the field of view of an image intensifier results in a magnified image but increases the dose to the patient over the smaller area.
- Standard broadcast television images consist of two interlaced fields that constitute a single frame. Sixty fields appear per second, and therefore there are 30 frames per second.
- Road mapping and frame hold systems can be used to reduce patient dose in fluoroscopy.
- Signal-to-noise ratio varies as the square root of the number of photons. Therefore it increases as patient dose is increased in a given imaging system.
- In digital systems, frame averaging can be used to reduce noise.
- Remasking and pixel shifting are possible remedies for motion artifacts in digital subtraction angiography.

REFERENCES

1. Chamberlain, W. Fluoroscopes and fluoroscopy. *Radiology* 1942; **38**:383.
2. Sturm, R., and Morgan, R. Screen intensification systems and their limitations. *AJR* 1959; **62**:617.
3. Niklas, W. Conversion efficiencies of conventional image amplifiers, in Janower, M. (ed.): *Technological Needs for Reduction of Patient Dosage from Diagnostic Radiology*. Springfield, IL, Charles C Thomas, 1963, p. 271.
4. International Commission on Radiological Units and Measurements: *Methods of Evaluating Radiological Equipment and Materials*. Recommendations of the ICRU, Report 10f. National Bureau of Standards Handbook 89, 1963, p. 3.
5. Gebauer, A., Lissner, J., and Schott, O. *Roentgen Television*. New York, Grune & Stratton, 1967.
6. Lin, P. P. Technical considerations of equipment selection and acceptance testing of cardiovascular angiography systems, in Seibert, J. A., Barnes, G. T., and Gould, R. G. *Specification Acceptance Testing and Quality Control of Diagnostic X-Ray Imaging Equipment*, Medical Physics Monograph No 20, Woodbury, NY, American Institute of Physics, 1994, p. 609.
7. Templeton, A., et al. Standard and high-scan line television systems. *Radiology* 1968; **91**:725.
8. Kell, R., Bedford, A., and Fredendall, G. A determination of optimum number of lines in a television system. *RCA Rev.* 1940; **5**:8.
9. Seibert, J. A. Improved fluoroscopic and cine-radiographic display with pulsed exposures and progressive TV scanning. *Radiology* 1986; **159**:277–278.
10. Pooley, R. A., McKinney, J. M., and Miller, D. A. The AAPM/RSNA physics tutorial for residents: digital fluoroscopy. *Radiographics* 2001; **21**(2): 521–534.
11. Hendee, W., and Chaney, E. Diagnostic x-ray survey procedures for fluoroscopic installations—Part I: Undertable units. *Health Phys.* 1975; **29**:331.
12. Whitaker, F., Addison, S., and Hendee, W. Diagnostic x-ray survey procedures for fluoroscopic installations—Part II: Automatic brightness controlled units. *Health Phys.* 1977; **32**:61.
13. International Commission on Radiation Units and Measurements. *Cameras for Image Intensifier Fluorography*, ICRU Report 15. Washington, D.C., 1969.
14. Addison, S., Hendee, W., Whitaker, F., et al. Diagnostic x-ray survey procedures for fluoroscopic installations —Part III: Cinefluorographic units. *Health Phys.* 1978; **35**:845.
15. Hale, J., et al. Physical factors in cinefluorography, in Moseley, R., Rust, J. (eds.). *The Reduction of Patient Dose by Diagnostic Radiologic Instrumentation*. Springfield, IL, Charles C Thomas Publishers, 1964, p. 78.
16. Henny, G. Dose Aspects of Cinefluorography, in Janower, M. (ed.). *Technical Needs for Reduction of Patient Dosage from Diagnostic Radiology*. Springfield, IL, Charles C Thomas, 1963, p. 319.
17. Carlson, C., and Kaude, J. Integral dose in 70-mm fluorography of the gastroduodenal tract. *Acta Radiol* 1968; **7**:84.
18. Olsson, O. Comparison of image quality: 35 mm, 70 mm and full size films, in Moseley, R., and Rust, J. (eds.): *Diagnostic Radiologic Instrumentation*. Springfield, IL, Charles C Thomas, 1965, p. 370.

CHAPTER

15

COMPUTED TOMOGRAPHY

OBJECTIVES

After completing this chapter, the reader should be able to:

- Explain the principles of x-ray transmission computed tomography.
- Compare the properties of x-ray projection images with x-ray CT images.
- Provide a brief history of the evolution of x-ray CT imaging.
- Describe different approaches to the reconstruction of CT images from projection measurements.
- Depict the configuration of spiral and "ultrafast" CT scanners.
- Outline the features of x-ray sources, detectors, collimators, and display systems used in x-ray CT.
- Characterize the relationship between CT numbers, linear attenuation coefficients, and physical densities associated with CT scanning.
- Identify various characteristics important in quality control of CT units.

INTRODUCTION

In conventional radiography, subtle differences of less than about 5 percent in subject contrast (i.e., x-ray attenuation in the body) are not visible in the image. This limitation exists for the following reasons:

1. The projection of three-dimensional anatomic information onto a two-dimensional image receptor obscures subtle differences in x-ray transmission through structures aligned parallel to the x-ray beam. Although conventional tomography resolves this problem to some degree, structures above and below the tomographic section may remain visible as "ghosts" in the image if they differ significantly in their x-ray attenuating properties from structures in the section.
2. Conventional image receptors (i.e., film, intensifying and fluoroscopic screens) are not able to resolve small differences (e.g., 2%) in the intensity of incident radiation.
3. Large-area x-ray beams used in conventional radiography produce considerable scattered radiation that interferes with the display of subtle differences in subject contrast.

To a significant degree, each of these difficulties is eliminated in computed tomography (CT). Hence, differences of a few tenths of a percent in subject contrast are revealed in the CT image. Although the spatial resolution of a millimeter or so provided by CT is notably poorer than that provided by conventional radiography, the superior visualization of subject contrast, together with the display of anatomy across planes (e.g., cross-sectional) that are not accessible by conventional imaging techniques, make CT exceptionally useful for visualizing anatomy in many regions of the body.

The term "tomography" is derived from the Greek word *tomos*, meaning "section."

Computed tomography was not the first x-ray method to produce cross-sectional images. In the late 1940s and 1950s, Takahaski in Japan published several papers describing the analog technique of transverse axial tomography. Takahashi's method was commercialized by Toshiba Inc. in the 1960s. The Toshiba product was usurped by computed tomography in the early 1970s.

William Oldendorf was a neuroscientist interested in improving the differentiation of brain tissue. In particular, he was searching for a better imaging method than pneumoencephalography for studying the brain. Many scientists believe he should have shared in the 1970 Nobel Prize in Medicine.

The 1970 Nobel Prize in Medicine was shared by two pioneers of computed tomography, Godfrey Hounsfield, an engineer with EMI Ltd., and Allen Cormack, a South African medical physicist.

Hounsfield was a radar instructor during World War II. Before developing the first x-ray CT scanner, he designed the first transistorized digital computer while working as an engineer at EMI Ltd.

HISTORY

The image reconstruction techniques used in CT were developed for use in radio astronomy,[1] electron microscopy,[2] and optics including holographic interferometry.[3–5] In 1961, Oldendorf explored the principle of CT with an apparatus using an ^{131}I source.[6] Shortly thereafter, Kuhl and colleagues developed emission and ^{241}AM transmission CT imaging systems and described the application of these systems to brain imaging.[7,8] In spite of these early efforts, CT remained unexploited for clinical imaging until the announcement by EMI Ltd. in 1972 of the first commercially available x-ray transmission CT unit designed exclusively for studies of the head.[9] The prototype for this unit had been studied since 1970 at Atkinson–Morley

Hospital in England, and the first commercial unit was installed in the United States in 1973. The same year, Ledley and colleagues announced the development of a whole-body CT scanner.[10] In 1974, Ohio Nuclear Inc. also developed a whole-body CT scanner, and clinical models of both units were installed in 1975. By 1977, 16 or so commercial companies were marketing more than 30 models of transmission CT scanners. Today, approximately 5000 CT units are installed in U.S. hospitals, at a cost of up to a million dollars or slightly more per unit.

Cormack was interested in the generation of cross-sectional displays of attenuation coefficients for use in treatment planning for radiation therapy. He published his results in the *Journal of Applied Physics*. He received only one reprint request. It was from members of the Swiss Avalanche Research Center, who were interested in possible use of his method to predict snow depths.

PRINCIPLE OF COMPUTED TOMOGRAPHIC IMAGING

In early CT imaging devices ("scanners") a narrow x-ray beam is scanned across a patient in synchrony with a radiation detector on the opposite side of the patient. If the beam is monoenergetic or nearly so, the transmission of x rays through the patient is

$$I = I_0 e^{-\mu x}$$

In this equation the patient is assumed to be a homogeneous medium. If the x-ray beam is intercepted by two regions with attenuation coefficients μ_1 and μ_2 and thicknesses x_1 and x_2, the x-ray transmission is

$$I = I_0 e^{-(\mu_1 x_1 + \mu_2 x_2)}$$

If many (n) regions with different linear attenuation coefficients occur along the path of x-rays, the transmission is

$$I = I_0 e^{-\sum_{i=1}^{n} \mu_i x_i}$$

where $-\sum_{i=1}^{n} \mu_i x_i = (\mu_1 x_1 + \mu_2 x_2 + \cdots + \mu_n x_n)$, and the fractional transmission I/I_0 is

$$e^{-\sum_{i=1}^{n} \mu_i x_i}$$

With a single transmission measurement, the separate attenuation coefficients cannot be determined because there are too many unknown values of μ_i in the equation. However, with multiple transmission measurements in the same plane but at different orientations of the x-ray source and detector, the coefficients can be separated so that a cross-sectional display of attenuation coefficients is obtained across the plane of transmission measurements. By assigning gray levels to different ranges of attenuation coefficients, a gray-scale image can be produced that represents various structures in the patient with different x-ray attenuation characteristics. This gray-scale display of attenuation coefficients constitutes a CT image.

In early (first-generation) CT scanners, multiple x-ray transmission measurements are obtained by scanning a pencil-like beam of x rays and an NaI detector in a straight line on opposite sides of the patient (Figure 15-1A). During this translational scan of perhaps 40 cm in length, multiple (e.g., 160) measurements of x-ray transmission are obtained. Next, the angular orientation of the scanning device is incremented 1 degree, and a second translational scan of 160 transmission measurements is performed. This process of translational scanning separated by 1-degree increments is repeated through an arc of 180 degrees. In this manner, 160 × 180 = 28,800 x-ray transmission measurements are accumulated. These measurements are transmitted to a computer equipped with a mathematical package for reconstructing an image of attenuation coefficients across the anatomic plane defined by the scanning x-ray beam.

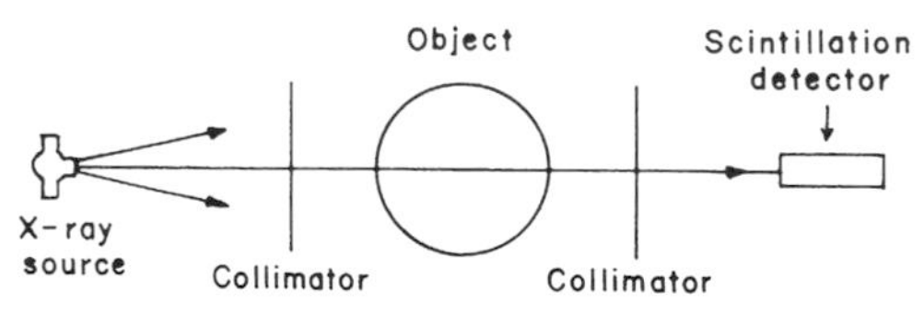

MARGIN FIGURE 15-1
In a first-generation computed tomographic scanner, x-ray transmission measurements are accumulated while an x-ray source and detector are translated and rotated in synchrony on opposite sides of the patient.

Hounsfield termed his technique "computerized transverse axial tomography." This expression was later abbreviated to "computerized axial tomography," and was referred to as "CAT scanning." After sufficient ridicule had been directed toward this acronym, the expression *computed tomography* (CT scanning) was adopted by major journals in medical imaging.

X-ray CT images are often described as "density distributions" because they provide a gray-scale display of linear attenuation coefficients that are closely related to the physical density of tissues.

The scanning technique employed by first generation CT scanners is referred to as "rectilinear pencil-beam scanning."

The first five clinical CT units marketed by EMI Ltd. were installed by early 1973 in London, Manchester, Glasgow, Rochester (MN), and Boston.

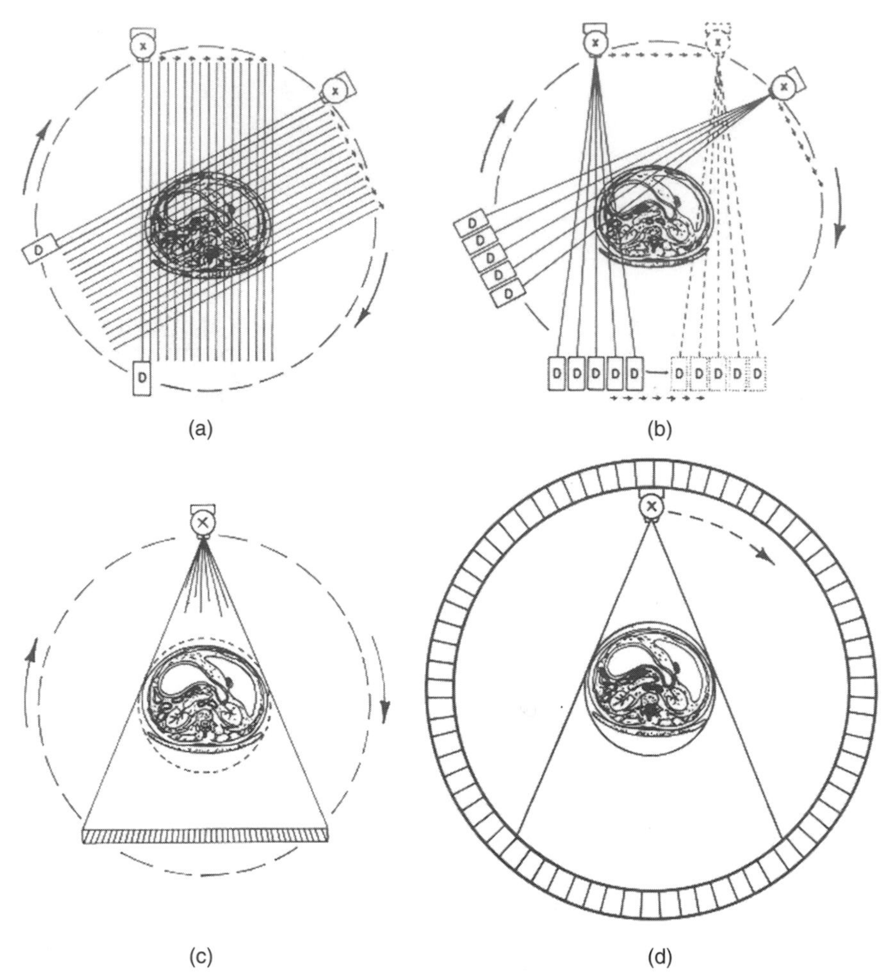

FIGURE 15-1
Scan motions in computed tomography. **A:** first-generation scanner using a pencil x-ray beam and a combination of translational and rotational motion. **B:** Second-generation scanner with a fan x-ray beam, multiple detectors, and a combination of translational and rotational motion. **C:** Third-generation scanner using a fan x-ray beam and smooth rotational motion of x-ray tube and detector array. **D:** Fourth-generation scanner with rotational motion of the x-ray tube within a stationary circular array of 600 or more detectors.

Backprojection is also known as the *summation method* or the *linear supposition method*.

Simple backprojection does not produce sharp and clear images, and it is not used in commercial CT scanners.

The integral equations approach to image reconstruction was first described in 1917 by the Austrian mathematician Radon. He developed his equations for studies of gravitational fields.[12]

As used in a mathematical sense, the word *filter* connotes a mathematical operation on a set of data. The convolution filter (also called a *kernel*) discussed here is applied to the data before they are backprojected to form an image.

RECONSTRUCTION ALGORITHMS

The foundation of the mathematical package for image reconstruction is the reconstruction algorithm, which may be one of four types.[11]

1. *Simple backprojection.* In this method, each x-ray transmission path through the body is divided into equally spaced elements, and each element is assumed to contribute equally to the total attenuation along the x-ray path. By summing the attenuation for each element over all x-ray paths that intersect the element at different angular orientations, a final summed attenuation coefficient is determined for each element. When this coefficient is combined with the summed coefficients for all other elements in the anatomic section scanned by the x-ray beam, a composite image of attenuation coefficients is obtained. Although the simple backprojection approach to reconstruction algorithms is straightforward, it produces blurred images of sharp features in the object.
2. *Filtered backprojection.* This reconstruction algorithm, often referred to as the *convolution* method, uses a one-dimensional integral equation for the

reconstruction of a two-dimensional image. In the convolution method of using integral equations, a deblurring function is combined (convolved) with the x-ray transmission data to remove most of the blurring before the data are backprojected. The most common deblurring function is a filter that removes the frequency components of the x-ray transmission data that are responsible for most of the blurring in the composite image. One of the advantages of the convolution method of image reconstruction is that the image can be constructed while x-ray transmission data are being collected. The convolution method is the most popular reconstruction algorithm used today in CT.

3. *Fourier transform*. In this approach, the x-ray attenuation pattern at each angular orientation is separated into frequency components of various amplitudes, similar to the way a musical note can be divided into relative contributions of different frequencies. From these frequency components, the entire image is assembled in "frequency space" into a spatially correct image and then reconstructed by an inverse Fourier transform process.
4. *Series expansion*. In this technique, variations of which are known as ART (algebraic reconstruction technique), ILST (interative least-squares technique), and SIRT (simultaneous iterative reconstruction technique), x-ray attenuation data at one angular orientation are divided into equally spaced elements along each of several rays. These data are compared with similar data at a different angular orientations, and differences in x-ray attenuation at the two orientations are added equally to the appropriate elements. This process is repeated for all angular orientations, with a decreasing fraction of the attenuation differences added each time to ensure convergence of the reconstruction data. In this method, all x-ray attenuation data must be available before reconstruction can begin.

A high-frequency convolution filter reduces noise and makes the image appear "smoother." A low-frequency filter enhances edges and makes the image appear "sharper." A low-frequency filter may be referred to as a "high-pass" filter because it suppresses low frequencies and allows high frequencies to pass.

Filtered backprojection removes the star-like blurring seen in simple backprojection. It is the principal reconstruction algorithm used in CT scanners.

The Fourier approach to image reconstruction is used commonly in magnetic resonance imaging, but seldom in CT scanning.

In 1958, the Ukrainian physicist Korenblyum[13] from the Kiev Polytechnic Institute reworked Radon's integral equations for application to fan-beam geometry. He verified his approach with images of a rotating body captured on film.

Series expansion (iterative reconstruction) techniques are not used in commercial CT scanners because the iteration cannot be started until all of the projection data have been acquired, causing a delay in reconstruction of the image.

SCAN MOTIONS

First- to Fourth-Generation CT Scanners

Early (*first-generation*) CT scanners used a pencil-like beam of x-rays and a combination of translational and rotational motion to accumulate the many transmission measurements required for image reconstruction (Figure 15-1A). Although this approach yields satisfactory images of stationary objects, considerable time (4 to 5 minutes) is required for data accumulation, and the images are subject to motion blurring. Soon after the introduction of pencil-like beam scanners, fan-shaped x-ray beams were introduced so that multiple measures of x-ray transmission could be made simultaneously (Figure 15-1B). Fan beam geometries with increments of a few degrees for the different angular orientations (e.g., a 30-degree fan beam and 10-degree angular increments) reduced the scan time to 20 to 60 seconds. Fan beam geometries also improved image quality by reducing the effects of motion. CT scanners with x-ray fan beam geometries and multiple radiation detectors constitute the *second generation* of CT scanners.

The third and fourth generations of CT scanners eliminate the translational motion of previous scanners and rely exclusively upon rotational motion of the x-ray tube and detector array (third generation Figure 15-1, C) or upon rotational motion of the x-ray tube within a stationary circular array of 700 or more detectors (fourth-generation scanner, Figure 15-1D). With these scanners, data accumulation times as short as 1 second are achievable.

The first CT scan of a patient was acquired at London's Atkinson-Morley Hospital in 1972.

In the mid-1970s more than 20 companies were developing x-ray CT scanners for the clinical market. Today, CT scanners are manufactured only by large international manufacturers of x-ray imaging equipment.

Because of their extended data acquisition and image reconstruction times, first- and second-generation CT scanners were limited principally to studies of the head and extremities where methods to immobilize anatomy could be employed.

Spiral CT Scanning

Several approaches to even faster CT scans have been pursued. Until recently, multiple scan sequences to produce contiguous image "slices" required that the x-ray tube stop its rotation and reverse its direction because the maximum extension of the

Development of a purely rotational CT scanner required a more complex reconstruction algorithm that accommodated a purely rotational geometry. This algorithm was developed in the mid-1970s by scientists at General Electric Medical Systems.

Spiral CT scanning is also known as *helical CT scanning*.

Spiral CT scanners today employ multiple detector rings to scan several slices through the body during each gantry rotation. These scanners are referred to as multislice CT scanners.

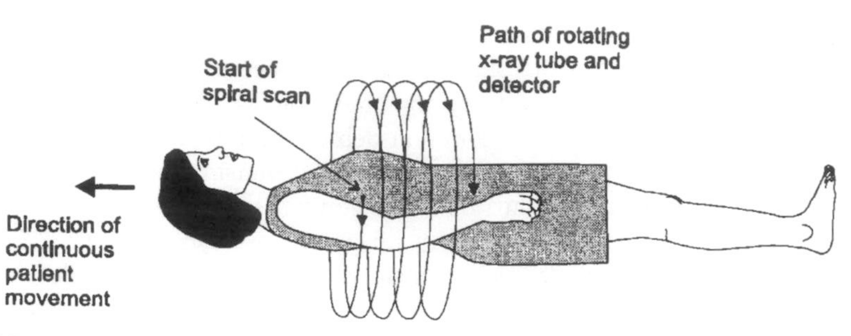

FIGURE 15-2
Spiral CT scanning. (From Rowlands, J.[14] Used with permission.)

The principal advantage of spiral CT is its ability to image a larger volume of tissue in a relatively short time. With spiral CT, for example, the entire torso can be imaged in a single breathold.

A pitch of one yields a contiguous spiral. A pitch of two yields an extended spiral. A pitch of $^1/_2$ yields an overlapping spiral.

Multislice CT scanners have many advantages, but also some disadvantages. Hundreds of CT images can be accumulated in a single study, resulting in high patient doses and massive amounts of digital imaging data.

high-voltage cables had been reached. Thus, a successive slice-by-slice accumulation technique was used to produced multislice images. In this technique, the total image acquisition time is significantly longer than the beam-on time because the table increments (moves) to the next slice location and the patient breathes between slices.

Spiral CT scanning was introduced in 1989 and is used almost universally today for third- and fourth-generation CT scanning. In this approach, image acquisition time is decreased significantly by connecting the tube voltage cables through a "slip ring" or sliding contact mounted on the rotating gantry of the unit. With slip ring technology, the x-ray tube rotates while the patient table moves without stopping. Hence, the patient is moved continuously through the gantry during the study, and the x-ray beam maps out a helical or spiral path in the patient, as depicted in Figure 15-2. Potential advantages of the spiral CT technique include a reduction of patient motion and a general increase in patient throughput.[15] A greater volume of the patient may be scanned during the passage of contrast media, permitting reduction in the volume of contrast needed. Also, the continuity of data along the axis of the patient (i.e., absence of gaps between scans) improves the quality of three-dimensional reconstruction.[16]

In single-slice CT scanning, *pitch* is defined as the *patient couch movement per rotation divided by the slice thickness*. In multislice CT, this definition is altered slightly to *patient couch movement per rotation divided by the beam width*. Low pitch (i.e., small increments of couch movement) yields improved spatial resolution along the long axis (Z axis) of the patient, but also results in higher patient doses and longer imaging times. For pitches greater than unity, the dose to the patient is less, but data must be interpolated so that resolution along the Z axis is preserved. Advantages of spiral over sectional computed tomography are listed in Table 15-1.

TABLE 15-1 Advantages of Spiral Compared with Conventional Computed Tomography

- Faster image acquisition
- Quicker response to contrast media
- Fewer motion artifacts
- Improved two-axis resolution
- Physiological imaging
- Improved coronal, sagittal, and 3D imaging
- Less partial volume artifact
- No misregistration

Ultrafast CT Scanners

Other approaches to fast CT scanning have involved radically different approaches to equipment design. In the late 1970s the first approach to subsecond CT scans was proposed by a group at the Mayo Clinic.[17] This approach, known as the dynamic spatial reconstructor (DSR), incorporated 28 gantry-mounted x-ray tubes over a 180-degree arc and used an equal number of image intensifier assemblies mounted on the opposite semicircle of the gantry. The entire assembly rotated about the patient at a rate of 15 rpm to provide 28 views every 1/60 second. Working models of the system were built for research,[18] but the technical complexity and cost prevented the DSR from being marketed commercially. A diagram of the DSR is shown in Figure 15-3.

Research on the dynamic spatial reconstructor has been discontinued.

Another approach to fast CT scanning eliminates mechanical motion entirely by converting the gantry of the unit into a giant x-ray tube in which the focal spot moves electronically about the patient.[20] This device, known as ultrafast CT (UFCT), cardiovascular CT (CVCT), or "cine CT," incorporates a semicircular tungsten x-ray target into the gantry.[21,22] A scanning electron beam with an energy of 130 keV is swept around the semicircular target so that the focal spot moves around the patient. A stationary semicircular bank of detectors records the x-ray transmission in a fashion

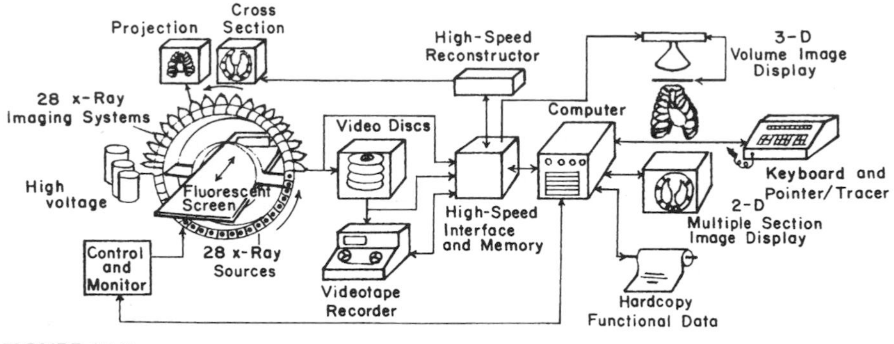

FIGURE 15-3
The dynamic spatial reconstructor. (From Robb, R., Lent, A., and Chu, A.[19] Used with permission.)

The UFCT is often referred to as electron-beam computed tomography [EBCT]. It was initially referred to as a cardiovascular CT scanner.

similar to that of a fourth-generation scanner. Because of the speed with which the electron beam may be steered magnetically, a scan may be accomplished in as little as 50 ms and repeated after a delay of 9 ms to yield up to 17 images per second.[23] By using four target rings and two detector banks, eight slices of the patient may be imaged without moving the patient. A diagram of a scanning electron beam CT scanner is shown in Figure 15-4.

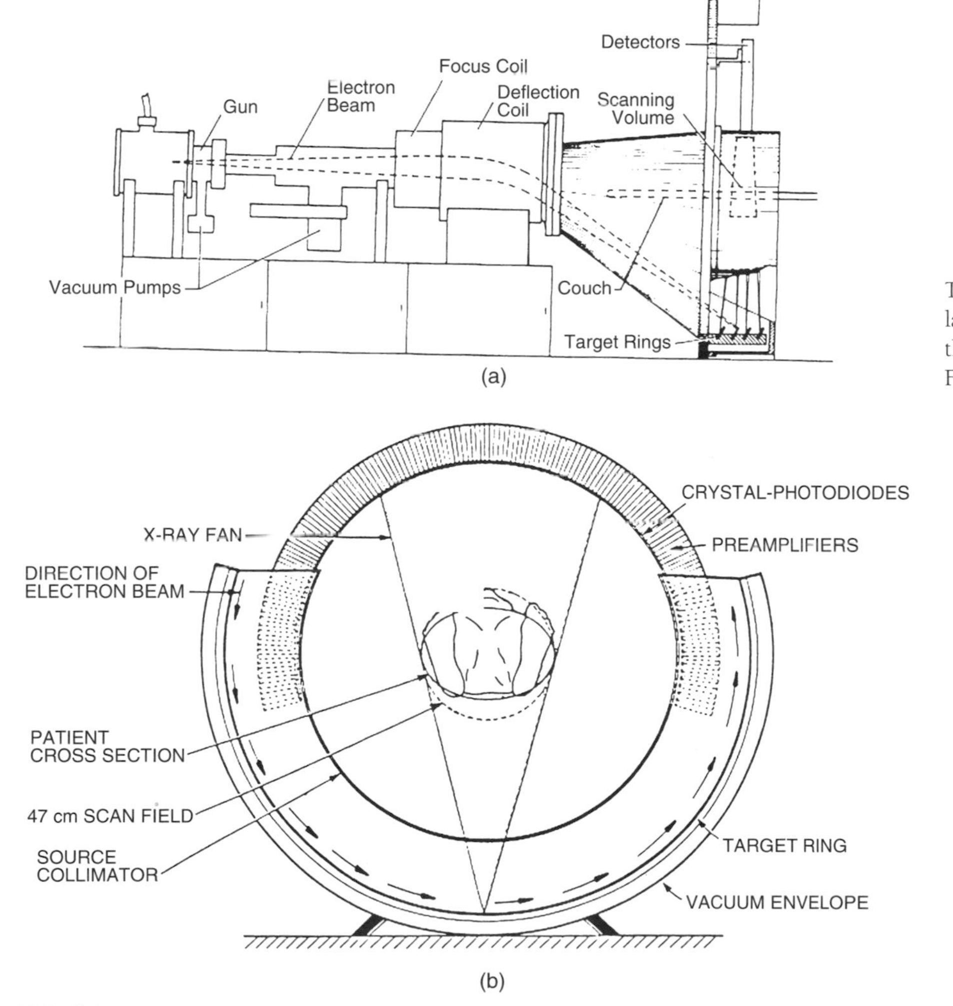

The UFCT unit was developed in the late 1970s by D. Boyd and colleagues at the University of California—San Francisco.

FIGURE 15-4
Longitudinal (**A**) and cross-sectional (**B**) view of the Imatron UFCT scanner.[24]

In CT units, the heel effect is eliminated by placing the anode–cathode axis of the x-ray tube at right angles to the long axis of the patient.

X-RAY SOURCES

Both stationary- and rotating-anode x-ray tubes have been used in CT scanners. Many of the translate–rotate CT scanners have an oil-cooled, stationary-anode x-ray tube with a focal spot on the order of 2 × 16 mm. The limited output of these x-ray tubes necessitates a sampling time of about 5 ms for each measurement of x-ray transmission. This sampling time, together with the time required to move and rotate the source and detector, limits the speed with which data can be accumulated with CT units using translational and rotational motion.

CT scanners employ compact, high-frequency x-ray generators that are positioned inside the CT gantry. In some units the generator rotates with the x-ray tube, while in others the generator is stationary.

To reduce the sampling time to 2 to 3 ms, newer CT units use 10,000 rpm rotating-anode x-ray tubes, often with a pulsed x-ray beam, to achieve higher x-ray outputs. To meet the demands of high-speed CT scanning, x-ray tubes with ratings in excess of 6 million heat units are becoming standard.

A special x-ray filter is used in CT to make the intensity of the x-ray beam more uniform. This filter is often referred to as "bow-tie filter."

COLLIMATION

Multislice CT scanning has placed added demands on the capacity of x-ray tubes to sustain high power levels over long periods of time.

After transmission through the patient, the x-ray beam is collimated to confine the transmission measurement to a slice with a thickness of a few millimeters. Collimation also serves to reduce scattered radiation to less than 1% of the primary beam intensity. The height of the collimator defines the thickness of the CT slice. This height, when combined with the area of a single picture element (pixel) in the display, defines the three-dimensional volume element (voxel) in the patient corresponding to the two-dimensional pixel of the display.

The voxel size is the major influence on image resolution in most CT units.

A voxel encompassing a boundary between two tissue structures (e.g., muscle and bone) yields an attenuation coefficient for the pixel that is intermediate between the values for the two structures. This "partial-volume artifact" may be reduced by narrowing the collimator to yield thinner slices. However, this approach reduces the number of x rays incident upon the detector. With fewer x rays interacting in the detector, the resulting signals are subject to greater statistical fluctuation and yield a noisier image in the final display.

CT units employ two types of collimators: source (prepatient) collimators to shape the x-ray beam and limit patient dose, and detector (postpatient) collimators to control slice thickness.

Reducing CT slice thickness yields the following:

- Decreased partial volume artifact
- Fewer x rays incident upon the detector
- Noisier images

X-RAY DETECTORS

To reduce the detector response time, all detectors used in CT are operated in current rather than pulse mode. Also, rejection of scattered radiation is accomplished with detector collimators rather than pulse height analyzers. Detectors for CT scanning, either gas-filled ionization chambers or solid-state detectors, are chosen for their detection efficiency, short response time, and stability of operation. Solid-state detectors include NaI (Tl), CaF, and CsI scintillation crystals; ceramic materials containing rare-earth oxides; and bismuth germanate (BGO) and cadmium tungstate [Cd WO_4] detectors chosen for their high detection efficiency and low fluorescence decay time. Gas-filled ionization chambers contain xenon pressurized up to 25 atm to improve their x-ray detection efficiency. With any detector, the stability of response from one transmission measurement to the next is essential for production of artifact-free images. With a rotational source and detector geometry, for example, detector instability gives rise to ring-shaped artifacts in the image. Minimum energy dependence of the detectors over the energy range of the x-ray beam is also important if corrections for beam hardening are to be applicable to all patient sizes and configurations.

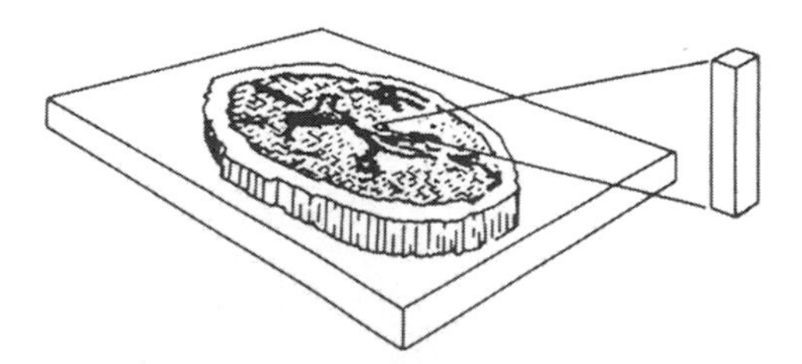

MARGIN FIGURE 15-2
A three-dimensional volume of tissue (voxel) displayed as a two-dimensional element in the CT image (pixel).

High-pressure xenon detectors provide detection efficiencies of about 50%. The detection efficiency of solid-state detectors used in CT is about 80%.

VIEWING SYSTEMS

The numbers computed by the reconstruction algorithm are not exact values of attenuation coefficients. Instead, they are integers termed CT numbers that are related to attenuation coefficients. On most CT units, the CT numbers range from −1000

TABLE 15-2 Electron Densities of Various Body Tissues[a]

Tissue	Electron Density (Electrons/cm^3)	Physical Density (g/cm^3)
Water	3.35×10^{-23}	1.00
Bone	3.72–5.59	1.2–1.8
Spleen	3.52	1.06
Liver	3.51	1.05
Heart	3.46	1.04
Muscle	3.44	1.06
Brain		
White matter	3.42	1.03
Gray matter	3.43	1.04
Kidney	3.42	1.05
Pancreas	3.40	1.02
Fat	3.07	0.92
Lung	0.83	0.25

[a] Data were calculated from the atomic composition and physical density.
Source: Geise and McCullough.[25] Used with permission.

In a CT image, higher CT numbers are brighter and lower CT numbers are darker.

for air to +1000 for bone, with the CT number for water set at 0. The relationship between CT number and linear attenuation coefficient μ of a material is

$$\text{CT number} = 1000\frac{(\mu - \mu_w)}{\mu_w}$$

where μ_w is the linear attenuation coefficient of water.

CT numbers normalized in this manner provide a range of several CT numbers for a 1% change in attenuation coefficient.

CT numbers are occasionally, but unofficially, referred to as *Hounsfield units*.

Linear attenuation coefficients of various body tissues for 60 keV x rays.[23]

Tissue	M (cm^{-1})
Bone	0.528
Blood	0.208
Gray matter	0.212
White matter	0.213
CSF	0.207
Water	0.206
Fat	0.185
Air	0.0004

A television monitor is used to portray CT numbers as a gray-scale visual display. This viewing device contains a contrast enhancement feature that superimposes the shades of gray available in the display device (i.e., the dynamic range of the display) over the range of CT numbers of diagnostic interest. Control of image contrast with the contrast enhancement feature is essential in x-ray CT because the electron density, and therefore the x-ray attenuation, are remarkably similar for most tissues of diagnostic interest. This similarity is apparent from the data in Table 15-2. The same cross-sectional CT data displayed at different settings of the "window" of the contrast enhancement control are illustrated in Figure 15-5. The viewing console of the CT scanner may contain auxiliary features such as image magnification, quantitative and statistical data display, and patient identification data. Also, many scanners permit the display of coronal and sagittal images by combining reconstruction data for successive slices through the body.

Filtering of lower-energy x rays from the x-ray beam as it penetrates the patient yields a beam of slightly higher energy in the center of the patient. This effect results in reduced attenuation coefficients in the center compared with the periphery. Hence, the center of the image contains pixels of reduced optical density. This effect is known as the "beam-hardening" artifact.

PATIENT DOSE

The radiation dose delivered during a CT scan is somewhat greater than that administered for an equivalent radiographic image. A CT image of the head requires a dose of about 1 to 2 rad, for example, whereas an abdominal CT image usually requires a dose of 3 to 5 rad. These doses would have to be increased significantly to improve the contrast and spatial resolution of CT images. The relationship between resolution and dose can be approximated as

Image storage devices for CT include magnetic tape and disks, digital videotape, and optical disks and tape.

$$D = a\left(\frac{s^2}{e^3 b}\right) \quad (15\text{-}1)$$

where D is the patient dose, s is the signal/noise ratio, e is the spatial resolution, b is the slice thickness, and a is a constant. From Eq. (15-1), the following are apparent:

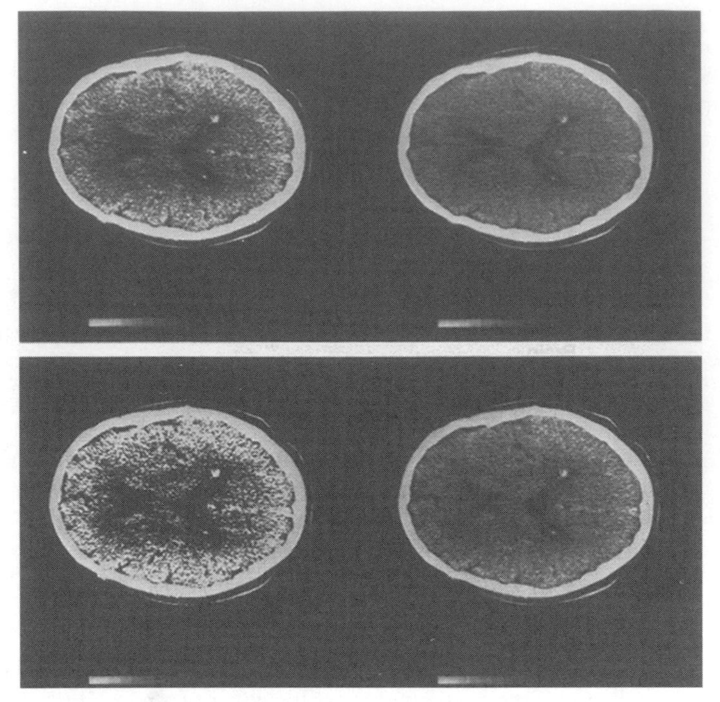

FIGURE 15-5
X-ray attenuation data at four positions of the window level of the contrast enhancement control.

1. A twofold improvement in the signal-to-noise ratio (contrast resolution) requires a fourfold increase in patient dose.
2. A twofold improvement in spatial resolution requires an eightfold increase in patient dose.
3. A twofold reduction in slice thickness requires a twofold increase in patient dose.

In multislice computed tomography, patient dose is described as the CT dose index (CTDI). When the distance that the patient moves between slices (the couch increment CI) equals the slice thickness ST, the CTDI equals the dose averaged over all slices (multislice average dose MSAD). When the couch increment is less than the slice thickness, the MSAD is the CTDI multiplied by the ratio of the slice thickness to the couch increment; that is,

$$\text{MSAD} = \text{CTDI}\left[\frac{\text{ST}}{\text{CI}}\right]$$

Patient dose decreases significantly outside of the slice. A conservative rule of thumb (i.e., an overestimate) is that the dose is 1% of the in-slice dose at an axial distance of 10 cm from the slice.

Techniques for measurement of dose parameters in CT have been described in detail by Cacak.[26]

In projection radiography, the dose is greatest where the x-ray beam enters the patient. In computed tomography the dose is relatively uniform across the section of tissue exposed to the x-ray beam, because the x-ray beam rotates around the patient during exposure.

QUALITY CONTROL

Many electronic components and massive amounts of data processing are involved in producing a CT image. A consequence of the separation between data acquisition and image display is the difficulty of observing and investigating imaging system problems through observation of the image alone. In such a complex system, image quality can be ensured only through prospective monitoring of system components and tests of

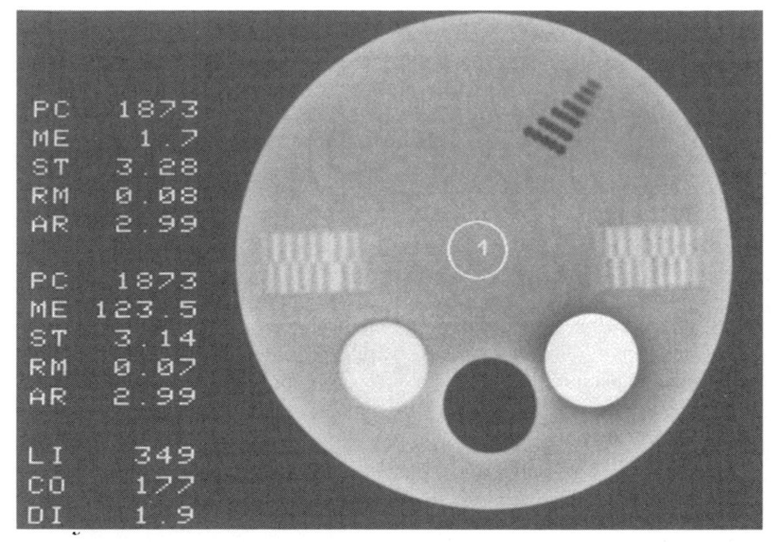

FIGURE 15-6
CT image of a quality control phantom. Image quality is evaluated by analysis of regions of interest and by visual inspection. The mean and standard deviation of pixel values in region 1 indicate CT number calibration, while comparison of region 2 with region 1 yields contrast information. The serrated patterns at 3 and 9 o'clock on the image indicate slice thickness and alignment. The rows of small dark circles (low CT number) at 1 o'clock is an indication of high contrast resolution. (Courtesy of Richard Geise, University of Minnesota Hospital; and Radiation Measurements Incorporated, Middleton, Wisconsin.)

overall system performance with standard phantoms. These measurements should be correlated with patient dose to ensure that the proper balance is maintained among variables that affect contrast, spatial resolution, image noise, and patient radiation dose.

Typical measurements of CT performance are given in Table 15-3, and examples are shown in Figure 15-6. The fundamental system performance indicators are CT number, resolution, noise, and patient dose.[26] The accuracy of CT numbers is measured by scanning a water-filled phantom at least monthly. The CT number for water should be zero over a 20-cm-diameter phantom, with a variation of less than 1 CT number. Deviation from the expected CT number of 0 for water at any energy is adjusted by applying a correction factor for the pixel value. Constancy of the value should be monitored with a daily scan.

TABLE 15-3 Common Quality Control Measurements for Computed Tomography

Measurement	*Frequency*
CT number	
Accuracy	Monthly
Constancy check	Daily
Noise	
Evaluation	Semiannually
Constancy check	Daily
Resolution	Monthly
Patient dose	Semiannually

A graph of CT number versus μ should yield a straight line passing through zero for water. This measure, known as the CT number linearity, is essential for quantitative computed tomography.

An overall check of system performance is obtained from semiannual measurements of CT image noise, defined as the standard deviation of CT numbers in a region of interest. Constancy of performance is checked by evaluation of the standard deviation in the daily water scan mentioned previously. Resolution is measured by scanning phantoms on a monthly basis. Of particular importance is low contrast resolution, which is a sensitive indicator of changes in component performance as they affect noise. Patient dose is evaluated semiannually. Specially designed ionization chambers[27] provide measurements from which the dose may be calculated for the exposure conditions (narrow beam, variable slice thickness) used in CT.[28] The values should agree with manufacturer's specifications to within 20%.

A variety of physical and mechanical factors such as patient couch positioning and indexing should be measured as part of a comprehensive quality control program. The performance of the hard-copy device and system monitors should be checked for distortion, brightness, contrast adjustment, and so on. The accuracy of image analysis features such as distance measurements and measurements of bone density should also be independently evaluated. Additional information on quality control in CT is available in the publications of a number of advisory groups and individuals.[29–32]

Reduction in CT number at the center of a water scan is termed "cupping." A heightened CT number at the scan center is know as "peaking." Either result compromises image quality.

SUMMARY

- X-ray transmission CT yields cross-sectional, sagittal, and coronal images with exquisite contrast resolution.
- CT imaging employs the principle of reconstructing images from measurements of x-ray transmission through the body.
- A variety of mathematical models are available for reconstructing x-ray images.
- Special demands are imposed on the x-ray sources and detectors used in CT imaging.
- Quality control and dose limitations are essential features of x-ray CT imaging.

PROBLEMS

15-1. Compare the spatial and contrast resolution of conventional radiographic and CT images.

15-2. List three reasons why contrast resolution is improved in CT imaging compared with conventional radiographic imaging.

15-3. Describe the mechanical features of the four generations of CT scanners.

15-4. Briefly describe the simple backprojection approach to image reconstruction, and explain how the convolution method improves these images.

15-5. What are the two major constraints on further reductions in CT scan time?

15-6. Explain the relationship between a voxel and a pixel.

15-7. What types of detectors are used in CT scanners?

15-8. Estimate the fetal dose 10 cm from a chest CT examination.

15-9. Define a CT (Hounsfield) unit, and explain the purpose of contrast enhancement in a CT viewing device.

15-10. Explain how image noise and patient dose in CT scanning are influenced by the signal-to-noise ratio, the size of each resolution element, and the slice thickness.

REFERENCES

1. Bracewell, R. Strip integration in radio astronomy. *Aust. J. Phys.* 1956; **9**:198.
2. DeRosier, D., and Klug, A. Reconstruction of three dimensional structures from electron micrographs. *Nature* 1968; **217**:130.
3. Rowley, P. Quantitative interpretation of three-dimensional weakly refractive phase objectives using holographic interferometry. *J. Opt. Soc. Am.* 1969; **59**:1496.
4. Berry, M., and Gibbs, D. The interpretation of optical projections. *Proc. R. Soc.* [A] 1970; **314**:143.
5. Webb, S. *From the Watching of Shadows*. New York, Adam Hilger, 1990.
6. Oldendorf, W. Isolated flying spot detection of radiodensity discontinuities—displaying the internal structural pattern of a complex object. *IRE Trans. Biomed. Electron.* 1961; **8**:68.
7. Kuhl D, Edwards R. Image separation radioisotope scanning. *Radiology* 1963; **80**:653.
8. Kuhl, D., Hale, J., and Eaton, W. Transmission scanning. A useful adjunct to conventional emission scanning for accurately keying isotope deposition to radiographic anatomy. *Radiology* 1966; **87**:278.
9. Hounsfield, G. Computerized transverse axial scanning (tomography): Part I. Description of system. *Br. J. Radiol.* 1973; **46**:1016.
10. Ledley, R., DiChiro, G., Lussenhop, A., et al. Computerized transaxial x-ray tomography of the human body. *Science* 1974; **186**:207.
11. Hendee, W. *Physical Principles of Computed Tomography*. Boston, Little Brown & Co., 1983.
12. Radon, J. Uber die Bestimming von Funktionen durch ihre integralwerte laengs gewisser Mannigfaltigkeiten (on the determination of functions from the integrals along certain manifolds). *Ber. Saechs. Akad. Wiss. Leipzig Math. Phys. Kl.* 1917, **69**:262.
13. Korenblyum, B., Tetelbaum, S., and Tyutin, A. About one scheme of tomography. *Bull. Inst. Higher Educ.—Radiophys.* 1958, **1**:151–157.
14. Rowlands, J. X-Ray imaging: Radiography, fluoroscopy, computed tomography, in Hendee, W. (ed.), *Biomedical Uses of Radiation. Part A: Diagnostic Applications*. Weinheim, Germany, Wiley-VCH, 1999.
15. Kalendar, W. A., Seissler, W., Klotz, E., et al. Spiral volumetric CT with single-breathhold technique, continuous transport, and continuous scanner rotation. *Radiology* 1990; **176**:181–183.
16. Hu, H. Multislice helical CT: Scan and reconstruction. *Med. Phys.* 1999; **26**:5.
17. Ritman, E. L., Robb, R. A., Johnson, S. A., et al. Quantitative imaging of the structure and function of the heart, lungs, and circulation. *Mayo Clin. Proc.* 1978; **53**:3–11.
18. Robb, R. A., Hoffman. E. A., Sinak, L. J., et al. High-speed three-dimensional x-ray computed tomography: The dynamic spatial reconstructor. *Proc. IEEE* 1983; **71**:308–319.
19. Robb, R. A., Lent, A. H., and Chu, A. A computer-based system for high-speed three-dimensional imaging of the heart and circulation: Evaluation of performance by simulation and prototype. *Proc. Thirteenth Hawaii Int. Conf. System Sci.* 1980; **III**:384–405.
20. Boyd, D. P., Lipton, M. J. Cardiac computed tomography. *Proc. IEEE* 1983: 198–307.
21. Linuma, T. A., Tateno, Y., Umegake, Y., et al. Proposed system for ultra-fast computed tomography. *J. Comput. Assist. Tomogr.* 1977; **1**:494–499.
22. Peschmann, K. R., Napel, S., Couch, J. L., et al. High-speed computed tomography: Systems and performance. *Appl. Opt.* 1985; **24**:4052–4060.
23. Seeram, E. *Computed Tomography*. Philadelphia, W. B. Saunders, 2001, p. 66.
24. Gould, R. G. Computed tomography overview and basics, in *Specification, Acceptance Testing and Quality Control of Diagnostic X-Ray Imaging Equipment*, Proceedings of the American Association of Physicists in Medicine, Vol. 2. Santa Cruz, Calif, July 1991, pp. 930–968.

25. Geise, R. A., and McCullough, E. C. The use of CT scanners in megavoltage photon-beam therapy planning. *Radiology* 1977; **124**:133–141.
26. Cacak, R. K. Measuring patient dose from computed tomography scanners, in Seeram, E. (ed.), *Computed Tomography*. Philadelphia, W. B. Saunders, 2001, pp. 199–208.
27. Moore, M. M., Cacak, R. K., and Hendee, W. R. Multisegmented ion chamber for CT scanner dosimetry. *Med. Phys.* 1981; **8**:640–645.
28. Shope, T. B., Gagne, R. M., Johnson, G. D. A method of describing the doses delivered by transmission x-ray computed tomography. *Med. Phys.* 1981; **8**:488–495.
29. National Council on Radiation Protection and Measurements. *Quality Assurance for Diagnostic Imaging Equipment*. Report 99. Bethesda, MD, NCRP, 1988.
30. American College of Medical Physics. *Radiation Control and Quality Assurance Surveys*, Report no. 1. Reston, VA, American College of Medical Physics, 1986.
31. Cacak, R. K. Design of a quality assurance program: CT scanners, in Hendee, W. R. (ed.), *Selection and Performance Evaluation of Radiologic Equipment*. Baltimore, Williams & Wilkins, 1985.
32. Cacak, R. K. Quality control for computed tomography scanners, in Seeram, E. (ed.), *Computed Tomography*. Philadelphia, W. B. Saunders, 2001, pp. 385–407.

CHAPTER

16

INFLUENCES ON IMAGE QUALITY

OBJECTIVES

After completing this chapter, the reader should be able to:

- Define the meaning of image unsharpness, and identify and explain the variables that contribute to it.
- Define the meaning of image contrast, and identify and explain the variables that contribute to it.
- Define the meaning of image noise, and identify and explain the variables that contribute to it.
- Provide examples of image distortion and artifacts.

INTRODUCTION

"A picture may be called a 'self image' when it is taken as a visual expression of its own properties, and a 'likeness' when it is taken as a statement about other objects, kinds of objects, or properties. The first conception, more elementary, can exist without the second; the second, more sophisticated, combines with the first."[1] Medical images are used as expressions of the second type of picture, and they combine features of the object with those of the recording process to create (more or less) a 'likeness' of the object.

In medical imaging, the word "image" usually connotes a pattern of visible light photons transmitted through a partially transparent film or emitted from a fluorescent screen.[2]

The purpose of the medical image may be to convey information that contributes to (1) detection of a disease or injury; (2) delineation of the nature and extent of disease or injury; (3) diagnosis of the underlying cause of the disease or injury; (4) guidance of treatment of the disease or injury; and (5) monitoring the treatment and its consequences to determine its effectiveness. The degree to which the image achieves its purpose, or demonstrates that no disease or injury is present, is described by the vague term *image quality*. In part, image quality connotes how clearly the image displays information about the anatomy, physiology, and functional capacity of the patient, including alterations in these characteristics caused by disease or injury. This component of image quality is referred to as the *clarity* of information in the image. Image quality depends on other features also, including whether the proper areas of the patient are examined, whether the correct images are obtained, and even whether a disease or injury is detectable by imaging.

Image clarity is a measure of how well information of interest is displayed in an image. Clarity is influenced by four fundamental characteristics of the image: unsharpness, contrast, noise, and distortion and artifacts. In any image, the clarity of information is affected by these image properties and how they interact with each other.[3,4]

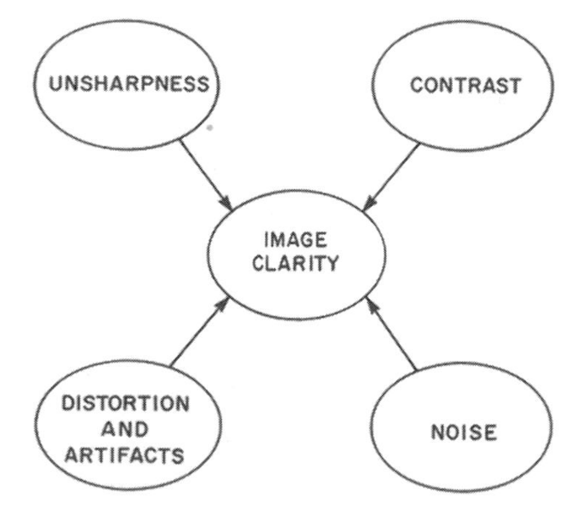

MARGIN FIGURE 16-1
Contributions to image clarity.

UNSHARPNESS

Every image introduces an element of "fuzziness" or blurring to well-defined (sharp) boundaries in the object (patient). This blurring is described as *unsharpness* and is depicted in Figure 16-1. Unsharpness (termed "blur" in some texts) is a measurable characteristic of images.

Unsharpness in an image is a consequence of four factors that contribute to image formation. These factors, considered collectively as the components of unsharpness, are geometric unsharpness U_g, subject unsharpness U_s, motion unsharpness U_m, and receptor unsharpness U_r. Under conditions where each of these factors contributes independently to the overall unsharpness of the image, the total unsharpness is

$$U_t = [(U_g)^2 + (U_s)^2 + (U_m)^2 + (U_r)^2]^{1/2}$$

A physician viewing an image with substantial unsharpness might comment that the image lacks detail. The term "detail" is a subjective reaction of the viewer to unsharpness present in an image.

Geometric Unsharpness

Geometric unsharpness is a direct consequence of the geometry of the image-forming process. In general, geometric unsharpness is influenced by the size of the radiation source and the distances between source and object (patient) and between object

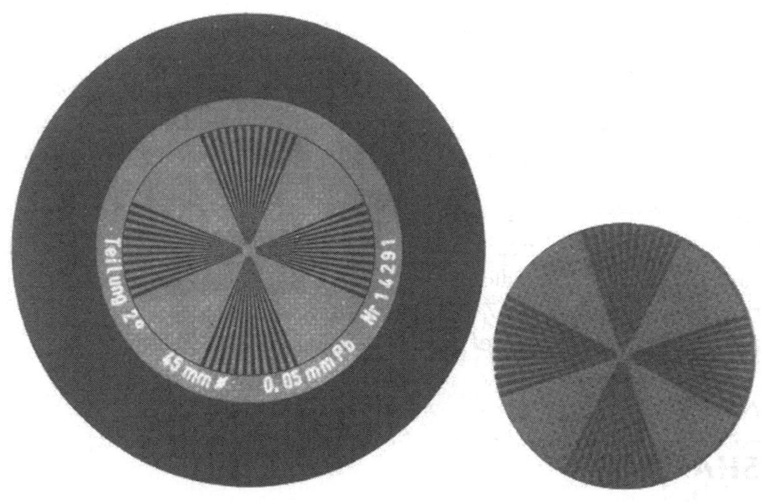

FIGURE 16-1
Image unsharpness is the blurring in an image of sharp edges in an object. The contact radiograph of a star test pattern on the left displays negligible unsharpness, while the x-ray image on the right reveals noticeable blurring. This is Figure 5-9, repeated here for convenience. (From Hendee, W. R., Chaney, E. L., and Rossi, R.P.[5] Used with permission.)

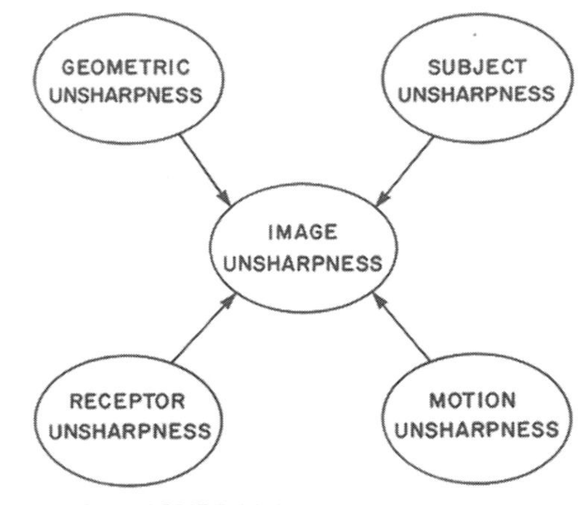

MARGIN FIGURE 16-2
Contributions to image unsharpness.

and image receptor. These influences are depicted in Figure 16-2 for the case of x-ray imaging. With minor modifications, they are applicable to all cases of radiologic imaging. In the upper left diagram of Figure 16-2, the radiation source is very small (i.e., a conventional focal spot of 0.6 to 1 mm). In this case, the sharp borders in the object are blurred over a finite region of the image. The image presents the borders with a degree of fuzziness that increases with the size of the focal spot. The blurring also increases with the distance of the image receptor from the object, as shown in the lower left diagram of Figure 16-2. By similar reasoning, the geometric unsharpness is reduced by moving the radiation source away from the object, thereby increasing the source-to-object distance (lower right diagram in Figure 16-2).

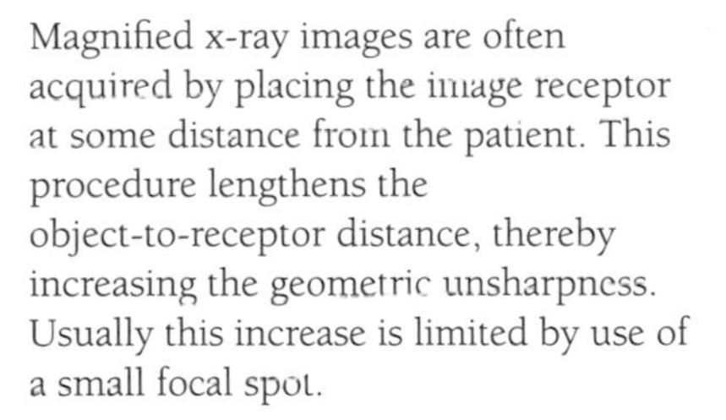
Magnified x-ray images are often acquired by placing the image receptor at some distance from the patient. This procedure lengthens the object-to-receptor distance, thereby increasing the geometric unsharpness. Usually this increase is limited by use of a small focal spot.

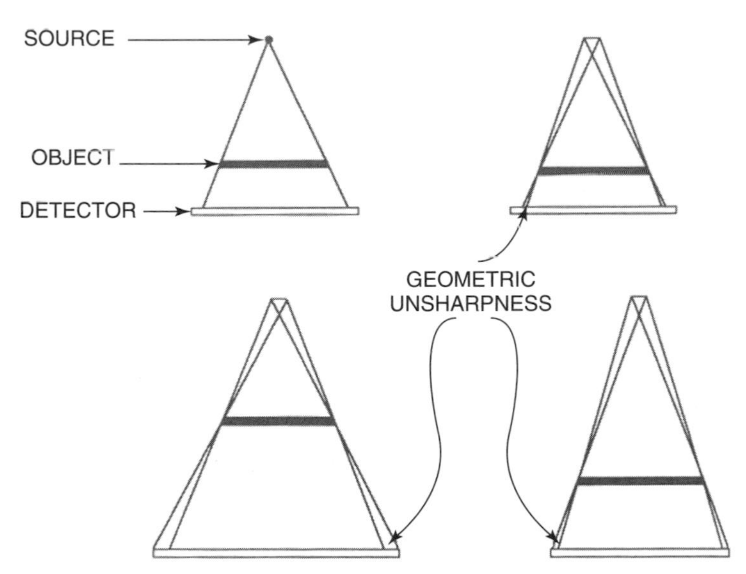

FIGURE 16-2
Geometric unsharpness. **Upper left:** Minimal unsharpness with a focal spot of infinitesimal size. **Upper right:** Unsharpness contributed by a focal spot of finite size. **Lower left:** Increased unsharpness caused by moving the receptor father from the object. **Lower right:** Reduction in unsharpness caused by moving the object farther from the radiation source and closer to the detector.

The relationship between the size of the radiation source and the distances between source and object and between object and image receptor (detector) can be expressed mathematically. The expression is

$$U_g = \frac{\text{(Focal spot size)(Object-to-receptor distance)}}{\text{(Source-to-object distance)}}$$

Geometric unsharpness sometimes is referred to as the "penumbra" or "edge gradient."

Compared with the cathode side, the anode side of the x-ray image reveals (a) reduced unsharpness because of the line-focus principle and (b) reduced x-ray intensity because of a greater heel effect.

Geometric unsharpness does not vary with the position of a boundary across an image. That is, objects near the edge of an image display the same geometric unsharpness as those near the center, provided that the projected size of the radiation source remains constant. This size is constant across the image in a direction perpendicular to the anode–cathode axis of the x-ray tube. In a direction parallel to the axis, however, the projected focal spot diminishes in size in a direction towards the anode. This effort, described earlier as the line focus principle, reduces the unsharpness on the anode side of the image compared with the cathode side.

Example 16-1

Determine the geometric unsharpness for a radiographic procedure with a 2-mm focal spot, object-to-receptor distance of 25 cm and source-to-receptor distance of 100 cm.

The source-to-object distance is 100 cm − 25 cm = 75 cm.

The image unsharpness is

$$U_g = \frac{\text{(Focal spot size)(Object-to-receptor distance)}}{\text{Source-to-object distance}}$$

$$= \frac{(2\text{ mm})(25\text{ cm})}{(75\text{ cm})}$$

$$U_g = 0.67\text{ mm}$$

Every imaging technology produces blurred images as a consequence of imaging features analogous to those described for radiography. For example, nuclear medicine images exhibit considerable unsharpness as a result of the finite diameter of the collimator holes and the thickness of the collimator. In computed tomography, geometric unsharpness is determined primarily by the size of the x-ray tube focal spot and the diameter and length of the collimators in front of the radiation detectors. In magnetic resonance imaging (MRI), image unsharpness is strongly influenced by the degree to which extracted signals can be localized to a specific region of tissue by the use of temporally and spatially varying magnetic fields. In ultrasound, image unsharpness is affected by many factors, including the width of the ultrasound beam and the distance between the transducer and reflecting boundaries separating different tissues in the body.

Subject Unsharpness

Not all structures within the patient present well-defined boundaries that can be displayed as sharp borders in the image. Often a structure is distinguishable anatomically from its surroundings only by characteristics that vary gradually over distance. Also, the shape of an object may prevent the projection of sharp boundaries onto the image receptor. This latter situation is depicted in Figure 16-3, where the object on the left presents sharp borders in the image and the object on the right is depicted with blurred edges. Image unsharpness that is contributed by the object is known as subject (object) unsharpness. It can be a result of the composition of the object, its shape, or a combination of both.

Subject unsharpness is sometimes referred to as "absorption unsharpness." Subject unsharpness may be described quantitatively with the expression

$$U_s = \frac{D_2 - D_1}{x}$$

where $D_2 - D_1$ is the difference in optical density or brightness over a boundary of linear dimension x[6].

Motion Unsharpness

Motion is often a major contributor to unsharpness in a radiologic image. Usually the motion occurs in the anatomic region of interest as a result of involuntary physiologic processes or voluntary actions of the patient. Motion causes boundaries in the patient to be projected onto different regions of the image receptor while the image is being

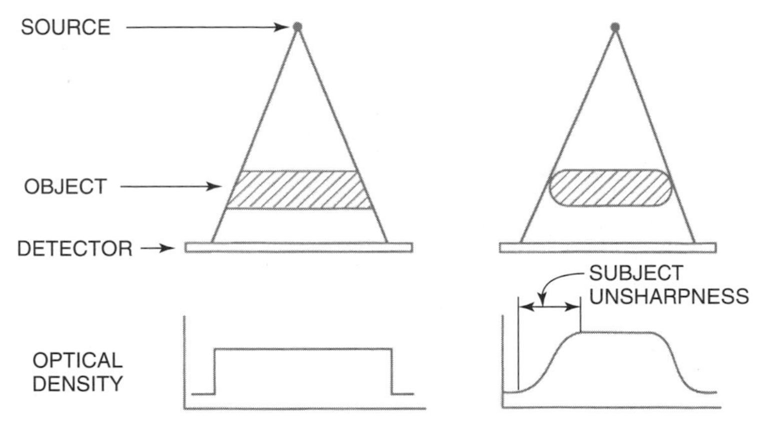

FIGURE 16-3
Subject unsharpness. Edges of the trapezoid on the left are parallel to the path of x-rays, and the resulting image reveals no subject unsharpness. The ellipsoid on the right yields an image of varying density from the edges to the center and presents the observer with substantial subject unsharpness.

formed. As a result, the boundaries are spread over a finite distance, and the resulting borders are blurred in the image.

Voluntary motion often can be controlled by keeping examination times short and asking the patient to remain still during the examination. However, these exhortations sometimes are ineffective, especially when the patient is an infant, demented, or in pain. In some cases, a family member can hold or soothe the patient and reduce voluntary motion during the examination. If the family member is exposed to radiation, protective garments should be provided to reduce the exposure. Occasionally, physical restraints and anesthetics may be required.

Typical Speeds of Anatomic Structures that Cause Blurring of the Image

Extremities	0.5–1 mm/sec
Head	1–2
Upper abdomen	20–40
Lower abdomen	15–30
Adult lungs	75–100
Infant lungs	150–200
Adult heart muscle	75–100
Infant heart muscle	150–200
Heart values	400–500
	~50 mi/hr

Involuntary motion is a more frequent problem and is especially troublesome when images of fast-moving structures such as the heart and great vessels are required. For many regions, motion can be "stopped" in the image by the use of very short examination times. In chest images, for example, examination times of a few milliseconds are used to gain a reasonable picture of the cardiac silhouette without the perturbing influence of heart motion. Short examination times are also desired in studies of the gastrointestinal tract to reduce motion unsharpness caused by peristalsis. In fact, short examination times are a general rule in radiography because voluntary and involuntary motion is invariably present to some degree when any anatomic region is being imaged.

The most frequent causes of motion unsharpness in x-ray images are heartbeat, the pulse, peristalsis, and the uncooperative patient.

Every image receptor requires a certain level of radiation exposure to produce an acceptable image. Some receptors are more sensitive to radiation than others and require less exposure. In radiography, for example, fast intensifying screens require less exposure than do detail screens, and shorter exposure times can be used. But fast screens produce more receptor unsharpness (discussed below), so the reduction in unsharpness achieved by using fast screens may be compromised by the greater receptor unsharpness contributed by the screens. Also, delivery of the required amount of radiation to a given set of screens must occur at higher intensity if short exposure times are used. Hence, higher tube currents are required with short exposure times so that the mA·s yields sufficient radiation to expose the screens properly. Higher tube currents (mA) require larger focal spots that produce greater geometric unsharpness in the image. In summary, every radiographic examination represents a trade-off among geometric, motion, and receptor unsharpness, with the goal of minimizing the total unsharpness in the image. This goal is achieved only by thoughtful selection of the tools and techniques of the examination based upon their characteristics and contributions to the image.

The radiographer should hold a patient during exposure only under extraordinary circumstances when no adult family member is available.

The evolution of x-ray tubes, including Coolidge's hot cathode, the rotating anode, and specially fabricated composites, reflects the nonending quest for shorter, higher-intensity exposures.

Receptor Unsharpness

The image receptor collects data generated during the imaging process and displays it as a gray-scale or color image. In some techniques such as conventional radiography,

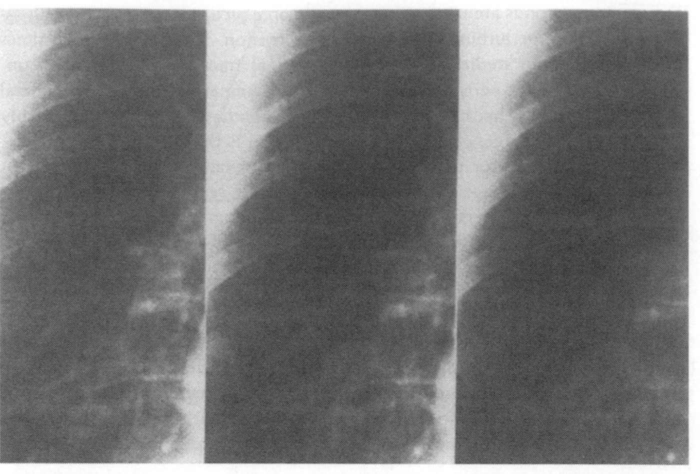

FIGURE 16-4
In this digital chest radiograph, multiple pulmonary nodules are displayed at three levels of spatial resolution. The finer display yields reduced receptor/display unsharpness. **Left:** 0.2 × 0.2-mm pixels. **Center:** 0.4 × 0.4-mm pixels. **Right:** 0.6 × 0.6-mm pixels. (From Foley, W. D.[8] Used with permission.)

In situations where image clarity is noise-limited, as discussed later in this chapter, receptor unsharpness sometimes can be used to reveal structures that are otherwise hidden (i.e., "buried in the noise"). The same effect can be achieved by defocusing the eyes slightly while viewing a noise-limited image.

The sensitivity of an x-ray receptor is the probability that an x ray will interact in the receptor. This probability is termed the *quantum detection efficiency n*.[7] This parameter is defined as $n = 1 - e^{-\mu x}$, where μ is the linear attenuation coefficient of the receptor material and x is the thickness of the receptor.

the receptor converts x-ray data to an image directly by the use of relatively simple devices such as intensifying screens and film. In other imaging methods the conversion process is more complex and employs a computer to form an image on a video display. In every display technique, no matter how simple, the image receptor inevitably adds unsharpness to the image. This contribution to image unsharpness is termed *receptor unsharpness*.

In radiography, receptor unsharpness is determined principally by the thickness and composition of the light-sensitive emulsion of the intensifying screens. These characteristics influence not only receptor unsharpness but also the sensitivity of the screens to x rays. With increasing thickness, for example, the sensitivity improves and the unsharpness increases. The choice of screens is, consequently, a trade-off between unsharpness introduced by the receptor and that resulting from motion caused by the finite time to record the imaging data.

In many imaging techniques, the technique of image display can influence the overall unsharpness of the image. In digital radiography, for example, the display device can provide different levels of unsharpness. In Figure 16-4 the same data are presented in three display formats. In the image on the left the data are presented in a fine matrix of 0.2 × 0.2-mm picture elements (pixels), and the image appears quite continuous and smooth. On the right the same data are displayed in a coarser matrix (0.6 × 0.6-mm pixels), and the image has a chunkier, rather "blocky" appearance. In this example, the coarser display adds significant unsharpness to the image.

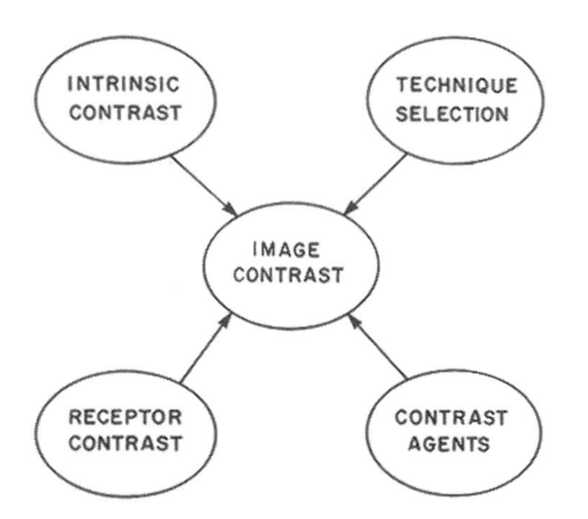

MARGIN FIGURE 16-3
Contributions to image contrast.

In medical imaging of soft tissues, contrast is frequently the image characteristic of greatest concern.

The contrast between adjacent regions in an image can be described quantitatively as $D_2 - D_1$, the difference in optical density or brightness between the two regions.

CONTRAST

Contrast is the second major feature of an image. This characteristic describes how well the image distinguishes subtle features in the object (patient). Image contrast is depicted in Figure 16-5. In diagnostic imaging, the contrast of an image is a product of complex interactions among the anatomic and physiologic attributes of the region of interest, the properties of the imaging method and receptor employed, and conscious efforts to influence both the intrinsic properties of the region and its presentation as an image. These interactions are characterized as four influences on image contrast, as depicted in the margin.

Intrinsic Contrast

The underlying philosophy of diagnostic imaging is that structures in the patient can be distinguished in an image because they differ in physical composition and

FIGURE 16-5
Different levels of image contrast. (From Hendee, W. R.[4] Used with permission.)

physiologic behavior. These differences are referred to as intrinsic (sometimes called "subject," "object," or "patient") contrast. Physical properties of the patient that contribute to intrinsic contrast are depicted in Table 16-1.

In radiography, intrinsic contrast is a reflection primarily of atomic number and physical density differences among different tissues. Some structures (e.g., breast) exhibit very subtle differences in composition and are said to have low intrinsic contrast. Other structures (e.g., chest) provide large differences in physical density and atomic composition and yield high intrinsic contrast.

In MRI, a number of tissue properties contribute to the imaging data, and the type of pulse sequence and its parameters are chosen to provide a desired level of contrast among tissues.

Imaging Technique

Every imaging application reflects a choice of a specific imaging technique and receptor among many alternatives to yield images of greatest potential to provide the desired information. For a specific imaging application, image contrast can be influenced by careful selection of the technique factors used to produce the image. In radiographic imaging of the breast, for example, subtle differences among tissues can be accentuated in the image by use of low peak kilovoltage and small amounts of beam filtration. These choices enhance the differential transmission of x rays through tissues that vary slightly in atomic composition and physical density. In chest radiography, relatively high peak kilovoltage and large amounts of beam filtration are chosen to suppress differential absorption of x rays in bone so that lesions of increased physical density can be detected in the lung parenchyma under the ribs. The choice of pulse sequence and other variables in MRI strongly influences the resulting contrast among structures in the image. This influence is shown in Figure 16-6 for the same anatomic region imaged by different pulse sequences.

Low kVp and small amounts of beam filtration yield a relatively "soft" x-ray beam with a significant likelihood of photoelectric interactions. High kVp and large amounts of beam filtration yield a relatively "hard" x-ray beam that interacts almost entirely by Compton interactions.

Contrast Agents

In some radiologic examinations a substance can be introduced into the body to enhance tissue contrast. This substance, termed a contrast agent (or contrast medium or dye), is selected to provide a signal different from that of the surrounding tissues. In angiography, a water-soluble agent containing iodine is injected into the circulatory system to displace blood and thereby enhance the contrast of blood vessels by increasing the attenuation of x rays impinging on the vessels. This technique can be further distinguished as arteriography or venography, depending on which side of

TABLE 16-1 Physical Properties that Influence Intrinsic Contrast for Various Imaging Techniques

Radiography	*Nuclear Medicine*	*Ultrasound*	*Computed Tomography*	*Magnetic Resonance*
Atomic number	Activity	Velocity	Physical density	Proton density
Physical density	Distribution	Physical density	Electron density	Relaxation times
Thickness	Thickness	Thickness	Flow	Flow

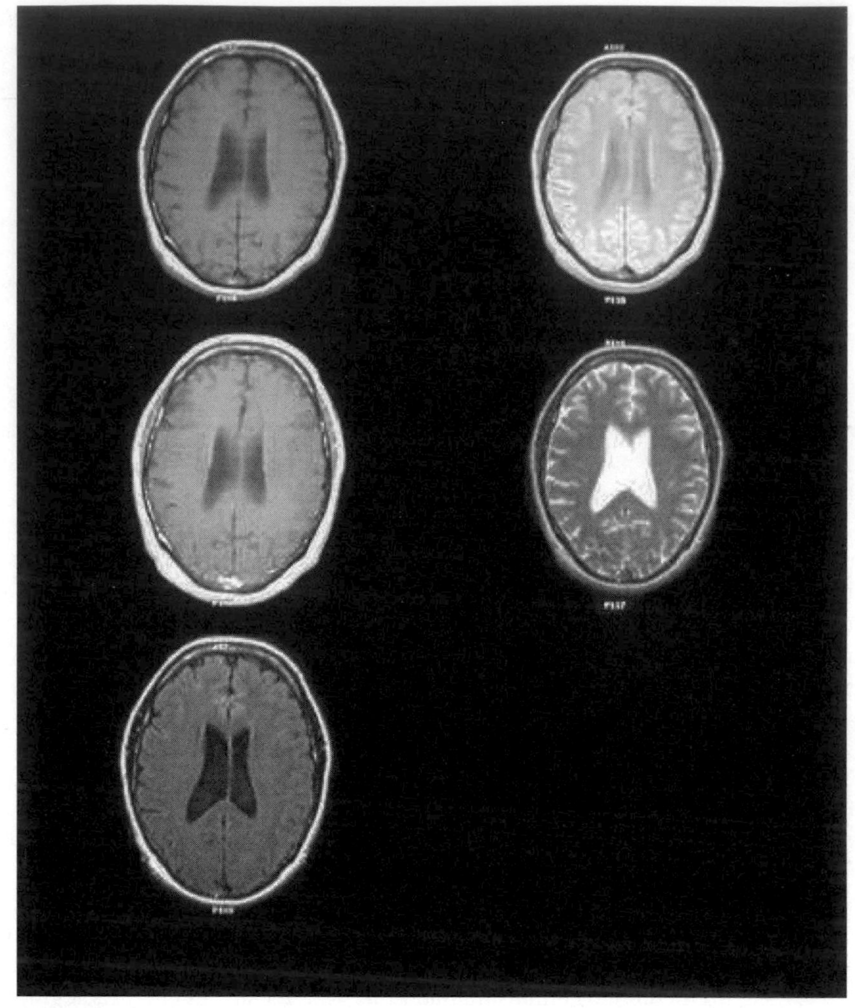

FIGURE 16-6
Representative magnetic resonance images of the normal brain
UL T_1 weighted image
ML T_1 weighted image with contrast
LL FLAIR (Fluid attenuated inversion recovery) image
UR Proton density image
LR T_2-weighted image
Source: Courtesy of David Daniels, M.D., Medical College of Wisconsin, Department of Radiology

the circulatory system the injection is made. The iodine attenuates x rays strongly because its K-absorption edge of 33 keV is ideally positioned with respect to the energy distribution of x rays typically employed for diagnosis.

Several other types of contrast agents are used in diagnostic imaging. For example, a contrast agent can be used to displace the cerebrospinal fluid in studies of the central nervous system. For studies of the spinal cord, the contrast agent may be "ionic" or "nonionic," and the procedure is termed "myelography." Contrast agents are useful in computed tomography as well as in radiography. Iodine-containing compounds are also helpful in studies of the kidneys and urinary tract, where the applications are known as urography. Compounds containing barium are frequently used for radiographic examination of the gastrointestinal tract, with the barium compound administered orally (barium swallows) or rectally (barium enemas). Solutions containing dissolved "microbubbles" of carbon dioxide are sometimes employed as contrast agents in ultrasound, and gadolinium and other substances are increasingly being used as contrast agents in MRI because of their influence on the relaxation constants of tissues.

New contrast agents, especially for MRI[10] and ultrasound,[11] are improving diagnostic imaging. With the arrival of nonionic agents for radiography and computed tomography, the risk of occasional reactions to contrast agents was reduced. These newer agents were initially more expensive and raised the issue of increased cost for a slight reduction in risk.[12] As the use of nonionic contrast agents has increased, the cost differential between ionic and nonionic agents has diminished and is no longer a major impediment to the use of nonionic agents.

Receptor Contrast

The contrast of an image is affected by the properties of the receptor used to form the image. In radiography, the choice of film has a substantial influence on image contrast. Observers who prefer a high-contrast image tend to select film that yields a steep characteristic (H–D) curve with high contrast and narrow latitude. Those who prefer less contrast and wider latitude of exposure choose a film with a more gradual slope for the characteristic curve.

When film is the image receptor, image contrast is highly dependent on the conditions under which the film is processed. A high level of quality control over film processing is essential to the production of acceptable film-based medical images.

Display of images on a video monitor adds flexibility in the choice of image contrast because the viewer can alter the receptor contrast. Just as with film, however, the dynamic range of the monitor is limited, and choosing a high-contrast setting limits the ability to display a wide range of exposures.

Increased flexibility in altering image contrast is one of the major advantages of digital imaging. When imaging data are available in digital form, the viewer can select any range of exposures for display in a gray-scale image. The range can be altered at will and the result captured on film if desired while the total range of data is preserved in computer storage. Shown in Figure 16-7 are displays at different contrast settings for the same set of computed tomographic data. Enhanced contrast and reduced exposure range are apparent in the images obtained with narrower windows.

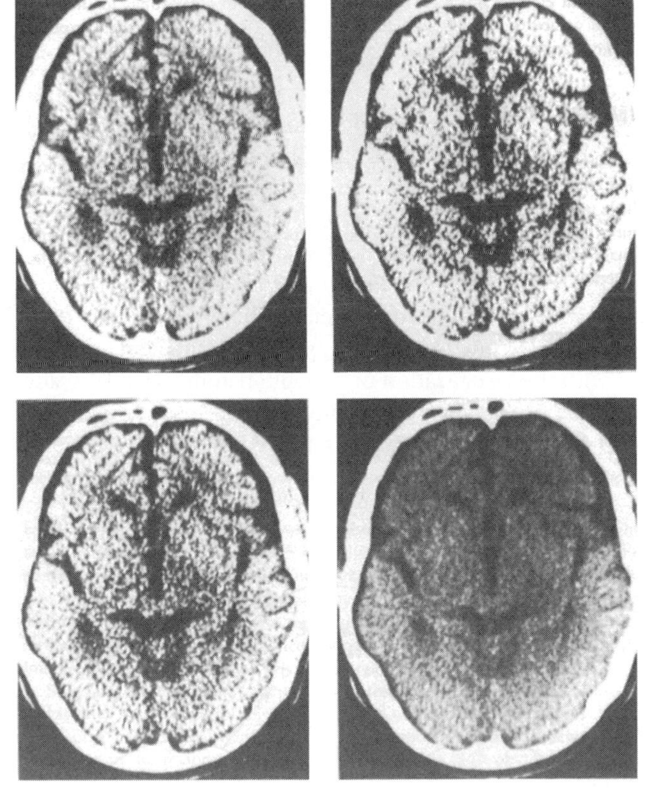

FIGURE 16-7
Computed tomographic data displayed with a constant window level but with different settings of window width. (From Hendee, W. R.[13] Used with permission.)

Noise in an imaging system can be demonstrated by exposing the system to a uniform fluence of radiation (e.g., x rays). The macroscopic variation in optical density or brightness about an average value is a result of the "noise" contributed by the imaging system.

IMAGE NOISE

Every radiologic image contains information that is not useful for diagnosis and characterization of the patient's condition. This information not only is of little interest to the observer, but often also interferes with visualization of image features crucial to the diagnosis. Irrelevant information in the image is defined as image noise. Image noise has four components: structure noise, radiation noise, receptor noise, and quantum mottle.[14]

Structure Noise

Information about the structure of the patient that is unimportant to diagnosis and characterization of the patient's condition is known as structure noise. For example, shadows of the ribs not only are irrelevant in chest radiographs taken to examine the lung parenchyma, but also can hide small lesions in the parenchyma that could be important to characterization of the patient's condition. However, rib shadows are important to the diagnosis of a cracked rib, and in this case the parenchyma and mediastinum could be characterized as structure noise. Structure noise is defined not only by the information present in the image but also by the reasons why the image was obtained in the first place.

Structure noise is one of the more disturbing features of radiologic images and is often responsible for missed lesions and undetected abnormalities. In radiography, patients frequently are positioned for examination so that structure noise is reduced in the region of interest in the image. The most obvious method of reducing structure noise is tomography, which blurs structures above and below the plane of interest within the patient. Tomographic images are obtained automatically in computed tomography and MRI, and the superior rendition of low-contrast structures by these technologies is attributable in part to their reduced level of structure noise.

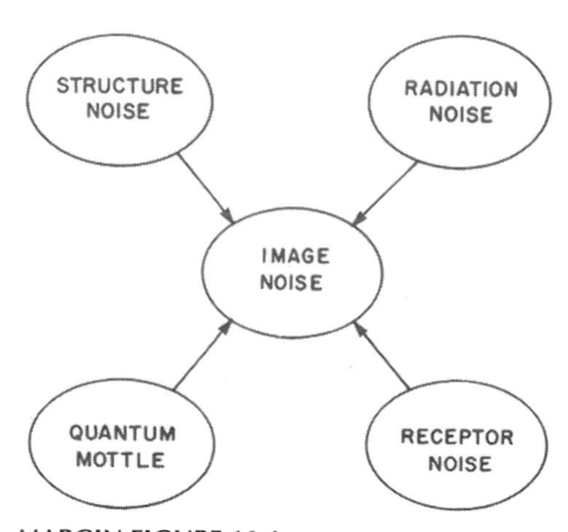

MARGIN FIGURE 16-4
Contributions to image noise.

Radiation Noise

Radiation noise is information present in a radiation beam that does not contribute to the usefulness of the image. For example, many x-ray beams exhibit a nonuniform intensity from one side to the other. This nonuniformity, termed the *heel effect,* is a result of the production and filtration of the beam in the anode of the x-ray tube. Although the nonuniformity provides information about these mechanisms, that information is irrelevant to diagnosis of the patient's condition. Similarly, the radiation beam exiting from the patient contains considerable scattered radiation. This radiation contains information about the patient, although in such a complicated form that it cannot be easily decoded. As it impinges on the image receptor, scattered radiation interferes with the visualization of patient anatomy. Hence, scattered radiation is radiation noise. In projection radiography, the contribution of scattered radiation to image noise is so severe that it must be removed by a grid to permit the structures of interest to be seen clearly.

The influence of radiation noise is often described by the signal-to-noise ratio (SNR), defined as:

$$\begin{aligned} \text{SNR} &= nN_0/[nN_0]^{1/2} \\ &= (nN_0)^{1/2} \end{aligned}$$

where n is the quantum detection efficiency and N_0 is the number of incident photons.

Radiation noise is present with every imaging technique, but is sometimes not very troublesome. In computed tomography, for example, the thin sections exposed during the acquisition of imaging data do not yield very much scattered radiation. In nuclear medicine, scattered radiation can be rejected by use of a pulse height analyzer because all scattered photons have energies less than the primary γ rays. In ultrasound, scattered radiation sometimes contributes to a "speckle pattern" in the image that aids in the diagnosis of certain conditions.

Receptor Noise

Most image receptors are not uniformly sensitive to radiation over their active surfaces. These receptors impose a pattern of receptor noise onto the image. In radiography,

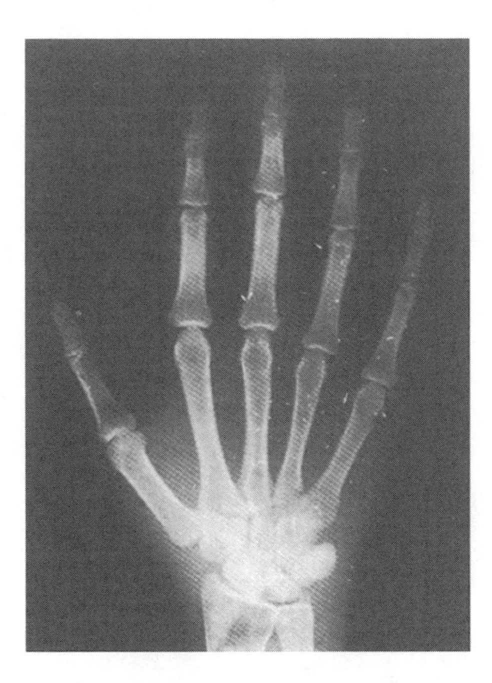

FIGURE 16-8
Contamination on an intensifying screen is a common contributor to receptor noise. (From Hendee, W. R.[4] Used with permission.)

intensifying screens and film are fairly uniform in their sensitivity to radiation, and their contribution to image noise is usually undetectable. However, contamination on a screen can add substantial noise to the image, as shown in Figure 16-8. In some imaging methods, receptor noise can be a major problem. In nuclear medicine, for example, unbalanced detectors can contribute significant receptor noise, as shown in Figure 16-9. In most scintillation cameras, minor variations in sensitivity among detectors are compensated automatically by computer monitoring and correcting. In MRI, surface coils contribute receptor noise through nonuniform patterns of sensitivity.[16]

Increasingly, imaging data are processed through computer algorithms to yield displayed images. Imperfections in the algorithms can produce imperfections in the images. These imperfections are an additional form of receptor noise. They are encountered most often with reconstruction algorithms used in CT, MRI, SPECT, and PET imaging.

Quantum Mottle

In many imaging methods, image noise is dominated by quantum mottle resulting from the finite number of information carriers (x rays, γ rays, etc.) used to create

FIGURE 16-9
Unbalanced detectors can produce considerable noise in nuclear medicine images. (From Adams, R.[15] Used with permission.)

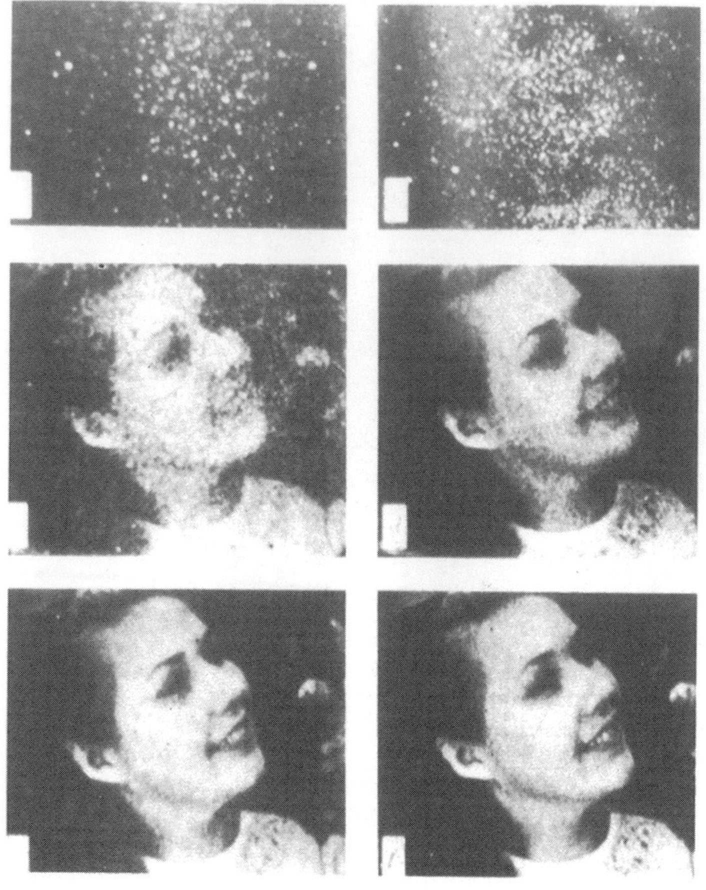

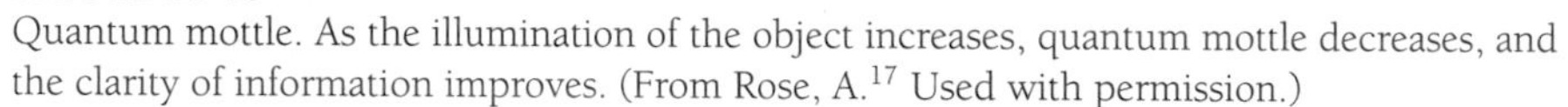

FIGURE 16-10
Quantum mottle. As the illumination of the object increases, quantum mottle decreases, and the clarity of information improves. (From Rose, A.[17] Used with permission.)

the image. Quantum mottle is illustrated in Figure 16-10. This contribution to image noise can be particularly disturbing in images of relatively low contrast. Quantum mottle can be reduced only by using more information carriers to form the image. In most cases, more information carriers are provided only at the expense of longer imaging times and increased radiation dose to the patient.

IMAGE DISTORTION AND ARTIFACTS

Distortion and artifacts are present to some degree in every diagnostic image. Image distortion is caused by unequal magnification of various structures in the image. In radiography, for example, structures near the intensifying screens are magnified less than those farther away, as shown in Figure 16-11. The experienced interpreter learns to use these differences in magnification to gain a perspective of depth among different structures in the image.

Magnification M in an image is the ratio of image to object size. In radiography, magnification can be computed as:

$$M = \frac{\text{Image size}}{\text{Object size}}$$

$$= \frac{\text{Source-to-receptor distance}}{\text{Source-to-object distance}}$$

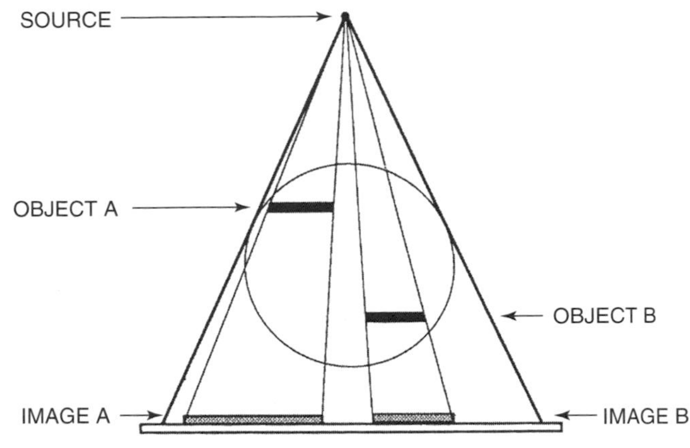

FIGURE 16-11
Distortion is the unequal magnification of different structures recorded in an image.

Artifacts can arise from so many causes that their complete description is not possible here. Examples include reverberation artifacts in ultrasound, streaks caused by moving structures of high density in computed tomography, and artifacts produced by nonuniformities in the magnetic field introduced by metallic structures (e.g., bridgework) in MRI.

Examples of artifacts are shown in Figure 16-12. Usually these artifacts are easily recognizable and do not interfere greatly with interpretation of the image.

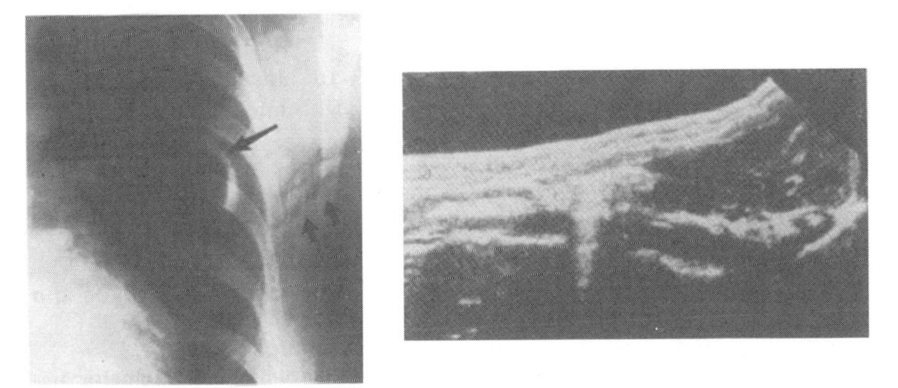

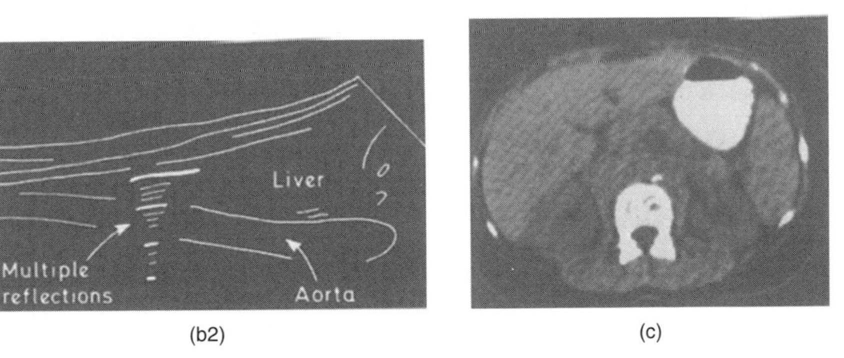

FIGURE 16-12
Examples of image artifacts. **A:** Radiographic artifacts caused by crinkling the film (*single arrow*) and exerting pressure on the film (*double arrows*). (From Sweeney, R. J.[18] Used with permission.) **B:** Reverberation artifacts cause by multiple reflections in ultrasonography. (From McDicken, W. R.[19] Used with permission.) **C:** Ring artifacts caused by imbalanced detectors in computed tomography. (From Morgan, C. L.[20] Used with permission.)

Example 16-2

For an image of 10 cm obtained at a source-to-receptor distance of 100 cm and an object-to-receptor distance of 20 cm, what is the size of the object?

The source-to-object distance is 100 cm – 20 cm = 80 cm.

$$M = \frac{\text{Image size}}{\text{Object size}} = \frac{\text{Source-to-receptor distance}}{\text{Source-to-object distance}}$$

$$\text{Object Size} = \text{Image Size}\left[\frac{(\text{Source-to-object distance})}{(\text{Source-to-receptor distance})}\right]$$

$$= 10\text{ cm}\left[\frac{80\text{ cm}}{100\text{ cm}}\right]$$

$$= 8\text{ cm}$$

SUMMARY

Image unsharpness is influenced by four factors:

- Geometric unsharpness
- Subject unsharpness
- Motion unsharpness
- Receptor unsharpness

Image contrast is influenced by four factors:

- Intrinsic contrast
- Image technique
- Contrast agents
- Receptor contrast

Image noise is influenced by four factors:

- Structure noise
- Radiation noise
- Receptor noise
- Quantum mottle

Image distortion is caused by unequal magnification of structure in the object. Image artifacts can arise from a multitude of conditions.

PROBLEMS

*16-1. Identify the four factors that influence the clarity of information in a diagnostic image.

*16-2. Identify the four contributions to image unsharpness.

*16-3. Estimate the total unsharpness in an image when the geometric unsharpness is 0.2 mm, motion unsharpness is 0.3 mm, subject unsharpness is 0.1 mm, and receptor unsharpness is 0.2 mm.

*16-4. For an object–receptor distance of 10 cm and a source–object distance of 90 cm, the geometric unsharpness is 0.2 mm. Determine the focal spot size.

*16-5. For a focal spot size of 1.0 mm and fixed source–object distance of 80 cm, what object–film distance would yield a geometric unsharpness of 0.3 mm?

*16-6. Identify the four factors that contribute to image contrast.

*16-7. Name the four contributions to image noise.

*For those problems marked with an asterisk, answers are provided on p. 492.

REFERENCES

1. Gregory, R. L. *The Oxford Companion to the Mind*. New York, Oxford University Press, 1987, p. 48.
2. Wohlbarst, A. B. *Physics of Radiology*. Norwalk, CT, Appleton & Lange, 1993, p. 148.
3. Sprawls, P., Jr. *Physical Principles of Medical Imaging*. Rockville, MD, Aspen Publications, 1987.
4. Hendee, W. R. Characteristics of the radiologic image, in Putman C. E, and Ravin, C. E. (eds.), *Textbook of Diagnostic Imaging*, 2nd edition. Philadelphia, W. B. Saunders , 1994.
5. Hendee, W. R., Chaney, E. L., and Rossi, R. P. *Radiologic Physics, Equipment and Quality Control*. St. Louis, Mosby–Year Book, 1977.
6. Seemann, H. E. *Physical and Photographic Principles of Medical Radiography*. New York, John Wiley & Sons, 1968.
7. Rowlands, J. A. X-Ray imaging: Radiography, fluoroscopy, computed tomography, in WR Hendee, *Biomedical Uses of Radiation, Part A—Diagnostic Applications*. New York: VCH Publishers (subsidiary of John Wiley & Sons), 1999, p. 100
8. Foley, W. D. The effect of varying spatial resolution on the detectability of diffuse pulmonary nodules. *Radiology* 1981; **141**:25–31.
9. Daniels, D. L., Haughton, V. M., and Naidich, T. P. *Cranial and Spinal Magnetic Resonance Imaging: An Atlas and Guide*. New York, Raven Press, 1987, pp. 6–7.
10. Birn, R. M., Donahue, K. M., and Bandettini, P. A. Magnetic resonance imaging: Principles, pulse sequences and functional imaging. In Hendee, W. R. (ed.), *Biomedical Uses of Radiation: Part A—Diagnostic Applications*. New York, VCH Publishers (subsidiary of John Wiley & Sons) 1999, pp. 431–548.
11. Kimme-Smith C. Ultrasound (including Doppler), in Hendee, W. R. (ed.), *Biomedical Uses of Radiation: Part A— Diagnostic Applications*. New York, VCH Publishers (subsidiary of John Wiley & Sons), 1999, pp. 381–418.
12. Lasser, E. C. , and Berry, C. C. Commentary. Nonionic vs ionic contrast agents. What does the data tell us? *AJR* 1989; **152**:945–947.
13. Hendee, W. R. *The Physical Principles of Computed Tomography*. Boston, Little, Brown & Co. 1983, p. 41.
14. Macovski, A. *Medical Imaging Systems*. Englewood Cliffs, NJ, Prentice-Hall, 1983, pp. 75–105.
15. Adams, R. The scintillation gamma camera, in Williams, L. E. (ed.), *Nuclear Medical Physics*, Vol. 2. Boca Raton, FL, CRC Press , 1987, p. 116.
16. Dendy, P. P., and Heaton, B. *Physics for Diagnostic Radiology*, 2nd edition. Philadelphia, Institute of Physics, 1999.
17. Rose, A. *Vision: Human and Electronic*. New York, Plenum, 1973. As reprinted in Hendee, W. R. Characteristics of the radiologic image, in Putman, C. E., and Ravin C. E. (eds.), *Textbook of Diagnostic Imaging*, 2nd edition. Philadelphia, W. B., Saunders, 1994.
18. Sweeney, R. J. *Radiographic Artifacts: Their Cause and Control*. Philadelphia, J. P. Lippincott, 1983, p. 46.
19. McDicken, W. R. *Diagnostic Ultrasound Principles and Use of Instruments*, 3rd edition. New York, John Wiley & Sons, 1981.
20. Morgan, C. L. *Principles of Computed Tomography*. Rockville, MD, Aspen Publishers, 1983, p. 284.

CHAPTER

17

ANALYTIC DESCRIPTION OF IMAGE QUALITY

Image clarity is an objective evaluation of the ability of the image to reveal fine anatomic detail.

Image quality is the subjective response of the observer to information in the image, and it incorporates not only the physical features of the image but also the perceptual properties of the observer responding to the information.

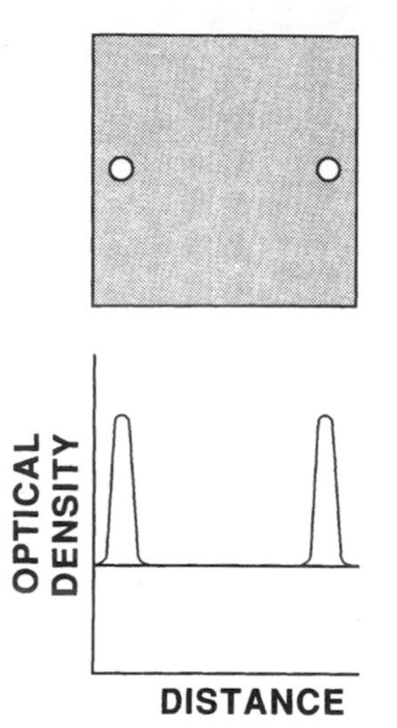

MARGIN FIGURE 17-1
When two point-like, high-contrast objects are widely separated they can be seen as separate entities in the image.

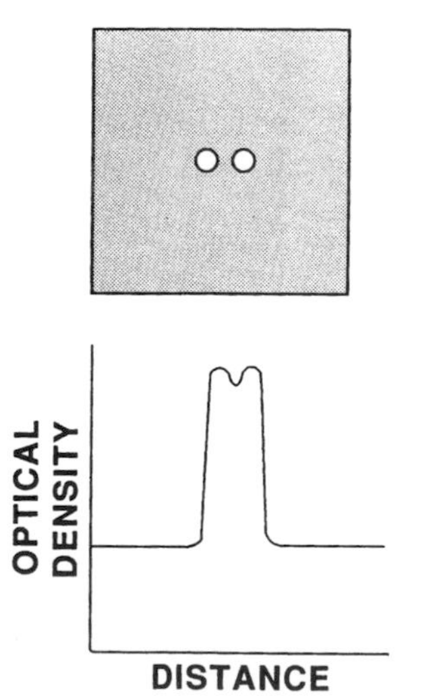

MARGIN FIGURE 17-2
When two point-like, high-contrast objects are close together, their unsharp boundaries overlap, and they are interpreted as a single object rather than as separate objects.

The distance between the objects required for their recognition as two separate entities is the *resolution distance*.

OBJECTIVES

After studying this chapter, the reader should be able to:

- Define and distinguish among the
  - Point response
  - Line response
  - Edge response
  - Contrast response
- Describe the properties and applications of the modulation transfer function.
- Discuss the principles of quantum levels, conversion efficiencies, and the quantum sink.

INTRODUCTION

The clarity of information in a medical image is a measure of how well the image conveys information about the patient. If the information is clear and unambiguous, the clarity is acceptable. If the information is blurred or distorted, or if the image lacks contrast or is too noisy, the clarity is compromised and the information is less helpful. In these situations, the image is said to be of "unsatisfactory quality." Image quality is a subjective expression that connotes the observer's reaction to the clarity of information in the image. Image clarity is discussed in this chapter. The perception characteristics of the observer and how they influence image interpretation are covered in the next chapter.

POINT RESPONSE

The clarity of information in an image is a measure of how well the image reveals fine structure in the object. One approach to evaluating this ability is to place two structures close together and determine how clearly they can be distinguished in the image. The structures should be very small so that they essentially constitute "point objects." Furthermore, they should differ substantially in composition from the surrounding medium so that they have high subject contrast. Pinpoint holes in an otherwise x-ray-opaque object constitute an excellent set of point objects.

When the point objects are far apart, they are easily distinguishable in the image. As they are moved closer, their unsharp boundaries begin to overlap in the image, and the contrast diminishes between the objects and their intervening space. Eventually the diminution in contrast is so severe that the objects lose their separate identity, and they begin to be seen as a single larger objective. A difference of 10% or more in optical density usually is required to yield enough contrast for the objects to be visually separable.

A simple but rather impractical way to describe the spatial resolution of an imaging system is by use of the point spread (PSF). The PSF is an equation that mathematically describes the image of an infinitesimal point-like object (see margin). A narrow, sharply peaked PSF indicates high spatial resolution, whereas a broad, flat PSF delineates low spatial resolution.

LINE RESPONSE

A hypothetical use of point-like objects is a convenient way to introduce the concept of spatial resolution and resolving power. However, this approach is impractical from an experimental point of view because the objects need to be infinitesimally small for the approach to be valid. But very small objects such as pinpoint holes transmit very

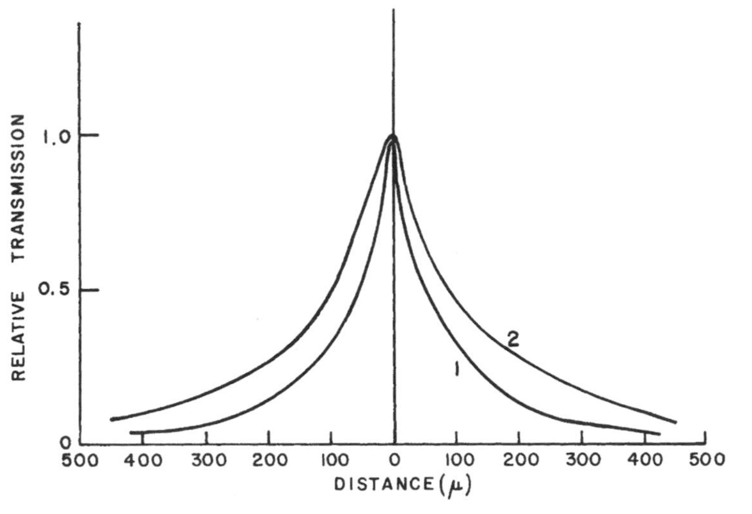

FIGURE 17-1
Two normalized graphs of a line source illustrate the blurring introduced by the imaging system. System 1, obtained with x rays detected with medium-speed intensifying screens, yields better spatial resolution than system 2, which employs fast intensifying screens.

The resolution distance is often said to reflect the *spatial resolution* of the imaging system. A system that yields excellent spatial resolution is said to have a high *resolving power*.

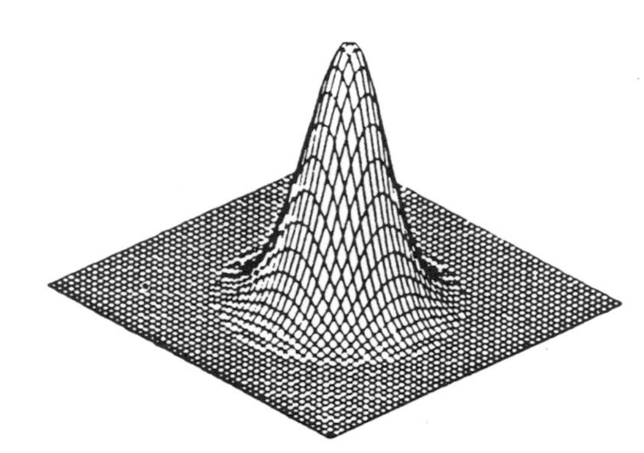

MARGIN FIGURE 17-3
The point spread function PSF (x, y).

few photons, so it is difficult to distinguish them from their surroundings. A better approach to describing spatial resolution is use of the "line response."

A line response is obtained by producing an image of a structure shaped as a line. A common object for producing a line response in radiography is an x-ray-opaque sheet that contains a narrow slit through which x rays are transmitted.[1] A narrow catheter containing radioactive material satisfies the same purpose in nuclear imaging. The image of the line source is blurred, as shown in Figure 17-1. The amount of blurring reflects the resolving power of the imaging system. The two graphs in the illustration depict two different imaging systems, with system 1 yielding images with greater spatial resolution.

Line objects have a specific advantage over point objects for determining the resolution characteristics of imaging systems. The advantage is that line objects transmit many more x rays, and therefore the precision of the evaluation is not restricted by the number of photons used in its determination. By use of a line object, the resolving power of an imaging system can be measured under conditions of high contrast with enough photons to eliminate quantum mottle as a significant factor affecting the accuracy of the determination.

A line response such as that depicted in Figure 17-1 can be described mathematically as well as displayed pictorially. The mathematical description is known as a line spread function (LSF).[2] The LSF is the PSF integrated over one dimension. It is widely used in diagnostic imaging as input data for computation of the modulation transfer function (MTF), discussed later in this chapter.

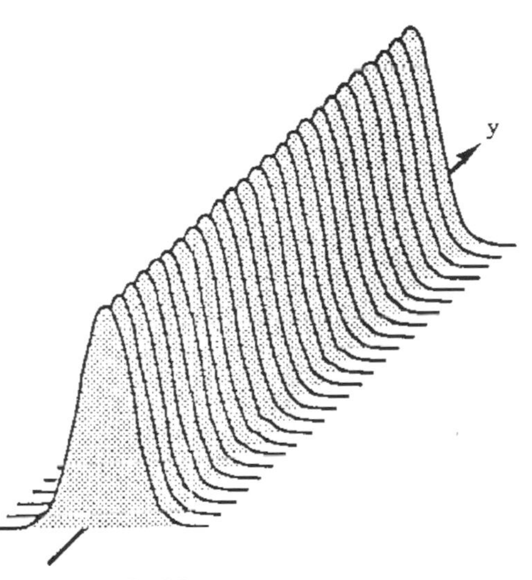

MARGIN FIGURE 17-4
The line spread function LSF (x). The derivative of the LSF is the PSF.

The edge response is a concept that is closely related to the line response. In this approach, the imaging system is presented with a source that transmits radiation on one side of the edge and attenuates it completely on the other side. A microdensitometric scan of the resulting image yields the edge response.[3]

If the spatial resolution of the image is the same at all locations and orientations, then the PSF and LSF are said to be *spatially invariant.*

CONTRAST RESPONSE

One approach to determination of the resolution characteristics of imaging systems is to use a test object containing a series of slits separated by x-ray-opaque spaces (Margin Figure 17-5A). The slits are widely separated on one end of the object and become increasingly close together toward the other end. The transmission of x rays through this object is depicted as a square-wave pattern of reduced spatial separation (i.e., higher spatial frequency) toward the end of the object where the slits

The edge spread function (ESF) is an equation that mathematically describes the shape of the edge response. The derivative of the ESF is the LSF.

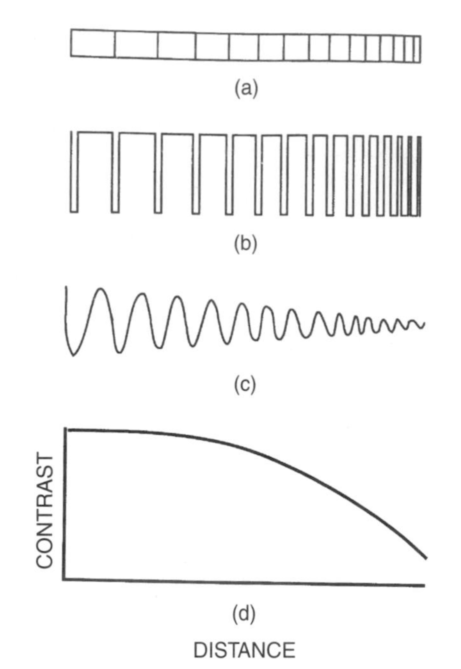

MARGIN FIGURE 17-5
Depiction of contrast response. **A:** Test object. **B:** X-ray transmission; **C:** X-ray image. **D:** Contrast response curve.

Actually, cycles and line pairs are not valid units, and the scientifically correct expression for spatial frequency is 1/(unit length) (e.g., 1/millimeter), just as the correct unit for temporal frequency is (1/unit time) and not (cycles/unit time).

The difference between subject contrast in the object and the contrast presented in the image is illustrated in Margin Figure 17-6 as a function of spatial frequency.

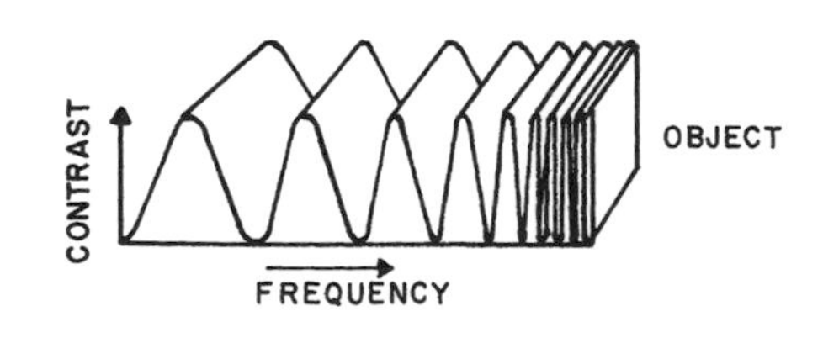

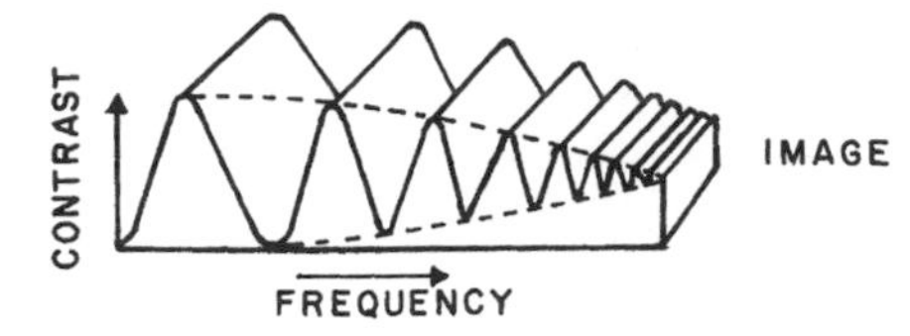

MARGIN FIGURE 17-6
Suppression of contrast in the image with increasing frequency of variation in the transmission of x rays through an object.

are closer together (Margin Figure 17-5B). The spatial frequency is commonly expressed in units of cycles or "line pairs" per unit length (e.g., cycles or line pairs per millimeter).

An image of transmitted x rays is unable to maintain the square-wave pattern because the imaging system introduces unsharpness that blurs the edges (Margin Figure 17-5C). This blurring does not interfere with appreciation of separate slits when the slits are widely separated. As the slits become closer together, however, they can no longer be distinguished as separate in the image. In this region of higher spatial frequencies, the images of the slits begin to run together, and the differences between the optical densities diminish. At even higher spatial frequencies, the image is a uniform grayness, and the individual-slits and intervening spaces are undetectable.

The contrast response curve (Margin Figure 17-5D) is a single curve that depicts the amplitude of response in the image as a function of spatial frequency. The amplitude of response is the variation in optical density between the image of a slit and the image of the adjacent x-ray-opaque space.

The contrast response curve can be described mathematically as well as pictorially. The mathematical description, known as the contrast response function, is an equation that describes the amplitude of response as a function of spatial frequency.

MODULATION TRANSFER FUNCTION

The contrast response curve and contrast response function have one significant limitation. They require a test object of alternating slits and spaces that increase in spatial frequency from one end to the other. For evaluation of a radiographic unit, frequencies as high as 10 to 15 cycles per millimeter may be required. Test objects with frequencies this high are difficult to manufacture and deploy, and they are not routinely used.

An x-ray beam transmitted through any object can be represented as a series of sine waves, with each wave possessing a characteristic frequency and amplitude. The transmitted x-ray beam is said to be modulated by the object because the distribution of amplitudes and frequencies is influenced by the object, and hence contains information about the features of the object. The function of an imaging system is to translate the modulation of the transmitted x-ray beam into a visible image and to reproduce the modulation faithfully so that features of the object are apparent. The ability of an imaging system to fulfill this responsibility is described by the modulation transfer function (MTF). That is, the MTF of an imaging system is a measure of how well sine waves that describe the transmission of x rays through an object are represented faithfully in the image.[4,5] The MTF for an imaging system can be calculated from the PSF or LSF measured with the test objects described earlier. A curve that represents the MTF as a function of spatial frequency is shown in Margin Figure 17-7.

In the illustration, the value of the MTF is unity (i.e., 100%) at low spatial frequencies, signifying that the imaging system reproduces low frequencies without distortion or loss of resolution. As the frequency increases, the MTF decreases until it reaches zero, signifying that the spatial frequencies are so high that the imaging system provides no reproduction at all.

The MTFs for two imaging systems are shown in MF 17-8. System 1 provides a higher MTF at every spatial frequency, reflecting greater spatial resolution for this imaging system.

The MTF is a useful descriptor of the spatial resolution of an imaging system. Other characteristics of the system are also important, including its ability to reveal subtle differences in subject contrast and to provide images with low levels of noise. The MTF provides information about the clarity of information in images furnished by an imaging system, but it is not a complete descriptor of image clarity.[5,6]

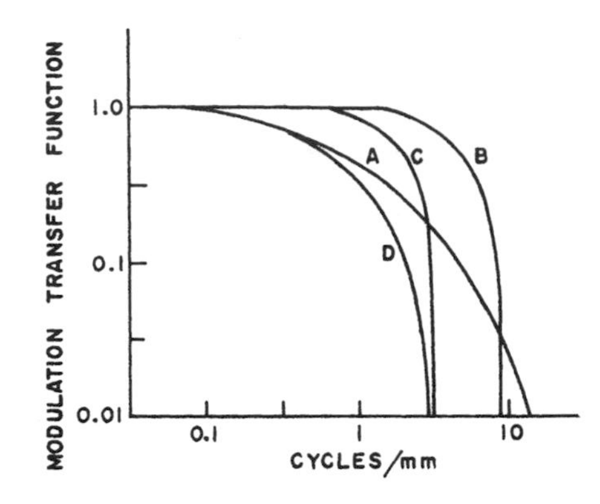

FIGURE 17-2
MTF curves for components of a radiographic imaging system and the composite MTF for the entire system. **A:** MTF for the screen–film combination. **B:** MTF for a 1-mm focal spot with 90 cm between the focal spot and the object and 10 cm between the object and film. **C:** MTF for 0.3-mm motion of the object during exposure. **D:** Composite MTF for the entire imaging system. (From Morgan, R.[7])

A test object of alternating slits and spaces is referred to as a "line pair phantom."

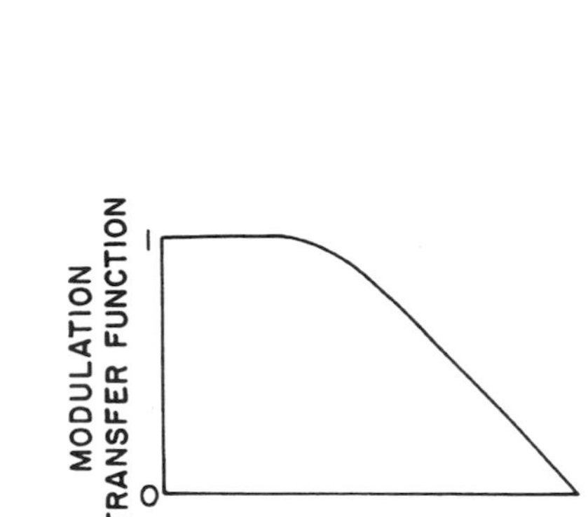

MARGIN FIGURE 17-7
Modulation transfer function (MTF) for a typical imaging system.

The MTF of any imaging system is a product of the MTFs of the individual components of the system. That is, the MTF of the complete system can be computed if the MTFs of the individual components are known. This principle is illustrated in Example 17-1.

Example 17-1

At a spatial frequency of 5 cycles per millimeter, the MTFs of components of an x-ray film–screen imaging system are focal spot, 0.9; motion, 0.8; and intensifying screen, 0.7. The MTF of the composite imaging system is $0.9 \times 0.8 \times 0.7 = 0.5$.

The frequency where the MTF is 0.1 is usually taken as the cutoff frequency; frequencies above this value are considered to be unrepresented in the image. The cutoff frequency sometimes is described as the resolving power of the imaging system.

Example 17-1 illustrates the rule that the MTF of a composite imaging system at any particular spatial frequency is always less than the MTF of the system component with the poorest resolution at that frequency. This rule is illustrated in Figure 17-2, where the composite MTF curve for a radiographic imaging unit is below the MTF curves for the components of the imaging system at every spatial frequency. This is shown also in the MTF curves for the fluoroscopic system shown in Figure 17-3.

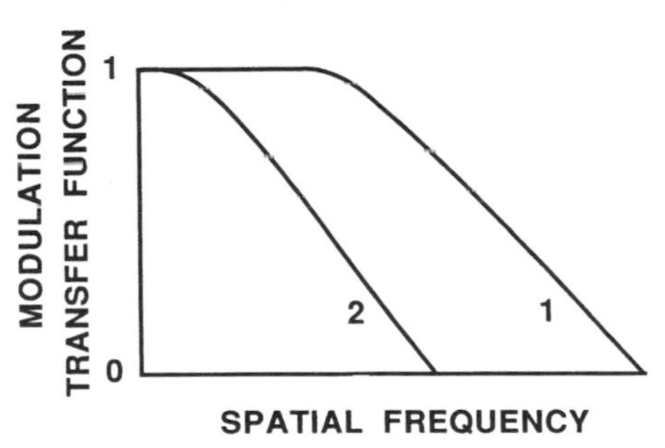

MARGIN FIGURE 17-8
MTF curves for two imaging system. Superior spatial resolution is provided by system 1.

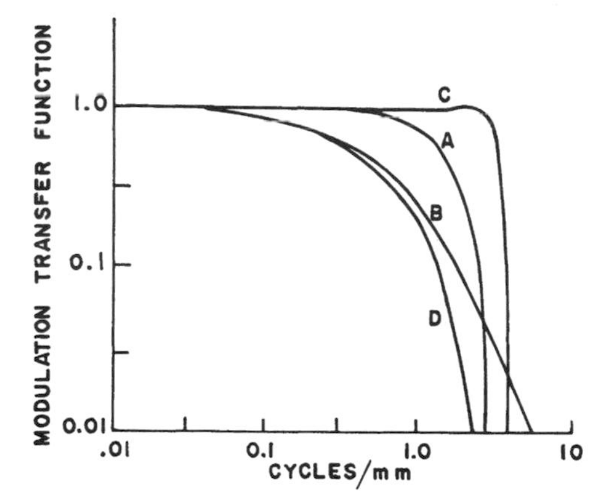

FIGURE 17-3
MTF curves for components of a fluoroscopic imaging system and the composite MTF for the system. **A:** MTF for a 1-mm focal spot with a ratio of 3 between the target–object distance and the object–screen distance. **B:** MTF for the image intensifier. **C:** MTF for a 525-line television system. **D:** composite MTF for the entire system. (From Morgan, R.[8])

The MTF is said to be the representation of the PSF or LSF in "frequency space."

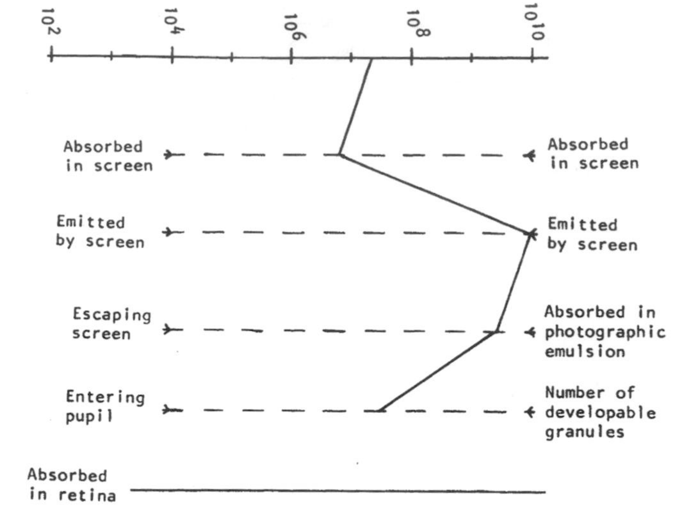

FIGURE 17-4
Quantum levels for stages of information conversion in film–screen radiography.

QUANTUM LEVELS AND CONVERSION EFFICIENCIES

The number of x rays or light photons per unit area may be described at each stage of transformation of information in an x-ray beam into a visual image. These numbers are termed *quantum levels*. For example, a typical radiograph may require an exposure of 3×10^7 x-ray photons per square centimeter to the intensifying screen (Figure 17-4). About 30% of the photons may be absorbed in the screen. For each x ray absorbed, 1000 photons of light may be liberated. About 50% of these light photons escape from the screen and interact in the emulsion of the x-ray film. In a typical photographic emulsion, 200 photons of light are absorbed for each photographic granule affected. In the data in Figure 17-4 the number of photons per unit area is lowest at the stage of absorption of x-ray photons in the intensifying screen. This stage is termed the *quantum sink*. The quantum sink identifies the stage of greatest influence of quantum mottle on image resolution.

SUMMARY

- The point spread function is a mathematical expression that describes the ability of an imaging system to reproduce a point object (point response).
- The line spread function is a mathematical expression that describes the ability of an imaging system to reproduce a line object (line response).
- The edge spread function is a mathematical expression that describes the ability of an imaging system to reproduce an edge (edge response).
- The contrast response function is a mathematical expression that describes the amplitude of response in the image as a function of spatial frequency (contrast response).
- The modulation transfer function is a mathematical expression that describes the ability of an imaging system to capture and display modulations present in the radiation beam from the object.
- The stage in the imaging process where the number of information carriers is lowest is the quantum sink.

PROBLEMS

*17-1. The space between two slits in an x-ray-opaque lead sheet can be narrowed to 0.2 mm before the slits are no longer distinguishable in an image. Determine the resolution distance and the cutoff frequency of the imaging system.

*17-2. In a radiographic unit, would the cutoff frequency be higher or lower on the anode side versus the cathode side of the x-ray beam? Why?

*17-3. A fluoroscopic imaging system has the following component MTFs at 3 line pairs per millimeter: focal spot, 0.9; motion, 0.7; input screen, 0.9; intensifier electronics, 1.0; output screen, 0.9; optics, 0.9; and television system, 0.9. What is the composite MTF of the system?

*For those problems marked with an asterisk, answers are provided on p. 492.

REFERENCES

1. Van Allen, W., and Morgan, R. Measurement of revolving powers of intensifying screens. *Radiology* 1946; **47**:166.
2. Swet, J. A., and Pickett, R. M. *Evaluation of Diagnostic Systems*. New York, Academic Press, 1982.
3. Hendee, W. R. Imaging science in medicine, in Hornak, J. (ed.), *Encyclopedia of Imaging Science and Technology*. New York: John Wiley & Sons, 2002.
4. Stone, J. M. *Radiation and Optics*. New York, McGraw-Hill, 1963.
5. Kijewski, M. F. Image quality, in Hendee, W. R. (ed.), *Biomedical Uses of Radiation*. Part A: Diagnostic Applications. Weinheim, Wiley VCH, 1999, pp. 573–614.
6. Rao, G. U. Measurement of modulation transfer functions, in Waggener, R. J., Keriakes, J. G., and Shalek, R. J. (eds.), *Handbook of Medical Physics*, Vol 2. Boca Raton, FL, CRC Press, 1982. pp. 159–180.
7. Morgan, R. The frequency response function: A valuable means of expressing the informational recording capabilities of diagnostic x-ray systems. *Am. J. Roentgenol. Radium Ther. Nucl. Med.* 1962; **88**:175–186.
8. Morgan, R. Physics of diagnostic radiology, in Glasser, O., et al. (eds.), *Physical Foundations of Radiology*, 2nd edition. New York, Harper & Row, 1961.

CHAPTER

18

VISUAL PERCEPTION

OBJECTIVES

After completing this chapter, the reader should be able to:

- Distinguish between detection, recognition and interpretation of information in an image.
- Describe the anatomy and physiology of the human eye.
- Explain various models for detecting visual signals in images.
- Delineate the concepts of visual acuity and contrast discrimination.
- Identify and explain various expressions for describing visual performance, including
  - Contrast-detail analysis
  - Conspicuity
  - Truth tables
  - Relative operating characteristics

INTRODUCTION

Visual images have always been essential to the efforts of humans to understand, characterize, and accommodate to their environment. Long before written communication, images were used to convey information and to express inner thoughts and desires.

Today, visual images serve in many ways to display and communicate complex information in a highly sophisticated world society. Economic transactions, transportation control, security and surveillance, and process control in industry depend on images, as do almost all interactions among individuals and groups. In the area of human services, medical diagnosis and treatment are among the specialties most dependent on images. Radiology is the medical discipline primarily responsible for producing and interpreting medical images. Consequently, an understanding of the detection and interpretation of visual information is an essential part of the radiologist's education.

The perception of visual information consists of the three sequential processes illustrated in Figure 18-1. The first process is the *detection* of visual signals by the observer (Figure 18-2). If visual information is not detected, it cannot be integrated into the perceptual process for its correct interpretation. Important information sometimes is overlooked in radiologic images for reasons that are not well understood. Experiments[4–7] have revealed that radiologists miss 20% to 30% of the visual clues present in diagnostic images. In 10% to 20% of instances, radiologists detect different signals (interobserver variation) in the same images. Even disagreement with a previous reading by the same radiologist (intraobserver variation) occurs in 5% to 10% of images. These undetected signals and inconsistencies reflect the tendency of all observers to miss important visual information present in images.

Even if visual signals are detected, they may be ignored because they are considered inconsequential. The tendency to dismiss information connotes a problem of *recognition*. The observer simply does not appreciate the importance of the information. Eye tracking experiments performed on radiologists reading diagnostic images have revealed that radiologists frequently look at visual clues that are important to the diagnosis but fail to incorporate them into the interpretive process. This failure indicates that the visual clues are detected but not recognized.

Visual signals may be detected and their importance recognized, but still the diagnosis reached by the radiologist may be incorrect. This error occurs in the *interpretation* phase of information gathering and processing (Figure 18-3), the third step in the model of radiologic diagnosis. Error at this level may reflect inadequate training, inexperience, prior conditioning, fatigue, poor judgment, or simply a difficult case in which the wrong option is chosen.

Paintings in caves of southern France and northern Spain are thought to reflect attempts by early humans to "visualize" events such as a successful hunt before embarking on the experience.

Visual perception has been the subject of philosophical speculation for centuries, beginning with the Greeks and probably before. Aristotle suggested that a unified perception of an object occurred through the exercise of the faculty of "common sense" that resides in the heart.[1]

Euclid and Ptolemy developed a geometric theory that suggested that the eye perceives an object by projecting something toward that object. Aristotle and the atomists believed the reverse. The argument was settled more than 800 years later when in 965 an Arabian scientist in Basra (now Iraq), Abu Ali al-Hasan Ibn al-Haytham, convinced some observers to stare at the sun until their eyes became damaged. This was accepted as evidence that something must enter the eye from the outside. While of questionable ethics, this event marked a transition in philosophy from complete faith in theory and conjecture to an evolving acceptance of the primacy of experimental evidence.

Three stages of visual perception:

- Detection
- Recognition
- Interpretation

The frequency of missed signals during interpretation of radiographs has not changed much in spite of major advances in equipment and technique.

Variations in Detection of Radiologic Information

Miss rate	20–30%
Interobserver variation	10–20%
Intraobserver variation	5–10%

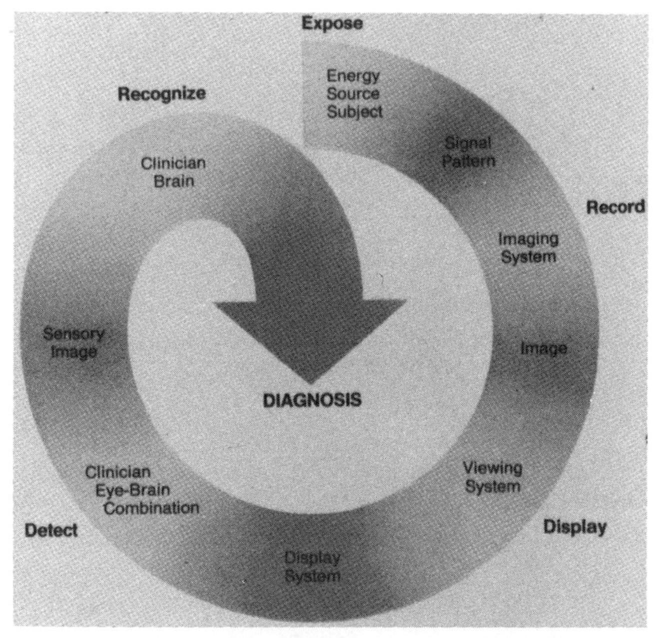

FIGURE 18-1
The perception of visual information. (From Jaffe, C. C. [2] Used with permission.)

Reasonably good models exist to describe the process of detecting visual signals. Paradigms to explain how these signals are recognized as normal and abnormal, important or unimportant, and expected or unexpected are more rudimentary. The greatest enigma of all is understanding how detected and recognized visual signals are woven into an interpretation that leads to a correct medical diagnosis of the patient's condition.[9]

FIGURE 18-2
Example of a detection challenge; Dalmation dog. (From Thurston, J. B., and Carraher, R. G.[3] Used with permission.)

MARGIN FIGURE 18-1
Example of recognition challenge; the observer should be able to recognize both a young and an old lady, although not simultaneously.

For centuries, philosophers assumed that visual artists (e.g., painters and sculptors) experienced a richly pictorial inner life of the mind that they were able to project into their art. In the last century, efforts of psychologists and psychiatrists to probe and characterize this inner life were largely discredited by logical positivists, behavior scientists, and some postmodernist philosophers who believed that people are biological machines that react to external stimuli rather than to thoughts, feelings, and intentions. Recent developments in philosophy, psychology, linguistics, and computer science are restoring interest in the inner life of the mind.

FIGURE 18-3
Example of interpretation challenge; moving water without origin. (From Escher, M. C.[8] Used with permission.)

HUMAN VISION

Images are the product of the interaction of the human visual system with its environment. The process of human vision is outlined here; more detailed treatments are available in texts on human physiology.

Anatomy and Physiology of the Eye

The human eye is diagramed in Figure 18-4. It is an approximate sphere that contains four principal features: the cornea, iris, lens, and retina. The retina contains photoreceptors that translate light energy into electrical signals that serve as nerve impulses to the brain. The other three components serve as focusing and filtering mechanisms to transmit a sharp, well-defined light image to the retina.

The wall of the eye consists of three layers (tunics) that are discontinuous in the posterior portion where the optic nerve enters the eye. The outermost tunic is a fibrous layer of dense connective tissue that includes the cornea and the sclera. The cornea comprises the front curved surface of the eye, contains an array of collagen fibers and no blood vessels, and is transparent to visible light. The cornea serves as a coarse focusing element to project light onto the observer's retina. The sclera, or white of the eye, is an opaque and resilient sheath to which the eye muscles are attached. The second layer of the wall is a vascular tunic termed the uvea. It contains the choroid, ciliary body, and iris. The choroid contains a dense array of capillaries that supply blood to the tunics. Pigments in the choroid reduce internal light reflection that otherwise would blur the images. The ciliary body contains the muscles that support and focus the lens. It also

The first known description of eye physiology was provided by Felix Plather of Basle. Others who contributed knowledge about how the human eye functions include Johanes Kepler, Rene Descartes, and Isaac Newton.

Although only about 24 mm in diameter, the eye contains several hundred million working parts that provide a vast range of responses to contrast, color, detail, form, and space. The eye is one of the most complex organs in the body.

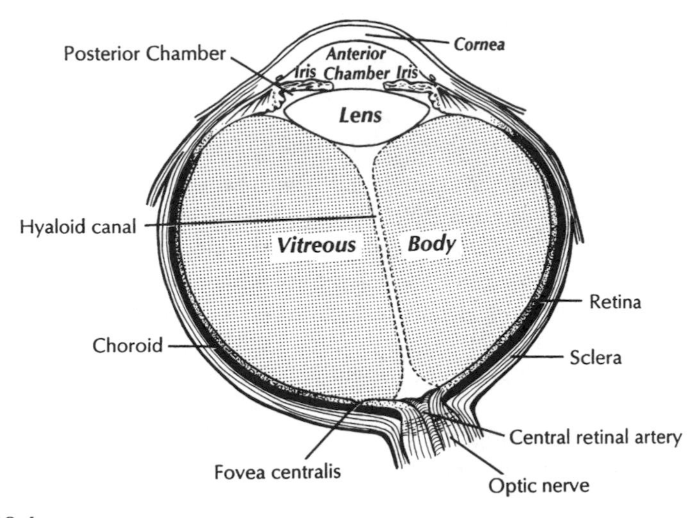

FIGURE 18-4
Horizontal section through the human eye.

contains capillaries that secrete fluid into the anterior segment of the eyeball. The iris is the colored part of the eye that has a central aperture termed the pupil. The diameter of the aperture can be altered through the action of muscles in the iris to control the amount of light entering the eye. The aperture can vary from about 1.5 mm to 8 mm.

The anterior and posterior chambers of the eye are filled with fluid. The anterior chamber contains aqueous humor, a clear plasma-like fluid that is continually drained and replaced. The posterior chamber is filled with vitreous humor, a clear viscous fluid that is not replenished. The cornea, the aqueous and vitreous humors, and the lens serve collectively as the refractive media of the eye.

The lens of the eye provides the fine focusing of incident light onto the retina. It is a convex lens that can change in thickness by action of the ciliary muscles. The index of refraction of the lens is close to that of the surrounding fluids in which it is suspended, so it serves as a fine-focusing adjustment to the coarse focusing function of the cornea. The process of accommodation by which near objects are brought into focus is achieved by contraction of the ciliary muscles. This contraction causes the elastic lens to bow forward into the aqueous humor, thereby increasing its thickness. Accommodation is accompanied by constriction of the pupil, which increases the depth of field of the eye.

With age the lens loses its flexibility and is unable to accommodate so that near objects can be focused onto the retina. This is the condition of presbyopia in which reading glasses are needed to supplement the focusing ability of the lens. Clouding of the lens with age causes a diminution in the amount of light reaching the retina. This condition is known as a lens cataract, which, when severe enough, makes the individual a candidate for surgical replacement of the lens, often with an artificial lens.

The innermost layer of the eye is the retina, which is composed of two components, an outer monolayer of pigmented cells and an inner neural layer of photoreceptors. Because considerable processing of visual information occurs in the retina, it often is thought of more as an outpost of the brain rather than as simply another component of the eye. There are no photoreceptors where the optic nerve enters the eye, creating a blind spot. Near the blind spot is the macula lutae, an area of about 3 mm^2 over which the retina is especially thin. Within the macula lutae is the fovea centralis, a slight depression about 0.4 mm in diameter. The fovea is on the optical axis of the eye and is the area where the visual cones are concentrated to yield the greatest visual acuity.

The retina contains two types of photoreceptors, termed rods and cones. Rods are distributed over the entire retina, except in the blind spot and the fovea centralis. The retina contains about 125 million rods, or about $10^5/mm^2$. Active elements in the rods

The cornea contributes about two-thirds of the refracting power of the eye (refracting power is the ability to change the direction of incident light). The remaining one-third is contributed principally by the surfaces of the lens.

Underwater vision without a mask is blurred and hazy because the eye is bathed by water of about the same refractive index as the cornea. Hence, the refracting power of the cornea is greatly diminished, and the retinal image is unfocused.

Accommodation of the lens is similar in function to the auto-focus capacity of a camera.

If an image is not projected in focus on the retina, the eye is said to be *ametropic*. There are three types of ametropia: myopia (short-sightedness), hyperopia (long-sightedness), and astigmatism (lack of symmetry on the cornea and lens surfaces).

Cataracts and other lens defects produce the vision problem known as aphakia.

The retina does not "record" an image projected onto it by the lens. Instead, it responds to the incident image by sending coded neural signals to the brain, where they are decoded to yield the sense of visual perception. Hence, the optics of the eye do not have to be as perfect as those of an optical instrument. Hermann von Helmholtz, the nineteenth-century physicist who developed the field of physiological optics, said that if an optician built so imperfect an optical instrument as the human eye, he would undoubtedly become bankrupt.[10]

(and in the cones as well) are replenished throughout an individual's lifetime. Rods have a low but variable threshold to light, and they respond to very low intensities of incident light. Vision under low illumination levels (e.g., night vision) is attributable almost entirely to the rods.

Rods contain the light-sensitive pigment rhodopsin (visual purple) which undergoes a chemical reaction (the rhodopsin cycle) when exposed to visible light. Rhodopsin consists of a lipoprotein called opsin and a chromophore (a light-absorbing chemical compound called 11-*cis*-retinal).[11] The chemical reaction begins with the breakdown of rhodopsin and ends with the recombination of the breakdown products back into rhodopsin. The recovery process takes 20–30 minutes, which is the time required to accommodate to low levels of illumination (dark adaptation). The rods are maximally sensitive to light of about 510 nm, in the blue-green region of the visible spectrum. Rods cannot discriminate different wavelengths of light, and vision under low illumination conditions is essentially "colorblind." More than 100 rods are connected to each ganglion cell, and the brain cannot discriminate among these photoreceptors to identify the origin of an action potential transmitted along the ganglion. Hence, rod vision yields relatively low visual acuity, in combination with a high sensitivity to low levels of ambient light.

The retina contains about 7 million cones, packed tightly in the fovea and diminishing rapidly across the macula lutae. The density of cones in the fovea is about 140,000/mm^2. Cones are maximally sensitive to light of about 550 nm, in the yellow-green region of the visible spectrum. Cones are much ($1/10^4$) less sensitive than rods to light, but in the fovea there is a 1:1 correspondence between cones and ganglions, so the visual acuity is very high. Cones are responsible for color vision through mechanisms that are imperfectly understood at present.

Properties of Vision

For two objects to be distinguished on the retina, light rays from the objects must define at least a minimum angle as they pass through the optical center of the eye. The minimum angle is defined as the visual angle. The visual angle, expressed in units of minutes of arc, defines the visual acuity of the eye. An individual with excellent vision and who is functioning in ideal viewing conditions can achieve a visual angle of about 0.5 minutes of arc, which is close to the theoretical minimum defined by the packing density of cones on the retina.[12]

The eye is extremely sensitive to small amounts of light. Although the cones do not respond to illumination levels below a threshold of about 0.001 cd/m^2, the rods are much more sensitive and respond to just a few photons. For example, as few as 10 photons can generate a visual stimulus in an area of the retina where the rods are present at high concentration.[13]

Differences in signal intensity that can just be detected by the human observer are known as just noticeable differences (JND). This concept applies to any type of signal, including light, that can be sensed by the observer. The smallest difference in signal that can be detected depends on the magnitude of the signal. For example, the brightness difference between one and two candles may be discernible, whereas the difference between 100 and 101 candles probably cannot be distinguished. This observation was quantified by the work of Weber, who demonstrated that the JND is directly proportional to the intensity of the signal. This finding was quantified by Fechner as

$$dS = k[dI/I]$$

where I is the intensity of stimulus, dS is an increment of perception (termed a limen), and k is a scaling factor. The integral form of this expression is known as the Weber–Fechner law:

$$S = k[\log I/I_0]$$

The Weber–Fechner law is similar to the expression for the intensity of sound in decibels, and it provides a connection between the objective measurement of sound intensity and the subjective impression of loudness.[14]

It is often said that rods encode the twilight picture, and cones encode the daylight picture.

The process of viewing with rods is known as "scoptopic" vision. Viewing with cones is known as "photopic" vision.

The physiologist Lord Adrian said that the "copy" furnished by the >100 million reporters (rods and cones) is severely edited and condensed, and only the more dramatic features are telegraphed in code from the retina to the central offices in the brain.[10]

One popular theory (the trichromatic theory) of color vision proposes that three types of cones exist, each with a different photosensitive pigment that responds maximally to a different wavelength (450 nm for "blue" cones, 525 nm for "green" cones, and 555 nm for "red" cones). The three cone pigments share the same chromophore as the rods; their different spectral sensitivities result from differences in the opsin component.

In a now famous prism experiment, Newton showed in 1666 that sunlight consists of a mixture of visible light rays, each with its own color. Goethe responded to this finding by stating that it was absurd to assert that mere mixing of rainbow colors could appear white, since white is without color.

The spectral response of human vision peaks near the center of the sun's spectrum under conditions of bright illumination. The peak response shifts toward the blue end of the visible spectrum at twilight, in order to match the spectrum of scattered light in the sky.

The detection threshold of the human eye is often said to be the light from a single candle on a clear, dark night at a distance of 30 miles.

An alternative expression for the relationship between stimulus intensity I and the visual perception S of the stimulus is known as Stevens' power law: $S = kI^n$, where k is a constant and n is the power.

In 1860 Gustav Fechner introduced the term *psychophysics* to characterize the study of the perceptions of stimuli and their relationship to the physical characteristics of the stimuli.

DETECTION OF VISUAL INFORMATION

Various models have been developed to describe how observers detect visual signals in images. These models can be categorized into three approaches to understanding visual detection: signal-to-noise, computational, and signal channel models.

Signal-to-Noise Model

The signal-to-noise model describes the ability of an observer to detect simple visual signals embedded in a noisy background. Noise in the image (e.g., quantum mottle, extraneous structures, and receptor artifacts) is referred to as external noise.[15] The model characterizes an "ideal observer" who detects signals with a certain likelihood whenever their amplitude differs from background by some prescribed threshold, usually a multiplicative factor of 2 to 5. The inability of a real observer to detect visual signals as well as the ideal observer is said to be caused by "internal noise" in the real observer. Sources of internal noise include distraction of the observer, preconditioning of the observer by the presence or absence of signals in previously viewed images, reduced intrinsic sensitivity to visual signals in some observers, and a myriad of other possible causes.

The performances of ideal and actual observers are compared in the margin for the simple case in which actual observers are asked to state whether a simple signal is present or absent in an image. The index of detectability on the ordinate (y axis) is the inverse of the size of subtle lesions that can be detected with reasonable assurance. As shown in the illustration, the performance of the human observer usually falls short of that for the ideal observer. This difference may be due in part to selection by the human observer of a visual threshold different from that assumed for the ideal observer. It may also be caused in part by internal noise (unpredictability) in the observer.

Wagner and Brown[16] have suggested that the performance of an imaging system can be characterized quantitatively by describing the ability of an ideal observer to detect simple visual signals provided by the system. They have proposed this approach for evaluation and comparison of imaging systems used in clinical radiology.

Both the Weber–Fechner law and the Stevens power law have been shown to be seriously deficient under certain experimental conditions. These findings reveal that the ability of humans to quantify sensations such as brightness and loudness is flawed. Consequently, we remain dependent on measuring devices such as light meters and speedometers.

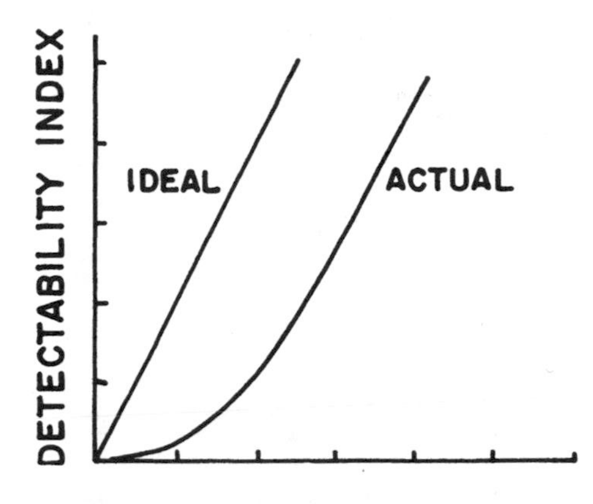

MARGIN FIGURE 18-2
The visual performance of actual and "ideal" observers in a lesion detectability experiment.

An experiment where the participant is required to select one of two choices is called a two-alternative, forced-choice test.

Computational Model

The computational model of visual detection has evolved from experience in designing computer programs to extract information from images. The model proposes that the first step in detecting visual information involves the rapid identification of brightness (density) changes in the image, together with properties such as orientation, location, depth, and spatial frequency of these changes. This "primal sketch" of the image is obtained by a simple psychophysical reaction to the image without use of *a priori* knowledge of the structure and function of the object represented in the image. That is, visual signals are extracted from the image without any assumptions about the nature of the image and the type of visual scene portrayed.[17] Initial applications of the computational model were promising; however, its development into a useful paradigm for visual detection has been disappointing. Probably our current understanding of psychophysical properties of human vision is simply too rudimentary to develop a satisfactory computational model for the visual detection process.[18]

Studies of the definition and communication of information, and its separation from background "noise," constitute the discipline of Information Theory. This discipline was founded in the late 1940s by the engineer Claude Shannon and the mathematician Norbert Wiener.

The computational model suggests that visual detection consists of two stages: (1) a preattentive (early warning) stage for almost instantaneous detection of textural changes in the environment and (2) an attentive stage for focusing attention on specific objects identified in the preattentive stage.[1]

Signal-Channel Model

In this model of visual detection, the human visual system is characterized as a series of parallel vision channels, with a specific frequency range of visual information assigned to each channel. Several studies suggest that the frequency range of each channel is on the order of ± 1 octave of spatial frequency information[19] and that the visual system as a whole functions essentially as a frequency analyzer.

Development of the signal channel model has been aided by the use of sinewave gratings, a series of light and dark bars repeated at increasing frequency (Figure 18-5).

Although the computational model is limited in its ability to fully characterize human vision, it has proven to be exceptionally useful in the design of reconnaissance systems and pattern recognition software programs in artificial intelligence.

A person who has excellent visual acuity (e.g., 20/20 with the Snellen test), but has impaired contrast vision, may be unable to see a truck in the fog or read a low-contrast x-ray image.

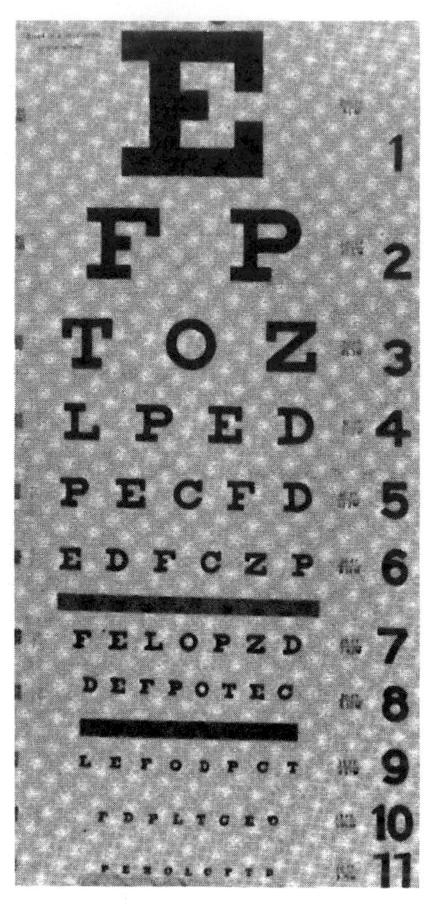

MARGIN FIGURE 18-3
Snellen eye test chart.

If the lettering on a Snellen chart is reversed (i.e., white letters on a black background), the ability of observers to recognize the letters from a distance is greatly impaired.

Magnification and minification lenses, along with viewing images from different distances, are useful techniques in radiology because they help match image detail to the spatial-frequency sensitivity of human vision (1–5 cycles per degree).

The ability of the human eye to filter spatial frequencies explains the usefulness of certain image-processing techniques such as edge enhancement, gray-level quantization, and color mapping.

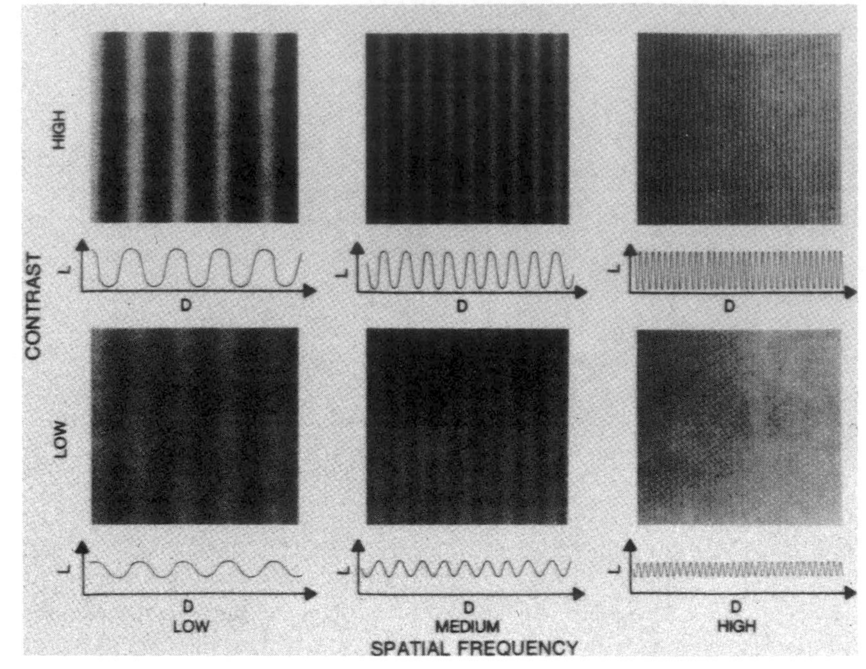

FIGURE 18-5
Sine-wave gratings used to determine the sensitivity to visual information of different spatial frequencies. (From Ginsburg, A. P.[21] Used with permission.)

Images of these gratings can be used as test objects to evaluate the ability of the observer to extract visual signals from images of increasing complexity.[20,21]

VISUAL ACUITY

Visual acuity describes the ability of an observer to extract detailed information from a small region of an image. In tests of visual acuity, high-contrast images are used to ensure that results are not compromised by the inability of the observer to distinguish subtle differences in brightness (density) in the image. Although visual acuity can be evaluated in several ways, the most common method is use of the Snellen eye chart shown in the margin. The Snellen chart consists of rows of letters that diminish in size from top to bottom. Results of Snellen tests are often expressed as the ratio of the maximum distance that an observer can just see an object divided by the distance that most people can see the same object. A measure of visual acuity of 20/40 simply means that the individual barely sees letters at 20 ft that the average person would see at 40 ft.

Persons are frequently tested for visual acuity, and corrective eyewear is designed principally to compensate for deficiencies in this property. Yet visual acuity is only part of the answer to whether people can detect visual information. Another important feature is the ability of the observer to distinguish subtle changes in brightness (density) in the image.

CONTRAST DISCRIMINATION

Contrast discrimination is the ability of the visual system to distinguish differences in brightness (density) in the image. This ability is described by a contrast sensitivity curve (Figure 18-6). This curve reveals the tendency of an observer to filter spatial frequencies so that coarse and fine details of the image are suppressed as compared with information of intermediate spatial frequencies.[22]

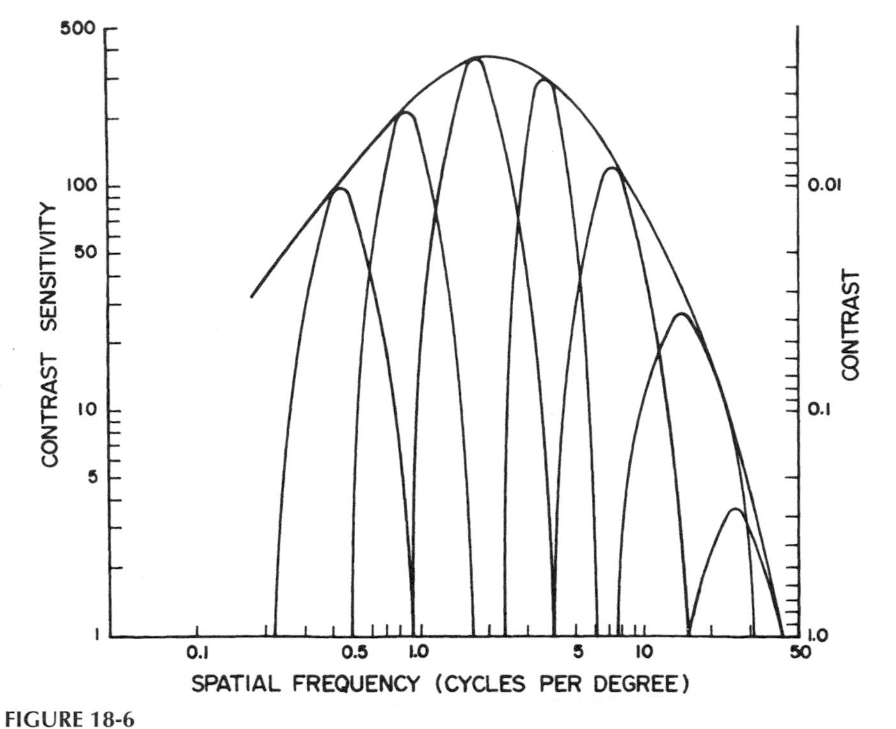

FIGURE 18-6
A contrast sensitivity curve. (From Ginsburg, A. P.[21] Used with permission.)

Contrast discrimination can be evaluated in different ways. One approach uses the vision contrast test system[23] depicted in Figure 18-7. Contrast discrimination is probably a more critical feature than visual acuity in determining how well the average person "sees." Determination of contrast discrimination has proved helpful in evaluating the extent of eye diseases such as cataracts, glaucoma, amblyopia, and the visual consequences of diabetes.

Vision loss at high frequencies, as detected with the Snellen test, is usually indicative of an optical (blurring) problem. Vision loss at middle and low frequencies, with vision at the higher frequencies intact, suggests a neuro-ophthalmological cause.

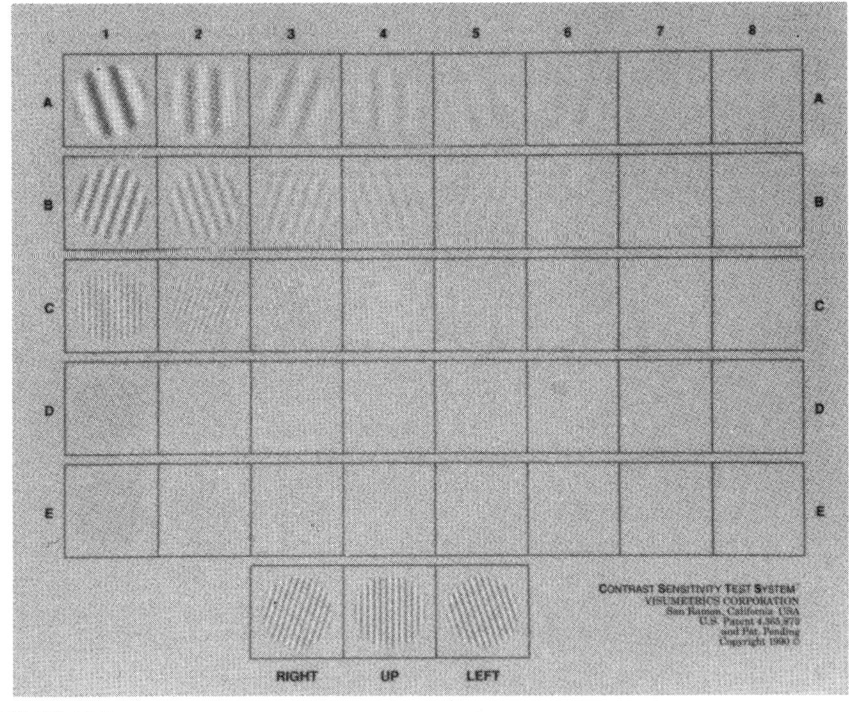

FIGURE 18-7
Vision contrast test system. (Courtesy of A. P. Ginsburg.)

Tests of contrast sensitivity have proven to predict the ability to detect and recognize letters, faces, aircraft in radar simulators, and road signs.[21]

Some authors have suggested that contrast discrimination should be evaluated whenever contact lenses and spectacles are fitted. A few[24] have even suggested that contrast discrimination might be a useful test in selecting candidates for occupations strongly dependent on image interpretation. Among these occupations might be aircraft pilots, radar operators, surveillance experts, and radiologists. Candidates could be selected by criteria that include matching the contrast discrimination characteristics of the individual to the contrast range and distribution presented in the images to be interpreted. In radiology, this suggestion has not been enthusiastically received.

RECOGNITION AND INTERPRETATION OF VISUAL INFORMATION

Radiologic diagnosis is a more complex process than simply detecting visual signals in a relatively uniform background. How these signals are recognized and interpreted and how this interpretation is woven into a radiologic diagnosis are poorly understood phenomena. Several vision scientists have proposed that an experienced observer first looks quickly at an image and compares it with a template of "remembered" images that represent normal structures similar to those in the image. In effect, the observer performs a quick mental subtraction to leave only those items in the perceived image that differ from the normal template. These differences are then examined more closely, and decisions are made about their significance. If the observer decides that they are acceptable deviations from normal, they receive no further attention. But if they are judged to be unusual, they are integrated into an interpretation of the image that reflects an abnormal condition in the patient. This "cognitive" model of image perception is a tempting explanation of the way visual information is detected, recognized, and interpreted. However, several experiments with human observers suggest that it is overly simplistic.[25–28]

Studies of visual search patterns reveal that observers inexperienced in the types of images presented (e.g., chest radiographs) tend to focus on the central part of the image, whereas more experienced observers employ a more circumferential and complete scanning pattern.

EXPRESSIONS OF VISUAL PERFORMANCE

Visual images are useful insofar as they convey aesthetically pleasing or helpful information to the observer. In the latter case, helpfulness is characterized by greater enlightenment and improved performance of the observer. Several approaches have been developed to express these characteristics in a quantitative manner. Some of the more common approaches are described in this section.

Contrast-Detail Analysis

The detection of simple objects in an image is a complex interplay of their size, brightness (density), and differences from background. This interplay can be represented in a contrast-detail diagram. In this approach, a test pattern is constructed that contains objects of different sizes and composition superimposed on a relatively uniform background. The objects of different composition are reflected as regions of different brightness (density) in the image. The observer is asked to identify the smallest object of each brightness or density that is visible in the image. Smaller objects must be displayed at greater brightness or density to be seen in the image.

Conspicuity

The detection of objects in an image is complicated by the presence of background information, especially when the background has an intrinsic structure of its own. In studies of this complicating factor, a description of the relationship of the objects and background is helpful. One useful description is conspicuity, defined as the contrast between the objects and background divided by the average complexity of

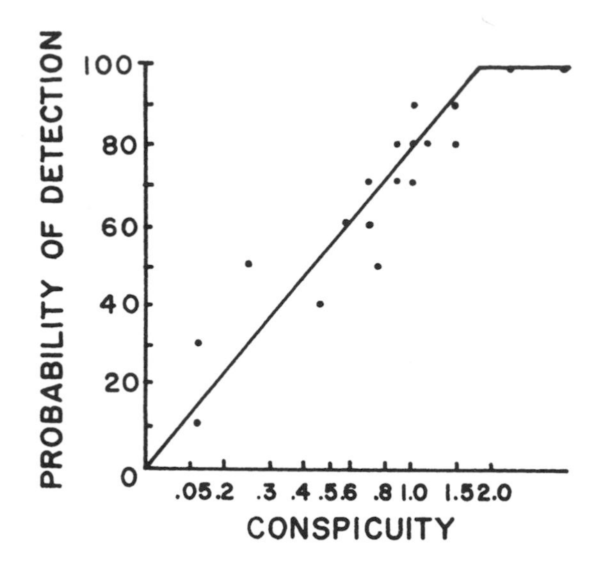

FIGURE 18-8
Probability of detecting solitary nodules on a chest film as a function of log (conspicuity). (From Revesz, G., Kundel, H. L., and Graber, M. A.[29] Used with permission.)

the background. An example of the influence of conspicuity on the detection of simple objects in an image is illustrated in Figure 18-8.

Truth Tables

An image may or may not portray a particular object, and an observer may or may not detect an object that is portrayed. That is, four responses are possible in the interpretation of an image.

1. True-positive (TP): Observer detects an object that is portrayed in the image.
2. False-positive (FP): Observer detects an object that is not portrayed in the image.
3. True-negative (TN): Observer does not detect an object, and no object is portrayed in the image.
4. False-negative (FN): Observer does not detect an object even though an object is portrayed in the image.

These responses are represented in the *truth table* shown in the margin.

Truth Table

	Object	
Response	*Present*	*Absent*
Positive	True-positive	False-positive
Negative	False-negative	True-negative

Truth tables require a binary response from the observer, either a positive or negative reaction. This is an example of a two-alternative, forced-choice experiment.

In radiology, a film (or examination) is reported as either positive or negative. The film is positive if an abnormality is detected, and it is negative if no abnormality is seen. An independent evaluation (e.g., a surgical biopsy) of the presence or absence of the abnormality in the patient yields the classifications of TP, FP, TN, and FN.

When an object is portrayed in an image, only two responses are possible: The object is either detected or not. That is, $p(\mathrm{TP}) + p(\mathrm{FN}) = 1$, where p represents the probability of each response when the object is present. Similarly, only two responses, FP and TN, are possible when no object is portrayed in the image. Therefore, $p(\mathrm{FP} + p(\mathrm{TN}) = 1$ when the object is absent. The sensitivity of an imaging procedure is defined as $p(\mathrm{TP})$, the probability of detecting an abnormality when it is present in the patient by examining the images provided by the procedure. The value of $p(\mathrm{TP}) = \mathrm{TP}/(\mathrm{TP} + \mathrm{FN})$, where TP denotes the number of true-positives in a series of responses, and (TP + FN) signifies the total number of responses when the object of interest is present in the images. The specificity is defined as $p(\mathrm{TN})$, the probability of a negative (normal) interpretation when no abnormality exists in the patient. Often the specificity is defined as $1 - p(\mathrm{FP})$. The value of $p(\mathrm{TN}) = \mathrm{TN}/(\mathrm{TN} + \mathrm{FP})$, where TN denotes the number of true-negatives in a series of responses, and (TN + FP)

The p(TP) and p(FP) values in an experiment may change depending on how willing or confident the observer is in judging an abnormality.

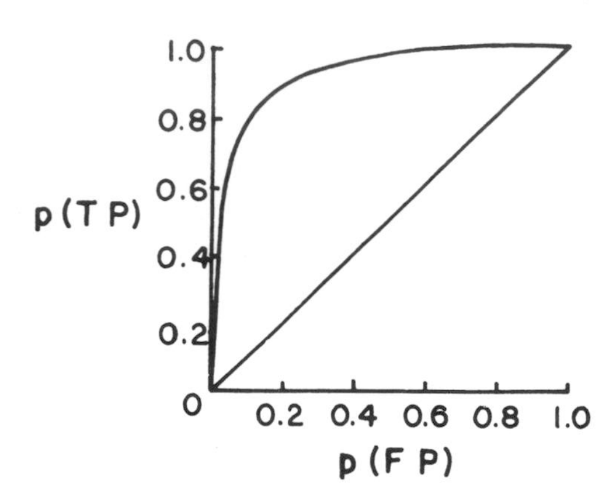

MARGIN FIGURE 18-4
A representative relative operating characteristics (ROC) curve.

The false-positive probability p(FP) in an ROC curve is frequently expressed as (1-specificity). The true-positive probability p(TP) is then described as the sensitivity.

"The main purpose in reporting an index of performance of a diagnostic system with a sample of cases, a sample of observers, and a sample of readings is not to tell the journal readership how well it performed in this particular sample that *was* studied, but to provide an estimate of how well it would perform 'on the average' in those similar cases and observers and readings that were *not* studied."[30]

TABLE 18-1 Expressions Important in Describing the Results of Diagnostic Tests

Expression	*Definition*[a]
Sensitivity	TP/(TP + FN)
Specificity	TN/(TN + FP)
False-positive ratio	FP/(FP + TN)
Positive predictive value	TP/(TP + FP)
Negative predictive value	TN/(FN + TN)
Prevalence	(TP + FN)/[TP + FN + FP + TN]

[a] TP, true-positive; FN, false-negative; TN, true-negative; FP, false-positive.

signifies the total number of responses when the object of concern is not present in the image.

Several other expressions are important to descriptions of the results of diagnostic tests. These expressions are summarized in Table 18-1.

Relative Operating Characteristics Analysis

The performance of observers who are examining images is often described by a relative operating characteristics (ROC) curve. In this approach, the p(TP) (the sensitivity) is plotted against p(FP). An ideal imaging procedure yields a p(TP) of unity and a p(FP) of zero. Data that fall high in the upper left quadrant of an ROC curve depict an accurate imaging procedure. Data that fall on a 45-degree diagonal represent pure guesswork by the observer.

The use of ROC curves is a powerful approach to analysis of the ability of an imaging system to yield definitive information about patients. However, ROC curves are tedious to generate because they must reflect an adequate sampling of all possible images and observers.

SUMMARY

- The three stages of visual perception are
  - Detection
  - Recognition
  - Interpretation
- The principal features of the human eye are the
  - Cornea
  - Iris
  - Lens
  - Retina
- Rod vision provides high sensitivity and low visual acuity.
- Cone vision provides low sensitivity and high visual acuity.
- Models that describe the detection of visual information include:
  - Signal-to-noise model
  - Computational model
  - Signal-channel model
- Visual acuity describes the ability for a observer to extract detailed information from an image. It can be evaluated with a Snellen chart.
- Contrast discrimination describes the ability of an observer to distinguish differences in brightness in an image. It is measured with a contrast discrimination chart.
- Expressions of visual performance include:
  - Contrast detail analysis
  - Conspicuity
  - Truth tables
  - ROC analysis

PROBLEMS

18-1. Distinguish between external noise in an image and internal noise in an observer.

*18-2. An individual receives a visual acuity score of 20/100 with a Snellen chart test. What does the score mean?

18-3. Explain the construction of a contrast-detail curve for an imaging system.

*18-4. Define the expressions "sensitivity" and "specificity" applied to an imaging system.

18-5. Describe the construction of an ROC curve.

*18-6. In screening x-ray mammography examinations, is a high value of sensitivity or specificity considered the more critical value? Why?

*For problems marked with an asterisk, answers are provided on p. 493.

REFERENCES

1. Hendee, W. R. Cognitive interpretation of visual signals, in Hendee, W., and Wells, P. (eds.), *The Perception of Visual Information*, 2nd edition. New York, Springer, 1997, pp. 149–175.
2. Jaffe, C. C. Medical imaging, vision, and visual psychophysics. *Med. Radiogr. Photogr.* 1984; **60**:1–48.
3. Thurston, J. B. and Carraher, R. G. *Optical Illusions and the Visual Arts*. New York, Van Nostrand Reinhold, 1986, p. 162.
4. Garland, L. H. On scientific evaluation of diagnostic procedures. *Radiology* 1949; **52**:309–329.
5. Yerushalemy, J. Reliability of chest radiography in the diagnosis of pulmonary lesions. *Am. J. Surg.* 1955; **80**:231–240.
6. Felson, B., Morgan, W. K., Bistow, L. J, et al. Observations on the results of multiple readings of chest films in coal miners' pneumoconiosis. *Radiology* 1973; **109**:19–23.
7. Kundel, H. L., Revesz, G. The evaluation of radiographic techniques by observer tests: Problems, pitfalls and procedures. *Invest. Radiol.* 1974; **9**:166–173.
8. Escher, M. C. *The Graphic Work of MC Escher,* 2nd edition. New York, Hawthorne Books, 1973.
9. Beam, C. A., Layde, P. M., and Sullivan, D. C. Variability in the interpretation of screening mammograms. *Arch. Intern. Med.* 1996; **156**:209–213.
10. Gregory, R. L. (ed.). *The Oxford Companion to the Mind*. New York, Oxford University Press, 1987.
11. Sharp, P. F., and Philips, R. Physiological Optics, in Hendee, W., and Wells, P. (eds.), *The Perception of Visual Information*, 2nd edition. New York, Springer, 1997, pp. 1–29.
12. Brown, B. H., Smallwood, R. H., Barber, D. C., Lawford, P. V., and Hose, D. R. *Medical Physics and Biomedical Engineering*. Philadelphia, Institute of Physics Publishing, 1999.
13. Stevens, S. S. The psychophysics of sensory function, in: Rosenblith, W. A. (ed.), *Sensory Communication.* Cambridge, MA, MIT Press, 1961, pp. 1–33.
14. Hendee, W. R. Imaging science in medicine, in Hornak, J. (ed.), *Encyclopedia of Imaging Science and Technology*. New York: John Wiley & Sons, 2002.
15. Green, D. M., and Swets, J. A. *Signal Detection Theory and Psychophysics*. New York, John Wiley & Sons, 1966.
16. Wagner, R. F., and Brown, D. G. Unified SNR analysis of medical imaging systems. *Phys. Med. Biol.* 1985; **30**:489–515.
17. Marr, D. *Vision*. New York, W. H., Freeman, 1982.
18. Ginsburg, A. P., and Chesters, M. S. Models of visual perception and their relationship to image interpretation, in Hendee, W., and Wells, P. N. T. (eds.), *Visualization Science in Engineering, Computing and Medicine*. Reston, VA, American College of Radiology, 1988.
19. Glezer, V. D., Ivanoff, V. A., Tscherback, T. A. Investigation of complex and hypercomplex receptive fields of visual cortex of the cat as spatial frequency filters. *Vision Res.* 1973; **13**:1875–1904.
20. Ginsburg, A. P. The visualization of diagnostic images. *Radiographics* 1987; **7**:1251–1260.
21. Ginsburg, A. P., and Hendee, W. R. Quantification of visual capability, in Hendee, W., and Wells, P. (eds.), *The Perception of Visual Information*, 2nd edition. New York, Springer, 1997, pp. 57–86.
22. Braddock, O. J., Sleigh, A. C. *Physical and Biological Processing of Images*. New York, Springer, 1983.
23. Ginsburg, A. P., Evans, D., Sekuler, R., et al. Contrast sensitivity predicts pilots' performance in aircraft simulators. *Am. J. Opt. Physiol. Optics* 1982; **59**:105–109.
24. Hendee, W. R. The perception of visual data. *AJR* 1989, **152**:1313–1317.
25. Kundel, H. L., Nodine, C. F., and Doi, K. Human interpretation of displayed images, in Hendee, W., and Wells, P. N. T. (eds.), *Visualization Science in Engineering, Computing and Medicine*. Boston, American College of Radiology, 1988.
26. Poggio, T., Torre, V., and Kock, C. Computational vision and regularization theory. *Nature* 1985; **317**:314–319.
27. Hendee, W. R., and Wells, P. N. T. *Perception of Visual Information*, 2nd edition. New York, Springer-Verlag, 1997.
28. Hendee, W. R. Image analysis and perception, in Putman, C. E., and Raven, C.E. (eds.), *Textbook of Diagnostic Imaging*. Philadelphia, W. B. Saunders, 1988.
29. Revesz, G., Kundel, H. L., and Graber, M. A. The influence of structured noise on the detection of radiologic abnormalities. *Invest Radiol* 1974; **9**:479–486.
30. Hanley, J. A. Receiver operating characteristic (ROC) methodology: The state of the art. *Crit. Rev. Diagn. Imaging* 1989; **29**:307–335.

CHAPTER

19

ULTRASOUND WAVES

OBJECTIVES

From studying this chapter, the reader should be able to:

- Explain the properties of ultrasound waves.
- Describe the decibel notation for ultrasound intensity and pressure.
- Delineate the ultrasound properties of velocity, attenuation, and absorption.
- Depict the consequences of an impedance mismatch at the boundary between two regions of tissue.
- Explain ultrasound reflection, refraction and scattering.

INTRODUCTION

Ultrasound is a mechanical disturbance that moves as a pressure wave through a medium. When the medium is a patient, the wavelike disturbance is the basis for use of ultrasound as a diagnostic tool. Appreciation of the characteristics of ultrasound waves and their behavior in various media is essential to understanding the use of diagnostic ultrasound in clinical medicine.[1–6]

In 1794, Spallanzi suggested correctly that bats avoided obstacles during flight by using sound signals beyond the range of the human ear.

HISTORY

In 1880, French physicists Pierre and Jacques Curie discovered the piezoelectric effect.[7] French physicist Paul Langevin attempted to develop piezoelectric materials as senders and receivers of high-frequency mechanical disturbances (ultrasound waves) through materials.[8] His specific application was the use of ultrasound to detect submarines during World War I. This technique, sound navigation and ranging (SONAR), finally became practical during World War II. Industrial uses of ultrasound began in 1928 with the suggestion of Soviet Physicist Sokolov that it could be used to detect hidden flaws in materials. Medical uses of ultrasound through the 1930s were confined to therapeutic applications such as cancer treatments and physical therapy for various ailments. Diagnostic applications of ultrasound began in the late 1940s through collaboration between physicians and engineers familiar with SONAR.[9]

"Piezo" is Greek for pressure. *Piezoelectricity* refers to the generation of an electrical response to applied pressure.

Pierre Curie used the piezoelectric properties of quartz crystals to construct a device to measure the small changes in mass that accompany radioactive decay. This work was done in collaboration with his wife Marie in her early studies of radioactivity.

The term "transducer" refers to any device that converts energy from one form to another (mechanical to electrical, electrical to heat, etc.). When someone asks to see a "transducer" in a radiology department, they will be shown ultrasound equipment. But strictly speaking, they could just as well be taken to see an x-ray tube.

WAVE MOTION

A fluid medium is a collection of molecules that are in continuous random motion. The molecules are represented as filled circles in the margin figure (Margin Figure 19-1). When no external force is applied to the medium, the molecules are distributed more or less uniformly (**A**). When a force is applied to the medium (represented by movement of the piston from left to right in **B**), the molecules are concentrated in front of the piston, resulting in an increased pressure at that location. The region of increased pressure is termed a *zone of compression*. Because of the forward motion imparted to the molecules by the piston, the region of increased pressure begins to migrate away from the piston and through the medium. That is, a mechanical disturbance introduced into the medium travels through the medium in a direction away from the source of the disturbance. In clinical applications of ultrasound, the piston is replaced by an ultrasound transducer.

As the zone of compression begins its migration through the medium, the piston may be withdrawn from right to left to create a region of reduced pressure immediately behind the compression zone. Molecules from the surrounding medium move into this region to restore it to normal particle density; and a second region, termed a *zone of rarefaction*, begins to migrate away from the piston (**C**). That is, the compression zone (high pressure) is followed by a zone of rarefaction (low pressure) also moving through the medium.

If the piston is displaced again to the right, a second compression zone is established that follows the zone of rarefaction through the medium, If the piston oscillates continuously, alternate zones of compression and rarefaction are propagated through the medium, as illustrated in **D**. The propagation of these zones establishes a wave disturbance in the medium. This disturbance is termed a *longitudinal wave* because the motion of the molecules in the medium is parallel to the direction of wave propagation. A wave with a frequency between about 20 and 20,000 Hz is a sound wave that is audible to the human ear. An infrasonic wave is a sound wave below 20 Hz; it is not audible to the human ear. An ultrasound (or ultrasonic) wave has a frequency greater than 20,000 Hz and is also inaudible. In clinical diagnosis, ultrasound waves of frequencies between 1 and 20 MHz are used.

As a longitudinal wave moves through a medium, molecules at the edge of the wave slide past one another. Resistance to this shearing effect causes these molecules to move somewhat in a direction away from the moving longitudinal wave. This transverse motion of molecules along the edge of the longitudinal wave establishes shear waves that radiate transversely from the longitudinal wave. In general, shear waves are significant only in a rigid medium such as a solid. In biologic tissues, bone is the only medium in which shear waves are important.

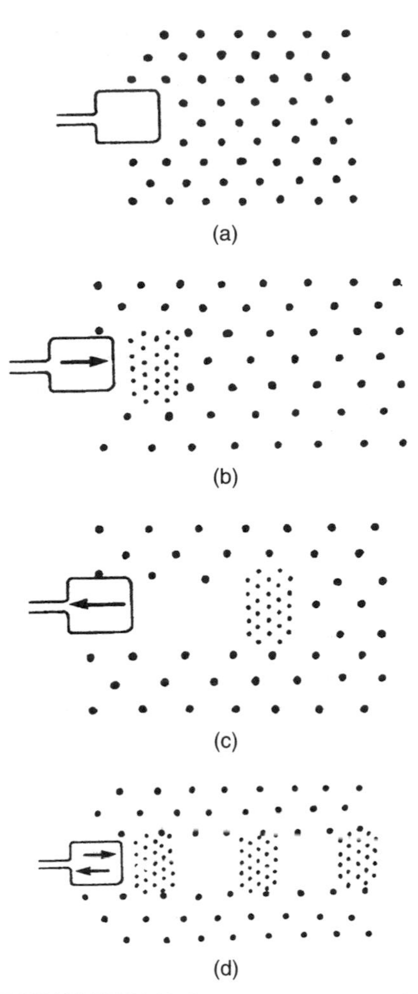

MARGIN FIGURE 19-1
Production of an ultrasound wave. **A:** Uniform distribution of molecules in a medium. **B:** Movement of the piston to the right produces a zone of compression. **C:** Withdrawal of the piston to the left produces a zone of rarefaction. **D:** Alternate movement of the piston to the right and left establishes a longitudinal wave in the medium.

WAVE CHARACTERISTICS

A zone of compression and an adjacent zone of rarefaction constitute one cycle of an ultrasound wave. A wave cycle can be represented as a graph of local pressure (particle density) in the medium versus distance in the direction of the ultrasound wave (Figure 19-1). The distance covered by one cycle is the wavelength of the ultrasound wave. The number of cycles per unit time (cps, or just sec^{-1}) introduced into the medium each second is referred to as the *frequency of the wave*, expressed in units of hertz, kilohertz, or megahertz where 1 Hz equals 1 cps. The maximum height of the wave cycle is the amplitude of the ultrasound wave. The product of the frequency (ν) and the wavelength (λ) is the velocity of the wave; that is, $c = \nu\lambda$.

In most soft tissues, the velocity of ultrasound is about 1540 m/sec. Frequencies of 1 MHz and greater are required to furnish ultrasound wavelengths suitable for diagnostic imaging.

When two waves meet, they are said to "interfere" with each other (see Margin). There are two extremes of interference. In constructive interference the waves are "in phase" (i.e., peak meets peak). In destructive interference the waves are "out of phase" (i.e., peak meets valley). Waves undergoing constructive interference add their amplitudes, whereas waves undergoing destructive interference may completely cancel each other.

During the propagation of an ultrasound wave, the molecules of the medium vibrate over very short distances in a direction parallel to the longitudinal wave. It is this vibration, during which momentum is transferred among molecules, that causes the wave to move through the medium.

Frequency Classification of Ultrasound

Frequency (Hz)	*Classification*
20–20,000	Audible sound
20,000–1,000,000	Ultrasound
1,000,000–30,000,000	Diagnostic medical ultrasound

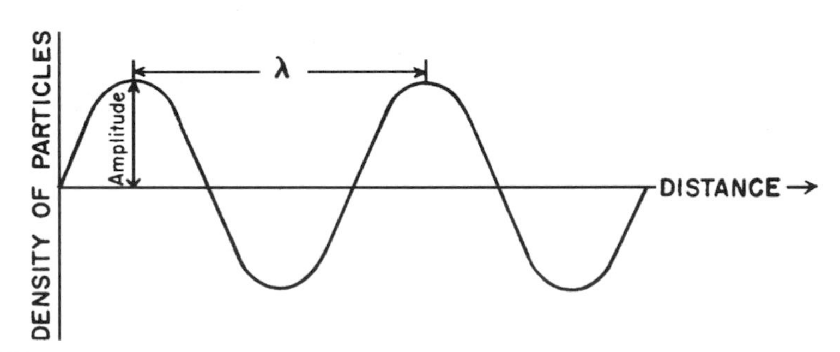

FIGURE 19-1
Characteristics of an ultrasound wave.

Ultrasound frequencies of 1 MHz and greater correspond to ultrasound wavelengths less than 1 mm in human soft tissue.

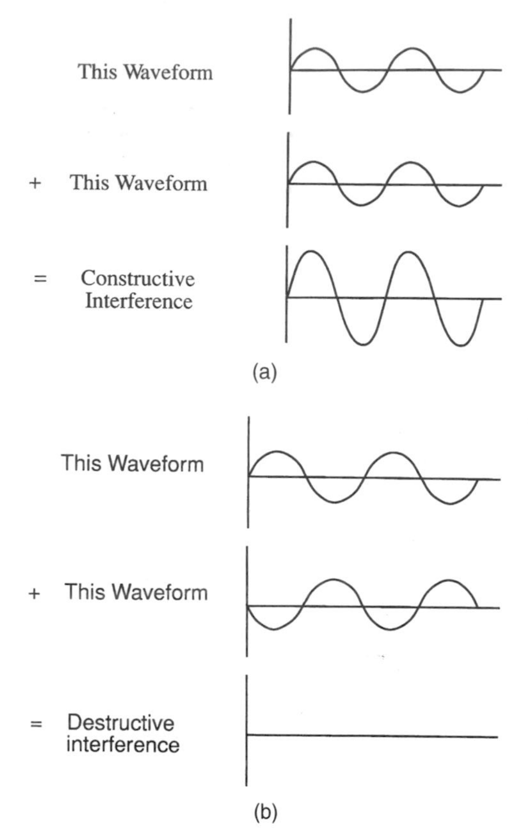

MARGIN FIGURE 19-2
Waves can exhibit interference, which in extreme cases of constructive and destructive interference leads to complete addition (**A**) or complete cancellation (**B**) of the two waves.

In the audible range, sound power and intensity are referred to as "loudness."

Pulsed ultrasound is used for most medical diagnostic applications. Ultrasound pulses vary in intensity and time and are characterized by four variables: spatial peak (SP), spatial average (SA), temporal peak (TP), and temporal average (TA).

Temporal average ultrasound intensities used in medical diagnosis are in the mW/cm^2 range.

TABLE 19-1 Quantities and Units Pertaining to Ultrasound Intensity

Quantity	*Definition*	*Unit*
Energy (E)	Ability to do work	joule
Power (P)	Rate at which energy is transported	watt (joule/sec)
Intensity (I)	Power per unit area (a), where t = time	$watt/cm^2$
Relationship	$I = \frac{P}{a} = \frac{E}{(t)(a)}$	

ULTRASOUND INTENSITY

As an ultrasound wave passes through a medium, it transports energy through the medium. The rate of energy transport is known as "power." Medical ultrasound is produced in beams that are usually focused into a small area, and the beam is described in terms of the power per unit area, defined as the beam's "intensity." The relationships among the quantities and units pertaining to intensity are summarized in Table 19-1.

Intensity is usually described relative to some reference intensity. For example, the intensity of ultrasound waves sent into the body may be compared with that of the ultrasound reflected back to the surface by structures in the body. For many clinical situations the reflected waves at the surface may be as much as a hundredth or so of the intensity of the transmitted waves. Waves reflected from structures at depths of 10 cm or more below the surface may be lowered in intensity by a much larger factor. A logarithmic scale is most appropriate for recording data over a range of many orders of magnitude. In acoustics, the decibel scale is used, with the decibel defined as

$$dB = 10 \log \frac{I}{I_0} \tag{19-1}$$

where I_0 is the reference intensity. Table 19-2 shows examples of decibel values for certain intensity ratios. Several rules can be extracted from this table:

- Positive decibel values result when a wave has a higher intensity than the reference wave; negative values denote a wave with lower intensity.
- Increasing a wave's intensity by a factor of 10 adds 10 dB to the intensity, and reducing the intensity by a factor of 10 subtracts 10 dB.
- Doubling the intensity adds 3 dB, and halving subtracts 3 dB.

No universal standard reference intensity exists for ultrasound. Thus the statement "ultrasound at 50 dB was used" is nonsensical. However, a statement such as "the returning echo was 50 dB below the transmitted signal" is informative. The transmitted signal then becomes the reference intensity for this particular application. For

TABLE 19-2 Calculation of Decibel Values From Intensity Ratios and Amplitude Ratios

Ratio of Ultrasound Wave Parameters	*Intensity Ratio (I/I_0) (dB)*	*Amplitude Ratio (A/A_0) (dB)*
1000	30	60
100	20	40
10	10	20
2	3	6
1	0	0
1/2	−3	−6
1/10	−10	−20
1/100	−20	−40
1/1000	−30	−60

audible sound, a statement such as "a jet engine produces sound at 100 dB" is appropriate because there is a generally accepted reference intensity of 10^{-16} W/cm^2 for audible sound.[10] A 1-kHz tone (musical note C one octave above middle C) at this intensity is barely audible to most listeners. A 1-kHz note at 120 dB (10^{-4} W/cm^2) is painfully loud.

The human ear is unable to distinguish a difference in loudness less than about 1 dB.

Because intensity is power per unit area and power is energy per unit time (Table 19-1), Eq. (19-1) may be used to compare the power or the energy contained within two ultrasound waves. Thus we could also write

$$\text{dB} = 10 \log \frac{\text{Power}}{\text{Power}_0} = 10 \log \frac{E}{E_0}$$

Ultrasound wave intensity is related to maximum pressure (P_m) in the medium by the following expression[11]:

$$1 = \frac{P_m^2}{2\rho c} \qquad (19\text{-}2)$$

where ρ is the density of the medium in grams per cubic centimeter and c is the speed of sound in the medium. Substituting Eq. (19-2) for I and I_0 in Eq. (19-1) yields

$$\text{dB} = 10 \log \frac{P_m^2/2\rho c}{(P_m^2)_0/2\rho c} = 10 \log \left[\frac{P_m}{P_{m_0}}\right]^2$$

$$= 20 \log \frac{P_m}{P_{m_0}} \qquad (19\text{-}3)$$

When comparing the pressure of two waves, Eq. (19-3) may be used directly. That is, the pressure does not have to be converted to intensity to determine the decibel value. An ultrasound transducer converts pressure amplitudes received from the patient (i.e., the reflected ultrasound wave) into voltages. The amplitude of voltages recorded for ultrasound waves is directly proportional to the variations in pressure in the reflected wave.

Calculation of Decibel Value from Wave Parameters

For X = Intensity in W/cm^2
= Power in watts
= Energy in joules
use $\text{dB} = 10 \log \frac{X}{X_0}$
For Y = Pressure in pascals or atmospheres
= Amplitude in volts
Use $\text{dB} = 20 \log \frac{Y}{Y_0}$

The decibel value for the ratio of two waves may be calculated from Eq. (19-1) or from Eq. (19-3), depending upon the information that is available concerning the waves (see Margin Table). The "half-power value" (ratio of 0.5 in power between two waves) is –3 dB, whereas the "half-amplitude value" (ratio of 0.5 in amplitude) is –6 dB (Table 19-2). This difference reflects the greater sensitivity of the decibel scale to amplitude compared with intensity values.

Ultrasound intensities may also be compared in units of nepers per centimeter by using the natural logarithm (ln) rather than the common logarithm (log) where

$$\text{Neper} = \ln (I/I_0)$$

ULTRASOUND VELOCITY

The velocity of an ultrasound wave through a medium varies with the physical properties of the medium. In low-density media such as air and other gases, molecules may move over relatively large distances before they influence neighboring molecules. In these media, the velocity of an ultrasound wave is relatively low. In solids, molecules are constrained in their motion, and the velocity of ultrasound is relatively high. Liquids exhibit ultrasound velocities intermediate between those in gases and solids. With the notable exceptions of lung and bone, biologic tissues yield velocities roughly similar to the velocity of ultrasound in liquids. In different media, changes in velocity are reflected in changes in wavelength of the ultrasound waves, with the frequency remaining relatively constant. In ultrasound imaging, variations in the velocity of ultrasound in different media introduce artifacts into the image, with the major artifacts attributable to bone, fat, and, in ophthalmologic applications, the lens of the eye. The velocities of ultrasound in various media are listed in Table 19-3.

In ultrasound, the term *propogation speed* is preferred over the term *velocity*.

The velocity of an ultrasound wave should be distinguished from the velocity of molecules whose displacement into zones of compression and rarefaction constitutes the wave. The molecular velocity describes the velocity of the individual molecules in the medium, whereas the wave velocity describes the velocity of the ultrasound wave

The velocity of ultrasound in a medium is virtually independent of the ultrasound frequency.

TABLE 19-3 Approximate Velocities of Ultrasound in Selected Materials

Nonbiologic Material	*Velocity (m/sec)*	*Biologic Material*	*Velocity (m/sec)*
Acetone	1174	Fat	1475
Air	331	Brain	1560
Aluminum (rolled)	6420	Liver	1570
Brass	4700	Kidney	1560
Ethanol	1207	Spleen	1570
Glass (Pyrex)	5640	Blood	1570
Acrylic plastic	2680	Muscle	1580
Mercury	1450	Lens of eye	1620
Nylon (6-6)	2620	Skull bone	3360
Polyethylene	1950	Soft tissue (mean value)	1540
Water (distilled), 25°C	1498		
Water (distilled), 50°C	1540		

The velocity of ultrasound is determined principally by the compressibility of the medium. A medium with high compressibility yields a slow ultrasound velocity, and vice versa. Hence, the velocity is relatively low in gases, intermediate in soft tissues, and greatest in solids such as bone.

through the medium. Properties of ultrasound such as reflection, transmission, and refraction are characteristic of the wave velocity rather than the molecular velocity.

ATTENUATION OF ULTRASOUND

As an ultrasound beam penetrates a medium, energy is removed from the beam by absorption, scattering, and reflection. These processes are summarized in Figure 19-2. As with x rays, the term *attenuation* refers to any mechanism that removes energy from the ultrasound beam. Ultrasound is "absorbed" by the medium if part of the beam's

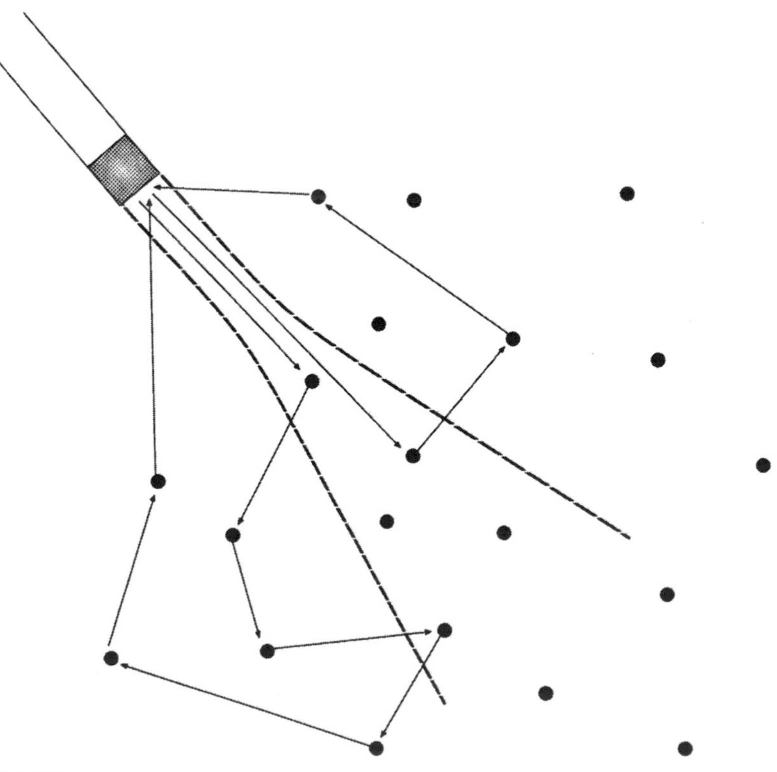

FIGURE 19-2
Constructive and destructive interference effects characterize the echoes from nonspecular reflections. Because the sound is reflected in all directions, there are many opportunities for waves to travel different pathways. The wave fronts that return to the transducer may constructively or destructively interfere at random. The random interference pattern is known as "speckle."

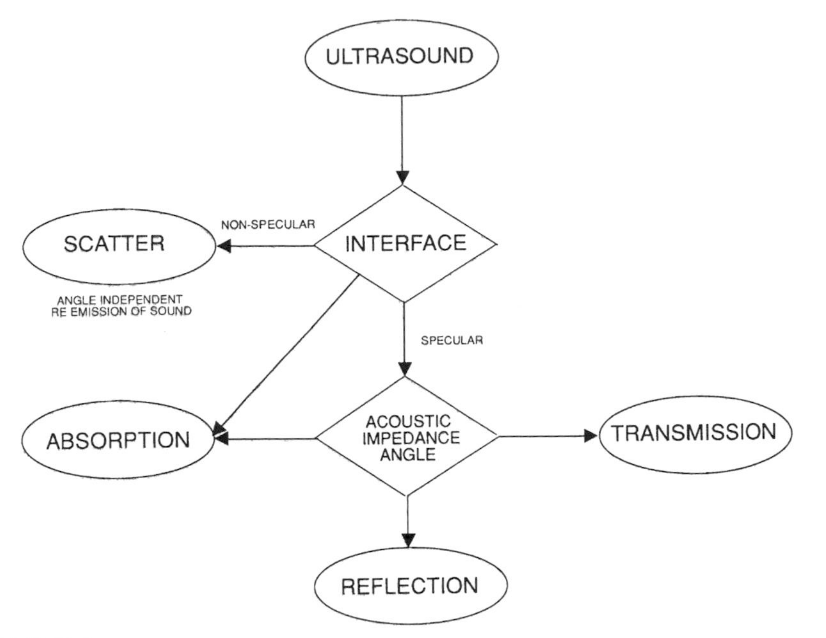

FIGURE 19-3
Summary of interactions of ultrasound at boundaries of materials.

energy is converted into other forms of energy, such as an increase in the random motion of molecules. Ultrasound is "reflected" if there is an orderly deflection of all or part of the beam. If part of an ultrasound beam changes direction in a less orderly fashion, the event is usually described as "scatter."

The behavior of a sound beam when it encounters an obstacle depends upon the size of the obstacle compared with the wavelength of the sound. If the obstacle's size is large compared with the wavelength of sound (and if the obstacle is relatively smooth), then the beam retains its integrity as it changes direction. Part of the sound beam may be reflected and the remainder transmitted through the obstacle as a beam of lower intensity.

If the size of the obstacle is comparable to or smaller than the wavelength of the ultrasound, the obstacle will scatter energy in various directions. Some of the ultrasound energy may return to its original source after "nonspecular" scatter, but probably not until many scatter events have occurred.

In ultrasound imaging, specular reflection permits visualization of the boundaries between organs, and nonspecular reflection permits visualization of tissue parenchyma (Figure 19-2). Structures in tissue such as collagen fibers are smaller than the wavelength of ultrasound. Such small structures provide scatter that returns to the transducer through multiple pathways. The sound that returns to the transducer from such nonspecular reflectors is no longer a coherent beam. It is instead the sum of a number of component waves that produces a complex pattern of constructive and destructive interference back at the source. This interference pattern, known as "speckle," provides the characteristic ultrasonic appearance of complex tissue such as liver.

The behavior of a sound beam as it encounters an obstacle such as an interface between structures in the medium is summarized in Figure 19-3. As illustrated in Figure 19-4, the energy remaining in the beam decreases approximately exponentially with the depth of penetration of the beam into the medium. The reduction in energy (i.e., the decrease in ultrasound intensity) is described in decibels, as noted earlier.

An increase in the random motion of molecules is measurable as an increase in temperature of the medium.

Contributions to attenuation of an ultrasound beam may include:

- Absorption
- Reflection
- Scattering
- Refraction
- Diffraction
- Interference
- Divergence

Reflection in which the ultrasound beam retains its integrity is said to be "specular," from the Latin for "mirror." Reflection of visible light from a plane mirror is an optical example of specular reflection (i.e., the original shape of the wave fronts).

An optical example of nonspecular reflection occurs when a mirror is "steamed up." The water droplets on the mirror are nonspecular reflectors that serve to scatter the light beam.

When scattering objects are much smaller than the ultrasound wavelength, the scattering process is referred to as Rayleigh scattering. Red blood cells are sometimes referred to as Rayleigh scatterers.

Example 19-1

Find the percent reduction in intensity for a 1-MHz ultrasound beam traversing 10 cm of material having an attenuation of 1 dB/cm. The reduction in intensity (dB) = (1 dB/cm) (10 cm) = −10 dB (the minus sign corresponds to a decrease in intensity

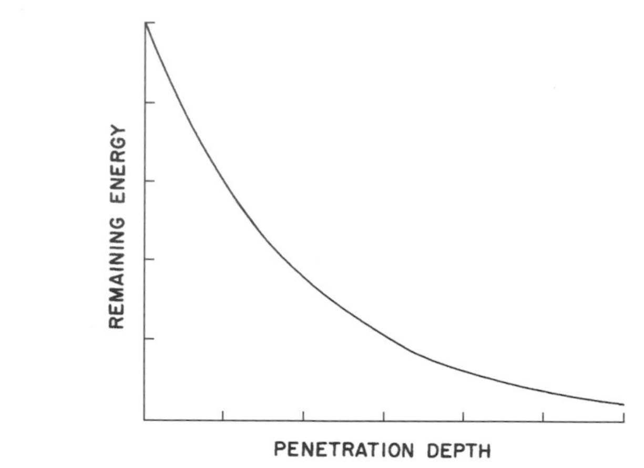

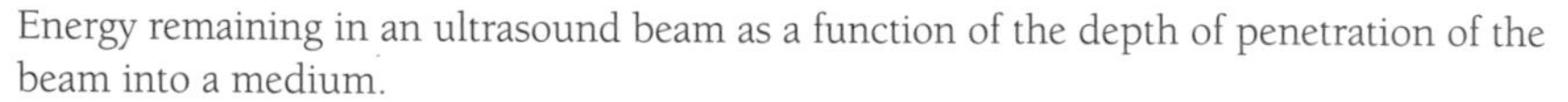

FIGURE 19-4
Energy remaining in an ultrasound beam as a function of the depth of penetration of the beam into a medium.

compared with the reference intensity, which in this case is the intensity of sound before the attenuating material is encountered).

$$\mathrm{dB} = 10 \log \frac{I}{I}$$

$$-10 = 10 \log \frac{I}{I_0}$$

$$1 = \log \frac{I_0}{I}$$

$$10 = \frac{I_0}{I}$$

$$I = \frac{I_0}{10}$$

There has been a 90% reduction in intensity. Determine the intensity reduction if the ultrasound frequency were increased to 2MHz.

Because the attenuation increases approximately linearly with frequency, the attenuation coefficient at 2 MHz would be 2 dB/cm, resulting in a –20 dB (99%) intensity reduction.

The attenuation of ultrasound in a material is described by the attenuation coefficient α in units of decibels per centimeter (Table 19-4). Many of the values in Table 19-4 are known only approximately and vary significantly with both the origin and condition of the biologic samples. The attenuation coefficient α is the sum of the individual coefficients for scatter and absorption. In soft tissue, the absorption

TABLE 19-4 Attenuation Coefficients α for 1-MHz Ultrasound

Material	*α (dB/cm)*	*Material*	*α (dB/cm)*
Blood	0.18	Lung	40
Fat	0.6	Liver	0.9
Muscle (across fibers)	3.3	Brain	0.85
Muscle (along fibers)	1.2	Kidney	1.0
Aqueous and vitreous humor of eye	0.1	Spinal cord	1.0
		Water	0.0022
Lens of eye	2.0	Caster oil	0.95
Skull bone	20	Lucite	2.0

coefficient accounts for 60% to 90% of the attenuation, and scatter accounts for the remainder.[11]

Table 19-4 shows that the attenuation of ultrasound is very high in bone. This property, along with the large reflection coefficient of a tissue-bone interface, makes it difficult to visualize structures lying behind bone. Little attenuation occurs in water, and this medium is a very good transmitter of ultrasound energy. To a first approximation, the attenuation coefficient of most soft tissues can be approximated as 0.9υ, where υ is the frequency of the ultrasound in MHz. This expression states that the attenuation of ultrasound energy increases with frequency in biologic tissues. That is, higher-frequency ultrasound is attenuated more readily and is less penetrating than ultrasound of lower frequency.

The energy loss in a medium composed of layers of different materials is the sum of the energy loss in each layer.

Attenuation causes a loss of signal intensity when structures at greater depths are imaged with ultrasound. This loss of intensity is compensated by increasing the amplification of the signals.

Differences in attenuation among tissues causes enhancement and shadowing in ultrasound images.

Example 19-2

Suppose that a block of tissue consists of 2 cm fat, 3 cm muscle (ultrasound propagated parallel to the fibers), and 4 cm liver. The total energy loss is

$$\begin{aligned}\text{Total energy loss} &= (\text{Energy loss in fat}) + (\text{Energy loss in muscle}) \\ &\quad + (\text{Energy loss in liver}) \\ &= (0.6\ \text{dB/cm})\ (2\ \text{cm}) + (1.2\ \text{dB/cm})\ (3\ \text{cm}) \\ &\quad + (09\ \text{dB/cm})\ (4\ \text{cm}) \\ &= 1.2\ \text{dB} + 3.6\ \text{dB} + 3.6\ \text{dB} \\ &= 8.4\ \text{dB}\end{aligned}$$

For an ultrasound beam that traverses the block of tissue and, after reflection, returns through the tissue block, the total attenuation is twice 8.4 dB or 16.8 dB.

REFLECTION

In most diagnostic applications of ultrasound, use is made of ultrasound waves reflected from interfaces between different tissues in the patient. The fraction of the impinging energy reflected from an interface depends on the difference in acoustic impedance of the media on opposite sides of the interface.

The acoustic impedance Z of a medium is the product of the density ρ of the medium and the velocity of ultrasound in the medium:

$$Z = \rho c$$

Acoustic impedances of several materials are listed in the margin. For an ultrasound wave incident perpendicularly upon an interface, the fraction α_R of the incident energy that is reflected (i.e., the reflection coefficient α_R) is

$$\alpha_R = \left(\frac{Z_2 - Z_1}{Z_2 + Z_1}\right)^2$$

where Z_1 and Z_2 are the acoustic impedances of the two media. The fraction of the incident energy that is transmitted across an interface is described by the transmission coefficient α_T, where

$$\alpha_T = \frac{4Z_1Z_2}{(Z_1 + Z_2)^2}$$

Obviously $\alpha_T + \alpha_R = 1$.

In this discussion, reflection is assumed to occur at interfaces that have dimensions greater than the ultrasound wavelength. In this case, the reflection is termed *specular reflection.*

Acoustic impedance may be expressed in units of rayls, where a rayl = 1 kg-m^{-2}-sec^{-1}.

The rayl is named for Lord Rayleigh [John Strutt (1842–1919), the 3rd Baron Rayleigh], a British physicist who pioneered the study of molecular motion in gasses that explains sound propogation.

Approximate Acoustic Impedances of Selected Materials

Material	*Acoustic Impedance (kg-m^{-2} -sec^{-1}) × 10^{-4}*
Air at standard temperature and pressure	0.0004
Water	1.50
Polyethylene	1.85
Plexiglas	3.20
Aluminum	18.0
Mercury	19.5
Brass	38.0
Fat	1.38
Aqueous and vitreous humor of eye	1.50
Brain	1.55
Blood	1.61
Kidney	1.62
Human soft tissue, mean value	1.63
Spleen	1.64
Liver	1.65
Muscle	1.70
Lens of eye	1.85
Skull bone	6.10

With a large impedance mismatch at an interface, much of the energy of an ultrasound wave is reflected, and only a small amount is transmitted across the interface. For example, ultrasound beams are reflected strongly at air–tissue and air–water interfaces because the impedance of air is much less than that of tissue or water.

Example 19-3

At a "liver–air" interface, $Z_1 = 1.65$ and $Z_2 = 0.0004$ (both multiplied by 10^{-4} with units of kg-m^{-2}-sec^{-1}).

$$\alpha_R = \left(\frac{1.65 - 0.0004}{1.65 + 0.0004}\right)^2, \qquad \alpha_T = \frac{4(1.65)(0.0004)}{(1.65 + 0.0004)^2}$$
$$= 0.9995 \qquad\qquad = 0.0005$$

Thus 99.95% of the ultrasound energy is reflected at the air–liver interface, and only 0.05% of the energy is transmitted. At a muscle ($Z = 1.70$)–liver ($Z = 1.65$) interface,

$$\alpha_R = \left(\frac{1.70 - 1.65}{1.70 + 1.65}\right)^2, \qquad \alpha_T = \frac{4(1.70)(1.65)}{(1.70 + 1.65)^2}$$
$$= 0.015 \qquad\qquad = 0.985$$

At a muscle–liver interface, slightly more than 1% of the incident energy is reflected, and about 99% of the energy is transmitted across the interface. Even though the reflected energy is small, it is often sufficient to reveal the liver border. The magnitudes of echoes from various interfaces in the body are described in Figure 19-5.

Because of the high value of the coefficient of ultrasound reflection at an air–tissue interface, water paths and various creams and gels are used during ultrasound examinations to remove air pockets (i.e., to obtain good acoustic coupling) between the ultrasound transducer and the patient's skin. With adequate acoustic coupling,

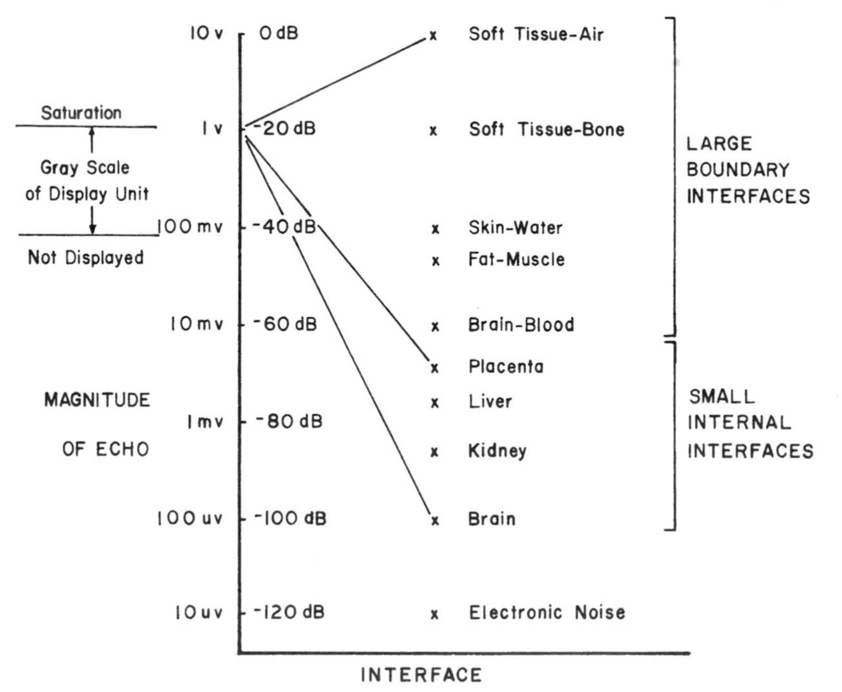

FIGURE 19-5
Range of echoes from biologic interfaces and selection of internal echoes to be displayed over the major portion of the gray scale in an ultrasound unit. (From Kossoff, G., et al.[12] Used with permission).

the ultrasound waves will enter the patient with little reflection at the skin surface. Similarly, strong reflections of ultrasound occur at the boundary between the chest wall and the lungs and at the millions of air–tissue interfaces within the lungs. Because of the large impedance mismatch at these interfaces, efforts to use ultrasound as a diagnostic tool for the lungs have been unrewarding. The impedance mismatch is also high between soft tissues and bone, and the use of ultrasound to identify tissue characteristics in regions behind bone has had limited success.

The discussion of ultrasound reflection above assumes that the ultrasound beam strikes the reflecting interface at a right angle. In the body, ultrasound impinges upon interfaces at all angles. For any angle of incidence, the angle at which the reflected ultrasound energy leaves the interface equals the angle of incidence of the ultrasound beam; that is,

$$\text{Angle of incidence} = \text{Angle of reflection}$$

In a typical medical examination that uses reflected ultrasound and a transducer that both transmits and detects ultrasound, very little reflected energy will be detected if the ultrasound strikes the interface at an angle more than about 3 degrees from perpendicular. A smooth reflecting interface must be essentially perpendicular to the ultrasound beam to permit visualization of the interface.

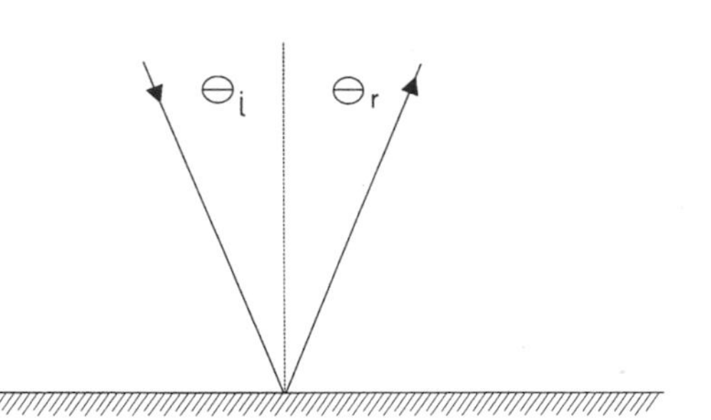

MARGIN FIGURE 19-3
Ultrasound reflection at an interface, where the angle of incidence θ_i equals the angle of reflection θ_r.

Two conditions are required for refraction to occur: (1) The sound beam must strike an interface at an angle other than 90°; (2) the speed of sound must differ on opposite sides of the interface.

REFRACTION

As an ultrasound beam crosses an interface obliquely between two media, its direction is changed (i.e., the beam is bent). If the velocity of ultrasound is higher in the second medium, then the beam enters this medium at a more oblique (less steep) angle. This behavior of ultrasound transmitted obliquely across an interface is termed *refraction*. The relationship between incident and refraction angles is described by Snell's law:

$$\frac{\text{Sine of incidence angle}}{\text{Sine of refractive angle}} = \frac{\text{Velocity in incidence medium}}{\text{Velocity in refractive medium}} \tag{19-4}$$

$$\frac{\sin \theta_i}{\sin \theta_r} = \frac{c_i}{c_r}$$

For example, an ultrasound beam incident obliquely upon an interface between muscle (velocity 1580 m/sec) and fat (velocity 1475 m/sec) will enter the fat at a steeper angle.

If an ultrasound beam impinges very obliquely upon a medium in which the ultrasound velocity is higher, the beam may be refracted so that no ultrasound energy enters the medium. The incidence angle at which refraction causes no ultrasound to enter a medium is termed the *critical angle* θ_c. For the critical angle, the angle of refraction is 90 degrees, and the sine of 90 degrees is 1. From Eq. (19-4),

$$\frac{\sin \theta_c}{\sin 90^\circ} = \frac{c_i}{c_r}$$

but

$$\sin 90^\circ = 1$$

therefore

$$\theta_c = \sin^{-1}[c_i/c_r]$$

where $\sin^{-1}$, or arcsin, refers to the angle whose sine is c_i/c_r. For any particular interface, the critical angle depends only upon the velocity of ultrasound in the two media separated by the interface.

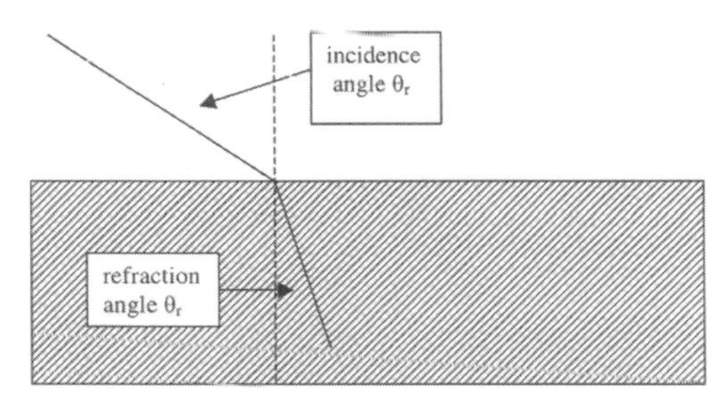

MARGIN FIGURE 19-4
Refraction of ultrasound at an interface, where the ratio of the velocities of ultrasound in the two media is related to the sine of the angles of incidence and refraction.

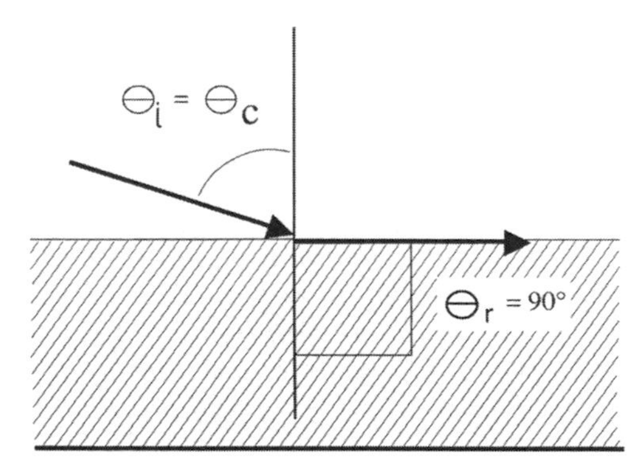

MARGIN FIGURE 19-5
For an incidence angle θ_c equal to the critical angle, refraction causes the sound to be transmitted along the surface of the material. For incidence angles greater than θ_c, sound transmission across the interface is prevented by refraction.

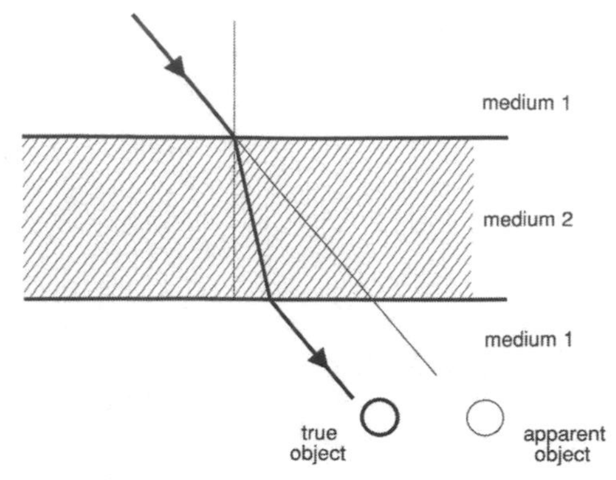

MARGIN FIGURE 19-6
Lateral displacement of an ultrasound beam as it traverses a slab interposed in an otherwise homogeneous medium.

Refraction is a principal cause of artifacts in clinical ultrasound images. In Margin Figure 19-6, for example, the ultrasound beam is refracted at a steeper angle as it crosses the interface between medium 1 and 2 ($c_1 > c_2$). As the beam emerges from medium 2 and reenters medium 1, it resumes its original direction of motion. The presence of medium 2 simply displaces the ultrasound beam laterally for a distance that depends upon the difference in ultrasound velocity and density in the two media and upon the thickness of medium 2. Suppose a small structure below medium 2 is visualized by reflected ultrasound. The position of the structure would appear to the viewer as an extension of the original direction of the ultrasound through medium 1. In this manner, refraction adds spatial distortion and resolution loss to ultrasound images.

Maximum ultrasound intensities (mW/cm^2) recommended by the U.S. Food and Drug Administration for various diagnostic applications. The values are spatial-peak, temporal-average (SPTA) values.[5,13]

Use	$(\text{Intensity})_{max}$
Cardiac	430
Peripheral vessels	720
Ophthalmic	17
Abdominal	94
Fetal	94

ABSORPTION

Relaxation processes are the primary mechanisms of energy dissipation for an ultrasound beam transversing tissue. These processes involve (a) removal of energy from the ultrasound beam and (b) eventual dissipation of this energy primarily as heat. As discussed earlier, ultrasound is propagated by displacement of molecules of a medium into regions of compression and rarefaction. This displacement requires energy that is provided to the medium by the source of ultrasound. As the molecules attain maximum displacement from an equilibrium position, their motion stops, and their energy is transformed from kinetic energy associated with motion to potential energy associated with position in the compression zone. From this position, the molecules begin to move in the opposite direction, and potential energy is gradually transformed into kinetic energy. The maximum kinetic energy (i.e., the highest molecular velocity) is achieved when the molecules pass through their original equilibrium position, where the displacement and potential energy are zero. If the kinetic energy of the molecule at this position equals the energy absorbed originally from the ultrasound beam, then no dissipation of energy has occurred, and the medium is an ideal transmitter of ultrasound. Actually, the conversion of kinetic to potential energy (and vice versa) is always accompanied by some dissipation of energy. Therefore, the energy of the ultrasound beam is gradually reduced as it passes through the medium. This reduction is termed *relaxation energy loss*. The rate at which the beam energy decreases is a reflection of the attenuation properties of the medium.

The effect of frequency on the attenuation of ultrasound in different media is described in Table 19-5.[14–18] Data in this table are reasonably good estimates of the influence of frequency on ultrasound absorption over the range of ultrasound frequencies used diagnostically. However, complicated structures such as tissue samples often exhibit a rather complex attenuation pattern for different frequencies, which probably reflects the existence of a variety of relaxation frequencies and other molecular energy absorption processes that are poorly understood at present. These complex attenuation patterns are reflected in the data in Figure 19-6.

American Institute of Ultrasound in Medicine Safety Statement on Diagnostic Ultrasound has been in use since the early 1950s. Given its known benefits and recognized efficacy for medical diagnosis, including use during human pregnancy, the American Institute of Ultrasound in Medicine hereby addresses the clinical safety of such use as follows:

> No confirmed biological effects on patients or instrument operators caused by exposure at intensities typical of present diagnostic ultrasound instruments have ever been reported. Although the possibility exists that such biological effects may be identified in the future, current data indicate that the benefits to patients of the prudent use of diagnostic ultrasound outweigh the risks, if any, that may be present.

TABLE 19-5 Variation of Ultrasound Attenuation Coefficient α with Frequency in Megahertz, Where α_1 Is the Attenuation Coefficient at 1 MHz

Tissue	*Frequency Variation*	*Material*	*Frequency Variation*
Blood	$\alpha = \alpha_1 \times v$	Lung	$\alpha = \alpha_1 \times v$
Fat	$\alpha = \alpha_1 \times v$	Liver	$\alpha = \alpha_1 \times v$
Muscle (across fibers)	$\alpha = \alpha_1 \times v$	Brain	$\alpha = \alpha_1 \times v$
Muscle (along fibers)	$\alpha = \alpha_1 \times v$	Kidney	$\alpha = \alpha_1 \times v$
Aqueous and vitreous humor of eye	$\alpha = \alpha_1 \times v$	Spinal cord	$\alpha = \alpha^1 \times v$
Lens of eye	$\alpha = \alpha_1 \times v$	Water	$\alpha = \alpha_1 \times v^2$
Skull bone	$\alpha = \alpha_1 \times v^2$	Caster oil	$\alpha = \alpha_1 \times v^2$
		Lucite	$\alpha = \alpha_1 \times v$

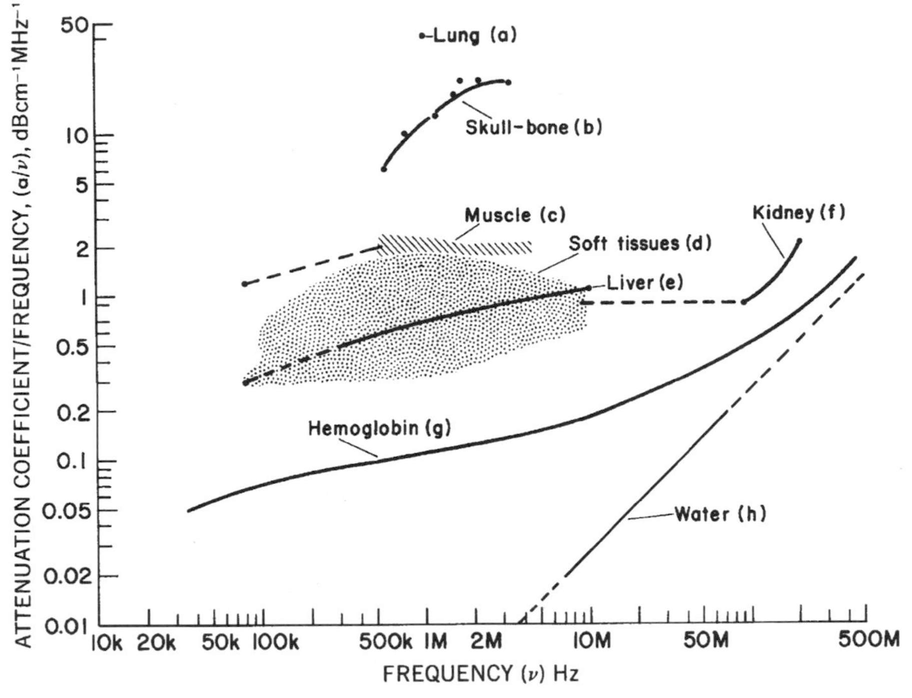

FIGURE 19-6
Ultrasound attenuation coefficient as a function of frequency for various tissue samples. (From Wells, P. N. T.[18] Used with permission.)

If gas bubbles are present in a material through which a sound wave is passing, the compressions and rarefactions cause the bubble to shrink and expand in resonance with the sound wave. The oscillation of such bubbles is referred to as *stable cavitation*. Stable cavitation is not a major mechanism for absorption at ultrasound intensities used diagnostically, but it can be a significant source of scatter.

If an ultrasound beam is intense enough and of the right frequency, the ultrasound-induced mechanical disturbance of the medium can be so great that microscopic bubbles are produced in the medium. The bubbles are formed at foci, such as molecules in the rarefaction zones, and may grow to a cubic millimeter or so in size. As the pressure in the rarefaction zone increases during the next phase of the ultrasound cycle, the bubbles shrink to 10^{-2} mm^3 or so and collapse, thereby creating minute shock waves that seriously disturb the medium if produced in large quantities. The effect, termed *dynamic cavitation*, produces high temperatures (up to 10,000°C) at the point where the collapse occurs.[19] Dynamic cavitation is associated with absorption of energy from the beam. Free radicals are also produced in water surrounding the collapse. Dynamic cavitation is not a significant mechanism of attenuation at diagnostic intensities, although there is evidence that it may occur under certain conditions.[20]

Dynamic cavitation is also termed "transient cavitation."

SUMMARY

- Properties of ultrasound waves include:
  - Compression and rarefaction
  - Requires a transmissive medium
  - Constructive and destructive interference
- The relative intensity and pressure of ultrasound waves are described in units of decibels.

- Ultrasound may be reflected or refracted at a boundary between two media. These properties are determined by the angle of incidence of the ultrasound and the impedance mismatch at the boundary.
- Energy may be removed from an ultrasound beam by various processes, including relaxation energy loss.
- The presence of gas bubbles in a medium may give rise to stable and dynamic cavitation.

PROBLEMS

19-1. Explain what is meant by a longitudinal wave, and describe how an ultrasound wave is propagated through a medium.

*19-2. An ultrasound beam is attenuated by a factor of 20 in passing through a medium. What is the attenuation of the medium in decibels?

*19-3. Determine the fraction of ultrasound energy transmitted and reflected at interfaces between (a) fat and muscle and (b) lens and aqueous and vitreous humor of the eye.

*19-4. What is the angle of refraction for an ultrasound beam incident at an angle of 15 degrees from muscle into bone?

19-5. Explain why refraction contributes to resolution loss in ultrasound imaging.

*19-6. A region of tissue consists of 3 cm fat, 2 cm muscle (ultrasound propagated parallel to fibers), and 3 cm liver. What is the approximate total energy loss of ultrasound in the tissue?

*For those problems marked with an asterisk, answers are provided on p. 493.

REFERENCES

1. Zagzebski, J. *Essentials of Ultrasound Physics*. St. Louis, Mosby–Year Book, 1996.
2. Wells, P. N. T. *Biomedical Ultrasonics*. New York, Academic Press, 1977.
3. McDicken, W. *Diagnostic Ultrasonics*. New York, John Wiley & Sons, 1976.
4. Eisenberg, R. *Radiology: An Illustrated History*. St. Louis, Mosby–Year Book, 1992, pp. 452–466.
5. Bushong, S. *Diagnostic Ultrasound*. New York, McGraw-Hill, 1999.
6. Palmer, P. E. S. *Manual of Diagnostic Ultrasound*. Geneva, Switzerland, World Health Organization, 1995.
7. Graff KF. *Ultrasonics: Historical aspects*. Presented at the IEEE Symposium on Sonics and Ultrasonics, Phoenix, October 26–28, 1977.
8. Hendee, W. R., and Holmes, J. H. History of Ultrasound Imaging, in Fullerton, G. D., and Zagzebski, J. A. (eds.), *Medical Physics of CT and Ultrasound*. New York: American Institute of Physics, 1980.
9. Hendee, W. R. Cross sectional medical imaging: A history. *Radiographics* 1989; **9**:1155–1180.
10. Kinsler, L. E., et al. *Fundamentals of Acoustics*, 3rd edition New York, John Wiley & Sons, 1982, pp. 115–117.
11. ter Haar GR. In CR Hill (ed): *Physical Principles of Medical Ultrasonics*. Chichester, England, Ellis Horwood/Wiley, 1986.
12. Kossoff, G., Garrett, W. J., Carpenter, D. A., Jellins, J., Dadd, M. J. Principles and classification of soft tissues by grey scale echography. *Ultrasound Med. Biol.* 1976; **2**:89–111.
13. Thrush, A., and Hartshorne, T. *Peripheral Vascular Ultrasound*. London, Churchill-Livingstone, 1999.
14. Chivers, R., and Hill, C. Ultrasonic attenuation in human tissues. *Ultrasound Med. Biol.* 1975; **2**:25.
15. Dunn, F., Edmonds, P., and Fry, W. Absorption and Dispersion of Ultrasound in Biological Media, in H. Schwan (ed.), *Biological Engineering*. New York, McGraw-Hill, 1969, p. 205
16. Powis, R. L., and Powis, W. J. *A Thinker's Guide to Ultrasonic Imaging*. Baltimore, Urban & Schwarzenberg, 1984.
17. Kertzfield, K., and Litovitz, T. *Absorption and Dispersion of Ultrasonic Waves*. New York, Academic Press, 1959.
18. Wells, P. N. T. Review: Absorption and dispersion of ultrasound in biological tissue. *Ultrasound Med Biol* 1975; **1**:369–376.
19. Suslick, K. S. (ed.). *Ultrasound, Its Chemical, Physical and Biological Effects*. New York, VCH Publishers, 1988.
20. Apfel, R. E. Possibility of microcavitation from diagnostic ultrasound. *Trans. IEEE* 1986; **33**:139–142.

CHAPTER

20

ULTRASOUND TRANSDUCERS

OBJECTIVES

After studying this chapter, the reader should be able to:

- Explain the piezoelectric effect and its use in ultrasound transducers.
- Characterize the properties of an ultrasound transducer, including those that influence the resonance frequency.
- Describe the properties of an ultrasound beam, including the Fresnel and Fraunhofer zones.
- Delineate the characteristics of focused ultrasound beams and various ultrasound probes.
- Identify different approaches to multitransducer arrays and the advantages of each.

INTRODUCTION

A transducer is any device that converts one form of energy into another. An ultrasound transducer converts electrical energy into ultrasound energy and vice versa. Transducers for ultrasound imaging consist of one or more piezoelectric crystals or elements. The basic properties of ultrasound transducers (resonance, frequency response, focusing, etc.) can be illustrated in terms of single-element transducers. However, imaging is often preformed with multiple-element "arrays" of piezoelectric crystals.

PIEZOELECTRIC EFFECT

The *piezoelectric effect* is exhibited by certain crystals that, in response to applied pressure, develop a voltage across opposite surfaces.[1–3] This effect is used to produce an electrical signal in response to incident ultrasound waves. The magnitude of the electrical signal varies directly with the wave pressure of the incident ultrasound. Similarly, application of a voltage across the crystal causes deformation of the crystal—either compression or extension depending upon the polarity of the voltage. This deforming effect, termed the *converse piezoelectric effect,* is used to produce an ultrasound beam from a transducer.

The piezoelectric effect was first described by Pierre and Jacques Curie in 1880.

The movement of the surface of a piezoelectric crystal used in diagnostic imaging is on the order of a few micrometers (10^{-3} mm) at a rate of several million times per second. This movement, although not discernible to the naked eye, is sufficient to transmit ultrasound energy into the patient.

An ultrasound transducer driven by a continuous alternating voltage produces a continuous ultrasound wave. Continuous-wave (CW) transducers are used in CW Doppler ultrasound. A transducer driven by a pulsed alternating voltage produces ultrasound bursts that are referred to collectively as pulse-wave ultrasound. Pulsed ultrasound is used in most applications of ultrasound imaging. Pulsed Doppler uses ultrasound pulses of longer duration than those employed in pulse-wave imaging.

Many crystals exhibit the piezoelectric effect at low temperatures, but are unsuitable as ultrasound transducers because their piezoelectric properties do not exist at room temperature. The temperature above which a crystal's piezoelectric properties disappear is known as the *Curie point* of the crystal.

A common definition of the efficiency of a transducer is the fraction of applied energy that is converted to the desired energy mode. For an ultrasound transducer, this definition of efficiency is described as the electromechanical coupling coefficient k_c. If mechanical energy (i.e., pressure) is applied, we obtain

$$k_c^2 = \frac{\text{Mechanical energy converted to electrical energy}}{\text{Applied mechanical energy}}$$

If electrical energy is applied, we obtain

$$k_c^2 = \frac{\text{Electrical energy converted to mechanical energy}}{\text{Applied electrical energy}}$$

Values of k_c for selected piezoelectric crystals are listed in Table 20-1.

Essentially all diagnostic ultrasound units use piezoelectric crystals for the generation and detection of ultrasound. A number of piezoelectric crystals occur in nature (e.g., quartz, Rochelle salts, lithium sulfate, tourmaline, and ammonium dihydrogen phosphate [ADP]). However, crystals used clinically are almost invariable man-made

TABLE 20-1 Properties of Selected Piezoelectric Crystals

Materials	*Electromechanical Coupling Coefficient (K_c)*	*Curie Point (°C)*
Quartz	0.11	550
Rochelle salt	0.78	45
Barium titanate	0.30	120
Lead zirconate titanate (PZT-4)	0.70	328
Lead zirconate titanate (PZT-5)	0.70	365

ceramic ferroelectrics. The most common man-made crystals are barium titanate, lead metaniobate, and lead zirconate titanate (PZT).

In some transducers of newer design, the piezoelectric ceramic is mixed with epoxy to form a *composite ceramic*. Composite ceramics have several performance advantages in comparison with conventional ceramics.[4]

TRANSDUCER DESIGN

The piezoelectric crystal is the functional component of an ultrasound transducer. A crystal exhibits its greatest response at the *resonance frequency*. The resonance frequency is determined by the thickness of the crystal (the dimension of the crystal along the axis of the ultrasound beam). As the crystal goes through a complete cycle from contraction to expansion to the next contraction, compression waves move toward the center of the crystal from opposite crystal faces. If the crystal thickness equals one wavelength of the sound waves, the compressions arrive at the opposite faces just as the next crystal contraction begins. The compression waves oppose the contraction and "dampen" the crystal's response. Therefore it is difficult (i.e., energy would be wasted) to "drive" a crystal with a thickness of one wavelength. If the crystal thickness equals half of the wavelength, a compression wave reaches the opposite crystal face just as expansion is beginning to occur. Each compression wave produced in the contraction phase aids in the expansion phase of the cycle. A similar result is obtained for any odd multiple of half wavelengths (e.g., $3\lambda/2$, $5\lambda/2$), with the crystal progressing through more than one cycle before a given compression wave arrives at the opposite face. Additional crystal thickness produces more attenuation, so the most efficient operation is achieved for a crystal with a thickness equal to half the wavelength of the desired ultrasound. A crystal of half-wavelength thickness resonates at a frequency ν:

$$\nu = \frac{c}{\lambda} = \frac{c}{2t}$$

where $\lambda = 2t$

The components of an ultrasound transducer include the

- Piezoelectric crystal
- Damping material
- Electrodes
- Housing
- Matching layer
- Insulating cover

The transducer described here is a "single-element" transducer. Such a transducer is used in some ophthalmological, m-mode, and pulsed Doppler applications. Most other applications of ultrasound employ one of the multielement transducers described later in this chapter.

Example 20-1

For a 1.5-mm-thick quartz disk (velocity of ultrasound in quartz = 5740 m/sec), what is the resonance frequency?

$$\nu = \frac{5740 \text{ m/sec}}{2(0.0015 \text{ m})} = 1.91 \text{ MHz}$$

High-frequency ultrasound transducers employ thin (<1 mm) piezoelectric crystals. A thicker crystal yields ultrasound of lower frequency.

To establish electrical contract with a piezoelectric crystal, faces of the crystal are coated with a thin conducting film, and electric contacts are applied. The crystal is mounted at one end of a hollow metal or metal-lined plastic cylinder, with the front face of the crystal coated with a protective plastic that provides efficient transfer of sound between the crystal and the body. The plastic coating at the face of the crystal has a thickness of $^1/_4\lambda$ and is called a *quarter-wavelength* matching layer. A $^1/_4\lambda$ thickness maximizes energy transfer from the transducer to the patient. An odd multiple of one-quarter wavelengths would perform the same function, but the greater thickness of material would increase attenuation. Therefore, a single one-quarter wavelength thickness commonly is used for the matching layer. The front face of the crystal is connected through the cylinder to ground potential. The remainder of the crystal is electrically and acoustically insulated from the cylinder.

With only air behind the crystal, ultrasound transmitted into the cylinder from the crystal is reflected from the cylinder's opposite end. The reflected ultrasound reinforces the ultrasound propagated in the forward direction from the transducer. This reverberation of ultrasound in the transducer itself contributes energy to the ultrasound beam. It also extends the pulse duration (the time over which the ultrasound pulse is produced). Extension of the pulse duration (sometimes called temporal pulse length) is no problem in some clinical uses of ultrasound such as continuous-wave and pulsed-Doppler applications. For these purposes, ultrasound probes with air-backed crystals may be used. However, most ultrasound imaging applications utilize short pulses of ultrasound, and suppression of ultrasound reverberation in the transducer is desirable. Suppression or "damping" of reverberation is accomplished by filling the transducer cylinder with a backing material such as tungsten powder embedded in epoxy resin. Sometimes, rubber is added to the backing to increase the absorption of ultrasound. Often, the rear surface of the backing material is sloped to prevent direct reflection of ultrasound pulses back to the crystal. The construction of a typical ultrasound transducer is illustrated in the margin. The crystal may be flat, as shown in the drawing, or curved to focus the ultrasound beam.

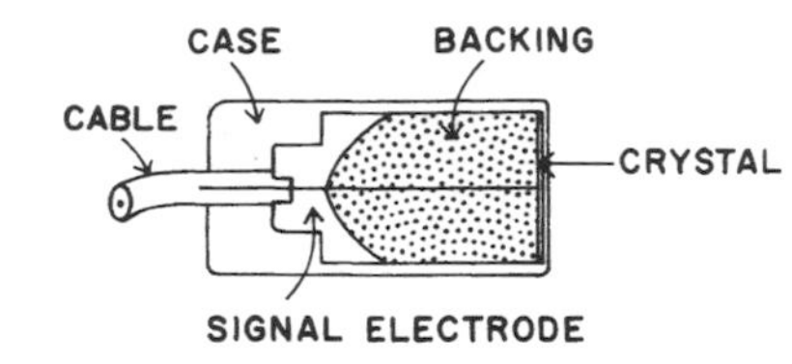

MARGIN FIGURE 20-1
Typical ultrasound transducer.

Damping the reverberation of an ultrasound transducer is similar to packing foam rubber around a ringing bell.

The few reverberation cycles after each voltage pulse applied to a damped ultrasound crystal is described as the *ringdown* of the crystal.

As a substitute for physical damping with selected materials placed behind the crystal, electronic damping may be used. In certain applications, including those that use a small receiving transducer, a resistor connecting the two faces of an air-backed crystal may provide adequate damping. Another approach, termed *dynamic damping*, uses an initial electrical pulse to stimulate the transducer, followed immediately by a voltage pulse of opposite polarity to suppress continuation of transducer action.

FREQUENCY RESPONSE OF TRANSDUCERS

Ultrasound frequencies from 2–10 MHz are used for most diagnostic medical applications.

An ultrasound transducer is designed to be maximally sensitive to ultrasound of a particular frequency, termed the *resonance frequency* of the transducer. The resonance frequency is determined principally by the thickness of the piezoelectric crystal. Thin crystals yield high resonance frequencies, and vice versa. The resonance frequency is revealed by a curve of transducer response plotted as a function of ultrasound frequency. In the illustration in the margin, the frequency response characteristics of two transducers are illustrated. The curve for the undamped transducer displays a sharp frequency response over a limited frequency range. Because of greater energy absorption in the damped transducer, the frequency response is much broader and not so sharply peaked at the transducer resonance frequency. On the curve for the undamped transducer, points ν_1 and ν_3 represent frequencies on either side of the resonance frequency where the response has diminished to half. These points are called the half-power points, and they encompass a range of frequencies termed the bandwidth of the transducer. The ratio of the "center," or resonance frequency ν_2, to the bandwidth $(\nu_3 - \nu_t)$ is termed the Q value of the transducer. The Q value describes the sharpness of the frequency response curve, with a high Q value indicating a sharply-peaked frequency response.

$$Q \text{ value} = \frac{\nu_2}{\nu_3 - \nu_1}$$

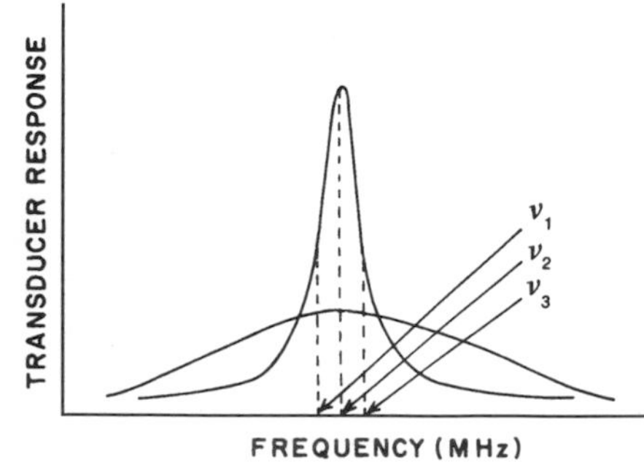

MARGIN FIGURE 20-2
Frequency–response curves for undamped (*sharp curve*) and damped (*broad curve*) transducers.

Transducers used in ultrasound imaging must furnish short ultrasound pulses and respond to returning echoes over a wide range of frequencies. For these reasons, heavily damped transducers with low Q values (e.g., 2 to 3) are usually desired.

Because part of the damping is provided by the crystal itself, crystals such as PZT (lead zirconate titanate) or lead metaniobate with high internal damping and low Q values are generally preferred for imaging.

The resonance frequency ν_2 is near the value posted on the imaging system for the transducer. For example, an individual "3.5"-MHz transducer may have an actual resonance frequency of 3.489 MHz.

The efficiency with which an ultrasound beam is transmitted from a transducer into a medium (and vice versa) depends on how well the transducer is coupled to the medium. If the acoustic impedance of the coupling medium is not too different from that of either the transducer or the medium and if the thickness of the coupling medium is much less than the ultrasound wavelength, then the ultrasound is transmitted into the medium with little energy loss. Transmission with almost no energy loss is accomplished, for example, with a thin layer of oil placed between transducer and skin during a diagnostic ultrasound examination. Transmission with minimum energy loss occurs when the impedance of the coupling medium is intermediate between the impedances of the crystal and the medium. The ideal impedance of the coupling medium is

Coupling of the transducer to the transmitting medium affects the size of the electrical signals generated by returning ultrasound pulses.

$$Z_{\text{coupling medium}} = \sqrt{Z_{\text{transducer}} \times Z_{\text{medium}}}$$

Two methods are commonly used to generate ultrasound beams. For continuous-wave beams, an oscillating voltage is applied with a frequency equal to that desired for the ultrasound beam. A similar voltage of prescribed duration is used to generate long pulses of ultrasound energy, as shown in the margin **(A)**. For clinical ultrasound imaging, short pulses usually are preferred. These pulses are produced by shocking the crystal into mechanical oscillation by a momentary change in the voltage across the crystal. The oscillation is damped quickly by the methods described earlier to furnish ultrasound pulses as short as half a cycle. The duration of a pulse usually is defined as the number of half cycles in the pulse with an amplitude greater than one fourth of peak amplitude. The effectiveness of damping is described by the pulse dynamic range, defined as the ratio of the peak amplitude of the pulse divided by the amplitude of ripples following the pulse. A typical ultrasound pulse of short duration is illustrated in the margin **(B)**.

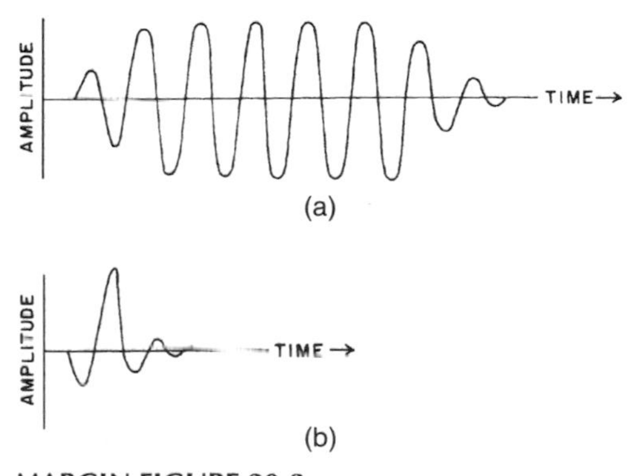

MARGIN FIGURE 20-3
Typical long **(A)** and short **(B)** ultrasound pulses.

ULTRASOUND BEAMS

Wave Fronts

The compression zones of an ultrasound wave are represented by lines perpendicular to the motion of the ultrasound wave in the medium. These lines are referred to as *wave fronts*. For an ultrasound source of large dimensions (i.e., a large-diameter transducer as compared with the wavelength), ultrasound wave fronts are represented as equally spaced straight lines such as those in Figure 20-1A. Wave fronts of this type are termed *planar wave fronts*, and the ultrasound wave they represent is termed a *planar* or *plane wave*. At the other extreme, an ultrasound wave originating from a source of very small dimensions (i.e., a point source) is represented by wave fronts that describe spheres of increasing diameter at increasing distance from the source. *Spherical wave fronts* from a point source are diagramed in Figure 20-1B.

Ultrasound from a point source creates spherical wave fronts. Ultrasound from a two-dimensional extended source creates planar wavefronts.

Sources of exceptionally small or large dimensions are not routinely used in diagnostic ultrasound. Instead, sources with finite dimensions are used. These sources

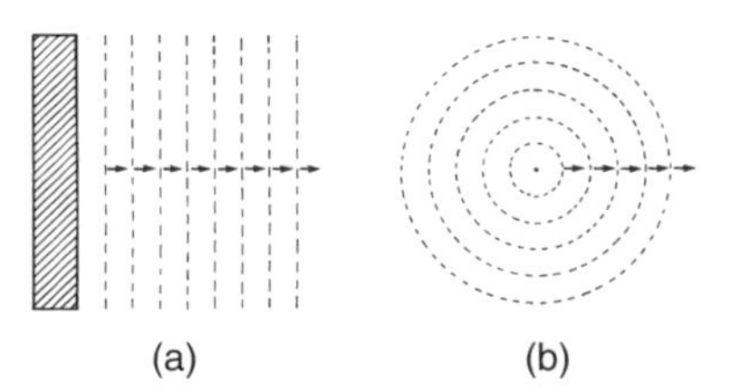

FIGURE 20-1
Ultrasound wave fronts from a source of large dimensions **(A)** and small dimensions **(B)**.

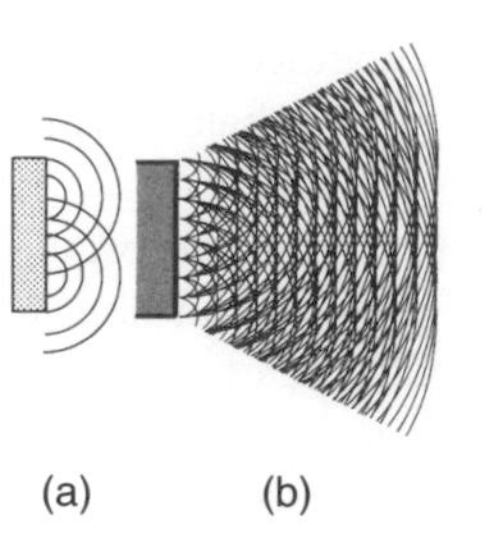

FIGURE 20-2
A: Ultrasound sources may be considered to be a collection of point sources, each radiating spherical wavelets into the medium. **B:** Interference of the spherical wavelets establishes a characteristic pattern for the resulting net wavefronts.

The principles of interference were described by Christian Huygens (1629–1695), the Dutch mathematician, physicist, and astronomer.

can be considered to be a collection of point sources, each radiating spherical wave fronts (termed wavelets) into the medium, as shown in Figure 20-2A. In regions where compression zones for one wavelet intersect those of another, a condition of constructive interference is established. With constructive interference, the wavelets reinforce each other, and the total pressure in the region is the sum of the pressures contributed by each wavelet. In regions where compression zones for one wavelet intersect zones of rarefaction for another wavelet, a condition of destructive interference is established. In these regions, the molecular density is reduced.

Ultrasound transducer rules: (1) The length of the near field increases with increasing transducer diameter and frequency; (2) divergence in the far field decreases with increasing transducer diameter and frequency.

With many spherical wavelets radiating from a transducer of reasonable size (i.e., the diameter of the transducer is considerably larger than the ultrasound wavelength), many regions of constructive and destructive interference are established in the medium. In Figure 20-2B, these regions are represented as intersections of lines depicting compression zones of individual wavelets. In this figure, the reinforcement and cancellation of individual wavelets are most noticeable in the region near the source of ultrasound. They are progressively less dramatic with increasing distance from the ultrasound source. The region near the source where the interference of wavelets is most apparent is termed the *Fresnel* (or *near*) *zone*. For a disk-shaped transducer of radius r, the length D of the Fresnel zone is

The length D of the Fresnel (near) zone can also be written as $D = d^2/4\lambda$, where d is the transducer diameter.

$$D_{\text{fresnel}} = \frac{r^2}{\lambda}$$

where λ is the ultrasound wavelength. Within the Fresnel zone, most of the ultrasound energy is confined to a beam width no greater than the transducer diameter. Beyond the Fresnel zone, some of the energy escapes along the periphery of the beam to produce a gradual divergence of the ultrasound beam that is described by

$$\sin\theta = 0.6\left(\frac{\lambda}{r}\right)$$

where θ is the Fraunhofer divergence angle in degrees (see margin). The region beyond the Fresnel zone is termed *the Fraunhofer* (or *far*) *zone*.

Rules for Transducer Design
For a given transducer diameter,

- the near-field length increases with increasing frequency;
- beam divergence in the far field decreases with increasing frequency

For a given transducer frequency

- the near-field length increases with increasing transducer diameter;
- beam divergence in the far field decreases with increasing transducer diameter.

Example 20-2

What is the length of the Fresnel zone for a 10-mm-diameter, 2-MHz unfocused ultrasound transducer?

A 10-mm-diameter transducer has a radius of 5 mm. A 2-MHz transducer has a λ of

$$\lambda = \frac{1540 \text{ m/sec}}{2 \times 10^6/\text{sec}} = 770 \times 10^{-6} \text{ m}$$
$$= 0.77 \text{ mm}$$
$$D_{\text{Fresnel}} = \frac{(5 \text{ mm})^2}{0.77 \text{ mm}}$$
$$= 32.5 \text{ mm}$$

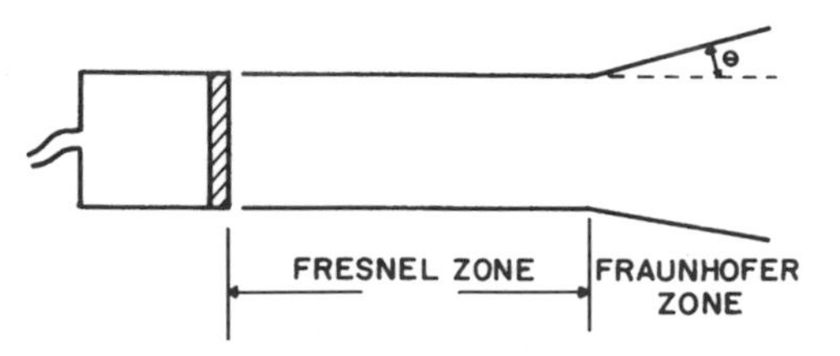

MARGIN FIGURE 20-4
Divergence of the ultrasound beam in the Fraunhofer region. Angle θ is the Fraunhofer divergence angle.

TABLE 20-2 Transducer Radius and Ultrasound Frequency and Their Relationship to Fresnel Zone Depth and Beam Divergence

Frequency (MHz)	*Wavelength (cm)*	*Fresnel Zone Depth (cm)*	*Fraunhofer Divergence Angle (degrees)*
Transducer radius constant at 0.5 cm			
0.5	0.30	0.82	21.5
1.0	0.15	1.63	10.5
2.0	0.075	3.25	5.2
4.0	0.0325	6.50	2.3
8.0	0.0163	13.0	1.1

Radius (cm)	*Fresnel Zone Depth (cm)*	*Fraunhofer Divergence Angle in Water (degrees)*
Frequency constant at 2 MHz		
0.25	0.83	10.6
0.5	3.33	5.3
1.0	13.33	2.6
2.0	53.33	1.3

For medical applications of ultrasound, beams with little lateral dispersion of energy (i.e., long Fresnel zones) are preferred. Hence a reasonably high ratio of transducer radius to wavelength (r/λ) is desired. This requirement can be satisfied by using ultrasound of short wavelengths (i.e., high frequencies). However, the absorption of ultrasound energy increases at higher frequencies, and frequencies for clinical imaging are limited to 2 to 20 MHz. At these frequencies a transducer radius of 10 mm or more usually furnishes an ultrasound beam with adequate directionality for medical use. The relationship of transducer radius and ultrasound frequency to the depth of the Fresnel zone and the amount of beam divergence is illustrated in Table 20-2.

Beam Profiles

The transmission and reception patterns of an ultrasound transducer are affected by slight variations in the construction and manner of electrical stimulation of the transducer. Hence, the exact shape of an ultrasound beam is difficult to predict, and beam shapes or profiles must be measured for a particular transducer. One approach to the display of ultrasound beam characteristics is a set of pulse-echo response profiles. A profile is obtained by placing an ultrasound reflector some distance from the transducer and scanning the transducer in a direction perpendicular to the axis of the ultrasound beam. During the scan, the amplitude of the signal induced in the transducer by the reflected ultrasound is plotted as a function of the distance between the central axis of the ultrasound beam and the reflector. A pulse-echo response profile is shown in Figure 20-3A, and a set of profiles obtained at different distances from the transducer is shown in Figure 20-3B.

In Figure 20-3A, locations are indicated where the transducer response (amplitude of the reflected signal) decreases to half (−6 dB) of the response when the transducer is aligned with the reflector. The distance between these locations is termed the response width of the transducer at the particular distance (range) from the transducer. If response widths are connected between profiles at different ranges (Figure 20-3C), 6-dB response margins are obtained on each side of the ultrasound beam axis. Similarly, 20-dB response margins may be obtained by connecting the $^1/_{10}$ amplitude (−20 dB) response widths on each side of the beam axis.

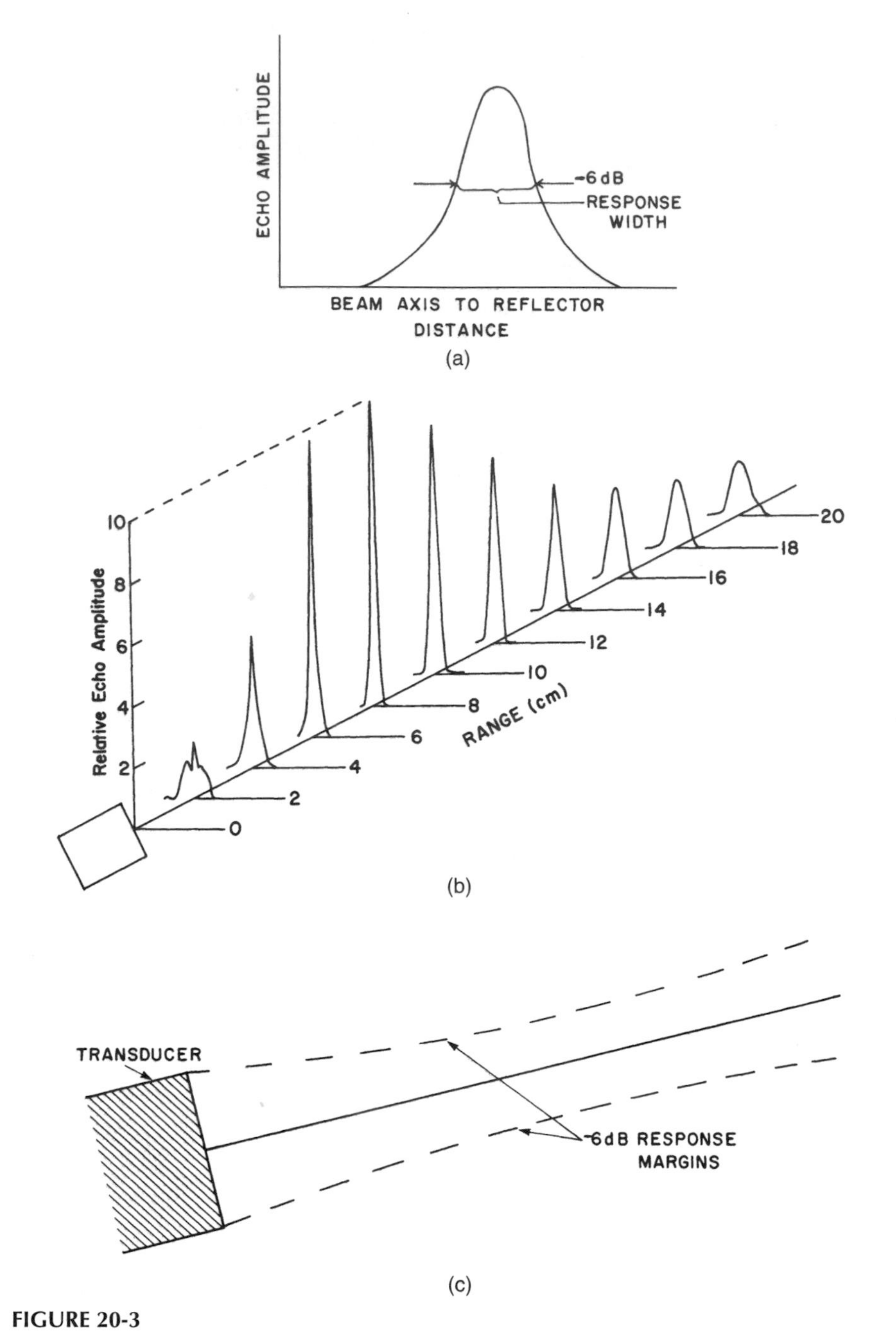

FIGURE 20-3
A: Pulse-echo response profile and the response width of a transducer. **B:** Set of response profiles along the axis of an ultrasound beam. **C:** Response margins of −6 dB along an ultrasound beam.

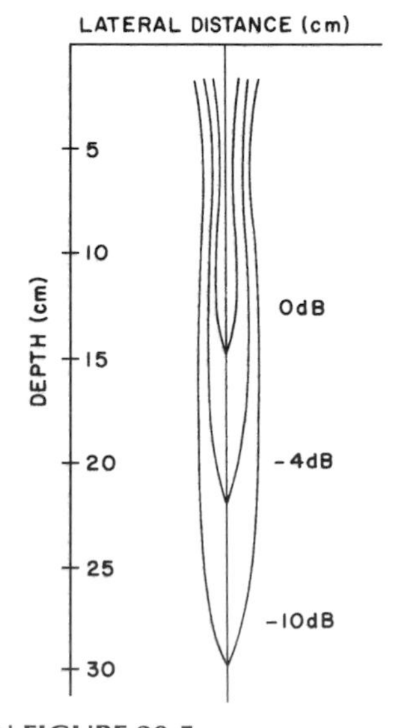

MARGIN FIGURE 20-5
Isoecho contours for a nonfocused transducer.

Response profiles for a particular transducer are influenced by several factors, including the nature of the stimulating voltage applied to the transducer, the characteristics of the electronic circuitry of the receiver, and the shape, size, and character of the reflector. Usually, the reflector is a steel sphere or rod with a diameter of three to ten times the ultrasound wavelength. Response profiles may be distorted if the receiver electronics do not faithfully represent low-intensity signals. Some older ultrasound units cannot accurately display echo amplitudes much less than $^1/_{10}$ (−20 dB) of the largest echo recorded. These units are said to have limited dynamic range.

Another approach to describing the character of an ultrasound beam is with isoecho contours. Each contour depicts the locations of equal echo intensity for the ultrasound beam. At each of these locations, a reflecting object will be detected with equal sensitivity. The approach usually used to measure isoecho contours is to place a small steel ball at a variety of positions in the ultrasound beam and to identify locations where the reflected echoes are equal. Connecting these locations with lines yields isoecho contours such as those in the margin, where the region of maximum sensitivity at a particular depth is labeled 0 dB, and isoecho contours of lesser intensity are labeled −4 dB, −10 dB, and so on. Isoecho contours help depict the lateral resolution of a transducer, as well as variations in lateral resolution with depth and with changes in instrument settings such as beam intensity, detector amplifier gain, and echo threshold.

Accompanying a primary ultrasound beam are small beams of greatly reduced intensity that are emitted at angles to the primary beam. These small beams, termed side lobes (see margin), are caused by vibratory modes of the transducer in the transverse plane. Side lobes can produce image artifacts in regions near the transducer, if a particularly echogenic material, such as a biopsy needle, is present.

Side lobes can be reduced further by the process of *apodization*, in which the voltage applied to the transducer is diminished from the center to the periphery.

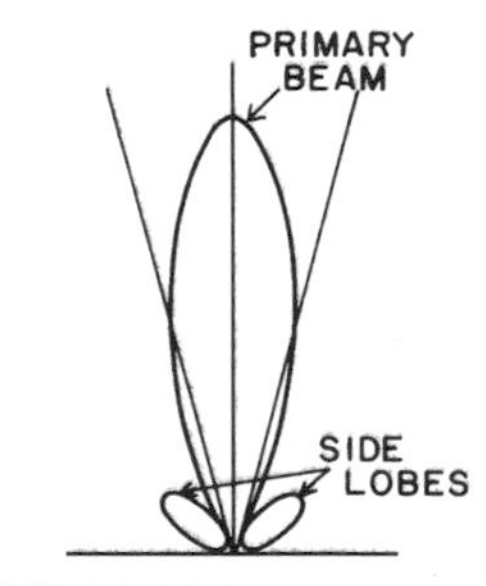

MARGIN FIGURE 20-6
Side lobes of an ultrasound beam.

The preceding discussion covers general-purpose, flat-surfaced transducers. For most ultrasound applications, transducers with special shapes are preferred. Among these special-purpose transducers are focused transducers, double-crystal transducers, ophthalmic probes, intravascular probes, esophageal probes, composite probes, variable-angle probes, and transducer arrays.

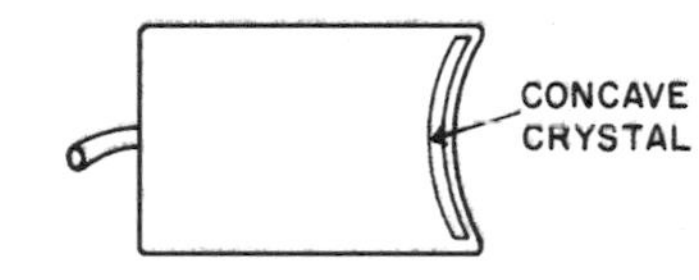

MARGIN FIGURE 20-7
Focused transducer.

Focused Transducers

A focused ultrasound transducer produces a beam that is narrower at some distance from the transducer face than its dimension at the face of the transducer.[5,6] In the region where the beam narrows (termed the focal zone of the transducer), the ultrasound intensity may be heightened by 100 times or more compared with the intensity outside of the focal zone. Because of this increased intensity, a much larger signal will be induced in a transducer from a reflector positioned in the focal zone. The distance between the location for maximum echo in the focal zone and the element responsible for focusing the ultrasound beam is termed the focal length of the transducer.

Often, the focusing element is the piezoelectric crystal itself, which is shaped like a concave disk (see figure in margin). An ultrasound beam also may be focused with mirrors and refracting lenses. Focusing lenses and mirrors are capable of increasing the intensity of an ultrasound beam by factors greater than 100. Focusing mirrors, usually constructed of tungsten-impregnated epoxy resin, are illustrated in the margin. Because the velocity of ultrasound generally is greater in a lens than in the surrounding medium, concave ultrasound lenses are focusing, and convex ultrasound lenses are defocusing (see margin). These effects are the opposite of those for the action of optical lenses on visible light. Ultrasound lenses usually are constructed of epoxy resins and plastics such as polystyrene.

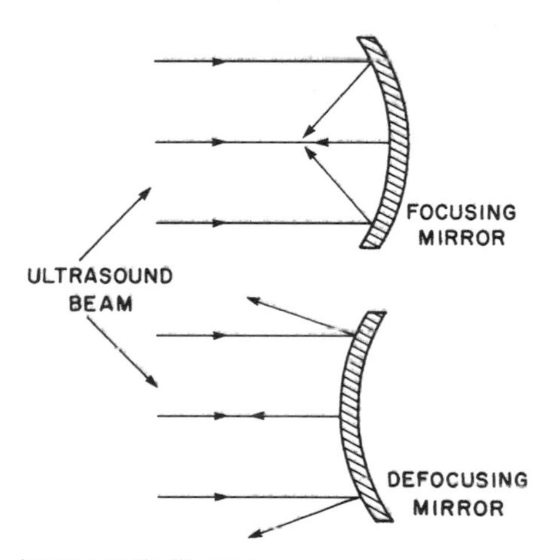

MARGIN FIGURE 20-8
Focusing and defocusing mirrors.

For an ultrasound beam with a circular cross section, focusing characteristics such as pulse-echo response width and relative sensitivity along the beam axis depend on the wavelength of the ultrasound and on the focal length f and radius r of the transducer or other focusing element. These variables may be used to distinguish the degree of focusing of transducers by dividing the near field length r^2/λ by the focal length f. For cupped transducer faces on all but weakly focused transducers, the focal length of the transducer is equal to or slightly shorter than the radius of curvature of the transducer face. If a planoconcave lens with a radius of curvature r is attached to the transducer face, then the focal length f is

$$f = \frac{r}{1 - c_M/c_L}$$

where c_M and c_L are the velocities of ultrasound in the medium and lens, respectively.

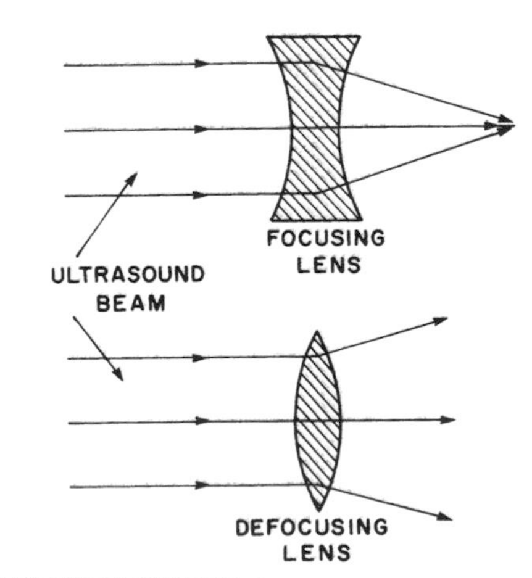

MARGIN FIGURE 20-9
Focusing and defocusing lenses.

The ratio f/d, where $d = 2r =$ the diameter of the transducer, is often described as the f-number of the transducer or other focusing element.

Degree of Focus of Transducers Expressed as a Ratio of the Near-field Length r^2/λ to the Focal Length f

Degree of Focus	*Near–Field Length* / *Focal Length*
Weak	$(r^2/\lambda)/f \leq 1.4$
Medium weak	$1.4 < (r^2/\lambda)/f \leq 6$
Medium	$6 < (r^2/\lambda)/f \leq 20$
Strong	$20 < (r^2/\lambda)/f$

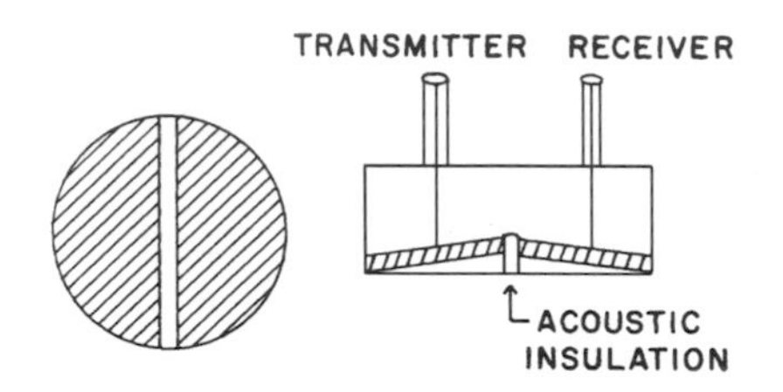

MARGIN FIGURE 20-10
Front (*left*) and side (*right*) views of a typical Doppler transducer.

MARGIN FIGURE 20-11
Linear array. (From Zagzebski, J. A.[4] Used with permission.)

The length of the focal zone of a particular ultrasound beam is the distance over which a reasonable focus and pulse-echo response are obtained. One estimate of focal zone length is

$$\text{Focal zone length} = 10\lambda\left(\frac{f}{d}\right)^2$$

where $d = 2r$ is the diameter of the transducer. These strongly focused transducers are also used for surgical applications of ultrasound where high ultrasound intensities in localized regions are needed for tissue destruction.

Doppler Probes

Transducers for continuous-wave Doppler ultrasound consist of separate transmitting and receiving crystals, usually oriented at slight angles to each other so that the transmitting and receiving areas intersect at some distance in front of the transducer (see margin). Because a sharp frequency response is desired for a Doppler transducer, only a small amount of damping material is used.

Multiple-Element Transducers

Scanning of the patient may be accomplished by physical motion of a single-element ultrasound transducer. The motion of the transducer may be executed manually by the sonographer or automatically with a mechanical system. Several methods for mechanical scanning are shown in Figure 20-4. The scanning technique that uses an automatic scanning mechanism for the transducer is referred to as *mechanical sector scanning*.

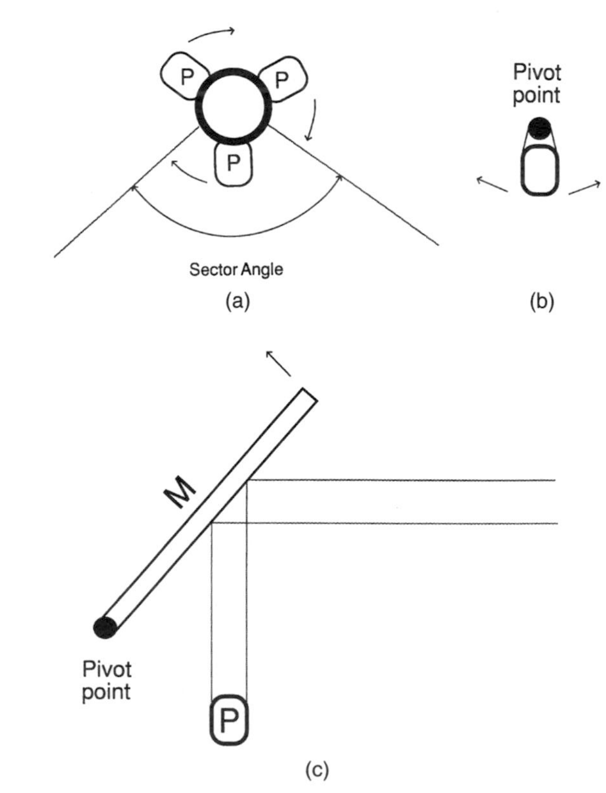

FIGURE 20-4
Three methods of mechanical scanning of the ultrasound beam. **A:** Multiple piezoelectric elements (P) mounted on a rotating head. One element is energized at a time as it rotates through the sector angle. **B:** Single piezoelectric element oscillating at high frequency. **C:** Single piezoelectric element reflected from an oscillating acoustic mirror (M).

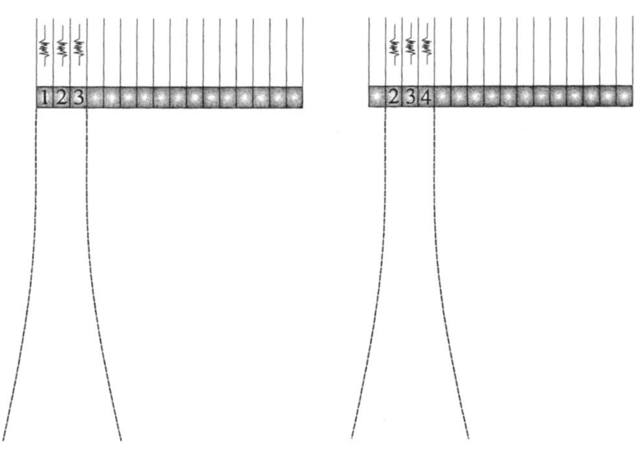

FIGURE 20-5
Electronic scanning with a linear switched array.

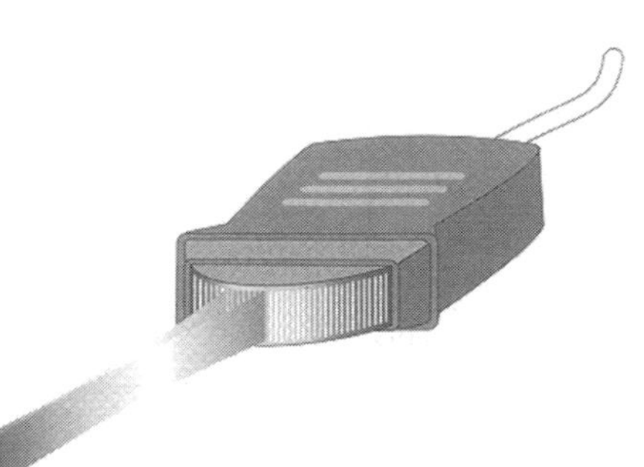

MARGIN FIGURE 20-12
Curved (curvilinear) array. (From Zagzebski, J. A.[4] Used with permission.)

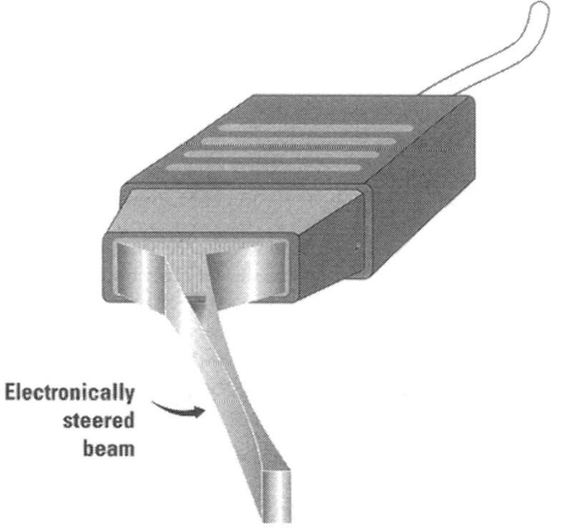

MARGIN FIGURE 20-13
Phased array. (From Zagzebski, J. A.[4] Used with permission.)

Alternatively, the ultrasound beam may be swept back and forth without the need for any mechanical motion through the use of transducer arrays. Transducer arrays are composed of multiple crystals that can change the direction or degree of focus of the ultrasound beam by timing the excitation of the crystals.

A linear (or curved) array of crystal elements is shown in Figure 20-5. The crystals, as many as 60 to 240 or more, are excited by voltage pulses in groups of three up to 20 or more. Each excitation of the crystal group results in a "scan line" (see next chapter). To obtain the succeeding scan line, the next crystal group is defined to overlap the first (e.g., scan line 1 is produced by crystals 1, 2, and 3, while scan line 2 is produced by crystals 2, 3, and 4, and so on). This pulsing scheme is referred to as a *linear switched array*. By sequentially exciting the entire array, an image composed of a number of scan lines (typically 64, 128, 256, etc.) is obtained.

Another method for scanning with a linear array uses *phased-array* technology. A phase array uses all (typically 128) of the elements of the array to obtain each scan line. By using slight delays between excitations of the elements, the beam may be "swept" to the left or right (Figure 20-6A). A variation in the time delay scheme in a phased array may also be used to focus the beam at various distances (Fig 20-6B) throughout each image. In this technique, called *dynamic focusing*, part of the image is acquired with the focal zone near the transducer face, and part is acquired with the focal zone farther from the transducer face. Thus, two or more images taken at different focal zones may be combined to produce a single image with better axial resolution over a broader range than is possible with a single-element transducer. *Dynamic focusing* is performed without moving parts by simply varying the timing of the excitation pulses.

The main distinction between the linear phased array and the linear switched array is that all of the elements of the phased array are used to produce each scan line while only a few of the elements of the switched array are used to produce each scan line. Linear phased arrays provide variable focus in only one dimension—that is, in the plane of the scan. Beam focus in the other dimension, the direction of slice thickness, is provided by acoustic lenses or by concavity of the elements (Figure 20-7).

Another type of array, the annular array, is capable of focusing in all planes perpendicular to the axis of the beam. The annular array (Figure 20-7) consists of a series of piezoelectric elements in the shape of concentric rings or annuli. The beam may be focused at various distances from the transducer face by varying the time delays among excitations of the rings.

Advantages of *transducer* arrays:

- They provide electronic beam steering.
- They permit dynamic focusing and beam shaping.

Transducer arrays employ the technique of a *dynamic aperture*, in which the diameter of the sensitive region of the array is expanded as signals are received from greater depths. This technique maintains a constant lateral resolution with depth.

Transducer Damage

Ultrasound transducers can be damaged in many ways.[4,7] The crystals are brittle, and the wire contacts on the crystals are fragile; hence transducers should be handled

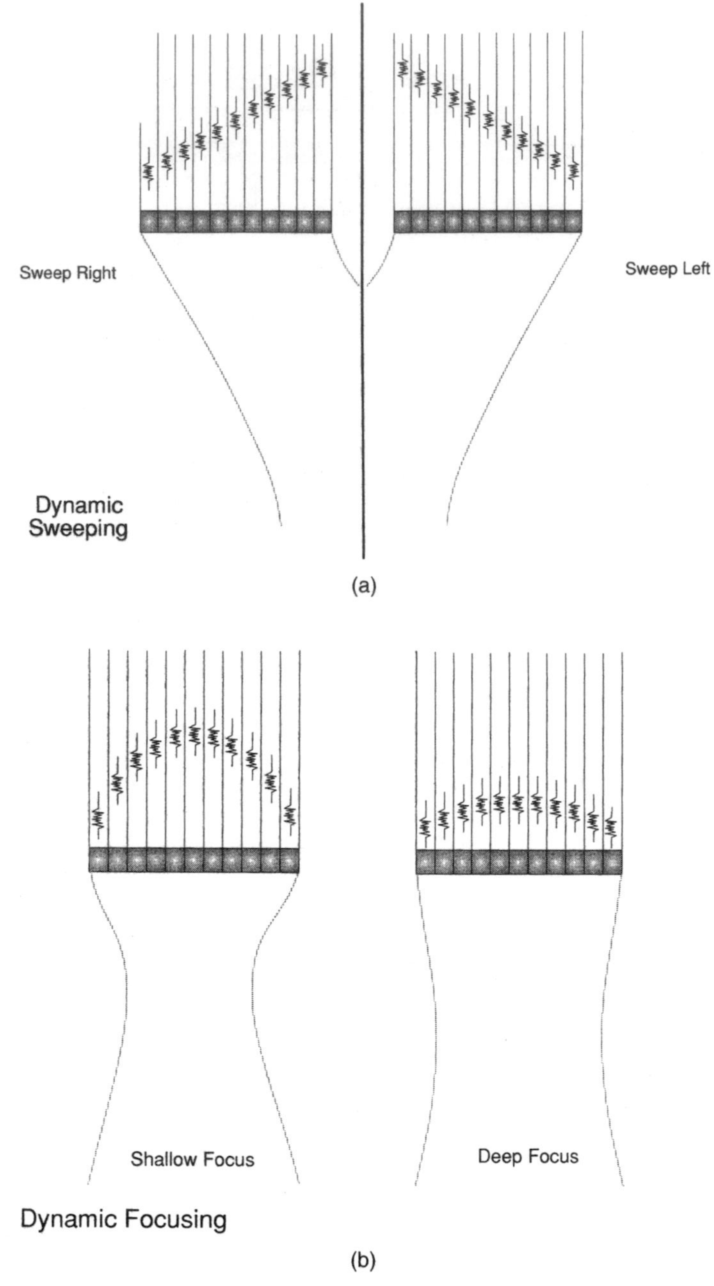

FIGURE 20-6
Electronic scanning with a linear phased array.

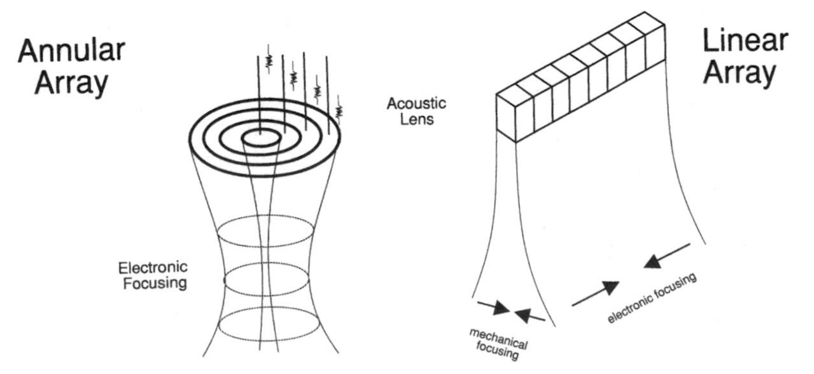

FIGURE 20-7
A linear array requires focusing in the "out-of-plane" dimension by mechanical means (lens or shape of the transducer), whereas an annular array focuses in all directions perpendicular to the axis of the beam.

carefully. Excessive voltages to the crystal should be avoided, and only watertight probes should be immersed in water. A transducer should never be heated to a temperature near the Curie point of the piezoelectric crystal. Dropping a transducer, twisting, bending, and crushing the transducer cables, and other examples of mishandling ultrasound transducers are frequent causes of dysfunction.

PROBLEMS

20-1. Explain what is meant by the piezoelectric effect and the converse piezoelectric effect.

*20-2. For a piezoelectric material with an ultrasound velocity of 6000 m/sec, what thickness should a disk-shaped crystal have to provide an ultrasound beam with a frequency of 2.5 MHz?

20-3. What is meant by damping an ultrasound transducer, and why is this necessary? What influence does damping have on the frequency response of the transducer?

*20-4. What is the estimated focal zone length for a 2-MHz ($\lambda = 0.075$ cm) focused ultrasound transducer with an f-number of 8?

20-5. A linear array can be electronically steering and focused. An annular array can be electronically focused. Can an annular array be electronically steered?

*For those problems marked with an asterisk, answers are provided on p. 493.

SUMMARY

- The active element of an ultrasound transducer is a crystal that exhibits the piezoelectric (receiver) and converse piezoelectric (transmitter) effect.
- The resonance frequency of a transducer is determined by the type of piezoelectric crystal and its thickness.
- The bandwidth of a transducer is expressed by its Q value.
- The Fresnel zone, the area of minimum beam divergence in front of the transducer, has an axial dimension determined by r^2/λ.
- The angle θ of divergence of the ultrasound beam in the Fraunhofer zone is determined by $\sin\theta = 0.6\,(\lambda/r)$.
- The shape of an ultrasound beam may be described by response profiles or isoecho contours.
- Real-time ultrasound scans may be obtained with mechanical sector scanners or multitransducer arrays.
- Multitransducer arrays include linear switched arrays, linear phased arrays, and annular arrays.

REFERENCES

1. McDicken, W. *Diagnostic Ultrasonics*. New York, John Wiley & Sons, 1976, p. 248.
2. Wells, P. *Biomedical Ultrasonics*. New York, Academic Press, 1977, p. 45.
3. Bumber, J. C., and Tristam, M. Diagnostic ultrasound, in Webb, S. (ed.), *Physics of Medical Imaging*. Philadelphia, Adam Hilger, 1988, pp. 319–388.
4. Zagzebski, J. A. *Essentials of Ultrasound Physics*. St Louis, Mosby–Year Book, 1996.
5. Kossoff, G. Improved techniques in ultrasonic echography. *Ultrasonics* 1972, **10**:221.
6. McElroy, J. Focused ultrasonic beams. *Int J. Nondestruc Testing* 1971; **3**:27.
7. Hendee, W. R., et al. Design of a quality control program, in Hendee, W. (ed.), *The Selection and Performance of Radiologic Equipment*. Baltimore, Williams & Wilkins, 1985, pp 163–208.

CHAPTER

21

ULTRASOUND INSTRUMENTATION

OBJECTIVES

After completing this chapter, the reader should be able to:

- Name three modes of ultrasound imaging.
- Determine the time required to obtain an ultrasound image.
- Define pulse repetition frequency and state the role it plays in frame rate.
- Name the main components of an ultrasound imaging system.
- Explain the preprocessing required to obtain digital data from ultrasound signals.
- Define the term *time gain compensation*.
- Explain how postprocessing is used to change the appearance of an ultrasound image.
- Describe several common artifacts in ultrasound images.
- List several common ultrasound quality control tests.

All approaches to obtaining pictorial information with ultrasound depend on echo-ranging, the principle that the time required for return of a reflected signal indicates distance. Strategies for displaying echo-range information—amplitude A, brightness B, and motion M—are discussed in this chapter.

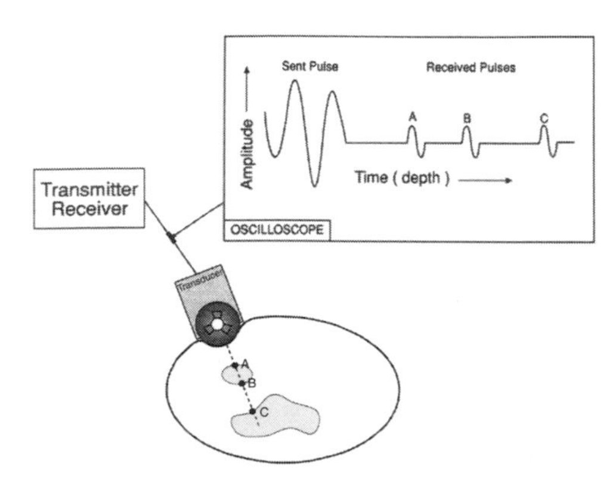

MARGIN FIGURE 21-1
A-mode (amplitude mode) of ultrasound display. An oscilloscope display records the amplitude of echoes as a function of time or depth. Points A, B, and C in the patient appear as peaks-A, B, and C in the A-mode display.

For each centimeter of depth, ultrasound travels 2 cm, 1 cm "out" for the transmitted pulse and 1 cm "back" for the reflected pulse. With the assumption of a speed of sound of 1540 m/sec for ultrasound in soft tissue, the time that corresponds to each centimeter of "depth" is

$$\frac{0.02\text{ m}}{1540\text{ m/sec}} = 0.000013\text{ sec} = 13\ \mu\text{sec}$$

The result is a conversion factor of 13 μsec/cm for diagnostic ultrasound that is used to convert time to distance and vice versa.

This conversion factor is appropriate for reflection imaging in soft tissue where a speed of sound of 1540 m/sec is assumed to be reasonably accurate.

A-mode displays of a transcranial scan (echoencephalography) were used until the late 1970s to detect shifts of the midline in the brain of newborns with suspected hydrocephaly. This technique has been replaced by B-mode real-time scanning, computed tomography and magnetic resonance imaging.

PRESENTATION MODES

A-Mode

In the A-mode presentation of ultrasound images, echoes returning from the body are displayed as signals on an oscilloscope. The oscilloscope presents a graph of voltage representing echo amplitude (hence the term "A-mode") on the ordinate, or y-axis, as a function of time on the abscissa, or x-axis. With the assumption of a constant speed of sound, time on the x-axis can be presented as distance from the ultrasound transducers (Margin Figure 21-1).

A-mode reveals the location of echo-producing structures only in the direction of the ultrasound beam. It has been used in the past to localize echo-producing interfaces such as midline structures in the brain (echoencephalography) and structures to be imaged in B-mode. A-mode displays are not found on most imaging systems used today. The concept of A-mode is however, useful in explaining how pixels are obtained from scan lines in B-mode imaging.

B-Mode

In B-mode presentation of pulse echo images (Margin Figure 21-2) the location of echo-producing interfaces is displayed in two dimensions (x and y) on a video screen. The amplitude of each echo is represented by the brightness value at the xy location. High-amplitude echoes can be presented as either high brightness or low brightness to provide either "white-on-black" or "black-on-white" presentations. Most images are viewed as white on black, so regions in the patient that are more echogenic correspond to regions in the image that are brighter (hence B for "brightness" mode).

B-mode images may be displayed as either "static" or "real-time" images. In static imaging the image is compiled as the sound beam is scanned across the patient, and the image presents a "snapshot" averaged over the time required to sweep the sound beam. In real-time imaging, the image is also built up as the sound beam scans across the patient, but the scanning is performed automatically and quickly, and one image follows another in quick succession. At image frequencies greater than approximately 24 per second, the motion of moving structures seems continuous, even though it may appear that the images are flickering (i.e., being flashed on and off). Images that are refreshed at frequencies greater than approximately 48 per second are free of flicker.

Real-time B-mode images are useful in the display of moving structures such as heart valves. They also permit the technologist to scan through the anatomy to locate structures of interest. Certain applications such as sequential slice imaging of organs are performed better with static imaging.

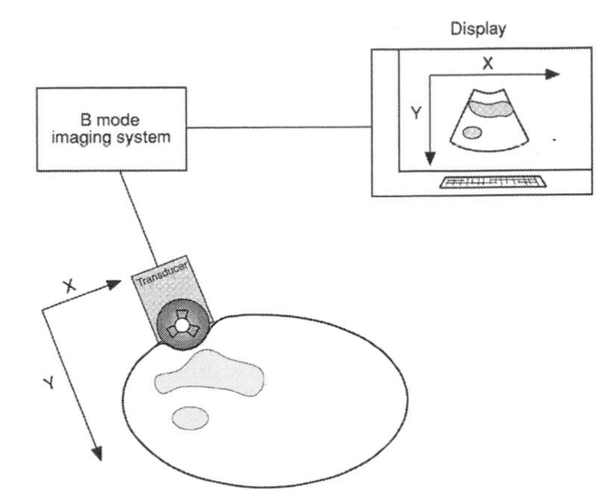

MARGIN FIGURE 21-2
B-mode, or brightness mode, ultrasound display. The amplitude of reflected signals is displayed as brightness at points of an image defined by their x- and y-coordinates.

M-Mode

The M-mode presentation of ultrasound images is designed specifically to depict moving structures. In an M-mode display, the position of each echo-producing interface is presented as a function of time. The most frequent application of M-mode scanning is echocardiography, where the motion of various interfaces in the heart is depicted graphically on a cathode-ray tube (CRT) display or chart recording.

In a typical M-mode display (Margin Figure 21-3), the depths of the structures of interest are portrayed as a series of dots in the vertical direction on the CRT, with the position of the transducer represented by the top of the display. With the transducer in a fixed position, a sweep voltage is applied to the CRT deflection plates to cause the dots to sweep at a controlled rate across the CRT screen. For stationary structures, the dots form horizontal lines in the image. Structures that move in a direction parallel to the ultrasound beam produce vertical fluctuations in the horizontal trace to reveal their motion. The image may be displayed on a short-persistence CRT or storage scope and may be recorded on film or a chart recorder. Modern systems digitize the information and display it on a digital monitor.

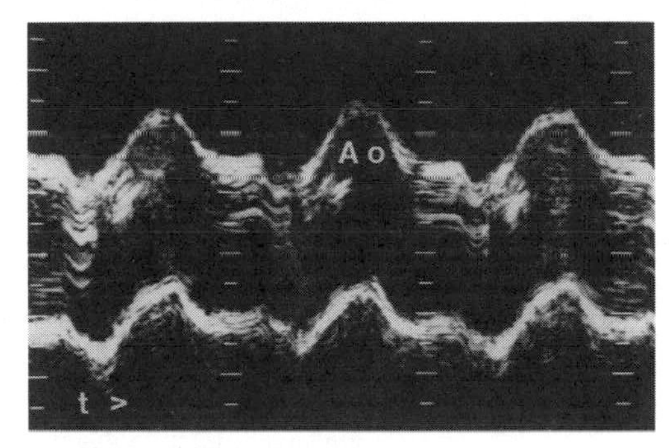

MARGIN FIGURE 21-3
A typical M-mode echocardiographic tracing.

TIME REQUIRED TO OBTAIN IMAGES

Image formation with ultrasound requires that echo information be received along discrete paths called "scan lines." The time required to complete each line is determined by the speed of sound. If all of the information must be received from one line before another can be initiated, a fundamental limit is imposed on the rate at which ultrasound images can be acquired (Margin Figure 21-4).

Each line of information is obtained as follows. First, an ultrasound pulse of several nanoseconds [called the pulse time (PT)] is sent. The transducer is then quiescent for the remainder of the pulse repetition period (PRP), defined as the time from the beginning of one pulse to the next. During the quiescent time, echoes returning from interfaces within the patient excite the transducer and cause voltage pulses to be transmitted to the imaging device. These "echoes" are processed by the device and added to the image only if they fall within a preselected "listen time." Acquisition of echoes during the listen time provides information about reflecting interfaces along a single path in the object—that is, the scan line (Margin Figure 21-5).

The PRP determines the maximum field of view (FOV), also known as depth of view (DOV)—that is, the length of the scan lines.

$$\text{PRP (sec)} = \text{FOV (cm)} \times 13 \times 10^{-6}\ \text{sec/cm} \tag{21-1}$$

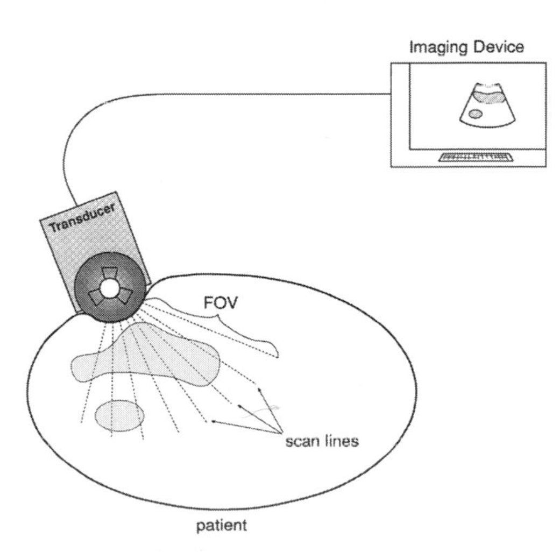

MARGIN FIGURE 21-4
A B-mode image consists of scan lines. The length of the scan lines determines the depth within the patient that is imaged [the field of view (FOV)].

Example 21-1

Find the maximum length of a scan line from an ultrasound unit having a PRP of 195 μs.

$$\begin{aligned}\text{Length of scan line} &= \frac{\text{PRP } (\mu\text{sec})}{13\ \mu\text{sec/cm}} \\ &= \frac{195\ \mu\text{sec}}{13\ \mu\text{sec/cm}} \\ &= 15\ \text{cm}\end{aligned}$$

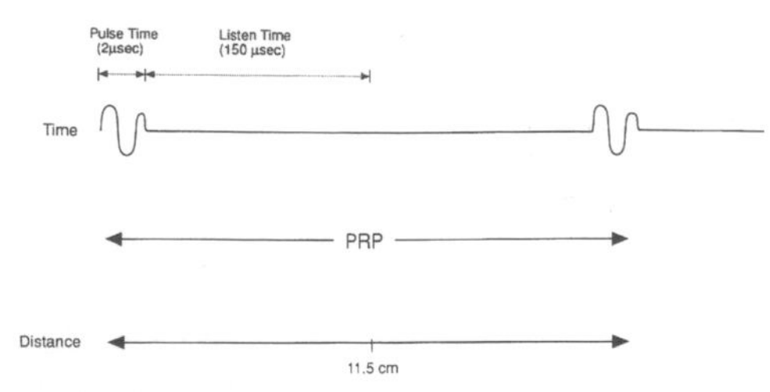

MARGIN FIGURE 21-5
Ultrasonic echoes from a single scan line in A- or B-mode imaging. In the example shown here, a 150-μsec "listen time" corresponds to a depth [field of view (FOV)] of 11.5 cm. The pulse time (or pulse duration) is short when compared with the listen time. The time from the beginning of one pulse to the beginning of the next is the pulse repetition period (PRP).

Note that the PRP determines the *maximum* depth of view. The depth of view selected on the imaging device could be smaller than PRP(μs)/13(μs/cm)μsec if the pulse listen time were shorter than the PRP.

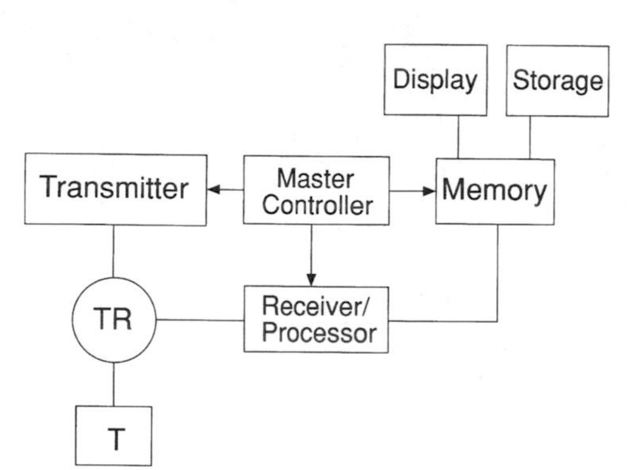

MARGIN FIGURE 21-6
Main components of an ultrasound B-mode imaging system.

Ultrasound instrumentation for medical imaging was developed after World War II when personnel, who were familiar with military electronics (signal generators, amplifiers) and the use of sonar, returned to civilian life and wanted to transfer these technologies to medicine. One of the first experimental scanners was built in Denver, CO in 1949 by Joseph Holmes and Douglas Howry. It employed an amplifier from surplus Air Force radar equipment, a power supply from a record player, and a gun turret from a B-29 bomber.

The pulse repetition frequency (PRF) is the inverse of the PRP:

$$\text{PRF} = \frac{1}{\text{PRP}}$$

where the PRF has units of pulses per second, inverse seconds, or hertz. The concept of PRF is of special interest in pulsed Doppler ultrasound (Chapter 22).

The frame time (FT) is the time required to obtain a complete image (frame) made up of multiple scan lines (N):

$$\text{FT} = \text{PRP} \times N \tag{21-2}$$

Obviously, the frame time can be shortened by obtaining two or more scan lines simultaneously. Some linear arrays are capable of providing "parallel" acquisition of image data. For the purpose of problems presented here, we will assume that scan lines are acquired sequentially (i.e., one after the other).

A quantity such as frame time is a "period." It is the time required to complete a periodic task. For any such quantity, there is the inverse, called a "frequency" or "rate," that depicts the number of such tasks that can be performed per unit time. So the rate at which images may be acquired, the frame rate (FR), is the inverse of the frame time.

$$\text{FR} = \frac{1}{\text{FT}}$$

By substituting from Eqs. (21-1) and (21-2), we obtain

$$\text{FR} = \frac{1}{13 \times 10^{-6} \times \text{FOV} \times N} \tag{21-3}$$

The frame rate has units of frames per second, inverse seconds, or hertz.

Example 21-2

For an ultrasound image with a 10-cm field of view (FOV) and 120 lines of sight, find the minimum pulse repetition period (PRP), the minimum frame time (FT), and the maximum frame rate (FR).

$$\begin{aligned}
\text{PRP} &= \text{FOV} \times 13 \times 10^{-6} \\
&= 10 \text{ cm} \times 13 \times 10^{-6} \text{ sec/cm} \\
&= 130 \times 10^{-6} \\
&= 130\ \mu\text{sec} \\
\text{FT} &= \text{PRP} \times N \\
&= 130 \times 10^{-6} \text{ sec} \times 120 \\
&= 15.6 \times 10^{-3} \text{ sec} \\
&= 15.6 \text{ msec} \\
\text{FR} &= \frac{1}{\text{FT}} \\
&= \frac{1}{15.6 \times 10^{-3} \text{ sec}} \\
&= 64 \text{ sec}^{-1} \\
&= 64 \text{ frames/sec} \\
&= 64 \text{ Hz}
\end{aligned}$$

An inverse relationship exists between the frame rate and each of the variables pulse repetition period, field of view, and number of scan lines. An increase in any of these variables decreases the frame rate in direct proportion.

SYSTEM COMPONENTS

The main components of an ultrasound B-mode imaging system are shown in Margin Figure 21-6. During a scan, the transmitter sends voltage pulses to the transducer. The pulses range from tens to hundreds of volts in amplitude. Returning echoes from the patient produce voltage pulses that are a few microvolts to a volt or so in amplitude. These small signals are sent to the receiver/processor for preprocessing functions such as demodulation, thresholding, amplification, time gain compensation, and detection, all described below. From the receiver/processor, signals are stored as digital values in random-access computer memory.

An automatic transmit/receive switch is provided to prevent large transmitter signals from finding their way into the sensitive receiver/processor. Failure of this switch could result in damage to the receiver/processor, as well as yield no image or one of diminished brightness. The latter condition would call for an increased gain setting beyond that normally used.

The functions of the components shown in Margin Figure 21-6 are coordinated by the master controller. The master controller provides reference signals against which the various components can time the arrival of echoes or keep track of which scan line is being processed.

For static B-mode imaging, a position-sensing device is added to Margin Figure 21-6. The device is attached to the transducer assembly to monitor the position and orientation of the transducer. Electronic components called "potentiometers" are used in the position-sensing device. When the potentiometer is turned, its resistance changes, producing a change in voltage in a position-sensing circuit. As the position-sensing device is moved, the potentiometers located at hinges provide voltages that are automatically converted to position coordinates (Margin Figure 21-7). For real-time B-mode imaging, the master controller encodes the echoes according to their location in space (Margin Figure 21-8). The echo signals are stored in memory in such a way that when read out as video signals, they appear on the display as brightness values in locations that are meaningful representations of their origins within the patient.

Ultrasound images are usually displayed on a CRT. M-mode data are usually recorded on a strip chart recorder. Static B-mode images usually are recorded on film with a hard-copy camera, and real-time B-mode images usually are preserved on videotape or disk.

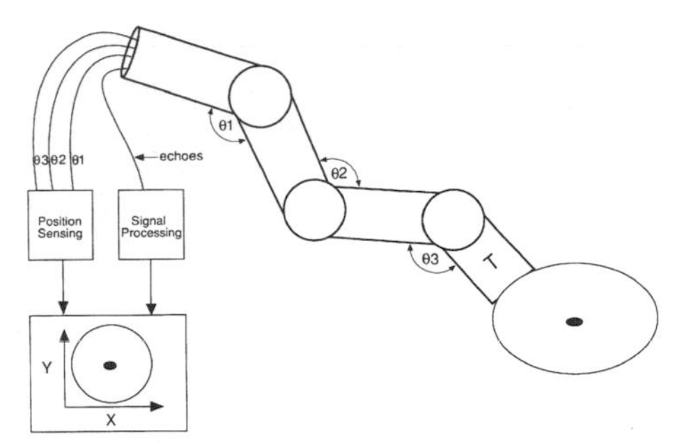

MARGIN FIGURE 21-7
The XY position—sensing circuits of a static B-mode scanner. Electric signals from potentiometers 1, 2, and 3 are proportional to the angles θ_1, θ_2, and θ_3 at the hinges of the articulated arm.

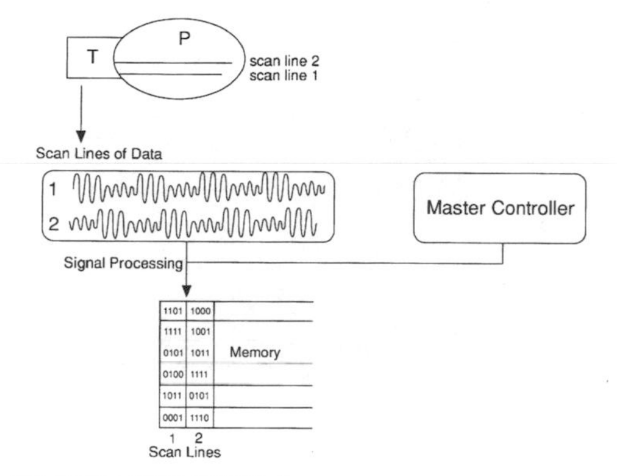

MARGIN FIGURE 21-8
In a real-time B-mode imaging system, spatial encoding of the echo signals is accomplished by storing scan-line data in memory in a methodical fashion.

SIGNAL PROCESSING

Returning echoes contain a tremendous amount of information about the patient. As in all imaging modalities, methods used to process the echoes determine the information content of the final image. In B-mode imaging, each location in the image is associated with one value of echo amplitude. This value is recorded in memory and translated into brightness on the display screen. Deriving a single value from each echo is the goal of signal processing.

Signal processing may be divided into two categories. If the processing schemes to be used are determined prior to scanning the patient, they are termed *preprocessing schemes*. Postprocessing refers to signal or image processing that alters stored values of data after they have been acquired, (i.e., after the patient has been scanned).

Preprocessing

The amplitude of an ultrasound echo is governed by the acoustic impedance mismatch at the interface where the echoes originated, the attenuation of intervening tissues, and the amplitude of the ultrasonic pulse that is sent out from the transducer. It is the presence and degree of acoustic impedance mismatch at interfaces that we wish to image. The attenuation by intervening tissues is usually considered an undesirable factor because it produces falloff of signal intensity with depth without yielding any useful information. This attenuation can be compensated for in the image by use of the

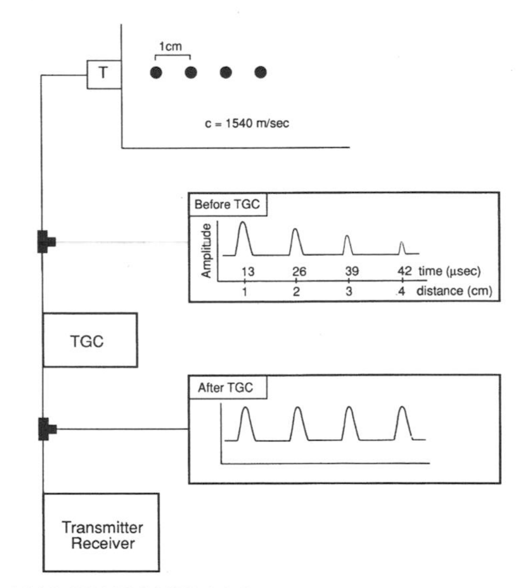

MARGIN FIGURE 21-9
Illustration of the time gain compensation (TGC) principle. Identical steel rods are imaged in a phantom containing an attenuating medium. The time at which echoes return to the transducer (T) depends only upon the speed of sound in the medium. TGC compensates for attenuation by varying the amplification as a function of time of echo reception.

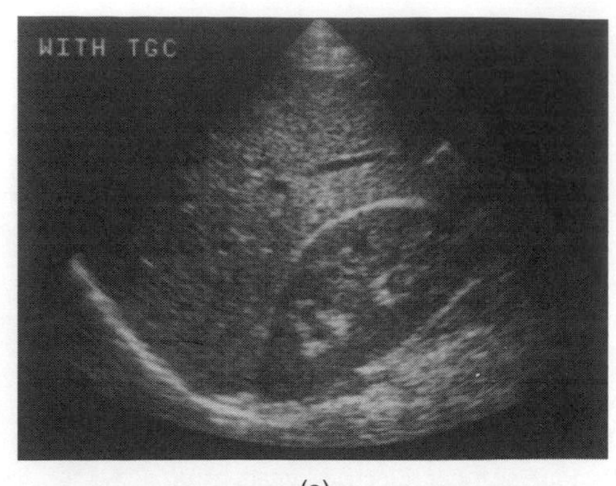

(a)

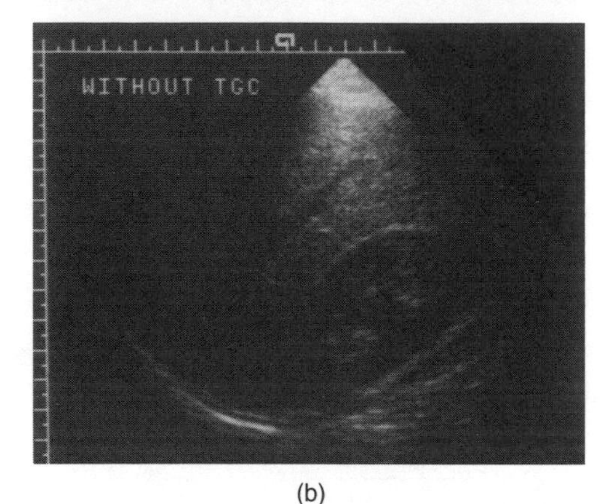

(b)

MARGIN FIGURE 21-10
Real-time B-mode images obtained with **(A)** and without **(B)** a properly adjusted TGC.

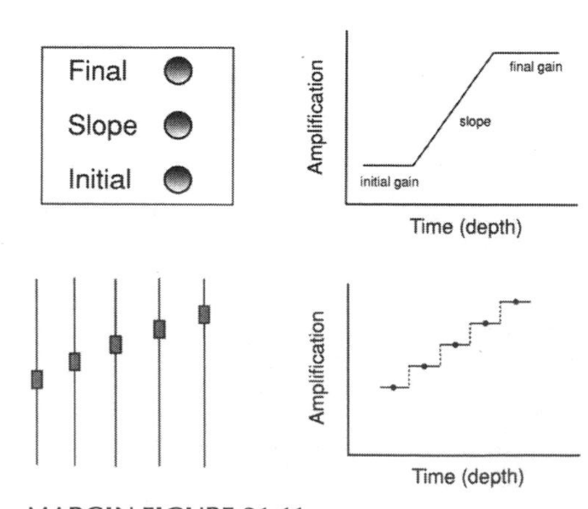

MARGIN FIGURE 21-11
Three-knob TGC control **(A)** and a linear potentiometer TGC control **(B)** along with the TGC curves produced.

time gain compensation (TGC) circuit, sometimes called the depth gain compensation (DGC) or time-varied gain (TVG) circuit.

The action of the TGC circuit is shown in Margin Figure 21-9. A plot of uncorrected or "raw" amplitudes or echoes (in the form of an A mode display) from a succession of equivalent reflectors in a homogeneous medium shows a predictable decrease in signal amplitude with depth. The key to compensating for this falloff is the relationship between the depth and time of arrival of the echo at the transducer. Echoes that return at longer times (and therefore from greater depth) receive greater amplification to compensate for increased attenuation by intervening soft tissue. The curve that describes the amplification or gain required to compensate for the attenuation is called the TGC curve. Margin Figure 21-10 shows images obtained with and without a properly adjusted TGC.

The attenuation of different body parts and variations from one patient to another render a single factory setting of the TGC impractical. The TGC must be adjusted for each patient and readjusted during the scan as different tissues are encountered. In practice, most ultrasound systems do not allow specification of a continuous curve to describe the TGC. Instead, a few controls allow reasonable flexibility in determining the curve. Two common types of TGC controls are (1) a series of linear potentiometers for discrete setting of time delays at various depths and (2) a three-knob control that allows adjustment of initial gain, slope, and far gain. The two types of TGC systems are illustrated in Margin Figure 21-11.

Other types of signal preprocessing commonly performed in ultrasound imaging include rejection, rectification, enveloping, and classification (Margin Figure 21-12). Rejection eliminates signals that are too large or too small to be of diagnostic value. All remaining signals are rectified and enveloped, a process sometimes referred to as "low-pass" filtering. Echo intensity may then be defined in a number of ways. Three typical methods are to classify echoes according to (1) the height of the peak value, (2) the area under the peak, or (3) the maximum rate of rise or slope of the echo.

A final preprocessing step involves assigning a value in computer memory that corresponds to signal intensity. Margin Figure 21-13 shows a preprocessing "map" that demonstrates how signal levels measured in decibels are converted into arbitrary numbers to be stored in memory. If the map is linear, for example, then a 10% increase in signal level would be stored as a 10% larger value in memory. By using a nonlinear map, it is possible to accentuate part of the range of echoes at the expense of another. In Margin Figure 21-13 the larger signals, from about 15 to 30 dB, are mapped into stored values of 50 to 256, while smaller signals (0 to 15 dB) are mapped into values of 0 to 50. Thus, the upper half of the range of echoes is accentuated because it takes up four-fifths of the stored values. Relationships among the stronger signals will be accentuated because the gray levels in the display will be taken up disproportionately by the stronger signals. Smaller signals will all appear very dark, with little differentiation. Therefore, the map illustrated in Margin Figure 21-13 provides high signal separation.

Postprocessing

The value stored for each location in the image (each picture element, or pixel) is eventually displayed on a monitor as a level of brightness. That is, the echo intensities are displayed as brightness values varying from black to white. As with preprocessing, one may choose a "map" for postprocessing that may be either linear or nonlinear. A nonlinear map can be designed to emphasize some parts of the range of stored values. In Margin Figure 21-14 a postprocessing map is shown that maps the lower stored values into most of the gray scale, thereby emphasizing differences among the lower values. Note that if the image were preprocessed with the map shown in Margin Figure 21-13 and then postprocessed with the map in Margin Figure 21-14, the result would be the same as if a linear map were used for both. That is, postprocessing can either "undo" or enhance the effects of preprocessing.

Other types of postprocessing features that may be present in an ultrasound device include zoom, region of interest, and digital filtering (see Chapter 10).

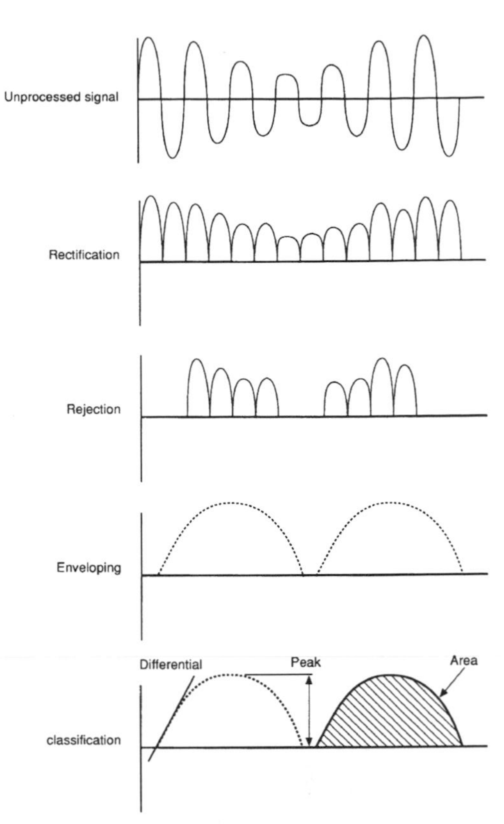

MARGIN FIGURE 21-12
Signal preprocessing in ultrasound.

DYNAMIC RANGE

The ratio of the largest to the smallest echoes processed by components of an ultrasound device is known as the dynamic range of the device. In general, the dynamic range decreases as signals pass through the imaging system (Margin Figure 21-15) because operations such as TGC and rejection eliminate small and large signals. Echoes returning from tissues can have dynamic ranges of 100 to 150 dB. The dynamic range in decibels may be easily converted to a ratio of amplitudes or intensities.

The terms "enveloping" and "low-pass filtering" are used interchangeably in signal processing. A low-pass filter removes higher frequencies but allows lower ones to pass through the circuit. If a complex signal (one consisting of varying amounts of many different frequencies) is sent through a low-pass filter, the resulting signal will look something like the outline or envelope of the original signal.

Example 21-3

Find the amplitude and intensity ratios corresponding to a dynamic range of 100 dB. For amplitudes,

$$\text{dB} = 20 \log \frac{A}{A_0}$$

can be rewritten as

$$\frac{\text{dB}}{20} = \log \frac{A}{A_0}$$

or

$$10^{\text{dB}/20} = \frac{A}{A_0}$$

So, for 100 dB we have

$$10^5 = 100{,}000 = \frac{A}{A_0}$$

The strongest echoes have 100,000 times the amplitude of the weakest echoes. Similarly, for intensity,

$$\text{dB} = 10 \log \frac{I}{I_0}$$

so that

$$10^{\text{dB}/10} = 10^{10} = 10{,}000{,}000{,}000 = \frac{I}{I_0}$$

The strongest echoes have 10^{10}, or 10 billion, times the intensity of the weakest echoes.

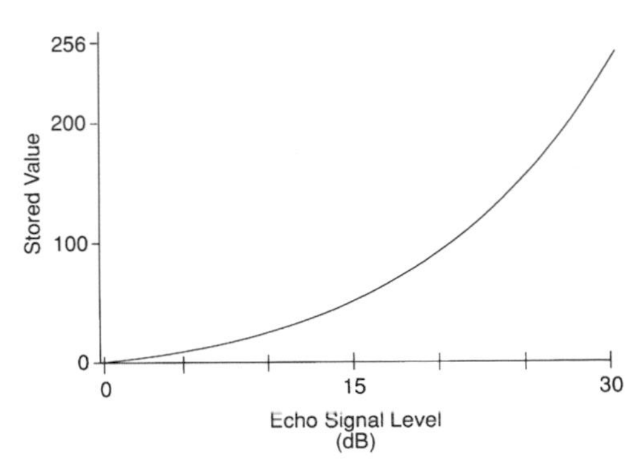

MARGIN FIGURE 21-13
Preprocessing "map." The graph shows how echo signal levels are converted to stored values. This particular example shows "high signal separation," with the upper half of the amplitude range taking up the upper 4/5 of the stored values.

Some of the dynamic range of echoes is caused by attenuation as the signals traverse several centimeters of tissue. The TGC circuit amplifies the weak echoes that arrive later, and rejection circuits eliminate low-amplitude noise. In this manner, the dynamic range is reduced, or "compressed," to about 40 dB in the receiver/processor. This dynamic range may be retained in memory while a range of about 20 dB is displayed on the system monitor.

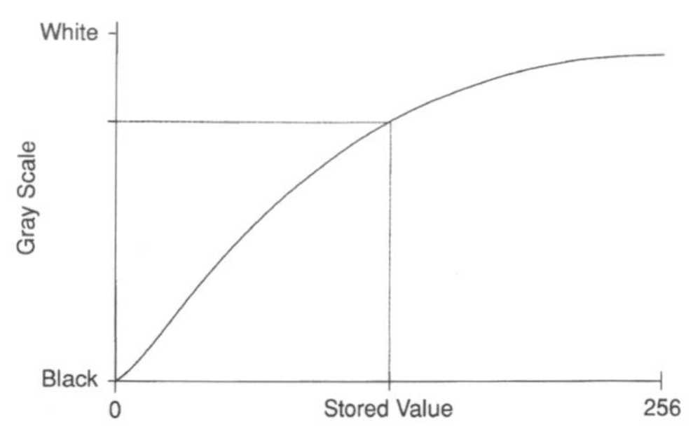

MARGIN FIGURE 21-14
Postprocessing "map." The graph shows how stored values from memory are converted to gray-scale values. This particular example shows "low signal separation," with the lower half of the stored values taking up the lower 4/5 of the gray scale (brightness values vary from black to white).

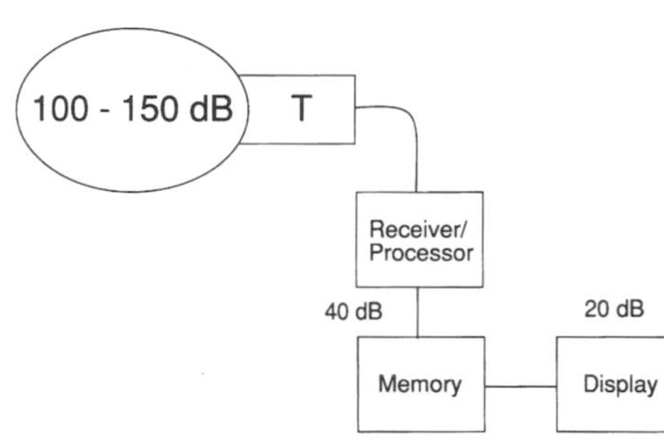

MARGIN FIGURE 21-15
Dynamic range of signals at different stages of the ultrasound imaging system.

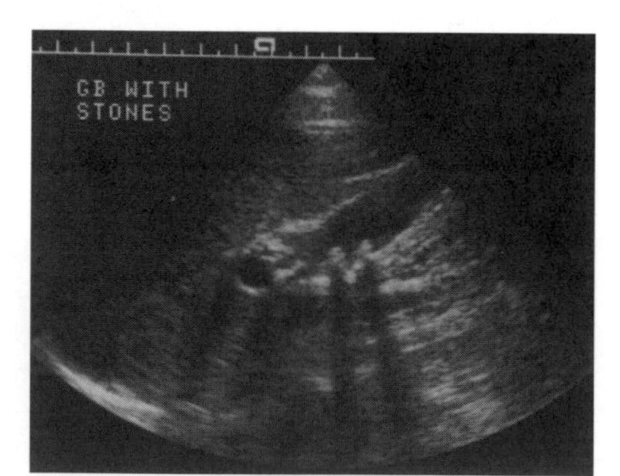

MARGIN FIGURE 21-16
Acoustic shadowing in an ultrasound image. Stones in a gallbladder reduce the transmission of sound. The display records lower amplitude echoes or "shadows" beyond the region of the stones.

ULTRASOUND IMAGE ARTIFACTS

Even though ultrasound systems produce digital images, it is impossible to identify pixel values in typical clinical images with specific fundamental properties of tissue in the way that x-ray CT numbers are associated with linear attenuation coefficients. Although pixel values in ultrasound images are influenced by fundamental properties such as acoustic impedance and physical density, each pixel value is determined by interrelationships among surrounding materials that are too complex to unfold quantitatively. However, these interrelationships sometimes create artifacts that are recognizable in a qualitative sense.

Speckle

The physical principles behind the phenomenon of speckle were described in Chapter 20. Although the presence of speckle in an image indicates the presence of features in the patient that are smaller than the wavelength of the ultrasound, there is no one-to-one correspondence between a dot in a speckle pattern and the location of a feature. The speckle pattern changes from frame to frame, even for stationary objects, and therefore is not a simple indicator of even the number of such objects. However, various attempts have been made to determine the presence or absence of disease in organs from statistical analysis of speckle patterns.[1]

Shadowing and Enhancement

If some object within the patient has a larger attenuation coefficient than the material that lies beyond it, then the settings of the TGC circuit that would provide appropriate compensation for normal tissue will undercompensate and cause the region beyond the object to appear less echogenic. This phenomenon is referred to as acoustic shadowing (Margin Figure 21-16). Similarly, if the object in the path of the ultrasound beam has a lower attenuation coefficient than its surroundings, acoustic enhancement may result.

Multiple Pathway

Various types of multiple-pathway artifacts occur in ultrasound images (see Chapter 20). When an echo returns to a transducer, the imaging device assumes that the sound traveled in a straight line following a single reflection from some interface in the patient. The scan converter then places the brightness value at an appropriate location in the image. If the actual path of the echo involved multiple reflections, the echo would take longer to return, and the scan converter would place the interface at a greater depth in the image.

Refraction

Refraction sometimes causes displacement of the sound beam as it crosses tissue boundaries. Because the scan converter assumes that the ultrasound travels in straight lines, refraction causes displacement errors in positioning reflective interfaces in the image. A refraction artifact is illustrated in Margin Figure 21-17.

QUALITY CONTROL

Frequent testing of the performance of the ultrasound imaging system by using phantoms whose properties are well understood can uncover problems that may be too subtle to discern in clinical images.[2] Standard phantoms are shown in Margin

Figure 21-18. Their properties and some protocols for typical performance tests are discussed below.

Shown in Margin Figure 21-18A is an example of a high-contrast test object for evaluation of ultrasound imaging systems. The test object mimics the speed of sound in soft tissue and contains a mixture of ethanol and water or NaCl and water. With this test object, the depth-measuring properties of ultrasound systems can be evaluated. "An equivalent speed of sound" also implies that the wavelength c/ν, near zone length r^2/λ for unfocused transducers, and the focal length for focused transducers are the same in the test medium as in soft tissue. Therefore, this phantom may be used to evaluate axial resolution at an appropriate depth.

The standard American Institute of Ultrasound in Medicine (AIUM) 100-mm test object (Margin Figure 21-18A) is an example of a high-contrast test object that employs a medium with a speed of sound equal to that of soft tissue. Reflected signals are produced by rods having diameters of 0.75 mm located at depths specified to an accuracy of ± 0.25 mm. The diameter and accuracy of location of the rods exceed the specifications for minimum resolution and distance for typical imaging systems.

A tissue-equivalent phantom (Margin Figure 21-18B) mimics more of the ultrasonic properties of human soft tissue as compared with the high-contrast test object. These properties may include scatter characteristics at typical frequencies as well as overall attenuation. The phantom provides an image with a texture pattern similar to that of liver parenchyma with a few low-contrast "cysts" in the phantom.

One of the first ultrasonic scanners for imaging the human body required that the patient be immersed in warm saline solution in a large metal tank in which a transducer rotated about the patient. Head scanning was not often performed with this technique. Shown here is the compound scanner (Somascope) developed by Howry and co-workers from a B-29 gun turret. (From *Am. J. Digest Dis.* 1963; **8**:12. Harper and Row, Hoeber Medical Division. Used with permission.)

Quality Control Tests

Dead Zone

A small portion of the image at the transducer–tissue interface is usually saturated with echoes because of the spatial pulse length of the transducer and multiple reverberations from the transducer–tissue interface. Examination of a group of high-contrast rods located near the surface of the phantom reveals the extent of this problem. The minimum depth at which a high-contrast rod can be resolved is usually quoted as the depth of the dead zone.

Axial Resolution

Any equipment problem that reduces the bandwidth or increases the spatial pulse length of the system interferes with axial resolution. Examples of such problems include cracks in the crystal materials and separation of the facing or backing material. Axial resolution is measured by attempting to resolve (produce a clear image of) rods that are closely spaced along the axis of the sound beam.

Lateral Resolution

There are several methods to evaluate lateral resolution, defined as resolution in a direction perpendicular to the axis of the ultrasound beam. A scan of rod groups that form a line parallel to the scanning surface can be used. Alternatively, rods that are perpendicular to the axis of the beam may be scanned. Because the lateral resolution *is* the width of the ultrasound beam, the apparent width of the rods scanned in this fashion (Margin Figure 21-19) yields the resolution.

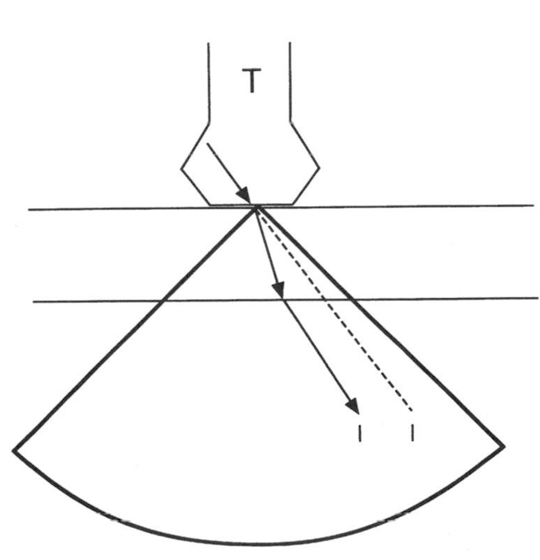

MARGIN FIGURE 21-17

Refraction artifact. The object on the left appears shifted to the right and at a slightly greater depth because of refraction in a layer having lower speed of sound than the rest of the tissue being imaged.

Distance Accuracy Measurement

Distance measurements play a key role in ultrasound dating of fetuses. Errors in distance calibration may lead to a change in the gestational age estimate that would adversely affect patient management. To perform the distance accuracy measurement, high-contrast rods with a separation of at least 10 cm are scanned. Distance

(a)

(b)

MARGIN FIGURE 21-18
Standard ultrasound phantoms. AIUM 100-mm test object **(A)** and a tissue-equivalent phantom **(B)**.

measurement options that are available on the imaging system (marker dots, calipers, etc.) are then used to determine the distance from the image. Because the rod spacings in the test object or phantom are known, the distance error may be reported as a percentage of the true value. For example, a measured distance of 10.5 for a known distance of 10.0 cm is reported as a 5% error. A reasonable goal for most imaging systems currently in use is an error of 2% or less measured over 10 cm.

Sensitivity and Uniformity

There are a number of ways to gauge the ability of an ultrasound system to record faint sources of echoes (sensitivity) and provide a consistent response to echoes (uniformity).[3] Rod groups both perpendicular and parallel to the scanning surface may be scanned, and the brightness and size of each rod may be compared. If a rod group perpendicular to the scanning surface is scanned in an attenuating phantom, then the TGC settings required to render a uniform rod image appearance may be recorded. Images obtained with those TGC settings may be reviewed periodically as a constancy check.

Display and Printing

The consistency of the recording medium may be evaluated with test patterns supplied by the manufacturer. Basically, the appearance of the "hard copy" should be the same as its appearance on the system monitor. This parameter should be measured qualitatively on all systems. More quantitative evaluations involving measurement of monitor luminance are possible[4] but not routine. If a test pattern is not available in vendor-supplied software, it is possible to use any image, such as a typical image of the tissue-equivalent phantom, to establish photographic consistency.

It is most important to separate performance of the image acquisition system from performance of the recording device. This is best accomplished by using a test pattern that is generated independently from the image acquisition system—that is, an "electronic" test pattern. An electronic test pattern that is frequently used for this purpose is a pattern that has been specified by the Society of Motion Picture and Television Engineers (SMPTE). The SMPTE pattern is shown in Margin Figure 21-20.

Frequency of Testing

The quality control tests outlined in this section should be performed monthly and after any service call that may affect the parameters measured.[5] More extensive tests may be performed by qualified personnel upon delivery of the unit.

PROBLEMS

21-1. If A, B, and M mode ultrasound display modes are described as X versus Y, what are the X and Y scales for each mode?

*21-2. How many scan lines may be obtained during a scan of a patient when the depth of view (DOV) is 8 cm if the total scan time is 13.4 msec? Assume sequential acquisition of scan lines.

*21-3. If the pulse repetition period is doubled, the field of view cut in half, and the number of scan lines doubled what is the effect on maximum allowable frame rate? Assume sequential acquisition of scan lines.

21-4. Describe the function of the master controller of a real-time ultrasound B-mode imaging system.

*21-5. What phenomenon is compensated for by the time gain compensation (TGC) circuit?

*For those problems marked with an asterisk, answers are provided on p. 493.

SUMMARY

- Display modes include
  - Amplitude
  - Brightness
  - Motion
- Frame rate is inversely proportional to depth of view, number of scan lines, and pulse repetition frequency.
- The main components of an ultrasound imaging system are
  - Transducer
  - Transmit/receive switch
  - Transmitter
  - Receiver/processor
  - Master controller
  - Memory
  - Display
  - Storage
- The time gain compensation system varies the amplification of return signals from each scan line in a variable manner as a function of time after the sent pulse.
- Preprocessing steps include
  - Rectification
  - Pulse height rejection
  - Enveloping (low-pass filtering)
  - Classification (e.g., differential, peak, area, alone or some combination)
- In preprocessing, echo signals are assigned to stored values according to according to a functional form or curve.
- In postprocessing, stored values are assigned to the gray scale according to a functional form or curve.
- Artifacts include
  - Speckle
  - Acoustic shadowing
  - Multiple pathway
  - Refraction
- Ultrasound quality control tests include
  - Dead zone
  - Axial resolution
  - Lateral resolution
  - Distance accuracy
  - Sensitivity and uniformity

(a) (b)

MARGIN FIGURE 21-19
Scan of a rod group in an ultrasound phantom demonstrates beam width (lateral resolution).

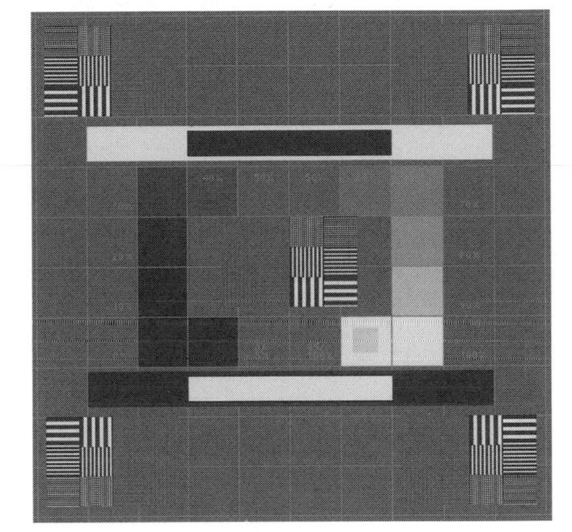

MARGIN FIGURE 21-20
The test pattern promulgated by the Society of Motion Pictures and Television Engineers (SMPTE), known colloquially as the "símtē pattern." This pattern, which tests image quality parameters such as high and low contrast resolution, distortion, and linearity, may be used as an electronic data file to test the fidelity of displays, hard-copy devices, or storage and transmission systems. (ref: Society of Motion Pictures and Television Engineers (SMPTE). Specifications for Medical Diagnostic Imaging Test Pattern for Television Monitors and Hard-Copy Recording Cameras. Recommended Practice RP 133-1986. *SMPTE J.* 1986; **95**:693–695.)

REFERENCES

1. Allison, J. W., et al. Understanding the process of quantitative ultrasonic tissue characterization. *Radiographics* 1994; **14**(5):1099–1108.
2. General Medical Physics Committee Ultrasound Task Group. *Pulse Echo Ultrasound Imaging Systems: Performance Tests and Criteria.* Medical Physics Publishing, Madison, WI, 1980.
3. Smith, S. W., and Lopez, L. A contrast-detail analysis of diagnostic ultrasound imaging. *Med. Phys.* 1982; **9**:4–12.
4. Ritenour, E. R., Sahu, S. N., Rossi, R. P., et al. Quantitative methods for hardcopy device adjustment, Paper 766-67. Presented at the Society of Photo-Optical Instrumentation Engineers, Medicine/Pattern Recognition, Newport Beach, CA, 1987.
5. Goodsit, M. M., et al. Real-time B-mode ultrasound quality control test procedures, Report of AAPM Ultrasound Task Group No 1, *Medical Phys.* 1998; **25**(8):1385–1406.

CHAPTER

22

DOPPLER EFFECT

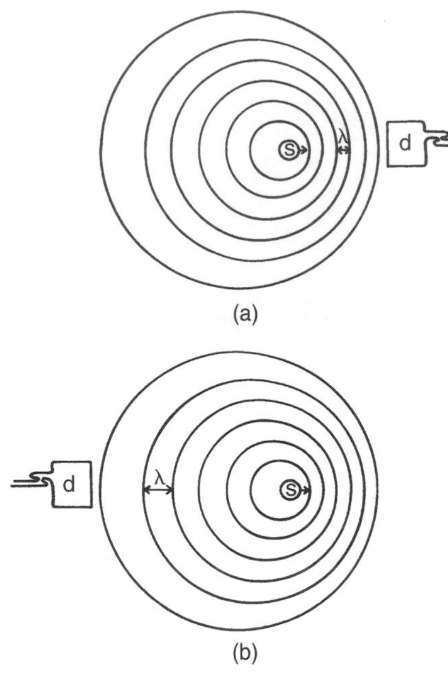

MARGIN FIGURE 22-1
Principles of Doppler ultrasound. **A:** Source moving toward a stationary detector d. **B:** Source moving away from a stationary detector d.

OBJECTIVES

After completing this chapter, the reader should be able to:

- Write the Doppler equation and use it to calculate the Doppler shift.
- Describe the effect of Doppler angle upon frequency shift and describe how error in estimation of the angle influences the accuracy of velocity estimation.
- Determine volumetric flow from average velocity and vessel cross-sectional area.
- Name the main components of a Doppler ultrasound imager.
- Calculate range gate size and depth from timing information.
- Describe the differences between information obtained from pulsed-wave versus continuous-wave Doppler.
- Describe the main features of a frequency spectrum and describe how it is used in constructing the Doppler spectral trace.
- Describe how color is assigned in color Doppler.
- Discuss some reasons for spectral broadening.

Ultrasound is used not only for display of static patient anatomy but also for identification of moving structures in the body. Approaches to the identification of moving structures include real-time pulse-echo imaging, motion mode (M-mode) display of reflected ultrasound pulses, and the Doppler-shift method. Discussed in this chapter are the basic principles of the Doppler-shift method. The Doppler method has a number of applications in clinical medicine, including detection of fetal heartbeat, detection of air emboli, blood pressure monitoring, detection and characterization of blood flow, and localization of blood vessel occlusions.[1,2]

ORIGIN OF DOPPLER SHIFT

When an emergency vehicle with its siren on passes someone on the street, a shift in frequency of one octave on the chromatic musical scale corresponds to a velocity difference between the siren and the listener of 40 miles per hour, two octaves corresponds to 80 miles per hour, and so on.

When there is relative motion between a source and a detector of ultrasound, the frequency of the detected ultrasound differs from that emitted by the source. The shift in frequency is illustrated in Margin Figure 22-1. In Margin Figure 22-1A, an ultrasound source is moving with velocity v_s toward the detector. After time t following the production of any particular wave front, the distance between the wave front and the source is $(c - v_s)t$, where c is the velocity of ultrasound in the medium. The wavelength λ of the ultrasound in the direction of motion is shortened to

$$\lambda = \frac{c - v_s}{\nu_0}$$

where ν_0 is the frequency of ultrasound from the source. With the shortened wavelength, the ultrasound reaches the detector with an increased frequency ν:

$$\nu = \frac{c}{\lambda} = \frac{c}{(c - v_s)/\nu_0}$$

$$= \nu_0\left(\frac{c}{c - v_s}\right)$$

That is, the frequency of the detected ultrasound shifts to a higher value when the ultrasound source is moving toward the detector. The shift in frequency $\Delta\nu$ is

$$\Delta\nu = \nu - \nu_0 = \nu_0\left(\frac{c}{c - v_s}\right) - \nu_0$$

$$= \nu_0\left(\frac{v_s}{c - v_s}\right)$$

If the velocity c of ultrasound in the medium is much greater than the velocity v_s of the ultrasound source, then $c - v_s \simeq c$ and

$$\Delta\nu = \nu_0\left(\frac{v_s}{c}\right)$$

A similar expression is applicable to the case in which the ultrasound source is stationary and the detector is moving toward the source with velocity v_d. In this case the Doppler-shift frequency is approximately

$$\Delta\nu = \nu_0\left(\frac{v_d}{c}\right)$$

where $c >> v_d$.

If the ultrasound source is moving away from the detector (Margin Figure 22-1B), then the distance between the source and a wavefront is $ct + v_s t = (c + v_s)t$, where t is the time elapsed since the production of the wavefront. The wavelength λ of the ultrasound is

$$\lambda = \frac{c + v_s}{\nu_0}$$

and the apparent frequency ν is

$$\begin{aligned}\nu &= \frac{c}{\lambda}\\ &= \frac{c}{(c + v_s)/\nu_0}\\ &= \nu_0\left(\frac{c}{c + v_s}\right)\end{aligned}$$

That is, the frequency shifts to a lower value when the ultrasound source is moving away from the detector. The shift in frequency $\Delta\nu$ is

$$\begin{aligned}\Delta\nu = \nu - \nu_0 &= \nu_0\left(\frac{c}{c + v_s}\right) - \nu_0\\ &= \nu_0\left(\frac{-v_s}{c + v_s}\right)\end{aligned}$$

where the negative sign implies a reduction in frequency. If the velocity c of ultrasound is much greater than the velocity v_s of the source, $c + v_s \simeq c$, and $\Delta\nu = \nu_0(-v_s/c)$. A similar expression is applicable to the case where the ultrasound source is stationary and the detector is moving away from the source with velocity v_d:

$$\Delta\nu = \nu_0\left(\frac{-v_d}{c}\right)$$

when $c >> v_d$.

If the source and detector are at the same location and ultrasound is reflected from an object moving toward the location with velocity v, the object acts first as a moving detector as it receives the ultrasound signal and then as a moving source as it reflects the signal. As a result, the ultrasound signal received by the detector exhibits a frequency shift (when $c >> v$)

$$\Delta\nu = 2\nu_0\frac{v}{c}$$

Similarly, for an object moving away from the source and detector, the shift in frequency $\Delta\nu$ is

$$\Delta\nu = 2\nu_0\left(\frac{-v}{c}\right)$$

where the negative sign indicates that the frequency of the detected ultrasound is lower than that emitted by the source.

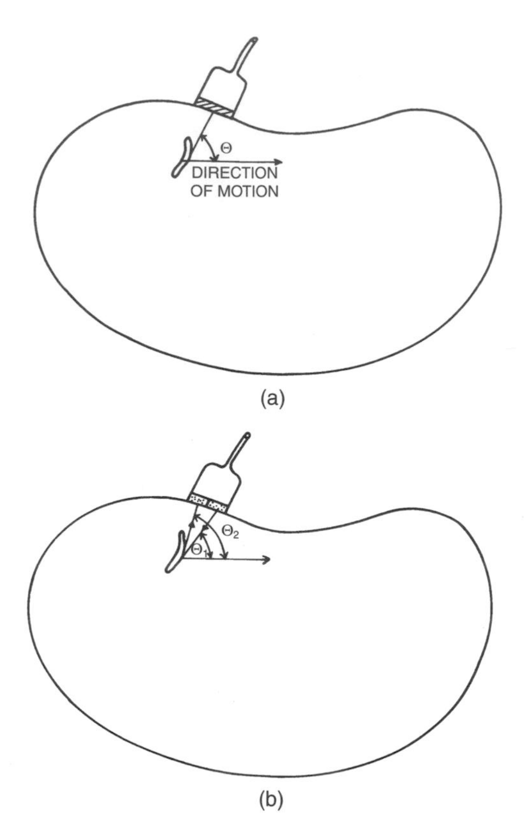

MARGIN FIGURE 22-2
A: Angle θ between an incident ultrasound beam and the direction of motion of the object. **B:** For a dual probe with separate transmitting and receiving transducers, the angle θ is the average of the angles θ_1 and θ_2 that the transmitted and detected signals make with the direction of motion.

The discussion above has assumed that the ultrasound beam is parallel to the motion of the object. For the more general case where the ultrasound beam strikes a moving object at an angle θ (Margin Figure 22-2A), the shift in frequency $\Delta\nu$ is

$$\Delta\nu = 2\nu_0\left(\frac{v}{c}\right)\cos\theta \tag{22-1}$$

A negative sign in this expression would imply that the object is moving away from the ultrasound source and detector and that the frequency of detected ultrasound is shifted to a lower value. The angle θ is the angle between the ultrasound beam and the direction of motion of the object. If the ultrasound source and detector are displaced slightly (Margin Figure 22-2B), then θ is the average of the angles that the transmitted and reflected beams make with the motion of the object. For the small displacement between ultrasound transmitter and detector commonly used in Doppler ultrasound units, this assumption contributes an error of only 2% to 3% in the estimated frequency shift.[3]

Example 22-1

A 10-MHz ultrasound is used for the detection of blood flow. For blood flowing at 15 cm/sec toward the ultrasound source, find the Doppler-shift frequency. Assume that the angle θ is zero so that $\cos\theta = 1$.

$$\begin{aligned}\Delta\nu &= 2\nu_0\left(\frac{v}{c}\right)\cos\theta \\ &= (2)(10\text{ MHz})\left(\frac{15\text{ cm/ sec}}{154{,}000\text{ cm/sec}}\right) \\ &= 1.9\text{ kHz}\end{aligned} \tag{22-1}$$

The audible frequency range is approximately 20 Hz to 20 kHz. Mechanical vibrations within this frequency range, if delivered with sufficient amplitude, are registered by the ear as audible sound. For physiologic flow rates the Doppler shift (the difference in frequency between sent and received signals) falls in this range and can be used to drive an audio speaker. The listener is not able to derive quantitative velocity information from the audible output but can appreciate changes in signal frequency and intensity. These changes may be caused by variations in the spatial location or angle of the probe with respect to the patient or by variations in the Doppler shift caused by temporal changes in flow within the patient during the cardiac cycle. As flow increases, the frequency of the sound (its "pitch") increases. The listener does not hear a pure single-frequency note but instead hears a complex assortment of frequencies similar to the sound of ocean waves breaking on the seashore. The assortment of frequencies, referred to as the spectral bandwidth of the Doppler signal, is discussed in detail later in this chapter.

Austrian physicist Christian Doppler (1803–1858) first described the effect that bears his name in a paper published in the proceedings of the Royal Bohemian Society of Learning in 1843. The paper's title translates into English as: "Concerning the colored light of double stars." He explained the shift in color (frequency of light) of stars as their velocities changed relative to earth. While this effect is too small to result in a detectable color shift in measurements available at that time, he demonstrated the reality of this effect in 1845 using sound waves. Two trumpeters who could maintain a steady note with perfect pitch were recruited. One performed at the side of a railway, and the other performed on a moving train. The difference in pitch (frequency of sound) was apparent to listeners.

The term involving $\cos\theta$ in the Doppler-shift equation [Eq. (22-1)] describes the variation in frequency shift as the sonation angle (the angle between the Doppler probe and the moving material) is varied. The shift is greatest when the longitudinal axis of the probe is oriented along the direction of motion. This occurs when the transmitted beam of sound is parallel (sonation angle = 0 degrees) or opposed to (sonation angle = 180 degrees) the direction of motion. Theoretically, no Doppler shift occurs when the probe is exactly perpendicular to the direction of motion. In practice, a small Doppler shift may be detected when the probe appears to be perpendicular to the direction of blood flow in a vessel. The small shift is due to several phenomena such as transverse components of flow in the vessel, detection of signals from side lobes, and so on.

The presence of the cosine term in the Doppler-shift equation complicates the direct measurement of velocity by the Doppler method. The only quantity that is measured directly by a typical Doppler ultrasound unit is Doppler shift. Determination of the velocity of blood in units such as centimeters per second requires an estimate of the sonation angle. Errors in the estimate of this angle (using on-screen cursors) limit the accuracy of velocity estimates. This error is more significant at larger angles (near 90 degrees) than at smaller angles.

An error in a sonographer's ability to estimate the Doppler angle in a real-time B-mode display affects the estimated value of blood velocity. The same error has a greater effect at angles near 90° than at angles near 0°. The error occurs because calculation of velocity from the Doppler frequency shift [Eq. (22-1)] involves multiplication by the cosine of the Doppler angle. This affect is appreciable even if the error is small.

If we assume an error of ± 3 degrees, then at an actual Doppler angle of 10° the sonographer's estimate will lead to a variation in the value of the cosine function from cos (7°) = 0.993 to cos (13°) = 0.974 where the true value of cos (10°) = 0.985. The error is less than ± 1%.

At an actual Doppler angle of 80° the same error of 3° leads to a variation in the value of the cosine function from cos (77°) = 0.225 to cos (83°) = 0.122 where the true value of cos (80°) = 0.174. Here, the error is approximately 30%!

Example 22-2

A 10-MHz ultrasound probe detects a 1.4-kHz Doppler shift at a sonation angle estimated as 45 degrees. Find the estimated velocity of blood flow and the percent error in the estimate if the angle of sonation is incorrect by as much as 3 degrees:

$$\Delta\nu = \frac{2\nu_0\, v \cos\theta}{c}$$

$$v = \frac{\Delta\nu c}{2\nu_0 \cos\theta}$$

$$= \frac{(1.4 \times 10^3\ \text{sec}^{-1})(1.54 \times 10^5\ \text{cm/sec})}{2(10 \times 10^6\ \text{sec}^{-1})} \frac{1}{\cos\theta}$$

$$= (10.8\ \text{cm/sec})\frac{1}{\cos\theta}$$

If $\theta = 45$ degrees

$$v = \frac{10.8}{0.707} = 15.3\ \text{cm/sec}$$

But if θ actually equaled 48 degrees instead of 45, then

$$v = \frac{10.8}{0.669} = 16.1\ \text{cm/sec}$$

The percent error is

$$\%\ \text{error} = \frac{16.1 - 15.3}{15.3}(100) = 5.5\%$$

The error is slightly smaller if θ were actually 42 degrees.

Determining the volumetric flow of blood in units of cubic centimeters per second requires an estimate of the area of the vessel as well as the sonation angle. Furthermore, it assumes measurement of average quantities that are not always valid. Volumetric flow Q is the product of the average velocity v and the cross-sectional area A of the vessel:

$$Q = vA \qquad (22\text{-}2)$$

In clinical measurements of volumetric flow, several sources of error reduce the accuracy below what would be expected based upon the intrinsic precision of Doppler-shift measurements. The measurement of vessel area is typically based upon measurement of vessel diameter in the image. However, the angle at which the image plane "cuts through" the vessel may cause either underestimation or overestimation of diameter. Also, the Doppler unit may not sample the average velocity in a vessel. Flow in a vessel (as in a mountain stream) does not have a uniform cross-section. It is usually greatest at the center and decreases to near zero at the vessel wall. Complex flow "profiles" are possible, particularly if a stenosis, bifurcation, or plaque formation is present. An accurate estimate of average velocity requires sampling of velocity at different radii within the vessel (at several points across the "stream"). It is possible to make some simplifying assumptions. For example, in laminar (nonturbulent) flow the average velocity is half the maximum value.[4]

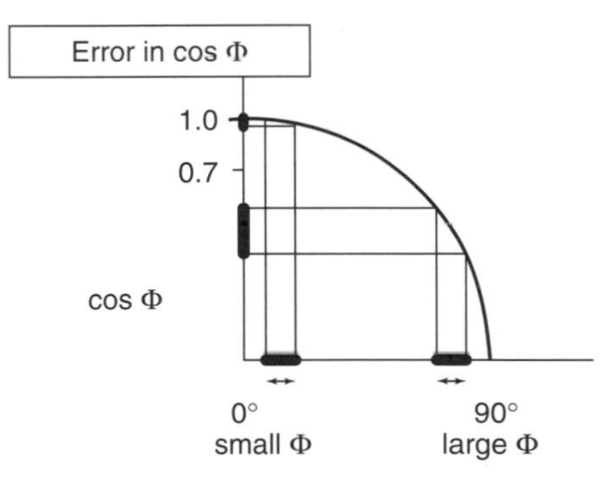

MARGIN FIGURE 22-3
Because of the dependence of the Doppler shift upon the cosine of the Doppler angle, calculation of the velocity of an object from an observation of its Doppler shift requires estimation of the Doppler angle. This estimate (usually acquired through observation of a freeze frame of a real-time B mode ultrasound image) could be off by a few degrees. In the graph shown here we see that if the few-degree error (horizontal axis) occurs at a small Doppler angle (close to parallel to the direction of flow), the error introduced by the cosine function (vertical axis) is relatively small. If the same few degree error occurs at a larger Doppler angle (close to perpendicular to the direction of flow), then the error introduced by the cosine function is greater.

Example 22-3

Find the volumetric flow in a 3-mm-diameter vessel in which the average velocity (v) of blood is estimated to be 10 cm/sec.

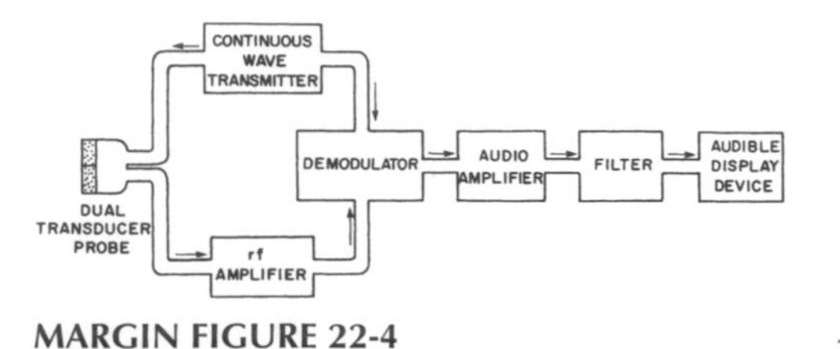

MARGIN FIGURE 22-4
Diagram of a continuous-wave Doppler unit.

First, find the cross-sectional area (A) of the vessel.

$$A = \pi\left(\frac{d}{2}\right)^2 = 3.14\left(\frac{0.3\ \text{cm}}{2}\right)^2$$
$$= 0.07\ \text{cm}^2$$

The volumetric flow (Q) is

$$Q = vA$$
$$= (10\ \text{cm/sec})(0.07\ \text{cm}^2)$$
$$= 0.7\ \text{cm}^3/\text{sec} \qquad \textbf{(22-2)}$$

A continuous-wave Doppler unit is diagramed in Margin Figure 22-4. The Doppler probe contains separate transmitting and receiving transducers. Application of a continuous electrical signal to the transmitting transducer produces a continuous ultrasound beam at a prescribed frequency. As the beam enters the patient, it is reflected from stationary and moving surfaces, and the reflected signals are returned to the receiver. Suppose that two signals are received, one directly from the transmitting transducer and the other from a moving surface. These signals interfere with each other to produce a net signal with a "beat frequency" equal to the Doppler-shift frequency. This Doppler-shift signal is amplified in the radio-frequency (RF) amplifier. The amplified signal is transmitted to the demodulator, where most of the signal is removed except the low-frequency beat signal. The beat signal is amplified in the audio amplifier and transmitted through the filter (where frequencies below the beat frequency are removed) to the loudspeaker or earphones for audible display. In practice, many moving interfaces reflect signals to the receiver, and many beat frequencies are produced.

Pulsed Doppler

Accurate estimates of the location of the source of a Doppler shift in a patient are difficult to achieve with continuous-wave Doppler. The source is obviously somewhere along the extended axis of the ultrasound transducer (direction of sonation), but exactly where is uncertain. To localize the source of the Doppler shift, a method known as range gating may be used. In this method the sending transducer is shut off, and a preset interval of time, the first range gate, is allowed to elapse before the return signal is processed to determine the frequency shift. The Doppler shift is then analyzed until a second range gate is encountered. With the assumption of an average speed of sound, the range gate settings define a specific Doppler sampling region over a particular range of depth. The time between the two gates determines the extent of the sampling region, and variation of the elapsed time before the first gate determines the sampling depth. With a speed of sound for soft tissue of 1540 m/sec, a range factor of 13 μs of elapsed time corresponds to 1 cm of depth. This range factor takes into account the "round trip" of 2 cm that sound travels (i.e., from the transducer to the reflector and back again) for each centimeter of depth.

Example 22-4

Range gates are set at 39 and 45 μsec with respect to a transmitted pulse of ultrasound. Determine the dimensions of the Doppler sampling region within soft tissue.

$$\frac{39\ \mu\text{sec (1st gate)}}{13\ \mu\text{sec/cm}} = 3\ \text{cm from the transducer}$$

$$\frac{45\ \mu\text{sec (2nd gate)}}{13\ \mu\text{sec/cm}} = 3.5\ \text{cm from the transducer}$$

The Doppler sampling region begins at a depth of 3 cm and extends to a depth of 3.5 cm. We could state that there is a 0.5-cm sampling region centered at a depth of 3.25 cm. The width of the sampling region is roughly equal to the ultrasound beam width at that depth.

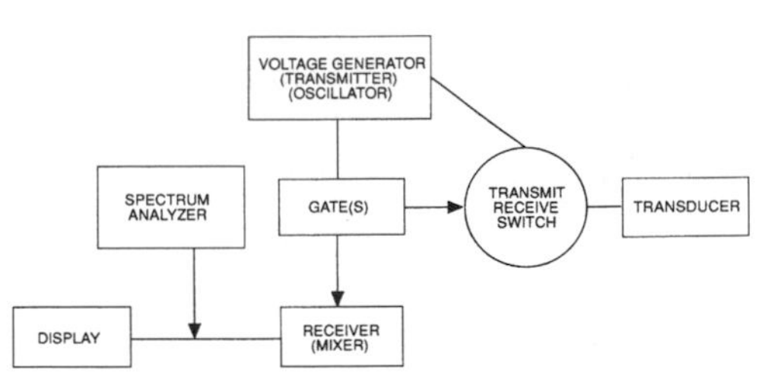

MARGIN FIGURE 22-5
Diagram of a pulsed Doppler unit.

In Doppler systems that are used in conjunction with imaging, the position of the sampling volume is usually indicated by cursors positioned on the display. The ultrasound computer performs the electronic equivalent of the calculations in Example 22-4 to determine where the cursors should appear on the image.

Range gating requires that the ultrasound be started and stopped (i.e., pulsed) during the signal acquisition process so that the time to echo may be determined unambiguously. The use of pulsed Doppler not only allows range gating but also permits imaging pulses to be interspersed with Doppler pulses. Thus a real-time B-mode image of the patient may be obtained from every other pulse. In continuous-wave Doppler there is, by definition, no time in which to send imaging pulses, because there is no gap in the Doppler process.

A simplified diagram of a pulsed Doppler ultrasound system is shown in Margin Figure 22-5. The main differences between the pulsed Doppler system in this illustration and the continuous-wave system of Margin Figure 22-4 are that the pulsed system includes range gates and a transmit–receive switch similar to that in B-mode systems. Units that intersperse Doppler and imaging pulses incorporate components found in standard B-mode systems with both Doppler and B-mode components connected to the voltage generator or oscillator so that the sending of Doppler and imaging pulses can be coordinated.

There are techniques for determining velocity in ultrasound images that do not use the Doppler frequency shift effect. In "time domain" analysis, the return signals along scan lines are compared in a process known as "autocorrelation analysis" to determine if a signal anywhere in the image in one frame appears at a different location in another frame. This information is then color-coded using any of the techniques that are used in color Doppler.

Spectral Analysis

The Doppler-shift signal contains a wealth of information. One of the fundamental challenges of Doppler ultrasound is to display that information in a useful fashion. Doppler shift in a blood vessel is produced by the flow of red blood cells (RBCs). However, within a single vessel there is an assortment (distribution) of RBC velocities. Also, the velocity of any given cell varies with time according to the cardiac cycle. Three methods for visual display of information are discussed here: frequency spectrum, spectral trace, and color mapping.

Frequency Spectrum

A frequency spectrum (Margin Figure 22-6) is a histogram in which the horizontal axis depicts Doppler shift frequency and the vertical axis shows the relative contribution of each frequency to the observed Doppler shift. For simplicity, the frequency spectrum can be thought of as a histogram of the relative distribution of RBC velocities in a blood vessel. The peak of the spectrum shows the Doppler shift associated with the greatest number of RBCs. The frequency spectrum is a "snapshot" depicting the conditions of flow in the sampling volume at a particular moment in time. To determine temporal variations, a display such as the flow plot is required.

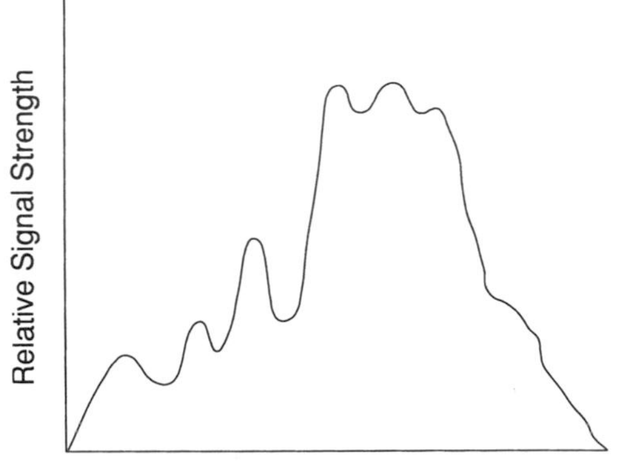

MARGIN FIGURE 22-6
Doppler-shift frequency spectrum showing the relative contribution of frequencies to the Doppler shift during a brief interval of time.

Spectral Trace

A spectral trace (Margin Figure 22-7) is a graph of frequency versus time that depicts the manner in which the value of Doppler shift varies with time. The value indicted may be the average, minimum, or maximum Doppler shift, and a distribution of values may be indicated by a vertical bar. The spectral trace may be displayed as a Doppler-shift frequency trace or as a velocity trace. The velocity trace is obtained by using the center frequency of the probe and the sonation angle measured by the operator to convert the Doppler-shift frequency scale a velocity scale (see Example 22-2). Audible sound may be broadcast by speakers as the trace is displayed in real time

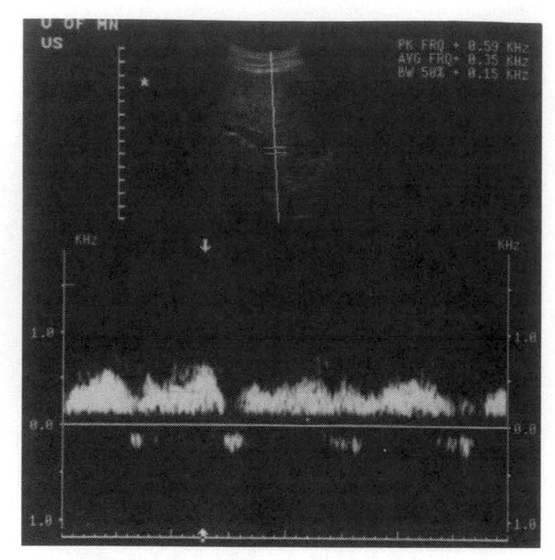

MARGIN FIGURE 22-7
Doppler spectral trace showing Doppler shift as a function of time. The distribution of Doppler shifts at any given time is the vertical "thickness" of the trace. Thus, the Doppler spectral trace shows changes in the frequency spectrum over time.

by using a "moving pen" or "strip chart" type of display. Cardiac information such as electrocardiographic (ECG) tracings may also be displayed to allow the viewer to coordinate features in the spectral trace with timing of the cardiac cycle. Some systems also allow the user to display a frequency spectrum for a selected moment of time during display of the spectral trace.

Color Mapping

Both frequency spectra and spectral traces may be displayed along with B-mode images by dividing the display screen into separate "windows" (compartments). Cursors on the image indicate the Doppler sampling volume. It is possible to indicate flow directly within the image by color-coding the pixels. In color mapping or color-flow Doppler, pixels in the B-mode image that correspond to regions in the patient where flow is present are assigned hues (colors such as red, blue, green) according to an arbitrary color scale. For example, red may indicate a large positive Doppler shift, orange a smaller positive shift, yellow a small negative shift, and so on.

Alternatively, the color scale may involve only two hues. One of the most common two-hue systems uses red for a positive Doppler shift and blue for a negative Doppler shift. The magnitude of the Doppler shift is encoded by the "saturation" of the color. Saturation is the degree to which the hue is present. A totally unsaturated color is white. It emits a broad distribution of all frequencies of visible light. Higher saturation implies that more of the frequencies in the spectrum are of a single frequency, interpreted by our retina as a purer color. The saturation encoding in color Doppler is usually arranged such that a greater shift (higher velocity either toward or away from the transducer) is encoded as a less saturated hue. For example, a small positive Doppler shift would be indicated by a dark red, whereas a larger positive Doppler shift would be indicated by a lighter, less saturated red.

In color Doppler, the most commonly used color scale codes flow away from the transducer as blue and flow toward the transducer as red. When such a "blue away red toward" (BART) display is used, it is sometimes necessary to remind observers that the colors are not necessarily correlated with the oxygenation (venous or arterial) status of the blood.

Whatever color scale is used in color Doppler, it is desirable to avoid color-coding the very slow velocities of tissues that occur in the patient as pliable tissues are pushed by the transducer and as the transducer itself is moved. Detection of such motions is not usually the purpose of the exam. Thus, most color scales do not assign color to very small Doppler shifts and display a black band at the center of the color scale indicator. The width of the black band may be changed by the operator to vary the range of velocities that will be color-coded.

The large variety of color scales that could be used in color Doppler, as well as the variation in perception of color from one individual to another, is an intriguing problem for optimization of visual perception.

Quantitative Use of Doppler

Various attempts have been made to correlate information obtained from Doppler displays, particularly spectral traces, with pathology. Simple measures such as the peak systolic or end-diastolic average velocities are probably too crude to differentiate normal and disease states, in part because of the wide range of normal values in a population and also because of the difficulties inherent in obtaining quantitative values for velocity. It is possible to construct ratios of features of spectral traces so that conversion factors such as the angle of sonation are the same for both features and therefore cancel out. With the symbol A used for peak systolic and B for end-diastolic frequency shift, some indices that have been used with varying degrees of success include A/B,[5] $(A - B)$/mean,[6] and $A - B/A$ (known as Pourcelot's or the resistivity index).[7] Still, attempts to provide definitive diagnoses on the basis of indices alone have not been generally successful. Such methods appear more useful for longitudinal studies than for specific diagnoses.[8]

LIMITATIONS OF DOPPLER SYSTEMS

Pulse Repetition Frequency

The use of pulsed Doppler limits the maximum velocity that can be detected. Each pulse reflected from the RBCs is a sample of the frequency shift. The greater the rate of sampling or pulsing, the better the measurement of the shift. Information theory suggests that an unknown periodic signal must be sampled at least twice per cycle to determine even rudimentary information such as the fundamental frequency.[4] Thus the rate of pulsing [the pulse repetition frequency (PRF)] of pulsed Doppler must be at least twice the Doppler-shift frequency produced by flow. If not, the frequency shift is said to be "undersampled"—that is, sampled so infrequently that the frequency reported by the instrument is erroneously low, an artifact known as "aliasing."

Example 22-5

A stenosis produces a high-speed jet in a vessel. The maximum velocity of RBCs in the stenosis is 80 cm/sec. Find the minimum PRF that must be used to avoid aliasing for pulsed Doppler at 7.5 MHz. Assume that the angle of sonation is 0 degrees.

The Doppler shift is

$$\begin{aligned} \Delta\nu &= \frac{2\nu_0\, v \cos\theta}{c} \\ &= \frac{2(7.5 \times 10^6 \text{ sec}^{-1})(80 \text{ cm/sec})(1)}{1.54 \times 10^5 \text{ cm/sec}} \\ &= 7.8 \text{ kHz} \end{aligned} \qquad \textbf{(22-1)}$$

The minimum PRF required to prevent aliasing is twice the Doppler-shift frequency, or 15.6 kHz.

In color Doppler, aliasing is characterized by colors from one end of the color spectrum appearing adjacent to colors from the opposite end. For example, an abrupt transition from blue to red with no dark band between the two colors signals aliasing. A dark band would indicate momentary cessation of flow which would have to occur if there was a change in flow direction.

There is a practical limit to the PRF of an ultrasound system. To obtain information at a given depth in the patient, the system must "wait and listen" over the time required for sound to travel from the transducer to that depth and back again. If an echo returns after another pulse has been sent, it is difficult to determine which travel time to use in determining the range. Because the travel time required for sound waves in soft tissue is 13 μsec/cm of depth, the maximum PRF for a given total depth of imaging [field of view (FOV)] can be computed.

Example 22-6

Find the field of view (FOV) if the PRF of 15.6 kHz calculated from Example 22-5 is used for imaging.

If the PRF is 15.6 kHz, then the wait-and-listen time between pulses, the pulse repetition period (PRP), is its reciprocal.

$$\begin{aligned} \text{PRP} = \frac{1}{\text{PRF}} = \frac{1}{15.6 \times 10^3 \text{ sec}^{-1}} &= 6.41 \times 10^{-5} \text{ sec} \\ &= 64.1\ \mu\text{sec} \end{aligned}$$

The field of view (FOV) is then

$$\text{FOV} = \frac{64.1\ \mu\text{sec}}{13\ \mu\text{sec/cm}} = 4.9 \text{ cm}$$

Clearly, the PRF required to measure the maximum flow in a high-speed jet is not routinely available in imaging systems because it would severely limit the field of view.

In echocardiography, the appearance of very rapidly moving blood that spurts from heart valves that do not close properly (known as "jets") may be enhanced in color flow Doppler. One technique that enhances jets is to color code signals with an extremely high Doppler shift in a color that is markedly different from the traditional red and blue (e.g., green or yellow). Then, when a jet appears in a real time image, a spurt of bright color is immediately apparent.

To avoid aliasing, most pulsed Doppler systems have a "high-PRF" option that permits operation at PRFs that may exceed those practical for imaging. An alternative is to switch to continuous-wave Doppler because it is not limited in the same way that pulsed Doppler is limited by the sampling process. In some cases, it may be possible to switch to a lower-frequency transducer so that the Doppler shift is lower. Several techniques to circumvent the PRF requirement have been proposed but are still in the research stages. These include pulse-coding schemes and the use of nonperiodic pulses.[9,10]

Spectral Broadening

There are many barriers to interpreting Doppler shift as a measure of a fundamental parameter such as velocity. Just as other imaging modalities are limited by noise, Doppler ultrasound is also limited by extraneous information or noise. This extraneous information may cause a "spectral broadening" of the range of Doppler-shift frequencies displayed by the unit. Spectral broadening compromises the interpretation of the width of the frequency spectrum as an indication of the range of RBC velocities. Spectral broadening can also be caused by other factors that may be lumped together as deterministic errors (see Chapter 11). These include the bandwidth or range of frequencies in the transmitted signal, the presence of adjacent vessels or flow within the fringes of the sampling volume, and tissue and wall motion (this low-velocity, low-frequency component is usually eliminated by electronic filtering known as "wall filtering").

Nondeterministic or random spectral broadening may be caused by electronic noise resulting from random fluctuations in electronic signals.

PROBLEMS

*22-1. Estimate the frequency shift for a 10-MHz ultrasound source moving toward a stationary detector in water ($c = 1540$ m/sec) at a speed of 5 cm/sec. Is the frequency shifted to a higher or lower value?

*22-2. Estimate the frequency shift for an object moving at a speed of 10 cm/sec away from a 10-MHz source and detector occupying the same location.

22-3. What is meant by a beat frequency?

22-4. Identify some of the major medical applications of Doppler systems.

*22-5. Find the minimum pulse repetition frequency required to prevent aliasing when measuring a velocity of 5 cm/sec with pulsed Doppler ultrasound at 5 MHz for a sonation angle of zero degrees. How would this PRF change if:
a. The angle of sonation was increased?
b. The center frequency was increased?

*For those problems marked with an asterisk, answers are provided on p. 493.

SUMMARY

- The Doppler equation is $\Delta\nu = 2\nu_0 \frac{v}{c} \cos\theta$.
- As the Doppler angle increases from zero to 90 degrees, the Doppler shift-decreases and errors in estimation of the angle cause greater inaccuracies in velocity estimation.
- Volumetric flow Q, average velocity v, and area A are related as follows:

$$Q = vA$$

- The main components of a Doppler ultrasound imager include
  - Transducer
  - Transmit/receive switch
  - Voltage generator (transmitter or oscillator)
  - Gate system

  - Receiver (mixer)
  - Spectrum analyzer
  - Display
- A Doppler spectral trace is a plot that shows the variation in time of a brightness modulated Doppler spectra.
- In color Doppler, the magnitude of frequency shift is indicated by saturation and the specific colors indicate direction of flow.
- Spectral broadening may indicate that a range of velocities exists in the sample volume. It may also be caused by
  - Bandwidth of the transmitted pulse
  - Signal from adjacent vessels
  - Tissue and vessel wall motion
  - Electronic noise

REFERENCES

1. Reid, J., and Baker, D. Physics and electronics of the ultrasonic Doppler method, in Böck, J., and Ossoining, K. (ed.), *Ultrasmographia Medica*, Vol 1. Proceedings of the First World Congress on Ultrasonics in Medicine and SIDUO III. Vienna, Verlag der Wiener Medizenischen Akademic/Vienna Academy of Medicine, 1971, p. 109.
2. McDicken, W. *Diagnostic Ultrasonics*. New York, John Wiley & Sons, 1976; p. 219.
3. Wells, P. The directivities of some ultrasonic Doppler peaks. *Med. Biol. Eng.* 1970; **8**:241.
4. Evans, D. H., McDicken, W. N., Skidmore, R., et al. *Doppler Ultrasound: Physics Instrumentation and Clinical Applications*. New York, John Wiley & Sons, 1989, p. 8.
5. Stuart, B., Drumm, J., Fitzgerald, D. E., et al. Fetal blood velocity waveforms in normal pregnancy. *Br. J. Obset. Gynaecol.* 1980; **87**:780–785.
6. Gosling, R. G., King, D. H., Newman, D. L., et al. Transcutaneous measurement of arterial blood velocity by ultrasound, in *Ultrasonics for Industry Conference Papers*. Guildford, England, IPC, 1969, pp. 16–32.
7. Pourcelot, L. Applications Cliniques de l'Examen Doppler Transcutane, in Peronneau, P. (ed.), *Velocimetrie Ultrasonore Doppler*. Paris, Institut National de la Santé et de la Recherche Médicale 34, 1974, pp. 780–785.
8. Taylor, K. J. W., and Holland, S. Doppler US, Part I. Basic principles, instrumentation, and pitfalls. *Radiology* 1990; **174**:297–307.
9. Bendick, P. J., and Newhouse, V. L. Ultrasonic random-signal flow measurement system. *J. Acoust. Soc. Am.* 1974; **56**:860–864.
10. Cathignol, D. N., Fourcade, C., and Chapelon, Y. Transcutaneus blood flow measurement using pseudorandom noise Doppler system. *IEEE Trans. Biomed. Eng.* 1980; **27**:30–36.

CHAPTER

23

FUNDAMENTALS OF MAGNETIC RESONANCE

OBJECTIVES

After completing this chapter, the reader should be able to:

- Explain the phenomenon of precession of nuclei in a static magnetic field.
- Describe the phenomenon of nutation and explain how it is responsible for causing variation in the strength of the MR signal.
- Trace the steps involved in reception of the MR signal beginning with insertion of the patient into the magnet.
- Solve the Larmor equation to determine resonance frequency.
- Give the quantum mechanical explanation for magnetic resonance.
- Describe the behavior of bulk magnetization of a sample in the presence of static and radio frequency magnetic fields.
- Define T1 and T2 relaxation.
- Give an equation for exponential relaxation of spins after application of a radio-frequency pulse has ended.
- Explain the behavior of relaxation times as the strength of the static magnetic field is increased.

Certain nuclei have magnetic properties as discussed in Chapter 2. By the use of magnetic fields and radio-frequency (RF) waves, these properties can be exploited to yield information about biological materials containing such nuclei.

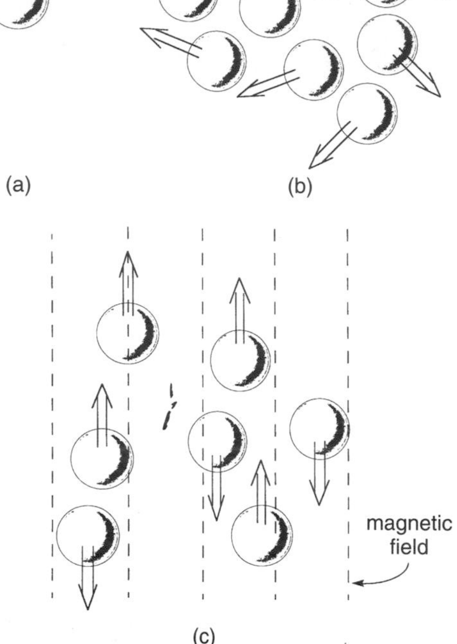

MARGIN FIGURE 23-1
A: Proton magnetic moment direction is indicated by arrow. **B:** In a typical material, magnetic moments are oriented randomly. **C:** If a magnetic field is applied, magnetic moments align themselves along the direction of the field. Note that some are parallel, while others are antiparallel.

In 1933, Stern and associates discovered that the proton had a magnetic moment and would therefore interact with a magnetic field.[1,2] Six years later, one of Stern's students, Rabi, designed an apparatus to examine the interaction of magnetic nuclei with time-varying magnetic fields.[3] He found that a certain frequency of the magnetic field was required to obtain the strongest interaction with an atom or molecule and that this frequency varied for each type of atom. Rabi demonstrated the principle of magnetic resonance (MR) and linked resonance frequencies to specific nuclei. He went on to investigate the magnetic forces exerted upon protons and neutrons. Rabi received the Nobel prize for this work in 1944.

INTERACTION OF NUCLEI WITH A STATIC MAGNETIC FIELD

This discussion of MR first presents the "classical interpretation" of the behavior of nuclear magnetic moments by using the hydrogen nucleus (i.e., a single proton) as a model. In the classical interpretation the position of the hydrogen nucleus can be specified with any desired degree of precision, and its movements are assumed to be continuous and completely predictable. Each proton behaves as a small magnet with a magnetic moment that has both magnitude and direction (Margin Figure 23-1). In a typical sample of hydrogen-containing material (such as the human body), the magnetic moments of the individual hydrogen nuclei are oriented in random directions. If a strong magnetic field is applied to the sample, the nuclei "align" their magnetic moments with the direction of the magnetic field in a manner similar to a compass needle aligned with the earth's magnetic field. The earth's magnetic field (0.5 gauss) is not strong enough to bring protons in a tissue sample into alignment. The field supplied by an MR system (e.g., 20,000 gauss) is strong enough to produce alignment.

ROTATION AND PRECESSION

In addition to aligning with a magnetic field, a magnetic moment also precesses about the field. Precession is easily demonstrated in rotating objects (another reason why we infer the property of "spin" for protons). A spinning top, for example, will "wobble" about a vertical axis defined by the earth's gravitational field (Margin Figure 23-2). This wobbling motion is precession.

Precession is a type of motion that is distinct from rotation. Rotation is the spinning of an object about its axis (an imaginary line through the center of mass of the object). The rapid spin of a top that causes its surface to blur is rotation. Precession is a "second-order" motion. It is the "rotation of a rotating object."

If you push against a ball, it will roll in the direction in which it was pushed. If you push against a gyroscope (a gyroscope is simply a top surrounded by a stationary framework), the gyroscope will move in a perpendicular direction (90 degrees) to the direction of the push. If you push it toward the north, it will move to the east or west depending on the direction of its spin (clockwise or counterclockwise) (Margin Figure 23-3).

In the macroscopic world, rotating objects have the property of angular momentum that causes them to behave as gyroscopes and tops. Nuclei react to forces in the microscopic world just as objects with angular momentum respond to forces in the macroscopic world. Protons and other subatomic particles are assumed to rotate about their axes and are described as having "spin."

Precession results from the interaction of forces with a rotating object. These forces and the resultant motions are diagramed in Margin Figure 23-4 for a gyroscope, where the force is gravity, and for a proton, where the force is a magnetic field.

The frequency f of precession of a proton in units of megahertz (10^6 cycles or rotations per second) depends upon its gyromagnetic ratio γ (in megahertz per tesla) and the strength B (in tesla, T) of the static magnetic field. This relationship is described by the Larmor equation.

$$f = \gamma B \tag{23-1}$$

Gyromagnetic ratios for a number of nuclei are given in Table 23-1.

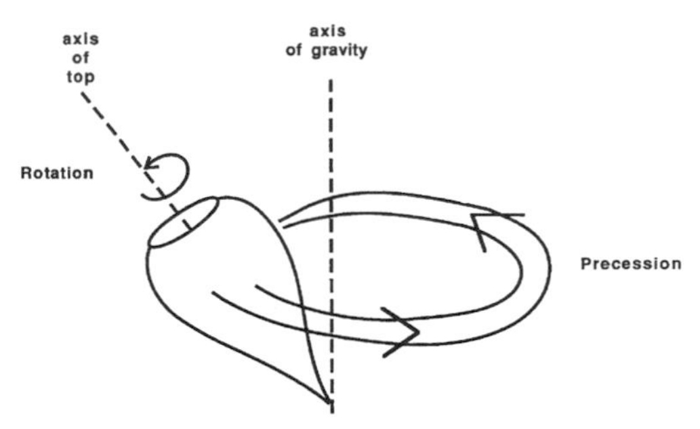

MARGIN FIGURE 23-2
Motions of a spinning top. Rotation or spin of the top about its own axis is first-order motion. Precession of the top about the vertical axis (axis of gravity) is second-order motion.

Example 23-1

Find the resonance frequency for protons in a 2-tesla magnetic field.

$$\begin{aligned} f &= \gamma B \\ &= \left(42.6 \frac{\text{MHz}}{\text{T}}\right)(2.0\text{T}) \\ &= 85.2 \text{ MHz} \end{aligned} \tag{23-1}$$

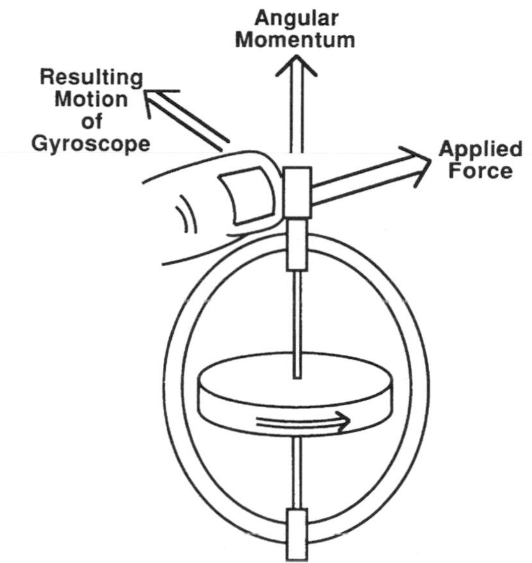

MARGIN FIGURE 23-3
Spinning object demonstrates a property known as angular momentum, shown as an arrow along the axis of the spinning object. When force is applied to an object having angular momentum, the resulting motion is at right angles to the force.

INTERACTION OF NUCLEI WITH A RADIO FREQUENCY WAVE: NUTATION

There is a "third-order" property of an object that is rotating (first-order property) and precessing (second-order property). This property, called nutation, is the result of forces that rotate *with* the precession of the object.

When force is applied to an object with angular momentum, the object tends to move at right angles to the force. If we try to speed up the precession of a gyroscope by using a finger to push it in a circle (Margin Figure 23-5), we will not affect the

TABLE 23-1 Magnetic Resonance Properties of Selected Nuclei

Nucleus	*Natural Abundance (%)*	*Gyromagnetic Ratio[a] (MHz/T)*	*Sensitivity[b]*
^{1}H	99.98	42.58	100.00
^{13}C	1.11	10.71	1.59
^{19}F	100.00	40.05	83.30
^{23}Na	100.00	11.26	9.25
^{31}P	100.00	17.23	6.63
^{39}K	93.10	1.99	0.05

[a] The gyromagnetic ratio listed here is the same value listed in some texts as $\gamma/2\pi$, where γ is the gyromagnetic ratio in units of radians per tesla. The convention in this text is to assume that γ is given in units of megahertz per tesla, with the factor of 2π already included.

[b] Relative sensitivity (signal strength as compared with hydrogen) for the same number of nuclei at constant field strength as a percentage of sensitivity for ^{1}H.

Source: Data from James, T. L. *Nuclear Magnetic Resonance in Biochemistry.* New York, Academic Press, 1975.

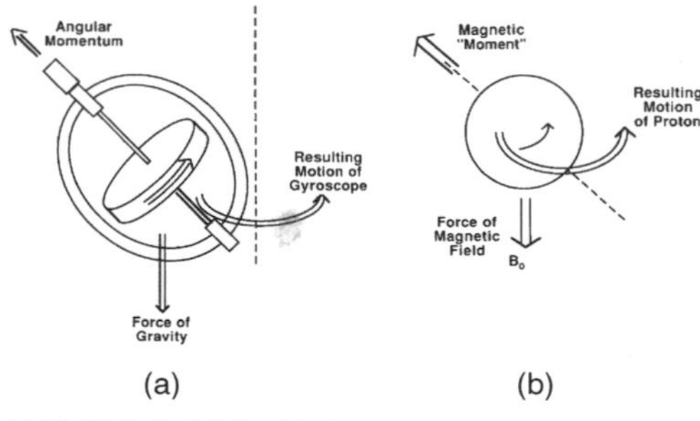

MARGIN FIGURE 23-4
Diagram of forces acting to cause precession. Angular momentum and gravity interact to cause precession of a gyroscope. Magnetic moment and a magnetic field interact to cause precession of a proton.

The frequency of precession is known as the *Larmor frequency.*

The emission of radio signals from nuclei was demonstrated independently by Bloch[4] and Purcell[5] in 1945. By the late 1940s, several groups had established that the electrical charges of electrons involved in chemical bonds caused small, precise shifts in the frequency of these signals. The frequency shifts were found to be characteristic for individual molecules. This finding identified nuclear magnetic resonance (NMR) as a method for identifying minute amounts of chemicals in unknown samples and led to the development of NMR in the 1950s as a standard tool for research in analytical chemistry.[6]

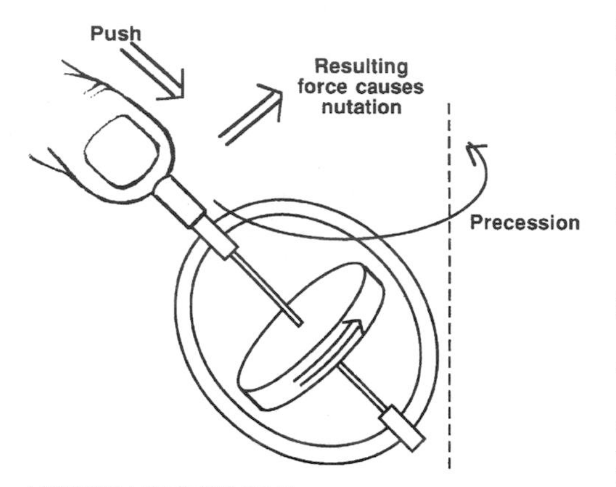

MARGIN FIGURE 23-5
Attemption to push a gyroscope in the direction of precession causes the gyroscope to change its angle of precession. Change in the angle of rotation is referred to as nutation.

Rotation: Cyclical motion of an object about an axis.
Precession: Compound motion of a rotating object about an axis other than its axis of rotation. The angle between the axis of rotation and the axis of precession is known as the *angle of precession*.
Nutation: Change in the angle of precession.

precessional speed. Instead, the *angle* of precession will change. That is, if we push in the direction of precession, the gyroscope precesses at a greater angle until it finally lies flat on the table. This result is best explained by a diagram that is drawn as if we were precessing along with the object. Such a "rotating-frame-of-reference" diagram is shown in Margin Figure 23-6. This diagram depicts a precessing object responding to a force by moving at a right angle to the force, thereby changing its *angle* of precession. This change in angle, called nutation, is a third-order circular motion (after rotation and precession).

In magnetic resonance of protons, the force that causes nutation is provided by a second magnetic field that varies with time. Colloquially, we could describe this field as follows: Each time the proton's magnetic moment "comes around" during precession, the second magnetic field is "switched on" (Margin Figure 23-7). As the magnetic moment rotates away, the magnetic field is "switched off." More precisely, the "ons" and "offs" are the peaks and valleys of a continuously varying magnetic field.

Example 23-1 shows that the Larmor equation [Eq. (23-1)] may be used to predict the frequency of precession of a proton in a magnetic field. The range of static field strengths employed in clinical MRI, 0.1 to 3.0 tesla, corresponds to precessional frequencies of 4.3 to 129 MHz for hydrogen nuclei. The time-varying magnetic field of an electromagnetic wave with a frequency in the megahertz range is a suitable source for inducing nutation. When this wave has a frequency that matches the precession of protons in a particular magnetic field, it is said to be in "resonance." This is the origin of the term *magnetic resonance*. Appropriate frequencies are in the FM radio portion of the electromagnetic spectrum. Thus, we refer to the use of RF pulses in MR.

INDUCTION OF A MAGNETIC RESONANCE SIGNAL IN A COIL

A changing magnetic field can induce a current in a loop of conducting wire. This principle is known as Faraday's law, or the law of electromagnetic induction. A proton has a magnetic moment and therefore acts as a small magnet. Protons that precess so that their magnetic fields intersect the plane of a nearby coil will induce an electrical current in the coil. This current is the MR "signal" induced in the receiver coil (Margin Figure 23-8).

The MR signal is greatest for precession of magnetic moments in a plane perpendicular to the plane of the receiver coil. The imaginary plane perpendicular to the receiver coil is known as the "transverse" or "*xy*" plane. If the magnetic moments are not precessing entirely within the transverse plane, a signal of reduced magnitude (i.e., a "weaker" signal) is induced in the coil. If the magnetic moments are perpendicular to the transverse plane (i.e., parallel to the static magnetic field), then no signal will be induced in the coil. In summary, it is the "component" of the magnetization in the transverse plane that is responsible for the MR signal.

The number of magnetic moments precessing near the coil also influences the strength of the MR signal. With other factors constant, the strength increases linearly with the number of magnetic moments. The magnetic moments that are used for MRI at the present time are those of protons. Signals from soft tissue in volumes on the order of a cubic millimeter can be resolved in MRI.

Example 23-2

Calculate the number of hydrogen nuclei in a cubic millimeter of water.

From the mass of a cubic millimeter of water, the number of water molecules and the number of protons can be computed.

The density of water is $1 g/cm^3$.

A cubic millimeter of water has a mass of

$$(1\ \text{mm}^3)\left(10^{-3}\frac{\text{cm}^3}{\text{mm}^3}\right)\left(1\frac{\text{g}}{\text{cm}^3}\right) = 0.001\ \text{g}$$

A gram molecular weight, or mole, of water is the mass in grams of pure water that contains Avogadro's number, 6.02×10^{23}, of molecules. The molecular weight of water is 18 g/mol. The number of water molecules in 0.001 g is

$$\frac{(0.001\text{ g})}{(18\text{ g/mol})}\left(6.02 \times 10^{23}\,\frac{\text{molecules}}{\text{mol}}\right) = 3.34 \times 10^{19}\text{ molecules.}$$

However, each molecule contains two hydrogen nuclei. Therefore, the total number of hydrogen nuclei in a cubic millimeter of water is

$$\left(2\frac{\text{H nuclei}}{\text{molecule}}\right)(3.34 \times 10^{19}\text{ molecule}) = 6.68 \times 10^{19}\text{ H nuclei}$$

Note that we do not count the eight protons in each oxygen atom in this calculation because those nuclei do not have the same Larmor frequency as hydrogen.

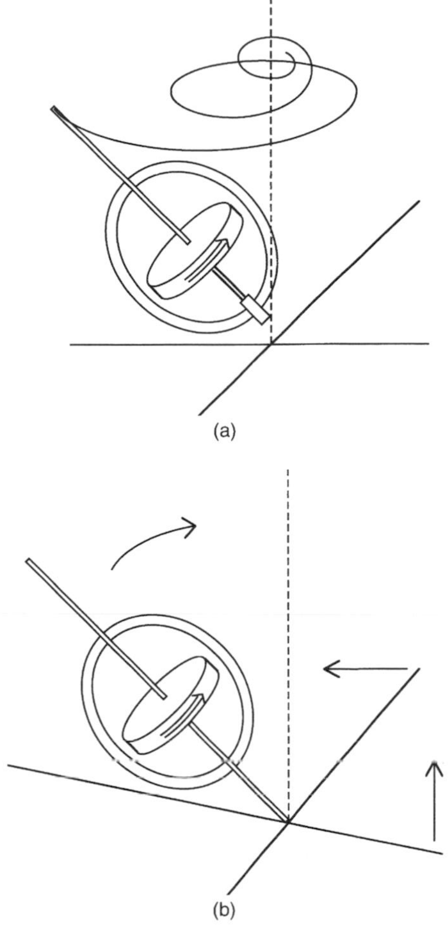

MARGIN FIGURE 23-6
Nutation of a gyroscope shown in a stationary frame of reference (**A**) and a rotating frame of reference (**B**). Rotating frame of reference shows motion of the gyroscope as if the observer were precessing along with the gyroscope.

The maximum MR signal that may be induced in a receiver coil depends linearly upon the number of protons available. Hence doubling either the volume or the proton density (protons per cubic millimeter) of the sample doubles the MR signal.

QUANTUM MECHANICAL INTERPRETATION

The preceding discussion follows the "classical interpretation" of the behavior of magnetic moments in the presence of magnetic fields and radio waves. This interpretation is useful for teaching, but it is not entirely correct. In the region of the atom, where distances are small and the "discreteness" of energy levels and angular momentum is apparent, it is not possible to predict the exact behavior of individual atoms with certainty.[7] Nuclei possess only certain discrete values of quantities such as angular momentum and energy; they cannot possess intermediate values of these variables. Properties such as angular momentum and energy are said to be quantized; that is, they are available only in specific values that are "allowed" by empirical rules. Therefore, the picture of smooth nutation of individual nuclei as if they were tiny tops gradually falling onto a tabletop is not strictly true.

Another important distinction must be made between the quantum mechanical and the classic interpretations of MR. In quantum mechanics, instead of radio "waves," we refer to photons or packets of electromagnetic energy. For example, in the quantum mechanical interpretation of a system of spins (such as protons) in a magnetic field, there are two possible "energy states," either parallel or antiparallel to the magnetic field. The protons having magnetic moments aligned with the magnetic field have slightly less energy (are in a lower energy state) than do protons with magnetic moments opposing the magnetic field (Margin Figure 23-9). A photon with an energy equal to the energy difference between the two states can promote, or "flip," protons from the lower to the higher energy state. In the classic model, the frequency of the radio wave that interacts in resonance is given by the Larmor equation [(Eq. (23-1)]. In the quantum mechanical model, the energy of the photon is given by the equation

$$E = hf \tag{23-2}$$

where E is the energy of the photon; h is Planck's constant, 4.14×10^{-15} eV-sec; and f is the Larmor frequency in Hertz, or cycles per second. This equation permits a calculation of the photon energy required to cause transitions between low and high energy states.

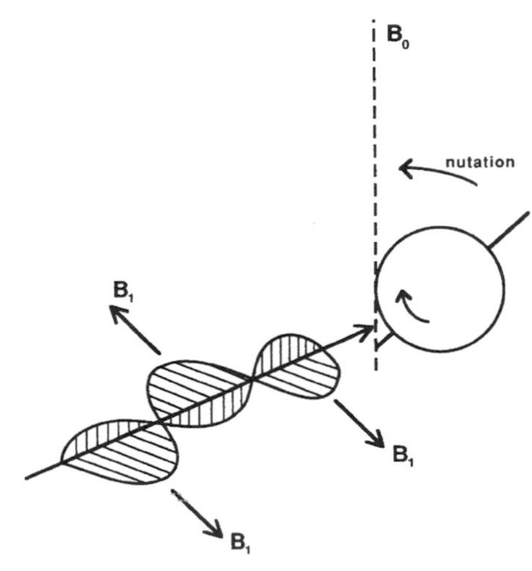

MARGIN FIGURE 23-7
Nutation of a precessing proton caused by application of an electromagnetic wave having intermittent magnetic field components (B_1). Motion is shown in the rotation frame of reference.

Example 23-3

Find the energy difference between low and high energy states of protons in a 1-tesla magnetic field.

There are a number of magnetic resonance phenomena. Magnetic resonance effects involving the spin and orbital magnetic moments of an atom's electrons are actually thousands of times more powerful than the nuclear magnetic resonance effect used in magnetic resonance imaging. To date, no application in imaging the human body has been found for electron resonance phenomena. The effect described in this chapter is more accurately known as nuclear paramagnetic resonance. If an accurate physical description was the only consideration, the imaging described in this chapter should be abbreviated NPMRI.

First, the precessional frequency of the protons is computed from the Larmor equation:

$$\begin{aligned} f &= \gamma B \\ &= \left(42.6 \frac{\text{MHz}}{\text{T}}\right)(1\text{T}) \\ &= 42.6 \text{ MHz} = 42.6 \times 10^6 \text{ sec}^{-1} \end{aligned} \tag{23-1}$$

Next the energy difference is computed from the Larmor frequency:

$$\begin{aligned} E &= hf \\ &= 4.14 \times 10^{-15} \text{ eV - sec} \left(42.6 \times 10^6 \text{ sec}^{-1}\right) \\ &= 1.76 \times 10^{-7} \text{ eV} \end{aligned} \tag{23-2}$$

Thus a photon energy of 1.76×10^{-7} eV is required. As expected, this photon is part of the RF portion of the electromagnetic spectrum.

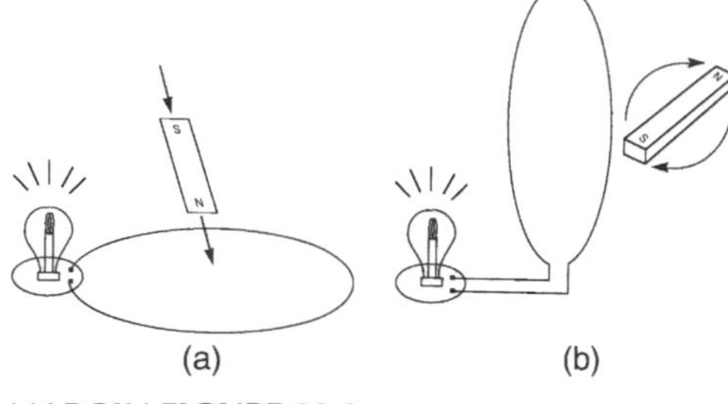

MARGIN FIGURE 23-8
A: Magnet that is in motion near a loop of conducting wire induces an electrical current in the wire, causing the light bulb to glow. **B:** Magnet that rotates near a loop of conducting wire induces an alternating current in the wire, causing the light bulb to glow (or flash on and off if the rate of rotation of the magnet is low).

BULK MAGNETIZATION

The classic and quantum mechanical interpretations of MR phenomena can be reconciled when a sample of macroscopic size consisting of millions of atoms is considered. In this case, the classical laws govern the composite, or "bulk," magnetization of the sample. It is a cornerstone principle of physics that as the quantum mechanical equations governing the behavior of individual atoms are averaged over "large" numbers of atoms, the classical equations of motion emerge. Thus, the statements about precession, nutation, and induction of signals in a coil are true of bulk magnetization. When a radio wave is applied to a sample at the resonance frequency, the bulk magnetization precesses at the Larmor frequency (Margin Figure 23-10). The angle of precession (nutation angle) is determined by the intensity of the radio wave and the amount of time over which it is applied.

When considering a collection of spins, as in a macroscopic sample, an important issue to include is whether the individual magnetic moments tend to "work together" (i.e., point in the same direction) or cancel each other (i.e., point in random directions). Before a radio wave is applied, the nuclei are precessing at nearly the same frequency, but their transverse components tend to cancel each other because they are rotating "out of phase." Hence the bulk magnetization is zero in the transverse plane. When the radio wave (RF pulse) is applied, the spins are brought into synchrony, and the individual nuclear magnetic moments reinforce each other to produce a strong bulk magnetic moment. The ability of an RF pulse to bring protons into phase is a consequence of applying a force that acts in resonance with a physical system. As an example, consider using a long board to simultaneously push several people who are on swings.* Assume that you will push at the resonance frequency. If the people are already swinging on their own in random phases as you approach them with the board to apply "pulses" to the system, you will bump some at just the right part of the swing cycle to give them an effective push. Others will be missed or will bump against the board at the wrong part of the cycle. If you persist over a number of cycles, those who are out of phase will be forced into phase, and finally the force will be applied to all components of the system in resonance. This technique is analogous to what happens when an RF pulse is applied to a system of precessing nuclear spins. As the nuclei begin to rotate in phase, the transverse component of the bulk magnetization becomes apparent.

Werner Heisenberg, in the early 1970s, reflected on discussions among physicists in the 1920s on the subject of using terms such as "energy," "magnetization," and "reality" when describing quantum systems:

"It was found that if we wanted to adapt the language to the quantum theoretical mathematical scheme, we would have to change even our Aristotelian logic. That is so disagreeable that nobody wants to do it: ... We have learned that language is a very dangerous instrument to use, and this fact will certainly have its repercussions in other fields, but this is a very long process which will last through many decades I should say."

(From Buckley, P., Peat, F. D., Peat, D. F. (eds.), *Glimpsing Reality: Ideas in Physics and the Link to Biology*, revised edition (March 1996), University of Toronto Press.

*In this physical system, resonance frequency is determined by the length of the ropes supporting the swings and the mass of the people.

TABLE 23-2 Relaxation Times (Mean ± SD) in Milliseconds for Various Tissues at 1 Tesla Static Magnetic Field Strength (42.6 MHz)

Tissue	*T1*			*T2*		
Skeletal muscle	732	±	132	47	±	13
Liver	423	±	93	43	±	14
Kidney[a]	589	±	159	58	±	24
Adipose	241	±	68	84	±	36
Brain						
Gray matter	813	±	138	101	±	13
White matter	683	±	116	92	±	22

[a] Average of medulla and cortex.

Source: Data from Bottomley, P. A., Foster, T. H., Argersinger, R. E., et al. *Med. Phys.* 1984; **11:** 524–547.

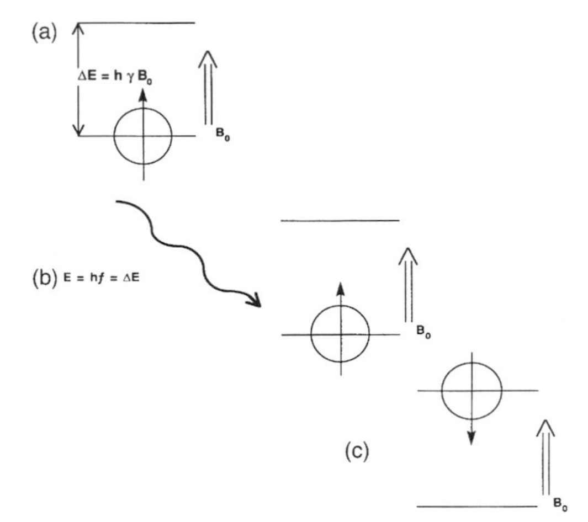

MARGIN FIGURE 23-9
In the quantum mechanical model of magnetic resonance, protons parallel to (aligned with) the magnetic field B_0 exist in a lower energy state than protons that are antiparallel. A photon having an energy equal to the difference in energy between the two states can change a proton in the lower energy state to the higher energy state.

RELAXATION PROCESSES: T1 AND T2

We have seen that when an RF pulse is applied to a sample, the bulk magnetization may be nutated into the *xy* plane and induce an MR signal in a receiver coil positioned perpendicular to the *xy* plane. When the radio wave is switched off, the signal decays away. This decay is the result of the return of protons to the state that existed before the radio wave was applied. This return is termed *relaxation* of the protons. There are two basic relaxation processes at work in the sample. Both processes account for the observed decay of the MR signal.

One relaxation process involves a return of the protons to their original alignment with the static magnetic field. This process, called *longitudinal* or *spin-lattice relaxation,* is characterized by a time constant T1 (Margin Figure 23-11). The term spin-lattice refers to the interaction of the protons (spins) with their surroundings (the "lattice" or network of other spins). This interaction causes a net release of energy to the surroundings as the protons return to the lower energy state of alignment.

The other relaxation process is a loss of synchrony of precession among the protons. Before a radio wave is applied, the precessional orientation of the protons is random.The application of a radio wave brings the protons into synchronous precession, or "in phase." When the radio wave is switched off, the protons begin to interact with their neighbors and give up energy in random collisions. In so doing, they revert to a state of random phase. As the protons revert to random orientation, the bulk signal decreases because the magnetic moments tend to cancel each other. This process is called *transverse* or *spin–spin relaxation* and is characterized by a time constant T2 (Margin Figure 23-11).

In a patient undergoing MRI, both longitudinal and transverse relaxation processes occur at the same time. The transverse (T2) relaxation time is always shorter than the longitudinal (T1) relaxation time. That is, magnetic moments dephase faster than they move into alignment with the static magnetic field (Margin Figure 23-11). For typical biological materials, T1 may be on the order of several hundred milliseconds while T2 is a few tens of milliseconds (Table 23-2).[16]

Relaxation (decay) of the MR signal is characterized by exponential expressions analogous to those used to describe radioactive decay, absorption of photons, and growth of cells in culture (see Appendix I). For longitudinal relaxation, the expression

$$S = S_0 e^{-t/T1} \quad (23\text{-}3)$$

reveals that the MR signal S decreases exponentially in time from the signal S_0 that was present immediately following an RF pulse. The value of S_0 is influenced by factors such as the number of protons in the sample, length of time that the radio wave was applied to the sample, sensitivity of the receiver coil, and overall sensitivity of the electronics.

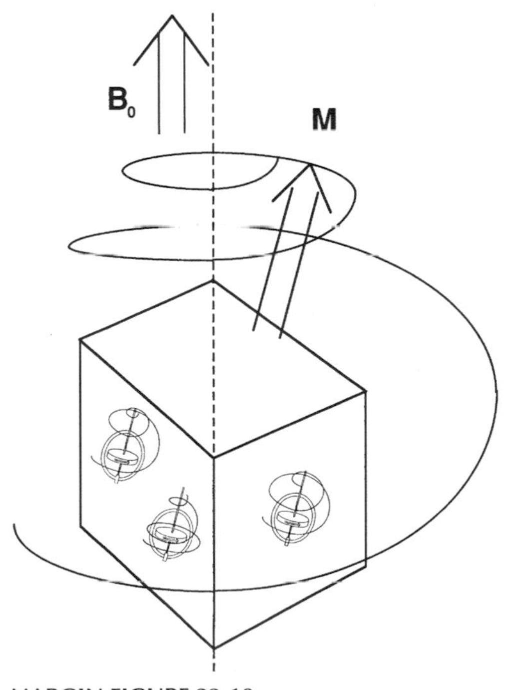

MARGIN FIGURE 23-10
Bulk magnetization of a sample of spins represents the net result of the precession and nutation of the individual spins within a sample.

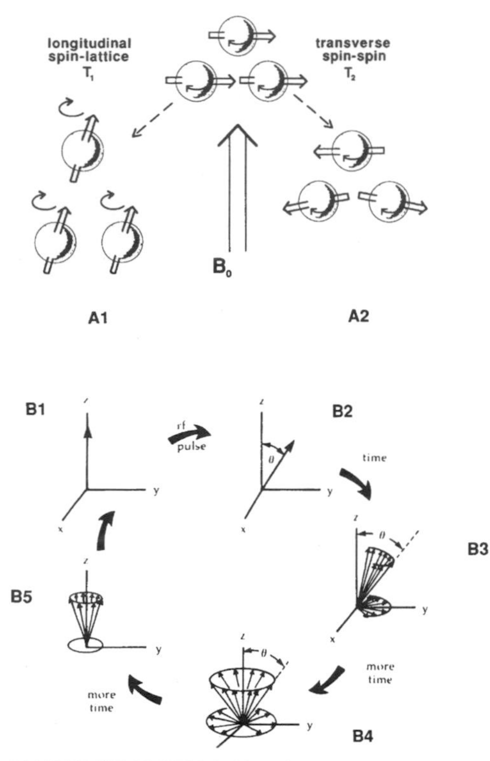

MARGIN FIGURE 23-11
A: Two relaxation processes in a sample of nuclear spins. There are several terms for each process. **A1:** Longitudinal relaxation occurs as the spins return to alignment with the static magnetic field, B_0. **A2:** Transverse relaxation occurs as the spins precess "out of phase." **B:** spin–spin relaxation diagramed in the rotating frame of reference. **B1:** Magnetic moment of the sample is aligned with the magnetic field. **B2:** Immediately after an RF pulse, the magnetic moment of the sample can be represented by a single vector. **B3:** As the magnetization vector begins to break up or dephase as a result of localized nonuniformities in the applied field, components of the vector begin to fan out in the *xy* plane. **B4:** When there are an equal number of components in all directions in the *xy* plane, the components cancel one another and the MR signal disappears. **B5:** As time passes, the cone representing the precessing but dephased magnetic moment continues to narrow because of spin–lattice relaxation. **B1:** Finally the magnetic moment once again is realigned with the applied field. (From Hendee, W. R., and Morgan, C. R. *West. J. Med.* 1984; **141**:491–500. Used with permission.)

Example 23-4

If a material has a longitudinal relaxation time constant T1 = 200 msec, what will the MR signal *S* be 200 ms after an RF pulse has flipped (nutated) the bulk magnetization into the transverse plane?

$$S = S_0 e^{-t/T1} \tag{23-3}$$

for t = T1 = 200 msec,

$$\begin{aligned} S &= S_0 e^{-200/200} \\ &= S_0 e^{-1} \\ &= S_0(0.37) \end{aligned}$$

This example illustrates that the longitudinal time constant T1 may be defined as the time required for the signal to decrease to 37% of its original value by the process of longitudinal (spin–lattice) relaxation.

For transverse relaxation, governed by the time constant T2, we obtain

$$S = S_0 e^{-t/T2} \tag{23-4}$$

where T2 is defined as the time required for the MR signal to decrease to 37% of its original value by the process of transverse (spin–spin) relaxation. Because T1 and T2 are independent processes both acting to reduce the signal S_0, Eq. (23-3) and (23-4) may be combined into

$$S = S_0 e^{-t/T1} e^{-t/T2} \tag{23-5}$$

Example 23-5

A sample of liver tissue has a longitudinal relaxation time constant T1 of 500 msec and a transverse relaxation time constant T2 of 40 ms. What will the MR signal be 40 msec after an RF pulse? 500 msec after?

$$S = S_0 e^{-t/T1} e^{-t/T2} \tag{23-5}$$

At t = 40 msec,

$$\begin{aligned} S &= S_0 e^{-40/500} e^{-40/40} \\ &= S_0(0.92)(0.37) \\ &= S_0(0.34) \end{aligned}$$

At 40 msec the signal is 34% of its value immediately following an RF pulse.

At t = 500 msec,

$$\begin{aligned} S &= S_0 e^{-500/500} e^{-500/40} \\ &= S_0(0.37)(3.7 \times 10^{-6}) \\ &= S_0(1.38 \times 10^{-6}) \end{aligned}$$

At 500 msec the signal is little more than a millionth of its value immediately after the RF pulse.

Note that during this time, both longitudinal and transverse relaxation processes act to reduce the magnitude of the MR signal. However, the effect of T2 dominates, particularly at longer time intervals. If T1 is long compared with T2 as in this example (and as is the case for many tissues of interest in MRI), the effect of T1 is overshadowed by the effect of T2. However, if T1 is about the same as T2, as for nonviscous liquids, then both relaxation processes are important influences on the MR signal.

In the next chapter; special RF pulsing techniques are introduced to vary the influence of T1 and T2 upon the MR signal that is finally recorded. With special pulsing techniques the MR signal may actually appear to increase in strength due to T1 relaxation. This is due to the fact that another RF pulse is applied just before the signal is received by the coil. Therefore the more relaxation that has occurred, the stronger the signal will be when flipped (nutated) by 90 degrees. Therefore, the RF signal that is measured after a 90 degree spin flip increases as the time between *successive* applied pulses is increased.

The influence of relaxation parameters on the MR signal is one of the central principles of tissue contrast in MRI. Any measurement technique that exploits differences in MR signal among volume elements (voxels) of the sample enhances contrast in the image. Just as contrast in computed tomography is influenced by factors such as electron density (no. of electrons per gram) and physical density (grams per cubic centimeter), contrast in MRI is influenced by differences in the relaxation parameters T1 and T2, and in the nuclear spin density, N(H) (number of spins per cubic centimeter). Measurement techniques that exploit differences in T1, T2, and N(H) among tissues are described in the next chapter.

The mathematical equations that describe relaxation of the MR signal have the same general form as the equations that describe attenuation of photons from an x-ray beam as it passes through a material, and the decay of activity of a radioactive sample over time. In these processes, a "constant" characterizes the behavior of photons of specific energies in specific materials (attenuation coefficient, μ) or of certain types of atoms (decay constant, λ). In magnetic resonance, the constants, 1/T1 and 1/T2, characterize relaxation phenomena of specific nuclides.

In 1971, Damadian used NMR to observe differences in magnetic properties between normal and cancerous tissues in rats.[9] These studies were extended to human tissue samples in 1973[10] and stimulated considerable interest in the use of NMR as a biochemical probe. Lauterbur proposed the first imaging technique in 1973 and named it zeugmatography, from the Greek *zeugma* for "that which joins together." In that year, Lauterbur published the first MR images.[11] In 1974, an image of a mouse was obtained by Mallard.[12] The first human anatomic images were published by several groups in 1977 and included a wrist by Hinshaw et al.,[13] a finger and various tissue phantoms by Mansfield and Maudsley,[14] and an in vivo scan of the human thorax by Damadian and colleagues.[15] The first MR image of the brain was obtained in 1980 at Nottingham by Holland et al.[16]

RELAXATION TIMES (T1 AND T2) FOR BIOLOGIC MATERIALS

Biologic materials may be characterized to some degree by their T1 and T2 values (Table 23-2). However, there are several difficulties. For example, the exact values are not unique to composite substances. A bulk sample may contain materials having a range of values for T1 and T2. The temperature of a sample also influences relaxation.

The rate of interaction among spins and their surroundings determines the rate at which the spins in a higher energy state dissipate energy to their surroundings. Molecules in a sample are in constant motion and rotate with frequencies that range from zero up to a maximum value determined by the temperature. Any magnetic moment is influenced by the rotation of nearby magnetic moments. Spins may change energy states from parallel to antiparallel alignment (T1), and they may also move out of phase with other spins (T2) as a result of these interactions. A number of other interactions, the discussion of which is beyond the scope of this text, also play a role.[7,17]

Longitudinal relaxation T1 has been shown to vary with magnetic field strength. Since magnetic field strength, along with the gyromagnetic ratio, determines the resonance frequency for a nucleus, T1 may be expressed as a function of resonance frequency. For a wide range of tissues, T1 for hydrogen can been approximated[8] as

$$\mathrm{T1} = \alpha f^{\beta} \tag{23-6}$$

where T1 is in milliseconds, f is the resonance frequency in hertz, and the parameters α and β vary with tissue type over a range of $\alpha = 0.5$ to 10 and $\beta = 0.2$ to 0.4. The transverse relaxation time T2 has been found to be independent of resonance frequency (magnetic field strength).

The properties of a material that give rise to observed values of T1 and T2 are complex. However, a theory of the contribution of molecular motion to relaxation was suggested by Bloembergen et al. in 1948.[17] An important characteristic of this theory is the correlation time of molecules. Molecules in liquids and semisolids are free to rotate. The rate at which they rotate is the rotational frequency. The inverse of the rotational frequency is the rotational period, also called the correlation time T_c (Margin Figure 23-12). Molecules that rotate more slowly have a greater probability of interacting with their neighbors. In such materials, relaxation can occur more rapidly, and the relaxation time constant is smaller. For T2 relaxation, the prediction is straightforward; a longer correlation time implies a faster transverse relaxation (shorter T2). For T1, however, there is a resonance phenomenon to consider. Molecules rotating at Larmor frequencies maximize their rate of interaction with their neighbors, just as the interaction of radio waves with magnetic moments is maximized at the Larmor

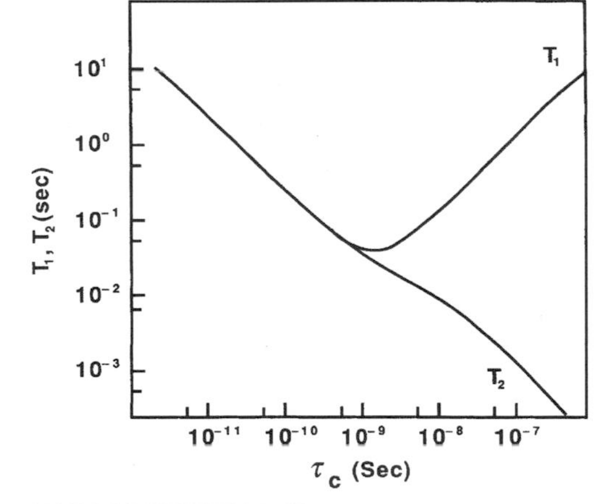

MARGIN FIGURE 23-12
Relaxation times T_1 and T_2 as a function of correlation time. Data for static magnetic field strength 1.4 tesla (proton resonance frequency, 60 MHz). (Redrawn from Martin, L., Delpuich, J. J., and Martin, G. J. *Practical NMR Spectroscopy*. London, Heydon, 1980.)

frequency. Thus a material having a correlation time that happens to coincide with the inverse of the Larmor frequency being used for MRI yields the minimum T1. Materials having smaller or larger values of correlation time have a larger T1. Furthermore, T1 depends on the Larmor frequency and increases as the magnetic field strength is increased. Thus, materials with correlation times near the resonance value for one field strength would have a short T1 at that field strength only. T2 is relatively unaffected by changes in resonance frequency and is therefore independent of field strength.

Correlation time analysis yields some straightforward predictions for T1 and T2 in biologic systems. Thermal motion is relatively fast in pure liquids such as water. Therefore, the correlation time is usually below the resonance value, and T1 is greater in water than in more viscous materials. Similarly, T1 in solids or anhydrous proteins is generally longer because the thermal motion is decreased and the correlation time is too long for resonance interactions. Because of the inverse relationship between T2 and correlation time, liquids tend to have longer T2s, and solids tend to have shorter T2s.

PROBLEMS

23-1. Show that the range of static magnetic field strengths 0.1 to 2.0 tesla corresponds to the resonance frequencies 4.3 to 86 MHz for hydrogen. By using data from Table 23-1, find the corresponding resonance frequencies for ^{19}F and ^{13}C.

*23-2 Find the number of ^{13}C nuclei in 1 mg of pure ^{13}C. By using the data in Table 23-1, determine the number of milligrams of ^{13}C needed to give the same signal as a cubic millimeter of water (see Example 23-2).

*23-3 A tissue T1 is characterized in Eq. (23-6) by $\alpha = 0.5$ and $\beta = 0.4$. What magnetic field strength is required to double the T1 for hydrogen that is obtained at a magnetic field strength of 0.35 tesla?

*23-4 For a material having a T2 of 80 msec 320 msec after a 90° pulse, what is the relative magnitude of the MR signal compared to the maximum signal.

*For those problems marked with an asterisk, answers are provided on p. 493.

SUMMARY

- Nuclei with magnetic moments will precess in a static magnetic field. This precession may be monitored by the signal it induces in a receiver coil.
- Larmor equation: $f = \gamma B$.
- Quantum mechanical equation: $E = hf$.
- T1 is also known as longitudinal or "spin–lattice" relaxation.
- T2 is also known as transverse or "spin–spin relaxation.
- The MR signal decreases in time after all RF pulsing has ceased, due to T1 and T2 relaxation.
- Relaxation properties of materials are determined by their correlation times, the period of rotation of the molecules.

REFERENCES

1. Frisch, R., and Stern, O. Über die magnetische ablenkung von Wasserstoff-Molekulen und das magnetische moment das protons I. *Z. Phys.* 1933; **85**:4–16.
2. Esterman, I., and Stern, O. Über die magnetische ablenkung von Wasserstoff-Molekulen und das magnetische moment das protons II. *Z. Phys.* 1933; **85**:17–24.
3. Rabi, I. I., et al. Molecular beam resonance method for measuring nuclear magnetic moments. *Physiol. Rev.* 1939; **55**:526–535.
4. Block, F. The principle of nuclear induction, in *Nobel Lectures in Physics: 1946–1962*. New York, Elsevier, 1964.
5. Purcell, E. M. Research in nuclear magnetism, in *Nobel Lectures in Physics: 1946–1962*. New York, Elsevier, 1964.
6. Rogers, E., Packard, E. M., and Shoolery, J. N. *The Origins of NMR Spectroscopy*. Palo Alto, CA, Varian Associates, 1963.
7. Baym, G. *Lectures on Quantum Mechanics*. Reading MA, W. A. Benjamin, 1974. This text is for the graduate physics student. A popular account of the principles of quantum mechanics may be found in Gamow, G. *Biography of Physics*. New York, Harper & Row, 1961.
8. Bottomley, P. A., Foster, T. H., Argersinger, R. E., et al. A review of normal tissue hydrogen NMR relaxation times and relaxation mechanisms from

1–100 MHz: Dependence on tissue type, NMR frequency, temperature, species, excision, and age. *Med. Phys.* 1984; **11**:425–448.
9. Damadian, R. Tumor detection by nuclear magnetic resonance. *Science* 1971; **171**:1151–1153.
10. Damadian, R., et al. Nuclear Magnetic resonance as a new tool in cancer research: Human tumors by NMR. *Ann. N. Y. Acad. Sci.* 1973; **222**:1048–1076.
11. Lauterbur, P. C. Image formation by induced local interactions: Examples employing nuclear magnetic resonance. *Nature* 1973; **242**:190–191.
12. Mallard, J., et al. Imaging by nuclear magnetic resonance and its bio-medical implications. *J. Biomed. Eng.* 1979; **1**:153–160.
13. Hinshaw, W. S., Bottomley, P. A., and Holland, G. N. Radiographic thin-section image of the human wrist by nuclear magnetic resonance. *Nature* 1977; **270**:722–723.
14. Mansfield, P., and Maudsley, A. A. Medical imaging by NMR. *Br. J. Radiol.* 1977; **50**:188–194.
15. Damadian, R., Goldsmith, M., and Minkoff, L. NMR in cancer: XVI. FONAR image of the live human body. *Physiol. Chem. Phys.* 1977; **9**:97–100.
16. Holland, G. N., Moore, W. S., and Hawkes, R. C. Nuclear magnetic resonance tomography of the brain. *J. Comput. Assist. Tomogr.* 1980; **4**:1–3.
17. Bloembergen, N., Purcell, E. M., and Pound, R. V. Relaxation effects in nuclear magnetic resonance absorption. *Physiol. Rev.* 1948; **73**:679–712.

CHAPTER

24

MAGNETIC RESONANCE IMAGING AND SPECTROSCOPY

OBJECTIVES

After completing this chapter, the reader should be able to:

- Explain the phenomenon of free induction decay.
- State the rationale for use of the spin-echo technique in imaging.
- Define T2*.
- Explain the pulsing and signal acquisition scheme used in the following pulse sequences:
  - Spin-echo
  - Carr–Purcell–Meiboom–Gill
  - Inversion recovery
  - Gradient-echo
- Chart the order of occurrence of RF pulses, gradient magnetic fields, and signal acquisition in spatially encoded spin-echo imaging.
- Explain how the two-dimensional Fourier transform (2DFT) method is used to construct the MR image.
- Calculate the time required to obtain an image using the spin-echo technique.
- List some motion suppression techniques.
- Give the physical principle that underlies the use of contrast agents in MR.
- Explain the physiological basis of functional MRI (fMRI).
- List reasons for tissue contrast, differences in MR signal among voxels, and how these differences are highlighted through the choice of pulse sequence parameters.
- Explain the chemical basis of MR spectroscopy and give some examples of nuclei that are studied.

OVERVIEW: MAGNETIC RESONANCE AS A PROBE OF THE BODY

The magnetic properties of nuclei described in Chapter 2 have significant applications in medical imaging and biochemical analysis. These applications are possible because the relaxation properties and the resonance frequency for a nucleus depend upon its environment. Factors such as the presence of chemical bonds, paramagnetic ions, and the rate of flow of fluids influence the magnetic resonance (MR) signal. Therefore, different regions of a biological sample produce different MR signals. While the reasons for these differences are often so complex that interpretation is difficult, their existence yields an intrinsic subject contrast between tissues. If the values of the MR signals can be mapped according to their spatial locations and can be encoded as brightness (gray scale) on a monitor, then an image may be produced. This is the method of magnetic resonance imaging (MRI). A plot of the magnitude of the MR signal as a function of frequency yields an MR spectrum. This latter approach is termed *magnetic resonance spectroscopy* (MRS).

The symbols B and H are used to indicate the presence of a magnetic field. While, for many purposes, they are used interchangeably, there are some differences that are important in the field of electromagnetism. H is the "magnetic field intensity." It may be thought of as an externally applied "magnetizing force" much as a "gravitational" force is said to act upon objects in gravitational field theory. B is the "magnetic induction field." It is the magnetic field that is actually induced in matter when that particular matter is placed in a given magnetic field intensity (H). In MR, we usually refer to the static (constant) magnetic field at the center of the MR scanner as the B_0 field (pronounced "bee-zero").

PULSE SEQUENCES

Both imaging and spectroscopy require the application of radiofrequency (RF) pulses to a sample so that coherent signals may be obtained from a receiver coil placed in close proximity to the sample. These pulses are applied in specific sequences to produce MR signals that yield information about the sample. All "pulse sequences" have in common the use of RF pulses to tip ("nutate") the bulk magnetization vector at some angle with respect to the static magnetic field.

Free Induction Decay

The simplest pulse sequence is free induction decay (FID) (Margin Figure 24-1). In FID, the magnetization is tipped (nutated) 90 degrees to orient the plane of precession of the protons perpendicular to the receiver coil. Such a pulse also has the effect of establishing "phase coherence" in the sample. That is, the nuclei are made to rotate together in phase. The signal is induced in the coil immediately after the 90-degree pulse. If another FID is performed, the magnitude of the resulting signal depends upon the amount of relaxation that has taken place since the last pulse.

There are two problems with using the FID sequence for imaging. First, it is difficult to record the MR signal immediately after the 90-degree pulse is transmitted. The transmitted pulse is orders of magnitude more intense than the received pulse. The presence of the large "transmitted" pulse induces a "ringing" in the receiver coil that requires at least a few milliseconds to dissipate. During this "ring-down" period the receiver coil is unable to respond to the signal induced by the rotating nuclei.

The second problem with the FID sequence is that the T2 relaxation of the tissue of interest is usually masked by another phenomenon that has the same effect upon the signal. The resonance frequency of a proton depends on the magnetic field to which it is exposed. Magnetic field inhomogeneities across the sample, particularly in imaging devices where the bore size is on the order of 100 cm, produce variations in the frequency of precession within the sample volume. These variations are indistinguishable from the "true" T2 signal decay produced as the protons interact with one another to go "out of phase." The reduction in magnitude of the FID is shown in Margin Figure 24-1 as the result of two processes:

- Dephasing produced by magnetic field inhomogeneities governed by a time constant T2* (pronounced "T two star")
- Dephasing produced by spin–spin interactions, governed by a time constant T2

Because the inhomogeneity phenomenon is a property of the MR system and not of the sample, we usually prefer to remove the T2* component of FID. The influence of magnetic field inhomogeneities may be reduced by applying a "rephasing pulse" described below in the spin-echo sequence.

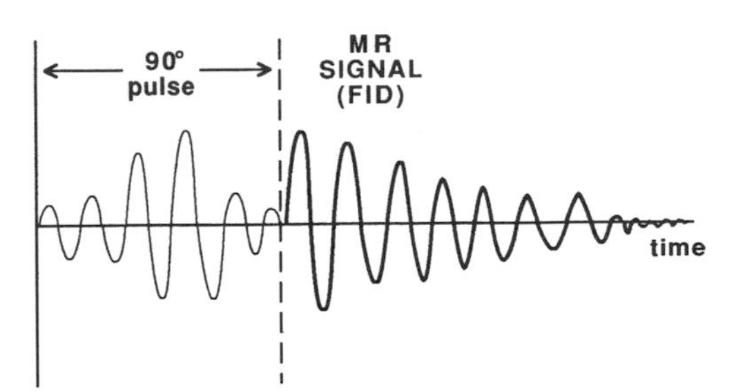

MARGIN FIGURE 24-1
The MR signal immediately following a 90-degree pulse decays rapidly due to the effects of magnetic field inhomogeneity. The MR signal for free induction decay (FID) is not shown to scale in this illustration. Note also that the 90-degree pulse and the received signal may occur in separate coils.

T2 is the dephasing of neighboring spins due to spin–spin interaction. The slightly different resonance frequencies among neighboring spins due to chemical shift, motion, and other effects causes the spins to loose phase coherence (stop precessing as a group with all spins pointing in the same direction).

T2* (pronounced "T two star") is dephasing that is produced by inhomogeneities of the static magnetic field within a sample. This effect results in a loss of phase coherence that mimics the "natural" or "true" spin–spin relaxation of a sample. Even in the best of MR system magnets the value of the T2* time constant for the magnet is much smaller than the value of T2 for most features of interest. Therefore, if uncorrected, the sample would relax (loose MR signal) so rapidly from T2* relaxation that the small differences in true T2 that produce contrast in the image would be completely overshadowed by T2*.

Spin Echo

As in FID, the spin-echo pulse sequence (Margin Figure 24-2) begins with an RF pulse that nutates the magnetization to some angle (not necessarily 90 degrees). After application of the pulse, the individual spins are in phase, or "coherent." With time, coherence (and therefore the MR signal) is reduced because of interactions among spins ("true" T2) and magnetic field inhomogeneities (T2*). The inhomogeneities may be as small as a few parts per million; nevertheless, the fact that the magnetic field is a little stronger in some regions of the sample than in others causes the nuclei to precess at slightly different rates and therefore to dephase.

To remove the effect of magnetic field inhomogeneities, a 180-degree "rephasing" pulse can be applied at some time T after the initial RF pulse. If the initial RF pulse produced nutation about the x-axis, the rephasing pulse is applied to produce nutation about the y-axis (Margin Figure 24-3). The effect of the rephasing pulse is to take spins that were precessing more rapidly and flip them behind the spins rotating at the average rate of precession. Similarly, spins that are rotating more slowly are automatically flipped "ahead" of the mean precession rate. The nuclei are still precessing at different rates because of inhomogeneities in the magnetic field of the system. After the rephasing pulse, the faster spins catch up with the average spins, and the slower spins are caught by the average spins. When this happens, the spins are said to have "rephased."

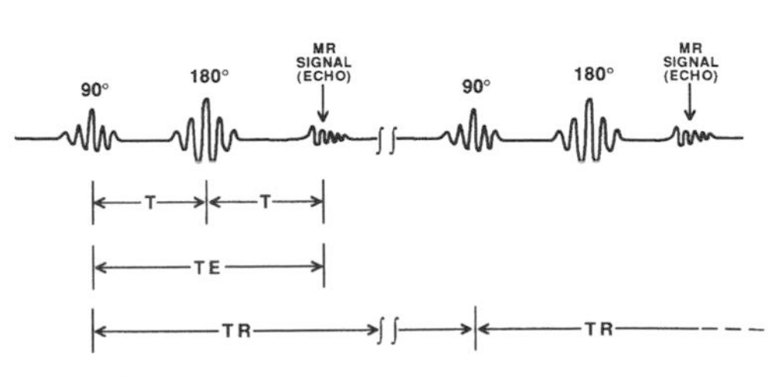

MARGIN FIGURE 24-2
The spin-echo pulse sequence.

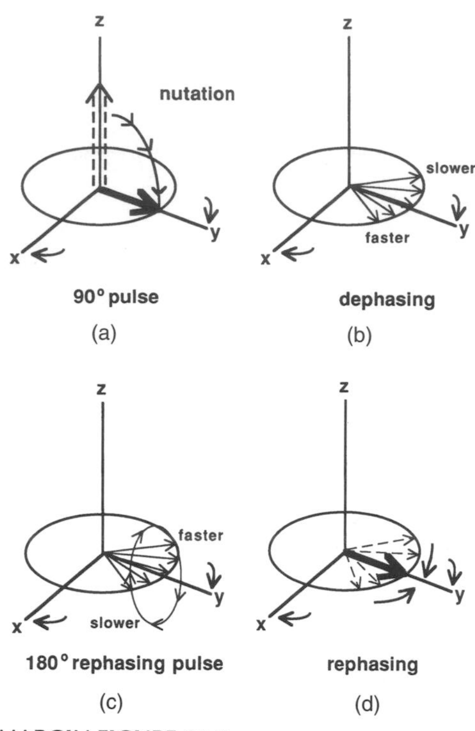

MARGIN FIGURE 24-3
In the spin-echo pulse sequence the 90-degree pulse **(A)** causes nutation of the bulk magnetization about the x-axis. Dephasing then occurs **(B)** as slower components of the magnetization lag and faster components advance. A 180-degree rephasing pulse **(C)** causes nutation about the y-axis and reverses the order of fast and slow components. The components then merge **(D)** to produce the maximum signal. Diagrams **A** to **D** are shown in the rotating frame of reference.

If dephasing progresses for a time T after the initial RF pulse, then a time T is required after application of the rephasing pulse for the spins to rephase. At the moment of exact rephasing, the signal will be strongest, since the effect of T2* is completely removed. The time after the initial RF pulse that the moment of exact rephasing occurs is referred to as TE (time to echo) and is equal to $2T$. Thus we choose TE by choosing T, the time between the application of the 90-degree pulse and the 180-degree pulse. The entire sequence is repeated at a time TR (repetition time) after the first 90-degree pulse. Typically, TR is at least 10 times TE. Obviously, TR must exceed TE.

Carr–Purcell–Meiboom–Gill

Carr and Purcell introduced the idea of using multiple rephasing pulses, with collection of the signal after each pulse.[1] Meiboom and Gill proposed applying nutation about an axis that is perpendicular to the axis of nutation for the original RF pulse (as described above for spin echo).[2] These techniques are used in the Carr–Purcell–Meiboom–Gill (CPMG) pulse sequence. Because the spins are rephased after TE, it is as if the experiment had started over with the initial 90-degree pulse. Therefore a second rephasing pulse delivered a time TE after the first produces a second rephasing, and another signal ("echo") is obtained. The diminution of signal between subsequent echoes is caused by T1 and T2 relaxation. The CPMG pulse sequence is diagramed in Margin Figure 24-4. In the CPMG pulse sequence the signal after three echoes spaced TE apart is similar to the signal that would be obtained in a single spin-echo sequence having a time to echo of 3TE. That is, signals may be measured at several different TE values (in this example, TE, 2TE, and 3TE) to yield significantly more information about the sample than from a single sequence. This is only true, however, if TR is much longer than the total TE for n pulses (nTE, where n is the number of 180-degree rephasing pulses in the sequence). In practice, the storage space and computer time required to process multiple echo images have caused them to be used less often than spin-echo images that may be tailored to bring out one or more of the factors that determine image contrast.

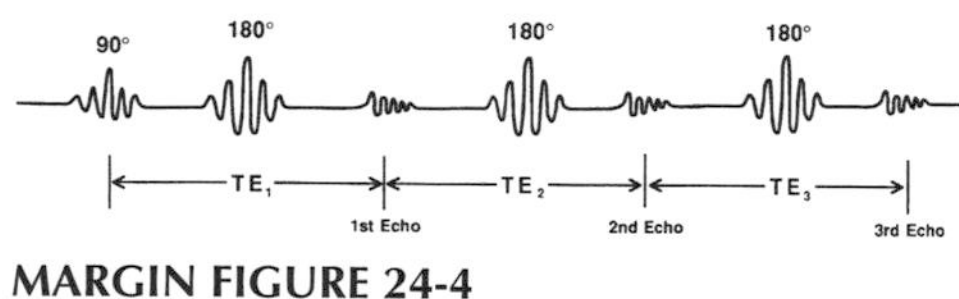

MARGIN FIGURE 24-4
The Carr–Purcell–Meiboom–Gill (CPMG) pulse sequence.

Inversion Recovery

Inversion recovery (Margin Figure 24-5) takes its name from an initial 180-degree pulse that inverts spins (nutates them 180 degrees). At a time TI (inversion time) after the initial 180-degree pulse, a 90-degree pulse is applied about the same axis. After the 90-degree pulse the inversion recovery sequence is the same as spin echo. At a time T after the 90-degree pulse, a 180-degree rephasing pulse is given. As in spin echo, this 180-degree pulse produces a rephased signal at a time $TE = 2T$ after the 90-degree pulse. As in all pulse sequences, the time between one complete pulse sequence and the next is defined as TR. The 180-degree pulse used at the beginning of inversion recovery allows more time for differences in longitudinal relaxation of spins between pulse sequences than the smaller (e.g., 90-degree) pulses used in spin echo. Thus, differences in T1 among tissues are, in principle, accentuated by the inversion recovery sequence.

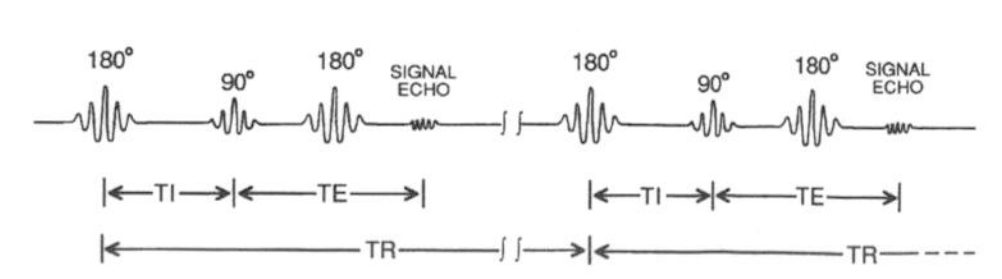

MARGIN FIGURE 24-5
The inversion recovery (IR) pulse sequence.

SPATIAL ENCODING OF MAGNETIC RESONANCE IMAGING SIGNAL

In Chapter 23, techniques for receiving and manipulating signals from bulk magnetic moments were discussed. To form an image or to determine the origin of spectral information, the signals must be related to their points of origin within a sample. This mapping or identification of signals with regions of space is called *spatial encoding*.

Sensitive-Point Method

The signal from each volume element ("voxel") in the patient could be received separately in sequence (i.e., one voxel at a time). One could then record the value of each signal in a separate storage location in the computer and reconstruct the image by displaying the values as a matrix of shades of gray. In practice, this method of data acquisition, called "sensitive-point" scanning,[3] is too time-consuming, and patient motion would be unacceptable. It is described here to illustrate several fundamental points about data acquisition and to emphasize the need for more complex but efficient methods of acquiring data.

If the magnetic field is made to vary gradually across the patient, then according to the Larmor equation the frequency of precession and therefore the resonance frequency of the protons will also vary. Suppose that the variation or "gradient" of the magnetic field is in the craniocaudal direction (Margin Figure 24-6); that is, suppose the magnetic field is stronger at the patient's head than at the feet. If an RF pulse of narrow bandwidth (a narrow range of frequencies) is sent into the patient while the gradient magnetic field is on, the pulse resonates with protons somewhere along the craniocaudal axis. Judicious selection of the gradient and the pulse frequency will result in resonance with protons in a selected transverse slice of the patient. The bandwidth or range of frequencies in the pulse will determine the thickness of the slice. Thus we have defined a plane within the patient from which we have placed the protons at some preselected orientation with respect to the plane of the receiver coils (e.g., with a 90-degree pulse). We could repeat the procedure in the other two orthogonal planes by using a dorsal–ventral magnetic field gradient to define a coronal plane and a left–right gradient to define a sagittal plane. The intersection of three planes is a point and yields a "sensitive point" for data acquisition.

In the sensitive point method, the "point" is actually a small volume element (voxel) of tissue with dimensions defined by the gradients and the frequency range of the RF pulse.

When RF pulses of a finite bandwidth are used, the intersection of the three orthogonal planes forms a volume element, or voxel, of time. We may "communicate" with (receive MR signals from) this voxel because the location has a uniquely defined resonance frequency. This spatial encoding procedure of activating three orthogonal magnetic field gradients is then repeated for each voxel in a slice to obtain data to be displayed as a matrix of picture elements ("pixels").

In practice, the sensitive-point technique has been implemented by using alternating gradient fields (Margin Figure 24-7). The effect of the alternating fields is to provide changing signals that can be recognized and discarded electronically at all locations except the "node" where the gradient is zero, leaving the spins to experience the static magnetic field of the imaging system. The location of the node can be defined by three orthogonal gradients and can be steered throughout the patient.

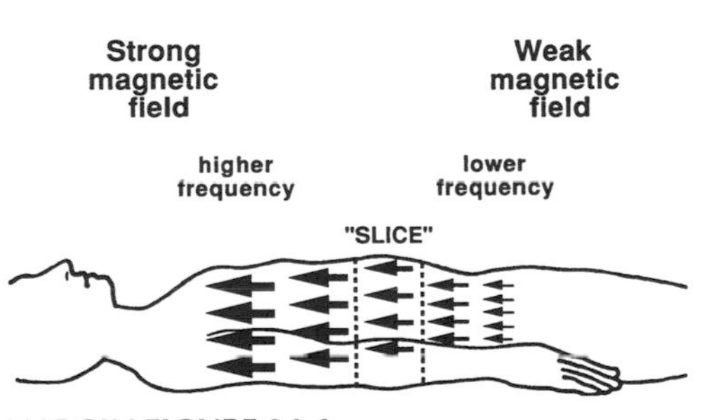

MARGIN FIGURE 24-6
A craniocaudal "gradient" or change in the magnetic field along the craniocaudal axis allows receiver coils to be "tuned in" to a slice.

Example 24-1

If it takes 300 msec to obtain a single voxel by using a sensitive-point technique, how much time is required to obtain five slices with a matrix size of 256 × 256?

$$\begin{aligned}\text{Imaging time} &= 256 \times 256 \times 5 \times 300 \times 10^{-3}\ \text{sec}\\ &= 98{,}300\ \text{sec}\\ &= 27.3\ \text{hr}\end{aligned}$$

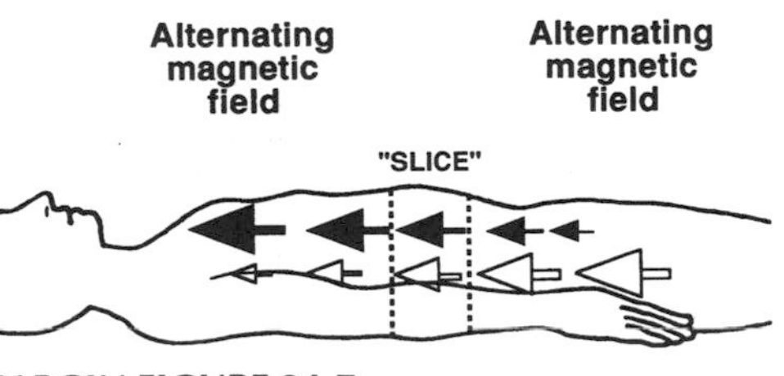

MARGIN FIGURE 24-7
The sensitive-point technique begins with the application of alternating magnetic fields. For part of a cycle the magnetic field increases from right to left as shown here. The direction of increase then reverses. The region defined as the "slice" maintains the same magnetic field value.

Two-Dimensional Fourier Transform

The sensitive-point technique makes use of "frequency encoding" to establish a unique resonance frequency for each voxel in the patient. The most obvious limitation of the sensitive-point technique is that excessive time is required to obtain enough information to construct an image. Fortunately, there are faster methods of spatial encoding. Signals may be encoded according to their phase. They may also be encoded by frequency a second time by using the Fourier transform. A number of alternate image acquisition schemes have been employed. These include sensitive-line,[4,5]

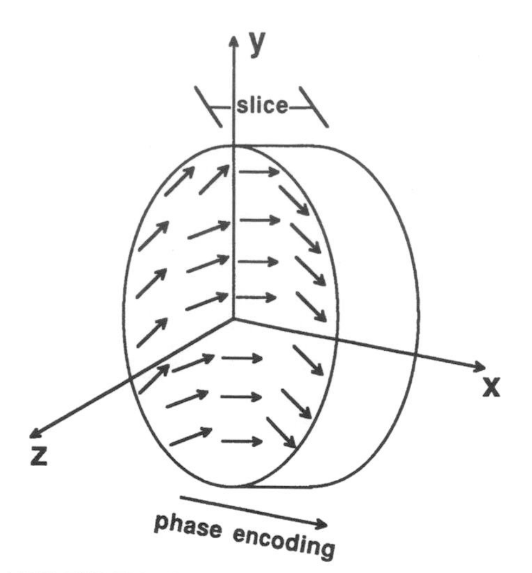

MARGIN FIGURE 24-8
The effect of the phase-encoding gradient is to vary the phase of precession of the protons along the x direction. The phase of precession of protons along the y direction is the same at any location along the x-axis.

sequential-plane,[6] and three-dimensional zeugmatography.[7] At the present time, the most commonly used method of spatial encoding in commercial MRI systems is the two-dimensional Fourier transform (2DFT).[8]

The first step in the 2DFT method of signal acquisition is activation of a slice-select gradient while a narrow-bandwidth RF pulse is sent. As discussed above, the presence of the slice-select gradient means that protons in a narrow section are in resonance and therefore nutated to a specified angle by the RF pulse. Protons elsewhere in the patient are relatively unaffected. Within the plane, however, the angle of nutation depends upon the magnitude and duration of the RF pulse. The location of the plane along the direction of the gradient is determined by the center frequency of the RF pulse, and the thickness of the section or "slice" is determined by the bandwidth of the RF pulse.

As in any spatial encoding scheme, the 2DFT requires that information about the location of the signal be provided along three orthogonal axes. The first axis is "slice," as defined above, which is usually identified as the "z direction" and is oriented along the cranial-quadrant axis of the patient. The other two axes, x and y, are identified with or "encoded" by the frequency and phase of the RF signals returning from the tissue. The second step of 2DFT is to turn on a phase-encoding gradient in a direction perpendicular to the slice-encoding gradient. This gradient is turned off before any RF pulses are applied. In Margin Figure 24-8 the phase-encoding direction is along the x-axis, but this alignment is arbitrary.

It may seem that turning the phase-encoding gradient on and off between pulses is a useless exercise. When it is on, protons at the end of the gradient where the magnetic field is stronger precess faster (at a higher frequency) than do protons at the other end, with intermediate values of precession frequency in between. When the gradient is turned off, protons within the slice revert to a common precession frequency. Thus the phase-encoding gradient has a very significant effect upon the MR signal (Margin Figure 24-8). When the phase-encoding gradient is on, it divides the slice into columns with different precessional frequencies. When the gradient is turned off, the protons revert to the precession frequency called for by the static magnetic field provided by the main system magnet, and no difference in frequency exists from one column to the next. However, protons in the columns that earlier experienced a higher frequency of precession rotated farther during the time that the phase-encoding gradient was on, and protons in the columns that experienced a lower frequency of precession did not rotate as far during the same interval. Thus when the gradient is turned off, the magnetization is locked in with protons in different columns precessing at different phases of rotation (i.e., 12 o'clock, 1 o'clock, etc.), thus giving rise to the term *phase encoding*. That is, the difference between columns is encoded by the phase of rotation.

The third step in 2DFT is to apply a frequency-encoding gradient at the time that the RF signal is received from the patient. The frequency-encoding gradient acts according to the same principle as the slice-select gradient or the phase-encoding gradient. When it is on, protons at one end of the slice precess faster than do protons at the other end. The "fast" and "slow" ends define an axis or direction. In Margin Figure 24-9 the frequency-encoding gradient is identified with the y-axis. The effect of this gradient is to divide the slice into rows. Thus, while the RF receiver is turned on, protons in different rows precess at different frequencies, and the rows are encoded by the frequency of rotation. Because the frequency-encoding gradient is turned on only during signal reception or "readout," it is often referred to as a "read gradient."

The fourth step of 2DFT is the application of the Fourier transform (see Appendix I). This approach permits a complex data set to be acquired in a single set of measurements over a relatively short period of time. The data are analyzed later with a mathematical transformation or mapping between Fourier "transform pairs." In MRI, there are two transformations (two-dimensional Fourier transformation). In step 3 above, the MR signal from all columns is read while the frequency-encoding gradient is on. Because of this gradient, the MR signal is a complex signal made up of the contributions from protons in voxels having different frequencies of precession

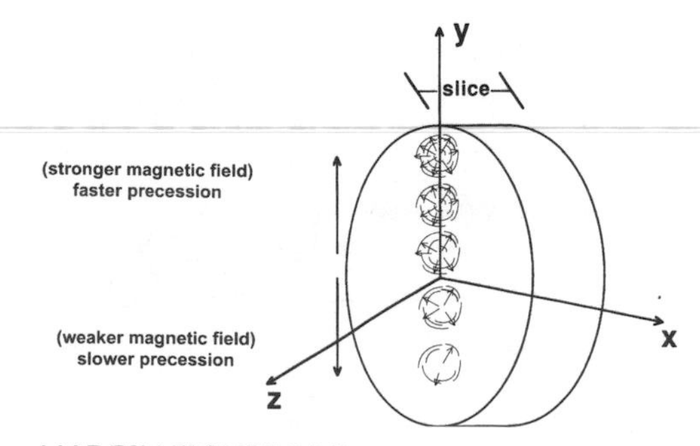

MARGIN FIGURE 24-9
The effect of the frequency-encoding gradient is to vary the rate or frequency of precession of protons along the y direction.

(Margin Figure 24-10). The Fourier transform determines the contributions that were made by different pixels by transforming the MR signal (acquired as a function of time) to its mathematical representation as a function of frequency. Thus, the Fourier transform pair in step 3 consists of time and frequency. The data from different rows (different frequencies of precession) are stored separately in preparation for the second Fourier transform.

The fifth step is to repeat steps 1 to 4 a number of times and average the result. This "signal averaging" reduces the contribution of noise to the final data.

At this point we might think of the 2DFT as having resulted in the encoding of information (the MR signal) from one slice of voxels. The slice-select gradient has defined the slice. The phase-encoding gradient has defined columns, and the frequency-encoding gradient has defined rows. However, it is not quite so simple. A single application of the phase-encoding gradient does not provide enough information to fully differentiate the columns within a slice. The essence of the Fourier transform method is that a complex signal can be analyzed to determine the contribution of different elements to the signal. It is thus necessary to acquire data for the entire slice before both Fourier transforms may be applied.

The sixth step of 2DFT is to change the magnitude of the phase-encoding gradient and then repeat steps 2 to 5. The number of times that this step is carried out determines the number of columns into which the slice will be divided—that is, the number of voxels in the phase-encoding direction. Recall that along a row the voxels have different phases (see Margin Figure 24-8). The MR signal that is received after application of each phase-encoding gradient yields additional information about the contribution of different protons along the row. That is, the row can be divided more finely (i.e., into more voxels) by exposing it to more gradients.

In the seventh and final step of the 2DFT method the second Fourier transform operates on the complex signal obtained after all of the phase-encoding steps have been completed. This second transformation makes use of the transform pair of phase and position. That is, the contributions from the different phase combinations obtained from the separate phase-encoding steps (rows) are determined by Fourier transformation of the complex signal made up of all of the data.

The steps involved in the 2DFT method may be summarized as follows:

1. Apply the slice-select gradient while RF pulses are transmitted.
2. Apply the phase-encoding gradient.
3. Apply the frequency-encoding gradient while the MR signal is received.
4. Fourier-transform the received MR signal.
5. Repeat steps 1 to 4 and average.
6. Repeat steps 2 to 5 with different phase-encoding gradients.
7. Fourier-transform the data from steps 1 to 6.

The 2DFT reduces scan time significantly compared with "sensitive-point" techniques. Mathematical analysis is required to determine the contribution of individual voxels within rows and columns to the total signal, but time spent in computation does not affect motion unsharpness in the final image. A diagram of the 2DFT method is shown in Margin Figure 24-11.

The acquisition time T_{aq} required to obtain an image by using 2DFT is determined by the number of signal averages N_a, the number of phase-encoding steps N_p, and the time that the imager has been programmed to wait to complete the pulse sequence that was selected, that is, TR (pulse repetition time). Thus

$$T_{aq} = N_a N_p \mathrm{TR} \tag{24-1}$$

Example 24-2

Find the total time required to obtain an image if four signal averages and 256 phase-encoding steps are used in a pulse sequence that requires 1500 msec to complete.

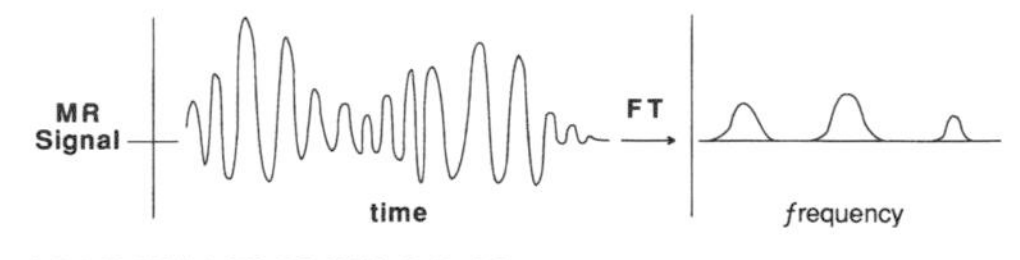

MARGIN FIGURE 24-10
The Fourier transform (FT) of the MR signal acquired during the application of the frequency-encoding or "read" gradient. The MR signal as a function of time is transformed mathematically into relative magnitudes at different frequencies.

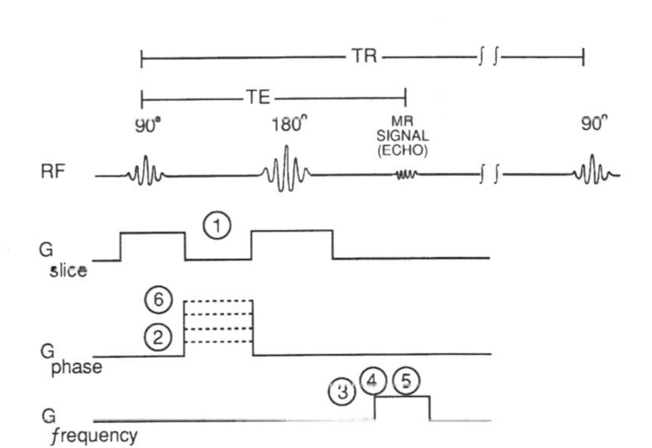

MARGIN FIGURE 24-11
The two-dimensional Fourier transform (2DFT) radio-frequency (RF) pulses and gradient magnetic fields (G) are shown on separate time lines illustrating the order in which events occur. The exact shapes of pulses and gradients are not represented. The steps involved in the entire 2DFT are as follows: (1) The slice select gradient (G_{slice}) is applied while RF pulses are sent; (2) a phase-encoding gradient (G_{phase}) is applied; (3) a frequency encoding gradient ($G_{\text{frequency}}$) is applied while an MR signal is received; (4) an MR signal is Fourier-transformed; (5) steps 1 to 4 are repeated, and the results are averaged; and (6) steps 2 to 5 are repeated with a different phase-encoding gradient (G_{phase}). A second Fourier transform is applied to the data set obtained in steps 1 to 6.

TABLE 24-1 MRI Motion Suppression Techniques

Technique	*References*
Breath holding, physical restraints	9,10
Respiratory gating	11,12
Data averaging	13–15
Respiratory ordering of phase encoding (ROPE)	16–18
Gradient-echo techniques	19–22
Gradient-moment nulling	23,24
Half-Fourier imaging	25
Snapshot	26

$$
\begin{aligned}
T_{aq} &= N_a N_p \mathrm{TR} \\
&= 4 \times 256 \times 1500 \times 10^{-3}\ \mathrm{sec} \\
&= 1536\ \mathrm{sec} \\
&= 25.6\ \mathrm{min}
\end{aligned}
$$

MOTION SUPPRESSION TECHNIQUES

Various schemes have been proposed for suppressing motion artifacts in MRI. Table 24-1 lists several of these techniques. Methods of motion suppression may be divided into three groups: techniques that suppress motion in the patient, those that make novel use of scan data to remove the effects of motion, and the use of faster pulsing schemes.

Suppression of Motion of Patient

Methods such as patient restraint, sedation, acquisition during breath holding, and gated acquisition are obvious extensions of techniques that have had some success in other imaging applications. Ultimately, breath holding is practical for most patients only if scan times are reduced to the order of seconds. Respiratory gating with a mechanical device to time data acquisition during one or more phases of the respiratory cycle has the undesirable feature of an increase in overall examination time. Also, the image is a composite, and transient physiologic phenomena will not be seen.

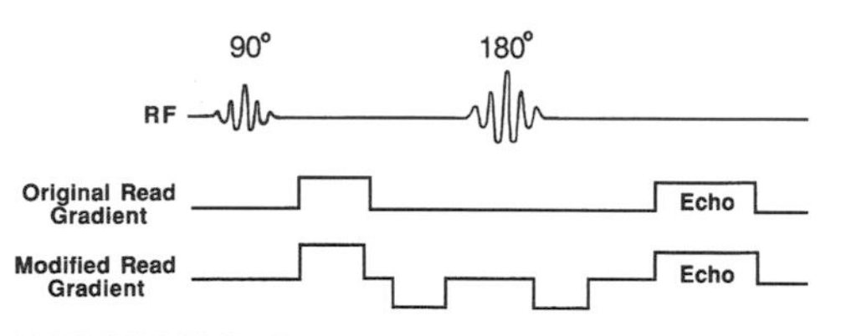

MARGIN FIGURE 24-12
A simple modification was made to the spin-echo imaging technique to rephase magnetization moving at constant speed along the readout direction. Two identical lobes were added on either side of the 180-degree refocusing RF pulse. The two lobes must have similar polarity because the 180-degree pulse reverses the orientation of the transverse magnetization. Consequently, the area under the readout gradient is left unchanged. The size of the new lobes can be adjusted to achieve similar dephasing for stationary tissue and tissue moving at a constant speed. (From Runge, V. M., and Wood, M. L. *Magn. Reson. Imaging* 1988; **6**:595–608. Used with permission.)

Removal of Effects of Motion

The order of phase-encoding steps may be selected according to the respiratory motion of the patient so that the end result is an image that appears to have been taken during a single respiration. There is some residual blurring with the respiratory-ordered phase-encoding technique (ROPE), but the reordering of data does not require intermittent scanning as in a gating technique.

Effects of motion that occur at a constant rate along the direction of a gradient may be removed by altering the gradient. In gradient-moment-nulling techniques, the shape of the gradient is changed over time (Margin Figure 24-12) so that symmetrical application of the gradient rephases spins that have moved a specified distance during a pulse sequence. In addition to allowing motion suppression in only the three gradient directions, gradient nulling places a limit upon minimum achievable TE because of the time required to reapply new gradients during a pulse repetition.

Fast Scanning Techniques

Two significant limitations on the time required for a pulse sequence are the repetition time (TR) required to allow recovery of longitudinal magnetization and the echo time (TE) mandated by use of a 180-degree rephasing pulse. Both of these limitations are addressed in gradient-echo imaging techniques.

As mentioned earlier, the purpose of the 180-degree pulse in a spin-echo sequence is to remove the effects of magnetic field inhomogeneities related to T2*. The gradient echo is another technique for dealing with this problem. In gradient-echo pulse sequences, while the phase-encoding gradient is on, the frequency-encoding gradient has a specified direction. For example, the magnetic field may be increased along the positive y-axis and decreased along the negative y-axis. After the phase-encoding gradient is switched off, the frequency-encoding gradient is reversed (e.g., decreased along the positive y-axis and increased along the negative y-axis). The effect of the reversal of gradient is the following: spins that experienced an increased field before reversal precess at a higher frequency and vice versa. Thus, a type of dephasing, albeit organized and directional, occurs. After gradient reversal, the faster spins become the slower spins and will be "caught" by the average spins. Thus, a type of rephasing occurs that results in a maximum signal that occurs at a predictable time after reversal of the frequency-encoding or read gradient (Margin Figure 24-13). In summary, the gradient echo does away with the 180-degree rephasing pulse by substituting a reversal of the magnitude of the frequency-encoding or read gradient.

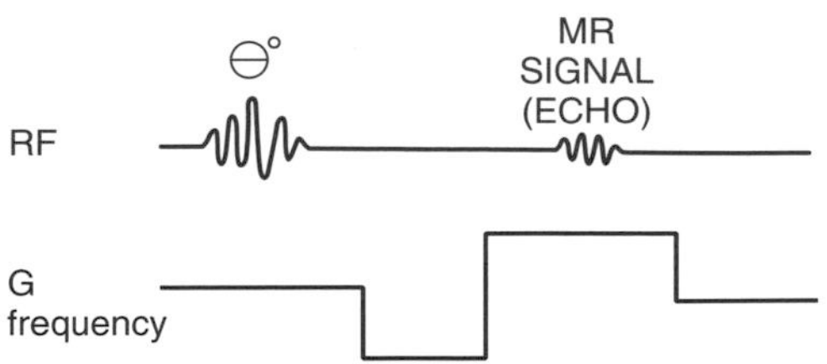

MARGIN FIGURE 24-13
Gradient-echo imaging technique. The radio-frequency pulse (RF) and frequency-encoding magnetic field gradient ($G_{\text{frequency}}$) are shown. Other gradients are similar to those used for spin-echo 2DFT image acquisition. The frequency-encoding gradient is reversed to bring about a rephasing of spins. This technique eliminates the 180-degree pulse used in spin echo. The RF pulse used in gradient-echo techniques is usually less than 90 degrees.

The gradient reversal does not accomplish all that the 180-degree rephasing pulse achieves because it does not remove the effects of magnetic field inhomogeneities. Therefore, the appearance of an image is influenced by T2* as well as T2. The effects of chemical shifts are also not removed by the gradient echo and can contribute to decreased signal intensity, especially at high field strengths. However, it performs the function of separating the echo, the maximum MR signal induced in the receiver coils, from the initial pulse. Because the initial part of the gradient, before the reversal, may be applied while the phase-encoding gradient is on, a reduction in TE is achieved when compared with the method of the 180-degree rephasing pulse.

The use of a gradient-echo technique allows a significantly shorter TR. An effect of shortening TR is that longitudinal magnetization does not have time to recover completely between pulse sequences. When complete recovery does occur, a 90-degree pulse provides maximum transverse magnetization and therefore a maximum induced signal in the receiver coils. During a pulse sequence in which TR is too short to allow complete recovery of longitudinal magnetization, however, a 90-degree pulse nutates the net magnetization past the transverse plane and provides a reduced signal. Thus, for techniques such as gradient echo where TR is shortened, flip angles of less than 90 degrees result in an improved signal-to-noise ratio.[27,28]

Examples of gradient-echo techniques include GRASS (gradient-recalled acquisition in the steady state),[29] FLASH (fast low-angle shot imaging),[30] and FISP (fast imaging with steady-state precession).[31]

Half-Fourier Technique

There is symmetry in the data produced by repeated application of the phase-encoding gradient (the phase-encoding "steps" of signal acquisition).[25] In theory, it is necessary to acquire only half of the steps, say those produced with a negative gradient, with the other half, the positive gradient steps, reproduced mathematically from those acquired. This half-Fourier technique saves imaging time, with the only drawback in theory being an increase in computer reconstruction time. In practice, however, use of the second half of the phase-encoding steps is a method of signal averaging (multiple measurement of the same data). Because the half-Fourier technique uses fewer data, the signal-to-noise ratio may be reduced.

Time-Varying Gradient

In standard imaging techniques the phase-encoding gradient is stepped through 128 or 256 discrete values, with the entire pulse sequence repeated for each value. In snapshot imaging, all the phase-encoded values are collected during a single RF pulsing sequence. Spatial encoding is accomplished by rapid oscillation of the frequency-encoding gradient interspersed with pulses of the phase-encoding gradient that occur as the x gradient returns to zero between oscillations. Rapid data sampling during each excursion of the frequency-encoding or read gradient results in the final signal acquisition. To date, snapshot techniques have yielded 64 × 128 images in 1/50th of a second.[26]

CONTRAST AGENTS

There are three basic classifications of materials with regard to magnetic properties: paramagnetic, diamagnetic, and ferromagnetic.

Electrons have spin and magnetic moments just as do nucleons. In fact, the magnetic moment of the electron is more than 100 times greater than the magnetic moments of nucleons. Their effect is less noticeable, however, because the magnetic moments of electrons in orbitals or shells surrounding the nucleus tend to "pair up"—for example, one electron having spin up, the other having spin down. In diamagnetic substances, virtually all electrons are paired, so interactions with external magnetic fields are minimized. In fact, diamagnetic substances actually repel an external magnetic field, albeit weakly. The repulsion is due to an induced magnetic field in the electron orbital. Such an induced field opposes the field that induced it (Lenz's law).[32] Approximately 99% of biologic tissues are diamagnetic. Because of their high natural abundance in body tissue, there is little current interest in the development of diamagnetic substances as contrast agents, although diamagnetic clay suspensions have been used for contrast in the gastrointestinal (GI) tract[33] by eliminating signal from the bowel contents.

Paramagnetic contrast agents contain unpaired electrons. They develop their own magnetic fields by tumbling in solution at some fraction of the Larmor frequency. The tumbling creates alternating magnetic fields that affect precessing nuclei. An interaction between the unpaired electrons and the precessing nuclei increases the rate of relaxation of neighboring protons, and the presence of variations in the magnetic field introduced by the paramagnetic agent increases dephasing. Thus, both T1 and T2 are shortened by the presence of a paramagnetic agent.

Only a small fraction, less than 0.5%, of the body's tissues are naturally paramagnetic. Paramagnetic agents may be added to enhance relaxation and provide a mechanism for contrast enhancement. Thus, paramagnetic contrast agents are not imaged directly in MRI. Instead, the presence of the agent influences the appearance of the image. One of the most widely used paramagnetic agents is gadolinium with diethylenetriamine pentaacetic acid as a chelating agent (Gd-DTPA).[34] An example of contrast enhancement by Gd-DTPA is shown in Margin Figure 24-14.

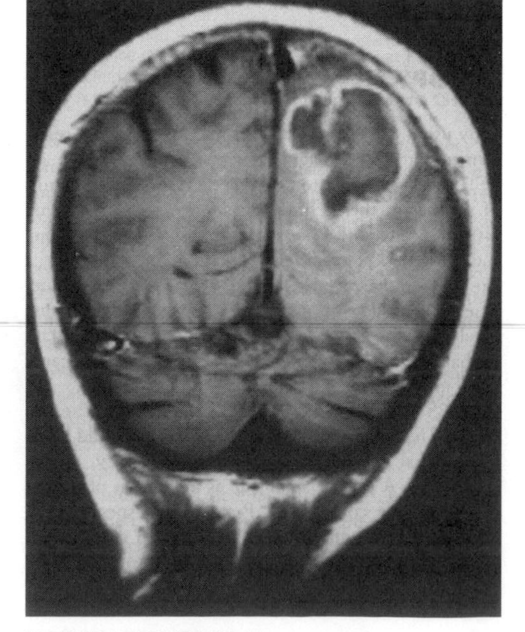

MARGIN FIGURE 24-14
An example of contrast enhancement by Gd-DTPA.

Ferromagnetic materials develop magnetic polarization (i.e., become magnetic) in the presence of an applied magnetic field. Ferromagnetism is a phenomenon in which hundreds to thousands of atoms having unpaired electrons group together into "domains." If many domains are aligned, then the material is said to be magnetized. If the alignment persists in the absence of an applied magnetic field, then the material is a permanent magnet. Because a large grouping of atoms is required for ferromagnetism, ferromagnetic contrast agents do not exist as solutions of metal ions. They are given in the form of compounds such as iron oxides (Fe_2O_3, Fe_3O_4). These agents selectively decrease T2 and cause a decrease in MR signal in conventional pulse sequences. Thus ferromagnetic MR contrast agents provide "negative" contrast.

In addition to diamagnetic, paramagnetic, and ferromagnetic mechanisms of contrast agent action in MR, other mechanisms are possible. Perfluorochemicals, substances in which all or part of the hydrogen has been replaced by fluorine, elicit no MR signal in conventional proton imaging because they contain no hydrogen. These substances have been used to provide negative contrast.[35] Similarly, the introduction of gas-evolving substances and of air itself into the body has been investigated.[36]

Functional MRI (FMRI)

Oxygenated blood, like most body tissues, is diamagnetic despite the presence of iron-containing hemoglobin. In oxygenated hemoglobin (oxyhemoglobin), the outer-shell electrons of the iron atom are transferred to oxygen molecules. In deoxygenated blood (deoxyhemoglobin), four of the six outer electrons of the heme ion are unpaired, placing the iron in a ferrous (Fe^{2+}) state that is paramagnetic. In the absence of neuronal activity, the presence of paramagnetic deoxyhemoglobin in red blood cells causes a susceptibility gradient (change in the effective magnetic field over a small region of space) that has the effect of shortening T2 and T2* in the surrounding tissue and blood plasma. These regions of magnetic field inhomogeneity produce measurable effects over distances of two or more times the radius of the vessel.[37]

It is known that neuronal activity causes a localized increase in blood flow, cerebral blood volume, and oxygen delivery.[38] However, oxygen extraction increases by a lesser amount, if at all.[39,40] The increased blood flow "drives out" the deoxyhemoglobin so that there is a net *decrease* in the concentration of deoxyhemoglobin in capillaries and venous circulation. The susceptibility gradient that exists between oxygenated blood and surrounding tissues in the *absence* of neuronal activity is reduced. Therefore, in the area of neuronal activity, less intravoxel dephasing occurs. The effect of this reduction in dephasing is a net increase in signal from tissue in the region of neuronal activation in T2- and T2*-weighted images. Such signal intensity changes are observable within seconds after the onset of stimulation of neuronal activity.[41] The time course of the rise of signal intensity is consistent with cerebrovascular transit times measured with radiotracer studies,[42] supporting the belief that the increase in signal intensity is related to the transition of relatively oxygenated blood through the capillary bed and into venuoles. Thus, the presence of a naturally occurring contrast agent (oxyhemoglobin) that responds to the spatial and temporal pattern of neuronal activation in the brain allows construction of a functional map of neuronal activity (see Margin Figure 24-15). The spatial resolution of this functional map is limited by the volume of the region of decreased oxyhemoglobin concentration. Temporal resolution is limited by the consistency of the several second lag time between onset of neuronal stimulation and the rise in MR signal.

Functional MRI (fMRI) has revolutionized the study of brain function. It has upgraded brain research from a field in which the approximate anatomical correlates of function were known through studies of individuals with neurological abnormalities or injuries, to a field in which it is possible to determine the specific locus of many neural activities in specific individuals.[44,45] This technique is replacing presurgical mapping[46–53] to identify the location of functional areas before tumor resection, and it is valuable in patient evaluation following stroke or other neurovascular events.[54–57] It has already led to major advances in understanding the linkage between structure and function in the cognitive sciences,[58–63] particularly in the study of learning and development.[64–66] Margin Figure 24-16 shows a functional MR study in which the areas associated with motor, memory, and visual processing are examined during successive tasks in which the subject moves an index finger in response to movement of a cursor on a screen, or simply watches the movement of the cursor. Included in the illustration is the image formed by subtraction of the two acquired images to isolate the role of vision in a simple task.

Magnetic Susceptibility is a property of matter that modifies an externally applied magnetic field so that an induced magnetic field appears in and around the matter. A related concept using a slightly different scale is **magnetic permeability.** A high value of susceptibility (or permeability) indicates that the material will "act like a magnet" when placed in a magnetic field. Such material is classified as **ferromagnetic.** Iron is an example of a ferromagnetic material. Some matter has negative magnetic susceptibility—it actually tries to repel a magnetic field (although the repulsive force is generally too weak to notice). Such materials are classified as **diamagnetic.** In fact, most biological materials, water, most organic molecules, and so on, are diamagnetic.

In MRI, variation in the actual magnetic field that is induced in a region of the patient is produced by variation in the susceptibility of adjacent structures. These **susceptibility gradients** between adjacent structures such as brain/sphenoid sinuses and CSF/vertebral bodies can cause artifacts and signal losses. Susceptibility gradients can also enhance the appearance of structures by changing local relaxation times in comparison with the surrounding materials where materials such as gadolinium are introduced as contrast agents.

Seiji Ogawa of Bell Labs was the first to propose the technique of functional MRI, also known as blood-oxygen-level-dependent (BOLD) imaging. He published his work on animal studies and suggested the application of this technique for imaging of brain function in humans in 1990. He collaborated with the Center for Magnetic Resonance Research at the University of Minnesota[43] to produce some of the earliest functional MR images made with human subjects.

TISSUE CONTRAST IN MAGNETIC RESONANCE IMAGING

Spatial encoding of MR signals is pointless if there are no differences in magnitude among the signals. Fortunately, there are substantial differences among human tissues in relaxation times (T1, T2) and proton density N(H). To illustrate concepts that

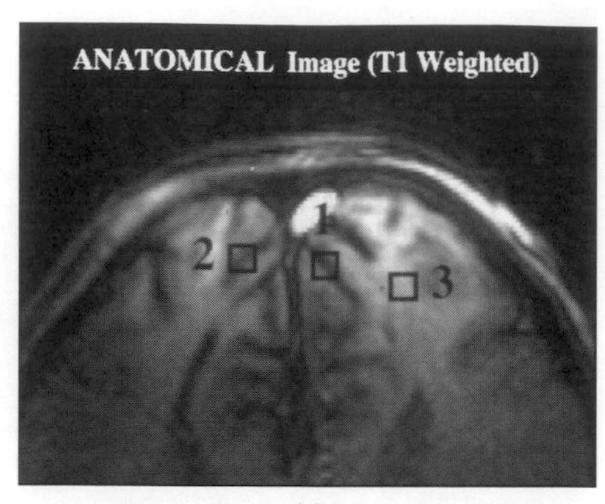

(a)

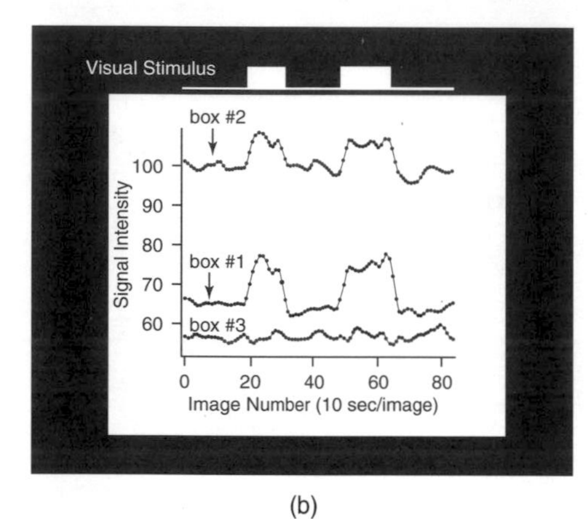

(b)

MARGIN FIGURE 24-15
An early indication of the utility of functional magnetic resonance imaging (FMRI) in mapping areas of brain function. A visual stimulus is shown to a research subject. In the T1-weighted MR image, areas 1 and 2 are known to be involved in processing visual stimuli and area 3 is not. A plot of FMRI signal intensity as a function of time shows that the signal increased during visual stimulus in the two areas associated with processing of visual stimuli, but did not increase in area 3.

are central to measurement of contrast in MRI, a rudimentary "patient" or sample containing only two materials is considered below, each having characteristic values of T1, T2, and $N(\mathrm{H})$. With fixed characteristics of the imaging system (e.g., field strength and receiver coil geometry) the imaging experiment consists of choosing operator-selectable parameters (TR, TE, TI, flip angle, etc.) that maximize the contrast or difference in signal between the two materials.

Spin Echo

The strength intensity S_{SE} of the MR signal during a spin-echo sequence is given by

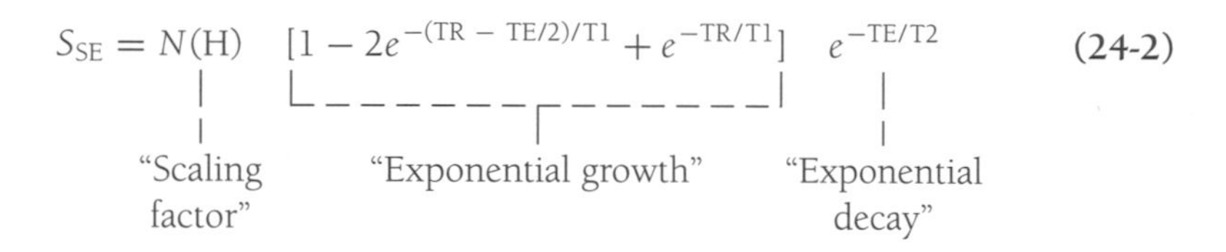

$$S_{\mathrm{SE}} = N(\mathrm{H}) \quad [1 - 2e^{-(\mathrm{TR}\,-\,\mathrm{TE}/2)/\mathrm{T1}} + e^{-\mathrm{TR}/\mathrm{T1}}] \quad e^{-\mathrm{TE}/\mathrm{T2}} \tag{24-2}$$

where $N(\mathrm{H})$ is the number of spins, T1 is the longitudinal relaxation time constant, T2 is the transverse relaxation time constant, TE is the time to echo (time between the 90-degree pulse and rephased signal or echo, which is twice the time between the 90- and 180-degree pulses), and TR is the pulse sequence repetition time.

The influence of T1 in this equation depends on the magnitude of TR. T1 and TR appear together in an "exponential growth" term. Spin density, $N(\mathrm{H})$, appears as a scaling factor or multiplier and has a linear effect upon signal strength. That is, if the spin density is doubled, the signal strength is doubled.

The time between pulse repetitions, TR, allows the bulk magnetization of a material to return to alignment with the axis of the static magnetic field (the z-axis). If alignment is complete, then the next 90-degree pulse nutates the bulk magnetization into the xy plane to yield a maximum signal. Margin Figure 24-17, A, where signal strength is plotted as a function of TR from Eq. (24-2) (assuming constant TE), shows the exponential growth of signal with TR. The magnitude of the maximum signal (when TR is infinitely long) is determined by the total number of spins contributing to the signal, that is, spin density $N(\mathrm{H})$. Also shown in Margin Figure 24-17A is the effect of T1. If the only difference between two materials is T1, then the material having the shorter T1 would produce a stronger signal at intermediate values of TR.

The time between the 90-degree pulse and the echo, TE, determines the degree to which dephasing of the spins, T2 relaxation, is allowed to reduce the transverse or "xy" component of the bulk magnetization. If the dephasing is allowed to progress to the point where the phase of spins is random, the signal becomes zero. Margin Figure 24-17B, where signal strength is plotted as a function of TE from Eq. (24-2) (assuming constant TR), shows the exponential decay of signal with TE. The magnitude of the maximum signal (when TE is zero) is determined by spin density, $N(\mathrm{H})$. Also shown in Margin Figure 24-17B is the effect of T2. Note that if the only difference between two materials were T2, then the material having the longer T2 would produce a stronger signal at intermediate values of TE.

If the goal of imaging were to maximize the signal from a particular tissue, longer TR values and shorter TE values would always be preferred for spin-echo imaging. To obtain an image, however, one must obtain signals from a variety of tissues and maximize the contrast, or relative signal difference, among tissues of interest. Margin Figure 24-17A shows that the difference between signals for two materials having different values of T1 (the vertical distance between the two curves at any TR) is maximum for TR near the average T1 of the two tissues. Similarly, Margin Figure 24-17B shows that the difference between signals for two materials having different values of T2 is maximum for TE near the average T2 of the two tissues.

TABLE 24-2 Rules of Thumb for Contrast in Spin-Echo Imaging

TR	*TE*	*Weighting*	*Stronger Signal Is Associated with*
Short (∼avg. T1)	Short	T1	Shorter T1
Long (>4 T1)	Short	*N*(H)	Greater *N*(H)
Long	Moderate (∼ avg. T2)	T2	Longer T2

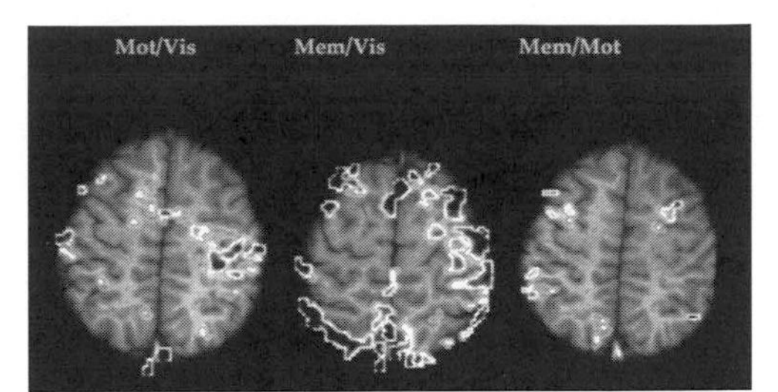

MARGIN FIGURE 24-16
A series of functional magnetic resonance (FMRI) images taken during a study of learning ability and its relation to short-term memory activity. In the first image, the subject is given a task that involves motor skills and the visual system (to press a key in response to the appearance of an object on a view screen). In the second, the subject is told to remember a sequence as different objects are shown on a screen. The third image is obtained by subtracting the previous two under the hypothesis that it represents what was common to both tasks. Although not definitive in itself, this study is one of a number of novel approaches to new areas of study in understanding brain function that has been made possible with FMRI.

Summary of Contrast in Spin-Echo Pulse Sequences

It is possible to generalize about the role of TR and TE in bringing out the contrast between tissues. Table 24-2 gives some rules of thumb for MR contrast. The terms "long, short, and moderate" are of course relative and must be defined in terms of the T1 and T2 of the materials of interest. Note that because the convention in imaging is to map higher signal values into brighter pixel values, the brighter pixels are correlated with shorter T1, longer T2, or higher spin density in spin-echo images.

The reader is cautioned that there are virtually no universally applicable rules for contrast optimization in MR. It is illustrative to consider the number of distinct cases that are possible when two materials that differ in one or more MR parameters [T1, T2, or *N*(H)], and T2 can be either greater than, less than, or the same. Margin Figure 24-18 shows the range of cases generated by selection of different values of TR and TE in a spin-echo sequence. By selecting different values of TR and TE, the operator can reverse contrast or eliminate it altogether.

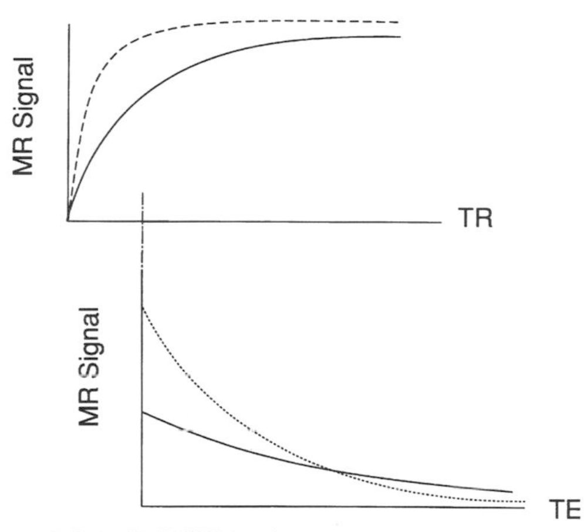

MARGIN FIGURE 24-17
Simulated MR signals from solution of the equation for spin echo [Eq. (24-2)] as TR and TE are varied. **A**: MR signal from two tissues with different T1 relaxation times (solid line, 600 msec; dashed line, 200 msec) as a function of pulse repetition time, TR, in spin-echo pulse sequence. TR varies from 0 to 2800 ms. Contrast (difference in MR signal, i.e., vertical distance between the curves) varies with TR. The two tissues shown here have the same spin density and therefore reach the same maximum signal value at long TR. **B**: MR signal from two tissues with different T2 relaxation times (solid line, 75 msec; dashed line, 40 msec) as a function of time to echo (TE) in a spin-echo pulse sequence. Contrast varies with TE. The two tissues shown here have the T1 and spin-density characteristics of the tissues in part **A** with a TR of 600 msec.

Inversion Recovery

The strength intensity S_{IR} of the MR signal during an inversion recovery pulse sequence is given by

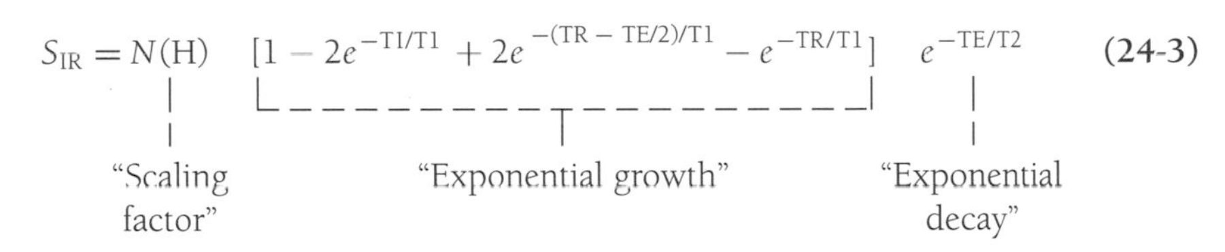

$$S_{IR} = N(H) \; [1 - 2e^{-TI/T1} + 2e^{-(TR - TE/2)/T1} - e^{-TR/T1}] \; e^{-TE/T2} \quad (24\text{-}3)$$

where *N*(H) is the number of spins, T1 is the longitudinal relaxation time constant, T2 is the transverse relaxation time constant, TE is the time to echo (time between the 90-degree pulse and the rephased signal or echo, which is twice the time between the 90- and 180-degree pulses), TR is the pulse sequence repetition time, and TI is the inversion time (time between the initial 180-degree pulse and the next 90-degree pulse).

The equation for inversion recovery is similar to that for spin echo. The difference is a more complicated exponential growth term for the inversion recovery equation. As with the spin-echo pulse sequence, TE influences the degree of T2 weighting. In inversion recovery, however, the degree of T1 weighting is influenced by both TR and TI. If TI is sufficiently short in comparison with T1 and TR, a negative value of the signal is obtained. Margin Figure 24-19 shows signal and contrast as a function of TI for signal-processing schemes that preserve the sign, positive or negative, of the signal ("signed reconstruction") and for those that do not ("absolute value" or "magnitude" reconstruction).

The same general rules for contrast optimization apply for inversion recovery as for spin echo, with the added factor of TI. Short TI enhances T1 by taking advantage of contrast reversal in magnitude reconstruction and is used in pulse sequences such as short TI inversion recovery (STIR).[67]

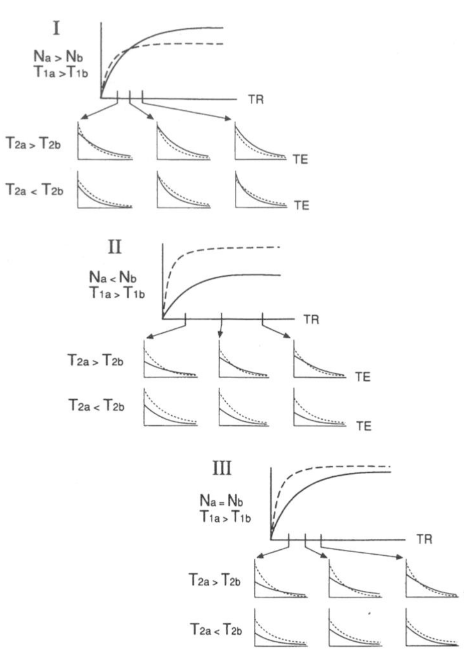

MARGIN FIGURE 24-18
Simulated MR signals showing contrast (vertical distance between curves) for spin-echo pulse sequences for two tissues having different relative values of spin density, N, longitudinal relaxation, T1, and transverse relaxation, T2. Tissue a (solid line) has a longer T1 than tissue b (dashed line). The 21 graphs in this figure show all possible cases for N and T2. Each set is plotted first as a function of TR and then as a function of TE for short, medium, and long TR. Set I begins with $N_a > N_b$, Set II with $N_a < N_b$, and Set III with $N_a = N_b$. Contrast varies as a function of repetition time, TR, and time to echo, TE, for each case. Note that a pulse sequence may be selected for any case such that the resulting MR signals provide positive contrast (signal a > signal b), negative contrast (signal a < signal b), or no contrast (signal a = signal b). Signal values were calculated from the equation for spin echo [Eq. (24-2)] by Manhot Lau, Ph.D., University of Minnesota Medical School.

The main determinants of contrast in MRI are:

- Spin density
- T1
- T2
- Flow
- Magnetic susceptibility

Gradient Echo

When gradient reversals replace 180-degree pulses in sequences where flip angles of less than 90 degrees are used, contrast weighting depends primarily upon flip angle and TE. In general, for flip angles greater than 45 degrees with short TE (less than the average T2), a T1 weighting is achieved. For flip angles of less than 20 degrees, weighting is shifted toward T2 and spin density.[19–22]

MR ANGIOGRAPHY

Selective visualization of blood vessels with MRI may be accomplished with or without contrast agents. In single slice imaging, protons that flow into the slice may not have received the RF pulses that prepare the in-slice protons for the normal imaging sequence. This phenomenon may be exploited to image flowing blood. For example, protons in blood components that were outside of the slice during the application of an initial 90-degree pulse in a spin-echo sequence have not had their bulk magnetic moments "saturated"—that is, flipped to the angle that will produce a maximum signal for that number of spins. In time-of-flight MR angiography (TOF-MRA), the spins entering the slice represent blood components that are unsaturated, and these spins are in alignment with the static magnetic field rather than at 90 degrees to it. T1-weighted images pick up the blood as a stronger signal than its surroundings so that the vessels appear white.

Because blood flow is rarely exactly perpendicular to the plane of slices of interest, various 2D and 3D TOF techniques are used where several slices (or an entire slab) of tissue are imaged together to provide a higher signal-to-noise ratio. During these techniques, a series of 90-degree pulses is used to saturate blood outside of regions of interest. Lower flip angles are also used to avoid oversaturating the signal prior to the end of the sequence.

Irrespective of how the information is obtained, once a volume of tissue has been imaged in MRA, single slice views or volume-rendered images may be obtained by using an algorithm that picks the strongest signal acquired along a particular projection angle. This technique is known as maximum-intensity projection, or MIP (see Margin Figure 24-20).

Contrast agents such as gadolinium may be used to further enhance the signal from the blood vessels. By shortening the T1 of the surrounding blood, the presence of gadolinium helps maintain the spins in their unsaturated state. However, gadolinium leaks quickly into surrounding tissues (except in the brain), so fast-scan techniques must be used. Many of the current generation fast-scan techniques push the envelope of acceptable levels of radiofrequency power levels (see Chapter 25). Thus, there is a trade-off between an increased signal-to-noise ratio and the desire to use more phase-encoding steps or number of slices.

SPECTROSCOPY

Spectroscopic applications of nuclear magnetic resonance (NMR) may be performed with some imaging systems. If performed in an MRI device, the technique is usually referred to as magnetic resonance spectroscopy (MRS). This technique has been used in analytical chemistry since the 1940s. MRS measures the differences in resonance frequencies among nuclei that occupy different positions in molecules. For example, protons in the CH_3 group of ethyl alcohol (Margin Figure 24-21) resonate at a slightly different frequency than do protons in the CH_2 group or protons in OH. The difference is only about $4 \times 10^{-4}\%$ (4 ppm) when the field strength is 1 tesla, much smaller than the differences in resonance frequency that are exploited for spatial encoding in MRI. Nevertheless, this difference can be measured.

In a typical MRS experiment, a radio-frequency wave is applied to the sample, and the signal is received. The naïve approach to spectroscopy would be to measure the amplitude of the signal, make a slight change in frequency, measure the amplitude again, and so on. A plot of amplitude as a function of frequency, the MR spectrum, could then be constructed. Peaks in the spectrum, frequencies at which a large amount of RF energy was absorbed and then returned by nuclei, correspond to the resonance frequencies of different nuclei. The resolution of the spectrum would be determined by the ability of the system to record the precise frequencies of radio waves and by the amount of time that is available for the experiment.

The time required for the spectroscopic approach described above is prohibitive. Instead of sending a radio wave with a single frequency, a radio wave containing a broad range of frequencies (a broad-bandwidth pulse) is sent. The return or re-emitted signal also contains a range of frequencies. The return signal may be Fourier-transformed to determine how much (what amplitude) of each frequency is contained within it. Thus, a single broadband pulse and receive sequence can replace the laborious set of measurements used in the more naïve approach. Some of the time that was "saved" is usually reinvested to improve resolution. The results of interrogation of the sample with hundreds of pulses are averaged to reduce the effects of noise.

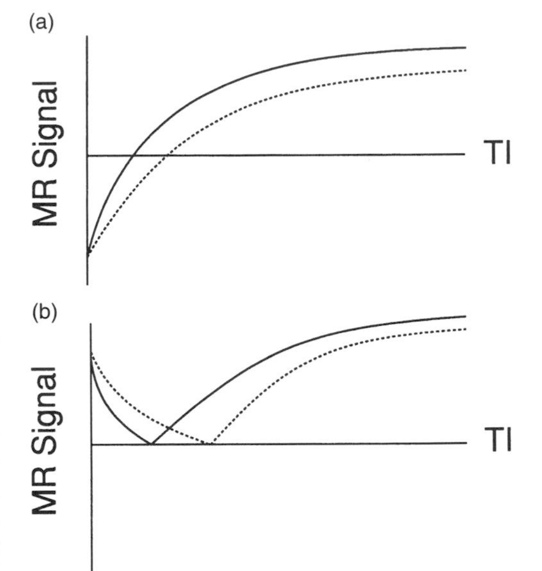

MARGIN FIGURE 24-19
Simulated MR signals from solution of the equation for inversion recovery [Eq. (24-3)] as inversion time, TI, is varied for two tissues. The longitudinal relaxation time T1 for the tissue represented by the solid line is shorter than T1 for the tissue represented by the dashed line. In **A** the sign of the signal (positive or negative) is preserved, while in **B** the absolute value (always positive) is retained. Spin density and T2 are equal.

The NMR spectrum for a molecule or compound is a unique, reproducible "signature" of great utility in determining the presence of unknown compounds in analytical chemistry. The origin of the unique pattern of chemical shifts for a molecule is the distortion of the magnetic field caused by the distribution of electrons that corresponds to each chemical bond. The effective field felt by each nucleus in the molecule is determined by the chemical bonds that surround it. Therefore, their resonance frequencies are shifted slightly. The individual protons within a chemical group experience slight differences in chemical shift as well because a hydrogen nucleus in the center of the group is influenced more strongly by the field of other hydrogen nuclei than is one at the periphery of the group. Therefore, a resonance peak for a chemical group is often split slightly into a number of equally spaced peaks, referred to collectively as a multiplet.

The chemical shifts discussed in this section so far are on the order of 10^{-7} tesla for a nominal magnetic field strength of 1 tesla. For spatial localization in imaging, gradient magnetic fields that produce differences of 10^{-4} tesla/cm are used. Therefore, the shift in resonance frequency produced by the magnetic field gradient would be 1000 times greater than the chemical shift and would completely overshadow it. One of the more significant differences between MRI and MRS is that during MRS no magnetic field gradients are applied. Furthermore, the homogeneity of the magnetic field over the MRS sample must be better than 1 part in 10 million.

According to the Larmor equation, resonance frequency depends upon the field strength applied to the sample. For this reason, chemical shifts are usually expressed as fractions, specifically as parts per million. That is, the resonance values are not expressed in terms of the absolute value of the resonance peaks f_P, but are described relative to some reference value of resonance frequency f_R and normalized to the nominal frequency of the system f_N according to the formula

$$\text{Chemical shift (ppm)} = \frac{f_R - f_P}{f_N}(10^6) \quad (24\text{-}4)$$

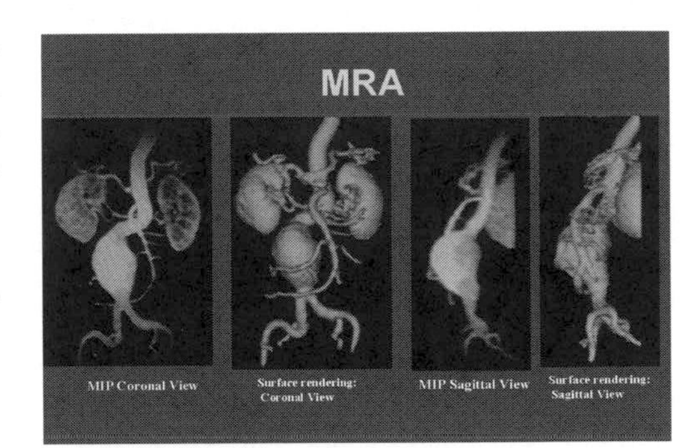

MARGIN FIGURE 24-20
Maximum Intensity Projection (MIP) and surface rendering of an aortic aneurysm from magnetic resonance image (MRI) data. (Courtesy of Chun Yuan, Ph.D., Department of Radiology, University of Washington.)

Example 24-3

An MRS system assumes a nominal frequency of exactly 300 MHz, the approximate resonance frequency for protons in a 7-tesla magnetic field. The peak against which other resonances are compared occurs at 300.004 MHz. Find the chemical shift in parts per million for a peak that occurs at a resonance frequency of 299.998 MHz.

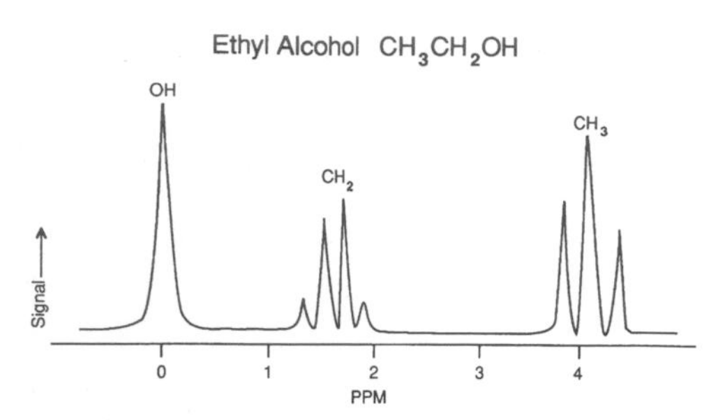

MARGIN FIGURE 24-21
Nuclear magnetic resonance (NMR) spectrum of ethyl alcohol at 1 tesla. The chemical shift is expressed as a fraction (parts per million) of the reference frequency. In this case the reference frequency is the resonance frequency of the proton in the OH group.

$$\begin{aligned}\text{Chemical shift (ppm)} &= \frac{f_R - f_P}{f_N}(10^6)\\ &= \frac{300.004\ \text{MHz} - 299.998\ \text{MHz}}{300.000\ \text{MHz}}(10^6)\\ &= 0.00002 \times 10^6\ \text{ppm}\\ &= 20\ \text{ppm}\end{aligned} \tag{24-4}$$

Other features of a spectrum, in addition to chemical shift, carry information about the sample. The area under a peak is proportional to the number of nuclei that contribute to the signal. The "line width" or full width at half-maximum (FWHM) amplitude is a measure of both the relaxation time in the sample and the degree of magnetic field inhomogeneity across the sample. Both factors tend to increase the range of frequencies gathered from equivalent nuclei during an experiment. A narrower line width is generally desirable, however, so that closely spaced peaks may be more readily resolved.

Nuclei Used for *In Vivo* Spectroscopy

Various nuclei may be examined with MRS. Table 23-1 shows the relative sensitivity for the same number of nuclei of interest.

^{31}P

High-energy phosphate reactions within the cell may be traced with the aid of ^{31}P spectroscopy. The phosphocreatine molecule is a storage site for high-energy phosphate groups used in metabolism and is often taken as the reference for chemical shift because its position does not change with pH.[68] The shift of inorganic phosphate with respect to phosphocreatine is then used as an indication of tissue pH. The relative heights of the inorganic and phosphocreatine peaks can indicate changes in the amount of substances necessary for cellular metabolism (Margin Figure 24-22). It is also thought that extracellular and intracellular components of the same metabolites differ in their chemical shifts. Therefore, compartmentalization studies, which are difficult to perform analytically, may be conducted spectroscopically.[69] Enzyme kinematics may be studied through selective excitation of individual peaks and measurement of the intensity and relaxation time of neighboring peaks.[70]

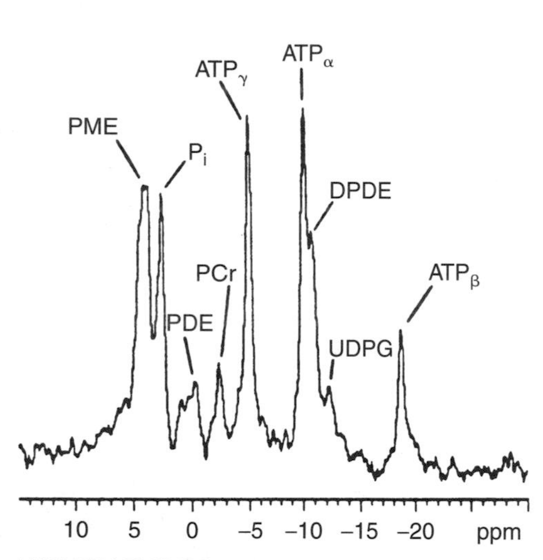

MARGIN FIGURE 24-22
^{31}P spectrum of normal liver obtained through the anterior abdominal wall of an intact rat at 7 tesla using multivoxel localization with subsequent liver voxel addition. The resonances for phosphomonoesters (PME) and inorganic phosphate (P_i) are prominent relative to the phosphocreatine (PCr) resonance, which indicates that a small amount of muscle tissue was present within the volume assayed. Also noted are well-resolved peaks for diphosphodiesters (DPDE), including uridine diphosphoglucose (UDPG). Phosphodiester (PDE) resonances are not prominent at this field strength. (Courtesy of Sally Weisdorf, M.D., University of Minnesota Medical School.)

^{1}H

Hydrogen or proton MR spectra may be used to observe metabolites such as lactate and glucose. An increase in the presence of lactate implies anaerobic conditions that may indicate ischemia or hypoxia. The peak from protons in water tends to obscure the peak from metabolites because it represents a 10,000-fold greater concentration of protons. However, suppression techniques are available to reduce the water peak through saturation (elimination of the net magnetization) of the water component or through selective excitation of the metabolites. The presence of the lipid peak also disrupts the observation of metabolites, which renders studies difficult unless performed in the brain where there is little detectable lipid.

^{23}Na, ^{39}K

Sodium and potassium are the major extracellular and intracellular cations, respectively. Studies of active transport mechanisms in nerve cells and other tissues are being pursued.[71–73]

^{13}C

The 1% isotopic abundance of ^{13}C renders observation somewhat difficult. However, spectra have been obtained that show promise as probes of brain proteins[74] and for the detection of glycogen.[75] Intermediary metabolism of compounds such as pyruvate, ethanol, or glycerol enriched with ^{13}C[76,77] has also been investigated.

^{19}F

The virtual lack of natural abundance of ^{19}F in the body precludes spectroscopy of that material as a component of tissue. However, this abundance does provide an opportunity for use of ^{19}F as a contrast agent. Fluorinated compounds have been used as blood substitutes.[78] Anesthetics containing fluorine have also been investigated and show promise as lipid-soluble compounds that may be suitable probes of membrane permeability.[79] One interesting feature of ^{19}F is the similarity of its gyromagnetic ratio to that of hydrogen (see Table 24-1). Thus a suitable concentration of ^{19}F may be useful for imaging in current scanners whose electronics are tuned for the reception of hydrogen.

CHEMICAL SHIFT IMAGING

In principle, the information obtained from the spectra of volume elements, or voxels, within a sample may be used to build up an image of the sample. In standard MRI, the spin density $N(\mathrm{H})$ and the relaxation parameters T1 and T2 determine the brightness of picture elements. These picture elements, or pixels, correspond to volume elements in the sample or patient. In chemical shift imaging (CSI), spectral information may be used directly to construct an image of a patient. A number of factors cause chemical shift imaging to be more difficult than proton imaging. In particular, the effects of magnetic field inhomogeneities tend to overshadow the small variations in resonance produced by the chemical shift.

Example 24-4

The separation of the main spectral peak due to water and the peak due to lipids (fat) is approximately 3 ppm. If the static magnetic field strength of an MRI unit is 1.5 tesla, what is the frequency shift in hydrogen between water and lipid?

$$\text{Chemical shift (ppm)} = \frac{f_{\mathrm{R}} - f_{\mathrm{P}}}{f_{\mathrm{N}}}(10^6) \qquad \textbf{(24-4)}$$

For hydrogen, with a gyromagnetic ratio of 42.6 MHz/tesla, a magnetic field strength of 1.5 tesla yields a nominal resonance frequency of

$$\begin{aligned} f &= \gamma B \\ &= (42.6\ \text{MHz/T})(1.5\text{T}) \\ &= 63.9\ \text{MHz} \\ &= f_{\mathrm{N}} \end{aligned}$$

Rewriting Eq. (25-4), we have

$$\begin{aligned} f_{\mathrm{R}} - f_{\mathrm{P}} = \text{Frequency shift} &= \frac{(\text{Chemical shift})(f_{\mathrm{N}})}{10^6} \\ &= \frac{(3)(63.9)}{10^6} \\ &= 0.000192\ \text{MHz} \\ &= 192\ \text{Hz} \end{aligned}$$

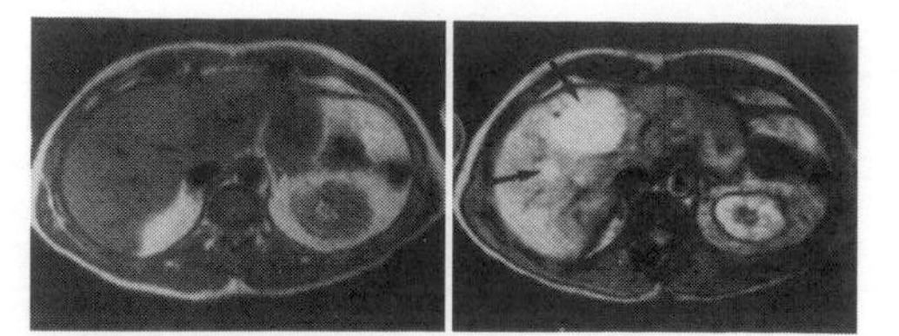

MARGIN FIGURE 24-23
Left: Conventional spin echo TR = 2000, TE = 30 shows only subtle difference in signal intensity between left and right hepatic lobes. **Right:** Chemical shift image (phase contrast "opposed phase" technique with same TR, TE as image on the left) shows high-intensity abnormalities consistent with metastatic cancer. Overall hepatic signal intensity is decreased due to diffuse fatty infiltration. The metastases do not contain MR observable fat, and their signal intensity relative to surrounding lever is increased. (From Stark, D. D., and Bradley, W. G. *Magnetic Resonance Imaging*. St. Louis, Mosby–Year Book, 1988, p. 240. Used with permission.)

Other factors, such as (a) variations in magnetic susceptibility of biologic materials within the body and (b) the presence of a strong water signal, render detection of small differences in resonance frequency problematic.

One of the main barriers to effective chemical shift imaging is that conventional 2DFT, the most widely used technique for spatial encoding of the MR signal, already uses frequency encoding for spatial location. Therefore, assignment of small variations in resonance frequency requires an alternative technique. Several options for signal localization have been considered.[80] One technique involves doing away with the frequency-encoding or read gradient and using only phase-encoding gradients stepped in both the x and y directions (Margin Figure 24-8). The obvious drawback of this technique is the need for $N \times N$ separate signal acquisitions, where N is the number of picture elements. Another CSI technique involves standard pulsing schemes for imaging with additional RF pulses.[81–83] The pulses for imaging are applied in the presence of a magnetic field gradient to allow spatial selection. Other pulses, applied in the absence of a magnetic field gradient, are used to excite certain lines in the spectrum of interest. To eliminate one or more lines from the spectrum that may obscure lines of interest, a "presaturation pulse" may be administered at the appropriate frequency to excite the unwanted line, followed by a dephasing gradient pulse to remove it from the image.[82,83]

A chemical shift image is shown in Margin Figure 24-23. At the present time, clinical applications of CSI have involved separation of fat and water components because they occur as relatively separable strong spectral lines. There is increasing interest in using spectral information from the phosphorous contained in various metabolites,[84] fluorine in blood-substitute compounds,[85] and intracellular and extracellular sodium.[80] Continued evolution of CSI shows considerable promise for the future.[86]

PROBLEMS

*24-1. A spin-echo pulse sequence has a TE of 100 msec. A new pulse sequence begins 800 msec after the MR signal or echo is received. Three signal averages and 256 phase encoding steps are used. Find the total image acquisition time.

*24-2. Assume that relaxation times for gray matter and white matter in the brain are as follows:

	T1	T2
Gray matter	800	100
White matter	700	90

In which one of the following pulse sequences would the white matter appear lighter than gray matter in the image?

a. TR = 750 msec, TE = 20 msec
b. TR = 2000 msec, TE = 750 msec
c. TR = 2000 msec, TE = 95 msec
d. TR = 1500 msec, TE = 200 msec

*24-3. The frequency shift for a spectral peak on an MR spectroscopy system is 200 Hz at a static magnetic field strength of 5.0 tesla. What is the frequency shift of the same peak at a static magnetic field strength of 7.0 tesla?

24-4. Describe the mechanism of contrast enhancement when gadolinium is used as a contrast agent for MRI.

24-5. Describe three motion suppression techniques for MRI that involve acquiring the image in a shorter time (i.e., "fast scan" techniques).

*For problems marked with an asterisk, answers are provided on p. 493.

SUMMARY

- Immediately after the application of the first 90-degree pulse in a sequence, the maximum MR signal is produced. However, free induction decay causes the signal to disappear rapidly.
- The spin-echo technique allows recovery of an MR signal at a preselected time (TE) and removes the effects of T2* relaxation.
- In spin-echo pulse sequences, the choice of TR primarily affects T1 contrast and the choice of TE primarily affects T2 contrast in the image.

- Materials having greater spin density tend to have a stronger MR signal for all values of TR and TE.
- Gradient magnetic fields are used to isolate a slice and then allow frequency and phase encoding of the signal.
- In two-dimensional Fourier transform image acquisition, the resulting image is a mapping of MR signal as a function of phase and frequency.
- Material may be classified according to their magnetic properties as
  - Diamagnetic
  - Paramagnetic
  - Ferromagnetic
- Motion suppression techniques include
  - Respiratory gating
  - Respiratory-ordered phase-encoding
  - Gradient moment nulling
  - Gradient echo techniques
  - Half-Fourier imaging
- The physiological basis of functional MRI (fMRI) is that neuronal activity in localized regions of the brain changes the level of blood oxygenation in those regions. Deoxyhemoglobin is paramagnetic, but oxyhemoglobin is diamagnetic. The increase in oxyhemoglobin concentration in capillaries and veins that follows neuronal activation alters T2 and T2* contrast in the region of neuronal activation.
- Spectroscopy is an analysis of the small (parts per million) shift in resonance frequency of nuclei as a result of surrounding chemical bonds.
- The chemical shift between water and lipids is large enough to affect frequency encoding of the MR signal and can affect the appearance of an MR image or be used as the basis of chemical shift imaging.

REFERENCES

1. Carr, H. Y., and Purcell, E. M. Effects of diffusion on free precession in nuclear magnetic resonance experiments. *Physiol. Rev.* 1954; **94**:630–638.
2. Meiboom, S., and Gill, D. Modified spin-echo method for measuring nuclear relaxation times. *Rev. Sci. Instrum.* 1958; **29**:688–691.
3. Hinshaw, W. S. Image formation by nuclear resonance: The sensitive point method. *J. Appl. Physiol.* 1976; **47**:3709–3721.
4. Andrew, E. R., and Bottomley, P. A., Hinshaw, W. S., et al. NMR images by the multiple sensitive pint method: Application to larger biological systems. *Phys. Med. Biol.* 1977; **22**:971–974.
5. Hinshaw, W. S. Spin mapping: The application of moving gradients to NMR. *Phys. Lett.* 1974; **48A**:87–88.
6. Mansfield, P., and Maudsley, A. A. Medical imaging by NMR. *Br. J. Radiol.* 1977; **50**:188–194.
7. Lauterbur, P. C. Image formation by induced local interactions: Examples employing nuclear magnetic resonance. *Nature* 1973; **242**:190–191.
8. Kumar, A., Welti, I., and Ernst, R. R. NMR Fourier zeugmatography. *J. Magn. Reson.* 1975; **18**:69–83.
9. Wood, M. L., and Henkelman, R. M. The magnetic field dependence of the breathing artifact. *Magn. Reson. Imaging* 1986; **4**:387–392.
10. Edelman, R. R., Hahn, P. F., Buxton, R., et al. Rapid MR imaging with suspended respiration: Clinical application in the liver. *Radiology* 1986; **161**:125–131.
11. Runge, V. M., Clanton, J. A., Partain, C. L., et al. Respiratory gating in magnetic resonance imaging at 0.5 tesla. *Radiology* 1984; **151**:521–523.
12. Ehman, R. L., McNamara, M. T., Pallack, M., et al. Magnetic resonance imaging with gating: Techniques and advantages. *AJR* 1985; **143**:1175–1182.
13. Stark, D. D., Wittenberg, J., Edelman, R. R., et al. Detection of hepatic metastases: Analysis of pulse sequence performance in MR imaging. *Radiology* 1986; **159**:365–370.
14. Axel, L., Summers, R. M., Kresel, H. Y., et al. Respiratory effects in two-dimensional Fourier transform MR imaging. *Radiology* 1986; **161**:795–801.
15. Stark, D. D., Hendrick, R. E., Hahn P. F., et al. Motion artifact suppression with fast spin-echo imaging. *Radiology* 1987; **164**:183–192.
16. Young, I. R. Special Pulse Sequences and Techniques, in Stark, D. D., Bradley, W. G. (eds.), *Magnetic Resonance Imaging*. St. Louis, Mosby–Year Book, 1988, pp. 84–106.
17. Bailes, D. R., Gilderdale, D. J., Bydder, G. M., et al. Respiratory ordering of phase encoding (ROPE): A method for reducing respiratory motion artifacts in MR imaging. *J. Comput. Assist. Tomogr.* 1985; **9**:835–838.
18. Haacke, E. M., and Patrick, J. L. Reducing motion artifacts in two-dimensional Fourier transform imaging. *Magn. Reson. Imaging* 1986; **4**:359–376.
19. Frahm, J., Haase, A., and Matthaei, D. Rapid NMR imaging of dynamic processes using the FLASH technique. *Magn. Reson. Med.* 1986; **3**:321.
20. Van der Meulen, P., Cuppen, J. J. M., and Groen, J. P. Very fast MR imaging by field echoes and small angle excitation. *Magn. Reson. Imaging* 1985; **3**:297–298.
21. Haase, A., Frahm, J., Matthaei, D., et al. FLASH imaging. Rapid NMR imaging using low flip-angle pulses. *J. Magn. Reson.* 1986; **67**:258–266.
22. Oppelt, A., Graumann, R., Barfub, H., et al. FISP—A new fast MRI sequence. *Electromedica* 1986; **54**:15–18.
23. Haacke, E. M., and Lenz, G. W. Improving MR image quality in the presence of motion by using rephasing gradients. *AJR* 1987; **148**:1251–1258.
24. Pattany, P. M., Phillips, J. J., Chiu, L. C., et al. Motion artifact suppression technique (MST) for MR imaging. *J. Comput. Assist. Tomogr.* 1987; **11**:369–377.
25. Feinberg, D. A., Crooks, L. E., Hoenninger, J. C., et al. Halving MR imaging time by conjugation: Demonstration at 3.5 kG. *Radiology* 1986; **161**:527–532.

26. Rzedzian, R. R., and Pykett, I. L. Instant images of the human heart using a new, whole-body MR imaging system. *AJR* 1987; **149:**245–250.
27. Farrar, T. C., Becker, E. D. *Pulse and Fourier Transform NMR*. New York, Academic Press, 1971.
28. Kumar, N. G., Karstaedt, N., and Moran, P. R. Fast scan 2DFT imaging at 0.15T by field echoes and small pulse angle excitations [Abstract]. *Magn. Reson. Imaging* 1986; **4:**108–109.
29. Glover, G. Unpublished data.
30. Haase, A., Frahm, J., Matthaei, D., et al. FLASH imaging: Rapid NMR imaging using low flip angle pulses. *J. Magn. Reson.* 1986; **67:**217–225.
31. Oppelt, A., Graumann, R., Barfub, H., et al. FISP—A new fast MRI sequence. *Electromedica* 1986; **54:**15–18.
32. Sears, F. W., and Zemansky, M. W. *University Physics*. Reading, MA, Addison-Wesley, 1970, pp. 471 and 487.
33. Listinsky, J. J., Bryant, R. G. Gastrointestinal contrast agents: A diamagnetic approach. *Magn. Reson. Med.* 1988; **8:**285–292.
34. Weinmann, H.-J., et al. Characteristics of gadolinium–DTPA complex: A potential NMR contrast agent. *AJR* 1984; **142:**619.
35. Mattrey, R. F., Hajek, P. C., Gylys-Morin, V. M., et al. Perfluoro chemicals as gastrointestinal contrast agents for MR imaging: Preliminary studies in rats and humans. *AJR* 1987; **148:**1259–1263.
36. Weinreb, J. C., Maravilla, K. R., Redman, H. C., et al. Improved MR imaging of the upper abdomen and glucagon and gas. *J. Comput. Assist. Tomogr.* 1984; **8:**835–838.
37. Orrison, W. W., Lewine, J. D., Sanders, J. A., and Hartshorne, M. F. *Functional Brain Imaging*. Mosby, New York, 1995, p. 242.
38. Fenstermacher, J. D. The flow of water in the blood–brain cerebrospinal fluid system. *Syllabus*, Society of Magnetic Resonance in Medicine FMRI Workshop, Arlington VA, 1993, pp. 9–17.
39. Fox, P. T., and Raichle, M. E. Focal physiological uncoupling of cerebral blood flow and oxidative metabolism during somatosensory stimulation in human subjects. *Neurobiology* 1986; **83:**1140–1144.
40. Belliveau, J. W., Kennedy, D. N., McKinstry, R. C., et al. Functional mapping of the human visual cortex by magnetic resonance imaging. *Science* 1991; **254:**716–719.
41. Kwong, K. K., Belliveau, J. W., Chesler, D. A., et al. Dynamic magnetic resonance imaging of human brain activity during primary sensory stimulation. *Proc. Natl. Acad. Sci.* USA 1992; **89:**5675–5679.
42. Grubb, R. L., Raichle, M. E., Eichling, J. O., et al. The effects of change in Pa-CO_2 on cerebral blood volume, blood flow and vascular mean transit time. *Stroke* 1974; **5:**630–639.
43. Ugurbil, K., Adriany, G., Andersen, G., Chen, W., Gruetter, R., Hu, X., Merkle, H., Kim, D. S., Kim, S. G., Strupp, J., Zhu, S. H., and Ogawa, S. Magnetic resonance studies of brain function and neurochemistry. *Annu. Rev. Biomed. Eng.* 2000; **2:**633–660.
44. Weng, X., Ding, Y. S., Volkow, N. D. Imaging the functioning human brain. *Proc. Natl. Acad. Sci. USA* 1999; 28; **96**(20):11073–11074.
45. George, J. S., Aine, C. J., Mosher, J. C., Schmidt, D. M., Ranken, D. M., Schlitt, H. A., Wood, C. C., Lewine, J. D., Sanders, J. A., and Belliveau, J. W. Mapping function in the human brain with magnetoencephalography, anatomical magnetic resonance imaging, and functional magnetic resonance imaging [Review] [90 refs]. *J. Clin. Neurophysiol.* 1995; **12**(5):406–431.
46. Schulder, M., Maldjian, J. A., Liu, W. C., Mun, I. K., Carmel, P. W. Functional MRI-guided surgery of intracranial tumors. *Stereotact. Funct. Neurosurg.* 1997; **68**(1–4 Pt 1):98–105.
47. Roux, F. E., Ranjeva, J. P., Boulanouar, K., Manelfe, C., Sabatier, J., Tremoulet, M., and Berry, I. Motor functional MRI for presurgical evaluation of cerebral tumors. *Stereotact. Funct. Neurosurg.* 1997; **68**(1–4 Pt 1):106–111.
48. Maldjian, J. A., Schulder, M., Liu, W. C., Mun, I. K., Hirschorn, D., Murthy, R., Carmel, P., and Kalnin, A: Intraoperative functional MRI using a real-time neurosurgical navigation system. *J. Comput. Assist. Tomogr.* 1997; **21**(6):910–912.
49. Witt, T. C., Kondziolka, D., Baumann, S. B., Noll, D. C., Small, S. L., Lunsford, LD: Preoperative cortical localization with functional MRI for use in stereotactic radiosurgery. *Stereotact. Funct. Neurosurg.* 1996; **66**(1–3):24–29.
50. Righini, A., de Divitiis, O., Prinster, A., Spagnoli, D., Appollonio, I., Bello, L., Scifo, P., Tomei, G., Villani, R., Fazio, F., and Leonardi, M. Functional MRI: primary motor cortex localization in patients with brain tumors. *J. Comput. Assist. Tomogr.* 1996; **20**(5):702–708.
51. Nitschke, M. F., Melchert, U. H., Hahn, C., Otto, V., Arnold, H., Herrmann, H. D., Nowak, G., Westphal, M., and Wessel, K. Preoperative functional magnetic resonance imaging (fMRI) of the motor system in patients with tumors in the parietal lobe. *Acta Neurochir.* 998; **140**(12):1223–1229.
52. Gering DT, Weber DM: Intraoperative, real-time, functional MRI. *J. Magn. Reson. Imaging* 1998; **8**(1):254–257.
53. Roux, F. E., Boulanouar, K., Ranjeva, J. P., Tremoulet, M., Henry, P., Manelfe, C., Sabatier, J., and Berry, I. Usefulness of motor functional MRI correlated to cortical mapping in Rolandic low-grade astrocytomas. *Acta Neurochir.* 1999; **141**(1):71–79.
54. Welch, K. M., Cao, Y., and Nagesh, V. Magnetic resonance assessment of acute and chronic stroke. [Review] [60 refs] *Prog. Cardiovasc. Dis.* 2000; **43**(2):113–134.
55. Neumann-Haefelin, T., Moseley, M. E., and Albers, G. W. New magnetic resonance imaging methods for cerebrovascular disease: emerging clinical applications [Review] [94 refs]. *Ann. Neurol.* 2000; **47**(5):559–570.
56. Benson, R. R., FitzGerald, D. B., LeSueur, L. L., Kennedy, D. N., Kwong, K. K., Buchbinder, B. R., Davis, T. L., Weisskoff, R. M., Talavage, T. M., Logan, W. J., Cosgrove, G. R., Belliveau, J. W., and Rosen, B. R. Language dominance determined by whole brain functional MRI in patients with brain lesions. *Neurology* 1999; **52**(4):798–809.
57. Dymarkowski, S., Sunaert, S., Van Oostende, S., Van Hecke, P., Wilms, G., Demaerel, P., Nuttin, B., Plets, C., and Marchal, G. Functional MRI of the brain: Localisation of eloquent cortex in focal brain lesion therapy. *Eur. Radiol.* 1998; **8**(9):1573–1580.
58. Teasdale, J. D., Howard, R. J., Cox, S. G., Ha, Y., Brammer, M. J., Williams, S. C., and Checkley, S. A. Functional MRI study of the cognitive generation of affect. *Am. J. Psychiatry* 1999; **156**(2):209–215.
59. Schlosser, R., Hutchinson, M., Joseffer, S., Rusinek, H., Saarimaki, A., Stevenson, J., Dewey, L., and Brodie, J. D. Functional magnetic resonance imaging of human brain activity in a verbal fluency task. *J. Neurol. Neurosurg. Psychiatry* 1998; **64**(4):492–498.
60. Muri, R. M., Iba-Zizen, M. T., Derosier, C., Cabanis, E. A., and Pierrot-Deseilligny, C. Location of the human posterior eye field with functional magnetic resonance imaging. *J. Neurol. Neurosurg. Psychiatry* 1996; **60**(4):445–448.
61. Binder, J. R., Rao, S. M., Hammeke, T. A., Frost, J. A., Bandettini, P. A., Jesmanowicz, A., Hyde, J. S. Lateralized human brain language systems demonstrated by task subtraction functional magnetic resonance imaging. *Arch. Neurol.* 1995; **52**(6):593–601.
62. Rosen, H. J., Ojemann, J. G., Ollinger, J. M., and Petersen, S. E. Comparison of brain activation during word retrieval done silently and aloud using fMRI. *Brain Cogn.* 2000; **42**(2):201–217.
63. Loring, D. W., Meador, K. J., Allison, J. D., and Wright, J. C. Relationship between motor and language activation using fMRI. *Neurology* 2000; **54**(4):981–983.
64. Seger, C. A., Poldrack, R. A., Prabhakaran, V., Zhao, M., Glover, G. H., and Gabrieli, J. D. Hemispheric symmetries and individual differences in visual concept learning as measured by functional MRI. *Neuropsychologia* 2000; **38**(9):1316–1324.
65. Reber, P. J., Stark, C. E., and Squire, L. R. Cortical areas supporting category learning identified using functional MRI. *Proc. Nat. Acad. Sci. USA* 1998; **95**(2):747–750.
66. Gaillard, W. D., Hertz-Pannier, L., Mott, S. H., Barnett, A. S., LeBihan, D., and Theodore, W. H. Functional anatomy of cognitive development: fMRI of verbal fluency in children and adults. *Neurology* 2000; **54**(1):180–185.
67. Young, I. R. Special Pulse Sequences and Techniques, in Stark, D. D., and Bradley, W. G., Jr. (eds.), *Magnetic Resonance Imaging*, Vol. 6. St Louis, Mosby–Year Book, 1988, pp. 84–107.
68. Moseley, M. E., Berry, I., Chew, W. M., et al. Magnetic resonance spectroscopy: Principles and potential applications, in Brant-Zawadzki, M., and Norman, D. (eds.), *Magnetic Resonance Imaging of the Central Nervous System*. New York, Raven Press, 1987, pp. 107–113.
69. Seeley, P. J., Busby, S. J. W., Gadian, DG: A new approach to metabolite compartmentation in muscle. *Biochem. Soc. Trans.* 1976; **4:**62–66.
70. Koretsky, A. P., and Weiner, M. W. Phosphorus-31 Nuclear Magnetic Resonance Magnetization Transfer Measurement of Exchange Reactions *in Vivo*, in James, T. L., and Margulis, A. R. (eds.), *Biomedical Magnetic Resonance*.

San Francisco, Radiology Research and Education Foundation, 1984, pp. 209–231.

71. Hilal, S. K., Maudsley, A. A., Simon, H. E., et al. *In vivo* NMR imaging of tissue sodium in the intact cat before and after acute cerebral stroke. *AJNR* 1983; **4**:245.
72. Moseley, M. E., Chew, W. M., and James, T. L. ^{23}Sodium magnetic resonance imaging, in James, T. L., and Margulis, A. R. (eds.), *Biomedical Magnetic Resonance*. San Francisco, Radiology Research and Education Foundation, 1984.
73. Adam, W. R., Koretsky, A. P., and Weiner, M. W. Measurement of tissue potassium *in vivo* using ^{39}K NMR. *Biophys. J.* 1987; **51**:265.
74. Matson, G. B., and Weiner, M. W. MR Spectroscopy *in vivo*: Principles, animal studies, and clinical applications, in Stark, D. D., Bradley W. G., Jr. (eds.), *Magnetic Resonance Imaging*. St. Louis, Mosby–Year Book, 1988, pp. 201–228.
75. Shulman, R. G., and Cohen, S. M. Simultaneous ^{13}C and ^{31}P NMR studies of perfused rat liver. *J. Biol. Chem.* 1983; **258**:14294.
76. Sillerud, L. O., and Shulman, R. G. Structure and metabolism of mammalian liver glycogen monitored by carbon-13 nuclear magnetic resonance. *Biochemistry* 1983; **22**:1087.
77. Koutcher, J. A., Burt, C. T., Lauffer R. B., et al. Contrast agents and spectroscopic probes in NMR. *J. Nucl. Med.* 1984; **25**:506–513.
78. Nunnally, R. L., Peshock, R. M., and Rehr, R. B. Fluorine-19 (^{19}F) NMR *in Vivo*: Potential for flow and perfusion [Abstract], in *Proceedings of the Society of Magnetic Resonance in Medicine. Second Annual Meeting*. San Francisco, Society of Magnetic Resonance in Medicine, 1983, p. 266.
79. Wyrwicz, A. M., Pszenny, M. H., Schofield, J. C., et al. Noninvasive observations of fluorinated anesthetics in rabbit brain by fluorine-19 nuclear magnetic resonance. *Science* 1983; **22**:428.
80. Axel, L. Chemical shift imaging, in Stark, D., Bradley, W. G., Jr. (eds.), *Magnetic Resonance Imaging*, Vol. 12. St. Louis, Mosby–Year Book, 1988, pp. 229–243.
81. Dumoulin, C. L. A method of chemical-shift-selective imaging. *Magn. Reson. Med.* 1985; **2**:583.
82. Haase, A., et al. ^{1}H NMR chemical shift selective (CHESS) imaging. *Phys. Med. Biol.* 1985; **30**:341.
83. Hall, L. D., Subramaniam, S., and Talagala, S. L. Chemical-shift-resolved tomography using frequency-selective excitation and suppression of specific resonance. *J. Magn. Reson.* 1984; **56**:275.
84. Maudsley, A. A., and Hilal, S. K. Field inhomogeneity correction and data processing for spectroscopic imaging. *Magn. Reson. Med.* 1985; **2**:218.
85. Parhami, P., and Fung, B. M. Fluorine-19 relaxation study of perfluoro chemicals as oxygen carriers. *J. Phys. Chem.* 1983; **87**:1928.
86. Runge, V. M., and Wood, M. L. Fast imaging and other motion artifact reduction schemes: A pictorial overview. *Magn. Reson. Imaging* 1988; **6**:595–608.

CHAPTER

25

MAGNETIC RESONANCE IMAGING: INSTRUMENTATION, BIOEFFECTS, AND SITE PLANNING

OBJECTIVES

After completing this chapter, the reader should be able to:

- List the components of an MRI system, state their principle of operation, and describe their contribution to the system.
- List some commonly encountered MRI artifacts, their causes, and their remedies.
- List typical quality assurance tests and state the frequency with which they should be performed.
- Give examples of bioeffects and explain the biophysical reason for each bioeffect.
- Discuss the factors that must be considered in MRI site planning.

MAIN SYSTEM MAGNET

In magnetic resonance imaging (MRI) and spectroscopy, the static magnetic field is supplied by the main system (static) magnet. This magnetic field defines the axis about which protons precess. The static field direction is also the reference axis for the nutation angle of the protons following the application of radio-frequency (RF) pulses. A range of field strengths is used for imaging, from millitesla up to 10 tesla in some research systems, with the bulk of clinical imaging conducted in the range of 0.1 to 3.0 tesla. Currently, it is not clear whether there is an optimum field strength for certain imaging situations. Some say that "more field strength is better" if the engineering challenges presented by high-field strength systems can be overcome. It is true that the RF signal increases with field strength.[1] T1 also increases with field strength, however, and the magnitude of T1 influences the selection of parameters for the pulsing scheme. These, in turn, influence image-degrading factors such as noise and patient motion. Furthermore, it is the relative contrast between two tissues, as compared with noise, that is of interest in imaging. For a given field strength, therefore, the optimization of pulse parameters depends upon the tissues of interest.[2]

Other challenges associated with the use of higher field strengths for MRI include difficulties in maintaining magnetic field homogeneity, increased RF heating, and chemical shift artifacts. The Larmor equation calls for higher RF frequencies as field strength is increased. Higher frequencies are more readily absorbed, which causes more RF heating of the patient. Chemical shift artifacts occur because higher field strengths shift the resonance frequencies of fat and water molecules. Because frequency encoding is used for spatial localization, the position of fat layers may be shifted in the image. One might infer from these challenges that lower field strengths will continue over the foreseeable future to be useful in magnetic resonance imaging.

There are two types of main system magnets: electromagnets and permanent magnets.

Electromagnets

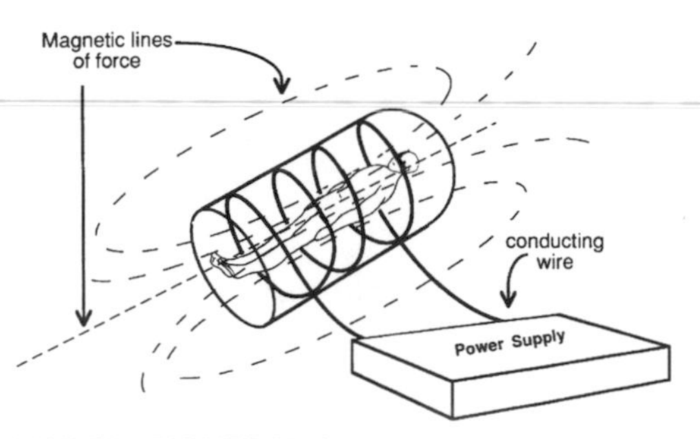

MARGIN FIGURE 25-1
A solenoidal electromagnet produces magnetic lines of force that are concentrated along the axis of the solenoid or coil.

An electromagnet is formed by twisting conducting wire in the shape of a coil or solenoid. Any wire that carries an electric current is surrounded by a magnetic field. When the conducting wire is twisted into the shape of a coil, the magnetic field from the coil loops reinforce each other to produce a magnetic field that is strongest in the coil center. Magnetic field "lines" are invisible lines in space that define the direction of force of the magnetic field upon a ferromagnetic object such as iron. The magnetic field lines at the center of a solenoidal electromagnet lie along the axis of the coil (Margin Figure 25-1).

Two types of electromagnets, resistive and superconductive, are used in MRI. The field strength of an electromagnet is limited by the amount of current that may be carried by the conducting wire. Some portion of the energy of an electric current is dissipated as heat in the wire because the conducting material resists the flow of electrons. Because of this heat, resistive-type MRI magnets must be water-cooled.

Attempting to force high currents through the wire produces excessive heat in the wire. For this reason, the maximum field attainable with a resistive MRI magnet is usually below 0.5 tesla. Power usage is also an issue; a typical resistive MRI system requires approximately 30 kW of power, the equivalent of 300 100-W light bulbs.

To achieve higher field strengths in magnets that are large enough for imaging human patients, conducting wire must be used in which there is no resistance. This is the property of superconductivity described in Chapter 2. Superconductive magnets allow higher currents and therefore higher field strengths. To maintain superconductivity, conductors must be kept near the temperature at which the element helium is a liquid (−269°C). This temperature is near absolute zero (0°K, or −273°C). Liquid helium is circulated around the superconducting wire in an insulating chamber called a dewar. A second insulating chamber, containing liquid nitrogen at −196°C, is used to help maintain the helium in its liquid state.

Clinical MRI using superconducting magnets is now routinely performed at magnetic field strengths of up to 2 tesla. Research with systems having bore sizes sufficiently large for humans is performed at up to 9 tesla. Small-bore research systems for animal imaging and spectroscopy exist at field strengths up to 10 tesla. There is hope that recent advances in the fabrication of ceramics that exhibit superconductivity at elevated temperatures may permit significant advances in magnetic resonance (MR) magnet technology in the not-too-distant future.

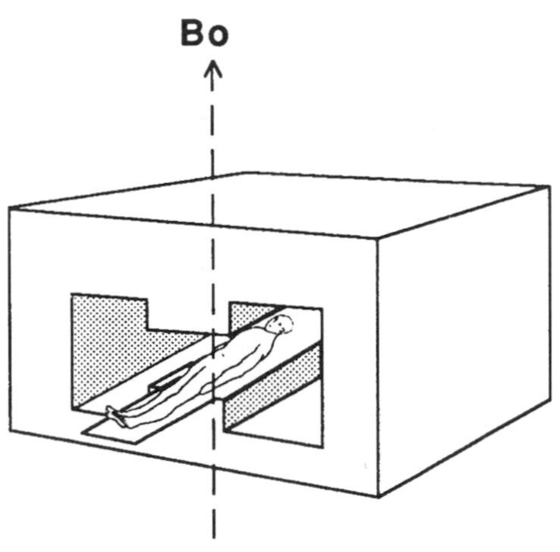

MARGIN FIGURE 25-2
Diagram of a permanent magnet MRI gantry. The orientation of the patient is shown in the static magnetic field (B_0) produced by the poles of the magnet.

1 gauss is equal to 0.1 millitesla.

Permanent Magnets

Permanent magnets are composed of ferromagnetic materials that maintain their magnetic properties with no additional input of energy. Large assemblies of ferromagnetic materials for clinical MRI systems have field strengths of up to 0.3 tesla. In such a system the patient is placed between the poles of the permanent magnet with the direction of magnetic field lines perpendicular to the axis of the patient (Margin Figure 25-2).

GRADIENT MAGNETIC FIELDS

The gradients required for spatial encoding of the MR signal are provided by gradient electromagnetic coils. The magnetic field generated by a gradient coil is substantially smaller than the field of the main system magnet. A typical rate of change of field strength is several gauss per centimeter over a distance of 100 cm. By applying the current in opposite directions to two loops on either side of the patient, magnetic fields are generated to provide a range of field strengths over the patient. These conducting loops are oriented to provide variations in magnetic field intensity along three orthogonal axes of the patient. An illustration of the orientation of the coils is shown in Margin Figure 25-3.

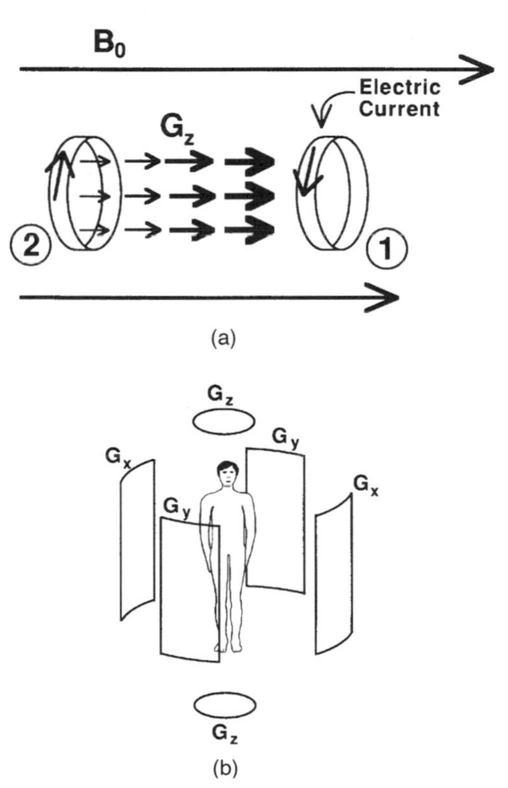

MARGIN FIGURE 25-3
A: Configuration of gradient coils in opposing current loops. The current in coil 1 augments the main-system magnetic field, B_0, while the current in coil 2 opposes the main-system magnetic field. The result is a gradient field, G_z, a field that varies with position along the z-axis. **B:** An illustration of the six coils that result in three gradient fields, G_x, G_y, G_z.

RADIO-FREQUENCY COILS

RF coils send and receive radio waves in a fashion similar to a radio antenna. The term "coil" is used rather than "antenna" when the source and receiver are very close, as is the case in MRI. When pulses are sent into the patient, the coil is the source, and the patient is the receiver. When the MR signal is read out, the patient is the source, and the coil is the receiver. In either case, the source and receiver are within a few coil diameters of each other.

Sending and Receiving Coils

The sending coil of an MRI system is usually built into the bore of the magnet (Margin Figure 25-4). Precise positioning of the patient within the sending coil is necessary

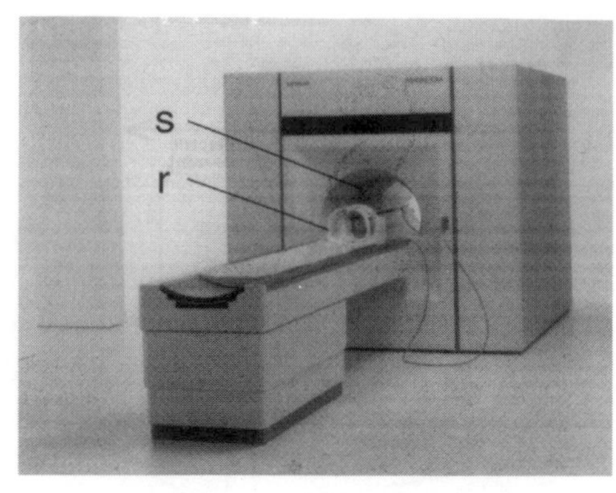

(a)

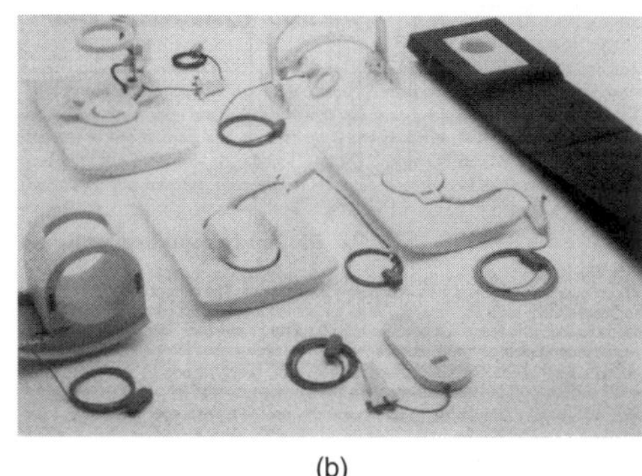

(b)

MARGIN FIGURE 25-4
A: The radio-frequency (RF) coil for sending is usually located within the bore of the magnet, while the RF coil for receiving is located on the patient table. **B:** An assortment of RF "surface coils" for imaging various body parts. (Courtesy of Siemens Medical Systems.)

to ensure that the anatomy of interest appears in the image. In most imaging units it is difficult to visualize patient position after the patient has entered the bore of the magnet. Therefore, a positioning system is used that involves the use of external markers or a patient table indexing scale.

Under some circumstances, the use of separate sending and receiving coils may be eliminated. The same coil may send and receive radio waves in the same fashion as radio antennas may be both senders and receivers. There is some interest in eliminating the in-bore sending coil to provide more room in the bore to help alleviate claustrophobic effects in patients. Alternately, deletion of the in-bore coil would allow a reduction in bore size that could resolve some of the design problems inherent in MRI magnet construction. The RF pulse produced in a small surface coil is less uniform than that provided by an in-bore solenoidal sending coil. It is possible to overcome this limitation with special pulsing techniques.[3]

Factors Affecting the Performance of Receiver Coils

Margin Figure 25-4 shows an assortment of receiver coils. The shape of the coil may be influenced by the body part being imaged. For small samples, a solenoidal coil completely surrounding the sample is often used.[4] For many parts of the human body, however, a surface coil is more practical. A surface coil is essentially a loop of conducting material, such as copper tubing, that is placed adjacent to the body part being imaged. The loop may form various shapes and be bent slightly to conform to a body part such as a joint.

A rule of thumb for surface coils is that the sensitivity decreases appreciably beyond a distance equal to the diameter of the coil.[5] Thus, positioning of the coil is an important determinant of performance. Surface coils, for example, are often positioned on the patient prior to entry of the patient into the magnet bore.

A fundamental limitation in the sensitivity of a receiver coil is the presence of spurious electrical signals (electronic noise) that do not contribute diagnostic information to the MR signal. A major source of noise is the movement of electrically charged molecules such as electrolytes within the patient. Other sources of noise produced in the coil itself are usually of smaller magnitude than patient noise.[6] Improvements in imaging performance are usually more achievable through increased signal rather than decreased noise.

Maximization of signal from a receiver coil requires "tuning" of the coil to a resonance frequency. The resonance frequency for a coil is the frequency at which transfer of energy from the patient to the coil is most efficient. Just as an organ pipe or a bell has a natural frequency of vibration (musical tone), an electronic component such as a coil has a natural electrical frequency. Operation of a coil at frequencies that are "off resonance" results in a reduction in signal strength.

The resonance frequency of a coil is determined by the geometry of the coil, including its length, diameter, number of loops, and so on. The resonance frequency may be adjusted by adding or subtracting capacitance, the ability of a circuit to store electrical charge. The resonance frequency of a coil changes when the coil is "loaded"—that is, when a patient is inserted into the magnet. The frequency of the coil may then have to be readjusted by changing the setting of variable or "tuning" capacitors. This adjustment may be an automatic part of signal optimization in the system, or the operator may have to make manual adjustments.

ELECTRONIC COMPONENTS

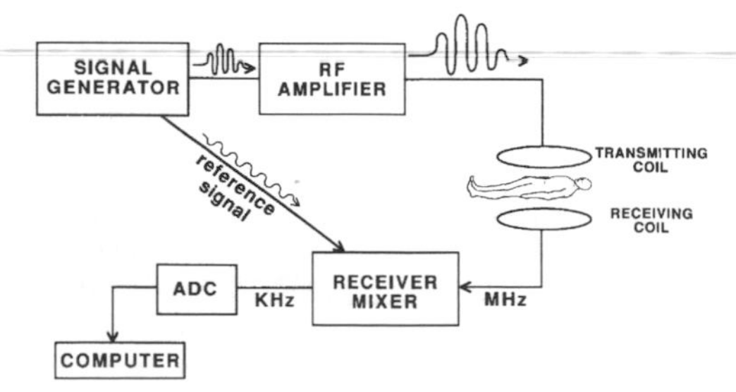

MARGIN FIGURE 25-5
Diagram of the main electronic components of an MRI system. The functions of the various components are described in the text.

The operation of an MRI system requires the use of several standard electronic components (Margin Figure 25-5). The transmitted radio waves originate in a device known as a signal generator. The signal generator's task is to produce a signal at a precise frequency and desired shape. The magnitude of the signal is then increased by an RF amplifier.

The receiver electronics include a mixer that combines the received signal (MR signal obtained from the patient by the receiver coil) with a reference signal at or near the Larmor frequency. The effect of mixing the signals is to subtract out the Larmor frequency, leaving only the small shifts in frequency. These frequency shifts are used for spatially encoding the MR signal. After mixing, a low-frequency (kilohertz instead of megahertz) MR signal remains. The purpose of mixing is to provide a low-frequency signal that can be digitized more precisely than a high-frequency signal. In a given cycle of the signal, the analog-to-digital converter (ADC) has more time to divide the signal into successive digital values to yield greater precision.

COMPUTER

As signals from the ADC are acquired, they are transferred to a temporary storage location in the computer as they are acquired. This storage area, called the buffer, allows the ADC to work at maximum speed. Additional memory, such as random access memory (RAM) in the computer, must be of sufficient size to permit storage of all pulse sequence software instructions. These instructions include the acquisition scheme [e.g., two-dimensional Fourier transform (2DFT)], number of slices, number of phase-encoding steps, and number of signal averages. An array processor is required for rapid reconstruction of images. Because the array processor must be able to operate on an entire data set, RAM must include space for a least several images. External storage of images may then be accomplished on magnetic tape or disk. Increasingly, both external and long-term storage are provided by optical disk systems.

ARTIFACTS

Artifacts in MR images may be grouped according to the system components most responsible for their occurrence (e.g., static magnetic field, gradient magnetic field, and RF artifacts). Other artifacts include motion artifacts (including flow), chemical shift, and foreign object artifacts.

Foreign object artifacts are sometimes referred to as "susceptibility" artifacts.

Foreign Objects

Most tissues in the body are diamagnetic and repel magnetic fields slightly. If paramagnetic or ferromagnetic objects, which attract magnetic fields, are present within tissue, they distort the magnetic field (Margin Figure 25-6) and may affect spatial encoding. For example, a ferromagnetic object produces a strong local magnetic field, as much as 100 times the static field of the magnet in its vicinity. This distortion of the magnetic field creates a signal void (dark region) surrounded by a white border at the location of the artifact. The extent of the distortion depends upon the size and composition of the object. Some common foreign objects that produce susceptibility artifacts include surgical clips, prosthetics, and shrapnel.

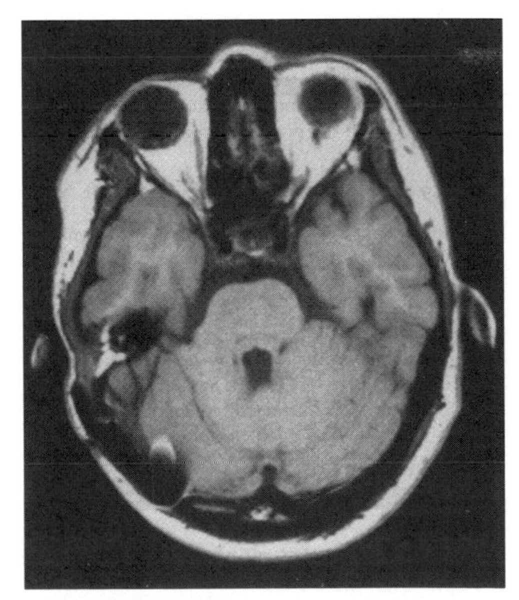

MARGIN FIGURE 25-6
Radio-frequency (RF) artifact caused by a metal object in the field of view.

Magnetic Field

Factors other than foreign objects may produce artifacts by causing abrupt changes in the static magnetic field across an image. For example, shim coils may be improperly adjusted, or inhomogeneities may arise at the edges of especially large fields of view. These effects cause distortion of the image due to disruption of spatial encoding.

Gradient Field

The usual spatial-encoding schemes assume that the gradient magnetic fields are linear. If nonlinearities exist along any of the gradient axes (slice, frequency, or

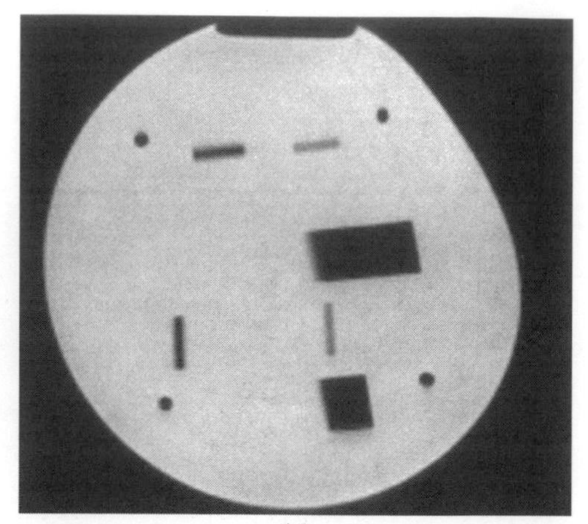

(a)

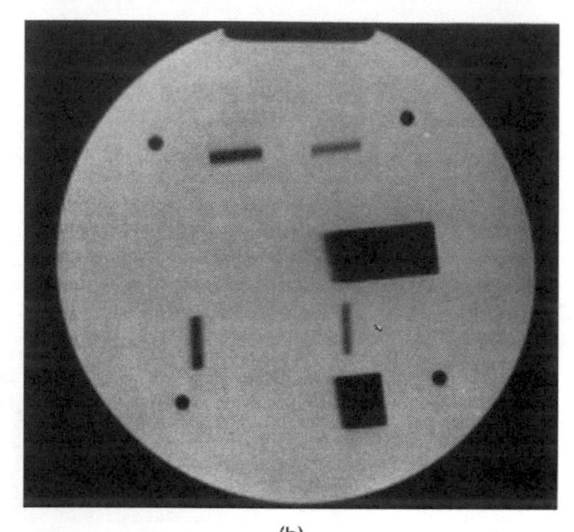

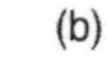

(b)

MARGIN FIGURE 25-7
Artifact caused by gradient nonlinearity. The artifact may be alleviated by switching the phase-encoding direction to coincide with the direction in which the gradient is most nonlinear.

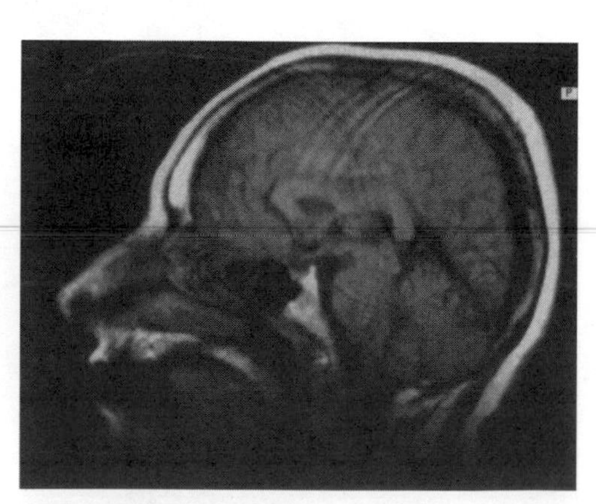

MARGIN FIGURE 25-8
Artifact caused by patient motion. The artifact is propagated along the direction in which phase encoding is performed.

phase-encoding directions), then variations in signal intensity will occur (Margin Figure 25-7). The phase-encoding process is the least sensitive of the three to small variations in linearity. Therefore, a standard technique for reducing the effect of gradient field nonlinearities is to choose the phase-encoding direction to coincide with the gradient direction with poorest linearity.

Radio Frequency

Radio waves from an outside source can interfere with signals received from the patient and be misinterpreted as pictorial information. These RF artifacts may appear as a line or band across the image in a direction perpendicular to the frequency-encoding direction. They represent RF "noise" as a single frequency or narrow range of frequencies. Artifacts may also affect the image in a more general but obviously nonphysiologic fashion such as a "herring bone" pattern.

Some RF-related problems in MR images are not caused by outside interference but instead by problems with internal components of the system such as malfunction of the RF transmitter, poor electrical connections, or failure of circuits associated with the receiver coil.

Motion

In MRI, patient motion causes artifacts that project through the image in the direction of phase encoding (Margin Figure 25-8). Phase encoding is usually performed over a number of phase-encoding steps, whereas frequency encoding takes place during a small number of pulse repetition periods. For this reason, spatial localization of signals in the phase-encoding direction is more easily disrupted than is localization along the frequency axis. Common sources of motion artifacts are gross patient movement, respiratory motion, cardiac motion, and flow. Flow artifacts may appear as decreased, increased, or variable changes in image brightness. Rapid flow may appear as decreased intensity because spins entering the slice during signal readout were not present during slice selection and do not provide a coherent signal. Less rapid flow may actually result in a signal increase because spins that were not saturated (nutated into the transverse plane) may be saturated by a subsequent pulse. Thus, high signal value in flowing material does not necessarily signify the effects of relaxation processes as would be the case for stationary material.

The flow phenomenon of enhanced signal during even-numbered echoes has been widely recognized.[7] In this artifact, seen in vessels exhibiting slow laminar flow, spins are not completely rephased by the initial 180-degree pulse. A subsequent 180-degree pulse completes rephasing, but the next is incomplete and so on. Much later echoes may fail to yield any signal as the spins leave the slice of interest.

Chemical Shift

Small differences in resonance frequency among different chemical environments of protons are not usually revealed in MR imaging. Chemical shifts produced by the screening effects of the electron "clouds" of nearby chemical bonds are usually on the order of tens of parts per million. Spatial localization using magnetic field gradients places signals in the appropriate pixels as a reflection of changes in field strength on the order of several parts per thousand—that is, one pixel out of the matrix size of 256 or 512. Thus chemical shifts do not usually affect the MR imaging process directly. An exception to this occurs when lipid-rich tissues are adjacent to tissues with high water content (Margin Figure 25-9). The lipid/water chemical shift is among the largest encountered in normal body tissues and may be detected as an artifact along the frequency-encoding direction. A region of increased signal intensity appears at one margin of the object, while a region of decreased signal appears at the opposite margin. The signal from lipid is shifted in the direction of decreasing magnetic field,

TABLE 25-1 Quality Assurance Tests for MRI Systems

Frequency	*Tests*
Daily	Resonance frequency
	Signal-to-noise ratio
Monthly	Image uniformity
	Linearity, hard copy
Yearly	Magnetic field homogeneity
	RF pulse shape and amplitude
	Gradient pulse shape and amplitude

while the signal from water is shifted in the opposite direction. If this shift exceeds the size of a pixel in the frequency-encoding dimension, then the signals from these substances are erroneously placed in the image matrix.

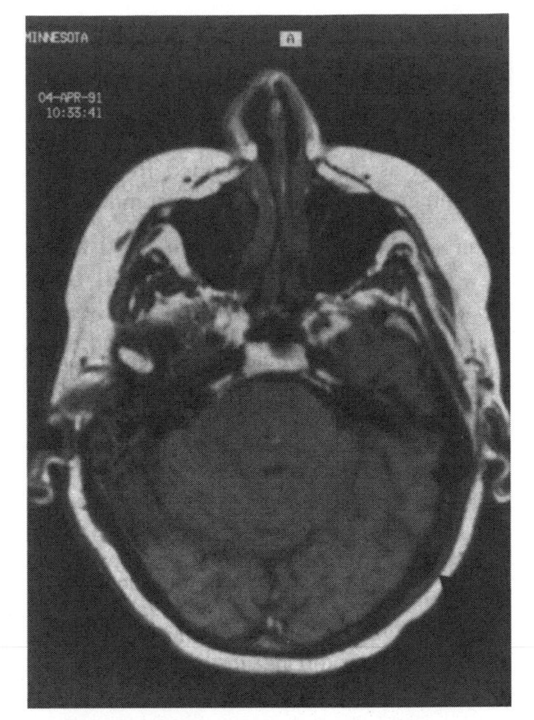

MARGIN FIGURE 25-9
A chemical shift artifact produced by the difference in resonance frequency between lipids and water. The frequency difference affects spatial encoding and causes the lipid signal to be placed in the image a short distance away from the water compartment of the same tissue.

QUALITY ASSURANCE

Quality assurance in MRI reflects the principles of image formation and display that determine the appearance of any medical image. Tests such as resolution, contrast, linearity, and sensitivity are a necessary part of evaluating equipment performance. The actual testing methods and the standards for performance are, of course, unique to MRI. There are two approaches to quality assurance measurements. Direct measurement of image quality through evaluation of end points such as imaging system uniformity is essential. Some imaging centers go further: they measure intermediate steps in the imaging process such as tip angle or gradient pulse shape to identify problems before they produce noticeable effects on image quality. A number of guides now exist for quality assurance in MRI.[8–10] These guides cover the range and recommended frequency of quality assurance measurements of these measurements. Table 25-1 summarizes some of these recommendations.

BIOEFFECTS

The US Food and Drug Administration has published guidelines for safe operating characteristics of MRI systems.[11] Manufacturers must follow these guidelines in order to obtain approval to market MRI devices. Exemptions to the guidelines require submission to the FDA of scientific evidence verifying the safety of patients if the guidance levels are exceeded. Under FDA guidance for clinical MRI, the maximum static magnetic field strength is 4 tesla and the maximum acoustic noise level (caused by thermal and mechanical stress in the magnet due to gradient switching) is 140 dB. The specific energy absorption rate and the time rate of change of the magnetic fields are addressed in three categories as "normal mode," "first level controlled," and "second level controlled." Normal mode refers to routine operation for patient studies. In first level controlled mode, a clear indication that the unit will operate in this mode must be visible to the operator and a positive action must be taken by the operator to initiate the scan. In second level controlled mode, security measures such as a key lock or a software password must be used. These parameters are described in Table 25-2.

At the present time, there is no conclusive evidence for adverse biologic effects in normal patients scanned in MRI systems using parameters that have been approved by the FDA. There are areas of concern, however, that are being monitored in a number of studies. Some of these studies are described in the following section. Areas of concern are categorized according to the main physical features of MRI, namely, the static magnetic field, time-varying magnetic field, and RF energy.

TABLE 25-2 FDA Safety Guidelines for Specific Absorption Rate (SAR) and Time Rate of Change of Magnetic Field (dB/dt) in Commercially Manufactured MRI Systems

	Watt/kg		
Specific Absorption Rate[a] *(SAR)*	*Normal Mode*	*First Level Controlled*	*Second Level Controlled*
Whole body, averaged over 15 minutes	1.5	4	10
Head, averaged over 10 minutes	≤3	None	≥3
Local tissue, averaged over any gram of tissue over 5 minutes			
Head or torso	≤8	None	≥8
Extremities	≤12	None	≥12
Time rate of change of magnetic field (*dB/dt*)	20 T/sec	Mild peripheral nerve stimulation[b]	Painful peripheral nerve stimulation[b]

[a] Alternatively, evidence may be submitted that RF energy levels will not cause more than 1°C rise in core body temperature, or localized temperatures in excess of 38°C in the head, 39°C in the trunk, and 40°C in the extremities.

[b] Determined by experimentation with at least 20 volunteer subjects.

A visual phenomenon may be produced if patients move their heads rapidly in a magnetic field above about 10 millitesla. This phenomenon, a visual sensation of flashing lights known as magnetophosphenes, is not thought to affect vision permanently and ceases immediately following magnetic field exposure. It was first discovered by d'Arsonval in 1896 and has been described in MRI systems since the mid-1980s.[12]

Static Magnetic Field

Some creatures such as mollusks and bees are known to be sensitive to magnetic fields. Such effects are caused by the presence of magnetic receptors in a sensory apparatus that contains small amounts of naturally occurring ferromagnetic materials. The migration patterns of birds, for example, are known to be affected by variations in the strength of the earth's magnetic field that are detected by receptors embedded in the cranium. Analogous structures have not been identified in humans. To date, the only significant biologic effect of static magnetic fields on higher organisms is the induction of electrical potentials on the order of millivolts due to motion of blood through a magnetic field. This is not known to have any adverse consequences.

Epidemiologic studies have been conducted in humans exposed to a wide range of static magnetic field strengths. Most of these studies are related more to fields experienced by workers in the control area or by technologists in the magnet room during patient setup than to the magnetic fields experienced by patients. Studies of workers who use electrolytic cells for chemical separation and experience fields typically below 15 millitesla, and of workers at national research laboratories who are exposed to field strengths of 0.5 millitesla for long periods of time or up to 2 tesla for short periods, have yielded no significant increases in disease rates.[13,14] Workers exposed to magnetic fields (less than 56 millitesla) in the aluminum industry have been found to be at greater risk for leukemia.[15,16] However, a wide spectrum of environmental contaminants is also present for these workers. At the present time, no causal relationships have been proposed.[17] The FDA has published clinical exposure guidelines for limiting human exposure to a maximum value of 4 tesla (Table 25-2). This upper bound represents the currently accepted limit for safe general usage, but is not thought to represent an absolute threshold of any kind.

Time-Varying Magnetic Fields

Electrical fields and currents may be induced in a conductor such as the human body when magnetic fields are changed (increased or decreased in magnitude) in the vicinity of the conductor. The principle behind this phenomenon, called Faraday's law, is the same as that responsible for induction of a current (the MR signal) in a receiver coil by precessing nuclei in the sample.

Small electrical currents are induced by time-varying gradient magnetic fields in MRI units. Gradient changes on the order of a few tesla per second are typical maximum values encountered during 2DFT clinical imaging techniques. These rates of change of the magnetic field induce currents on the order of a few tens of milliamperes per square meter. To put this figure into perspective, one should note that currents of up to 10 mA/m^2 are routinely present in brain and heart tissue.

Some fast imaging schemes may lead to larger induced current densities. The number of gradient-switching operations per unit time is not the only determinant in this regard. The instantaneous rate of change of the magnetic field during any single gradient-switching operation determines the maximum induced current, while the total number of such operations determines the duration.

Induced current densities of 10 to 100 mA/m^2 have been shown to produce permanent changes in cell function if the exposure is chronic. Examples of these changes include cell growth rate, respiration and metabolism, immune response, and gene expression. A bibliography and summary of these studies is given in work by Tenforde and Budinger.[17]

Whether deleterious or beneficial, the occurrence of biologic effects at current densities exceeding 10 mA/m^2 suggests the need for further research. Various mechanisms have been proposed for the cell response to induced electrical fields and their associated currents at levels encountered by MRI patients. These models suggest that membrane effects are plausible,[19] even at electrical field levels that are below the random background electronic "noise" at the cell surface.[20,21]

Current densities above 1000 mA/m^2 (1 A/m^2) produce irreversible deleterious effects such as cardiac fibrillation, stimulation of seizures, and respiratory tetanus. These current densities are orders of magnitude above those induced by clinical MRI systems used at the present time. It is difficult to produce such current densities in biologic specimens without the implantation of electrodes directly into the tissue.

Rates of change of gradient field strength used for imaging are the same order of magnitude as the changing magnetic field component of extremely low frequency (ELF) electromagnetic fields such as those associated with municipal transmission and distribution electrical lines. Such effects have received attention as a result of epidemiologic studies that suggest weak links with deleterious health effects. While laboratory studies involving human volunteers have failed to demonstrate any deleterious effects, more than a dozen epidemiologic studies during the decade 1981 to 1990 suggested an increased probability of cancer for individuals living near electrical power lines.[22–27] Residential power lines operate at frequencies of 60 Hz, which qualifies as ELF, and induce currents in humans that range from 10^{-3} to 1 mA/m^2. Although some studies contain serious design flaws such as unmatched controls, failure to establish a dose–response relationship, failure to control for the presence of known carcinogens, and so on, some suggestive data remain. The electrical field strengths at issue are the same order of magnitude as those experienced by users of commonly used appliances and devices such as electric blankets, so attempts to limit exposure would require significant changes in lifestyle for large segments of the population and significant design changes for many appliances, home wiring, and so on. In any case, the duration of exposure for individuals in these studies exceeds the duration of exposure of MRI patients by many orders of magnitude, so it is safe to conclude that MRI exposures do not make a significant contribution to population-wide ELF exposure at the present time.

In addition to deleterious effects, induced electrical currents have been associated with at least one beneficial effect. Current densities of the order of 20 mA/m^2 have been used clinically to stimulate union of bone fractures. However, there is some controversy regarding the efficacy of this technique.[18]

A reversible and apparently innocuous phenomenon related to induced current densities above 10 mA/m^2 is the magnetophosphene phenomenon. This phenomenon was discussed previously in relation to motion of the patient's head in a static magnetic field. Changes in the magnetic field over time have the same result when the patient's head is stationary. These phenomena have been associated with rates of change of the magnetic field below 2 tesla/sec at repetition rates of 15 magnetic field pulses per second,[12] well within the operating parameters of current MRI systems.

Radio-Frequency Electromagnetic Waves

Radio waves are capable of heating biologic materials. Heating occurs because the conductivity of tissues supports the transfer of energy from the radio wave to induced electrical currents in the patient. The energy dissipated in the patient from resistance to the induced currents leads ultimately to heating of tissues in the patient. This energy dissipation is described in terms of the specific absorption rate (SAR) in units of watts per kilogram. For MRI systems, the phenomenon of tissue heating in the patient is of concern during the sending part of the imaging cycle. The SAR from MRI depends upon the particular pulse sequence used (the shape of the pulses, fraction of time that power is applied, etc.). It also depends upon the frequency of the radio waves and the size, density, and electric conductivity of the patient. This parameter has been calculated for models of the human head and torso[28–30] and has been measured for cases of clinical interest.[31,32] The maximum temperature rise resulting from a given SAR may be estimated from

$$T = \frac{\text{SAR } t}{C_{\text{H}}} \tag{25-1}$$

where T is the temperature rise in degrees Celsius, SAR is the specific absorption rate in watts per kilogram, t is the duration of exposure in seconds, and C_H is the heat capacity of the sample in joules per kilogram-degree Celsius. A typical value of heat capacity for human soft tissue is 3470 J/(kg-°C). Equation (25-1) yields a maximum estimate of temperature rise because it does not take into account effects such as conduction (vascular flow through a volume), convection (mixing that spreads out the effects of heating in a liquid), or radiation (heat energy given off in the form of infrared radiation). These effects act to reduce heating in most clinical situations below the amount calculated by Eq. (25-1).

Example 25-1

Find the maximum temperature rise produced during an MRI session in which a SAR of 3 W/kg is applied for 5 minutes.

$$\begin{aligned} T &= \frac{\text{SAR } t}{C_{\text{H}}} \\ &= \frac{3 \text{ W/kg } (5 \text{ min})(60 \text{ sec/min})}{3470 \text{ J/(kg-°C)}} \\ &= 0.26°\text{C} \end{aligned} \tag{25-1}$$

The watt, the unit for power, equals 1 J of energy per second—that is, the rate of energy deposition. Therefore, all units except degrees Celsius cancel in Eq. (25-1).

It is generally agreed that heating of tissues in healthy humans is not of concern when the rise in core body temperature is less than 1°C. Cell killing is associated with temperatures above 42°C, approximately 5°C above core body temperature. There is some disagreement as to the biological effects of heating between these extremes.

Biologic effects of RF exposure that are not directly related to heating and that may occur at low power levels have been suggested on a theoretical basis and have been reviewed extensively.[34] One of the most widely studied nonthermal effects is the alteration of calcium ion binding to nerve cell surfaces. However, specific waveforms not routinely encountered in MRI are required to elicit this effect.

It is known that many of the body's autonomic response mechanisms are reversible.[33] These mechanisms include the neuroendocrine and cardiovascular systems. Thus elicitation of a biologic response is not necessarily evidence of deleterious biologic effects. In addition to an overall temperature rise, there is some concern that "hot spots" could occur. If, for example, pathways of electrical current become narrowed in tissue with little vascularity, local heating effects may occur, although global models would not predict it. The presence of foreign materials such as ferromagnetic prostheses complicates the estimate of local heating and is considered later in this chapter.

Epidemiologic studies of groups such as military and civilian personnel exposed to radar equipment, and U.S. embassy employees exposed to microwave surveillance devices, have failed to produce conclusive evidence of harm.[35]

Federal guidelines for RF exposure in humans for noninvestigational purposes are given in Table 25-2. It should be noted that these guidelines are based predominantly upon models for RF-induced heating in biologic materials. At present, the FDA approves scanning protocols for commercially available MRI systems on an individual basis. Most systems in clinical use have software indications of estimated SAR and do not allow operation of the system if guidelines are exceeded.

SITE PLANNING

As with all major imaging systems, the acquisition of an MRI unit involves analysis and preparation of the site. In the case of MRI, the major factors to consider are (1) the effects of ferromagnetic materials upon the homogeneity of the magnetic field in the bore of the magnet, (2) the effects of the static magnetic field upon metal objects in the scan room, (3) the effects of the static magnetic field outside of the scan room, (4) RF interference with the MRI scan produced by equipment outside of the scan room, and (5) RF interference with external communications produced by the MRI unit.

Site features that may complicate an MRI installation such as insufficient space, proximity of sensitive electronic components, and so on, are hallmarks of typical hospitals. A conceptually simple alternative to retrofitting existing space is construction of new space physically separate from the main hospital site. Although a number of severe problems were encountered in early installations, a great deal of experience has been accumulated. Vendors of MRI systems provide site planning services, and a number of general references are now available.[36] With foresight and planning, conversion of existing facilities is usually accommodated. Some of the major features of site selection and preparation are discussed in this section.

Magnetic Field Homogeneity

The presence of ferromagnetic materials in the vicinity of an MRI system may affect magnetic field homogeneity. Homogeneity is adjusted through the use of shim coils, electromagnets usually mounted on the ends of the main magnet bore that can be moved slightly with positioning screws to produce small variations in magnetic field strength. The effects of moving ferromagnetic materials such as elevators or laundry transport systems cannot be canceled by a static shimming system. At present there is no system to dynamically adjust magnetic field homogeneity. Therefore, the main system magnet should be located a distance away from large moving metallic objects. The minimum tolerable distance depends upon the requirements for magnetic field homogeneity. These requirements are more severe for large-bore, high-field strength magnets.

Radio-Frequency Interference

The presence of sources of RF energy external to the MRI system can interfere with MR signals from the patient and compromise the raw data accumulated during a scan. When the image is reconstructed, artifacts may occur because the amplitude of MR signals is increased at the frequencies of the interfering signals. When the image is reconstructed by a technique such as the Fourier transform, the interfering signals appear on the image as bands or other regular patterns. Some interference with the signals sent into the patient may also occur, but this effect is smaller because the magnitude of the signals sent into the patient is much greater than the magnitude of the signals received from the patient.

Sources of RF interference or electronic noise commonly found in the hospital include fluorescent lights, power supplies, electrical power lines, sprinkler systems, and communications networks. The connections between the MR scanner and the computer, RF power supplies, and gradient power supplies may also be significant sources of electronic noise.

The basic plan of RF shielding is to surround the magnet with an electrically conductive enclosure. The enclosure isolates the scanner from unwanted signals conducted through metallic objects and from waves radiated through space. Continuity of the enclosure ensures a pathway for conduction of unwanted electric signals around the magnet so that they do not enter the bore and interfere with reception of signals from the patient. Penetration of RF waves into an enclosure is determined in part by the size of openings in the enclosure. Holes in the RF enclosure are generally kept smaller than one-half wavelength of the smallest interference wavelengths that can be tolerated. For example, the viewing window is usually covered with fine copper mesh screen having hole diameters of a millimeter or less.

When conduits or ventilation ducts must pass through the enclosure, waveguides are used to reduce unwanted RF penetration. A waveguide is constructed by breaking the conduit, inserting a plastic sleeve, and covering the outside of the sleeve with

Electromagnetic field theory, pioneered by British scientist Michael Faraday in the 1800s, predicts that a conducting surface will not be penetrated by electromagnetic radiation so long as there are no holes that are larger than the wavelength of the radiation. This means that for the relatively long wavelength radiation (on the order of meters) of concern in magnetic resonance imaging, a cage with bars spaced no wider than a fraction of a meter apart will shield the MR scanner from unwanted interference. Such a cage, known as a Faraday cage, can reduce the level of radiofrequency interference by as much as 100 dB. Faraday cages are not usually required in MR site planning but may be necessary in areas of unusual radio-frequency interference. The principle of the Faraday cage explains why you are much safer during a thunderstorm when in a closed car than when standing outside or even in most buildings. This affect is often misattributed to insulation provided by the tires. Tires do not provide enough insulation to prevent the extremely high voltages associated with a lightning strike from producing electrical currents. The car is safer because it acts as a Faraday cage. The electromagnetic energy can strike the surface of the car, but will not penetrate to the interior.

It is recommended that all patients undergo screening prior to admittance into the magnetic resonance scanner. The screening should check for the presence of foreign materials within the patient that may produce an adverse effect in the presence of static or changing magnetic fields. In addition, the FDA recommends particular caution with regard to the following patients[11]:

a. Patients with greater than normal potential for cardiac arrest.
b. Patients who are likely to develop seizures or claustrophobic reactions.
c. Decompensated cardiac patients, febrile patients, and patients with impaired ability to perspire.
d. Patients who are unconscious, heavily sedated or confused, and with whom no reliable communication can be maintained.
e. Examinations that are carried out at a room temperature above 24°C or at a relative humidity above 60%.

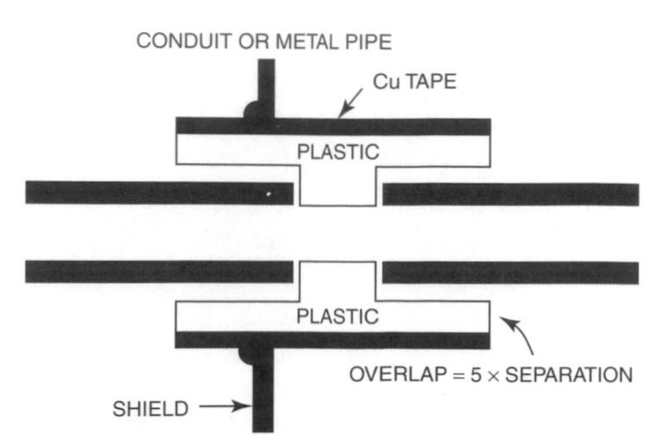

MARGIN FIGURE 25-10
Construction of a waveguide to allow conduit or a communication aperture for the MR scan room. (From AAPM. *Site Planning for Magnetic Resonance Imaging Systems*, Report to AAPM NMR Task Group No. 2, Report No. 20. New York, American Institute of Physics, 1987. Used with permission.)

tape that is electrically conductive (Margin Figure 25-10). A small gap is maintained between the conductive tape and the conduit so that there is no continuous path for conduction of electricity along the conduit.

Patient and Personnel Safety

The presence of a high-strength magnetic field imposes limitations on the use of MRI and restrictions on the immediate surroundings of the scanner. Metal objects are attracted to the magnet of an MRI unit. The force with which they are attracted depends upon the type of material, the total mass, and the shape of the object. For example, rod-shaped objects aligned with the magnetic field lines are attracted with greater force than are spherical objects of the same material. At a distance d of several times the size of the magnet, the force varies approximately as $1/d^3$. The force is quite sensitive to small changes in distance, so attempts to "test" the pull of the magnet by moving an object close to the magnet may result in an unexpectedly sudden onset of a substantial force. One benchmark for safety is the distance at which the pull of the magnet overcomes the pull of gravity and renders the object weightless (Margin Figure 25-11). The pull of the earth's magnetic field, approximately 0.05 millitesla, is small when compared with the gravitational attraction for objects of appreciable mass.

Patients must be examined for the presence of ferromagnetic objects such as surgical clips, shrapnel, prostheses, and so on. The three issues of concern for such objects are (1) torque or force upon the object within the patient, (2) heating of the material, and (3) distortion of the image. The presence of ferromagnetic objects within the patient may not contraindicate MRI. However, it is standard practice to interview the patient prior to a scan so that decisions may be made. For example, most institutions have established protocols for exclusion of patients who have ferromagnetic materials in the eye. In patient populations such as sheet metal workers or combat veterans who are suspected to be at risk, these protocols may include pre-MRI x rays to detect metallic objects.

Implants such as surgical clips, even though they may be made from stainless steel, may not be affected by magnetic fields because there are several different types of surgical-grade steel.[37–41] The size and location of the clips are also important. In the absence of specific information, the most prudent course is to assume that the implant materials will be affected.

Prostheses such as plates in the skull, hip replacements, and other types of hardware that are firmly attached to bone are of little concern with respect to torque. However, they are of greater concern with respect to RF heating. As with direct RF heating of tissues, a temperature rise of less than 1°C for objects within the body is of little concern. The degree of heating depends upon the pulse sequence used, and is influenced by the mass of the implant. For example, at field strengths below 2 tesla, heating of single hip replacements is not usually significant, but double hip replacements may be of concern.[42]

At the present time, the presence of a cardiac pacemaker mandates that the patient be excluded from MRI.[43,44] Torque on the pacemaker is usually unacceptable within the bore of the magnet at moderate field strengths. Moreover, the RF fields and gradient switching can act to divert the programming circuitry of the pacemaker into an asynchronous mode. Therefore, individuals with pacemakers are usually excluded even from the control area of the scanner.

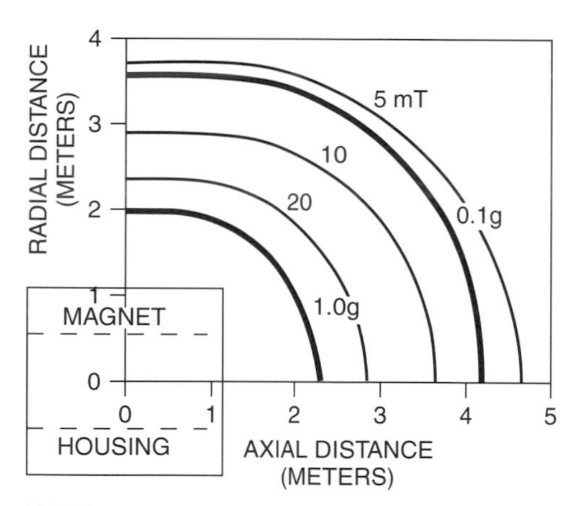

MARGIN FIGURE 25-11
Bold lines show the locations at which an iron object is pulled toward a 1.0-tesla MR system magnet with a force equal to the weight of the object (1.0 g) and with a force equal to one tenth the weight of the object (0.1 g). The magnet is assumed to contain no active shielding, and the iron is assumed to be in the form of a rod oriented along the lines of magnetic force, the shape and position that maximizes attraction. (From AAPM. *Site Planning for Magnetic Resonance Imaging Systems*, Report to AAPM NMR Task Group No. 2, Report No. 20. New York, American Institute of Physics, 1987. Used with permission.)

SUMMARY

- The main components of an MRI system are
  - Main system (static field) magnet
  - Gradient coils
  - Radio-frequency coils

  - Transmitter
  - Mixer
  - Analog-to-digital converter
  - Computer
- MRI system magnets that produce the static magnetic field are categorized as
  - Electromagnets
    - Resistive
    - Superconducting
  - Permanent
- MRI artifacts include
  - Foreign objects
  - Magnetic field inhomogeneities
  - Gradient nonlinearity
  - Radio-frequency interference
  - Motion
  - Chemical shift
- Quality assurance tests for MRI systems include
  - Daily tests
    - Resonance frequency
    - Signal-to-noise
  - Monthly tests
    - Image uniformity
    - Linearity
    - Hard copy
  - Yearly tests
    - Magnetic field homogeneity
    - RF pulse shape and amplitude
    - Gradient pulse shape and amplitude
- Bioeffects that are possible in MRI include
  - Movement of ferromagnetic objects into the magnet from the external environment
  - Movement of metallic devices, foreign bodies, or implants within the patient
  - Disruption of implanted devices
  - RF heating
  - Induction of electrical currents within the patient
- In site planning for MRI systems, the following must be considered:
  - Exclusion of metallic objects from the scan room
  - RF interference from the outside world and from the scanner
  - Cryogen ventilation in the event of a quench of a superconducting system
  - Methods to control traffic (i.e., excluding individuals wearing pacemakers)

REFERENCES

1. Hoult, D. I., and Lauterbur, P. C. The sensitivity of the zeugmatographic experiment involving human samples. *J. Magn. Reson.* 1979; **34**:425.
2. Crooks, L. E., Arakawa, M., Hoenninger, J., et al. Magnetic resonance imaging: Effects of magnetic field strength. *Radiology* 1984; **151**:127.
3. Ugurbil, K., Garwood, M., Rath, A. R., et al. Amplitude and frequency/phase modulated refocusing pulses that induce plane rotations even in the presence of inhomogenous fields. *J. Magn. Reson.* 1988; **78**:472.
4. Hayes, C. E., Edelstein, W. A., and Schenck, J. F. Radiofrequency coils, in Thomas, S. R., and Dixon, R. L. (eds.), *NMR in Medicine, American Association of Physicists in Medicine*, Medical Physics Monograph No. 14. New York, American Institute of Physics, 1986, pp. 142–165.
5. Hyde J. S., Froncisz, W., Jesmanowicz, A., et al. Planar-pair local coils for high-resolution magnetic imaging, particularly of the temporomandibular joint. *Med. Phys.* 1986; **13**:1–7.
6. Kneeland, J. B. Instrumentation, in Stark, D. D., and Bradley, W. G. (eds.), *Magnetic Resonance Imaging*. St. Louis, Mosby–Year Book, 1988, pp. 56–65.
7. Bradley, W. G., Waluch, V., Lai, K. S., et al. The appearance of rapidly flowing blood on magnetic resonance images. *AJR* 1984; **143**:1167–1174.
8. Knowles, R. J., and Markisz, J. A. *Quality Assurance and Image Artifacts in Magnetic Resonance Imaging*. Boston, Little, Brown & Co, 1988.
9. Dixon, R. L. *MRI: Acceptance Testing and Quality Control*, Proceedings of an AAPM Symposium, April 6–8, 1988, Bowman Gray School of Medicine. Madison, WI, Medical Physics Publishing Corp., 1988.
10. American College of Medical Physics. *Radiation Control and Quality Assurance Surveys: Magnetic Resonance Imaging, A Suggested Protocol*, Report No. 5. New York, American College of Medical Physics, 1989.
11. Guidance for the submission of premarket notifications for Magnetic Resonance Diagnostic Devices, Food and Drug administration, Center for Devices

and Radiologic Health, Computed Imaging Devices Branch, Office of Device Evaluation, FDA, Rockville, MD, www.fda.gov/chrh/ode/mri340.pdf, November 14, 1998.

12. Budinger, T. F., Cullander, C., and Bordow, R. Switched magnetic field thresholds for the induction of magnetophosphenes. Presented at the Third Annual Meeting of the Society of Magnetic Resonance in Medicine, New York, August 13–17, 1984.
13. Marsh, J. L., Armstrong, T. J., Jacobson, A. P., et al. Health effect of occupational exposure to steady magnetic fields. *Am. Indust. Hyg. Assoc. J.* 1982; **43**:387.
14. Budinger, T. F., Bristol, K. S., Yen C. K., et al. Biological effects of static magnetic fields. Presented at the Third Annual Meeting of the Society of Magnetic Resonance in Medicine, NewYork, August 13–17, 1984.
15. Milham, S. Mortality from leukemia in workers exposed to electrical and magnetic fields. *N. Engl. J. Med.* 1982; **307**:249.
16. Rockette, H. E., and Arena, V. C. Mortality studies of aluminum reduction plant workers: Potroom and carbon department. *J. Occup. Med.* 1983; **25**:549.
17. Tenforde, T. S., and Budinger, T. F. Biological effects and physical safety aspects of NMR imaging and *in vivo* spectroscopy, in Thomas, S. R., and Dixon, R. L. (eds.), *NMR in Medicine: The Instrumentation and Clinical Applications*, Medical Physics Monograph No. 14. American Institute of Physics, 1986, pp. 493–548.
18. Bassett, C. A. L. Pulsing electromagnetic fields: A new approach to surgical problems, in Buchwald, H., and Varco, R. L. (eds.), *Metabolic Surgery*. New York, Grune & Stratton, 1978, pp. 255–306.
19. Luben, R. A., Cain, C. D., Chen, M. C.-Y., et al. Effects of electromagnetic stimuli on bone and bone cells *in vitro*: Inhibition of responses to parathyroid hormone by low-energy low-frequency fields. *Proc. Natl. Acad. Sci.* USA 1982; **79**:4180.
20. Hoult, D. I., and Lauterbur, P. C. The sensitivity of the zeugmatographic experiment involving human samples. *J. Magn. Reson.* 1979; **34**:425–433.
21. Weaver, J. C., and Astumian, R. D. The response of living cells to very weak electric fields: The thermal noise limit. *Science* 1990; **247**:459–462.
22. Roberts, N. J., Jr., Michaelson, S. M., and Lu, S. T. The biological effects of radiofrequency radiation: A critical review and recommendations. *Int. J. Radiat. Biol.* 1986; **50**:379–420.
23. McDowall, M. E. Mortality of persons resident in the vicinity of electricity transmission lines. *Br. J. Cancer* 1986; **53**:271–279.
24. Skeikh, K. Exposure to electromagnetic fields and the risk of leukemia. *Arch. Environ. Health* 1986; **41**:56–63.
25. Easterly, C. E. Cancer linked to magnetic field exposure: A hypothesis. *Am. J. Epidemiol.* 1981; **114**:169–174.
26. Savitz, D. A., and Calle, E. E. Leukemia and occupational exposure to electromagnetic fields: Review of epidemiologic surveys. *J. Occup. Med.* 1987; **29**:47–51.
27. Elder, J. A. Radiofrequency radiation activities and issues: A 1986 perspective. *Health Phys.* 1987; **53**:607–611.
28. Brown, R. W., Haacke, E. M., Martens, M. A., et al. A layer model for RF penetration, heating, and screening in NMR. *J. Magn. Reson.* 1988; **80**:225–247.
29. Bottomley, P. A., and Andrew, E. R. RF magnetic field penetration, phase shift and power dissipation in biological tissue: Implications for NMR imaging. *Phys. Med. Biol.* 1978; **23**:630–643.
30. Bottomley, P. A., Redington, R. W., Edelstein, W. A., et al. Estimating radiofrequency power deposition in body NMR imaging. *Magn. Reson. Med.* 1985; **2**:336–349.
31. Roschmann P. Radiofrequency penetration and absorption in the human body: Limitations to high-field whole body nuclear magnetic resonance imaging. *Med. Phys.* 1987; **14**:922–932.
32. Shuman, W. P., Haynor, D. R., Guy, A. W., et al. Superficial and deep-tissue temperature increases in anesthetized dogs during exposure to high specific absorption rates in a 1.5T MR imager. *Radiology* 1988; **167**:551–554.
33. Parker, L. N. Thyroid suppression and adrenomedullary activation by low-intensity microwave radiation. *Am. J. Physiol.* 1973; **224**:1388.
34. Postow, E., and Swicord, M. L. Modulated Field and "Window" Effects, in Polk, C., and Postow, E. (eds.), *Handbook of Biological Effects of Electromagnetic Fields*. Boca Raton, FL, CRC Press, 1986.
35. Tenforde, T. S., and Budinger, T. F. Biological Effects and Physical Safety Aspects of NMR Imaging and *in Vivo* Spectroscopy, in Thomas, S. R., and Dixon, R. L. (eds.), *NMR in Medicine: The Instrumentation and Clinical Applications*, Medical Physics Monograph No. 14. New York, American Institute of Physics, 1986, pp. 521–522.
36. AAPM. *Site Planning for Magnetic Resonance Imaging Systems*, Report of AAPM NMR Task Group No. 2, AAPM Report No. 20. New York, American institute of Physics, 1987.
    See also MR equipment vendors (e.g., Siemens Medical Systems Inc., Iselin, NJ; General Electric, Milwuakee, WI; Picker International, Cleveland, Ohio; Philips Medical Systems, Shelton, CT; Elscint Ltd., Boston, MA).
37. Barrafato, D., and Helkelman, M. Magnetic resonance imaging and surgical clips. *Can. J. Surg.* 1984; **27**:509.
38. Dujovny, M., Kossowski, N., Kossowsky, R., et al. Aneurysm clip motion during magnetic resonance imaging: In vivo experimental study with metallurgical factor analysis. *Neurosurgery* 1985; **17**:543–548.
39. Finn, E. J., DeChiro, G., Brooks, R. A., et al. Ferromagnetic materials in patients: Detection before MR imaging. *Radiology* 1985; **156**:139–141.
40. Laakman, R. W., Kaufman, B., Han, J. S., et al: MR imaging in patients with metallic implants. *Radiology* 1985; **157**:711–714.
41. New, P. F. J., Rosen, B. R., Brady, T. J., et al. Potential hazards and artifacts of ferromagnetic and nonferromagnetic surgical and dental materials and devices in nuclear magnetic resonance imaging. *Radiology* 1983; **147**:139–148.
42. Davis, P. L., Crooks, L., Arakawa, M., et al. Potential hazards in NMR imaging: Heating effects of changing magnetic fields on small metallic implants. *AJR* 1981; **137**:857–860.
43. Pavlicek, W., Geisinger, M., Castle, L., et al. The effects of nuclear magnetic resonance on patients with cardiac pacemakers. *Radiology* 1983; **157**:149–153.
44. Erlebacher, J. A., Cahill, P. T., Pannizzo, F., et al. Effect of magnetic resonance imaging on DDD pacemakers. *Am. J. Cardiol.* 1986; **57**:437–440.

CHAPTER

26

EXPERIMENTAL RADIOBIOLOGY

OBJECTIVES

After studying this chapter, the reader should be able to:

- Describe models for the direct and indirect effects of radiation.
- Characterize the components of a cell survival curve.
- Explain the effects of oxygen, LET, dose, dose rate, and fractionation on cell survival studies.
- Provide examples and proposed mechanisms of action of representative radioprotecting and radiosensitizing agents.
- Distinguish between somatic and genetic effects of radiation.

INTRODUCTION

An "almost instantaneous" interaction is one that occurs in less than 10^{-18} seconds.

A free radical is an electrically neutral chemical species that has an unpaired electron and is chemically very reactive. Electrons possess the quality "spin," and electrons tend to pair together, with one spin "up" and one spin "down." The presence of an unpaired electron in a free radical is an unstable situation that is remedied when the free radical either acquires or shares an electron from another molecule. Thus free radicals act as strong oxidizing agents and disrupt chemical bonds in the process.

Direct effects of radiation are caused by electrons released during interactions of radiation directly in biological material. Indirect effects of radiation are produced by free radicals formed as radiation interacts in biological material.

For direct effects, the relationship between dose and number of entities (molecules, cells, etc.) affected is expressed as $E = E_0(1 - e^{-kD})$, where E is the number affected, E_0 is the number of entities present, D is the absorbed dose, and k is a constant called the "inactivation constant." The value $1/k$ is the dose required to leave only 37% ($1/e$) of the entities unaffected. When the entities are cells, $1/k$ is termed the *mean lethal dose*.

Living organisms contain large amounts of water in both free and bound molecular states. Water plays a central role in radiation interactions within biological systems. During these interactions, water molecules are ionized by removal of electrons, leaving positively charged molecules (H_2O^+). The released electrons combine almost instantaneously with neutral water molecules to form negatively charged molecules (H_2O^-). Almost instantaneously means less than 10^{-18} sec, the shortest measurable time interval. The positively and negatively charged water molecules are unstable, dissociating into ions (H^+ and OH^-) and *free radicals* ($H^\bullet$ and $OH^\bullet$).

Free radicals in biologic specimens tend to react within 10^{-5} sec. They react among themselves and with water, organic molecules, oxygen, and their own reaction products. The products of free radical reactions include poisons such as hydrogen peroxide (H_2O_2). Of particular interest is the interaction of free radicals with oxygen, which serves to prolong the presence of free radicals. When a free radical $X^\bullet$ interacts with molecular oxygen O_2, the product is a longer-lived free radical.

$$X^\bullet + O_2 \rightarrow XO_2^\bullet$$

Oxygen serves to heighten the effects of free radicals by allowing the radicals to "survive" long enough to migrate farther from their origin, thus expanding the potential for biologic damage.

Because they are produced at one or more steps removed from the initial ionizing event, radiobiologic effects of free radicals are termed *indirect effects*. Effects produced directly by electrons set in motion as radiation interacts in a material, such as disruption of chemical bonds in macromolecules, are termed direct effects. These effects are included in the summary of radiation interactions depicted in Figure 26-1.

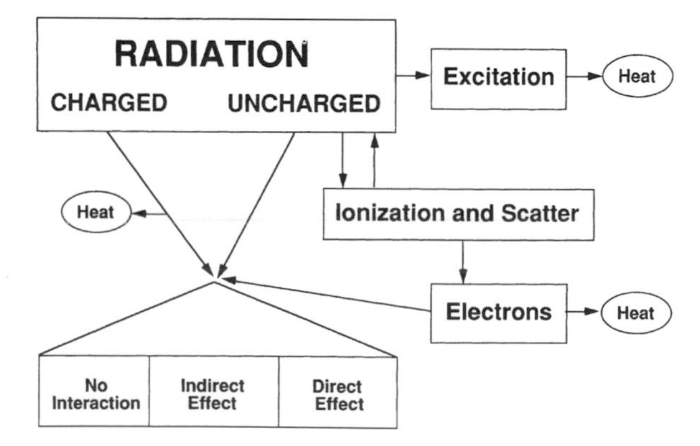

FIGURE 26-1
As radiation (charged particles or uncharged particles such as photons) enters matter, various outcomes are possible, including no interaction, an indirect effect (free radical production), a direct effect (disruption of chemical bonds in biomolecules), and the deposition of heat in the material.

INTERACTIONS AT THE CELL AND TISSUE LEVELS

Radiation is energy in motion. Radiation damage is the consequence of transfer of sufficient energy from radiation to tissue to produce a biological effect. The mechanism of damage is important. Although heat is always produced during radiation interactions, the mechanism for radiation bioeffects is definitely not heating of tissue. Calculation shows that a dose of over 4 kGy is required to raise the average temperature of water (or soft tissue) by 1°C. If heating were the mechanism of radiation bioeffects, the effects would be negligible below a dose of several kilogray, because temperature elevations on the order of 1°C or less are overshadowed by normal metabolic fluctuations in mammalian organisms.

The effects of radiation are traceable to the disruption of chemical bonds. Because several repair mechanisms exist for important macromolecules and also because the concentration of molecules often exceeds the minimum concentration required for cellular function, massive amounts of chemical disruption would be necessary to cause discernible effects. There is, however, an exception to this conclusion, because there exists a critical site of radiation damage within cells. That site is DNA.

DNA is a critical site for direct radiation damage in a cell because copies of the DNA molecule are not present.

The impact of radiation on DNA affects the ability of the cell to function and reproduce. Alteration of most biological molecules simply reduces the number of those molecules available for chemical reactions. However, alteration of DNA causes information to be "read" erroneously by the cell so that it loses or changes its normal function. If the cell in question is a germ cell, the transfer of information to the next generation of the organism may be compromised.

Ionizing radiation is particularly effective in disrupting DNA because it can produce damage in DNA components (the base, phosphate, and deoxyribose fractions), all within a relatively small region.

Through its ability to produce local, multiply damaged sites in DNA, ionizing radiation can induce the same amount of cell killing as aflatoxin, benzopyrene, and other agents known to create DNA damage.[1]

CELL SURVIVAL STUDIES

Cells may be grown in culture, surviving on nutrients provided by growth media and replicating on a time scale that reflects the growth kinetics of the specific cell line. The effects of various agents on cell survival under controlled laboratory conditions can be studied in these cell cultures. Direct observation of metabolic processes in cells is difficult; a more practical end point for observation is cell death (or survival). However, the definition and detection of cell death is also challenging. A more easily recognized end point is loss of reproductive capacity. After exposure to an agent, cell survival can be evaluated in terms of the fraction of cells that are able to replicate as evidenced by their ability to form colonies in culture.

The technique of using cultured cells to study radiation effects was developed in the 1950s by Puck and his colleagues at the University of Colorado.[2]

A graph showing the surviving fraction (defined in terms of reproductive capacity) of a cell population as a function of dose of an agent is referred to as a *cell survival curve*. It is conventional to show the surviving fraction on a log scale. Most cell survival curves have at least a portion that is a straight line, which suggests an exponential equation to describe the curve.

The inability of a cell to divide and form a colony is referred to as "reproductive death." A cell that experiences reproductive death may still be "metabolically alive."

A typical cell survival curve for x rays is shown in Figure 26-2. By definition, the fraction of surviving cells is unity when the dose is zero. As the dose is increased, the surviving fraction may remain equal to or near 1. Small doses of an agent may have no effect on reproductive capacity; or if cells are damaged, the damage may be repaired prior to the next replication. This *repair of sublethal damage* contributes to the shoulder observed on cell survival curves for x rays.

The exponential nature of cell killing is reminiscent mathematically of radioactive decay, the attenuation of x rays, and the relaxation of nuclear spins.

Beyond the shoulder, the cell survival cure displays a linear portion. Here the surviving fraction N/N_0 is described by the equation

$$\frac{N}{N_0} = e^{-D/D_0} \tag{26-1}$$

When the dose D equals D_0, the mean lethal dose, the surviving fraction is e^{-1}, or 0.37.

The mean lethal dose is defined as the dose required to reduce the "survival" to e^{-1} or 37% along the straight line portion of the survival curve. D_0 is 1 to 2 Gy for most mammalian cells exposed to x rays.

Douglas Lea was one of the pioneers in mathematical modeling of cellular radiation injury. His book *Actions of Radiation on Living Cells* (Cambridge University Press) is a classic in the radiation biology literature.

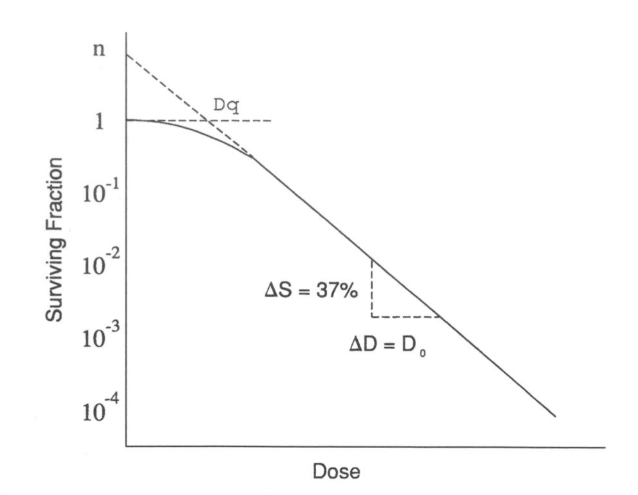

FIGURE 26-2
Representation of a cell survival curve illustrating extrapolation number n, quasithreshold dose D_q, and mean lethal dose, D_0.

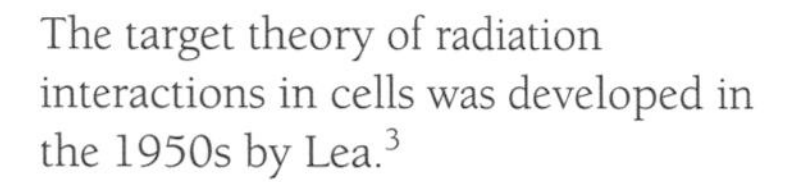

The target theory of radiation interactions in cells was developed in the 1950s by Lea.[3]

Mathematical models of radiation-induced loss of cellular reproductive capacity contain several assumptions:

- Loss of reproductive capacity is a multistep process.
- Absorption of energy in a critical volume(s) of the cell is the first step.
- Energy absorption creates molecular lesions in the cell.
- These lesions cause the loss of the cell's ability to carry out normal DNA replication and cell division.[6]

Three types (lines) of mammalian cells have been widely employed in tissue-culture studies of radiation effects. These lines are HeLa cells derived from a patient with cervical carcinoma, CHO cells derived from a Chinese hamster ovary, and T1 cells with a human kidney origin.

The width of the shoulder on a cell survival curve also reflects the number of sensitive targets that must be inactivated in the cell before the cell survival curve begins to decrease exponentially (i.e., reach the straight-line portion of a semilog plot). A "single-hit, multitarget" model of cell survival implies that there is no repair of sublethal damage and that only a single damaging "event" is necessary to inactivate each sensitive target within the cell. The number of targets that must be inactivated to kill the cell is found by extrapolating the straight-line portion of the cell survival curve back to the vertical axis. The value of the surviving fraction at the intersection with the vertical axis is the *extrapolation number n,* representing the average number of sensitive targets per cell. A wider shoulder yields a point of interaction higher on the vertical axis, implying that more targets must be inactivated. The extrapolation number n for mammalian cells exposed to x rays is highly variable and ranges from 1 (i.e., no shoulder, one target per cell) up to 20. The variability reflects the error inherent in extrapolation as well as the variability in cell types.

In actual cell survival curves, the shoulder probably indicates the effects of repair enzymes as well as the number of targets. It may not be possible to identify and specify the characteristics of sensitive targets within cells by using survival curves alone.[4] Models that are more complex than the multitarget single-hit model have been proposed.[5]

Another measure of the shoulder of a cell survival curve is the *quasithreshold dose* D_q. If the surviving fraction of cells is unity up to a certain dose and decreases exponentially beyond that dose, then such a dose would be a true threshold, a dose below which there is no cell killing. The quasithreshold dose, while not implying a true threshold, characterizes the end of the shoulder. It may be estimated as the dose at which a horizontal line, at the level of a surviving fraction of 1, crosses the extrapolated straight-line portion of the cell survival curve. It may also be shown[7] that

$$D_q = D_0 \ln(n)$$

Given the previous estimates for D_0 and n, D_q for mammalian cells ranges from 0 to 6 Gy.

MODIFICATION OF CELLULAR RESPONSES

Factors that influence the effects of radiation upon cells may be characterized by their effects upon cell survival curves. These factors include the presence of oxygen, linear energy transfer (LET) of the radiation, dose rate, fractionation, phase of the cell cycle, and presence of radiation modifiers.

Oxygen

Oxygen influences the indirect effects of ionizing radiation by modifying free radical interactions. Hence, the partial pressure of oxygen in the vicinity of the cell modifies the cell sensitivity. Two cell populations with different partial pressures of oxygen are illustrated in the margin. Population A is fully oxygenated and population B is less oxygenated. At a given dose beyond the quasithreshold dose, the surviving fraction is lower for population A. For a specific biologic effect at a particular partial pressure of oxygen, the oxygen enhancement ratio (OER) is defined as

$$\text{OER} = \frac{\text{Dose required with no oxygen present}}{\text{Dose required at given partial pressure of oxygen}}$$

The OER does not exceed 3 for mammalian cells and increases little for partial pressures above 1 atm.

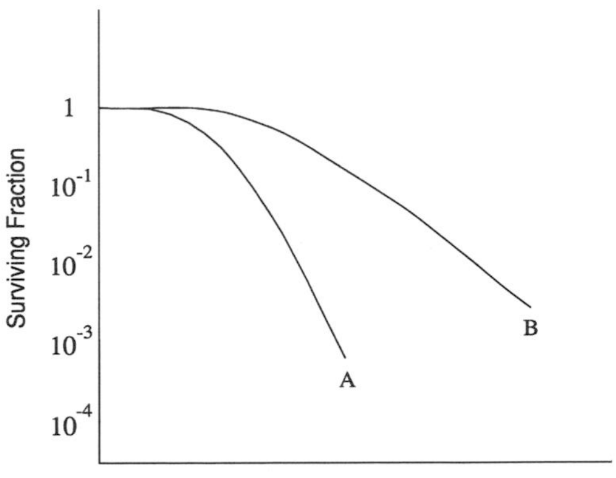

MARGIN FIGURE 26-1
Cell survival curves showing the effect of oxygenation. The cell population represented by curve A is fully oxygenated, with population B less oxygenated.

Linear Energy Transfer

LET is a measure of the density of ionizing events in the medium. As the LET is increased, the width of the shoulder of cell survival curves tends to decrease, reflecting an increase in damage per unit dose. Types of ionizing radiation that deposit more energy over shorter spatial intervals have a higher probability of damaging DNA within the cell, compared with radiations in which ionization is spread over larger distances. Therefore, damage is increased for a given dose if LET is increased.

The width of a single strand of DNA is on the order of a few thousandths of a micrometer. Neutrons, having an LET of 20 keV/μm, exhibit a higher probability of interacting more than once in the thickness of a DNA molecule, compared with x rays with an LET of less than 1 keV/μm. Cell populations exposed to low- and high-LET radiation are compared in the margin.

Higher-LET radiation produces relatively more damage by direct effects than by indirect effects. Thus, the oxygen effect, basically a result of indirect effects, is diminished for higher-LET radiation. Also, the disappearance of the shoulder for higher-LET radiation implies that effects related to repair of sublethal damage also diminish.

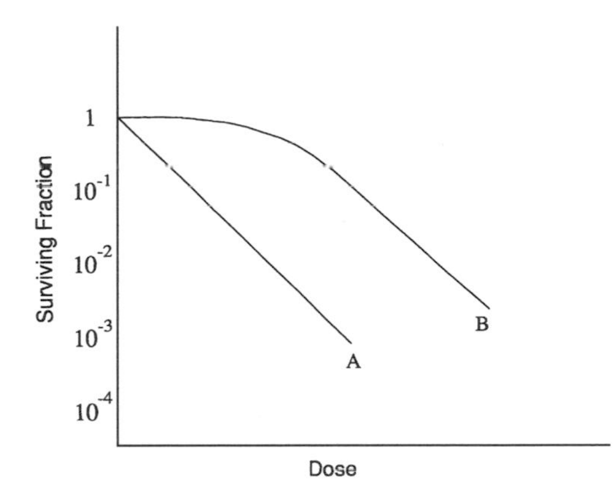

MARGIN FIGURE 26-2
Cell survival curves showing the effect of linear energy transfer (LET). The LET of the radiation is lower for B than for A. The shoulder disappears as LET increases.

Dose Rate and Fractionation

Compared with radiation delivered at a higher dose rate, radiation delivered at a lower rate may produce less cell killing even though the total doses are identical. Also, radiation delivered at a high rate but in fractions or discrete doses separated in time may be less damaging to cells than the same amount of radiation delivered in a single session. Repair of sublethal damage is the mechanism underlying dose rate and fractionation effects. An increase in dose rate may produce lesions at a frequency that exceeds the capacity of the cell for repair. The aggregate amount of damage produced in a short, high dose-rate fraction may be repairable if a suitable interval is allowed before additional irradiation. The effect of increasing dose rate upon cell survival is shown in the margin.

In Margin Figure 26-4, the cell survival curve for treatments given in a single dose is shown as curve abc, while the curve for an identical cell population is which a recovery period is allowed is curve abd. A recovery period of 10 to 20 hours is required for a detectable modification in mammalian cell survival through fractionation. This minimum period between fractions reflects the timing of events in the mammalian cell cycle.

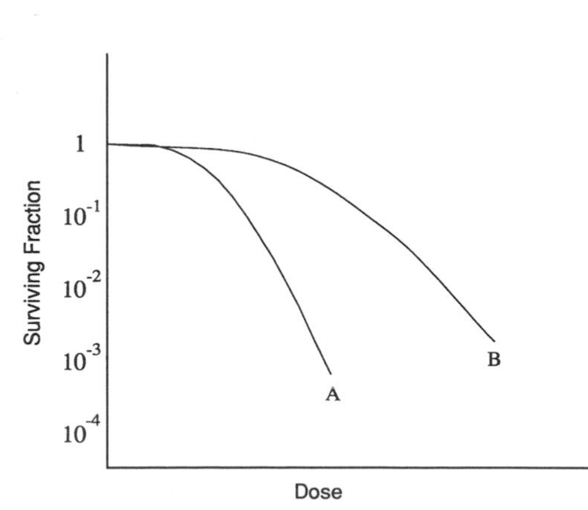

MARGIN FIGURE 26-3
Cell survival curves showing the effect of dose rate. The radiation is delivered at a higher dose rate for A than for B.

Cell Cycle

Figure 26-3 shows a typical cell cycle for somatic mammalian cells grown in culture. The cell cycle is divided into two parts, interphase and mitosis. During interphase, which occupies the majority of time in the cell cycle for somatic cells, the amount of cellular DNA is doubled as replicas of the chromosomes are synthesized. Interphase

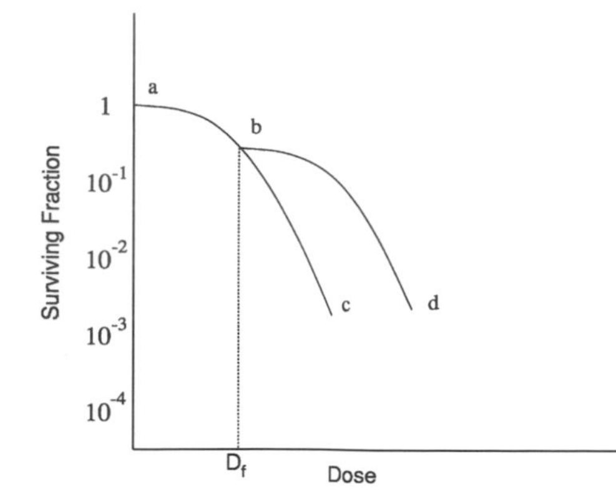

MARGIN FIGURE 26-4
Cell survival curves showing the effect of fractionated treatment. Cell populations were irradiated in single doses at all points along curve abc. To construct curve abd a recovery period of 10 to 20 hours was allowed for any dose greater than D_r, the total dose fraction.

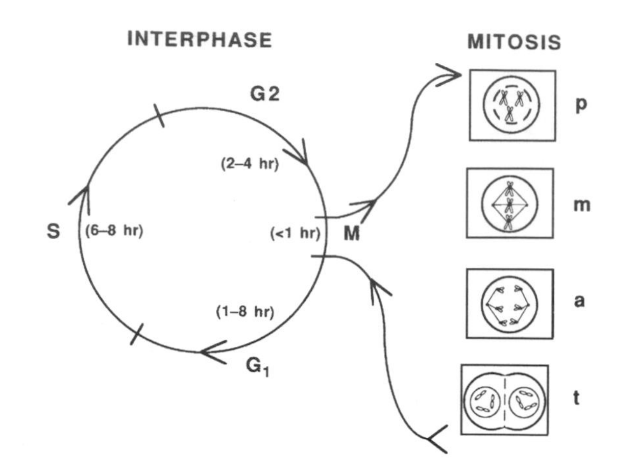

FIGURE 26-3
The cell cycle in mammalian cells is divided into interphase and mitosis. During interphase, DNA synthesis S is separated from mitosis M by gaps G_1 and G_2 during which normal activity of the cell is maintained. In mitosis or fission the nuclear membrane disappears in prophase p, a mitotic spindle forms in metaphase m, chromosomes separate in anaphase a, and nuclear and cellular membranes reappear to complete the separation of two progeny cells in telophase t.

is subdivided into three stages: G_1, or first gap, lasting 1 to 8 hours; S or DNA synthesis, lasting 6 to 8 hours; and G_2, or second gap, lasting 2 to 4 hours. During mitosis, the actions leading to cell division (fission of the cell) take place during a time period of approximately 1 hour. These actions are classified into four stages. During the first stage of mitosis, *prophase,* the nuclear membrane disappears. In the second stage, *metaphase,* a mitotic spindle forms, and the chromosomes become attached to the spindle. During the third phase, *anaphase,* the chromosomes migrate along the spindle to opposite sides of the cell. During the last stage, *telophase,* nuclear and cell membranes reappear, and two cells are present where formerly there was only one. The length of time spent by particular cells in each stage of the cell cycle, and the total length of the cell cycle, affect the radiation sensitivity of the cell.[8]

The mitotic spindle is a framework for coordinated movement of the chromosomes.

The radiation sensitivity of cells varies in different stages of the cell cycle. In general, mammalian cells are most resistant in late S phase and most sensitive in late G2 and mitosis.

Split-Dose Survival Curve

In general, radiation is less damaging to cells if given in discrete doses or fractions rather than all at once. Fractionation techniques are used in radiation therapy to provide normal cells time for recovery. Tumor cells often lack many of the functional capabilities of the cells from which they are derived. When the repair capabilities of tumor cells are below those of normal cells, fractionated radiation therapy treatments give normal cells an advantage in repairing sublethal damage. Therefore, fractionation tends to increase the ratio of damage to tumor cells compared with damage to normal cells.[9,10] A split-dose survival curve is shown in the margin. The shape of this curve illustrates many of the radiobiology principles discussed in this section. An initial dose fraction is given at time zero. The curve illustrates the effect upon survival when the second dose fraction is given at different times after the first fraction. At the time of the initial dose fraction, radiation encounters cells in all stages of division. Because the sensitivity of cells varies with the stage of the cell cycle, the cells that survive the first dose fraction are predominantly those that happened, by chance, to be in the radioresistant S phase. Thus, the second dose fraction encounters a cell population that is not randomly distributed. Stage I in the figure represents repair of sublethal damage between the dose fractions. Stage II represents the effect of waiting to administer the second dose fraction until the cells that were in S phase after the first dose fraction have progressed into the more sensitive G2 and M stages. Stage III represents the effect of repopulation; after the length of a cell cycle, the survivors of the first dose fraction divide. Thus the surviving fraction actually increases during this stage because new cells are produced between fractions.

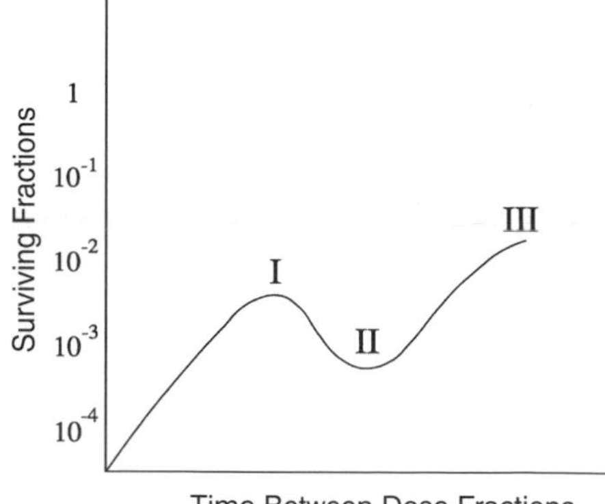

MARGIN FIGURE 26-5
A typical split-dose survival curve showing cell survival as a function of time allowed between doses. Stages I, II, and III are discussed in the text.

Radioprotectors and Radiosensitizers

Chemical agents may be used to modify the response of cells to ionizing radiation. These agents generally influence the indirect effects of radiation—that is, the effects of free radicals. Because of the speed with which free radical damage is produced, the agents must be present during irradiation. It is more difficult to modify the direct effects of ionization. Since the direct effects of radiation dominate as LET is increased, high-LET radiation effects are usually not modified significantly by administration of chemical agents.

Agents that reduce the effects of radiation upon cells are termed *radioprotectors*. Substantial interest in radioprotectors has arisen because of their potential applications in military and space programs as well as for medical uses such as protection of normal tissues during radiation therapy. One group of researchers has synthesized over 2000 potentially radioprotective compounds in the last 40 years.[11] Toxicity, unproven effectiveness of radioprotectors at low doses, and a lack of evidence for bioeffects of ionizing radiation at low doses make the use of radioprotectors in diagnostic radiology impractical.

Mechanisms of action for radioprotectors include inactivation of free radicals by preferential binding or scavenging, induction of hypoxia to reduce the sensitizing presence of oxygen, and occupation of sensitive sites in biomolecules by forming covalent bonds that are less easily disrupted by free radicals. Some protection against both direct and indirect effects has also been achieved by blocking cells in the G_1, phase of the cell cycle, thereby allowing more time for the action of repair enzymes before the S phase where DNA is replicated. The efficacy of radioprotectors is measured by the dose-modifying factor (DMF), where

Sulfhydril components such as cysteine and cysteamine are good free-radical scavengers.

Epinephrine and carbon monoxide are often used to induce hypoxia.

$$\text{DMF} = \frac{\text{Dose to produce an effect with the radioprotector}}{\text{Dose to produce an effect without the radioprotector}}$$

The amount of a radioprotector that can be administered is usually limited by its toxic effect on cells.

DMFs do not usually exceed 2. The theoretical limit of a DMF is 3 because removal of all indirect effects of radiation still leaves approximately one-third of the damage due to direct effects. The DMF varies with cell type, stage of the cell cycle, and LET of the radiation.

Radiosensitizers are agents that increase the effects of radiation on cells. Radiosensitizers are most applicable in radiation therapy, where distinction must be made between "apparent" sensitizers and "true" sensitizers. Apparent radiosensitizers are actually toxic agents that are particularly effective in cases where radiation alone is less effective. True radiosensitizers such as analogues of pyrimidines (5-bromodeoxyuridine [5-BUDR] and 5-iododeoxyuridine [5-IUDR]) and purines (6-mercaptopurine) promote both direct and indirect effects of radiation by inhibiting repair enzymes and weakening the sugar–phosphate bond of the DNA molecule.

Examples of apparent radiosensitizers include: methotrexate and 5-fluorouracil (5-FU), which tend to kill cells in the radioresistant S phase; and electron-affinic compounds such as nitroimidazoles, which are toxic to hypoxic cells.

5-Bromodeoxyuridine (5-BUDR) and 5-iododeoxyuridine (5-IUDR) are radiosensitizing pyrimidine analogues, and 6-mercaptopurine is a radiosensitizing purine analogue.

The direct effects of radiation may also be enhanced through the use of compounds containing boron to enhance the effects of neutron radiation therapy.[12] Low-energy neutrons (0.025 eV, or "thermal neutrons") have an unusually high probability of interaction with boron compared with elements that are normally present in the body. When a neutron interacts with ^{10}B, the unstable nuclide ^{11}B is formed. This nuclide fissions to produce α-particles that deliver a high absorbed dose in the immediate vicinity of the boron-containing compound. Although boron neutron capture therapy (BNCT) has been pursued experimentally over several decades, clinical applications have proven unsuccessful to date.

BNCT as a method to enhance radiation effects has been revisited several times over the past few decades. Each of these renewed efforts has failed to date to produce a clinically useful approach to radiation therapy.

ANIMAL STUDIES

Knowledge of the basic interactions of radiation at the cellular level is the foundation for understanding the bioeffects of radiation exposure. However, predictions based upon simple cell models must be tested against actual exposure of organisms. Radiation effects involve complicating factors such as the presence of co-carcinogens and the actions of the immune system. These effects must be explored at a level of system

The word "somatic" is derived from the Greek *somatikos,* meaning "pertaining to the body (soma)."

The word "genetic" is derived from the Greek *gennan,* meaning "pertaining to the gene (gennan)."

complexity above that of the cell. Both somatic (from the Greek *soma,* "of the body") and genetic effects are of interest.

Somatic Effects

Studies of animal survival following large doses of ionizing radiation have demonstrated a range of lethal doses that vary with the species and the irradiation conditions. A widely used measure of lethality is the dose required to kill 50% of the test animals within 30 days. This dose is called the $LD_{50/30}$. Among mammalian species the $LD_{50/30}$ varies from about 3 to 7 Gy, with a trend toward lower $LD_{50/30}$ for larger mammals. Death at these dose levels is caused by depletion of hematopoietic precursor cells, resulting in death by infection and bleeding due to a lack of mature erythrocytes, leukocytes, and platelets in the circulating blood supply. Larger species tend to have fewer hematopoietic system stem cells per unit body weight and are therefore more susceptible to these effects of radiation.[11,13,14]

Biological end points other than death have been studied extensively in animal populations. Tumor systems have been examined with regard to the effects of radiation therapy alone or in combination with other agents.[15] Normal tissue response has also been investigated in vivo by using end points such as skin reactions,[16] breathing rate (for lung tissue),[17] neurologic tests (for the spinal cord),[18] and sterilization (for gonadal irradiation).[19]

The induction of cancer (carcinogenesis) is an often-studied end point for radiation effects in exposed animal populations. These studies have revealed three distinct stages of carcinogenesis: initiation, promotion, and progression.[20]

Genetic Effects

In any animal population, there is a natural rate of spontaneous mutation. Because genetic variation is one of the factors that drives natural selection, it might be suggested that an increase in mutation rate could be beneficial to a population. However, there is no generally ccepted evidence of this suggestion in experimental populations. In fact, it is generally assumed that in the absence of other selection pressures, an increase in the background mutation rate beyond a certain point would be deleterious.

William Bateson first used the word "genetics" in 1905 to define the discipline of variation and heredity.

One measure of the genetic impact of an agent is the doubling dose. This dose is the amount of an agent required to double the spontaneous mutation rate in a population. Experiments suggest that the spontaneous mutation rate in most mammalian species is doubled by an absorbed dose to the germ cells of 1 Gy.[21] Some epidemiologic studies of human populations suggest that the doubling dose is slightly higher for humans.

Mutations in somatic cells are referred to as "somatic mutations." Mutations in germ cells are referred to as "genetic mutations."

Genetic effects have been studied extensively in a number of species. Dominant mutations may be demonstrated in first-generation offspring and identified within two generations, so studies may be carried out in a reasonable length of time in rodents and other populations that multiply rapidly.[22,23] Longer-term multigenerational studies have been conducted since the 1920s in short-lived species such as the fruit fly, *Drosophila melanogaster*.[21,24] By the early 1950s, there was considerable evidence that radiation-induced mutation rates were higher in rodents than in the fruit fly. Furthermore, it was found that mutation incidence in *Drosophila* varied linearly with dose and appeared to have no threshold. Moreover, there appeared to be no dose rate effect, suggesting that damage caused by exposure of germ cells is cumulative. If these effects occur in humans, then even a small amount of exposure per year, when summed over a long career, would lead to an appreciable risk of mutation.

Radiation-induced genetic effects in mice include

- Dominant lethal mutations
- Reciprocal gene translocations, sex chromosome losses and gains
- Recessive lethal mutations
- Recessive mutations causing morphological and biochemical traits
- Dominant mutations affecting histocompatibility genes
- dominant mutations causing
  - Skeletal abnormalities and cataracts
  - Preweaning mortality
  - Congenital abnormalities
  - Heritable tumors[25]

To answer some of the questions raised by the *Drosophila* data, a series of large-scale experiments (involving millions of mice, dubbed the "mega-mouse" experiments) were carried out in the early 1950s. These experiments were designed to examine the genetic effects of ionizing radiation upon specific gene loci in mice. These loci determine observable traits such as coat color and ear size[26–28] and can therefore be followed in experimental populations. One of the most important conclusions of the experiments was that for genetic effects, dose rate is an extremely important variable. At very low dose rates, no significant genetic effects were found, even when the total dose exceeded several gray. These results tempered the concern over occupational exposure of germ cells in humans where the dose rate is low.

Another result of the mega-mouse experiments was that delay of conception reduced the rate of genetic malformations in offspring. Thus, for gonadal doses large enough to create concern about genetic effects, (above 0.1 Gy to the gonads from radiation therapy or prolonged fluoroscopic procedures, for example), delay of conception for up to 6 months following exposure could be helpful in preventing genetic effects.[11]

Japanese A-bomb survivors and their descendants have been studied for more than 50 years to identify radiation-induced genetic effects in humans. Of the eight indicators of genetic abnormalities monitored, none has shown a statistically significant increase in the exposed population compared with controls.[29]

CONCLUSIONS

- Radiation acts directly on sensitive structures in cells (direct action) and through creation of sensitive intermediates (indirect action) to produce biologic damage.
- DNA (deoxyribonucleic acid) is a major sensitive site in the cell for the direct action of ionizing radiation.
- The loss of reproductive capacity of cells is a major way to describe cell "death" caused by ionizing radiation. This description is usually through the use of cell survival curves.
- Characteristics of a cell survival curve include the extrapolation number n, quasithreshold dose D_q, and mean lethal dose D_0.
- Factors that modify the radiation response of cells include the presence of oxygen, linear energy transfer (LET) of the radiation, total dose, dose rate, and dose fractionation.
- Mammalian cells are most sensitive when irradiated in the late G_2 and M phases of the cell cycle, and they are least sensitive in the S phase.
- Certain chemical substances can either heighten or suppress the response of cells to radiation. The action of these substances, known as radiosensitizing and radioprotecting agents, is described by the dose modifying factor (DMF).
- Actions of radiation directly on cells of the body other than germ cells are related to the somatic effects of radiation exposure.
- Actions of radiation on cells, giving rise to biological effects in future generations, are described as the genetic effects of radiation exposure.
- The doubling dose is the dose of ionizing radiation needed to double the spontaneous mutation rate in a biological organism.

REFERENCES

1. Ward, J. F., Limoli, C. L., Calabro-Jones, P., et al. Radiation vs. chemical damage to DNA, in Nygaard, D. F., Simic, M., and Cerutti, P. (eds.), *Anticarcinogenesis and Radiation Protection*. New York, Plenum, 1988.
2. Puck, T. T., Morkovin, P. I., Marcus, S. J., et al. Action of x rays on mammalian cells II. Survival curves of cells from normal human tissue. *J. Exp. Med.* 1957; **106**:485.
3. Lea, D. E. *Actions of Radiation on Living Cells,* 2nd edition. New York, Cambridge University Press, 1955.
4. Haynes, R. H. The interpretation of microbial inactivation and recovery phenomena. *Radiat. Res. Suppl.* 1966; **6**:1.
5. Chapman, J. D. Biophysical models of mammalian cell inactivation by radiation, in Meyn, R. E., and Withers, H. R. *Radiation Biology in Cancer Research*. New York, Raven Press, 1980, p. 21.
6. Alpen, E. L. *Radiation Biophysics,* 2nd edition. San Diego, CA, Academic Press, 1998.
7. Johns, H. L., and Cunningham, J. R. *The Physics of Radiology,* 4th edition. Springfield, IL, Charles C Thomas Publishers, 1983, p. 680.
8. Sinclair, W. K., Morton, R. A. X-ray sensitivity during the cell generation cycle of cultured Chinese hamster cells. *Radiat. Res.* 1966; **29**:450–474.
9. Strandqvist, M. Studien ber ūdie kumulative Wirkung der Röntgenstrahlen bei Fraktionierung. *Acta Radiol. Suppl.* (Stockh.) 1944:55.
10. Ellis, F. Dose, time and fractionation: A clinical hypothesis. *Clin. Radiol.* 1969; **20**:1.
11. Hall, E. J. *Radiobiology for the Radiologist,* 3rd edition Philadelphia, J. B. Lippincott, 1988.
12. Barth, R. F., and Soloway, A. H. Boron neutron capture therapy of cancer. *Cancer Res.* 1990; **50**:1061–1070.
13. Vriesendorp, H. M., and vanBekkum, D. W. Role of total body irradiation in conditioning for bone marrow transplantation. *Haematol. Bluttbansfus.* 1981; **25**:349–364.
14. Lusbaugh, C. C. Reflections on Some Recent Progress in Human Radiobiology, in Augenstein, L. G., Mason, R., and Zelle, M. (eds.), *Advances in Radiation Biology*. New York, Academic Press, 1969, pp. 277–314.
15. Looney, W. B., Trefil, J. S., Hopkins, H. A., et al. Solid tumor models for the assessment of different treatment modalities: Therapeutic strategy for sequential chemotherapy with radiotherapy. *Proc. Natl. Acad. Sci.* USA 1977; **74**:1983–1987.
16. Fowler, J. F., Denekamp, J., Page, A. L., et al. Fractionation with x-rays and neutrons in mice: Response of skin and C3H mammary tumors. *Br. J. Radiol.* 1972; **45**:237–249.
17. Travis, E. L., Vojnovic, B., Davies, E. E., et al. A plethysmographic method for measuring function in locally irradiated mouse lung. *Br. J. Radiol.* 1979; **52**:67–74.
18. Van der Kogel, A. J. Mechanisms of Late Radiation Injury in the Spinal Cord, in Meyn, R. E., and Withers, H. R. (eds.), *Radiation Biology in Cancer Research*, New York, Raven Press, 1980, pp. 461–470.

19. Morgan, K. Z., and Turner, J. E. *Principles of Radiation Protection*. New York, John Wiley & Sons, 1967, p. 422.
20. Hendee, W. R. Radiation Carcinogenesis, in Hendee, W. R., and Edwards, F. M. (eds.), *Health Effects of Exposure to Low-Level Ionizing Radiation*. Philadelphia, Institute of Physics Publishing, 1996, pp. 103–111.
21. BEIR V, National Academy of Sciences, Committee on the Biological Effects of Ionizing Radiations. *Health Effects of Exposure to Low Levels of Ionizing Radiation,* Report No. 5. Washington, D.C., National Academy Press, 1980.
22. Selby PB, Selby PR. Gamma-ray-induced dominant mutations that cause skeletal abnormalities in mice. I. Plan, summary of results and discussion. *Mutat. Res.* 1977; **43**:357–375.
23. Selby, P. B. Genetic Effects of Low-Level Irradiation, in Fullerton, G. D., Kopp, D. T., and Webster, E. W. (eds.), *Biological Risks of Medical Irradiations*. New York, American Institute of Physics, 1980, pp. 1–20.
24. Muller, H. J. The effects of x-radiation on genes and chromosomes. *Science* 1928; **67**:82.
25. Sankaranarayanan, S. Radiation Mutagenesis in Animals and Humans and the Estimation of Genetic Risk, in Hendee, W. R., and Edwards, F. M. (eds.), *Health Effects of Exposure to Low-Level Ionizing Radiation*. Philadelphia, Institute of Physics Publishing, 1996, pp. 113–167.
26. Russell, W. L. X ray induced mutations in mice. *Cold Spring Harbor Symp. Quant. Biol.* 1951; **16**:327–336.
27. Russell, W. L., Russell, L. B., and Kelly, E. M. Radiation dose rate and mutation frequency. *Science* 1958; **128**:1546–1550.
28. Russell, W. L., and Kelly, E. M. Mutation frequencies in male mice and the estimation of genetic hazards of radiation in men. *Proc. Natl. Acad. Sci.* USA 1982; **79**:542–544.
29. Neel, J. V., and Schull, W. J. (eds.), *The Children of Atomic Bomb Survivors: A Genetic Study*. Washington, D.C., U.S. National Academy of Science, 1991.

CHAPTER

27

HUMAN RADIOBIOLOGY

OBJECTIVES

After completing this chapter, the reader should be able to:

- Define the term "stochastic effects" and name the three effects that have been identified as stochastic radiation effects.
- Name the subcategories of the acute radiation syndrome (ARS).
- Define the term "collective dose equivalent."
- Define the term "effective dose."
- List the sources of natural and manmade background radiation and name the largest contributors to each.
- List 10 examples of human populations that are or have been exposed to unusual levels of radiation.
- Discuss current dose-effect models and the factors that influence them.
- Give an estimate of the risk per unit effective dose of fatal cancer due to a single exposure to ionizing radiation.

STOCHASTIC EFFECTS OF RADIATION

It is common knowledge that the incidence of leukemia among atomic bomb survivors in Hiroshima and Nagasaki began to increase a few years after the explosions. This increase has been associated with exposure to ionizing radiation produced by the blasts. Leukemia is one example of a long-term consequence of radiation exposure.

Radiation effects that appear several years or more after individuals are irradiated have several characteristics in common:

- The probability of occurrence of the effect (i.e., the fraction of persons in a population who exhibit the effect) increases with dose.
- The severity of the effect in a single individual is unrelated to the magnitude of the dose.
- There is no threshold below which it can be said with certainty that the effect will not occur.

Clearly, these effects are a result of radiation activation of mechanisms within the body that produce disease. A stochastic process (disorder or radiation effect) is one in which the probability of occurrence, but not the severity of the disorder, is related to the dose of radiation. The fact that severity is unrelated to dose suggests that radiation triggers a disease process whose ultimate course is determined by other factors in the individual.

One of the difficulties in providing precise estimates of the role of an agent such as radiation in producing stochastic effects is that these effects also occur in the absence of irradiation. That is, there is a "natural incidence" of the effects due to causes other than the agent under consideration. The three stochastic effects of radiation that have been identified to date are:

- Carcinogenesis, the induction of cancer
- Teratogenesis, the induction of birth defects by irradiation of the fetus
- Mutagenesis, the induction of genetic disorders in future generations by irradiation of germ cells (sperm and ova) prior to conception

In one case of fluoroscopic overexposure of the patient's skin, the patient's arms were at her side during a cardiac catheter ablation procedure in which the x-ray tube was repeatedly maneuvered such that the right humerus was plainly visible in the center of the beam. In addition, a spacer had been removed from the x-ray tube housing, allowing the tube to be much closer than usual to the x-ray tube. It has been estimated that the total dose to the arm exceeded 25 Gy. *Source:* Wolff D, Heinrich KW. Strahlenschaden der haut nach herzkatheterdiagnostik und -therapie 2 kasuistiken. *Hautnah Derm.* 1993; **5**:450–452.

As of 2001, 71 cases of radiation injury to the skin are known, with most associated with cardiac catheterization. Even assuming that many have gone unreported, the true number is undoubtedly a very small fraction of the approximately 30,000 interventional procedures performed yearly in the United States.

NONSTOCHASTIC EFFECTS OF RADIATION

Most of the "immediate" effects of radiation—that is, those that occur within several weeks of radiation exposure—are associated with dose levels far above those delivered in modern diagnostic radiology. One set of such effects is known coloquially as "radiation sickness" or, more formally, as acute radiation syndrome (ARS).

The acute radiation syndrome is divided into three subcategories: gastrointestinal (GI), hematopoietic, and cerebrovascular syndromes. These subcategories reflect both the relative sensitivities of various organ systems to radiation and the time required to produce effects on the overall health of the organism.

Hematopoietic Syndrome

The stem cells of the hematopoietic system, precursors of mature blood cells that are found principally in the bone marrow, can be inactivated in significant numbers at doses of a few gray. Loss of the precursor population may not pose a significant threat to the individual until it is time for the precursor cells to reach maturity. The body may then lose the ability to combat infection. Thus, a latency period may occur for a week to a month, during which attempts may be made to reestablish the bone marrow population through transplants or other techniques.

Gastrointestinal Syndrome

The cells associated with the GI tract, particularly epithelial cells that line the intestinal surfaces, are particularly susceptible to radiation damage. Following an exposure of several gray or more, diarrhea, hemorrhage, electrolyte imbalance, dehydration, etc., occur as a consequence of cell damage.

Cerebrovascular Syndrome

At doses above 50 gray, damage to the relatively radiation insensitive neurologic and cardiovascular compartments is severe enough to produce an almost immediate life-threatening syndrome. Death in this syndrome is brought about by destruction of blood vessels in the brain, fluid buildup, and neuronal damage. Death may occur in a few days or, at higher doses (greater than 100 Gy), within hours.[1] Because of the involvement of brain and nervous system tissues, the cerebrovascular syndrome (CVS) is sometimes referred to as central nervous system syndrome. The hematopoietic and GI systems are also severely damaged at this dose level but would not contribute to the cause of death in such a short time.

Fertility and sterility are also affected at high dose levels. Temporary sterility may occur in the male for testis irradiation above 0.15 Gy. Permanent sterility results from doses exceeding 3.5 Sv.[2] In the female, temporary sterility has been observed for ovarian doses above 0.65 Gy, and permanent sterility for doses above 2.5 Gy.[3] These thresholds are for single instantaneous exposures. Doses required to produce temporary or permanent sterility are increased by orders of magnitude if delivered either in fractions or at lower dose rates.

Because of the radionuclides contained in our bodies, our cells are irradiated by 30 million radioactive decays per hour. An ionizing interaction occurs in every cell in our body at least once a year.

One nonstochastic effect that occurs within weeks of radiation exposure that may be encountered in diagnostic radiology is the skin effect. In 1994, the FDA issued an "advisory," a document intended to warn health care practitioners that certain fluoroscopic procedures (cardiac ablation, coronary angioplasty, percutaneous transluminal angioplasty, and a few others) had been found, in rare instances, to produce skin effects.[4] In some cases, this was reddening of the skin that eventually ceased. However, in some cases, the skin became necrotic and multiple skin grafts were required. All were associated with the use of "high level control" mode, a mode of fluoroscopy in which a dose of as much as 10 times higher than normal is used to decrease noise so that moving catheters may be seen without exceeding the patient's level of tolerance for contrast material. In most cases the total fluoroscopy "beam on" time was excessive, due to complications or the need for multiple procedures. One factor that has caused this fairly recent phenomenon is the ability of modern x-ray tubes to tolerate greater heat loads so that procedures can go on much longer than a decade ago. Table 27-1 gives the doses required to obtain various skin reactions. At typical

TABLE 27-1 Radiogenic Skin Effects[a]

Skin Effect	*Approximate Single-Dose Threshold (Gy)*	*Approximate Time to Onset*	*Peak*
Transient erythema	2	a few hours	24 hr
Temporary epilation	3	3 weeks	NA
Main erythema	6	10 days	—
Dry desquamation	10	4 weeks	5 weeks
Moist desquamation	15	4 weeks	5 weeks
Late erythema	15	6–10 weeks	NA
Dermal necrosis	18	>10 weeks	NA
Ulceration	20	>6 weeks	NA

[a] Data from Wagner, L. K., Eifel, P. J., and Geise, R. A. Potential biological effects following high x-ray dose interventional procedures. *J. Vascular Interventional Radiol.* 1994; **5**:71–84.

fluoroscopic dose rates (~0.02 Gy/min at the entrance surface), erythema would probably not occur unless 1.7 hours of fluoroscopy beam-on time were exceeded. However, at high-level dose rate (~0.20 Gy/min), only 10 minutes is required.

Radiogenic skin injury from this source is especially challenging in terms of medical management. Some patient's reports of skin problems were initially treated as nonspecific dermatitis or some other nonradiogenic disorder because the injury was so far removed in time from the fluoroscopic procedure and dermatologists were unfamiliar with fluoroscopically induced skin damage. It is assumed that many cases of reversible skin changes have gone unreported.

Because of the potential for serious harm, the FDA recommends that patient dose be monitored if a procedure has the potential for delivering more than 1 Gy. Following the procedure, some indication of the specific areas of skin and the doses estimated for those areas is to be entered into the patient's chart. Some indication is also to be included regarding the types of skin effects that might be seen.

Although the majority of procedures for which skin effects have been demonstrated have been cardiac, neurointerventional procedures also involve patient doses that would put them within the criteria for monitoring and including notes in the patient's record. A review of over 500 neurointerventional procedures at a major medical center where the x-ray equipment was equipped with an automated dosimetry system demonstrated that 6% of embolization procedures and 1% of cerebral angiograms resulted in skin doses exceeding the erythema threshold of 6 Gy.[5]

DOSIMETRY IN INDIVIDUALS AND POPULATIONS

The dosimetry of ionizing radiation as it concerns the individual patient or irradiated specimen has been discussed elsewhere (Chapter 7). Estimation of radiation risk to individuals within populations requires some additional terminology. At the core of this terminology is the assumption that the effects of radiation at low doses are stochastic (probabilistic for the occurrence of effect, with the severity of effect not related to dose). As far as the individual is concerned, an increase in dose increases the probability of occurrence of an adverse effect. In terms of dose to a population, when more individuals receive a given dose, a greater number exhibit an adverse effect. The collective dose equivalent (CDE), defined as the sum of the dose equivalents for individuals in a population, has units of person-sievert (or person-rem). The average dose equivalent (ADE), which is the CDE divided by the total number of individuals in a population, has units of sievert (or rem).

The modifiers "committed" and "effective" are sometimes added to CDE and ADE. Committed dose refers to the dose delivered by radioactive materials deposited internally within the body. Such materials deliver radiation over a period of time determined by the biological and physical half-lives of the radionuclide. Thus, by summing (integrating) the dose delivered during a series of time intervals, the total commitment of dose over the life span of an individual may be determined.

To describe the overall harm to an organism, one must consider the varying degrees of damage caused by the irradiation of different organs.[6] Harm, in this context, refers to the risk of cancer and hereditary effects. As an index of overall harm, the effective dose (ED) concept uses weighting factors for each organ. These weighting factors, which reflect the radiosensitivity of the organ and its importance to the organism as a whole, are applied to doses delivered to each organ. For example, the weighting factor for radiation delivered to the breast (0.15) is higher than the weighting factor for a dose delivered to the thyroid (0.03). This concept was discussed extensively in Chapter 6.

Immediate effects of radiation exposure are also referred to as *early effects;* delayed effects are frequently termed *late effects* of radiation exposure.

BACKGROUND RADIATION

All of the earth's inhabitants are exposed to background radiation. Some of this radiation is man-made, such as radiation used in medical applications, and some is "natural." Natural sources include cosmic rays, terrestrial, and internal. Man-made radiation includes medical x-rays, medical nuclear procedures, consumer products, industrial sources, and some miscellaneous sources of radiation. A typical U.S. resident receives a total of 3.6 mSv (360 mrem) in one year from all sources of background radiation.[7] The actual background encountered by any one individual varies significantly, depending upon where he or she lives, the food that is consumed, the radon levels in the home, and so on.[8]

Natural Background

Cosmic rays have always bombarded the earth. A typical U.S. resident receives 0.29 mSv (29 mrem) from cosmic rays. The earth's atmosphere provides some shielding from cosmic rays. This shielding is reduced at higher elevations, and the cosmic ray dose is increased. Inhabitants of Denver, Colorado, at an elevation of 1600 meters (1 mile), receive 0.50 mSv/yr (50 mrem/yr) from cosmic rays, while those in Leadville, Colorado, at an elevation of 3200 meters (2 miles), receive 1.25 mSv (125 mrem). The average value of 0.29 mSv reflects the distribution of the U.S. population living at different elevations.

Terrestrial background originates from radioisotopes that are found everywhere in our surroundings. All elements found in nature have radioactive isotopes, many of which are also present in the environment. The exact composition of soil influences the local terrestrial background, because the minerals present determine which elements are most abundant. Terrestrial background sources are categorized as "primordial" if their half-lives are the same order of magnitude as the presumed lifetime of the earth (4.5×10^9 years). That is, these sources were present when the earth was formed and there is no way to replenish them in nature. Two isotopes of uranium, ^{238}U and ^{235}U, and one of thorium ^{232}T, give rise to three different decay series. In each of these series, the radioactive nuclide decays to another stable isotope of bismuth or lead. Seventeen other nuclides are primordial, but are not part of a decay series. Of these nonseries radionuclides, ^{40}K and ^{87}Rb make the greatest contribution to background dose.

The ICRP, NCRP, and BEIR Committee are voluntary efforts, and their recommendations are purely advisory. However, in the United States, NCRP recommendations are frequently codified into radiation regulations by federal and state agencies, including the Nuclear Regulatory Commission and the Environmental Protection Agency.

Internal background is the dose imposed by the isotopes contained in our bodies. A small percentage of the potassium in the human body is ^{40}K. This radioactive nuclide emits both locally absorbed beta radiation and more penetrating gamma radiation. Similarly, ^{14}C, which comprises a small percentage of the carbon atoms found in organic molecules throughout our bodies, contributes to a total dose of 0.39 mSv/yr (39 mrem/yr) to the average U.S. resident from internal background.

Most texts consider radon exposure to be a part of natural background, although it could be argued that there is a component of human activity involved in exposure to radon. As part of the ^{238}U decay series, radon (specifically, ^{222}Rn, referred to hereafter as simply radon) is significant because it is an alpha emitter that exists as an inert gas.

It is estimated that, anywhere on earth, a million-ton sample of the earth's crust would contain four tons of uranium (also, for comparison, 70 tons of copper and 10 lbs of gold).

Since it is inert, radon generated by decay of ^{226}Ra at some depth in the soil does not bind chemically with other elements. Instead, it percolates up to the surface to escape into the atmosphere. Being heavier than most constituents of the atmosphere, it tends to remain at lower elevations. Although minable deposits of uranium ore are primarily associated with granite rock formations, uranium is found everywhere in the earth's crust. Ninety-nine percent of the uranium found in nature is ^{238}U. Thus, the air we breathe anywhere on earth contains some amount of radon.

Radon itself is not particularly hazardous, when inhaled, because it does not react and in most cases is simply exhaled. A more significant concern is that two of the decay products of radon (^{218}Po and ^{214}Po), delivered to the air by decaying radon, are not inert. These products adhere to dust particles in the air, which may then be inhaled into the lung. The two alpha emitting isotopes, ^{218}Po and ^{214}Po, account for 85% of the dose to the lung from "radon;" two other beta and gamma emitting isotopes in the series, ^{214}Pb and ^{214}Bi, account for most of the remaining 15%. A typical U.S. resident receives an annual dose equivalent of 2.0 mSv (200 mrem) from radon and its decay products.[7]

The exact value of dose equivalent received by an individual is most heavily influenced by the location where the person spends his or her time—buildings. With a combination of an unusually strong source (e.g., an unfinished crawl space below a house in a region with a high concentration of uranium in the soil) and air flow (e.g., radon buildup due to the lack of air exchange in an energy-efficient house, or radon buildup because of a "chimney effect" in a drafty house), some homes can have radon levels that are significantly elevated. Although the risk to any one individual is very small, many people (i.e., the entire population) are exposed to some level of radon. By applying the linear, no-threshold model of radiation-induced cancer, one can estimate that a significant number of lung cancers may be caused each year by long-term exposure to high radon levels. Therefore, the U.S. Environmental Protection Agency recommends that radon concentrations be measured in buildings and that remedial action be taken if the radon concentrations exceed 4 pCi/l of air. This level of radon has been estimated to yield (a) a lung dose in a typical adult of 0.05 mGy (5 mrad) and (b) an increase in the risk of lung cancer of 2% for each year of exposure.

Stochastic radiation effects are those effects for which the probability of occurrence of the effect (as opposed to the severity of the effect) increases with dose. Although at small doses the probability of occurrence may be very small, it is presumed to never reach zero. That is, it is assumed that there is no threshold dose. Stochastic effects are sometimes called *probabilistic effects*.

Man-Made Background

Various human activities add to the annual radiation background. Chief among these is diagnostic radiology (0.39 mSv, 39 mrem), and second is nuclear medicine (0.14 mSv, 14 mrem).[7] Radiation received by patients in radiation therapy is not counted in man-made background, because the intention is to track radiation doses that are associated only with stochastic effects. The large doses received in radiation therapy are associated with both deterministic and stochastic effects. This omission has little impact on the average level of man-made background, because the number of persons who receive radiation treatment is a very small fraction of the U.S. population.

Of course, in any given year an individual may or may not undergo a medical diagnostic study. The actual dose received by an individual from a medical procedure can be estimated from clinical information such as the measured output of the x-ray tube, patient thickness, technique factors, and so on. The concern that is addressed by calculating the population average of 0.39 mSv per person per year is the collective equivalent dose received by the entire population, since it is the collective dose that is responsible for population bioeffects such as increases in somatic and genetic mutation rates.

Various consumer products emit small amounts of radiation.[9] Some examples include exit signs that contain ^{3}H, a low-energy beta emitter, and smoke detectors that contain ^{241}Am, an alpha emitter. Luminous dials on watches, clocks, and instruments contained ^{226}Ra at one time but, since 1968, have been completely replaced by

^{3}H and ^{147}Pm, both of which are low-energy beta emitters. The low-energy beta particles emitted by these substances are absorbed in the instrument components and provide negligible amounts of radiation dose to their owners. Increasingly, liquid crystal displays and light-emitting diodes are replacing the use of radioactive materials in luminous displays. Collectively, consumer products are estimated to contribute approximately 0.1 Sv (10 mrem) to the yearly dose from man-made background radiation.

Nonstochastic radiation effects are those effects that exhibit a dose threshold. After the threshold is exceeded, the severity of the effect increases as the dose increases. Nonstochastic effects are sometimes called *deterministic effects*.

Other contributions to man-made radiation include (a) aspects of the nuclear fuel cycle, ranging from mining to transportation and to waste storage, (b) the occupational exposure that a relatively small fraction of the population receives, and (c) fallout from nuclear weapons tests. These components are thought to account for approximately 0.02 mSv (2 mrem) per year.

The United States and other countries detonated over 500 nuclear weapons above ground during the time period 1945–1963. This activity had a measurable effect upon natural background radiation levels. In a nuclear explosion, new elements are created through the processes of nuclear fission and fusion. These elements are injected into the upper atmosphere by the force of the blast, where they drift over large distances before eventually returning to earth. The United States and the former Soviet Union ceased above-ground testing in 1963. A few countries conducted above-ground tests after that time, but by 1980 all above-ground testing had stopped. Individuals who were alive during the time period 1945–1980 received radiation doses from inhalation and ingestion of some atoms of radioactive material as a result of these tests. Calculation of this dose involves complicated functions of rate of uptake, rate of decay of the nuclides in individuals, rate of insertion of new isotopes into the environment, and other factors. The United Nations Scientific Committee on the Effects of Atomic Radiation (UNSCEAR) has estimated that an individual who was born before 1945 has received a total of 4.5 mSv (450 mrem) until now because of nuclear testing. The current dose rate is estimated to be 0.01 mSv/yr (1 mrem/yr).[10]

Radon was first identified as a significant source of background exposure when a worker reported to the Limerick Nuclear Power Plant in Pennsylvania for his first day of work in 1984. Radiation detectors were located at all exits from the plant to detect any contamination resident upon workers as they left. This worker set off radiation detectors when he entered the plant. Subsequent investigation revealed an unusually high concentration of uranium near the foundation of his home.

HUMAN POPULATIONS THAT HAVE BEEN EXPOSED TO UNUSUAL LEVELS OF RADIATION

Atom Bomb Survivors

The atomic bomb blasts at Hiroshima and Nagasaki left a population of nearly 280,000 persons who survived the immediate effects of the blast but were subjected to the long-term consequences of exposure to ionizing radiation.[11–17] This population represents the highest CDE for any large population to date and is the most extensively studied of all groups of exposed individuals. Of the survivors, dosimetry estimates are available for approximately 80,000. Of that group, 42,000 are estimated to have received greater than 0.005 Gy (0.5 rad), with the rest serving as controls. Cancer incidence and the effects of radiation exposure *in utero* have been documented in this population. Much of the risk estimate data discussed elsewhere in this chapter is based upon studies of atom bomb survivors.

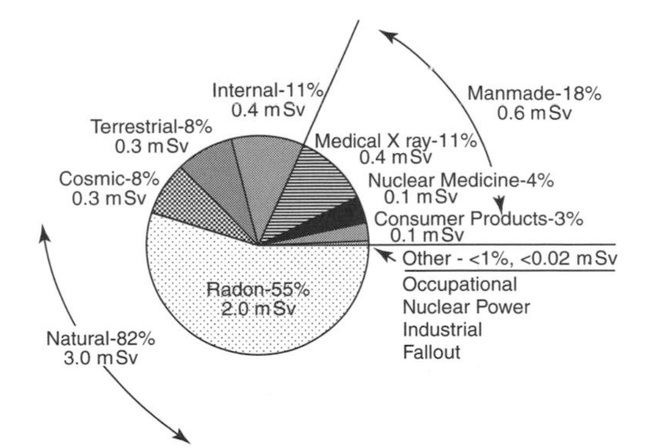

MARGIN FIGURE 27-1
A typical U.S. citizen's yearly background radiation equivalent dose, broken down into various natural and manmade sources. (From National Council on Radiation Protection and Measurements (NCRP) Report No. 93, *Ionizing Radiation Exposure of the Population of the United States*, NCRP, Bethesda, MD, 1987.)

Early Radiologists, Nurses, and Technologists

Radiologists in the early part of the twentieth century often worked with x-ray equipment that lacked safety features such as lead shielding and collimation.[18–20] Safety standards did not become commonplace until after World War II. Although it is difficult to reconstruct dosimetry for individuals, this group of radiation workers has been studied to demonstrate the need for current safety standards. The incidence of leukemia, for example, has been found to be elevated for radiologists who experienced a significant part of their careers prior to World War II.[21]

Uranium and Other Miners

It has been known for many years that the "mountain sickness" reported in miners as early as the fifteenth century was probably lung cancer.[22–25] The excess rates of cancer were probably initiated by inhalation of particulates with attached radioactive atoms such as the decay products of radon. Current estimates of radon hazards in the home have arisen in part from studies of uranium miners.

Radium Dial Painters

About 1% of radium dial painters who worked in the industry between 1913 and 1925 died of bone and sinus cancer. This is 33% higher than the rate found in control populations. In 1925, the industry passed a rule stating that the paint brush containing the radium paint shall never be placed in the mouth. Cancer rates for workers who started work after 1925 have been found to be equal to those of control groups.

In the period from 1915 to 1930, approximately 1500 workers, mostly women, were employed to paint watch dials with a luminous compound containing the α-emitter radium (^{226}Ra).[26–30] This activity was conducted before modern safety standards for handling radioactive materials were commonplace. Many of the workers ingested or absorbed radium compounds through work habits such as using the mouth to bring the tip of paint brushes to a point. Excess numbers of carcinomas of the paranasal sinuses and mastoid air cells have been observed in this population of workers. These cancers are thought to have arisen from irradiation of the tissue surrounding air cavities containing radon gas (^{222}Rn) and its decay products, and from direct irradiation of underlying bone by the radium, a chemical analogue of calcium that is taken up by bone and incorporated into the skeleton. Of the radium dial painters, approximately 35 cases of osteogenic sarcomas and sinus carcinomas have been identified over a follow-up period exceeding 50 years.

Radiation Therapy Patients

It is difficult to determine if cancer patients who have been treated with radiation are at increased risk for cancers occurring specifically as a result of the treatment. Studies have been attempted, but are hampered by the existence of proven cancers in those populations and the fact that second cancers are more prevalent inherently in patients who have had an earlier cancer. However, a number of benign disorders have been treated in the past with radiation at dose levels of several gray and above. These include tinea capitis (ringworm), thymic enlargement, ankylosing spondylitis, post-partum mastitis, and menorrhagia. Positive correlations between radiation exposure and several types of subsequent cancer, particularly skin cancer, have been shown in these populations.[31–36]

Diagnostic Radiology Patients

Radiation-induced changes in the vascular endothelium are thought to be the underlying cause of most nonstochastic radiation effects.

In the past, some diagnostic studies were performed on patients with techniques that yielded exposures that were orders of magnitude greater than those for similar procedures today.[37–43] For example, fluoroscopy was used from 1930 to 1954 to monitor pneumothoraces induced as part of the treatment for tuberculosis. Radiation doses of up to 6 Gy from this procedure have been associated with an increased incidence of breast cancer.

Most population studies of patients that have received doses on the order of those used currently have failed to provide statistically significant evidence of harmful effects. For example, a study of more than 10,000 ^{131}I thyroid examinations performed in Sweden in the 1950s failed to show an increase subsequent in thyroid cancer.[39]

Some marginal associations between diagnostic radiation exposure and adverse health effects remain. Retrospective studies of leukemia patients have examined the evidence of a link between adult-onset myelogenous leukemia and x-ray examinations. Some correlations have been found, particularly for individuals who experienced multiple radiographic examinations. There is also some evidence that *in utero* exposure increases the risk of childhood leukemia. A study of prenatal x-ray exposure in asymptomatic twin pregnancies (a technique that was used until the early 1970s to determine fetal position, resulting in a fetal dose of a few rem) demonstrated an increased incidence of childhood leukemia for the exposed twins by 1 chance in 2000. However, questions remain regarding these studies related to the possibility of statistical bias, inadequate dosimetry, and statistical power.

Nuclear Weapons Tests

In addition to the Japanese A-bomb survivors, a number of individuals were exposed to radiation as a result of the detonation of nuclear weapons.[44–46] These individuals include military personnel who conducted maneuvers in the vicinity of detonations, and residents of areas "downwind" of test sites. Epidemiologic studies among these groups are difficult because of inexact dosimetry. Also, organ doses for most of these are estimated to be below 50 mGy (5 rem). Thus it is not surprising that there is little statistically significant correlation between radiation exposure and adverse health effects in most of these groups.

Adverse health effects have been detected in individuals whose organ doses may have been higher than 50 mGy. These individuals include some residents near the Nevada test site, and Marshall Islanders in the Pacific who were accidentally exposed to fallout from a thermonuclear test explosion in 1954. Ingestion of ^{131}I and external exposure have been associated with an increased rate of nodule formation and cancer in the thyroids of the Marshall Islanders.

Residents of Regions with High Natural Background

The presence of naturally occurring radioactive materials in the soil has resulted in increased natural background radiation by inhabitants in many parts of the world.[47–52] Some of the most extensive epidemiologic studies of native populations have examined the effects of monazite, a mineral containing thorium, uranium, and radium. The global average natural background is approximately 1 mGy/yr, excluding exposure from radon and radon progeny in the lungs. Because of monazite, the natural background for the 12,000 inhabitants of Guarapari, Brazil is 6.4 mGy/yr, for the 70,000 individuals in the Kerala Coast of India it is 3.8 mGy/yr, and for the Guangdong Province of the People's Republic of China it is 3.5 mGy/yr. Studies of these populations have repeatedly shown an increased frequency of chromosomal aberrations. However, a heightened incidence of genetic disorders has not been unequivocally demonstrated. In the United States, the relatively smaller range of natural background radiation, coupled with the mobility of the population, has created difficulties for studies of the effects of background radiation upon regional morbidity and mortality. Studies have shown, as expected, no correlation.

Monazite sand formations in Guarapari, Brazil were expected to produce elevated natural background levels for inhabitants of that region due to a higher concentration of thorium, uranium, and radium than is found in most soils. The amount of this radiation actually received would depend upon how much time people spent outdoors, what kind of food they ate, and so on. Dr. Thomas L. Cullen, a physicist and Jesuit priest at the Catholic University of Rio de Janeiro, wanted to measure background radiation levels in this population. He needed to convince a large sample of the rural population to wear radiation dosimeters for three months as they went about their normal lives. Reasoning that the majority of inhabitants were devout Catholics, he had hollow medals of the Virgin Mary filled with thermoluminescent dosimetry powder. After they were blessed by Dr. Cullen, he was able to obtain the cooperation of 115 people. The medals were opened three months later and the powder was returned to Dr. Cullen. The subjects were allowed to keep the medals. This statistical sample was still somewhat biased as most non-Catholics in the region refused to wear the medals.
Source: John Cameron, *Health Physics Society Newsletter,* vol. xxix, No. 3, p. 11, March 2001.

Air and Space Travel: Encountering Cosmic Rays and Solar Flares

Cosmic rays are composed of various types of radiation that are emitted into space throughout the universe by supernovae, quasars, nebulae, and so on. The earth's atmosphere, along with the magnetic fields that surround the planet, act to shield the surface of the earth from most of the cosmic rays. On the surface of the earth, cosmic rays are nearly 100% protons. At sea level, the equivalent annual dose from cosmic rays is approximately 0.3 mSv (30 mrem). This figure approximately doubles for every mile above sea level. The dose also increases with latitude, because the earth's magnetic field "steers" charged particles toward the poles. A flight in a typical commercial airliner results in an equivalent dose rate of approximately 0.005 to 0.01 mSv/hr (0.5 to 1 mrem/hr), depending upon altitude and latitude. Outside of the earth's magnetic field, cosmic rays are 80% protons, 12% alpha particles, and 8% ions of practically all of the naturally occurring elements, with the greatest contribution in dose coming from iron ions. Such "heavy" ions ($Z_{iron} = 56$) have extremely high LET.

Sunspots are dark spots that occasionally appear on the surface of the sun. They indicate regions of increased electromagnetic activity and are sometimes responsible for ejecting particulate radiation into space. This ejected radiation is called a solar flare. Solar radiation normally constitutes a small fraction of the dose from cosmic radiation on earth. However, the solar contribution to the cosmic ray background increases during periods of high sunspot activity. A person flying 10 hours aboard a commercial aircraft during a period of normal sunspot activity will receive a radiation dose equivalent that is approximately equivalent to that received from a single chest x-ray examination. During a solar flare, this cosmic-ray dose can be 10 to as much as 100 times greater. Still, this increase in radiation exposure carries an immeasurably small health risk. The Atomic Bomb survivor studies revealed no excess cancers in subjects who received less than about 35 rem (350 millisievert).

The publication of *Silent Spring* by Rachel Carson in 1962 reinforced the public's awareness about the potential health consequences of environmental agents such as fallout from nuclear weapon blasts (atomic tests) and pesticides.

In space, the same rules for dose from cosmic rays apply as for earthbound individuals: Higher elevation and higher latitude (more northerly) means higher dose from cosmic rays and solar flares. In addition, astronauts may encounter the Van Allen radiation belts. These belts are clouds of charged particles (mostly electrons and protons) that collect in specific, stable patterns around the earth. They are not perfect spherical shells because of the motion of the earth, the effect of the earth's magnetic field, and the effect of the solar "wind," streams of charged particles that emanate from the sun. The Van Allen belt comes closest to the earth in a region over the South Atlantic. There are actually several "belts"; but the main one, in terms of radiation dose, begins at a height of approximately 18,000 km and extends to a height of approximately 80,000 km.

Space travel is associated with a risk of dangerously high radiation exposure. The astronauts who traveled to the moon in the 1970s carried radioprotective drugs. These drugs, while able to reduce the effects of radiation by almost a factor of three (so that three times as much radiation would be required to produce the effect once the drug had been taken), were also quite toxic. Among other side effects, they reduce the blood pressure to dangerously low levels. If a large solar flare had occurred during a lunar mission, they would have taken the drug, because the amount of radiation they would have received could conceivably have been lethal. *Source:* Hall, E. J., *Radiobiology for the Radiologist,* Lippincott, New York, pp. 204–205.

The Space Shuttle usually flies to minimize radiation dose by avoiding the poles and the South Atlantic region and by keeping the altitude below 500 kilometers (∼300 miles). During a typical mission of the Space Shuttle, the astronauts receive approximately 1 mSv (100 mrem) per day.

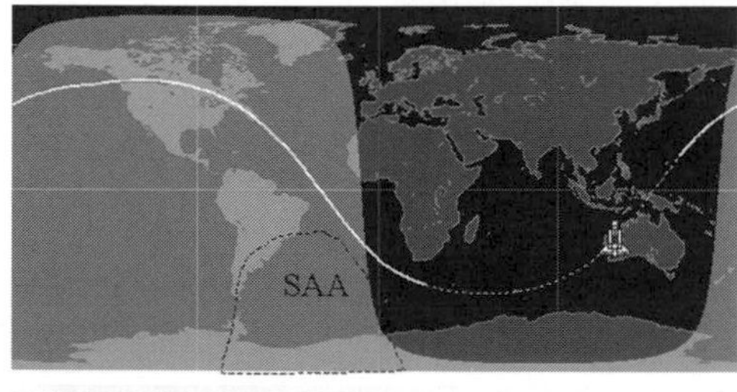

MARGIN FIGURE 27-2
The Space Shuttle usually flies in such a way that radiation dose is minimized, by avoiding the poles and the South Atlantic region and keeping the altitude below 500 kilometers (∼300 miles). During a typical mission of the Space Shuttle, the astronauts receive approximately 1 mSv (100 mrem) per day. This plot of the orbit of the Space Shuttle Discovery on March 9, 2001 shows that the South Atlantic Anomaly (SAA), a region in which the Van Allen radiation belts exist at lower altitude, has been avoided.

Nuclear Accidents

A few persons have received large doses of radiation in the course of accidents involving critical masses of fissionable materials or accidentally discarded radioactive sources.[53–55] Two nuclear power reactor accidents in recent years that have generated considerable publicity actually produced very different outcomes. The Three-Mile Island (TMI) reactor accident in March 1979, although it had the potential for severe impact, actually resulted in negligible exposure of employees and persons living near the reactor. The average dose to the population of 2 million persons living within 50 miles of TMI was 20 μSv (2 mrem). No detectable increase in cancer deaths is expected for this population, and no adverse health effects, with the exception of anxiety and possibly stress disorders, have been identified in this population. The Chernobyl nuclear reactor accident in April 1986, on the other hand, resulted in whole-body doses exceeding 1 Gy (100 rad) to over 200 workers. Thirty workers who received over 4 Gy (400 rad) subsequently died as a result of a combination of radiation exposure and fire-related injuries. There was a significant increase above background radiation to more than 75 million Soviet residents, with the highest exposure occurring in a population of 272,800 living within 200 miles of the reactor. More than 100 million curies of radioactivity was released from the reactor. Human exposures were caused primarily by the release of approximately 17 million curies of ^{131}I and 2 million curies of ^{137}Cs. For comparison, 15 curies of ^{131}I were released in the TMI accident. The radioactive release from Chernobyl resulted in a total collective dose of 55,000 person-sievert (5.5 million person-rem) to the population of 272,800 persons, yielding an average dose of 0.2 Sv (20 rem). In some individuals, thyroid doses caused by the ingestion of milk could have exceeded several hundred rem. The long-term health consequences in the population are not yet fully known.

DOSE-EFFECT MODELS

Many factors influence the effects of radiation upon humans. It is tempting to speculate that if a complete list of all possible "lifestyle" factors such as socioeconomic status, diet, exposure to chemicals at work and in the home, and detailed biochemical and genetic information were available for an individual, one could then compute with reasonable certainty the probability of occurrence of a particular radiation-induced stochastic health effect for that individual. Even if complete knowledge were available, however, the computed result would still be a probability of effect because the nature of radiation interactions and their effect on tissues are probalistic in nature. Furthermore, to make this calculation one would need to know the effects of each of the lifestyle and biochemical factors and whether they operate independently or in concert (synergistically) with other factors in the list.

At the present time, knowledge concerning the role of many of the factors listed above is incomplete. Undoubtedly, major advances will occur in the near future in the understanding of fundamental biochemical mechanisms such as the role of tumor promoters and inhibitors, co-carcinogenic interactions, and the role of genetics. For the present, estimates of the radiation risk to individuals are based primarily upon epidemiologic evidence. From this evidence, efforts have been directed to determine the role of several factors, principally dose, dose rate (including dose-modifying factors described in greater detail below), age, time since exposure, and the gender of exposed individuals. The role of these factors in risk estimation is discussed below.

Responses to concerns over radioactive fallout in the 1950s and early 1960s led to an effort to reduce exposures from medical radiation. One consequence of this effort was passage of the Radiation Control for Health and Safety Act of 1968. This act established manufacturing performance standards for radiation-emitting devices.

The Effects of Dose

The most fundamental variable in risk estimation is the amount (dose) of radiation to which the individual is exposed. If an epidemiologic study fails to show a relationship between dose and effect for an agent such as radiation, then there is little justification for ascribing the effect to the agent. However, if there is a relationship and the epidemiologic data are sufficiently robust, then a mathematical expression for the relationship between dose and effect can be determined.

The fitting of a mathematical expression to experimental dose–effect data is an interesting undertaking but is of little value unless it leads to other insights. Two goals of this effort are (1) the development of a theory that leads to an understanding of mechanisms and (2) extrapolation of risk into dose regimens that, for one reason or another, cannot be measured. The second goal has attracted considerable attention over a number of years. Because the stochastic effects of radiation all occur naturally in any human population, it is difficult to identify the magnitude of radiation effects at low dose levels (levels at which the effects of radiation are well below the natural rates of occurrence of the effects). Given the statistical fluctuation of the natural incidence of stochastic effects in a population, the additional contribution of radiation exposure at low levels of dose may not be directly observable in an epidemiologic study. It is precisely at these levels that there is the greatest interest in estimating risk, because the low-dose region encompasses diagnostic radiology, occupational, and background radiation exposures. Thus there is a great interest in choosing the proper functional form for a graph of stochastic effects versus radiation dose.

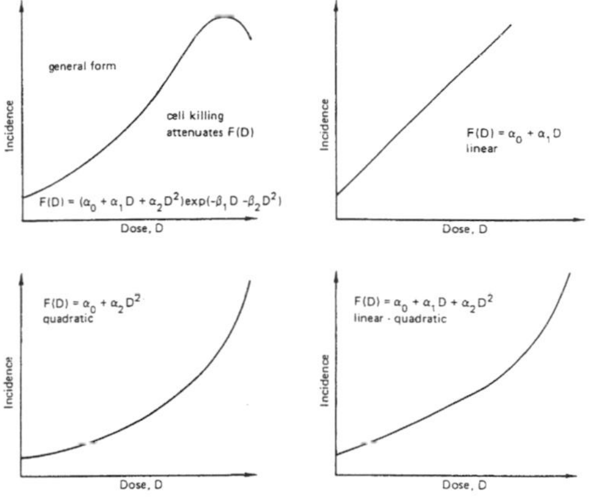

MARGIN FIGURE 27-3

Mathematical forms of dose–effect models for radiation. (From Committee on the Biological Effects on Ionizing Radiation III. *The Effects on Populations of Exposure to Low Levels of Ionizing Radiation.* Washington, D.C., National Academy Press, 1980, p. 28. Used with permission.)

Generic dose–effect curves are shown in Margin Figure 27-3, and actual dose–effect curves are provided in Margin Figure 27-4. The generic curves demonstrate features that have been identified in many epidemiologic studies. The curves contain a region characterized by an increase in biologic effect as dose is increased. But at even higher doses, the effect actually diminishes because of cell killing. For reasons described earlier, there are no data at very low dose levels.

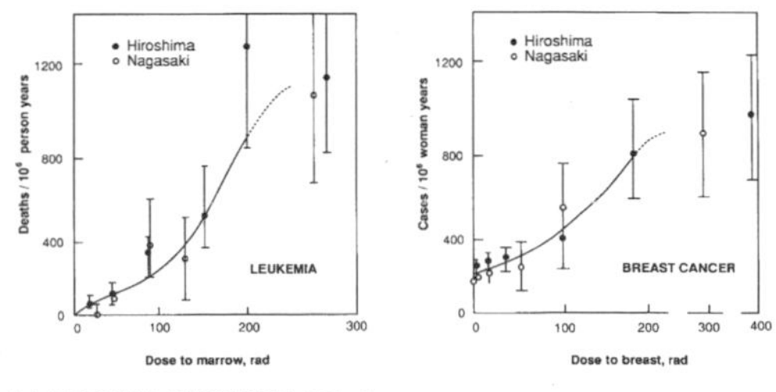

MARGIN FIGURE 27-4

Dose–effect data from atom bomb survivors. (From Straume, T., Dobson, R. L. *Health Phys.* 1981; **41**:666–671. Used with permission.)

A generally accepted model for the magnitude of radiation effects in a population, $F(D)$, as a function of dose, D, is the equation

$$F(D) = (a_0 + a_1 D + a_2 D^2 + \cdots + a_n D^n)e^{(-b_1 D - b_2 D^2)} \tag{27-1}$$

Polynomial Exponential

This equation is the product of two functional forms, a polynomial and an exponential. The polynomial is a simple functional form that contains easily interpretable parameters. The coefficients $a_0, a_1, \ldots$ are selected, usually by computer search, to provide the best fit to the data. The coefficient a_0 is the constant term, or y-axis intercept, that represents the natural incidence of the disorder (i.e., incidence at zero dose). The coefficient a_1 is the linear coefficient that characterizes the slope of the curve. The "higher-order" coefficients a_2, a_3, and so on, determine curvature and other detailed elements of the curve.

The exponential portion of Eq. (27-1) has little effect upon the curve if the dose D is small. At high doses (large values of D), the exponential portion illustrates the decrease in stochastic effects due to cell killing. That is, at high doses there are fewer stochastic consequences of radiation exposure because the nonstochastic effects take precedence. Because we are primarily interested in the lower-dose portion of dose–effect curves, examination of the dose–effect curve will be restricted to the polynomial portion.

For the purpose of radiation protection, controlled areas are areas under the direct supervision of a radiation safety officer (RSO), sometimes called a radiation protection officer (RPO). An RSO must be able to document extensive training and experience in radiobiology and radiation safety. The Nuclear Regulatory Commission (NRC) also requires that this individual must have administrative authority of a level that would allow him or her to halt unsafe programs or practices.

In general, the polynomial portion of Eq. (27-1) can be shaped to fit the data of a typical dose-effect curve. For a particular data set, a polynomial with the number of coefficients equal to the number of data points can fit the data exactly. However, this is not necessarily desirable. Because there is uncertainty in each data point, the "true" curve would not oscillate to fit each actual point, but would take a straighter path through the points and require only lower-order terms.

There is considerable controversy about which polynomial coefficients should be used to fit epidemiologic data related to the general dose–effect relationship for radiation exposure. The reason for the controversy is that although various coefficients may fit the data at higher doses, the predictions at lower doses, where there are no data (or less reliable data), are different. The controversy centers around whether the polynomial should be linear (coefficients a_0 and a_1 only), linear quadratic (a_0, a_1, and a_2), or pure quadratic (a_0, a_2 only). Fits to epidemiologic data tend to extrapolate to lower doses in such a way that the pure linear fit predicts the highest effect, pure quadratic predicts the lowest, and the linear-quadratic predicts an intermediate effect. The issue of the best model for dose–effect estimates at low doses is highly significant, because the predictions vary by as much as an order of magnitude at low doses. Currently, there is growing support for using the linear dose–response model for most forms of cancer and for genetic effects, and the linear quadratic model for leukemia.

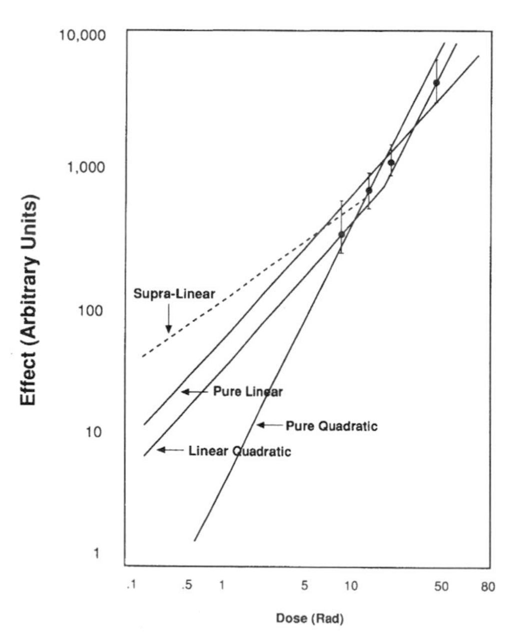

MARGIN FIGURE 27-5
Extrapolations of dose–effect curve fits to a low-dose regimen (below the range of reliable data) yield predictions that differ by an order of magnitude or more. (From Edwards, F. M. Dose–Effect Models for Radiation Exposure, in Hendee, W. R. (ed.), *Health Effects of Low-Level Radiation*. East Norwalk, CT, Appleton–Century–Crofts, 1984, p. 87. Used with permission.)

The Effects of Dose Rate

From basic experiments involving cell cultures and laboratory animals, it is known that dose rate plays a role in determining biologic effects. In the range of total doses over which an effect is observed, the effect often decreases if the dose rate is lowered. Data in human populations support this concept.[56,57] Indeed, dose rate and dose fractionation are two commonly used principles of radiation therapy. Tissue repair mechanisms generally are not as robust in tumors as in normal tissue. Fractionated radiation therapy treatments are designed to provide normal tissue a chance for repair without significantly reducing the damage to tumor tissue. At low doses where only stochastic processes are expected, there is some uncertainty about the role of dose rate.

The dose rate effectiveness factor (DREF) is defined as

$$\frac{\text{Total dose to elicit an effect at the reference dose rate}}{\text{Total dose to elicit an effect at the dose rate of interest}} = \text{DREF}$$

For example, assume that a given biologic effect is elicited at a particular reference dose rate when the total dose is 1.5 Gy. If the same radiation is delivered at a dose rate with a DREF of 3, then a total dose of only 0.5 Gy would be required to elicit the same effect. Estimates for DREF in humans at low total doses are usually in the range from 2 to 3, although some estimates yield a value as high as 10.[56]

Refinement of dose–effect models will continue in at least three areas, all of which will ultimately lead to more precise risk estimates. The first is better epidemiologic data—that is, epidemiologic studies with less uncertainty in each data point, thereby restricting the range of polynomials that fit the data. The second is extension of data into lower doses. The third is a greater understanding of fundamental mechanisms of radiation injury.

FACTORS THAT INFLUENCE DOSE–EFFECT MODELS

The effects of radiation are known to be modified by a number of factors. These effects can be expressed independently as a function of each of the variables. For example, age of the individual at the time of exposure is known to be important in estimating cancer risk. Furthermore, the time that has elapsed since a single exposure also influences the probability of occurrence of an effect. Table 27-2 gives the excess cancer deaths in a population of 100,000 exposed to 0.1 Sv (10 rem) according to the age at which the exposure occurred. For two reasons, the risk of cancer is greater when exposure occurs at an earlier age. First, if radiation exposure occurs at an earlier age, more years of life remain in which cancer may occur. Therefore, the lifetime risk of cancer is greater. Second, for most types of cancer the probability of cancer induction is actually greater on a yearly basis for exposures that occur at an earlier age.[57–58]

Regardless of the age at irradiation, a latency period, an interval of time that must elapse before the rate of occurrence of a radiation-induced effect rises above the natural rate of occurrence of that disorder in a population, always occurs before appearance of a stochastic radiation-induced effect. The latency period varies with different disorders. For radiation-induced leukemia, the latency period may be as short as 2 years, whereas for solid tumors it may be more than 20 years and may vary

The NCRP recommends that astronauts receive no more than 0.5 Sv per year and no more than 0.25 Sv per month. The limit for exposure during a career as an astronaut depends upon the age of first exposure and gender. It varies from 1.0 Sv for females first exposed at age 25 to 4.0 Sv for males first exposed at age 55. The extra lifetime risk of fatal cancer for the maximum dose limit is 4.0 Sv × 4% = 16%, meaning that the total lifetime risk of cancer for this astronaut is, theoretically, twice the typical risk of dying of cancer from all causes, which for a typical U.S. citizen is 16%.

A typical adult contains:
a total of ~9000 Bq of radioactivity
This includes:

4000 Bq of ^{14}C
3750 Bq of ^{40}K
600 Bq of ^{3}H
2 Bq of ^{226}Ra

and smaller amounts of ^{87}Rb, ^{238}U, $^{228,230,232}Th$, $^{220,222}Rn$, ^{210}Pb, and ^{210}Po.

TABLE 27-2 Cancer Excess Mortality by Age at Exposure and Site for 100,000 Persons of Each Age Exposed to 0.1 Sv (10 rem)

Age at Exposure (yr)	*Total*	*Leukemia*	*Nonleukemia*	*Breast*	*Respiratory*	*Digestive*	*Other*
Males							
5	1276	111	1165		17	361	787
15	1144	109	1035		54	369	612
25	921	36	885		124	389	372
35	566	62	504		243	28	233
45	600	108	492		353	22	117
55	616	166	450		393	15	42
65	481	191	290		272	11	7
75	258	165	93		90	5	—
85	110	96	14		17	—	—
Average[a]	770	110	660		190	170	300
Females							
5	1532	75	1457	129	48	655	625
15	1566	72	1494	295	70	653	476
25	1178	29	1149	52	125	679	293
35	557	46	511	43	208	73	187
45	541	73	468	20	277	71	100
55	505	117	388	6	273	64	45
65	386	146	240	—	172	52	16
75	227	127	100	—	72	26	3
85	90	73	17	—	15	4	—
Average	810	80	730	70	150	290	220

[a] Averages are weighted for the age distribution in a stationary population having U.S. mortality rates and have been rounded to the nearest 10.

Source: Committee on the Biological Effects of Ionizing Radiations. *Health Effects of Exposure to Low Levels of Ionizing Radiation,* Report No.5. Washington, D.C., National Academy Press, 1990, p. 175. Used with permission.

with the age of the individual at the time of exposure.[59] The latency period is also probabilistic, with values quoted for given disorders representing population averages.

There is some disagreement about the relationship between radiogenic stochastic effects and the natural incidence of the effects as a function of time after exposure. Two models have been proposed to describe this relationship. The absolute-risk model assumes that a constant increase in incidence of the effect takes place at all times after the latency period. The relative-risk model assumes that the radiation-induced incidence of an effect varies with the natural incidence of the effect at any time after the latency period. Thus, as natural incidence increases with age for many cancers, radiogenic cancer incidence increases also. Both models allow for a smooth increase in incidence of the effect during the latency period.

You eat approximately 40 Bq of radioactivity per year.

One of the current challenges of epidemiology is to determine which model best fits the available data. At the present time, both models fit the data reasonably well. Other factors being equal, the relative-risk model tends to predict a greater lifetime risk of cancer, because the incidence of disorders such as cancer increases with age. That is, the relative-risk model predicts that the number of cancers appearing later in life is greater than would be estimated from the absolute-risk model. Data concerning the largest and most exhaustively studied population, the atomic bomb survivors, were reviewed by the US National Research Council after 46 years (1945 to 1990) of study. Both absolute- and relative-risk models fit the data for different cancers in that population. However, the relative-risk model appears to yield more accurate estimates for most cancers, and it is the only model used in the (BEIR V) report from the National Research Council.[60]

Because of radioactivity present naturally in your body, you give yourself a dose equivalent of about 0.50 mSv (50 mrem) each year to your bone marrow,
1.9 mSv (190 mrem) to your bone cells (osteocytes),
and
0.35 mSv (35 mrem) to your soft tissues [the total effective dose equivalent is 0.4 mSv (40 mrem)].

ESTIMATING RISKS OF RADIATION: BEIR REPORT

A number of scientific and advisory groups have published risk estimates for ionizing radiation. According to current NCRP and ICRP risk estimates, the risk of fatal cancer is 4% per Sv (4.0×10^{-4} per rem). For radiographers, radiologists, and medical physicists, the annual dose limit is 0.05 Sv (50 mSv, 5 rem) and the cumulative limit is 10 mSv times the age in years. Therefore, a 55-year-old radiographer should have received a total occupational exposure of no more than 550 mSv (0.55 Sv, 55 rem). This maximum level of exposure is associated with an estimated $0.55 \times 4\% = 2.2\%$ increase in risk of fatal cancer.

One of the most recent and widely quoted reports on risk estimation is Report No. 5 of the National Academy of Sciences Committee on the Biological Effects of Ionizing Radiation (BEIR V). Table 27-3 summarizes the risk estimates for cancer from that publication.

The BEIR V report uses the linear dose–effect model for all cancers other than leukemia, along with the linear quadratic model for leukemia. The effects of age at irradiation and time elapsed since irradiation are considered with latency period beginning 2 years after irradiation for leukemia and 10 years after irradiation for other cancers. In keeping with recent trends in stochastic effects among atom bomb survivors, the relative-risk model of cancer incidence as a function of age is used.

The BEIR V report estimates (see Table 27-3) that if a population of 100,000 received a single exposure to all body organs of 0.1 Sv (10 rem), then there would be an extra (i.e., radiation-induced or radiogenic excess) 770 cases of cancer if the population were male and 810 if the population were female. Given the 90% confidence limits for these estimates (540 to 1240 for males, 630 to 1160 for females), it is reasonable to adopt an average of 800 cancers for a population of 100,000 with a reasonable distribution of males and females.

Many factors enter into the average estimate of radiation risk. The estimate of 800 excess cases per 100,000 per 0.1 Sv is the result of a calculation in which excess fatal cancers were estimated for single exposures occurring at different ages of individuals. The individuals are members of a hypothetical population having the same age distribution as the U.S. population. The results of these calculations are then averaged to arrive at the figure 800.

TABLE 27-3 Excess Cancer Mortality Estimates and Their Statistical Uncertainty: Lifetime Risks per 100,000 Exposed Persons[a]

	Male			*Female*		
	Total	*Nonleukemia[b]*	*Leukemia[c]*	*Total*	*Nonleukemia*	*Leukemia*
Single exposure to 0.1 Sv (10 rem)	770	660	110	810	730	80
90% confidence limits	540–1240	420–1040	50–280	630–1160	550–1020	30–190
Normal expectation	20,510	19,750	760	16,150	15,540	610
% of normal	3.7	3.3	15	5	4.7	14
Total years of life lost	12,000			14,500		
Average years of life lost per excess death	16			18		
Continuous lifetime exposure[d] to 1 mSv/yr (0.1 rem/yr)	520	450	70	600	540	60
90% confidence limits	410–980	320–830	20–260	500–930	430–800	20–200
Normal expectation	20,560	19,760	790	17,520	16,850	660
% of normal	2.5	2.3	8.9	3.4	3.2	8.6
Total years of life lost	8100			10,500		
Average years of life lost per excess death	16			18		
Continuous exposure[d] to 0.01 Sv/yr (1 rem/yr) from age 18 until age 65	2880	2480	400	3070	2760	310
90% confidence limits	2150–5460	1670–4560	130–1160	2510–4580	2120–4190	110–910
Normal expectation	20,910	20,140	780	17,710	17,050	650
% of normal	14	12	52	17	16	48
Total years of life lost	42,200			51,600		
Average years of life lost per excess death	15			17		

[a] Based on an equal dose to all organs and the committee's preferred risk models—estimates are rounded to nearest 10.

[b] Sum of respiratory, breast, digestive, and other cancers.

[c] Estimates for leukemia contain an implicit dose rate reduction factor.

[d] A dose rate reduction factor has not been applied to the risk estimates for solid cancers.

Source: Committee on the Biological Effects of Ionizing Radiation. *Health Effects of Exposure to Low Levels of Ionizing Radiation*, Report No. 5, Washington, D.C., National Academy Press, 1990, p. 172. Used with permission.

Another approach to risk estimation is the specification of risk to an individual who has been irradiated. One could state that the risk to an individual who has received 0.1 Sv (10 rem) in a single whole-body exposure is approximately 800 chances per 100,000 of acquiring a fatal radiation-induced cancer. This number, which may also be expressed as 8 chances in 1000, 8×10^{-3}, or 0.8%, applies only to a 0.1-Sv exposure. For higher exposures, the linear dose–effect model allows increasing the risk in proportion to the dose (e.g., doubling the risk as the dose is doubled). This is appropriate up to the dose level where incidence levels off, which is beyond dose levels for which data have been obtained from epidemiologic studies. In the BEIR V committee, actual data above 4 Sv (400 rem) are not used.[61] The BEIR committee's model, if extrapolated to larger doses, corresponds to an 8% risk of fatal cancer per sievert. This value is greater than the NCRP and ICRP estimate of 4% per sievert, although the confidence limits of these estimates (Table 27-3) overlap. Many epidemiologists and actuarial specialists summarize the various estimates with an easily remembered "rule of thumb" of "5% increase in fatal cancer per sievert."

One may also scale or extrapolate the risk estimates downward in proportion to the dose. However, caution should be used for the reasons cited earlier. Dose levels in the epidemiologic studies on which risk estimates are based are above 0.1 Sv. Using the linear dose–effect model to extrapolate to lower dose levels should be considered to be simply an estimate, quite possibly an overestimate, of risk.

The natural incidence of fatal cancers is also given in Table 27-3. Out of a population of 100,000 typical U.S. citizens followed until death has occurred, it is expected that 16,000 to 20,000 of the deaths would be due to cancer. Thus 0.1 Sv (10 rem) to a population would be estimated to increase the number of fatal cancers by about 4%. Average years of life lost per radiogenic cancer and total years of life lost in the population are also shown.

Example 27-1

Find the risk of contracting a fatal cancer from a single, whole-body dose of 0.02 Sv (2 rem), and compare this risk with the natural incidence of fatal cancer.

By extrapolation from Table 27-3, since 0.02 Sv is 20% of 0.1 Sv, the risk is

$$\frac{0.2(770)}{100{,}000} = 1.54 \times 10^{-3} \text{ for males}$$

and

$$\frac{(0.2)(810)}{100{,}000} = 1.62 \times 10^{-3} \text{ for females}$$

This represents a fraction of the naturally occurring cancers in a population as follows:

$$\frac{(0.2)(770)}{20{,}510} = 7.5 \times 10^{-3} = 0.75\% \text{ for males}$$

and

$$\frac{0.2(810)}{16{,}150} = 10.0 \times 10^{-3} = 1\% \text{ for females}$$

For several years before BEIR V, a range of 1×10^{-4} to 2×10^{-4} per rem (0.01 Sv) was widely used as an overall estimate of radiation risk. There are several reasons for the increase to the currently accepted value of 8×10^{-3} per 0.1 Sv. One reason is that as more years of follow-up have been accumulated in groups such as the atom bomb survivors, it has been found that the number of tumors, particularly solid tumors, has increased beyond earlier predictions. This finding is in keeping with the predictions of the relative-risk model.

A second reason for the increase is that dose estimates for atom bomb survivors have been redone[62,63] from information that had previously been classified. It is now thought that the γ-ray doses in both Hiroshima and Nagasaki were lower than previously believed and that the neutron component of the dose was much less significant. Therefore, the radiogenic cancers seen in those populations may have been caused by smaller doses than had been previously assumed.

The BEIR V report also gives estimates of the risk of the continuous exposure. Table 27-3 shows risks calculated for (a) a continuous lifetime exposure to all body organs of 1 mSv/yr (0.1 rem/yr) and (b) a continuous exposure to all body organs of 10 mSv/yr (1 rem/yr) delivered between ages 18 and 65. The former simulates an upper level for exposure of the general public, and the latter simulates an upper level for occupational exposure. These estimates are obtained by summing the age-specific incidence data for doses delivered over time. They indicate that continuous lifetime exposure yields a risk of fatal cancer that is 2% to 3% of the normal incidence and that occupational exposure yields a risk that is 14% to 17% of the normal incidence. However, these estimates should be interpreted as overestimates because the BEIR V committee chose not to apply a dose rate effectiveness factor (DREF) to the data. As was discussed earlier, the inclusion of a DREF would be expected to lower risk estimates by a factor of 2 to 3.[64]

The BEIR V report also contains an extensive compilation of organ-specific models for cancer risk. These data are more appropriate for computing risk to a patient in radiology than is the whole-body risk estimate of 800 cases of fatal cancer per 100,000 for each 0.1 Sv, because almost all diagnostic studies involve only partial-body exposure. As with whole-body irradiation, caution should be used in applying risk estimates to dose levels below those for which epidemiologic data exist.

The National Academy of Sciences is revising the BEIRV report on the Health Effects of Exposure to Low Levels of Ionizing Radiations. As with previous reports, panels of experts are reviewing the worldwide literature on the subject and are conducting their own analyses of various predictive models. The new report, is expected to include new information on:

- Mechanisms of the cellular and molecular events involved in the neoplastic process
- The role of genetics on radiation susceptibility as revealed in recent studies involving genetically engineered mice
- New evidence for specific molecular events that affect the shape of the dose–effect curve at low doses. These effects include:
  - DNA repair
  - Signal transduction
  - Chromosomal instability
  - Genetic adaptation
- An assessment of the current status of all models of carcinogenesis, including the possibility of thresholds, adaptive responses, and hormetic effects.

The National Academy of Sciences Report No. 6 of The Committee on Biological Effects of Ionizing Radiations (BEIR VI) was on the subject of the health effects of exposure to radon. The new report on the Health Effects of Exposure to Low Levels of Ionizing Radiations will be available from the National Academy Press (books.nap.edu) sometime after this text has gone to press.

Because of the radionuclides you contain within you body, if you sleep with someone (or somehow spend eight hours very close to someone each day) you give them (and they give you) an extra 0.001 mSv (0.1 mrem).

Effects on Embryo

Irradiation of the embryo is estimated by the BEIR V Committee to increase the incidence of mental retardation by 4% for a 0.1-Sv (10 rem) dose delivered during weeks 8 to 15 of gestation. As gestational age increases beyond week 15, the risk decreases. No statistically significant increase is noted for irradiation prior to week 8 or beyond week 25.

There are also data to support the hypothesis that irradiation of the embryo increases the risk of childhood (conception to the age of 15 years) cancer. Both leukemia and nervous system tumors have been linked to radiation exposure *in utero*. A risk of approximately 200 excess fatal cancers per 10,000 person-Gy is cited in BEIR V.

Mental retardation is a potential bioeffect of radiation exposure *in utero*. The risk of occurrence is greatest for exposures between the eighth and the fifteenth week of gestation. It is during this period that significant neurological organogenesis (differentiation of gray and white matter) occurs.

Genetic Effects

At the present time, there are no statistically significant human epidemiologic data from which information on the heritable effects of radiation can be derived. Risk estimates given by the BEIR V committee are based on chromosomal aberration data from humans coupled with studies of other mammals, principally mice. These estimates are summarized in Table 27-4. BEIR V estimates that the doubling dose for humans, the dose at which the natural mutation rate is doubled, is 1 Gy (100 rad). The risk of severe genetic damage appearing in the next few generations is 0.8% per Sv (0.8×10^{-4}per rem).

Human populations, fortunately, have not been exposed to high doses in sufficiently large numbers (the average dose to the atom bomb survivors was approximately 50 Gy) to provide a definitive demonstration of genetic risk. This provides evidence that, in all likelihood, the current estimates of genetic risk are conservative.

SOURCES OF INFORMATION

A number of scientific organizations have examined the biologic effects of ionizing radiation. These groups have reviewed the scientific literature and in many cases made risk estimates or recommendations concerning radiation safety. A listing is shown in Table 27-5 of organizations whose reports should be available in any major library or on the World Wide Web.

TABLE 27-4 Estimated Induced Mutation Rates per Rad (Primarily Mouse)

Genetic End Point, Cell Stage, Sex	*Low-LET Radiation Exposure*		
	High Dose Rate	*Low Dose Rate*	*Fission Neutrons (Any Dose Rate)*
Dominant lethal mutations			
Postgonial, male	10×10^{-4}/gamete	5×10^{-4}/gamete	75×10^{-4}/gamete
Gonial, male	10×10^{-5}/gamete	2×10^{-5}/gamete	40×10^{-5}/gamete
Recessive lethal mutations			
Gonial, male	1×10^{-4}/gamete		
Postgonial, female	1×10^{-4}/gamete		
Dominant visible mutations			
Gonial, male	2×10^{-5}/gamete		
Skeletal	5×10^{-7}/gamete		
Cataract	$5–10 \times 10^{-7}$/gamete		
Other	$5–10 \times 10^{-7}$/gamete	1×10^{-7}/gamete	25×10^{-7}/gamete
Postgonial, female	$5–10 \times 10^{-7}$/gamete		
Recessive visible mutations (specific locus tests)			
Postgonial, male	65×10^{-8}/locus		
Postgonial, female	40×10^{-8}/locus	$1–3 \times 10^{-8}$/locus	145×10^{-8}/locus
Gonial, male	22×10^{-8}/locus	7×10^{-8}/locus	125×10^{-8}/locus
Reciprocal translocations			
Gonial, male			
Mouse	$1–2 \times 10^{-4}$/cell	$1–2 \times 10^{-5}$/cell	$5–10 \times 10^{-4}$/cell
Rhesus	2×10^{-4}/cell		
Marmoset	7×10^{-4}/cell		
Human	3×10^{-4}/cell		
Postgonial, female			
Mouse	$2–6 \times 10^{-4}$/cell		
Heritable translocations			
Gonial, male	4×10^{-5}/gamete		
Postgonial, female	2×10^{-5}/gamete		
Congenital malformations			
Postgonial, female	2×10^{-4}/gamete		
Postgonial, male	4×10^{-5}/gamete		
Gonial, male	$2–6 \times 10^{-5}$/gamete		
Aneuploidy (trisomy)			
Postgonial, female			
Preovulatory oocyte	6×10^{-4}/cell		
Less mature oocyte	6×10^{-5}/cell		

Source: Committee on the Biological Effects of Ionizing Radiations: Health Effects of Exposure to Low Levels of Ionizing Radiation, Report No. 5 (BEIR V), Washington, D.C., National Academy Press, 1990, p. 100. Used with permission.

TABLE 27-5 Reports and Recommendations of Scientific Committees on Biological Effects of Ionizing Radiation

UNSCEAR	United Nations Scientific Committee on the Effects of Atomic Radiation
BEIR	National Academy of Sciences—National Research Council, Committee on the Biological Effects of Ionizing Radiation, Washington, DC
NIH	National Institutes of Health Ad Hoc Working Group to Develop Radioepidemiological Tables; reports available from U.S. Government Printing Office
ICRP	International Commission on Radiological Protection; reports available through Pergamon Press, New York
NCRP	National Council on Radiation Protection and Measurement, Bethesda, MD
NRC	United States Nuclear Regulatory Commission, Washington D.C.

SUMMARY

- Stochastic effects are effects for which the probability of occurrence (and not the severity) is proportional to dose. The three stochastic effects that have been identified with radiation are:
  - Carcinognesis
  - Teratogenesis
  - Mutagenesis
- The subcategories of acute radiation syndrome (ARS) are:
  - Hematopoietic
  - Gastrointestinal
  - Cerebrovascular
- Collective dose equivalent (CDE), the sum of the dose equivalents experienced by all the individuals in a population, has units of person-sievert (person-rem).
- Effective dose uses weighting factors to convert the equivalent dose delivered to specific body regions into a measure of overall harm to the organism and to future generations through genetic defects.
- Sources of natural background radiation, from largest to smallest are:
  - Radon
  - Internal
  - Terrestrial
  - Cosmic
- The yearly total equivalent dose from natural background is 3.0 mSv (300 mrem).
- Sources of man-made background radiation, from largest to smallest, are:
  - Medical x rays
  - Nuclear medicine
  - Consumer products
  - Other sources (occupational, nuclear power, industrial, fallout from nuclear weapons exploded in the atmosphere since 1945)
- The yearly total equivalent dose from man-made background is 0.6 mSv.
- Populations that have been exposed to unusual levels of radiation include:
  - Atom bomb survivors
  - Early radiologists, nurses, technologists (prior to World War II)
  - Uranium and other miners
  - Radium dial painters
  - Radiation therapy patients
  - Early diagnostic radiology patients
  - Populations exposed to fallout from nuclear weapons tests
  - Residents of regions with high natural background
  - Individuals participating in air and space travel
  - Victims of nuclear accidents
- A reasonable and widely accepted mathematical form that fits currently epidemiological data is the linear, nonthreshold model for most forms of cancer and the linear-quadratic, nonthreshold model for leukemia.
- A widely accepted "rule of thumb" for estimating the risk of fatal cancer per unit effective dose due to a single exposure to ionizing radiation is "5% per sievert."

REFERENCES

1. Statkiewicz-Sherer, M. A., Visconti, P. J., and Ritenour, E. R. *Radiation Protection in Medical Radiography,* 4th edition. Chicago, Mosby, 2001.
2. International Commission on Radiological Protection (ICRP): *Nonstochastic Effects of Ionizing Radiation,* ICRP Publication 41. Oxford, Pergamon Press, 1984.
3. United Nations Scientific Committee on the Effects of Atomic Radiation (UNSCEAR): *Ionizing Radiation: Sources and Biological Effects,* Report E.82.IX.8. New York, United Nations, 1982.
4. "U.S. Food and Drug Administration: FDA Advisory: Avoidance of Serious X-Ray Induced Skin Injuries During Fluoroscopically-Guided Prodecures," Center for Devices and Radiological Health, U.S. Food and drug Administration, 8 September 1994.

5. O'Dea, J. O., Geise, R. A., and Ritenour, E. R. The potential for radiation-induced skin damage in interventional neuroradiological procedures: A review of 522 cases using automated dosimetry. *Med. Phys.* 1999; **26**(9):2027–2033.
6. Recommendations of the International Commission on Radiological Protection: *International Commission on Radiological Protection,* Publication 26. New York, Pergamon Press, 1977.
7. National Council on Radiation Protection and Measurements (NCRP) Report No. 93, Ionizing Radiation Exposure of the Population of the United States, Bethesda, MD, NCRP, 1987.
8. National Council on Radiation Protection and Measurements (NCRP) Report No. 94, Exposure of the Population in the United States and Canada from Natural Background Radiation, Bethesda, MD, NCRP, 1987.
9. National Council on Radiation Protection and Measurements (NCRP) Report No. 95, Radiation Exposure of the US Population from Consumer Products and Miscellaneous Sources, Bethesda, MD, NCRP, 1987.
10. United Nations Scientific Committee on the Effects of Atomic Radiation, Ionizing Radiation: Sources and Biological Effects, No.E.82.IX.8., 06300P (United Nations, New York), 1982.
11. Committee on the Biological Effects of Ionizing Radiations (BEIR V): *Health Effects of Exposure to low Levels of Ionizing Radiation*. National Research Council. Washington, D.C., National Academy Press, 1990, p. 162.
12. Shimizu, Y., Kato, H., Schull, W. J., et al. Life Span Study Report 11, Part 1. Comparison of Risk Coefficients for Site-Specific Cancer Mortality Based on the DS86 and T65DR Shielded Kerma and Organ Doses. Technical Report RERF TR 12-87. Hiroshima Radiation Effects Research Foundation, 1987.
13. Tokunaga, M., Land, C., Yamamoto, T., et al. Incidence of female breast cancer among atomic bomb survivors, Hiroshima and Nagasaki, 1950–1985. *Radiat. Res.* 1987; **112**:243–272.
14. Beebe, G., Kato, H., and Land, C. Mortality experience of atomic bomb survivors, 1950–1974 Life Span Study Report 8, Radiation Effects Research Foundation, 1978.
15. Shigematsu, I. The 2000 Sievert Lecture—Lessons from atomic bomb survivors in Hiroshima and Nagasaki. *Health Phys.* 2000; **79**(3):234–241.
16. Ishimaru, T., Cihak, R. W., Land, C. E., et al. Lung cancer at autopsy in A-bomb survivors and controls, Hiroshima and Nagasaki, 1961–1970 II. Smoking, occupation, and A-bomb exposure. *Cancer* 1975; **36**:1723–1728.
17. Jablon, S., and Kato, H. Childhood cancer in relation to prenatal exposure to atomic bomb radiation. *Lancet* 1970; **2**:1000–1003.
18. Seltser, R., and Sartwell, P. E. The influence of occupational exposure on the mortality of American radiologists and other medical specialists. *Am. J. Epidemiol.* 1965; **81**:2–22.
19. Matanoski, G., Seltser, R., Sartwell, P., et al. The current mortality rates of radiologists and other physician specialists. Deaths from all causes and from cancer. *Am. J. Epidemiol.* 1975; **101**:188–198.
20. Matanoski, G., et al. The current mortality rates of radiologists and other physician specialists. Specific causes of death. *Am J Epidemiol* 1975; **101**:199–210.
21. Ulrich, H. The incidence of leukemia in radiologists. *N. Engl. J. Med.* 1946; **234**:45–46.
22. Lorenz, E. Radioactivity and lung cancer, a critical review of lung cancer in the miners of Schneeberg and Joachimsthal. *J. Natl. Cancer Inst.* 1944; **5**:1–13.
23. Whittemore, A. S., and McMillan, A. Lung cancer mortality among U.S. uranium miners. A reappraisal. *J. Natl. Cancer Inst.* 1983; **71**:489–499.
24. Archer, V. E., Wagoner, J. K., Lundin, F. E., Jr. Lung cancer among uranium miners in the United States. *Health Phys.* 1973; **25**:351–371.
25. Committee on the Biological Effects of Ionizing Radiations (BEIR IV). *Health Risks of Radon and Other Internally Deposited Alpha-Emitters*. National Research Council, Washington, D.C., National Academy Press, 1988, pp. 445–488.
26. Martland, H. S. The occurrence of malignancy in radioactive persons. *Am. J. Cancer.* 1931; **15**:2435–2516.
27. Spiers, F. W., Lucas, H. F., Rundo, J., et al. Leukemia incidence in the U.S. dial workers. *Health Phys.* 1983; **44**:65–72.
28. Stebbings, J. H., Lucas HF, Stehney AF: Mortality from cancers of major sites in female radium dial workers. *Am. J. Ind. Med.* 1984; **5**:435–459.
29. Rowland, R. E., Stehney, A. F., and Lucas, H. F., Jr. Dose-response relationships for female radium dial workers. *Radiat. Res.* 1978; **76**:368–383.
30. Polednak, A. P., Stehney, A. F., Rowland RE: Mortality among women first employed before 1930 in the U.S. radium dial-painting industry. *Am. J. Epidemiol.* 1978; **107**:179–195.
31. Darby, S. G., Doll, R., and Smith, P. G. Paper 9. Trends in long-term mortality in ankylosing spondylitis treated with a single course of x-rays. Health effects of low dose ionizing radiation. *London,* BNES, 1988.
32. Shore, R. E., Hildreth, N., Woodard, E., et al. Breast cancer among women given x-ray therapy for acute postpartum mastitis. *J. Natl. Cancer Inst.* 1986; **77**:689–696.
33. Court Brown, W. M., and Doll, R. Mortality from cancer and other causes after radiotherapy for ankylosing spondylitis. *Br. Med. J.* 1965; **2**:1327–1332.
34. Hempelmann, L., Hall, W., Phillips, M., et al. Neoplasms in persons treated with x-ray in infancy: Fourth survey in 20 years. *J. Natl. Cancer Inst.* 1975; **55**:519–530.
35. Modan, B., Ron, E., and Werner, A. Thyroid cancer following scalp irradiation. *Radiology* 1977; **123**:741–744.
36. Smith, P., and Doll, R. Late effects of x-irradiation in patients treated for metropathia haemorrhagica. *Br. J. Radiol.* 1976; **49**:224–232.
37. Boice, J. Jr., Rosenstein, M., and Trout, E. D. Estimation of breast doses and breast cancer risk associated with repeated fluoroscopic chest examinations of woman with tuberculosis. *Radiat. Res.* 1978; **73**:373–390.
38. Gibson, R., Graham, S., Lilienfeld, A., et al. Irradiation in the epidemiology of leukemia among adults. *J. Natl. Cancer Inst.* 1972; **48**:301–311.
39. Holm, L. E., Lundell, G., and Walinder, G. Incidence of malignant thyroid tumors in humans after exposure to diagnostic doses of iodine-131. Retrospective cohort study. *J. Natl. Cancer Inst.* 1980; **64**:1055–1059.
40. Harvey, G., Boice, J., Honeyman, M., et al. Prenatal x-ray exposure and childhood cancer in twins. *N. Engl. J. Med.* 1985; **312**:541–545.
41. Sherman, G. J., Howe, G. R., Miller, A. B., et al. Organ dose per unit exposure resulting from fluoroscopy for artificial pneumothorax. *Health Phys.* 1978; **35**:259–269.
42. Gunz, F., and Atkinson, H. Medical radiation and leukemia: A retrospective survey. *Br. Med. J.* 1964; **1**:389.
43. Gibson, R., Graham, S., Lilienfeld, A., et al. Irradiation in the epidemiology of leukemia among adults. *J. Natl. Cancer Inst.* 1972; **48**:301–311.
44. Committee on the Biological Effects of Ionizing Radiations (BEIR V). *Health Effects of Exposure to Low Levels of Ionizing Radiation*. National Research Council. Washington, D.C., National Academy Press, 1990, pp. 375–377.
45. Conard, R. A., et al. March 1958 Medical survey of Rongelap people. Brookhaven National Laboratory, Report 534 (T-135), 1959.
46. Conard, R. A., and Hicking, A. Medical findings in Marshallese people exposed to fallout radiation. *JAMA* 1965; **192**:457.
47. Barcinski, M. A., Abreu, M. D. C. A., De Almeida, J. C. C., et al. Cytogenetic investigation in a Brazilian population living in an area of high natural radioactivity. *Am. J. Hum. Genet.* 1975; **27**:802–806.
48. Kochupillan, N., Verma, I. C., Grewal, M. S., et al. Down's syndrome and related abnormalities in an area of high background radiation in coastal Kerala. *Nature* 1976; **262**:60–61.
49. Tao, Z., and Wei, L. An epidemiological investigation of mutational diseases in the high background radiation area of Yangjiang, China. *J. Radiat. Res.* 1986; **27**:141–150.
50. Jablon, S., and Bailar, S. C., III: The contribution of ionizing radiation to cancer mortality in the United States. *Prevent. Med.* 1980; **9**:219–226.
51. Pochin, E. E. Problems involved in detecting increased malignancy rates in the areas of high natural radiation background. *Health Phys* 1976; **31**:148–154.
52. Frigerio, N. A., and Stowe, R. S. Carcinogenic and genetic hazard from background radiation, in *Biological and Environmental Effects of Low-Level Radiation*. Vienna, International Atomic Energy Agency, 1976, pp. 385–393.
53. Vargo, G. J. A brief history of nuclear criticality accidents in Russia—1953–1997. *Health Phys* .1999; **77**(5):505–511.
54. Hess, D. B., and Hendee, W. R. Radiation and Political Fallout of Three Mile Island, in Hendee, W. R. (ed.), *Health Effects of Low-Level Radiation*. East Norwalk, CT, Appleton–Century–Crofts, 1984, pp. 247–258.
55. Normal, C., and Dickson, D. The aftermath of Chernobyl. *Science* 1986; **233**:1141–1143.
56. Committee on the Biological Effects of Ionizing Radiations (BEIR V): *Health Effects of Exposure to Low Levels of Ionizing Radiation*. National Research Council. Washington, D.C., National Academy Press, 1990, p. 23.
57. National Council on Radiation Protection and Measurements. *Influence of Doses and Its Distribution in Time on Dose-Response Relationships for Low LET*

Radiations, Report No. 64. Bethesda, MD, National Council on Radiation Protection and Measurements, 1980.

58. Committee on the Biological Effects of Ionizing Radiations (BEIR V). *Health Effects of Exposure to Low Levels of Ionizing Radiation*. National Research Council. Washington, D.C., National Academy Press, 1990, p. 267.
59. United Nations Scientific Committee on the Effects of Atomic Radiation (UNSCEAR). *Sources and Effects of Ionizing Radiation,* Report E. 77 IX. New York, United Nations, 1977.
60. Committee on the Biological Effects of Ionizing Radiations (BEIR V). *Health Effects of Exposure to Low Levels of Ionizing Radiation*. National Research Council. Washington, D.C., National Academy Press, 1990, pp. 1–8.
61. Committee on the Biological Effects of Ionizing Radiations (BEIR V). *Health Effects of Exposure to Low Levels of Ionizing Radiation*. National Research Council. Washington, D.C., National Academy Press, 1990, p. 165.
62. Radiation Effects Research Foundation. *U.S.–Japan Joint Reassessment of Atomic Bomb Radiation Dosimetry in Hiroshima and Nagasaki—Final Report,* Vol. 1. Hiroshima, Japan, 1987.
63. Radiation Effects Research Foundation: *U.S.–Japan Joint Reassessment of Atomic Bomb Radiation Dosimetry in Hiroshima and Nagasaki—Final Report,* Vol. 2. Hiroshima, Japan, 1988.
64. Hendee, W. R. Estimation of radiation risks: BEIR V and its significance for medicine. *JAMA* **268**(5):620–624, 1992.

CHAPTER

28

PROTECTION FROM EXTERNAL SOURCES OF RADIATION

OBJECTIVES

After completing this chapter, the reader should be able to:

- List the effective dose limits recommended by the NCRP.
- Discuss special regulations that concern pregnant workers.
- Convert monitor badge readings to effective dose.
- List some regulatory and some advisory groups in the field of radiation protection.
- Give an estimate for the radiation dose associated with radiologic procedures.
- Describe the methods used to calculate the amount of shielding required for various types of radiologic equipment.
- Explain the difference between
  - Primary and secondary barriers
  - Scatter and leakage exposure
  - Controlled and uncontrolled areas
- Write the fundamental equation used to calculate barrier thickness and explain each of its terms.
- Define the terms
  - Population dose
  - Genetically significant dose
  - Gonadal dose

For a few years after their discovery, x rays and radioactive materials were used with little regard for their biological effects. After a few years, the consequences of careless handling and indiscriminate use of radiation sources became apparent. These consequences included severe burns, epilation, and, later, leukemia and other forms of cancer in persons who received high exposures to radiation. Persons affected included many who pioneered the medical applications of ionizing radiation.

Because ionizing radiation had proved beneficial to humans in many ways, the question to be answered was whether individuals in particular, and the human race in general, could enjoy the benefits of ionizing radiation without suffering unacceptable consequences in current or future generations. This question is often described as the problem of *risk versus benefit*. To reduce the risk of using ionizing radiation, advisory groups were formed to establish upper limits for the exposure of individuals to radiation. Advisory groups have reduced the upper limits recommended for radiation exposure many times since the first limits were promulgated. These reductions reflect the use of ionizing radiation by a greater number of persons, the implications of new data concerning the sensitivity of biological tissue to radiation, and improvements in the design of radiation devices and in the architecture of facilities where persons use radiation devices.

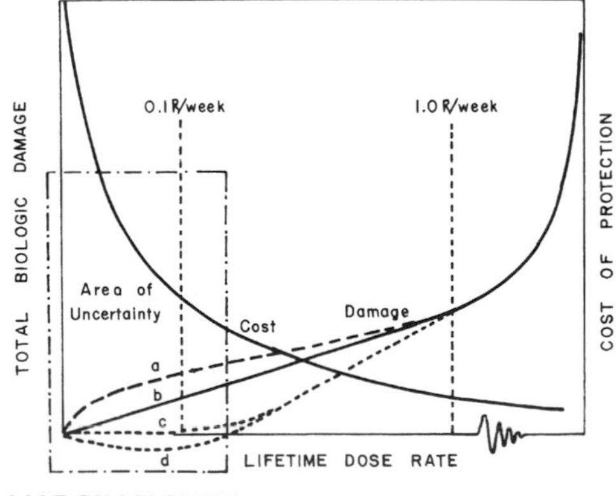

MARGIN FIGURE 28-1
Total biologic damage to a population expressed as a function of the average dose rate to individuals in the population. The cost of protection is reduced as greater biologic damage is tolerated. (From Claus, W. The Concept and Philosophy of Permissible Dose to Radiation, in Claus, W. (ed.), *Radiation Biology and Medicine*. Reading, MA, Addison-Wesley, 1958. Used with permission.)

The philosophy underlying the control of radiation hazards is a *philosophy of risk*. With this philosophy, advisory groups attempt to establish upper limits for radiation exposure that minimize the hazard to individuals and to the population but do not interfere greatly with the beneficial uses of radiation.[1] This philosophy of risk is depicted in Margin Figure 28-1. The total biological damage to a population is expressed as the sum of individual effects, such as reduced vitality, morbidity, shortened life span, and genetic damage, that may result from receipt of some average dose rate over the lifetime of each individual in the population. The total biological damage is assumed to increase gradually as the average dose rate increases to a value of perhaps 0.01 Sv (1 rem) per week. Damage is assumed to increase more rapidly beyond this dose rate. Although the exact shape is unknown for the curve of total biological damage versus dose rate, the curve in Margin Figure 28-1 is probably a reasonable estimate. The region of the curve enclosed within the rectangle labeled the "Area of Uncertainty" is the region of greatest interest to persons concerned with radiation protection; however, this part of the curve is also the region for which few data are available. As indicated by curve c, the damage may remain at zero for average dose rates below some threshold value. Conceivably, the curve for total biological damage may fall below the axis near the origin (curve d), which suggests that very low dose rates are beneficial, a concept known as radiation hormesis. Limited experimental data support this hypothesis. Perhaps the curve rises rapidly at the origin, as suggested in curve a. The data are meager for the biological consequences of radiation exposure at low dose rates, and curve b is usually assumed to be correct. This curve suggests that the total biological damage to a population is linearly proportional to the average dose rate down to a dose rate of zero for individuals in the population. This suggestion is referred to as the *linear nonthreshold hypothesis of radiation injury*.

Adequate data are not available for the area of uncertainty depicted in Margin Figure 28-1. Consequently, the cost of radiation protection must be balanced against unknown biological effects that might result if adequate protection were not provided. The cost of protecting individuals from ionizing radiation (e.g., by shielding, remote control techniques, monitoring procedures, and personnel restriction) is plotted as a function of the average dose rate acceptable to the population. The cost increases from almost zero with no restrictions on radiation exposure, to a cost that approaches infinity if the population desires no exposure to radiation. Somewhere within the area of uncertainty, an upper limit must be accepted for the exposure of persons to radiation. This limit should be recognized as an upper limit for radiation exposure that provides a risk acceptable to the population, without depriving the population of benefits derived from the judicious use of ionizing radiation. In practice, the exposure of individuals to radiation should be kept as low as reasonably achievable.

REGULATORY AUTHORITY FOR RADIATION PROTECTION

Various agencies play some role in the limitation of radiation dose to the general public and to occupationally exposed individuals. There are two types of agencies, advisory and regulatory. Advisory agencies assemble teams of experts in epidemiology, public health, radiobiology, and other fields to publish analyses of the world's literature on radiation bioeffects and to recommend guidelines, limits, and procedures that would lead to the most effective use of radiation for the public good while minimizing the hazard. Examples of advisory agencies include:

National Council on Radiation Protection and Measurements (NCRP)
International Commission on Radiological Protection (ICRP)
United Nations Scientific Committee on the Effects of Atomic Radiation (UNSCEAR)
National Academy of Sciences (NAS)
Conference of Radiation Control Program Directors (CRCPD)

Regulatory agencies make laws, inspect facilities, review records, levy fines, and issue and rescind licenses. The rules that they promulgate generally follow the guidelines of the advisory groups, usually with some time delay. Examples of regulatory agencies (and their jurisdictions) include:

Nuclear Regulatory Commission (NRC)—radioactive materials used or produced in nuclear reactors (source and by product radioactive material).
State Health Departments—x-ray producing equipment and, in some states, radioactive materials.
Environmental Protection Agency (EPA)—radioactive materials and sources that affect the environment.
Food and Drug Administration (FDA)—manufacture of X-ray producing equipment.

State Health Departments in some states have established their own regulations for radiation protection that are equivalent to or stricter than those of the NRC. Through

The responsibility for recommending limits for radiation exposure has been delegated to advisory groups composed of persons experienced in the use of ionizing radiation. In 1922, the American Roentgen Ray Society and the American Radium Society began a study of acceptable levels of radiation exposure. In 1928, the International Congress of Radiology formed the International Commission on Radiological Protection, referred to commonly as the ICRP. This organization is recognized as the international authority for the safe use of sources of ionizing radiation. The U.S. Advisory Committee on X-Ray and Radium Protection, known now as the National Council on Radiation Protection and Measurements (the NCRP), was formed in 1929 to interpret and implement recommendations of the ICRP for the United States.

TABLE 28-1 Effective Dose Limits Recommended by the NCRP

Occupational exposures	
Effective dose limits	
(a) Annual	50 mSv
(b) Cumulative	10 mSv × age
Equivalent dose limits for tissues and organs (annual)	
(a) Lens of eye	150 mSv
(b) Skin, hands, and feet	500 mSv
Guidance for emergency occupational exposure	500 mSv
Public exposures (annual)	
Effective dose limit, continuous or frequent exposure	1 mSv
Effective dose limit, infrequent exposure	5 mSv
Equivalent dose limits for tissues and organs	
(a) Lens of eye	15 mSv
(b) Skin, hands, and feet	50 mSv
Education and training exposures (annual)	
Effective dose limit	1 mSv
Equivalent dose limit for tissues and organs	
(a) Lens of eye	15 mSv
(b) Skin, hands, and feet	50 mSv
Embryo–fetus exposures (monthly)	
Equivalent dose limit	0.5 mSv

an agreement with the NRC, these states, known as agreement states, have accepted authority to regulate radioactive byproduct material within their borders.

There is no one set of regulations for radiation protection. The regulations that apply in a particular facility depend upon what state it is in, whether it is a federal facility, whether it produces emissions that fall under the purview of the EPA, and so on. Many of the agreement states do have similar regulations, however. The Conference of Radiation Control Program Directors is an organization that is composed of State Health Department representatives. They produce a document, known as the Suggested State Regulations for the Control of Radiation, that is often adopted with minor changes by individual states. Also, the U.S. Code of Federal Regulations, although technically binding only upon federal facilities and non-agreement states, is often adopted in whole or in part by individual states.

To reduce confusion and redundancy, the dose limits and radiation safety procedures presented in this chapter are those that are recommended by the NCRP and the ICRP. Those recommendations are at any time similar to or slightly ahead of implementation schedules of regulatory authorities and therefore represent the most up-to-date-information.

EFFECTIVE DOSE LIMITS

For the purpose of developing recommendations for the upper limits of exposure of personnel to ionizing radiation, the ICRP and NCRP assume that a population is divided into two main groups: occupationally exposed individuals and members of the general public.[2,3]

Radiation Dose to Occupationally-Exposed Persons

For persons whose occupation requires exposure to radiation, recommendations concerning limits to the effective dose* (E) are provided in Table 28-1. In the past, the expression maximum permissible dose (MPD) was used to describe regulations regarding personnel exposure. This expression has now been replaced by the term "effective dose limit." Effective dose limits do not include exposures from natural

*Effective dose is defined in Chapter 6.

background radiation or medical exposures when the worker is a patient. Medical exposures involve risk/benefit decisions regarding the patient, and are fundamentally the prerogative of the patient and or family, with guidance from the physician. However, guidelines for such decisions have been promulgated by the U.S. Food and Drug Administration,[4] the U.S. Nuclear Regulatory Commission, and the Environmental Protection Agency.[5]

The effective dose limit for occupationally exposed individuals is 50 mSv (5 rem) per year. As defined in Chapter 6, calculation of effective dose involves the use of both tissue-weighting factors and radiation weighting factors to arrive at a measure of overall somatic and genetic detriment. Because some organs have a low tissue-weighting factor, it is possible that deterministic effects could occur in such an organ while the effective dose to the individual is below the limit. Therefore, separate limits are set for the lens of the eye (150 mSv) and localized areas of the skin, hands, and feet (500 mSv). These limits are expressed as equivalent doses rather than effective doses because the intention is not to determine the overall somatic and genetic detriment caused by the organ dose, but to specify the organ dose itself. The lens of the eye, skin, hands, and feet are the only organs for which the special deterministic dose limits apply. All other organs are considered by the NCRP to be adequately protected by the effective dose limit.[6]

Most occupationally exposed persons are able to keep their exposures far below the effective dose limit. To ensure that the lifetime risk of these individuals is acceptable, the NCRP recommends that the cumulative lifetime occupational effective dose in millisieverts should not exceed ten times the age of the worker in years. For example, a 50-year-old worker exposed to the yearly effective dose limit each year since beginning work at the age of 18 years would have received a total of $50\,(50 - 18) = 1600$ mSv (160 rem). This exposure violates the cumulative lifetime limit of 10 times the age, that is, 500 mSV (50 rem). The purpose of the cumulative limit is to control the lifetime total effective dose as well as the yearly effective dose, and to encourage ongoing evaluation of this number to keep career exposures to a "reasonable" level. In short, the intent is to keep actual yearly effective doses well below the yearly limits.

A number of special cases arise when exposure limits are applied to particular individuals. These special cases include the following:

1. **The Pregnant Worker and Protection of the Embryo–Fetus.** An equivalent dose limit of 0.5 mSv per month is recommended for the fetus once the pregnancy is known. This amount does not include medical exposure or natural background radiation.

When a pregnant worker notifies her employer of her pregnancy, the radiation safety officer should review her badge readings over recent months to estimate what the fetal dose would have been during that time period. Because of the margin of safety usually employed in designing the workplace to limit the worker's actual exposure, there is seldom any need to reassign a worker or make any change in her activities. To reassure all parties during a pregnancy, some institutions routinely issue extra monitoring devices that are read more frequently or have an audible alarm if a preset dose threshold is reached.

Because of the legal outcome of "right to work" lawsuits brought on behalf of female workers in industries where hazardous materials may be encountered (e.g., lead and other heavy metals), most radiation safety regulations require a woman to notify her employer that she is pregnant before any special actions are taken. The institution does not have the authority to make changes in activities, responsibilities, record keeping, monitoring, or counseling unless or until the employee notifies the employer that she is pregnant.

There are no separate specific controls for occupationally-exposed women who are not known to be pregnant. Although reports of the United Nations Scientific Committee on the Effects of Atomic Radiation[7] estimate that the lifetime cancer risk from radiation received during the fetal stage is two to three times that of the adult (i.e., approximately 10% per Sv), the effective dose experienced by the fetus in a

In 1991 the ICRP revised their recommendations on radiation protection to incorporate recent information from the U.S. National Academy of Sciences and from the United Nations Scientific Committee on the Effects of Atomic Radiation. This revision involved a reanalysis of the risk of cancer and genetic effects. The reanalysis was possible because of declassification of atomic bomb yield data and because of the appearance of cancers in the atomic bomb survivor population that occurred later in the life of the individuals who were exposed to unusual amounts of radiation 46 years earlier. The main change was the reformulation of tissue weighting factors (see Chapter 6). Their recommendations are considered more inclusive not only because of the new data, but also because they took into account the risk of severe hereditary effects in all future generations (previous formulations had concerned only the first two generations) and the risk of nonfatal cancers. However, the most recent federal regulations adopted by the Nuclear Regulatory Commission in 1991 had already been through lengthy revision cycles, including public comment periods. These regulations were adopted using older terminology and weighting factors. At the present time, when a radiation safety officer of a facility calculates dose from personal dosimeter records, the results are expressed in terms of the older "effective dose equivalent" system which uses earlier versions of weighting factors and an equation that is slightly different from Eq. (28-9) and (28-10). Because of the large margins of safety that are built into regulations, the discrepancy has little effect on actual safety concerns. However, it is expected that there will be revisions in the future that bring Federal and State regulations into alignment with recommendations of the scientific advisory groups.

controlled occupational environment would be small. In the unlikely scenerio of a fetus receiving as much as 50 mSv prior to discovery of a pregnancy, the lifetime cancer risk to the offspring would be estimated to be at most (0.05 Sv) (10% Sv^{-1}) = 0.5% lifetime risk.

2. **Education and Training Exposures.** Exposure of individuals, even if under the age of 18, is permitted for purposes of education and training. This includes demonstrations for K-12 classrooms as well as for radiologic technology students who have not been formally employed.

It is recommended that education and training exposures be limited to an effective dose of less than 1 mSv per year. This limit is intended to count as part of the 5 mSv per year limit for "infrequent exposure" of members of the general public discussed in a later section. The equivalent dose to the lens of the eye is limited to 15 mSv annually, and the equivalent dose to the skin, hands, and feet is limited to 50 mSv annually.

3. **Emergency Occupational Exposure.** Many emergency situations may be handled within occupational limits for radiation exposure. In cases of extreme emergencies such as lifesaving events, however, there may be occasions when the risk versus benefit analysis favors a one-time overexposure of individuals to achieve a greater good. In such situations, most advisory groups recommend using older volunteers because most stochastic effects of radiation exposure are more likely to occur if the exposure is received at an earlier age.

In an emergency situation, the National Council on Radiation Protection and Measurements recommends that measures be taken to ensure that the emergency exposure is less than 500 mSv. It is further recommended that the equivalent dose to the skin be kept below 5 Sv. The latter requirement applies to exposures from low-energy photons or beta particles during lifesaving actions. In such circumstances, the tolerance dose for skin damage could be reached before the 500 mSv effective dose for the whole body was exceeded.

If exposures exceed the limits for an individual by a small amount, the expectation of harm to that individual is small. Nevertheless, this occurrence should be investigated so that the reason for exceeding the limit may be ascertained. Questions such as whether there was a lapse in institutional safety procedures, and what will be done to prevent this occurrence in the future, must be answered. To facilitate investigation of problem areas before significant exposures occur, institutions set their own action levels. Action levels are levels that are well below effective dose limits but higher than exposures necessary to perform routine occupational tasks. Many institutions set action levels at one-tenth of the effective dose limits, and they monitor the equivalent monthly or quarterly readings of personnel exposures. If the action level is exceeded but the effective dose limit has not been reached, an internal investigation is conducted without necessarily altering the individual's work conditions.

The guiding philosophy of radiation protection is that it is not desirable for personnel to routinely reach an effective dose limit. When designing facilities and in developing policies and procedures, the goal is to ensure that exposures of personnel are maintained as low as reasonably achievable (ALARA). While the definition of what is "reasonable" may vary from one situation to another, the principle of ALARA mandates consideration of measures beyond those strictly required to maintain effective dose limits. For example, the cost of adding extra shielding in the walls of examination rooms, or the modification of policies to provide further restrictions on exposure time or increased distance from radioactive sources, may be part of an ALARA program.[2,8]

Radiation Dose to Members of the Public

Personal monitors are not feasible for members of the public exposed occasionally to radiation. The effectiveness of protective measures for these persons is evaluated indirectly by sampling the air, water, soil, and other elements of the environment. The expected dose for members of the public is computed from analyses of the environment and from knowledge of the living habits of the members. The effective dose limit for members of the public excludes the radiation dose contributed by background radiation and by medical exposures, and in general it is one-tenth of the whole-body effective dose limit for occupationally exposed persons. That is, the effective dose limit for a whole-body dose is 5 mSv/yr (0.5 rem/yr) for occasional exposure of members of the public. If the exposure is frequent or continuous, the limit is lowered to 1 mSv/yr (0.1 rem/yr).

Radiation exposure limits for the general public do not include medical exposures. That is, the 10 to 50 mSv (1 to 5 rem) received from a computed tomographic (CT) scan does not "exceed" the dose limit for the patient. Such an exposure has nothing to do with the limits discussed here. Medical procedures are performed on the basis of a risk/benefit analysis for the individual and are considered to be separate from the coincidental radiation exposures discussed in this section.

Population Exposure, Gonadal Dose, and Genetically Significant Dose

In addition to recommending limits for the exposure of individuals, scientific advisory groups periodically estimate the actual impact of radiation exposure upon a population of persons. For this purpose, the *collective effective dose* is defined as the average dose estimated for individuals in the population multiplied by the total number of individuals in the population. The current estimate of collective effective dose for the U.S. population is 835,000 person-Sv. This estimate is the product of the average annual background of 3.6 mSv times the U.S. population, which, at the time of the most recent estimate, was 230,000,000.

The exposure of gonadal tissue in a large population is of concern because of the possibility of disrupting a significant fraction of the gene pool. The approximate dose to gonadal tissue that is delivered by various medical procedures is given in Table 28-2. In a medical facility, the output of each radiation-producing device should be measured yearly, and these measurements should be used to compute patient doses, including gonadal doses for each medical procedure in which the device is used. A prediction of the "genetic impact" of such exposures is based upon estimates of the

TABLE 28-2 Typical Gonad Doses from Various Radiographic Examinations

	Gonad Dose (mrad)[a]	
X-Ray Examination	*Male*	*Female*
Skull	<1	<1
Cervical spine	<1	<1
Full-mouth dental	<1	<1
Chest	<1	<1
Stomach and upper gastrointestinal	2	40
Gallbladder	1	20
Lumbar spine	175	400
Intravenous urography	150	300
Abdomen	100	200
Pelvis	300	150
Upper limb	<1	<1
Lower limb	<1	<1

[a] For some radiologic examinations the female gonad dose is greater than the dose received by the male because the female reproductive organs are located within the pelvic cavity, unlike the male reproductive organs, which are located outside and below the pelvic cavity. The distribution of biologic tissue overlying the ovaries also affects the dose received for a given radiologic examination.

Source: From Ballinger, P. W. *Merrill's Atlas of Radiographic Positions and Radiologic Procedures,* Vol. 1, 8th edition. St. Louis, Mosby, 1995.

fraction of the population that undergoes diagnostic and nuclear medicine procedures each year, in addition to all other sources of exposure. Furthermore, the impact is weighted by the age and sex distribution of the population to take into account the expected number of future children that would be produced by each category. The *genetically significant dose* (GSD) is a quantity that is calculated in this manner. It is the dose that, if received equally by all members of the population, would result in the same genetic impact as the uneven distribution of gonadal exposures that actually occurs. Current estimates reveal a GSD of the U.S. population of 1.3 mSv annually. Diagnostic x-ray exposure accounts for 0.2 mSv and nuclear medicine accounts for 0.02 mSv of the total. The rest is attributable to background radiation exposure.

SAFETY RECOMMENDATIONS FOR SOURCES OF X AND γ RADIATION

Using recommendations of the ICRP for guidance, members of the NCRP have suggested a number of guidelines for evaluating the safety of x-ray equipment and sources of γ radiation. A complete description of these recommendations is available in a report of the NCRP.[9]

PROTECTIVE BARRIERS FOR RADIATION SOURCES

All occupations are associated with some risk of morbidity and mortality. Occupations in trade, service, and government are considered, in an actuarial sense, to be "safe" and are associated with risks of fatality of the order of 1 in 10,000 per year. Occupations such as construction, mining, and agriculture are associated with risks that are 3–5 times higher and are considered to be "more hazardous," although still within a safe range. Because of the assumption that the risk of stochastic effects following exposure to ionizing radiation is nonthreshold, radiation safety regulations cannot be set so low that there is no theoretical risk. The National Council on Radiation Protection and Measurements uses a process of recommending limits based upon the goal of maintaining the risk of genetic effects, as well as fatal and nonfatal cancers, at the same level as that of "safe" industries. Estimates of the NCRP show that the lifetime risk from an average worker's exposures over a working career is approximately 1 in 10,000. A worker who receives the maximum cumulative limit of 10 mSv/yr over an entire career would have a risk of approximately 5 in 10,000.

Occupational exposure should be monitored for each individual radiation worker. Any area where a yearly whole-body effective dose of 15 mSv (1.5 rem) or more could conceivably be delivered to an individual should be considered a controlled area and supervised by a radiation protection officer. Persons working in a controlled area and who are exposed to radiation should carry one or more personal monitors for the measurement of radiation dose (e.g., a film badge, thermoluminescent dosimeter, or pocket ionization chamber), and access to the area should be restricted.

The walls, ceiling, and floor of a room containing an x-ray machine or radioactive source may be constructed to permit the use of adjacent rooms when the x-ray machine is energized or the radioactive source is exposed. With an effective dose limit of 50 mSv/yr (5 rem/yr) for occupational exposure, the maximum permissible average dose equivalent per week equals 1.0 mSv (0.1 rem). Therefore, a dose of 1.0 mSv/wk (0.1 rem/wk) (or less) is used when computing the thickness of radiation barriers required for controlled areas. For areas outside the supervision of a radiation safety officer, a value of 0.1 mSv (0.01 rem) per week is usually used. Because the conversion from roentgens to rads is nearly 1 for x and γ rays (see Chapter 6) and because the quality factor is also 1 for these radiations, the maximum permissible exposure is often taken as 2.5×10^{-5} C/(kg-wk) (0.1 R/wk) or 2.5×10^{-6} C/(kg-wk)(0.01 R/wk) for computation of the barrier thicknesses for sources of x and γ rays. In a shielded room, *primary barriers* are designed to attenuate the primary (useful) beam from the radiation source, and *secondary barriers* are designed to reduce scattered and leakage radiation from the source.

Protection from Small Sealed γ-Ray Sources

The exposure X in roentgens* to an individual in the vicinity of a point source of radioactivity is

$$X = \frac{\Gamma_\infty At}{d^2}(B) \tag{28-1}$$

where X is the exposure in roentgens, Γ_∞ is the exposure rate constant in (R-m^2)/(hr-Ci) at 1 m, A is the activity of the source in curies, t is the time in hours spent

*Shielding calculations are usually performed by using traditional units (e.g., roentgen, rad, rem) at the present time.

in the vicinity of the source, d is the distance from the source in meters, and B is the fraction of radiation transmitted by a protective barrier between the source and the individual. The exposure may be reduced by (1) increasing the distance d between the source and the individual, (2) decreasing the time t spent in the vicinity of the source, and (3) reducing the transmission B of the protective barrier (Margin Figure 28-2).

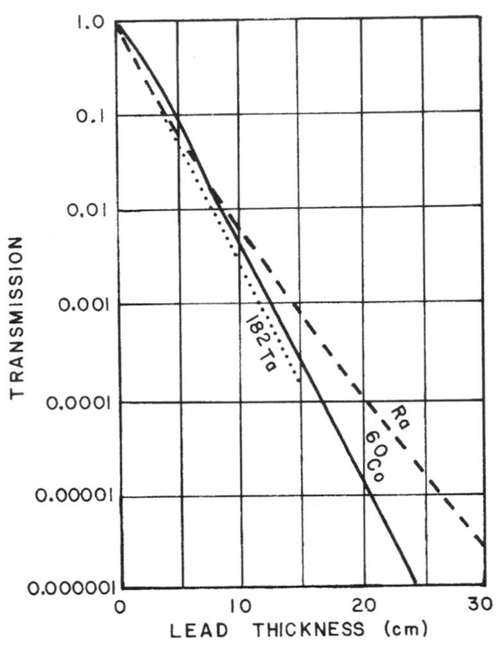

MARGIN FIGURE 28-2
Transmission through lead of γ rays from radium, ^{60}Co, and ^{182}Ta. (From International Commission on Radiological Protection. *Report of Committee III on Protection Against X Rays Up to Energies of 3 MeV and Beta and Gamma Rays from Sealed Sources*, ICRP Publication No. 3. New York, Pergamon Press, 1960. Used with permission.)

Example 28-1

What is the thickness of lead that reduces the exposure to 0.1 R/wk for a person at a distance of 1 m from a 100 mg radium source encapsulated with 0.5 mm Pt(Ir) for 40 hr/wk? The exposure rate constant Γ_∞ is 0.825(R-m^2)/(hr-Ci) for radium filtered by 0.5 mm Pt(Ir).

The 100-mg radium source is 0.1 g, which for radium is approximately 0.1 Ci.

$$X = \frac{\Gamma_\infty At}{d^2}(B)$$

$$B = \frac{Xd^2}{\Gamma_\infty At}$$

$$= \frac{(0.1)(1)^2}{(0.825)(0.1)(40)}$$

$$= 0.03 \tag{28-1}$$

From Margin Figure 28-2, about 7 cm of lead is required for the radiation barrier to yield a transmission of 0.03 (3%).

Primary Barriers for X-Ray Beams

For an individual some distance from an x-ray unit, the maximum expected exposure to radiation depends on the following:

1. Emission rate of radiation from the unit. As a general rule, a diagnostic x-ray unit operating at 100 kVp furnishes an exposure rate of less than 2.58×10^{-4} C/(kg-min) at 1 m for each milliampere of current flowing through the x-ray tube. That is, the exposure rate is $\leq 2.58 \times 10^{-4}$ C/(kg-mA-min) or ≤ 1 R/(mA-min) at 1 m.
2. Average tube current and operating time of the x-ray unit per week. The product of these variables has units of milliampere-minutes per week and is termed the *work load* W of the x-ray unit. Typical work loads for busy radiographic installations are described in Table 28-3.
3. Fraction of operating time that the x-ray beam is directed toward the location of interest. This fraction is termed the *use factor* U. Typical use factors are listed in Table 28-4.
4. Fraction of operating time that the location of interest is occupied by the individual. This fraction is termed the *occupancy factor* T. Typical occupancy factors are listed in Table 28-5. For controlled areas, the occupancy factor always is 1.
5. Distance d in meters from the x-ray unit to the location of interest because the exposure decreases with the inverse square of the distance ($1/d^2$).

Radiation protection may be reduced to three fundamental principles: time, distance, and shielding. Radiation exposure may be reduced by decreasing the exposure time, increasing distance from the source of radiation, or providing more shielding between the source and exposed individuals.

Occupancy factors are always equal to one for controlled areas because it is assumed that if a given worker is not in one controlled area during a workday, he or she is likely to be in another controlled area.

The maximum expected exposure to an individual at the location of interest may be estimated as

$$\begin{array}{l}\text{Maximum}\\\text{expected}\\\text{exposure}\end{array} = \frac{(1\text{R-m}^2/\text{mA-min})(\text{mA-min/wk})\left(\begin{array}{c}\text{Use}\\\text{factor}\end{array}\right)\left(\begin{array}{c}\text{Occupancy}\\\text{factor}\end{array}\right)}{d^2} \tag{28-2}$$

$$= \frac{WUT}{d^2}$$

TABLE 28-3 Typical Weekly Work Loads for Busy Installations

Diagnostic	*Daily Patient Load*	*Weekly Work Load (W), mA-min*[a]		
		100 kV or less	*125 kV*	*150 kV*
Chest (14 × 17:3 films per patient, no grid)	60	150	—	—
Cystoscopy	8	600	—	—
Fluoroscopy including spot filming	24	1500	600	300
Fluoroscopy without spot filming	24	1000	400	200
Fluoroscopy with image intensification including spot filming	24	750	300	150
General radiography	24	1000	400	200
Special procedures	8	700	280	140

[a]Peak pulsating x-ray tube potential.

Source: National Council on Radiation Protection and Measurements. *Structural Shielding Design and Evaluation for Medical Use of X Rays and Gamma Rays of Energies Up to 10 MeV*. Recommendations of the NCRP, Report No. 49. Washington, D.C., 1976.

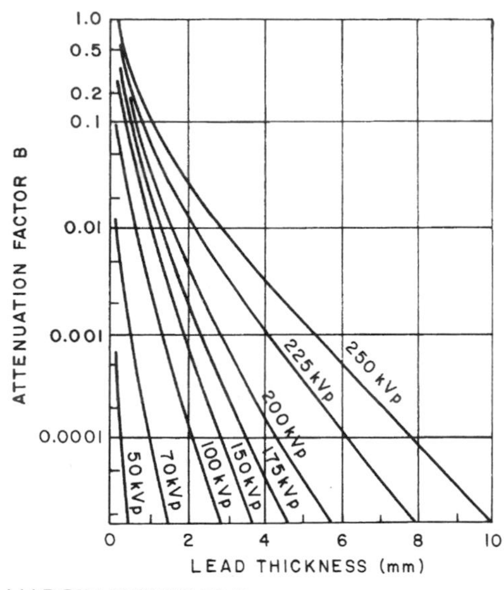

MARGIN FIGURE 28-3
Attenuation in lead of x rays generated at tube voltages from 50 to 250 kVp. The curves were obtained with a half-wave-rectified x-ray generator and with a 90-degree angle between the electron beam and the axis of the x-ray beam. The inherent filtration was 3 mm of aluminum for the 150-kVp to 250-kVp curves and 0.5 mm of aluminum for the other curves. X-ray beams generated with a constant tube voltage require barriers 10% thicker than those indicated by the curves. (From International Commission on Radiological Protection. *Report of Committee III on Protection Against X Rays Up to Energies of 3 MeV and Beta and Gamma Rays from Sealed Sources*, ICRP Publication No. 3. New York, Pergamon Press, 1960. Used with permission.)

This exposure can be reduced to a maximum permissible level X_p (e.g., 0.1 or 0.01 R/wk) by introduction of an attenuation factor B into Eq. (28-2).

$$X_p = \frac{WUT}{d^2}(B)$$

The attenuation factor B is

$$B = \frac{X_p d^2}{WUT} \tag{28-3}$$

With the attenuation factor B and the curves in Margin Figure 28-3, the thickness of concrete or lead may be determined that reduces the exposure rate to an acceptable level. Listed in Table 28-6 are the thicknesses of lead that are equivalent to concrete walls of different thicknesses.[10] Half-value layers for other commonly used shielding materials are given in Table 28-7.

Example 28-2

A diagnostic x-ray generator has a busy work load of 1000 mA-min/wk at 100 kVp. The x-ray tube is positioned 4.5 m from a wall between the radiation room and a

TABLE 28-4 Use Factors Recommended by the ICRP

Full use ($U = 1$)	Floors of radiation rooms except dental installations, doors, walls, and ceilings of radiation rooms exposed routinely to the primary beam.
Partial use ($U = \frac{1}{4}$)	Doors and walls of radiation rooms not exposed routinely to the primary beam; also, floors of dental installations.
Occasional use ($U = \frac{1}{16}$)	Ceilings of radiation rooms not exposed routinely to the primary beam. Because of the low use factor, shielding requirements for a ceiling are usually determined by secondary rather than primary beam considerations.

Source: International Commission on Radiological Protection. *Report of Committee III on Protection Against X Rays Up to Energies of 3 MeV and Beta and Gamma Rays from Sealed Sources*. ICRP Publication No. 3. New York, Pergamon Press, 1960. Used with permission.

TABLE 28-5 Occupancy Factors Recommended by the ICRP

Full occupancy ($T = 1$)	Control spaces, offices, corridors, and waiting spaces large enough to hold desks, darkrooms, workrooms and shops, nurse stations, rest and lounge rooms used routinely by occupationally exposed presonnel, living quarters, children's play areas, and occupied space in adjoining buildings.
Partial use $\left(T = \frac{1}{4}\right)$	Corridors too narrow for desks, utility rooms, rest and lounge rooms not used routinely by occupationally exposed personnel, wards and patients' rooms, elevators with operators, and unattended parking lots.
Occasional occupancy $\left(T = \frac{1}{16}\right)$	Closets too small for future occupancy, toilets not used routinely by occupationally exposed personnel, stairways, automatic elevators, pavements, and streets.

Source: International Commission on Radiological Protection. *Report of Committee III on Protection Against X Rays Up to Energies of 3 MeV and Beta and Gamma Rays from Sealed Sources,* ICRP Publication No. 3. New York, Pergamon Press, 1960. Used with permission.

radiologist's office. The wall contains 3 cm of ordinary concrete. For a use factor of 0.5, what is the thickness of lead that must be added to the wall?

$$B = \frac{X_p d^2}{WUT} \tag{28-3}$$

Because the radiologist's office is a controlled area for occupationally exposed persons, X_p — 0.1 R/wk, T — 1, and

$$\begin{aligned} B &= \frac{0.1d^2}{WUT} \\ &= \frac{(0.1)(4.5)^2}{(1.000)(0.5)(1)} \\ &= 0.004 \end{aligned}$$

From the 100-kVp curve in Margin Figure 28-3, the thickness of lead required is 1.0 mm. A 3-cm thickness of concrete is approximately equivalent to 0.4 mm of lead

TABLE 28-6 Thicknesses of Lead Equivalent to Thicknesses of Ordinary Concrete (Density, 22 g/cm³) at Selected Tube Voyages for Broad X-Ray Beams

Thickness of Concrete (cm)	*Lead Equivalent (mm) for X-Rays Generated at the Following Voltages (kVp)*										
	50	*75*	*100*	*150*	*200*	*250*	*300*	*400*	*500*	*1000*	*2000*
5	0.4	0.5	0.6	0.5	0.5	0.6	0.8	1.1	1.6	4.0	6
10	0.9	1.2	1.4	1.2	1.2	1.7	2.2	3.0	3.9	8.6	13
15	1.4	2.0	2.4	1.9	2.1	3.0	3.8	5.4	7.1	13	22
20	2.0	2.8	3.4	2.7	2.9	4.4	5.8	8.5	11	21	31
25	2.5	3.6	4.4	3.4	3.8	5.8	7.9	11	15	29	40
30	3.1	4.3	5.4	4.2	4.7	7.3	10	14	19	37	49
35	—	—	—	5.1	5.6	8.6	12	18	24	45	58
40	—	—	—	—	—	—	—	21	28	54	67
45	—	—	—	—	—	—	—	24	33	62	76
50	—	—	—	—	—	—	—	—	37	71	85
60	—	—	—	—	—	—	—	—	46	88	103
75	—	—	—	—	—	—	—	—	60	112	130
90	—	—	—	—	—	—	—	—	—	138	159

Source: Appleton, G., and Krishmamoorthy, P. *Safe Handling of Radioisotopes; Health Physics Addendum.* Vienna, International Atomic Energy Agency, 1960. Used with permission.

TABLE 28-7 Half-Value Layers for Commonly Encountered Shielding Materials[a]

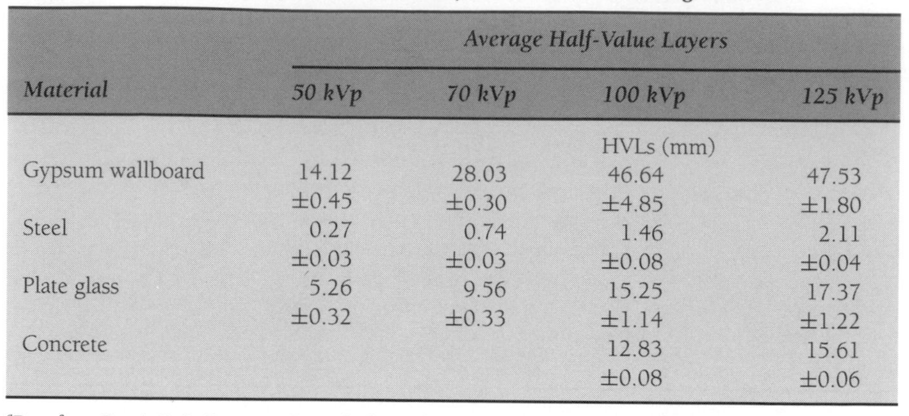

	Average Half-Value Layers			
Material	*50 kVp*	*70 kVp*	*100 kVp*	*125 kVp*
			HVLs (mm)	
Gypsum wallboard	14.12 ±0.45	28.03 ±0.30	46.64 ±4.85	47.53 ±1.80
Steel	0.27 ±0.03	0.74 ±0.03	1.46 ±0.08	2.11 ±0.04
Plate glass	5.26 ±0.32	9.56 ±0.33	15.25 ±1.14	17.37 ±1.22
Concrete			12.83 ±0.08	15.61 ±0.06

[a]Data from Rossi, R. P., Ritenour, R., and Christodoulou, E. Broad beam transmission properties of some common shielding materials for use in diagnostic radiology. *Health Physics* 1991; **61**(5):601–608.

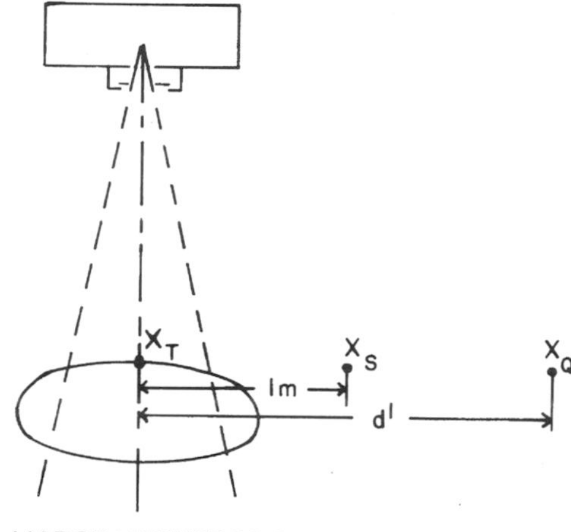

MARGIN FIGURE 28-4
Geometry for determining the radiation exposure X_Q at a distance d' from a scatterer where the exposure is X_T. The exposure X_S at a distance of 1 m from the scatterer is 1/1000 X_T.

(Table 28-6). Consequently, at least 0.6 mm (or 1/32 in.) of lead should be added to the wall.

Secondary Barriers for Scattered X-Radiation

Radiation may be scattered to a location that is not exposed to the primary x-ray beam. The amount of scattered radiation varies with the intensity of radiation incident upon the scatterer, the area of the x-ray beam at the scatterer, the distance between the scatterer and the location of interest, and the angle of scattering of the x rays. For a scattering angle of 90 degrees with respect to a diagnostic x-ray beam 400 cm^2 in cross-sectional area that is incident upon a scatterer, the exposure rate 1 m from the scatterer is less than 1/1000 of the primary beam exposure rate at the scatterer (Margin Figure 28-4). At other distances from the scatterer, the exposure rate varies as $1/d^2$.

Over a week, the exposure at the scatterer (the patient) is

$$\text{Exposure at scatterer} = \frac{WUT}{d^2} = \frac{W}{d^2}$$

where U and T are 1 because the beam is always directed at the patient and a patient is always present. At some location a distance d' from and at right angles to the primary beam, the exposure due to scattered radiation is

$$\text{Exposure per week from scattered radiation} = \frac{(W/d^2)(1/1000)(F/400)}{(d')^2} \tag{28-4}$$

where 1/1000 is the ratio of scattered radiation at 1 m to primary radiation at the scatterer, F is the area of the x ray beam at the scatterer, $F/400$ corrects the ratio 1/1000 for more or less scattered radiation from fields larger or smaller than 400 cm^2, and $(d')^2$ corrects the scattered radiation for the inverse square falloff of exposure with distance. The use factor is always 1 for scattered radiation. With insertion of an occupancy factor T for the location of interest, the exposure per week to an individual from scattered radiation may be estimated as

$$\text{Exposure to an individual due to scattered radiation (R/wk)} = \frac{(W/d^2)(1/1000)(F/400)(T)}{(d')^2} \tag{28-5}$$

To reduce the exposure to an acceptable level X_p, a secondary barrier may be interposed between the scatterer and the location of interest. The barrier introduces an attenuation factor B into Eq. (28-5):

$$X_p = \frac{(W/d^2)(1/1000)(F/400)(T)}{(d')^2}(B)$$

The attenuation factor B is

$$B = \frac{X_p(d')^2}{(W/d^2)(1/1000)(F/400)(T)} \quad \text{(28-6)}$$

In diagnostic radiology, x-ray beams are almost always generated at tube voltages less than 500 kVp. Radiation scattered from these beams is almost as energetic as the primary radiation, and the barrier thickness for scattered radiation is determined from curves for the primary radiation.

William Herbert Rollins, a Boston dentist, was the first to suggest the use of the intensifying screen in 1902 as a means of reducing patient dose. Such screens were not commonly used until the 1940s. Among other contributions that Rollins made to radiation safety practices were lead shielding, collimators, and the use of increased kVp to decrease patient skin dose.

Example 28-3

A diagnostic x-ray generator has a busy workload of 3000 mA-min/wk at 100 kVp and an average field size of 20 × 20 cm. The x-ray tube is positioned 1.5 m from a wall between the radiation room and a urology laboratory with uncontrolled access. The wall contains 3 cm of ordinary concrete. For an occupancy factor of 1, what thickness of lead should be added to the wall if the primary beam is not directed toward the wall?

$$B = \frac{X_p(d')^2}{(W/d^2)(1/1000)(F/400)(T)} \quad \text{(28-6)}$$

If the radiation scattered to the wall originates within a patient positioned 1 m below the x-ray tube, then

$$B = \frac{(0.01)(1.5)^2}{[3000/(1)^2](1/1000)(400/400)(1)}$$
$$= 0.0075$$

where the maximum permissible exposure is 0.01 R/wk for persons in the urology laboratory. From the 100-kVp curve in Margin Figure 28-3, the thickness of lead required is 0.8 mm for an attenuation of 0.0075 at 100 kVp.

Secondary Barriers for Leakage Radiation

Radiation escaping in undesired directions through the x-ray tube housing is termed *leakage radiation*. According to the ICRP, this radiation should not exceed 0.1 R/hr at 1 m at the highest rated kilovolt-peak for the x-ray tube and at the highest milliamperage permitted for continuous operation of the x-ray tube. To determine the maximum tube current for continuous operation, the anode thermal characteristics chart for the x-ray tube must be consulted (Margin Figure 28-5). From this chart, it is possible to determine the maximum heat units* per second that the x-ray tube can tolerate during continuous operation. By dividing this value by the maximum rated kilovolts peak for the x-ray tube, the maximum milliamperage for continuous operation is computed. At this milliamperage value, the exposure rate due to leakage radiation should not exceed 0.1 R/hr at 1 m from the x-ray tube.

To determine the maximum exposure per week at 1 m due to leakage radiation, the workload in mA-min/wk is divided by the maximum milliamperage for continuous operation. The quotient is the effective operating time W' of the x-ray unit for purposes of computing the exposure due to leakage radiation. The exposure at a location of interest a distance d from the x-ray tube is

$$\text{Exposure due to leakage radiation} = \frac{(0.1)(W')}{60d^2}$$

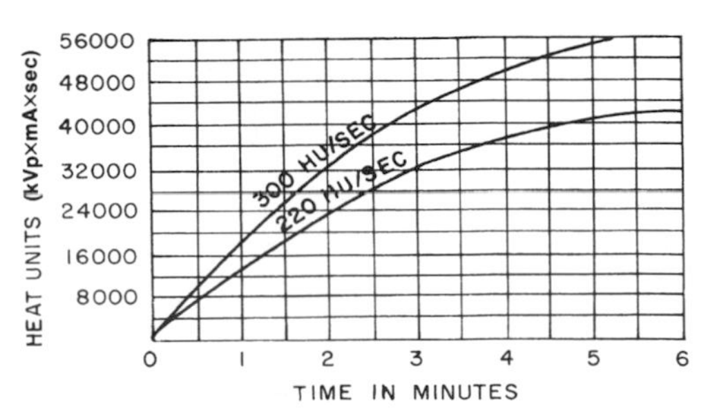

MARGIN FIGURE 28-5
A typical anode thermal characteristics chart.

*Heat units are discussed in Chapter 5.

where W' has units of minutes per week and 60 is a conversion from minutes to hours. The use factor for leakage radiation is always 1 because leakage radiation is emitted in all directions whenever the x-ray tube is used. To determine the exposure to an individual due to leakage radiation, an occupancy factor T for the location of interest should be added to the equation:

$$\text{Exposure to individual from leakage radiation} = \frac{0.1W'T}{60d^2} \tag{28-7}$$

By adding an attenuation factor B to the equation, leakage radiation may be reduced to the permissible level X_p:

$$X_p = \frac{0.1W'T}{60d^2}(B)$$

The attenuation factor B is

$$B = \frac{60X_p d^2}{0.1W'T} \tag{28-8}$$

Because leakage radiation is at least as energetic as the primary beam, the thickness of a barrier that provides the attenuation factor B is determined from curves in Margin Figure 28-3 for the primary x-ray beam.

A barrier designed for primary radiation provides adequate protection against scattered and leakage radiation. For a barrier designed for scattered and leakage radiation only (i.e., a secondary barrier), the thickness is computed separately for each type of radiation. If the barrier thickness required for scattered radiation differs by more than three half-value layers from that required for leakage, the greater thickness is used for the barrier. If the thicknesses computed for scattered and leakage radiation differ by less than three half-value layers, then the thickness of the barrier is increased by adding one half-value layer to the greater of the two computed thicknesses. Half-value layers for diagnostic x-ray beams are listed in Table 28-8.[11]

X rays were discovered in November 1895. By March 1896 there were already reports of adverse symptoms related to x-ray exposure. In late 1896, a physician purposely exposed the little finger of his left hand to x rays on repeated occasions over several days and wrote about the pain, swelling, and stiffness that resulted. The episode that finally convinced most experts that x rays are associated with adverse bioeffects was the death in 1902 of Clarence Dally, one of Thomas Edison's assistants. Dally's job was to energize x-ray tubes as they were finished to determine if their output was sufficient for imaging. The pattern of skin cancer matched the parts of his body that were habitually close to the x ray tubes. After this incident, Edison ceased all research involving x rays.

Source: Brown, P., *American Martyrs to Science Through the Roentgen Rays*. Charles C Thomas, Springfield, IL, 1936.

Example 28-4

For the x-ray tube with a maximum rated kilovolt-peak of 100 and the anode thermal characteristics chart shown in Margin Figure 28-5, determine the maximum milliamperage for continuous operation and the thickness of lead that should be added

TABLE 28-8 Half-Value Layers in Millimeters of Lead and in Centimeters and Inches of Concrete for X-Ray Beams Generated at Different Tube Voltages

	Half-Value Layer						
Attenuating Material	*50 kVp*	*70 kVp*	*100 kVp*	*125 kVp*	*150 kVp*	*200 kVp*	*250 kVp*
Lead (mm)	0.05	0.15	0.24	0.27	0.29	0.98	0.9
Concrete (in.)	0.17	0.33	0.6	0.8	0.88	1.0	1.1
Concrete (cm)	0.43	0.84	1.5	2.0	2.2	2.5	2.8

	Half-Value Layer					
Attenuating Material	*300 kVp*	*400 kVp*	*500 kVp*	*1,000 kVp*	*2,000 kVp*	*3,000 kVp*
Lead (mm)	1.4	2.2	3.6	7.9	12.7	14.7
Concrete (in.)	1.23	1.3	1.4	1.75	2.5	2.9
Concrete (cm)	3.1	3.3	3.6	4.4	6.4	7.4

Source: National Council on Radiation Protection and Measurements. *Medical X-Ray and Gamma-Ray Protection for Energies Up to 10 MeV*. Recommendations of the NCRP, Report No. 34, Washington, D.C., NCRP, 1970.

to the wall described in Example 28-3 to protect against both scattered and leakage radiation.

From Margin Figure 28-6, the maximum anode thermal load that can be tolerated by the x-ray tube is 300 heat units (HU) per second.

$$\text{Maximum mA for continuous operation of maximum kVp} = \frac{300 \text{ HU/sec}}{100 \text{ kVp}} = 3 \text{ mA} \qquad (28\text{-}8)$$

$$\text{Effective operating time } W' = \frac{3000 \text{ mA-min/wk}}{3 \text{ mA}} = 1000 \text{ min/wk}$$

$$B = \frac{60 X_p d^2}{0.1 W' T} = \frac{60(0.01)(1.5)^2}{0.1\,(1000)\,(1)} = 0.0135$$

From Margin Figure 28-3, the thickness of lead required to protect against leakage radiation is 0.6 mm Pb. From Example 28-3, the thickness of lead required to protect against scattered radiation is 0.8 mm Pb. The difference between these thicknesses is less than three times the half-value layer of 0.24 mm Pb for a 100-kVp x-ray beam (see Table 28-8). Hence, the total thickness of lead required to shield against both scattered and leakage radiation is (0.8 + 0.24) = 1.04 mm Pb. The concrete in the wall provides shielding equivalent to 0.4 mm Pb, so 0.64 mm, or 1/32 in., of lead must be added to the wall.

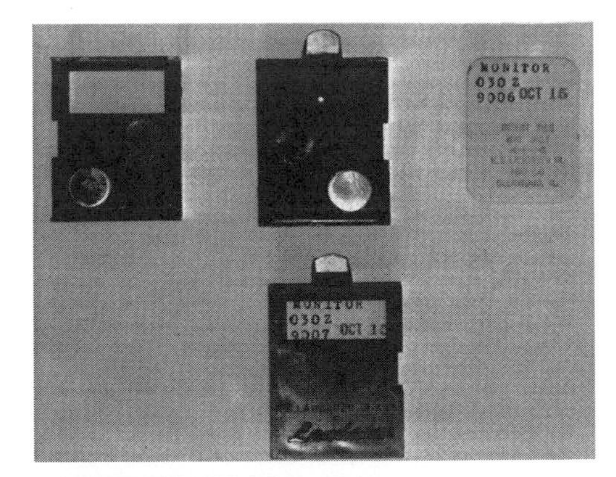

MARGIN FIGURE 28-6
Film badge for measuring the exposure of individuals to x and γ radiation and to high-energy β radiation. (Courtesy of R. S. Landauer, Jr., and Co.)

Special Shielding Requirements for X-Ray Film

X-ray film is especially sensitive to radiation, and film storage areas and bins must often be shielded to a greater extent than areas occupied by individuals. For film storage areas, shielding computations are identical with those described above, except that maximum permissible exposures X_p must be reduced to much lower values. As a rule of thumb, the exposure should be limited to no more than 1 mR for the maximum storage time of the film (e.g., 2 weeks).

Changes in Shielding Recommendations

At the present time, the National Council on Radiation Protection and Measurements is in the process of revising its 26-year-old recommendations on shielding design for diagnostic radiology. The changes are expected to include:

- Revisions of workload estimates that take into account the range of kVps that are currently used.
- Leakage and scatter calculations will be modeled more explicitly with regard to values found in modern equipment. The present rule of adding a half-value layer if the leakage and scatter barrier requirements are similar will be abandoned, and both sources of exposure will be modeled together.
- Currently, occupancy factors are not allowed to be lower than 1/16. The revised guidelines will allow for spaces such as closets, stairways, storage facilities, and so on, to be assigned occupancy factors as low as 1/40.
- Current guidelines do not take into account some attenuating material that is present. These materials include the image receptor assembly and the patient table. In the new guidelines, the attenuating properties of these materials will be considered.
- Use factors will be revised to reflect current practices.

The average glandular dose during mammography is 2.5–3.0 mSv for a 4.5-cm compressed breast composed of 50% fat and 50% glandular tissue, two-view study (craniocaudal and oblique).

Source: McLelland, R. et al. *AJR* 1991; **157**:473–479.

Estimation of Patient Dose in Diagnostic Radiology

The radiation dose delivered to a patient during a radiographic procedure may be estimated with reasonable accuracy with the rule of thumb of 1 R/mA-min at 1 m from a diagnostic x-ray unit. Frequently, this estimate is needed when a patient is discovered shortly after an examination to have been in an early stage of pregnancy when the x-ray procedure was performed. In this case, an estimate of fetal dose is required. A simple procedure for obtaining a rough estimate of fetal dose is given in Example 28-5.

Example 28-5

A patient receives 2 minutes of fluoroscopy (80 kVp, 3 mA) and four spot films (80 kVp, 30 mA-s) over the 20-cm thick lower abdominal region. The x-ray tube-to-tabletop distance is 0.5 m for the x-ray unit used for the examination. A short time later it is determined that the patient was 3 weeks pregnant at the time of the examination. Estimate the dose to the fetus.

The rule of thumb of 1 R/mA-min yields $[1 \text{ R/mA-min}/(0.5)^2]$, or 4 R/mA-min, at a distance of 0.5 m. The exposure at the skin delivered during fluoroscopy is

$$\text{Fluoroscopic exposure} = \left(4\frac{\text{R}}{\text{mA-min}}\right)(3 \text{ mA})(2 \text{ min})$$
$$= 24 \text{ R}$$

The exposure at the skin delivered by the spot films is

$$\text{Spot film exposure} = \frac{(4 \text{ R/mA-min})(30 \text{ mA-sec/film})(4 \text{ films})}{60 \text{ sec/min}}$$
$$= 8 \text{ R}$$

The total exposure at the skin is 24 R + 8 R = 32 R. For average-size patients, the transmission of the x-ray beam through the abdomen is roughly 0.1% to 0.5% of the incident exposure. A transmission of 0.4% means that the patient provides 8 HVL's of attenuation. With the assumption that the fetus is positioned midway in the abdomen, shielding equivalent to 4 HVL's (6.2% transmission) is provided to the fetus. With a conservative estimate of 7%, the exposure to the fetus is

$$\text{Exposure to fetus} = (32 \text{ R})(0.07)$$
$$= 2.2 \text{ R}$$

The absorbed dose in milligray is approximately equal numerically to ten times the exposure in roentgens, so the absorbed dose to the fetus is roughly 22 mGy (2.2 rad).

Values of entrance surface absorbed dose (skin dose, ESAD) are given in Table 28-9, and multiplicative factors that may be used to convert ESAD to fetal dose are given in Table 28-10. Typical bone marrow doses for radiographic exams are given in Table 28-11. For actual patient and fetal dosimetry, a qualified medical physicist can determine a more accurate estimate of dose based upon specific output measurements from the source and data such as patient thickness, technique factors, and so on.

AREA AND PERSONNEL MONITORING

The radiation exposure must be known for persons working with or near sources of ionizing radiation. This knowledge may be obtained by periodic measurements of exposure rates at locations accessible to the individuals. Furthermore, persons working with or near sources of ionizing radiation should be equipped with personnel monitors

TABLE 28-9 Mean Entrance Surface Absorbed Dose (ESAD) to Soft Tissue

Examination (Projection)[a]	*ESAD*
Film/screen	
Chest (P/A)	0.20 mGy
Skull (lateral)	1.93 mGy
Abdomen (kidneys, ureters, bladder) (A/P)	6.27 mGy
Retrograde pyelogram (A/P)	4.98 mGy
Thoracic spine (A/P)	2.85 mGy
Lumbar spine (A/P)	6.41 mGy
Full spine (A/P)	2.03 mGy
Feet (D/P)	0.70 mGy
Non-film/screen	
Fluoroscopy	<100 mGy min^{-1}
Cineradiography[b]	250 mGy min^{-1}
Digital radiography	2.5 mGy/image
Computed tomography	10–50 mGy

[a] P/A refers to posterior (back) to anterior (front) beam projection (i.e., the x-ray beam enters the posterior portion of the body and exits anteriorly). A/P refers to anterior to posterior beam projection. D/P refers to dorsal (top) to pedal (bottom of foot) beam projection.

[b] 0.14 mGy/frame, 30 frame s^{-1}.

Source: Ritenour, E. R., Geise, R. A., Radiation Sources: Medicine, in Hendee, W. R., and Edwards, F. M. (eds.), *Health Effects of Exposure to Low-Level Ionizing Radiation. Philadelphia,* Institute of Physics Publishing, 1996, p. 440.

TABLE 28-10 Typical Fetal Dose Factors as a Function of Entrance Surface Absorbed Dose (ESAD)[a]

X-Ray Exam	*Fetal Dose Factor mGy/[mGy^{ESAD}]*
Skull	<0.01
Cervical	<0.01
Limb	<0.01
Chest	0.03
Stomach and upper GI	0.30
Lumbar spine	2.90
Abdomen	3.05
Pelvis	3.40

[a] Calculated from data found in Ballinger, P. W. *Merrill's Atlas of Radiographic Positions and Radiologic Procedures,* Vol. 1. 8th edition, St Louis, Mosby, 1995.

that reveal the amount of exposure the individuals have received. The monitor used most frequently is the film badge containing two or more small photographic films enclosed within a light-tight envelope. The badge is worn for a selected interval of time (1 to 4 weeks) at some convenient location on the clothing over the trunk of the body. During fluoroscopy, most persons wear the film badge under a lead apron to monitor the exposure to the trunk of the body, although some state regulations require a second badge on or near the collar to measure the dose to organs (e.g., thyroid gland) not shielded by the apron. After the desired interval of time has elapsed, the film is processed, and its optical density is compared with the optical density of similar films that have received known exposures. From this comparison, the radiation exposure may be determined for the film badge and, supposedly, for the individual wearing the film badge. Small metal filters mounted in the plastic holder for the film permit some differentiation between different types of radiation and different energies of x rays that contribute to the exposure (Margin Figure 28-6). Types of radiation that may be monitored with a film badge include x and γ rays, high-energy electrons, and neutrons. Special holders have been designed for wrist badges and ring badges that are used to estimate the exposure to limited regions of the body. Although film badges furnish a convenient method of personnel monitoring, the difficulties of accurate dosimetry with photographic film limit the accuracy of measured exposures. Other methods for personnel dosimetry include pocket ionization chambers and thermoluminescent and photoluminescent dosimeters.

Personnel monitors such as badges do not provide a complete dosimetric picture of individual exposure. However, effective dose may be estimated from them. They are usually calibrated to estimate dose at a depth of 10 mm in soft tissue. This concept is referred to in federal radiation safety regulations as *deep dose equivalent*.

Calculation of effective dose from badge readings is of particular concern for personnel who perform fluoroscopy on a regular basis and who may approach equivalent dose limits. An extensive review of the relationships among effective does equivalent, x-ray tube output, apron thickness, and geometric considerations in neurointerventional fluoroscopy environments[12] has yielded the following equation[13]:

$$E = 0.5H_W + 0.025H_N \tag{28-9}$$

TABLE 28-11 Typical Bone Marrow Doses for Various Radiographic Examinations

X-Ray Examination	*Mean Marrow Dose (mrad)*
Skull	10
Cervical	20
Chest	2
Stomach and upper gastrointestinal	100
Gallbladder	80
Lumbar spine	60
Intravenous urography	25
Abdomen	30
Pelvis	20
Extremity	2

Source: Ballinger, P. W. *Merrill's Atlas of Radiographic Positions and Radiologic Procedures,* Vol. 1, 8th edition. St Louis, Mosby, 1995.

where E is effective dose, H_W is the deep dose equivalent obtained from a badge worn at the waist under an apron, and H_N is the deep dose equivalent obtained from a badge worn at the neck outside the apron. If a badge is worn at the neck only, the effective dose may be estimated from

$$E = 0.048H_N \tag{28-10}$$

Equations (28-9) and (28-10) are intended to provide an estimate of effective dose that is two to three times the actual effective dose received by the badged worker.

PROBLEMS

*28-1. What is the maximum permissible accumulated effective dose equivalent for a 30-year-old occupationally exposed individual?

*28-2. A nuclear medicine technologist received an estimated 200 mrem whole-body dose equivalent during the first 2 months of pregnancy. According to the NCRP, what is the maximum permissible dose equivalent over the remaining 7 months of pregnancy?

*28-3. The exposure rate constant Γ_∞ is 0.22 R-m^2/hr-Ci for ^{131}I. What is the exposure over a 40-hour week at a distance of 2 m from 200 mCi of ^{131}I?

*28-4. For a workload of 750 mA-min/wk for a dedicated 125-kVp chest radiographic unit, determine the shielding required behind the chest cassette at a distance of 6 ft from the x-ray tube if an office with uncontrolled access is behind the cassette.

*28-5. For the chest radiographic unit in Problem 28-4, determine the shielding required to protect against radiation scattered to a wall 2 m from and at right angles to the chest cassette if the area behind the wall has uncontrolled access with an occupancy of one.

*28-6. Repeat the computation in Problem 28-5 if film with a turnover time of 2 weeks is stored behind the wall.

*28-7. For the x-ray tube described by the anode thermal characteristics chart in Margin Figure 28-5, determine the shielding required for leakage radiation at a wall 2 m from the x-ray tube if the area behind the wall has uncontrolled access with an occupancy of one. The x-ray tube is rated at 125 kVp maximum.

*28-8. For the wall in Problem 28-7, 1.0 mm of Pb is required to shield against scattered radiation. What thickness of shielding is required to protect against both scattered and radiation leakage?

*For those problems marked with an asterisk, answers are provided on p. 493.

SUMMARY

- The dose limits recommended by the NCRP include:
  - Stochastic limits in terms of effective dose
  - Organ limits in terms of equivalent dose
- The United Nations Scientific Committee on the Effects of Atomic Radiation estimates that the lifetime cancer risk from radiation received during the fetal stage is two to three times that of the adult (i.e., approximately 10% per Sv).
- After a pregnancy is declared by a worker, an equivalent dose limit of 0.5 mSv per month is recommended for the fetus.
- An estimate of occupational effective dose E can be derived from monitor badge deep dose equivalent readings at the waist, H_W, and at the neck, H_N by the following equation:

$$E = 0.5H_W + 0.025H_N$$

If only the neck badge is worn, then the estimate is

$$E = 0.048H_N$$

- Groups that provide recommendations and information on the subject of radiation protection include:
  - National Council on Radiation Protection and Measurements (NCRP)
  - International Commission on Radiological Protection (ICRP)

- United Nations Scientific Committee on the Effects of Atomic Radiation (UNSCEAR)
- National Academy of Sciences (NAS)
- Conference of Radiation Control Program Directors (CRCPD)

- Groups that have legal authority to regulate manufacture and use of sources of ionizing radiation include:
  - Nuclear Regulatory Commission (NRC)
  - State Health Departments
  - Environmental Protection Agency (EPA)
  - Food and Drug Administration (FDA)
- The ***collective effective dose*** is defined as the average dose estimated for individuals in the population times the total number of individuals in the population. It is used to predict the magnitude of stochastic effects in a population.
- The ***genetically significant dose is*** an estimate of the "genetic impact" of radiation exposures to a population. It is based upon estimates of the fraction of the population that undergo diagnostic and nuclear medicine exams each year, in addition to all other sources of background exposure. It is weighted by the age and sex distribution of the population to take into account the expected number of children that would be produced by each category during the remainder of their lifetimes.
- The NCRP has based its process of recommending limits on radiation exposure with the goal of maintaining the risk of genetic effects, fatal cancers, and nonfatal cancers at the same level as that found in "safe" industries. Their estimates show that the lifetime risk from an average worker's exposure values over a working career is approximately 1 in 10,000. Maximum exposure levels would result in a lifetime risk of 5 in ten thousand.
- The attenuation factor that must be provided by shielding material is calculated from knowledge of workload W, use factor U, occupancy factor T, and distance-d as follows:

$$B = \frac{X_p d^2}{WUT}$$

REFERENCES

1. Claus, W. The Concept and Philosophy of Permissible Dose to Radiation, in Claus, W. (ed.), *Radiation Biology and Medicine*. Reading, MA, Addison-Wesley, 1958, p. 389.
2. National Council on Radiation Protection and Measurements: *Limitation of Exposure to Ionizing Radiation,* Report No. 116. Bethesda, Maryland, National Council on Radiation Protection and Measurements, 1993.
3. International Commission on Radiological Protection: *1990 Recommendations of the International Commission on Radiological Protection,* ICRP Publication No. 60, Annals of the ICRP 21. Elmsford, New York, Pergamon Press, 1991.
4. Food and Drug Administration: *Recommendations for Evaluation of Radiation Exposure from Diagnostic Radiology Examinations,* HHS Publication (FDA) 85-8247. Springfield, VA, National Technical Information Service, 1985.
5. Environmental Protection Agency. Radiation protection guidance to federal agencies for diagnostic x rays. *Fed. Reg.* 1978; **43**(22):4377.
6. National Council on Radiation Protection and Measurements: *Limitation of Exposure to Ionizing Radiation,* Report No. 116. Bethesda, Maryland, National Council on Radiation Protection and Measurements, 1993, p. 36.
7. United Nations Scientific Committee on the Effects of Atomic Radiation. *Genetic and Somatic Effects of Ionizing Radiation,* Report to the General Assembly with Annexes. United Nations Publications, New York, 1986.
8. International Commission on Radiological Protection. *Cost–Benefit Analysis in the Optimization of Radiation Protection,* ICRP Publication No. 37. New York, Pergamon Press, 1977.
9. National Council on Radiation Protection: *Medical X-Ray, Electron Beam and Gamma-Ray Protection for Energies up to 50 MeV (Equipment Design, Performance and Use)*. Recommendations of the NCRP, Report No. 102. Washington, D.C., 1989.
10. Appleton, G., and Krishnamoorthy, P. *Safe Handling of Radioisotopes: Health Physics Addendum*. Vienna, International Atomic Energy Agency, 1960.
11. *Structural Shielding Design and Evaluation for Medical Use of X Rays and Gamma Rays of Energies Up to 10 MeV*. Recommendations of the NCRP, Report No. 49. Washington, D.C., 1976.
12. Faulkner, K., and Marshall, N. W. The relationship of effective dose to personnel and monitor reading for simulated fluoroscopic irradiation conditions. *Health Phys.* 1993; **64**:502–508.
13. Webster, E. W. EDE for exposure with protective aprons. *Health Phys.* 1989; **56**:568–569.

CHAPTER

29

PROTECTION FROM INTERNAL SOURCES OF RADIATION

OBJECTIVES

After studying this chapter, the reader should be able to:

- Define the expressions committed dose equivalent, annual limit on intake, and derived air concentration.
- Distinguish between effective half-life for uptake and effective half-life for elimination, and identify the influencing variables on each.
- Identify the factors that influence the dose rate to an organ from an internally deposited radionuclide.
- Define the absorbed fraction, the absorbed dose constant, cumulative activity, and mean absorbed dose per cumulative activity.
- Explain the concept of "locally absorbed radiation" and the types of radiation it encompasses.
- Perform a few simple internal dose computations.

INTRODUCTION

Radioactive substances can enter the body accidentally by mechanisms of absorption, inhalation and ingestion.

Hazards associated with internal sources of radiation are best controlled by minimizing the absorption, inhalation, and ingestion of radioactive materials. Protection standards for internal sources of radiation are based on the concept of *committed dose equivalent* (CDE).

COMMITTED DOSE EQUIVALENT

The *residence time* of a radioactive nuclide in an organ is the time that would be required to deliver the radiation dose to an organ if the initial activity in the organ remained constant until it had been completely eliminated, at which time it would decrease precipitously to zero.

The radiation dose to an organ from a radioactive nuclide deposited in the body is delivered over a period of time as the nuclide is eliminated from the body. The time required for the nuclide to be eliminated by biological processes or radioactive decay depends on the physical characteristics of the nuclide and its chemical form. To cover a wide range of radiopharmaceuticals the International Commission on Radiological Protection (ICRP) has defined[1] the *committed dose equivalent* (CDE) for an organ as the dose accumulated in the organ over a 50-year period. The CDE is obtained by integrating (summing) the dose equivalent rate over time. The dose equivalent rate in an organ is determined from knowledge of the spatial distribution of the nuclide, the type of radiation emitted, and the range of the radiation in tissue. Finally, the committed effective dose equivalent (CEDE) (see Chapter 28) may be calculated by multiplying the CDE by the weighting factor for the organ. The committed effective dose equivalent for the entire body is determined by summing the CEDE for all organs containing the radioactive nuclide.

To prevent workers from accumulating internal nuclides that would exceed an acceptable CDE, limits have been set for the amount of various nuclides that may be taken into the body during any 1 year. These annual limits on intake (ALI) are based upon mathematical models for retention and excretion of radionuclides from the adult human body.[2,3] Some examples of ALIs are given in the margin.

Annual Limits on Intake for Selected Radionuclides

Radionuclide	*GBq (yr)*
Hydrogen 3	1.8
Carbon 14	0.088
Phosphorus 32	0.036
Technetium 99m	5.3
Iodine 131	0.0018
Xenon 133	8.8

The workplace where radioactive materials are handled must be designed so that workers do not exceed the intake and dose limits. It is not practical to monitor the actual uptake in workers continuously for all nuclides that may be present in the workplace. It is also not consistent with the principles of as low as reasonably achievable (ALARA) to base safety procedures solely upon a response to accidental exposures. It is more appropriate to design the workplace so that concentrations of potential contaminants are kept well below levels that could lead to values near the ALI. To aid in this approach to radiation protection, the derived air concentration (DAC) is defined as the in-air concentration of a radionuclide that would exceed the ALI if breathed by a reference human[4] for a year. That is, the (DAC) is defined as

$$\text{DAC} = \frac{\text{ALI}}{2.4 \times 10^9 \text{ ml}} \quad (29\text{-}1)$$

where the denominator (2.4×10^9 ml) is the average volume of air intake by a worker during 1 year. This air intake is based on the assumption that the worker is present in the workplace for fifty 40-hour weeks per year and that he or she breathes at a rate of 2×10^4 ml/min.

DACs for the workplace may be determined for any nuclide. Similar procedures may be used for other uptake pathways such as food or drinking water.[4]

The DAC is similar to but supersedes an older concept of maximum permissible concentration (MPC). See reports of the National Council on Radiation Protection and Measurement (NCRP) and ICRP[1,2] for a comparison of these concepts.

Derived Air Concentrations for Selected Radionuclides (Occupational)

Radionuclide	*Bq/m³(×10⁻⁴)*
Hydrogen 3 (HTO water)	74
Carbon 14 ($^{14}CO_2$ gas)	330
Phosphorus 32	0.74
Iodine 131	0.074
Xenon 133	370

ESTIMATING INTERNAL DOSE

The concentration of a radioactive nuclide in a particular organ changes with time after intake of the nuclide into the body. The change in concentration reflects not only radioactive decay of the nuclide but also the influence of physiologic processes that move chemical substances into and out of organs in the body. The effective half-life for uptake or elimination of a nuclide in a particular organ is the time required for the nuclide to increase (uptake) or decrease (elimination) to half of its maximum concentration in the organ. The *effective half-life for uptake* (T_{up}) of a radioactive nuclide is computed from the half-life T_1 for physiologic uptake, excluding radioactive decay, and the half-life $T_{1/2}$ for radioactive decay. The *effective half-life for elimination* (T_{eff}) of a radioactive nuclide from an organ is computed from the half-life T_b for physiologic elimination, excluding radioactive decay, and the half-life for radioactive decay. Expressions for the effective half-life for elimination and for uptake are stated in Eqs. (29-2) and (29-3):

The Medical Internal Radiation Dose [MIRD] committee of the Society of Nuclear Medicine recommends that the term "half-life" should be used to describe radioactive decay of a nuclide and that the term "half-time" should be used to describe biological elimination (i.e., biological half-time) and the net effect of both radioactive decay and biological elimination (i.e., effective half-time). This recommendation has not been widely adopted and is not used in this text.

Decay constant for elimination of nuclide = Decay constant for physiologic elimination + Decay constant for radioactive decay

$$\lambda_{eff} = \lambda_b + \lambda_{1/2}$$

In computations of internal dose, $\lambda_{1/2}$ and $T_{1/2}$ are often expressed as λ_p and T_p to represent physical (radioactive) decay

The decay constant $\lambda_{1/2}$ equals $0.693/T_{1/2}$. Similarly, $\lambda_b = 0.693/T_b$ and $\lambda_{eff} = 0.693/T_{eff}$, where T_b is the half-life for physiologic elimination, excluding radioactive decay, and T_{eff} is the effective half-life for elimination:

$$\frac{0.693}{T_{eff}} = \frac{0.693}{T_b} + \frac{0.693}{T_{1/2}}$$

$$\frac{1}{T_{eff}} = \frac{1}{T_b} + \frac{1}{T_{1/2}}$$

$$T_{eff} = \frac{T_b T_{1/2}}{T_b + T_{1/2}} \quad (29\text{-}2)$$

The effective half-life T_{eff} for elimination of a radioactive nuclide in an organ is the time required for the activity to diminish to half by both physical decay and biological elimination.

An expression for the effective half-life T_{up} for uptake may be derived by a similar procedure. The effective half-life T_{up} for uptake of a radioactive nuclide in an organ is the time required for the activity to increase to half of its maximum value.

$$T_{up} = \frac{T_1 T_{1/2}}{T_1 + T_{1/2}} \quad (29\text{-}3)$$

where T_1 is the half-life for biologic uptake, excluding radioactive decay.

T_{eff} is always less than either T_b or $T_{1/2}$ (however, $T_{eff} \simeq T_b$ when $T_{1/2}$ is very long, and $T_{eff} \simeq T_{1/2}$ when T_b is very long).

Example 29-1

^{35}S in soluble form is eliminated from its critical organ, the testes, with a biologic half-life of 630 days. The half-life is 87 days for radioactive decay of ^{35}S. What is the effective half-life for elimination of this nuclide from the testes?

$$T_{eff} = \frac{T_b T_{1/2}}{T_b + T_{1/2}} = \frac{(630 \text{ days})(87 \text{ days})}{(630 \text{ days}) + (87 \text{ days})} = 76 \text{ days}$$

T_{up} is always less than either T_1 or $T_{1/2}$ (except $T_{up} \simeq T_1$ when $T_{1/2}$ is very long, and $T_{up} \simeq T_{1/2}$ when T_1 is very long).

Example 29-2

^{123}I is absorbed into the thyroid with a half-life of 5 hours for physiologic uptake. The half-life for radioactive decay is 13 hours for ^{123}I. What is the effective half-life for uptake of ^{123}I into the thyroid?

$$T_{up} = \frac{T_1 T_{1/2}}{T_1 + T_{1/2}} = \frac{(5 \text{ hr})(13 \text{ hr})}{(5 \text{ hr}) + (13 \text{ hr})} = 3.6 \text{ hr}$$

The metabolism, biodistribution, and excretion of pharmaceuticals is different in children compared with adults. Various approaches based on surface area and body weight have been recommended to adjust for the differences between children and adults. One set of recommendations for the fraction of the adult dose of a radiopharmaceutical that should be administered to children has been developed by the Paediatric Task Group of the European Association of Nuclear Medicine.[6] These recommendations are shown below:

Fraction of Adult Doses for Pediatric Administration

Weight in kg (1b)	*Fraction*
3 (6.6)	0.10
4 (8.8)	0.14
8 (17.6)	0.23
10 (22.0)	0.27
12 (26.4)	0.32
14 (30.8)	0.36
16 (35.2)	0.40
18 (39.6)	0.44
20 (44.0)	0.46
22 (48.4)	0.50
24 (52.8)	0.53
26 (57.2)	0.56
28 (61.6)	0.58
30 (66.0)	0.62
32 (70.4)	0.65
34 (74.8)	0.68
36 (79.2)	0.71
38 (83.6)	0.73
40 (88.0)	0.76
42 (92.4)	0.78
44 (96.8)	0.80
46 (101.2)	0.83
48 (105.6)	0.85
50 (110.0)	0.88

RADIATION DOSE FROM INTERNAL RADIOACTIVITY

Occasionally, radioactive nuclides are inhaled, ingested, or absorbed accidentally by individuals working with or near radioactive materials. Also, radioactive nuclides are administered orally, intravenously, and occasionally by inhalation, for the diagnosis and treatment of patient disorders. The radiation dose delivered to various organs in the body should be estimated for any person with a substantial amount of internally deposited radioactive nuclide.

Standard Man

To compute the radiation dose delivered to an organ by radioactivity deposited within the organ, the mass of the organ must be estimated. The ICRP has developed a "standard man," which includes estimates of the average mass of organs in the adult.[5] These estimates are included in Table 29-1.

Measurement of Internal Dose

Body burdens for γ-emitting isotopes may be measured with a whole-body counter. Shown in Figure 29-1 is a whole-body counter that includes a large NaI(T1) crystal and a 512–channel analyzer. The person whose body burden is to be counted sits for a few minutes in a tilted chair directly under the crystal. The room is heavily shielded to reduce background radiation.

The pulse height spectrum in Figure 29-2 was obtained during a 40-minute count of a nuclear medicine technologist. The percent abundance is 0.012 for ^{40}K in natural potassium, and a photopeak indicating the presence of this nuclide occurs in whole-body pulse height spectra for all persons. The ^{137}Cs photopeak reflects the accumulation of this nuclide from radioactive fallout of nuclear explosions. The low-energy photopeak represents ^{99m}Tc, which was probably deposited internally as the technologist "milked" a ^{99m}Tc generator or worked with a ^{99m}Tc solution.

Various "bioassay" techniques exist to measure the actual amount of radioactive material present in an individual. These techniques include sampling urine,

TABLE 29-1 Average Mass of Organs in the Adult Human Body

Organ	*Mass (g)*	*Percentage of Total Body*
Total body[a]	70,000	100
Muscle	30,000	43
Skin and subcutaneous tissue[b]	6,100	8.7
Fat	10,000	14
Skeleton		
Without bone marrow	7,000	10
Fed marrow	1,500	2.1
Yellow marrow	1,500	2.1
Blood	5,400	7.7
Gastrointestinal tract	2,000	2.9
Contents of GI tract		
Lower large intestine	150	
Stomach	250	
Small intestine	1,100	
Upper large intestine	135	
Liver	1,700	2.4
Brain	1,500	2.1
Lungs (2)	1,000	1.4
Lymphoid tissue	700	1.0
Kidneys (2)	300	0.43
Heart	300	0.43
Spleen	150	0.21
Urinary bladder	150	0.21
Pancreas	70	0.10
Salivary glands (6)	50	0.071
Testes (2)	40	0.057
Spinal cord	30	0.043
Eyes (2)	30	0.043
Thyroid gland	20	0.029
Teeth	20	0.029
Prostate gland	20	0.029
Adrenal glands or suprarenal (2)	20	0.029
Thymus	10	0.014
Ovaries (2)	8	0.011
Hypophysis (pituitary)	0.6	8.6×10^{-6}
Pineal gland	0.2	2.9×10^{-6}
Parathyroids (4)	0.15	2.1×10^{-6}
Miscellaneous (blood vessels, cartilage, nerve, etc.)	390	0.56

[a] Does not include contents of the gastrointestinal tract.

[b] The mass of the skin alone is about 2,000 g.

Source: International Commission on Radiological Protection.[5] Used with permission.

The U.S. Nuclear Regulatory Commission (NRC Regulatory Guide 8.20) requires bioassays (thyroid uptake measurements in the case of radioiodine) of individuals working with volatile iodine radionuclides when the level of radioactivity handled exceeds the following amounts:

Open bench:	1 mCi (37 MBq)
Fume hood:	10 mCi (370 Bq)
Glove box:	100 mCi (3.7 GBq)

feces, and hair, and in postmortem samples they include direct counting of internal organs.

Internal Absorbed Dose

Consider a homogeneous mass of tissue containing a uniform distribution of a radioactive nuclide. The instantaneous dose rate to the tissue depends on three variables:

1. The concentration C in becquerels per kilogram of the nuclide in the tissue.
2. The average energy $\bar{E}$ in MeV released per disintegration of the nuclide.
3. The fraction φ of the released energy that is absorbed in the tissue.

The fraction ϕ is known as the *absorbed fraction*. Its value for any particular tissue or organ depends on the type and energy of the radiation and the size, shape, and location of the tissue or organ with respect to the radiation source.

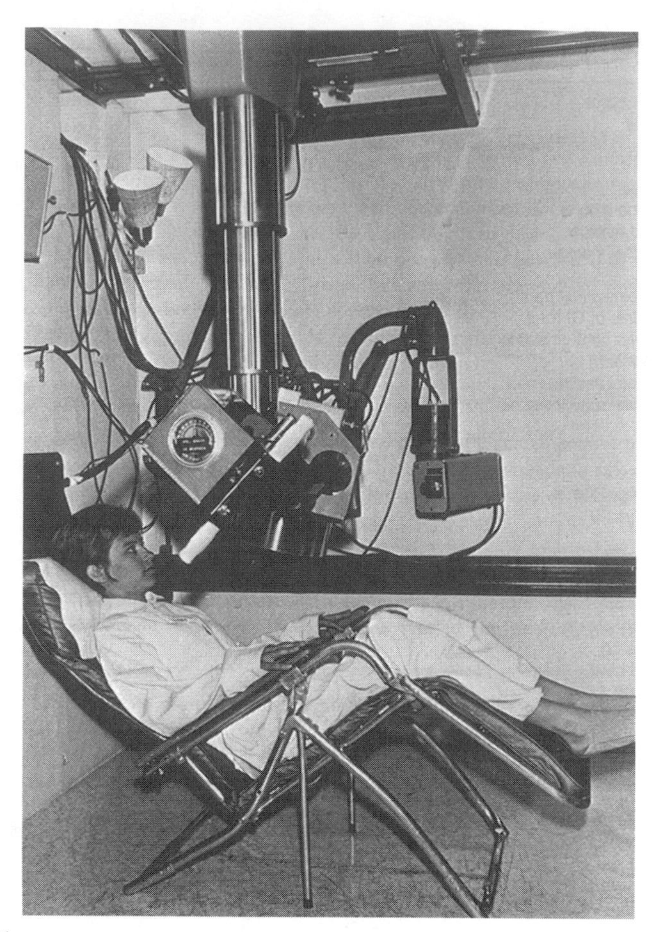

FIGURE 29-1
Detector assembly and tilting chair for a whole-body counter. (Courtesy of Colorado Department of Health.)

The average energy $\bar{E}$ released per disintegration may be estimated as

$$\bar{E} = n_1 E_1 + n_2 E_2 + n_3 E_3 + \cdots$$

where $E_1, E_2, E_3, \ldots$ are the energies of different radiations contributing to the absorbed dose and $n_1, n_2, n_3, \ldots$ are the fractions of disintegrations in which the corresponding radiations are emitted. The average energy $\bar{E}_{\text{abs}}$ absorbed per disintegration is

$$\bar{E}_{\text{abs}} = n_1 E_1 \varphi_1 + n_2 E_2 \varphi_2 + n_3 E_3 \varphi_3 + \cdots$$

In internal dose computations, an organ or volume of tissue for which the radiation dose is to be estimated is referred to as the *target organ*. The volume of tissue or organ from which the radiation is emanating is known as the *source organ*. Often the target and source organ are identical.

where $\varphi_1, \varphi_2, \varphi_3, \ldots$ are the fractions of the emitted energies that are absorbed (i.e., the absorbed fractions). A shorthand notation for the addition process is

$$\begin{aligned}\bar{E}_{\text{abs}} &= n_1 E_1 \varphi_1 + n_2 E_2 \varphi_2 + n_3 E_3 \varphi_3 + \cdots \\ &= \sum n_i E_i \varphi_i\end{aligned}$$

where $\sum n_i E_i \varphi_i$ indicates the sum of the products $n E \varphi_i$ for each of the "i" radiations. The instantaneous dose rate $\dot{D}$ may be written as

Often the rate of change (e.g., the radiation dose rate) in a quantity (e.g., the radiation dose) is depicted by a "dot" (e.g., $\dot{D}$) above the symbol for the quantity. This nomenclature is followed in the present chapter.

$$\begin{aligned}\dot{D}(\text{Gy/sec}) &= C(\text{Bq/kg})\bar{E}_{\text{abs}}(\text{MeV/dis})(1.6 \times 10^{-13}\ \text{J/MeV}) \\ &= 1.6 \times 10^{-13} C \sum n_i E_i \varphi_i \end{aligned} \qquad \textbf{(29-4)}$$

By representing $1.6 \times 10^{-13}\ n_i E_i$ by the symbol Δ_i, Eq. (29-3) may be written

$$\dot{D}(\text{Gy/sec}) = C \sum \Delta_i \phi_i \qquad \textbf{(29-5)}$$

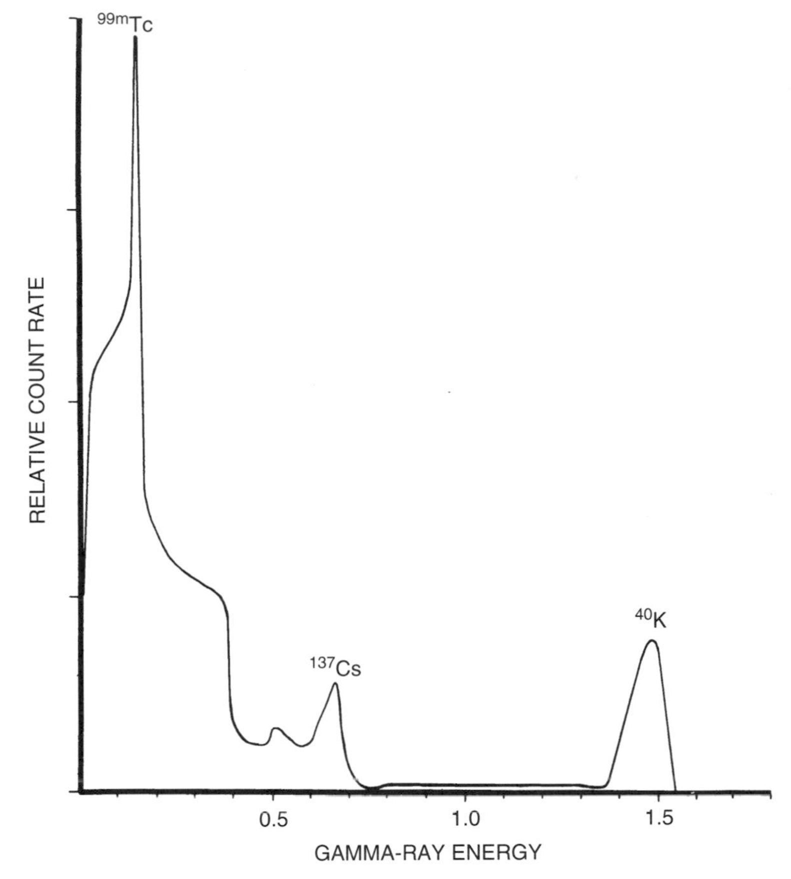

FIGURE 29-2
Pulse height spectrum furnished by a whole-body counter. (Courtesy of Colorado Department of Health.)

where Δ_i is the mean energy emitted per nuclear transition in units of Gy-kg/Bq-sec.

Locally absorbed radiations are those that are absorbed within 1 cm of their origin. For these radiations, the absorbed fraction is usually taken as 1. Locally absorbed radiations include (1) α-particles, negatrons, and positrons; (2) internal conversion electrons; (3) Auger electrons; and (4) x and γ rays with energies less than 11.3 keV.[7] Characteristic x rays considered to be locally absorbed radiations include K-characteristic x rays from elements with $Z < 35$, L-characteristic x rays from elements with $Z < 85$, and M-characteristic x rays from all elements.

Radiations that are not absorbed locally have an absorbed fraction less than 1. Generally, these radiations are x and γ rays with energies of 11.3 keV or greater. Illustrated in Table 29-2 are absorbed fractions for 140-kev γ rays from ^{99m}Tc deposited in various organs of the standard man.[8] More complete data on absorbed fractions, as well as extensive listings of absorbed dose constants, are available in publications of the Medical Internal Radiation Dose (MIRD) committee, Society of Nuclear Medicine (1850 Samuel Morse Dr. Reston, VA 20190). Included in the MIRD compilations are absorbed fractions for organs at some distance from an accumulation of radioactivity. For these situations, the absorbed fraction is zero for locally absorbed radiations and less than 1 for more penetrating radiations.

The instantaneous dose rate varies with changes in the concentration C of the nuclide in the mass of tissue. If the nuclide is eliminated exponentially, then

$$C = C_{\max}\lfloor e^{-\lambda_{\text{eff}}t}\rfloor$$
$$= C_{\max}[e^{-0.693t/T_{\text{eff}}}]$$

The symbol Δ_i is referred to as the *absorbed dose constant* for a specific transition in a particular radioactive nuclide. In some publications, Δ_i is termed the *equilibrium dose constant*.

X and γ rays with energies below 11.3 keV have a probability >95% of being absorbed in 1 cm of muscle.

For locally absorbed radiations, $\phi = 1$ when the radioactivity is in the target organ, and $\phi = 0$ when the radioactivity is outside the target organ.

TABLE 29-2 Absorbed Fractions for 140-keV γ Rays from ^{99m}Tc Deposited in Various Organs of the Standard Man

Source and Target Organ	*Absorbed Fraction*	*Source and Target Organ*	*Absorbed Fraction*
Bladder	0.102	Lungs	0.421×10^{-1}
Brain	0.139	Ovaries	0.181×10^{-1}
Gastrointestinal tract (stomach)	0.863×10^{-1}	Pancreas	0.379×10^{-1}
Gastrointestinal tract (small intestine)	0.128×10^{-1}	Spleen	0.660×10^{-1}
Gastrointestinal tract (upper large intestine)	0.710×10^{-1}	Testicles	0.389×10^{-1}
Gastrointestinal tract (lower large intestine)	0.569×10^{-1}	Thyroid	0.280×10^{-1}
Kidneys	0.614×10^{-1}	Total body	0.306×10^{-1}
Liver	0.134×10^{-1}		

Source: Snyder, W., et al.[9] Used with permission.

If the assumption of exponential elimination is too great a simplification, then the elimination of the nuclide from the tissue or organ must be described by a more complex expression.

Often an organ can be considered as several compartments, with different concentrations of the radionuclide in each compartment. The mathematics of describing elimination of a radionuclide from such an organ is termed "compartmental kinematics."

The mathematics of internal dose computations presented here are greatly simplified. A more complete analysis would require the use of differential and integral calculus.

where t is the time elapsed since the concentration was at its maximum value C_{max}. With exponential elimination,

$$\dot{D}(\text{Gy/sec}) = C_{max}[e^{-0.693t/T_{eff}}] \sum \Delta_i \varphi_i$$

For a finite interval of time t that begins at $t = 0$ when $C = C_{max}$, the cumulative absorbed dose D is

$$D(\text{Gy}) = C_{max}(1.44T_{eff})(1 - e^{-0.693t/T_{eff}}) \sum \Delta_i \varphi_i \tag{29-6}$$

with t and T_{eff} expressed in seconds and with $1.44T_{eff}$ termed the average life for elimination of the nuclide.

The total absorbed dose during complete exponential elimination of the nuclide from the mass of tissue is obtained by setting $t = \infty$ in Eq. (29-6). The exponential term $e^{-0.693t/T_{eff}}$ approaches zero, and

$$D(\text{Gy}) = C_{max}(1.44T_{eff}) \sum \Delta_i \varphi_i \tag{29-7}$$

That is, the total cumulative dose depends on the three variables that affect the dose rate (concentration, energy released per disintegration, and fraction of the released energy that is absorbed), together with the time (represented by T_{eff}) that the nuclide is present in the tissue.

If the effective half-life for uptake of a radionuclide is short compared with the effective half-life for elimination, Eq. (29-6) may be used to estimate the total dose delivered from the moment the nuclide is introduced into tissue until the time it is eliminated completely. If the uptake of the nuclide must be considered, then it may be reasonably accurate to assume that the uptake is exponential. If elimination of the nuclide is also exponential, then the concentration C at some time t is

$$C = C_{max}(e^{-0.093t/T_{eff}} - e^{-0.693t/T_{up}})$$

From this expression and Eq. (29-7), an expression may be derived for the total cumulative absorbed dose:

$$D(\text{Gy}) = C_{max}(1.44T_{eff}) \left(1 - \frac{T_{up}}{T_{eff}}\right) \sum \Delta_i \varphi_i \tag{29-8}$$

with T_{up} and T_{eff} expressed in seconds. If $T_{eff} \geq 20T_{up}$, the effective half-life T_{up} for uptake (i.e., the term $[1 - T_{up}/T_{eff}]$) may be neglected with an error no greater than 5% in the total cumulative absorbed dose.

A simplification of the MIRD approach to absorbed dose computations has been prepared for radionuclides absorbed in specific internal organs. In this approach, the quotient of $\sum \Delta_i \varphi_i$ divided by the mass of the organ containing the radionuclide is recognized as having a specific value for a particular nuclide, source organ (organ containing the nuclide), and target organ (organ for which the absorbed dose is to be computed). This specific value is symbolized as S, the mean absorbed dose per unit cumulative activity. The absorbed dose may be computed as

$$D(\text{Gy}) = \bar{A}S \tag{29-9}$$

where $\bar{A}$ is the cumulative activity in the organ. For exponential elimination and short uptake times, $\bar{A}$ is simply $A_{max}T_{avg}$, where A_{max} is the maximum activity in the organ and $T_{avg} = 1.44T_{eff}$. Further information on the "S method" of internal dose computation is available in MIRD pamphlet 11, and the MIRD Primer (both available from the Society of Nuclear Medicine). Values of S for ^{99m}Tc in selected organs are shown in Tables 29-3 and 29-4.

TABLE 29-3 Values of $S\left(\frac{\text{Gy}}{\text{MBq-sec}}\right)$ for ^{99m}Tc (Source Organ: Liver)

Target Organ	*S*
Liver	3.5×10^{-9}
Ovaries	3.4×10^{-11}
Spleen	6.9×10^{-11}
Red bone marrow	1.2×10^{-10}

Source: Snyder, W., et al. "*S*," *Absorbed Dose per Unit Cumulated Activity for Selected Radionuclides and Organs*, MIRD pamphlet 11. Reston, VA, Society of Nuclear Medicine. Used with permission.

Example 29-3

Thirty-seven megabecquerels (1 mCi) of $Na_2{}^{35}SO_4$ is given orally to a 70-kg man. Approximately 0.5% of the activity is absorbed into the blood, and 50% of the absorbed activity is rapidly concentrated in the testes. What are the instantaneous dose rate, the dose delivered over the first week, and the total dose delivered to the testes?

From Table 29-1, the mass of the testes is 40 g in the standard man. For ^{35}S, only locally absorbed radiation is emitted, with an average energy of 0.049 MeV per disintegration. The total activity reaching the testes is 37 MBq (0.0025) = 0.093 MBq. The instantaneous dose rate $\dot{D}$ is

$$\begin{aligned}\dot{D} &= 1.6 \times 10^{-13} C \sum n_i E_i \varphi_i \\ &= 1.6 \times 10^{-13}\left(\frac{9.3 \times 10^4 \text{ Bq}}{4 \times 10^{-2} \text{ kg}}\right)(0.049 \text{ Mev}) \\ &= 1.8 \times 10^{-8} \text{Gy/sec}\end{aligned} \tag{29-4}$$

From Example 29-1, the effective half-life is 76 days for elimination of ^{35}S from the testes. Assuming rapid uptake and exponential elimination, the dose delivered over the first week is

$$\begin{aligned}D &= C_{max}(1.44T_{eff})(1 - e^{-0.693t/T_{eff}}) \sum \Delta_i \varphi_i \\ &= \frac{9.3 \times 10^4 \text{ Bq}}{4 \times 10^{-2} \text{ kg}} 1.44(76 \text{ days})(24 \text{ hr/day})(3600 \text{ sec/hr}) \\ &\quad (1 - e^{-0.693(7/76)})(1.6 \times 10^{-13})(0.049 \text{ MeV}) \\ &= 0.011 \text{Gy}\end{aligned} \tag{29-6}$$

TABLE 29-4 Values of S $\left(\frac{\text{Gy}}{\text{MBq-sec}}\right)$ for ^{99m}Tc

	Source Organs			
Target Organ	*Cortical Bone*	*Trabecular Bone*	*Total Body*	*Bladder*
Red bone marrow	3.1×10^{-10}	6.8×10^{-10}	2.2×10^{-10}	1.7×10^{-10}

Source: Snyder, W., et al. "*S*," *Absorbed Dose per Unit Cumulated Activity for Selected Radionuclides and Organs*, MIRD pamphlet 11. Reston, VA, Society of Nuclear Medicine. Used with permission.

The total dose delivered to the testes during complete elimination of the nuclide is

$$
\begin{aligned}
D &= C_{max}(1.44T_{eff}) \sum \Delta_i \varphi_i \\
&= \frac{9.3 \times 10^4 \text{ Bq}}{4 \times 10^{-2} \text{ kg}} 1.44(76 \text{ days})(24 \text{ hr/day}) \\
&\quad (3{,}600 \text{ sec/hr})(1.6 \times 10^{-13})(0.049 \text{ MeV}) \\
&= 0.17 \text{ Gy}
\end{aligned}
\qquad (29\text{-}7)
$$

Example 29-4

Estimate the absorbed dose to the liver, ovaries, spleen, and bone marrow from the intravenous ingestion of 37 MBq (1 mCi) of ^{99m}Tc-sulfur colloid that localizes uniformly and completely in the liver (Table 29-3). Assume instantaneous uptake of the colloid in the liver and an infinite biologic half-life.

$$
\begin{aligned}
\bar{A} &= 1.44 T_{eff} A_{max} \\
&= 1.44(6 \text{ hr})(3600 \text{ sec/hr})(37 \text{ MBq}) \\
&= 1.2 \times 10^6 \text{ MBq-sec} \\
D &= \bar{A} S \\
D_{L_i} &= (1.2 \times 10^6 \text{ MBq-sec}) \left(3.5 \times 10^{-9} \frac{\text{Gy}}{\text{MBq-sec}}\right) \\
&= 4.2 \times 10^{-3} \text{ Gy} \\
D_{Ov} &= (1.2 \times 10^6 \text{ MBq-sec}) \left(3.4 \times 10^{-11} \frac{\text{Gy}}{\text{MBq-sec}}\right) \\
&= 4.1 \times 10^{-5} \text{ Gy} \\
D_{sp} &= (1.2 \times 10^6 \text{ MBq-sec}) \left(6.9 \times 10^{-11} \frac{\text{Gy}}{\text{MBq-sec}}\right) \\
&= 8.3 \times 10^{-5} \text{ Gy} \\
D_{BM} &= (1.2 \times 10^6 \text{ MBq-sec}) \left(1.2 \times 10^{-10} \frac{\text{Gy}}{\text{MBq-sec}}\right) \\
&= 1.4 \times 10^{-4} \text{ Gy}
\end{aligned}
$$

Example 29-5

Estimate the absorbed dose to the bone marrow from an intravenous injection of 37 MBq (1 mCi) of ^{99m}Tc labeled to a compound that localizes 35% in bone ($\bar{A} = 6.0 \times 10^4$ MBq-sec) with the remaining activity excreted in the urine ($\bar{A} = 7.2 \times 10^5$ MBq-sec for the bladder) (Table 29-4). For the activity in bone, assume that half is in cortical bone and half is in trabecular bone.

$$
\begin{aligned}
D &= \bar{A} S \\
D_{BM} &= 0.5(4.0 \times 10^5 \text{ MBq-sec})(3.1 \times 10^{-10}) \\
&= 6.2 \times 10^{-5} \text{ Gy from activity in cortical bone} \\
&= 0.5(4.0 \times 10^5 \text{ MBq-sec})(6.8 \times 10^{-10}) \\
&= 13.7 \times 10^{-5} \text{ Gy from activity in trabecular bone} \\
&= (6.0 \times 10^4 \text{ MBq-sec})(2.2 \times 10^{-10}) \\
&= 1.2 \times 10^{-5} \text{ Gy from activity in the total body}
\end{aligned}
$$

$$= (7.2 \times 10^4 \text{ MBq-sec})(1.7 \times 10^{-10})$$
$$= 1.3 \times 10^{-5} \text{ Gy from activity in the bladder}$$

The total dose to the bone marrow is

$$(D_{BM})_{total} = (6.2 + 13.7 + 1.2 + 1.3) \times 10^{-5} \text{ Gy}$$
$$= 22.4 \times 10^{-5} \text{ Gy} = 224 \ \mu\text{Gy}$$

RECOMMENDATIONS FOR SAFE USE OF RADIOACTIVE NUCLIDES

Procedures to minimize the intake of radioactive nuclides into the body depend on the facilities available within an institution and vary from one institution to another. A few guidelines for the safe use of radioactive materials are included in publications by individuals[10,11] and by advisory groups such as the ICRP and NCRP.[12–18]

SUMMARY

- The committed dose equivalent (CDE) is the dose accumulated in an organ over a 50-year period.
- The derived air concentration (DAC) and annual limit on intake [ALI] for a radioactive nuclide are values computed so that the CDE for an individual does not exceed accepted limits.
- The expressions T_{up} and T_{eff} related to the uptake and elimination of a radioactive nuclide in an organ are computed from the half-life for physiologic uptake (T_1), the half-life for physiologic elimination (T_b), and the half-life for radioactive decay ($T_{1/2}$).
- Factors that influence the dose rate from an internally deposited radionuclide include the concentration C, average energy E per disintegration, and absorbed fraction.
- The radiation dose rate to an organ may be written

$$\dot{D}(\text{Gy/sec}) = C \sum \Delta_i \varphi_i$$

  where Δ_i is the absorbed dose constant and φ_i is the absorbed fraction.
- Locally absorbed radiations include: α-particles, negatrons, and positrons; internal conversion electrons; Auger electrons; and very low energy x and γ rays.
- Most dose computations are performed with the sometimes oversimplified assumption of exponential elimination of the radionuclide from an organ.
- The MIRD committee of the Society of Nuclear Medicine has developed a simplified approach to internal dose computations that uses two variables, namely, the cumulative activity $\tilde{A}$ and the mean absorbed dose per unit cumulative activity S.

REFERENCES

1. International Commission on Radiological Protection. *Recommendations of the International Commission on Radiological Protection,* ICRP Publication No. 26. Elmsford, NY, Pergamon Press, 1977.
2. National Council on Radiation Protection and Measurements. *General Concepts for the Dosimetry of Internally Deposited Radionuclides,* Report No. 84. Washington D.C., NCRP, 1985.
3. International Commission on Radiological Protection. *Limits for Intakes of Radionuclides by Workers,* ICRP Publication No. 30. Elmsford, NY, Pergamon Press, 1980.
4. Environmental Protection Agency. *Limiting Values of Radionuclide Intake and Air Concentration, and Dose Conversion Factors for Inhalation, Submersion and Ingestion.* Federal Guidance Report No. 11, USEPA Report EPA-520/1-88-020. Washington, D.C., September 1988.
5. International Commission on Radiological Protection. *Reference Man: Anatomical Physiological and Metabolic Characteristics,* Report No. 23. Elmsford, NY, Pergamon Press, 1975.
6. Saha, G. *Physics and Radiobiology of Nuclear Medicine,* 2nd edition. New York, Springer-Verlag, 2001.

7. Hendee, W. *Radioactive Isotopes in Biological Research*. New York, John Wiley & Sons, 1973.
8. International Commission on Radiological Protection. *Report of the Task Group on Reference Man,* ICRP Publication No. 23. New York, Pergamon Press, 1975.
9. Snyder, W. S., Fisher, H. L., Jr., Ford, M. R., and Warner, G. G. Estimates of absorbed fractions for monoenergetic photon sources uniformly distributed in various organs of a heterogeneous phantom. *J. Nucl. Med.* 1969; **10**(Suppl. 3):7–52.
10. Owunwanne, A., Patel, M., and Sadek, S. *The Handbook of Radiopharmaceuticals*. New York, Chapman & Hall, 1995.
11. Moore, M., and Hendee, W. *Radionuclide Handling and Radiopharmaceutical Quality Assurance, Workshop Manual*. Washington, D.C., Bureau of Radiological Health, USDHEW-FDA, 1977.
12. Shapiro, J. *Radiation Protection,* 3rd edition. Cambridge, MA, Harvard University Press, 1990.
13. International Commission on Radiological Protection. *Report of Committee V (1953–62): Handling and Disposal of Radioactive Materials in Hospitals and Medical Research Establishments,* ICRP Publication No. 5. New York, Pergamon Press, 1964.
14. International Commission on Radiological Protection. *Protection of the Patient in Radionuclide Investigations,* ICRP Publication No. 17. New York, Pergamon Press, 1971.
15. International Commission on Radiological Protection. *Radiation Doses to Patients from Radiopharmaceuticals,* ICRP Publication No. 80. New York, Pergamon Press, 1999.
16. National Council on Radiation Protection and Measurements. *Precautions in the Management of Patients Who have Received Therapeutic Amounts of Radionuclides. Recommendations of the NCRP,* Report No. 37. Washington, D.C., NCRP, 1970.
17. National Council on Radiation Protection and Measurements. *Protection Against Radiation from Brachytherapy Sources. Recommendations of the NCRP,* Report No. 40. Washington, D.C., NCRP, 1972.
18. National Council on Radiation Protection and Measurements. *Radiation Protection of Allied Health Personnel,* Report No. 105. Bethesda, MD, NCRP, 1989.

CHAPTER

30

FUTURE DEVELOPMENTS IN MEDICAL IMAGING

OBJECTIVES

From review of this chapter, the reader should be able to:

- Describe the potential and challenges of several new imaging technologies.
- Define the purpose of an Information Management and Communications Systems (IMACS).
- Explain some of the challenges to implementation of an IMACS.
- Delineate the ultimate advantage of radiologists over other medical specialists in the interpretation of medical images.

INTRODUCTION

The evolution of medical imaging has been nothing short of spectacular over the past three decades. It is hard to imagine that 30 years ago technologies such as x-ray computed tomography, functional nuclear medicine, real-time, gray-scale, and Doppler ultrasonography, single-photon and positron emission computed tomography, magnetic resonance imaging, interventional angiography, and digital radiography were unavailable in the clinical setting. Because of these advances, it is tempting to focus backward on how much has happened, as if the major changes have now all occurred, and we can finally sit back, catch our breath, and fine-tune medical imaging to make it work even better. That posture, however, is unrealistic. The evolution of medical imaging is the product of a revolution in the way fundamental science is being applied to clinical medicine. The revolution is still occurring, and anyone who thinks it is not is in jeopardy of being bypassed by events yet to come.

NEW IMAGING TECHNOLOGIES

It is difficult to appreciate the possibility and potential of imaging technologies that have not yet been invented or evaluated clinically. Several additional technologies are being explored, and in some cases their potential usefulness is promising. A few of these technologies are described briefly here. Also included are features of existing imaging methods that need further development.

In several cases, the new technologies are functionally rather than anatomically based, and the underlying morphology is not directly obvious by viewing the images. This feature should not be surprising; several clinical techniques today, including phase studies in nuclear cardiology, Doppler methods in ultrasonography, and functional magnetic resonance imaging, do not directly reveal anatomic information. In fact, the movement away from direct anatomic correlation in medical imaging may be a liberating influence on further substantial growth of the discipline.

Electrical Source Imaging

Nerve impulses are electrical currents that are propagated in neural tissue. Accompanying these currents are electrical fields that emanate from the tissues and permeate the surrounding environment. *Electrical source imaging* (ESI) is the technique of using electrical field measurements to construct maps of the underlying electrical activity. ESI is an extension of electroencephalography (EEG) and electrocardiography (ECG or EKG) that permit identification of sources of intense electrical activity in the brain and heart. ESI could improve the diagnosis of certain abnormalities such as epilepsy and disorders in impulse conduction in the heart, and it could prove useful in guiding surgical procedures and monitoring the effectiveness of drug treatments. The technique has several limitations that have impeded the migration of ESI into the clinical setting. Among these limitations are:

- Severe absorption and distortion of electrical fields in bone (e.g., skull), which reduces the usefulness of the technique in brain studies
- Difficulties in identifying and localizing multiple sources of electrical activity from measurements of electrical fields at a distance (the so-called "inverse problem")
- Relatively poor spatial resolution caused in part by the problems identified above
- Uncertainty in optimizing the number and placement of surface electrodes for electrical field measurements
- Complexities introduced when the electrical activity varies in time.

At the present time, electrical source imaging (essentially functional maps of electrical activity) is considered an experimental technique that may have some future potential for imaging in the clinical arena.[1]

Magnetic Source Imaging

Electrical fields produced by nerve impulses are strongly absorbed by bone. Attempts to detect them with devices positioned around the head are compromised by the very small signals induced by the weak electrical fields emerging through the skull. Magnetic fields are not so strongly attenuated by the skull, however, and their intensities can be mapped with special devices known as SQUIDs (superconducting quantum interference devices). These maps can then be used to compute the origin and intensity of the neural activities that created the measured fields. These activities may have occurred spontaneously or may have been evoked in the patients in response to applied stimuli.

Magnetic source imaging (MSI), sometimes referred to as magnetoencelphalographic imaging when used to study the brain and magnetocardiographic imaging when applied to the heart, is being explored as an approach to clinical evaluation of a number of abnormal conditions, including epilepsy, migraine headache, and diabetic coma. It may also be useful for studying the normal response of the brain under a variety of conditions, including auditory, olfactory, and visual stimuli. The technique shows promise as an investigational tool for studying the physiology of neural tissue and the cognitive properties of the brain.[2]

Electrical Impedance Imaging

Electricity can be conducted through biologic tissues. Different tissues exhibit different resistances to the flow of electrical current. Furthermore, tissues contain bound as well as free electrical charges and can act as a dielectric when a voltage is applied across them. Differences in the electrical conductivity (electron flow) and permittivity (potential difference) that occur among tissues when a voltage is applied across them offer opportunities for producing images. Although the technology is still in the experimental stage, recent results show promise, and enthusiasm is moderately high for the eventual evolution of the approach into a clinically useful tool. One promising feature is that certain tissues exhibit large differences in electrical resistance (impedance). This difference suggests that excellent contrast resolution might be achievable among tissues if ways could be developed to produce impedance images with satisfactory spatial resolution.[3]

Measurements of electrical impedance can be obtained by applying a voltage across two electrodes on opposite sides of a mass of tissue. These measurements can be misleading, however, because current flow is not confined exclusively to straight-line paths between the electrodes. Instead, the current can follow paths that bulge significantly beyond the boundaries of the electrodes. Hence, the measured resistance to current flow is not simply an indication of the electric impedance of the tissue confined to the volume between the electrodes.

Combining resistance measurements obtained at various orientations can be used to reconstruct impedance tomography images in a manner similar to x-ray CT.

However, the multipath diffusion of current compromises the spatial resolution of the images. This problem has been addressed by designing the recording electrode as a matrix of small electrodes, with those at the periphery serving as guard electrodes and those in the center used to measure the current flow. Although this approach reduces distortions caused by assuming straight-line paths of current flow between electrodes, it is far from perfect.

In a slightly different approach to impedance tomography, an electrical current is impressed across tissue between two electrodes, and the distribution of electrical potential is measured with additional electrodes positioned at various locations on the surface of the tissue. Variations in the measured potentials from those predicted by assuming a uniform tissue composition can be used to construct images. These images reveal the distribution of electrical impedance over the mass of tissue defined by the electrodes. The technique is referred to as applied potential tomography (APT). The acquisition of APT data is very fast (a second or so). However, the spatial resolution of the images has been rather crude to date. On the other hand, the contrast resolution is exquisitely sensitive to very small changes in tissue composition. The method may eventually prove to be useful for dynamic studies and physiologic monitoring. A discussion of APT by Webb[4] includes a review of the literature of potential applications, including cardiac output measurements, pulmonary edema monitoring, and tissue characterization in breast cancer.

Microwave Computed Tomography

Various biological tissues exhibit large differences in their absorption characteristics for microwave radiation. These differences are especially notable between fat and other soft tissues, which raises the possibility of microwave imaging as a way to detect the presence of cancer in the fat-containing breasts of older women. This possibility has been explored experimentally by Cacak et al.[5] Rao et al.[6] have designed an experimental microwave scanner using 10.5-GHz, 10-mW microwaves and a translate–rotate geometry for producing CT images. The long wavelength of microwaves limits the spatial resolution of images reconstructed with this form of electromagnetic radiation. Clinical applications of microwave imaging, including microwave CT, remain to be explored.

Light Imaging

Soft tissues of the body are partially transparent to radiation in the visible and near-infrared portions of the electromagnetic spectrum. Images produced by transmitted visible and IR light have been explored for their possible use in examining tissues of moderate mass. Such tissues might include the female breast, testes, and neonatal brain. In applications to breast imaging, benign cysts are revealed as regions of increased brightness (increased light transmission), while darker areas (reduced light transmission) may indicate cancerous lesions. The technique, referred to as transillumination or diaphonography, has been proposed as a screening method for the detection of breast cancer. However, high false-positive and false-negative rates detract from the usefulness of transillumination as a screening tool. Also, the ability of this technique to reveal small lesions is compromised by the scattering of light and the resulting deterioration of spatial resolution. Although transillumination has a few advocates, most investigators who have attempted to use it remain unconvinced of its ultimate clinical value.

Recently efforts to develop light imaging have focussed on one or more challenges:

- Better understanding of tissue properties that govern light diffusion and scattering in tissue
- Improved ways to distinguish transmitted from scattered light
- Use of excited fluorochromes that target specific tissues and emit radiation of specific wavelengths under illumination

The third challenge listed above, if met and solved, raises the possibility of a new technique for functional imaging at the molecular level. It may be possible to tag a substance with a "marker" that emits visible light only when the substance undergoes a specific chemical reaction in the body. Thus, levels of activity involving, for example, minute amounts of specific enzymes could be detected.

PHASE-CONTRAST X-RAY IMAGING

Conventional x-ray images exhibit contrast that depicts differences in x-ray absorption among various tissue constituents in the path of the x-ray beam. These images provide excellent visualization of tissues with significantly different absorption characteristics resulting from differences in physical density and atomic number. When these differences are slight, however, conventional x-ray imaging methods are limited. Mammography is one example where conventional x-ray imaging methods are challenged.

In addition to absorption differences, x-rays also experience phase shifts during transmission through materials. At x-ray energies employed in mammography, the phase shifts may exceed the absorption differential by as much as 1000 times. Hence, it is possible to observe phase contrast in the image when absorption contrast is undetectable. Three methods to detect phase differences are conceivable.[7] They are (1) x-ray interferometry, (2) diffraction-enhanced imaging, and (3) phase-contrast radiography.

For phase-contrast x-ray imaging, a spatially coherent source of x rays is required. To date, synchrotron radiation sources have been employed. These sources are impractical for routine clinical use. Several efforts are underway to develop small x-ray tubes with a microfocal x-ray source to provide spatial coherence, and with enough x-ray intensity to achieve reasonable exposure times.[8] If and when such tubes become available, phase-contrast x-ray imaging may prove useful for soft-tissue imaging.[9]

INFORMATION MANAGEMENT AND COMMUNICATION

Medicine today is experiencing a severe case of information overload. Nowhere are the symptoms of this condition more apparent than in medical imaging. A rapid growth in imaging technologies and an explosion in diagnostic information about patients have occurred over the past three decades. Yet the methods to use these technologies and handle the information they generate are only now beginning to change.

Most x-ray imaging studies are still interpreted from hard-copy (film) images. These images are handled in file rooms in much the way they were managed years ago. At that time the film file was viewed as the weak link in the flow of patient information from radiology to the clinical services that need it. In most institutions, not much is different today except that the problem has worsened.

The efficient management and communication of diagnostic information is essential to a high-quality service for medical imaging. Several institutions are meeting this challenge by installing a digitally based information management and communications system (IMACS), sometimes referred to as a digital imaging network (DIN), that integrates a picture archiving and communications system (PACS) with a radiology information service (RIS) and a hospital information system (HIS). In most cases the installation is implemented in stages, with linkage of intrinsic digital imaging technologies such as nuclear medicine, CT, computed and digital radiography, and magnetic resonance imaging as a first step in the process.

Current efforts to improve the acquisition and delivery of imaging information by installation of an IMACS are handicapped by several technical and performance-related challenges. Some of these challenges are described below.

Cost

An IMACS represents a major financial investment that is not recoverable by additional fees for radiologic services. Charges for imaging services cannot be increased simply because the services are delivered through an IMACS rather than by traditional methods. Some persons have suggested that over time the cost of an IMACS installation may be partially offset by savings in film and personnel costs. However, it is doubtful that the full financial cost of investment will ever by recovered. Hence an IMACS must be viewed in part as an investment in improved consultative services and patient care. This point of view may be a "hard sell" in institutions that are already experiencing financial difficulties. On the other hand, the "business as usual" alternative to IMACS encourages clinical services to develop their own imaging capabilities.

Soft-Copy Viewing

Radiologists and referring physicians are accustomed to viewing "hard-copy" images on film. Interpretation from "soft-copy" images on video monitors is not an easy adjustment. The difficulty of this transition is exemplified in the continued use today of hard-copy images for final diagnoses in CT and MRI. New video monitors with 2000 × 2000 pixel spatial resolution and 12-bit contrast resolution remove the technical limitations present in early monitors and also remove most of the technical reasons for continued reliance on film for interpretation of studies. Still, it is difficult to change work habits, and dependence on film will probably persist—at least for a while.

Component Incompatibilities

Anyone attempting to initiate even a partial IMACS is familiar with the inadequate standards for interfaces among different components of the system. Maintaining the integrity of electronic signals across these interfaces is a challenge that often has to be solved locally. Professional and vendor organizations are addressing this problem, and progress is being made. Still, complete acquiescence to a single input/output configuration has not been achieved.

Data Storage Requirements

A medical imaging facility produces massive amounts of data each day. Technologies for storing and retrieving these data in a dependable and rapid manner have not yet been perfected. Laser-driven optical disks arranged in a "juke-box" configuration are generally accepted as the solution to this problem, and the capacity of these systems continues to improve. Additional improvements are required, however, before the systems are fully able to function in the manner required for a complete IMACS operation.

Workstation Design

Radiologists seldom arrive at a diagnosis by viewing one or two images. This is especially true if an abnormal condition is suspected from the patient's symptoms or physical examination, or if a suspicious area is detected in an image. Instead, images of past examinations and those from multiple current studies are frequently used to provide more information about the patient. A digital workstation must provide simultaneous access to a number of images and permit the radiologist to navigate through current and past studies in an effort to arrive at a diagnosis. These requirements introduce several technical challenges to the design of workstations, as well as logistical challenges for the persons who use them.

Previous Records

A medical imaging facility is not only a place where imaging data are acquired, images are interpreted, and consultations are held. It also is a major repository of patient information acquired during previous imaging studies. In planning a transition to an IMACS, decisions must be reached about what should be done with the existing film file. Converting all of the file to digital format is an expensive and time-consuming process. Retaining the file in analog form means, however, that the department must function in parallel as both a digital and analog facility. This problem is another justification for changing to a complete IMACS over several years rather than precipitously.

Turf Battles

Radiology's traditional turf is gradually being eroded as other medical specialties incorporate imaging technologies and interpretations into their services (and charges) to patients. As IMACS increases the immediate availability of images to nonradiologists, it enhances the possibilities for interpretation of these images by primary-care specialists without the input of radiologists. This capability has the potential for abusing radiology as a laboratory service for the production of images rather than utilizing it is as a professional consultative service for patient diagnoses. The only effective deterrent to this possibility is continued vigilance and documentation that radiologists interpret images significantly more accurately and completely than do nonradiologists. The threat of medical malpractice and the costs of insuring against it help to preserve radiology's turf.

Real-Time Demands

Currently most radiologic facilities attempt to provide typed reports to referring physicians within 24 to 72 hours of an examination. They also offer personal consultations upon request when a critical decision about patient care is required. Some facilities have added telephone access to the dictated but untyped reports of radiologists in an effort to improve services and to relieve the demand for personal consultations. With IMACS, however, radiologic images could be immediately available on the imaging network for access by nonradiologists who wish to interpret images and make decisions that impact on patient care. The only effective way to counter this temptation will be for radiologists to furnish their interpretive services in "real time."

The problems and challenges presented by IMACS are significant issues. But there is no real alternative to the implementation of IMACS over the next few years if medical institutions are to survive the ever-growing burden of information overload. And there are some added benefits to IMACS, not the least of which is the opportunity for radiologists to reorient themselves toward patient care as true medical consultants. In the long run this advantage by itself could offset the expense and inconvenience of changing to an IMACS method of data management.

TECHNOLOGY ASSESSMENT

For years critics have complained that early studies of new imaging procedures rely more on anecdotal reports and professional opinions than on rigorous methods of evaluation. They have suggested that the accuracy, specificity, and sensitivity of new diagnostic technologies should be determined before the technologies are widely incorporated into clinical practice and before decisions are made about reimbursement for their use.

Some healthcare savants believe that evaluation criteria such as accuracy, sensitivity, and specificity are only intermediate measures that may be related to, but not necessarily indicative of, the usefulness of diagnostic technologies. They emphasize the importance of outcome measures, such as morbidity, mortality, and quality of life, as the preferred ways to judge the effectiveness of a technology. These effects are exceedingly difficult to quantify for diagnostic technologies such as new imaging techniques. Furthermore, their growing importance as criteria for technology assessment, along with the increasing interest in tying coverage and reimbursement to such criteria, could have a decidedly negative influence on the development and diffusion of diagnostic technologies in medicine.[10] Those interested in the continued growth and evolution of medical imaging might be well advised to begin addressing the applicability and implications of outcome measures to diagnostic technologies.

TECHNICAL EXPERTISE IN RADIOLOGY

The 1990s are an exciting and challenging time in medical imaging. The technology of the discipline will continue to evolve over the next several years, and imaging will almost certainly maintain its position at the leading edge of the technological revolution occurring in medicine. This technology, however, will significantly increase the pressure on imaging services to operate in a "real-time, on-line" fashion with respect to requests for procedural and interpretive services. Certainly a higher level of technical performance will be demanded from those in imaging both in the conduct of procedures and in their interpretation. This demand will be satisfied only if imaging physicians and scientists have a solid understanding of the fundamental sciences that underlie their specialty. They also must demonstrate that understanding by responding in a knowledgeable and timely fashion to requests for imaging services.

The physical principles of imaging are among the fundamental sciences that every imaging physician and scientist must understand well to interface effectively with the technological and clinical demands of medical imaging. Provision of this understanding is the purpose of this text. It is only through such understanding that medical imaging will continue to progress as one of the most sophisticated specialties of medicine.

SUMMARY

- Several potential imaging technologies are being explored. They include:
  - Electrical source imaging
  - Magnetic source imaging
  - Electrical impedance imaging
  - Microwave computed tomography
  - Visible light imaging
  - Phase-contrast x-ray imaging
- An Information Management and Communications System (IMACS) integrates PACS, RIS, and HIS.
- Challenges to implementing an IMACs include:
  - Cost
  - Soft-copy viewing
  - Component incompatibilities
  - Data storage requirements
  - Workstation design for viewing multiple images
  - Management of previous records
  - Turf battles
  - Demands for real-time radiologic services
- The effective assessment of emerging medical technologies, including imaging technologies, presents major conceptual and financial challenges.
- The ultimate advantage of radiologists in medical imaging is a thorough understanding of the science of the discipline.

REFERENCES

1. National Research Council, Institute of Medicine of the National Academy of Sciences. *Mathematics and Physics of Emerging Biomedical Imaging*. Washington, D.C., National Academy Press, 1996, pp. 133–142.
2. Hari, R. The role of MEG in cognitive neuroscience. *Second Heidelberg Conference on Dynamic and Functional Radiologic Imaging of the Brain*. Heidelberg, Germany, October 1998.
3. Barber, D., Brown B., and Freeston, I. Imaging spatial distributions of resistivity using applied potential tomography. *Electron Lett.* 1983; **19**:933–935.
4. Webb, S. *The Physics of Medical Imaging*. Philadelphia, Adam Hilger Publications, 1988, pp. 509–523.
5. Cacak, R. K., Winans, D. E., Edrich, J., and Hendee, W. R. Millimeter wavelength thermographic scanner. *Med. Phys.* 1981; **8**:462–465.
6. Rao, P., Santosh, K., and Gregg, E. Computed tomography with microwaves. *Radiology* 1980; **135**:769–770.
7. Fitzgerald, R. Phase-sensitive x-ray imaging. *Physics Today* July 2000, 23–26.
8. Wilkins, S. W., Gureyev, T. E., Gao, D., and Pogany, A., and Stevenson, A. W. Phase-contrast imaging using polychromatic hard X-rays. *Nature* 1996; **384**: 335–338.
9. Pisano, E. D., Johnston, R. E., Chapman, D., Geradts, J., Iacocca, M. V., Livasy, C. A., Washburn, D. B., Sayers, D. E., Zhong, Z., Kiss, M. Z., and Thomlinson, W. C. Human breast cancer specimens: diffraction-enhanced imaging with histologic correlation—improved conspicuity of lesion detail compared with digital radiography. *Radiology* 2000; **214**(3):895–901.
10. Hendee, W. R. New Imaging Techniques, in Bragg, D. G., Rubin, P., and Hricak, H. (editors), *Oncologic Imaging,* second edition. Philadelphia: W. B. Saunders, 2002.

APPENDIX

I

REVIEW OF MATHEMATICS

A. Algebraic formulas:

If $a = b + c$, then solve for b, c (answer: $b = a - c, c = a - b$)

If $a - b = c$, then solve for b (answer: $b = a - c$)

B. Ratio and Proportion

1. Comparison (ratio):

$$\frac{\text{Dog weight}}{\text{Cat weight}} = \frac{48\ \text{lb}}{16\ \text{lb}} = 3 \text{ or } \frac{3}{1}$$

$$\frac{1\ \text{hr}}{1\ \text{day}\ 2\ \text{hr}} = \frac{1}{24 + 2} = \frac{1}{26} \text{ (must have same units throughout)}$$

$$\frac{6}{3} = \frac{2}{1} = 2 \text{ (fractions may sometimes be reduced)}$$

2. Proportion:

$$\frac{a}{b} = \frac{c}{d} \quad a{:}b{::}c{:}d \qquad ad = bc$$

Example: Solve for the value of x:

$$3{:}5{::}x{:}20 \quad \frac{3}{5} = \frac{x}{20} \qquad 5x = 3(20)$$

$$5x = 60$$

$$x = 12$$

3. Direct proportion:

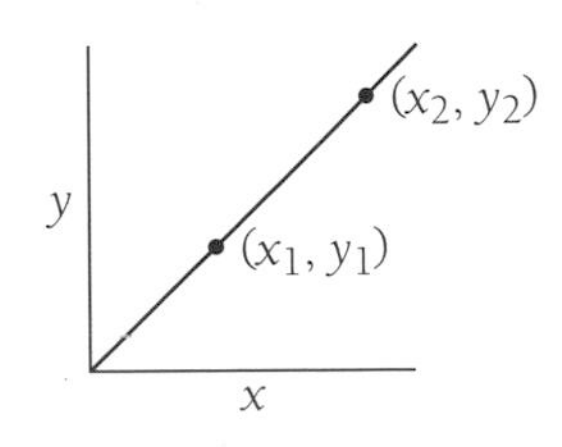

$$y = bx \qquad\qquad \frac{y_1}{y_2} = \frac{x_1}{x_2}$$

4. Inverse proportion:

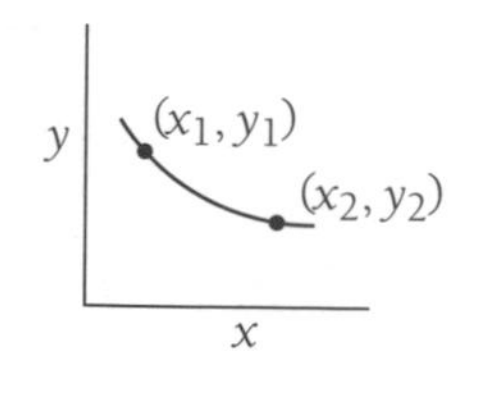

$$y = \frac{b}{x} \qquad \frac{y_1}{y_2} = \frac{x_2}{x_1}$$

C. *Powers* (exponents)

1. Powers of *any* number a^x (x = exponent; a = base):

$2^3 = 2 \times 2 \times 2 = 8$

$3^3 = 3 \times 3 \times 3 = 27$

2. Multiplication of powers:

$2^3 \times 2^2 = (2 \times 2 \times 2) \times (2 \times 2) = 2^5 = 32$

Add exponents

$(a^x)(a^y) = a^{x+y}$

3. Division:

$$\frac{2^3}{2^2} = \frac{2 \times 2 \times 2}{2 \times 2} = 2^1 = 2$$

Subtract exponents:

$$\frac{a^x}{a^y} = a^{x-y}$$

4. Negative exponents: Suppose we have

$$\frac{2^2}{2^3} = \frac{2 \times 2}{2 \times 2 \times 2} = \frac{1}{2}$$

Or by the rule above

$$= 2^{2-3} = 2^{-1} = \frac{1}{2}$$

So

$$2^{-3} = \frac{1}{2 \times 2 \times 2}; a^{-x} = \frac{1}{a^x}$$

5. Zero exponent: Suppose we have

$$\frac{5^2}{5^2} = \frac{5 \times 5}{5 \times 5} = 1$$

Or by the rule above

$= 5^{2-2} = 5^0$

So

$a^0 = 1$

6. Powers of powers:

$(4^2)^3 = (4 \times 4) \times (4 \times 4) \times (4 \times 4)$

$= 4^6$

Multiply exponents

$(a^x)^y = a^{xy}$

7. Powers of 10:
   Since our number system is *based* on 10, it is quite convenient to express any number as a power of 10.
   Note first that all the rules for any base a apply to base 10:
   $10^0 = 1; 10^2 \times 10^3 = 10^5; (10^2)^3 = 10^6$
   $$10^{-2} = \frac{1}{10^2}$$
   $$\frac{10^5}{10^6} = 10^{-1}; \frac{10^7}{10^{-3}} = 10^{7-(-3)} = 10^{10}$$
8. Powers of any number (scientific notation):
   $$751{,}239 = 7.51239 \times 100{,}000$$
   $$= 7.51239 \times 10^5$$
   $1000 = 10^3$ (3 zeros)
   $100 = 10^2$
   $10 = 10^1$
   $1 = 10^0$
   $0.1 = 10^{-1}$
   $0.01 = 10^{-2}$
   $0.001 = 10^{-3}$ (2 zeros + 1 decimal point)
   Rewriting numbers in scientific notation. Why do it? Because of numbers like
   $0.0000000000000003 = 3 \times 10^{-16}$
   $5.98 \times 10^3 = a \times b$ (product of a and b must remain the same; if a increases, b must decrease)
   59.86×10^2
   0.5986×10^4
9. Multiplication:
   $$(3 \times 10^6)\,(4 \times 10^2) = (3)\,(4) \times (10^6)(10^2)$$
   $$= 12 \times 10^8$$
   $$(a \times 10^x)(b \times 10^y) = (a)(b) \times 10^{x+y}$$
10. Division:
   $$\frac{6 \times 10^5}{3 \times 10^4} = \frac{6}{3} \times \frac{10^5}{10^4} = 2 \times 10^1$$
   $$\frac{a \times 10^x}{b \times 10^y} = \frac{a}{b} \times 10^{x-y}$$
11. Addition or subtraction:
   $(5 \times 10^5) - (4 \times 10^4) =$
   (must make powers of 10 equal before adding or subtracting numbers in scientific notation)
   $(50 \times 10^4) - (4 \times 10^4) = 46 \times 10^4$
   $(a \times 10^x) - (b \times 10^x) = (a - b) \times 10^x$

D. Logarithms:
   If $N = a^x$, then we say that "x is the log to the base a of N" and "x is the power to which a must be raised to equal N."

Common logarithm - base 10 - $\log_{10}$
Q. What is the common log of 1000?
A. $1000 = 10^3$
$\log_{10} 1{,}000 = 3$
Q. How do you find log 37?
A. Use a calculator: 1.57
$10^{1.57} = 37$
Rules:

$$\log(a)(b) = \log a + \log b$$

$$\log \frac{a}{b} = \log a - \log b$$

E. "e," "Euler's number" or the exponential quantity, the base of natural logarithms:

$$e = 2.7182818\cdots = \lim_{n\to\infty}\left(1 + \frac{1}{n}\right)^n$$

for example,

$$n = 10;\ \left(1 + \frac{1}{10}\right)^{10} = 2.594$$

$$n = 100;\ \left(1 + \frac{1}{100}\right)^{100} = 2.705$$

e^x is a function F that is a solution to the differential equation:

$\frac{dF}{dx} = F$ (The rate of change of the function F depends linearly upon the value of F.)

Graphs of the exponential function:

Exponential decay   Exponential growth

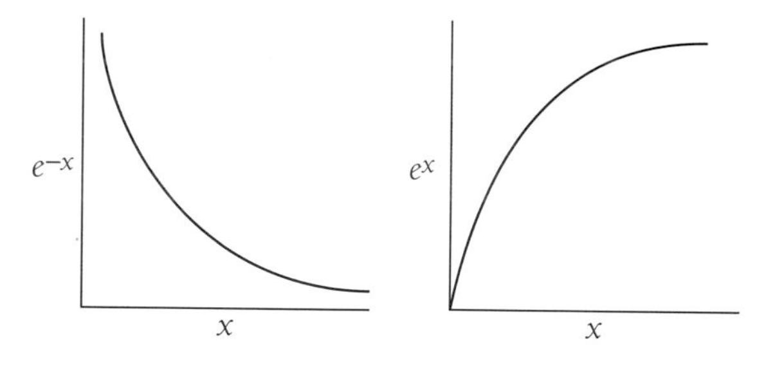

Special (useful) property: Half-value = x_H

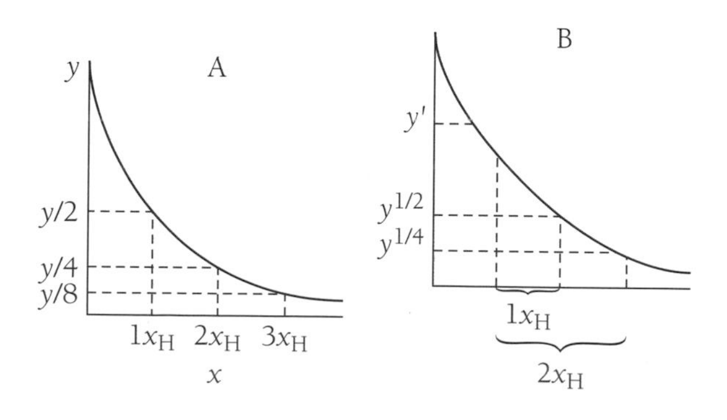

Pick any value of the function (e.g., the value y or the value y') move a distance x_H along the x-axis, and the value of e^{-x} will be cut in half.

You do not have to start at $x = 0$, but can start anywhere. Moving a distance x_H along the x-axis reduces any y-axis value by a factor of 2.
"e" is the *base* of the natural logarithms.
If $N = e^x$, then we say that "x is the natural log of N." "x is the power to which e must be raised to obtain N"
or
$x = \ln N$; "x is the natural log of N."
If follows that $\ln e^x = x$ (i.e., the power to which e must be raised to obtain e^x is obviously x).
To eliminate exponentials from equations, take the natural log (ln) of both sides:
$e^x = 7$ (To solve this, we must obtain x alone. So take the natural log of both sides.)
$\ln e^x = \ln 7$
$x = \ln 7$ (The equation is "solved" because we now know what x is equal to. If we want a numerical answer, we can use a calculator to find that $\ln 7 = 1.945$.)
Note also that

$$e^{-1} = 0.36788 \cdots \cong 0.37 = 37\%$$

How do you find the half-value, x_H?

Note that if $0.5 = e^{-x_H}$

$\ln(0.5) = \ln(e^{-x_H})$

$-0.693 = -x_H$

then $0.693 = x_H$

So $e^{-0.693} = \frac{1}{2}$ (similarly we can show that $e^{+0.693} = 2$).

So we can "measure" x in units of x_H by writing

$$e^{-0.693\left(\frac{x}{x_H}\right)} = [e^{-0.693}]^{\left(\frac{x}{x_H}\right)} = \left[\frac{1}{2}\right]^{\left(\frac{x}{x_H}\right)}$$

because when $x = 1x_H$, $e^{-0.693\left(\frac{x_H}{x_H}\right)} = e^{(-0.693)(1)} = \left(\frac{1}{2}\right)^1 = \frac{1}{2}$

$$x = 2x_H,\ e^{-0.693\left(\frac{2x_H}{x_H}\right)} = e^{(-0.693)(2)} = \left(\frac{1}{2}\right)^2 = \frac{1}{4}$$

$$x = 3x_H,\ e^{-0.693\left(\frac{3x_H}{x_H}\right)} = e^{(-0.693)(3)} = \left(\frac{1}{2}\right)^3 = \frac{1}{8}$$

$$x = 4x_H, \quad \cdot \quad \cdot \quad \cdot \quad = \frac{1}{16}$$

$$x = 5x_H, \quad \cdot \quad \cdot \quad \cdot \quad = \frac{1}{32}$$

We make use of a characteristic parameter α such that $e^{-\alpha x}$, where $\alpha = 0.693/x_H$. Table I-1 shows some characteristic exponential parameters used in radiologic physics and their corresponding half-values.

TABLE I-1 Characteristic Exponential Parameters Used in Radiology

Application	*Exponential Parameter*	*Half-Value*	*Equation*
Radioactive decay	Decay constant λ	Half-life $T_H = 0.693/\lambda$	$A = A_0e^{-\lambda t}$
Attenuation of photons	Attenuation coefficient μ	Half-value layer $X_H = 0.693/\mu$	$N = N_0e^{-\mu X}$
Magnetic resonance	Inverse of relaxation times $1/T_1, 1/T_2$	Relaxation times $(0.693)T_1, (0.693)T_2$	$S = S_0e^{-t/T_2}$

APPENDIX

II FOURIER TRANSFORM

The Fourier transform (FT) is a mathematical technique that allows a complex mathematical function to be broken down into a series of simpler functions. Although its precise mathematical statement is, admittedly, esoteric, the fundamental principle of the FT is used routinely in radiology. We may state the FT process as a mathematical rewriting (transformation) of a function A, having variables x, y, and so on, into a function B having variables p, q, and so on. Symbolically,

$$(\mathrm{FT})\{A[x,y]\} = B[p,q]$$

In the two examples that follow, FT is used to convert data from a form in which it is easier to acquire into a form in which it is easier to comprehend.

DOPPLER ULTRASOUND

The Doppler-shift signal is an electrical "function" that could be represented as a graph of voltage versus time. We know that it is a complex signal because it is the result of echoes from red blood cells traveling at different velocities. If all of the red blood cells were traveling at the same velocity (and if the signal sent into the patient were of a pure, single ultrasound frequency), the return signal would be a sine wave having a single frequency, or number of peaks per second. The return signal is, however, not a simple sine wave. It shows peaks of different amplitude at different spacings, which implies that it is composed of various amounts of different frequencies. The FT can be used to determine how much (what amplitude) of each frequency of sine wave must be added together to account for the signal that we have obtained from the patient. We then know what fraction of each frequency, and therefore of each velocity, that is represented in the signal. With this information, we may construct a graph of fraction of the signal versus frequency shift, known as a power spectrum. Thus, the frequency shift S is transformed into the power spectrum P. The FT pairs in this example are

$$(\mathrm{FT})\{S(\text{voltage, time})\} = P\ (\text{fraction of total signal, frequency})$$

The Doppler-shift equation may then be used to convert the frequency scale to a velocity scale.

MAGNETIC RESONANCE

The magnetic resonance signal is a complex radio-frequency signal that may be represented as a graph of voltage versus time. Gradient magnetic fields are applied to the patient so that the magnetic resonance signal may be spatially encoded. The effect of these fields is to cause the resonance frequency to vary across a sample. Therefore, signals originating from one end of the sample have frequencies that are slightly different from signals from the other end. The complex magnetic resonance signals may be mathematically transformed to determine the fraction of different amounts of signal at different frequencies across a sample. Thus, the magnetic resonance signal M is transformed into the image profile I. The Fourier transform pairs in this example are

$$\text{(FT)}\ \{M(\text{voltage, time})\} = I\ (\text{fraction of total signal, frequency}).$$

The frequency scale may then be converted into a distance scale by using the Larmor equation to convert frequency to magnetic field strength and dividing the result by the gradient strength in millitesla per meter.

APPENDIX

III

MULTIPLES AND PREFIXES

Multiple	*Prefix*	*Symbol*
10^{24}	yotta	Y
10^{21}	zetta	Z
10^{18}	exa	E
10^{15}	peta	P
10^{12}	tera	T
10^{9}	giga	G
10^{6}	mega	M
10^{3}	kilo	k
10^{-1}	deci	d
10^{-2}	centi	c
10^{-3}	milli	m
10^{-6}	micro	μ
10^{-9}	nano	n
10^{-12}	pico	p
10^{-15}	femto	f
10^{-18}	atto	a
10^{-21}	zepto	z
10^{-24}	yocto	y

APPENDIX

IV

MASSES IN ATOMIC MASS UNITS FOR NEUTRAL ATOMS OF STABLE NUCLIDES AND A FEW UNSTABLE NUCLIDES

Element	*Mass No.*	*Atomic Mass (AMU)*	*Element*	*Mass No.*	*Atomic Mass (AMU)*
O^n	1[a]	1.008665	$_{12}Mg$	23[a]	22.994125
$_{1}H$	1	1.007825		24	23.990962
	2	2.014102		25	24.989955
	3[a]	3.016050		26	25.991740
$_{2}He$	3	3.016030	$_{13}Al$	27	26.981539
	4	4.002603	$_{14}Si$	28	27.976930
	6[a]	6.018893		29	28.976496
$_{3}Li$	6	6.015125		30	29.973763
	7	7.016004	$_{15}P$	31	30.973765
	8[a]	8.022487	$_{16}S$	32	31.972074
$_{4}Be$	7[a]	7.016929		33	32.971462
	9	9.012186		34	33.967865
	10[a]	10.013534		36	35.967090
$_{5}B$	8[a]	8.024609	$_{17}Cl$	35	34.968851
	10	10.012939		36[a]	35.968309
	11	11.009305		37	36.965898
	12[a]	12.014354	$_{18}Ar$	36	35.967544
$_{6}C$	10[a]	10.016810		38	37.962728
	11[a]	11.011432		40	39.962384
	12	12.000000	$_{19}K$	39	38.963710
	13	13.003354		40[a]	39.964000
	14[a]	14.003242		41	40.961832
	15[a]	15.010599	$_{20}Ca$	40	39.962589
$_{7}N$	12[a]	12.018641		41[a]	40.962275

Element	Mass No.	Atomic Mass (AMU)	Element	Mass No.	Atomic Mass (AMU)
	13[a]	13.005738		42	41.958625
	14	14.003074		43	42.958780
	15	15.000108		44	43.955490
	16[a]	16.006103		46	45.953689
	17[a]	17.008450		48	47.952531
$_{8}$O	14[a]	14.008597	$_{21}$Sc	41[a]	40.969247
	15[a]	15.003070		45	44.955919
	16	15.994915	$_{22}$Ti	46	45.952632
	17	16.999133		47	46.951769
	18	17.999160		48	47.947951
	19[a]	19.003578		49	48.947871
$_{9}$F	17[a]	17.002095		50	49.944786
	18[a]	18.000937	$_{23}$V	48[a]	47.952259
	19	18.998405		50[a]	49.947164
	20[a]	19.999987		51	50.943962
	21[a]	20.999951	$_{24}$Cr	48[a]	47.953760
$_{10}$Ne	18[a]	18.005711		50	49.946055
	19[a]	19.001881		52	51.940514
	20	19.992440		53	52.940653
	21	20.993849		54	53.938882
	22	21.991385	$_{25}$Mn	54[a]	53.940362
	23[a]	22.994473		55	54.938051
$_{11}$Na	22[a]	21.994437	$_{26}$Fe	54	53.939617
	23	22.989771		56	55.934937
	57	56.935398		95	94.905839
	58	57.933282		96	95.904674
$_{27}$Co	59	58.933190		97	96.906022
	60[a]	59.933814		98	97.905409
$_{28}$Ni	58	57.935342		100	99.907475
	60	59.930787	$_{44}$Ru	96	95.907598
	61	60.931056		98	97.905289
	62	61.928342		99	98.905936
	64	63.927958		100	99.904218
$_{29}$Cu	63	62.929592		101	100.905577
	65	64.927786		102	101.904348
$_{30}$Zn	64	63.929145		104	103.905430
	66	65.926052	$_{45}$Rh	103	102.905511
	67	66.927145	$_{46}$Pd	102	101.905609
	68	67.924857		104	103.904011
	70[a]	69.925334		105	104.905064
$_{31}$Ga	69	68.925574		106	105.903479
	71	70.924706		108	107.903891
$_{32}$Ge	70	69.924252		110	109.905164
	72	71.922082	$_{47}$Ag	107	106.905094
	73	72.923463		109	108.904756
	74	73.921181	$_{48}$Cd	106	105.906463
	76	75.921406		108	107.904187
$_{33}$As	75	74.921597		110	109.903012
$_{34}$Se	74	73.922476		111	110.904189
	76	75.919207		112	111.902763
	77	76.919911		113	112.904409
	78	77.917314		114	113.903361
	80	79.916528		116	115.904762
	82	81.916707	$_{49}$In	113	112.904089
$_{35}$Br	79	78.918330		115[a]	114.903871
	81	80.916292	$_{50}$Sn	112	111.904835

Element	*Mass No.*	*Atomic Mass (AMU)*	*Element*	*Mass No.*	*Atomic Mass (AMU)*
$_{36}$Kr	78	77.920403		114	113.902773
	80	79.916380		115	114.903346
	82	81.913482		116	115.901745
	83	82.914132		117	116.902959
	84	83.911504		118	117.901606
	86	85.910616		119	118.903314
$_{37}$Rb	85	84.911800		120	119.902199
	87[a]	86.909187		122	121.903442
$_{38}$Sr	84	83.913431		124	123.905272
	86	85.909285	$_{51}$Sb	121	120.903817
	87	86.908893		123	122.904213
	88	87.905641	$_{52}$Te	120	119.904023
$_{39}$Y	89	88.905872		122	121.903066
$_{40}$Zr	90	89.904700		123	122.904277
	91	90.905642		124	123.902842
	92	91.905031		125	124.904418
	94	93.906314		126	125.903322
	96	95.908286		128	127.904476
$_{41}$Nb	93	92.906382		130	129.906238
$_{42}$Mo	92	91.906811	$_{53}$I	127	126.904470
	94	93.905091	$_{54}$Xe	124	123.906120
	126	125.904288		160	159.925202
	128	127.903540		161	160.926945
	129	128.904784		162	161.926803
	130	129.903509		163	162.928755
	131	130.905086		164	163.929200
	132	131.904161	$_{67}$Ho	165	164.930421
	134	133.905398	$_{68}$Er	162	161.928740
	136	135.907221		164	163.929287
$_{55}$Cs	133	132.905355		166	165.930307
$_{56}$Ba	130	129.906245		167	166.932060
	132	131.905120		168	167.932383
	134	133.904612		170	169.935560
	135	134.905550	$_{69}$Tm	169	168.934245
	136	135.904300	$_{70}$Yb	168	167.934160
	137	136.905500		170	169.935020
	138	137.905000		171	170.936430
$_{57}$La	138[a]	137.906910		172	171.936360
	139	138.906140		173	172.938060
$_{58}$Ce	136	135.907100		174	173.938740
	138	137.905830		176	175.942680
	140	139.905392	$_{71}$Lu	175	174.940640
	142	141.909140		176[a]	175.942660
$_{59}$Pr	141	140.907596	$_{72}$Wf	174	173.940360
$_{60}$Nd	142	141.907663		176	175.941570
	143	142.909779		177	176.943400
	144[a]	143.910039		178	177.943880
	145	144.912538		179	178.946030
	146	145.913086		180	179.946820
	148	147.916869	$_{73}$Ta	181	180.948007
	150	149.920915	$_{74}$W	180	179.947000
$_{62}$Sm	144	143.911989		182	181.948301
	147[a]	146.914867		183	182.950324
	148	147.914791		184	183.951025
	149	148.917180		186	185.954440
	150	149.917276	$_{75}$Re	185	184.953059

Element	*Mass No.*	*Atomic Mass (AMU)*	*Element*	*Mass No.*	*Atomic Mass (AMU)*
	152	151.919756		187[a]	186.955833
	154	153.922282	$_{76}$Os	184	183.952750
$_{63}$Eu	151	150.919838		186	185.953870
	153	152.921242		187	186.955832
$_{64}$Gd	152	151.919794		188	187.956081
	154	153.920929		189	188.958300
	155	154.922664		190	189.958630
	156	155.922175		192	191.961450
	157	156.924025	$_{77}$Ir	191	190.960640
	158	157.924178		193	192.963012
	160	159.927115	$_{78}$Pt	190[a]	189.959950
$_{65}$Tb	159	158.925351		192	191.961150
$_{66}$Dy	156	155.923930		194	193.962725
	158	157.924449		195	194.964813
	196	195.964967		205	204.974442
	198	197.967895	$_{82}$Pb	204	203.973044
$_{79}$Au	197	196.966541		206	205.974468
$_{80}$Hg	196	195.965820		207	206.975903
	198	197.966756		208	207.976650
	199	198.968279	$_{83}$Bi	209	208.981082
	200	199.968327	$_{90}$Th	232[a]	232.038124
	201	200.970308	$_{92}$U	234[a]	234.040904
	202	201.970642		235[a]	235.043915
	204	203.973495		238[a]	238.050770
$_{81}$Tl	203	202.972353			

[a] Unstable nuclide.

Source: From Weidner, R., and Sells, R. *Elementary Modern Physics*, 2nd edition. Boston, Allyn & Bacon, 1968. Used by permission.

ANSWERS TO SELECTED PROBLEMS

Chapter 2

2-1. 8; 16; 0.1369 amu; 127.5 MeV; 8.0 MeV/nucleon
2-2. 15.999 amu
2-4. Tungsten, 58, 220 eV; hydrogen, 10.1 eV
2-5. 0.51 MeV
2-6. 2.37×10^{24}, 0.85 g
2-7. 15.2×10^{15} J; 3.63×10^{12} kcal

Chapter 3

3-1. 4.6 hr; 6.9 hr; 156 hr
3-2. 3.5×10^{-7}; 6.6×10^{15}; 1.8×10^{-6}
3-4. 36.7%
3-6. 2.4×10^{19} sec^{-1}; 0.124 Å
3-8. 3.14 hr
3-10. 6.15×10^{14}; 9.2×10^{-8} g
3-11. 30.3 mCi
3-12. 1.09 days; 4100 Ci; 16.4 Ci
3-13. ^{74}As, positron decay and/or electron capture; ^{76}As, negatron decay
3-14. 80

Chapter 4

4-1. 2.031 keV/cm
4-2. 51,699 IP/cm
4-4. 0.59
4-5. 0.94 cm
4-7. 12 keV; 35 keV
4-9. 138 keV; 12 keV; decreased
4-10. 865 keV

Chapter 5

5-2. 3.12×10^{17}; 5000 J/sec
5-4. 11.5 degrees
5-5. No; yes
5-6. No; 1 should be eliminated
5-7. 3 per min
5-8. 250 KeV; 0.023; 0.05 A
5-9. 2 mm
5-10. 9.6 mm

Chapter 6

6-1. 1.06×10^{9} MeV/(m^2-sec); 1.06×10^{10} meV/m^2
6-2. 0.68 J/kg
6-3. 363 J/m^2; 2.27×10^{15} photons/m^2
6-4. 2.4×10^{11}
6-5. 7.4
6-6. 1.53×10^{-10} A
6-7. 52 picofarads
6-8. 270 R
6-11. 10^{-4} J/g; 10^{-3} J
6-12. 0.85
6-13. 1,500 mrem
6-14. 0.014°C
6-15. 1.44×10^{18}
6-17. (a) 1.5 mm A1; (b) 2.25 mm A1; (c) 0.7

Chapter 8

8-1. (a) No; (b) yes, yes; (c) no
8-2. No
8-4. 4.25×10^{6}
8-5. 0.6 mV

Chapter 9

9-1. 3.6×10^{10}
9-2. 10,500; 2000
9-3. 1.60; 60%
9-6. (a) 75 mCi/mL; (b) 0.064
9-9. 2.76 MeV = photopeak; 2.25 MeV = single-escape peak; 1.74 MeV = double-escape peak; 1.38 MeV = photopeak; 0.87 MeV = single-escape peak; 0.51 MeV = annihilation peak; 0.20 MeV = backscatter peak
9-11. About 580
9-14. 220 MCi
9-15. 3400 yr
9-16. 2.84×10^5 Ci/g

Chapter 10

10-1. 5120 bits = 5×2^{10}
10-2. 503,316,480 bits = $8 \times 60 \times 2^{20}$
62,914,560 bytes = 60×2^{20}
10-3. 16,777,216 bits = $32 \times 0.5 \times 2^{20}$
2,097,152 bytes = $4 \times 0.5 \times 2^{20}$
10-4. 20,480 images
10-5. 31.25 mV
10-6. PROM
10-7. 1.8 (line pairs)/mm < 2.5 (line pairs)/mm, does not meet the standard
10-8. 4.8 Mbit

Chapter 11

11-1. a. 6 cpm, 1.67%
b. 4.7 cpm, 3.60%
c. 8 cpm, 3.32%
11-2. a. The precision is described by the 2-mm standard deviation. Accuracy relates the researcher's data to the "true" answer determined by some independent method, which in this case is the study of excised bone. The degree of accuracy is therefore given by the 2% difference between the means of the two methods of bone measurement. The results are biased because the original measurements are consistently larger than the true value.
b. Both mean and standard deviation are "scaled" by the same proportion (see Table 11-3). Corrected mean = (0.98)(5.5) = 5.4 cm. Corrected standard deviation = (0.98)(2 mm) = 1.96 mm = 0.02 cm.
Note: The corrected values are given with the same number of significant figures as the original data, and the same units (cm) are used for both.
11-3. t value = 2.31; $P = 0.022$

Chapter 12

12-1. 59%
12-2. 5500 mL
12-4. 43%
12-5. (a) ^{241}Am; (b) ^{197}Hg

Chapter 13

13-1. 10% and 3%
13-2. 126.5 mR

Chapter 14

14-1. 3240
14-3. Yes
14-4. 9 MHz
14-6. 1.8×10^{-3} c/(kg-min), 7.0 roentgen/min
7.1×10^{-3} c/(kg-min), 27.7 roentgen/min

Chapter 16

16-1. Unsharpness, contrast, noise, distortion and artifacts
16-2. Geometric, subject, motion, receptor, unsharpness
16-3. 0.4 mm
16-4. 1.8 mm
16-5. 24 cm
16-6. Subject, technique, contrast agents, receptor
16-7. Structure noise, radiation noise, receptor noise, quantum mottle

Chapter 17

17-1. 0.2 mm; 2.5 line pairs/mm
17-2. Higher; projected focal spot is smaller
17-3. 0.41

Chapter 18

18-2. The individual sees at 20 ft what an ordinary person sees at 100 ft.
18-4. Sensitivity $= p(\text{TP}) = \text{TP}/(\text{TP} + \text{FN})$
Specificity $= p(\text{TN}) = \text{TN}/(\text{TN} + \text{FP})$
18-6. Specificity, because a false-negative means that a tumor that should have been detected is missed.

Chapter 19

19-2. -13 dB
19-3. (a) $\alpha_R = 0.011$ $\alpha_T = 0.989$
(b) $\alpha_R = 0.011$ $\alpha_T = 0.989$
19-4. 33.4 degrees
19-6. 6.9 dB

Chapter 20

20-2. 1.2 mm
20-4. 48 cm

Chapter 21

21-2. 128
21-3. $\frac{1}{4}$
21-5. Attenuation

Chapter 22

22-1. 325 Hz; higher
22-2. 1.3 kHz
22-5. 650 Hz; a, decrease; b, increase

Chapter 23

23-2. 7.63×10^{18}
23-3. 1.98 tesla
23-4. 1/16 of the maximum signal

Chapter 24

24-1. 11.52 min
24-2. a
24-3. 280 Hz

Chapter 28

28-1. 30 rem
28-2. 300 mrem
28-3. 440 mR
28-4. 2.0 mm Pb
28-5. None
28-6. 0.4 mm Pb
28-7. 0.7 mm Pb
28-8. 1.27 mm Pb

INDEX